Geography and Development

A World Regional Approach

The cover illustration is a portion of the
photograph on the following page showing the
eastern United States from Boston,
Massachusetts, to Norfolk, Virginia, taken at an
altitude of 568 miles (914 kilometers). Three
photos—green, red, and infrared—were seen and
recorded separately by the Earth Resources
Technology Satellite 1, then combined at
NASA's Goddard Space Flight Center.

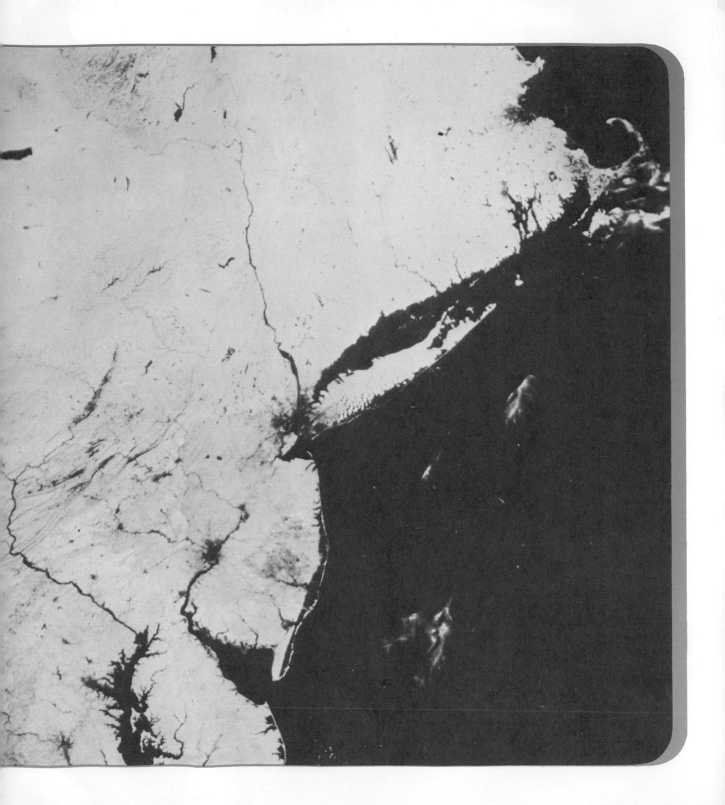

Geography and Development

A World Regional Approach

Contributors

Leonard Berry

Louis De Vorsey, Jr.

James S. Fisher

Don R. Hoy

Douglas L. Johnson

Clifton W. Pannell

Roger L. Thiede

Jack F. Williams

Edited by

DON R. HOY

Macmillan Publishing Co., Inc.
NEW YORK

Collier Macmillan Publishers
LONDON

Macmillan Publishing Co., Inc.
866 Third Avenue, New York, New York 10022

Collier Macmillan Canada, Ltd.

Library of Congress Cataloging in Publication Data

Main entry under title:

geography and development.

    Includes bibliographies and index.
    1. Economic development—Addresses, essays, lectures. 2. Economic history—Addresses, essays, lectures. 3. Underdeveloped areas—Addresses, essays, lectures. 4. Geography, Economic—Addresses, essays, lectures. I. Hoy, Don R.
HD82.G39   330.9   76–842
ISBN 0-02-357400-3

Printing: 1 2 3 4 5 6 7 8   Year: 8 9 0 1 2 3 4

Preface

I̲t̲ ̲i̲s̲ ̲a̲n̲ increasingly trite but accurate truism that the world is getting smaller. Not only are we constantly bombarded with news of places such as Kuwait, India, Vietnam, Switzerland, and Japan, but events in these and other nations also influence our daily lives. The quadrupling of crude oil prices by the Organization of Petroleum Exporting Countries (OPEC) awakened the public to the degree we rely on other nations to supply our needs. A coffee failure in Brazil, the development of a new high-yielding variety of wheat, the discovery of a chemical process for making plastic, and a myriad of other events all materially affect our life styles. As our population grows and our standards of living improve, the level of international interdependence also increases. Consequently it behooves us to know something about the world in which we live, for without knowledge we can exert little influence on our fate.

Our concern is knowledge about economic achievement. There is a great disparity in material well-being among the world's societies. With a smaller world brought about by better communication and transportation, knowledge of how others live is spread rapidly. Those economically less fortunate usually wish to emulate their materially richer neighbors but often are frustrated in their attempts by cultural, economic, and political constraints. The disparity in economic achievement is widening, and the social and political ramifications are manifold. The reasons for this disparity provide the focus and theme of this book: the processes of economic development, broadly defined.

Economic development is basically a Western societal concept. As we shall see, Western ideas of materialism are spreading widely throughout the world. To support development, modern technology has been accepted by other cultures. Penetration of Western ideas and technology into other cultures often has led to disruption and conflict. Not only are traditional economic patterns altered but also new modes of behavior and interpersonal relations are established. Few nations have "Westernized" completely, and cultural pluralism has resulted.

To appreciate economic development, we must examine the process from four principal points of view. The first point of view is the people: their number, growth rates, and distribution. Improved sanitation and medical science have greatly lowered the death rate and have caused a rapid rate of population growth—so great that the appellation "population explosion" has been coined. Where birth rates have not dropped to a comparable level, advances in economic development are dissipated among an increasing number of people. Population characteristics are outlined in Chapter 1 and examined in more detail in the treatment of each of the world's major regions.

The second point of view is the natural environment, which provides the stage and materials used in economic activities. Some environments are rich in resources mankind can use for economic gain. A well-watered alluvial plain with a long growing season provides many opportunities for productive agriculture. Similarly, highly mineralized areas with easily extractable ores can support other means of livelihood. Conversely, areas with steeply sloping land, thin soils, or moisture deficiency are relatively poor for cropping and present obstacles to development. Chapter 2 gives an overview of the various elements of the natural environment, and particular environments are treated in the various regional chapters.

The third point of view is culture. The way in which a society organizes itself in terms of belief systems, customs, and life styles greatly influences the direction and degree of economic development. One aspect of Western culture is materialism. The Puritan work ethic is an example of our material bent with wealth an index of success. Although not Western in tradition or culture, Japan has a similar work goal. Some other cultures do not place so high a priority on the material advantages of life. The wandering nomad cannot accumulate many possessions and still practice a migratory existence. Of more direct influence is the character of economic organization. Some cultures (including some Western cultures) have an economic structure ill-designed to use modern technology effectively or have such a rigid social and economic system that the human resource base is constrained. Other more flexible systems are able to adapt to new ideas and accept technology relatively easily. The elements of culture, and particularly economic organization, are the subject of Chapter 3. Culture and economic opportunities and constraints receive special attention in the regional chapters.

The fourth point of view is that of history. That the past is a key to the present is well demonstrated through a study of the evolution of economic activity and the world's various cultures. Economic development is not a short-term process. The necessary foundations or prerequisite conditions for development of most nations that are now undergoing rapid development or that have attained a high level of economic well-being were laid decades and even centuries ago. The cornerstones of Europe's Industrial Revolution, which began in the mid-eighteenth century, were formed during the preceding Renaissance period with roots going back to Roman and Greek times and even earlier. A more recent example is the case of Taiwan where many of the foundations for the island's recent remarkable growth were laid in the early part of this century. The historical perspective is a theme that runs throughout our discussion.

We begin our discussion with a unit of four chapters that provide a background of ideas and knowledge developed more fully in succeeding chapters. Chapter 1 presents an overview of population and the resource concept. Chapter 2 views the natural environment primarily from the resource standpoint. Chapter 3 is an appraisal of the elements of culture focused mainly on characteristics influencing development. Chapter 4 presents the rich and poor regions of the world based on some measures of economic well-being along with some characteristics of rich and poor regions and some theories of development.

The development theme permits a broad twofold division of the world into regions with relatively rich populations and those inhabited by relatively poor people. The rich regions are considered first and comprise Anglo-America, Western Europe, Eastern Europe and the Soviet Union, Japan, and Australia–New Zealand. The poor regions are Latin America, Africa and the Middle East, South Asia, China and its neighbors, and Southeast Asia.

Each regional unit is organized along the same lines to include a historical perspective and an examination of the physical basis for development; culture; economic structure; and present patterns, trends, and prospects of development. Emphasis, however, varies from region to region. The high standard of living in Anglo-America is viewed in the light of a varied and bountiful physical resource base. Poverty pockets and cultural conflicts are also indicated. Western Europe has a greater historical stress to explain the region's multiplicity of nation-states and present advanced technologic attainment. The Eastern European and the Soviet Union chapters contain an analysis of Communist development theory and appraisal of the Soviet Union's drive for rapid economic growth. Japan's status in the rich world is unique, and we must view Japanese growth, in spite of a poor natural endowment, from the perspective of a blend of Western and local culture traits.

In the poor world, Latin America's status is viewed from the perspective of cultural pluralism, societal attitudes to the resource base, and rapid population growth. The African and Middle Eastern region, truly diverse, has a recent history of colonialism, and the many newly independent nations struggle for self-identity that is expressed in many ways. The Arab-Israeli conflict and the oil-rich states present another aspect of the development process. Finally, in Monsoon Asia (South Asia, China and its neighbors, and Southeast Asia), emphasis is on the origin of different cultures and the relationship of economic organization to other aspects of culture. Attention is paid especially to the roles of the Monsoon wind system and religion in South Asia and the contrasts between traditional and Communist China.

Several aids are presented to assist the reader. A definition or explanation of many terms is given in the Glossary. An Appendix provides a ready reference of selected national data useful in comparing nations and regions. At the end of each chapter is a short annotated list of readings for those interested in exploring a bit further a particular idea or who wish more background information. Maps, graphs, tables, and photographs illustrate many of the ideas and points under discussion and are keyed to the textual material. Finally, in this age of interdependence and standardization, the move toward the metric system is recognized by presenting material in both the English and metric systems of measurement. Conversion factors for some of the more widely used measures are

- acres to hectares, multiply acres by .4047
- centimeters to inches, multiply centimeters by .3937
- cubic feet to cubic meters, multiply cubic feet by .0283
- cubic meters to cubic feet, multiply cubic meters by 35.3145
- feet to meters, multiply feet by .3048
- gallons (U.S.) to liters, multiply gallons by 3.7853
- hectares to acres, multiply hectares by 2.4710
- inches to centimeters, multiply inches by 2.54
- kilograms to pounds, multiply kilograms by 2.2046
- kilometers to miles, multiply kilometers by .6214
- liters to gallons, multiply liters by .2642
- meters to feet, multiply meters by 3.2808
- miles to kilometers, multiply miles by 1.6093
- square kilometers to square miles, multiply square kilometers by .3861
- square miles to square kilometers, multiply square miles by 2.59

The book is geared primarily for undergraduate college students and is organized for a variety of curriculum and teaching approaches. The development theme provides a means to divide the world into two or more parts for treatment in a course sequence. A two-course sequence, for example, could emphasize rich nations in one course and poor in another. Treatment by continental or major regional divisions is also feasible. Furthermore, the large number of chapters allows for differing course structures. For a five-hour semester course, all chapters can be used without overly long reading assignments. For a five-hour quarter course or a three-hour semester course, chapter selection can stress specific teaching strategies. If present conditions and trends are to be emphasized, Chapters 1, 4, 7, 8, 11, 12, 13, 16, 18, 19, 21, 22, 23, 30, 31, 34, 35, and 37 are appropriate. On the other hand, a blend of chapters can be assembled to provide a more varied coverage. One example, after the introductory chapters, is to examine Anglo-America (5–8), then the following: Europe (11–13); the Soviet Union (14 and 16); Japan (17); Australia–New Zealand (19); Latin America (20 and 21); Africa and the Middle East (24 and 27); and Monsoon Asia (28–30, 32, 34, and 36). Other chapter groupings are equally feasible which stress the historical perspective, the relationship of the natural environment to the development process, or a case-study approach.

Finally, acknowledgments are in order. The authors thank the students who, by their questions and interest, proved the approach useful and relevant to the understanding of many of the world's problems and prospects. Special thanks are given to those whose questioning demanded fuller explanation and required the authors to explain their ideas more clearly. We also appreciate the valuable criticisms and suggestions offered by many fellow geographers who read various portions of the manuscript. Finally, a number of individuals and organizations, both domestic and foreign, have graciously provided photographs and information vital to the completion of our task.

D. R. H.

Contents

Part Five   Japan and Australia–New Zealand

Jack F. Williams

Part Six   Latin America

Don R. Hoy

Part Seven   Africa and the Middle East

Douglas L. Johnson and Leonard Berry

Part Eight   Monsoon Asia

Clifton W. Pannell

List of Maps and Charts

List of Tables

Some Basic Concepts and Ideas

James S. Fisher
Don R. Hoy

PART ONE

People and Resources

1

A FEW YEARS ago Karl Sax, an authority on population, stated there are more people alive today than have ever died. Sax made two observations: first, for most of mankind's existence, the world's population has been relatively small; and second, there has been lately a rapid increase in population. The dramatic increase in population numbers within the last 100 years has been so great that the phrase "population explosion" has been applied. An increase in population, if rapid and sustained, places a strain on the productive capacity of a society to fulfill the material needs and aspirations of its members. To counterbalance the increased demand, society's productive capacity must be expanded by development of new resources and accelerated production of present resources. Even if production capacity matches population growth, diverse heretofore minor problems like pollution, interpersonal and group rivalries, and occasional scarcity of goods may be magnified to major and serious proportions. New situations may challenge basic and traditional societal values and ways of life. Society itself may collapse or be drastically altered.

People

Throughout most of human history, population growth has been negligible (Figs. 1–1a and 1–1b). At the dawn of the Agricultural Revolution some 7,000 to 10,000 years ago, the world's population probably numbered some 5 million. Plant cultivation and animal domestication, however, heralded a long and sustained period of population growth. Growth, at first confined to the areas of agricultural innovation, later spread worldwide with the diffusion of agriculture (Fig. 1–2). In a few places, such as Australia, the diffusion process was delayed until the coming of European colonists. Today, only in some of the harsher environments, polar zones, and remote dry lands, do a small and dwindling number of people live by the age-old hunting, fishing, and gathering occupations.

Impact of Agricultural Revolution

The Agricultural Revolution led to many fundamental changes in living patterns. No longer were people dependent solely upon nature, for domestic crops and animals far surpassed their wild cousins in utility. Productive capacity was greatly expanded, and many more people could live close together at a higher level of living. Gone were the days of the isolated extended family: four or five adults with their children, ranging over tens of square miles and having little contact with others. Families lived permanently in one place close to others. Villages grew, and a new social order was created to resolve the increasing conflicts brought about by more people living together.

All in all, life was easier and more secure. For some, substantial houses were built where formerly crude huts or caves served as temporary lodgings. Many tools and other large and small luxuries—a chair, table, and a bed—were acquired which previously were impractical because of the migratory nature of life. Indeed, materialism had its true beginnings with the development of sedentary (settled) agriculture. Possessions could be accumulated and passed on to

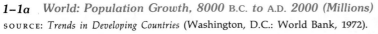

1-1a *World: Population Growth, 8000 B.C. to A.D. 2000 (Millions)*
SOURCE: *Trends in Developing Countries* (Washington, D.C.: World Bank, 1972).

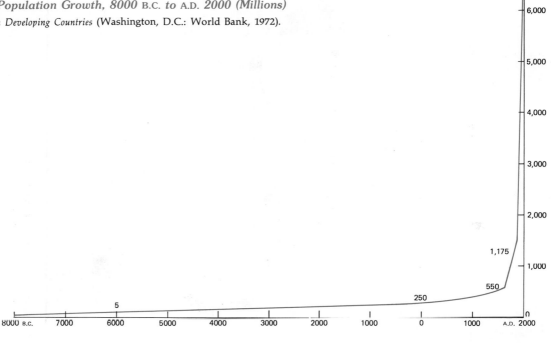

1-1b *World: Population Growth by Area, 1650–2000 (Millions)*
SOURCE: *Trends in Developing Countries* (Washington, D.C.: World Bank, 1972).

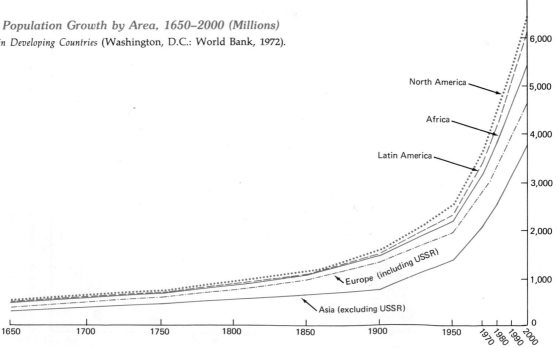

3

1–2 World: The Spread of Agriculture

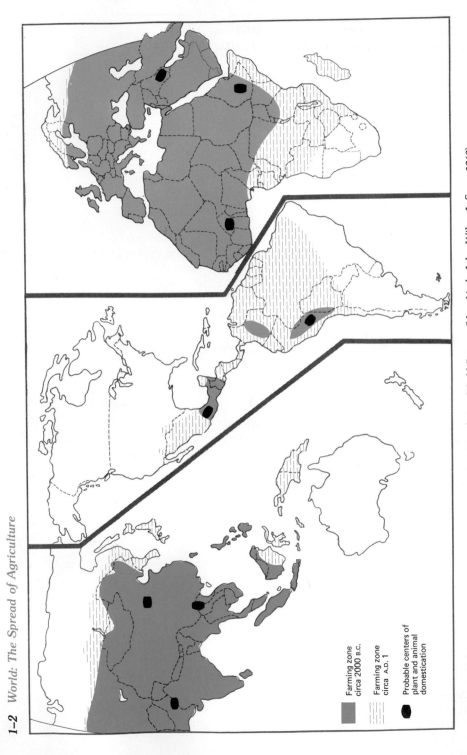

Farming zone
circa 2000 B.C.

Farming zone
circa A.D. 1

Probable centers of
plant and animal
domestication

SOURCE: Adapted from Glenn T. Trewartha, *A Geography of Population: World Patterns* (New York: John Wiley & Sons, 1969).

new generations. A sense of security increased too as production increased. Surpluses could be stored for an emergency. Agriculture, despite weather unpredictability, was more reliable than complete dependence on nature. Protection from hostile groups was also an attribute of village life. Permanent fortifications and many defenders tended to discourage outside attempts to capture and control the agriculturist. Defense was not always successful; for example, in recorded times the hunting and gathering Plains Indians of the United States, after obtaining the horse from the Spanish, were more than a match for the sedentary agricultural Indians along the eastern margin of the Plains. Similarly the Roman legions eventually failed to defend their agricultural tributary regions from the more primitive northern and eastern tribes.

Increased population, production, and village life led to the establishment of numerous organizations. A political system was derived to settle disputes, govern, and lead collective action in warfare and such public works as irrigation, drainage, and roads. Initial political organization was probably informal, but in some areas it gradually evolved into a highly complex system as duties expanded. In a similar manner, the formation of a priestly class helped formalize religion. Religious leaders frequently were the holders of both philosophical and practical knowledge, often playing a role in village life as medical men and weather forecasters. In the Mayan civilization of southern Mexico and Guatemala, for example, the priests developed an agricultural calendar that gave the times for cultivation of crops. This calendar, based on the progression of the sun, planets, and stars, predicted the beginning of wet and dry seasons. Increased production per worker provided more than the family unit needed. A portion of the labor force was freed not only for government and religious activities but also for other nonagricultural activities such as pottery making, metallurgy, and weaving.

A.D. 1 to 1650

By the beginning of the Christian era the Old World's population had grown to about 250 million. Most of these people lived within three great empires: the Greco-Roman (53 to 55 million) around the fringe of the Mediterranean Sea and northward into Europe; the Chin and Han dynasties (70 to 75 million) of China extending into Southeast Asia; and the Mauryan (100 to 140 million) of northern India. In these empires the simpler political, economic, and social organizations of the agricultural village vied with the more complex, integrating structure of the empire and the newly created city. Urbanism in all its facets became a way of life.

Urban life meant specialization of labor. The city dweller became dependent on the farmer for food and fiber. To a large degree the farmer's products were tribute and rent payments, but he did obtain manufactured products and silver or gold in small quantities. The city was the focal point of national life. The riches of the countryside and tribute from afar were concentrated in the city. Like a magnet, it drew people to its core. Arts flourished, and education was available to those who could pay. The best lawyers and doctors practiced there. Government employed increasing numbers to construct, maintain, and supply sewage lines, roads and streets, and drinking water. Fire and police protection was needed. In all, the activities of these empire cities were not much different from those of a modern metropolis.

The empire and the city fostered regional specialization. Each part of the empire, rather than producing all of its own needs, traded what it produced in surplus and received in turn items it did not possess or could only produce with difficulty. To be sure, regional specialization was not fully developed, but the concept was recognized and used. Rome, for example, exported Arretine pottery, wines and oils, metalware, glass, and perfume; gold too was an important export. In return Rome imported wheat from Egypt and North Africa, cattle and hides from Sicily, metals and livestock from Spain, slaves and fur from Germany, and even rare Indian spices and gems.

In contrast to the Old World, the New World's population probably numbered only about 10 million at the beginning of the Christian era. Originally centered in Mesoamerica (Central Mexico to Honduras) and in the Peruvian-Bolivian Andes, the Agricultural Revolution had not spread widely, perhaps because of a later start and more recent appearance of mankind. In any case the northern limit of agriculture approximated the United States-Mexican border. In South America, eastern and southern Brazil, Uruguay, most of Paraguay, Argentina, and Chile were outside the agricultural sphere. In Mesoamerica, the Aztec and Mayan civilizations were still in the formative process, as was the Inca empire in Peru. Urbanization was just beginning.

By the end of the first 1,650 years of the Christian era, the world's population had grown to over 500 million. Most of this growth was in the preexisting centers and the gradual expansion of the centers' populations in formerly sparsely settled areas. Productive capacity expanded with improved technology and new resources. Urbanization became more pronounced, although agriculture remained the base of livelihood.

Growth was by no means uniform in time. Diseases of epidemic proportion decimated many areas. Bubonic plague destroyed millions in Europe and in northern Africa. In the New World the introduction of smallpox, measles, and other Eurasian-African diseases may have resulted in a mortality rate of 50 to 90 percent of the Indian population. Some authorities estimate the New World's population in A.D. 1500 at 100 million; by 1650, the population may have dropped to only 15 million to 25 million.

Regionally, population growth also varied. Northern and Eastern Europe experienced considerable growth. The area around the Mediterranean Sea and into the Middle East grew more slowly, and in some places actually declined. South and East Asia probably contained nearly one-half the world's total population with the greatest increases in China and southern Asia.

1650 to Present

From A.D. 1650 to the present, the world's population has increased more and more rapidly. It took 1,650 years for the population to double from 250,000,000 to 500,000,000. By 1850, the population had doubled again and was estimated at 1,175,000,000. Within the succeeding 100 years, the population doubled a third time to 2,400,000,000. By 1985, the population may again double, approaching 5,000,000,000, one billion more people than at present. During this relatively short period a second revolution, the Industrial Revolution, produced an impact on mankind the results of which are still incomplete. The revolution is continuing and not yet applied everywhere.

There are now approximately 4,000,000,000 people

City	1800	1850	1900	1950	1975
New York	.1	.7	3.4	9.6	11.6
London	.8	?	1.8	8.3	7.5
Tokyo	1.0	2.3	4.5	8.4	11.5
Paris	.5	1.1	2.7	6.3	8.2
Moscow	.3	.4	1.0	5.4	7.3
Buenos Aires	.04	.1	.8	5.2	8.4
Shanghai	.3	.4	.9	5.0	10.8
Chicago	—	.03	1.7	5.1	7.0
Calcutta	.6	?	.8	4.6	7.1
Berlin	.2	.4	1.9	3.4	3.3
Mexico City	.1	.3	.3	3.0	8.6
Rio de Janeiro	.04	.3	.8	3.0	7.2
Detroit	—	.02	.3	3.0	4.2
Cairo	.3	.4	.6	2.3	5.0
Washington, D.C.	.003	.04	.3	1.5	2.9

Table 1–1 *Growth of Selected Cities (in millions)*

SOURCES: W. S. Woytinsky and E. S. Woytinsky, *World Population and Production: Trends and Outlook* (New York: Twentieth Century Fund, 1953); *World Population Situation in 1970* (New York: United Nations, 1970); *Demographic Yearbook* (New York: United Nations, 1974); and estimates of national governments.

in the world, and prospects are for a continued rapid increase in the foreseeable future. As we shall see, however, a controversy exists whether the "population explosion" poses a threat to mankind.

Industrial Revolution

The Industrial Revolution is characterized by a countless number of innovations. These range in complexity from the paper clip to interplanetary flight. The age of invention apparently arrived, and each new idea seemed to spawn many others. Muscle power from people and animals was replaced by inanimate power: the steam engine, water turbine, and the internal-combustion engine. In the agricultural sector the use of the tractor and its attachments has made the farmer so productive in some parts of the world that only a fragment of the labor force is needed to supply an abundance of food. In the United States, for example, farmers now account for only 3 to 5 percent of the labor force. Not all of this productivity can be attributed to the tractor; other scientific advances, such as improved higher-yielding seed, application of fertilizer, herbicides and insecticides, have contributed substantially.

Inanimate energy (coal, water, wind, oil) greatly facilitated the growth of cities. More raw materials from agriculture, mining, and forestry combined with new energy sources spurred industrialization. Especially since the eighteenth century, energy and manufacturing innovations have made the factory worker many times more productive. Manufactured products have become cheaper and more readily available. Craftsmen and small guilds have gradually given way to the modern factory where the worker is primarily a machine tender. Other factors also encouraged city growth. Cities from small to large became service, financial, educational, governmental, wholesale, and retail centers. From 1800 to 1975 many cities grew several times over (Table 1-1).

As the Industrial Revolution continues and its effect spreads, urbanization is expected to increase. The revolution, which began primarily in Western Europe, moved quickly to Anglo-America and other areas where European colonists settled. It moved more slowly into Eastern Europe, the Soviet Union, southern Europe, and Japan. Since the end of World War II, industrialization has become a major force nearly everywhere.

Distribution and Density

A map of the distribution and density (number of people per unit area) of the world's population shows a strong tie with the past (Fig. 1-3). Three principal centers of dense population are readily apparent: Indian subcontinent, eastern China and adjacent areas, and Europe. China and India represent old areas of large populations based on an early start in the Agricultural Revolution and in empire building. In these areas agriculture and village life are important facets of society. At the same time, however, the modern city with its service-manufacturing functions is also present. There are numerous cities of over 1 million inhabitants, a few of which are shown in Table 1-1. Population density in India and China varies considerably, usually in association with the relative productivity of the land. On the coastal and river plains of rich alluvial soils and abundant water, rural densities of 3,000 to 4,000 per square mile (1,158 to 1,544 per square kilometer) are not uncommon. Away from the well-watered lowlands, densities diminish but still may be in the range of 250 to 750 people per square mile (97 to 290 per square kilometer). At least one-half of the world's population lives in southern and eastern Asia, the same proportion of world population as at the beginning of the Christian era. Much of Asia's growth has come in the last century.

High population density in Europe is traced from technologic developments in the Middle East which were adopted by the Greeks and Romans and further expanded by the Industrial Revolution. The increase of the European population in the north and west is readily associated with the shift in the center of the technological revolution and political strength. Europe's population density is high but significantly less than in the Indian and Chinese areas. Moreover, the agricultural village and agriculture itself are overshadowed by the modern metropolis and manufacturing.

Secondary centers of high population density are more numerous. The northeast quarter of the United States and adjacent Canada are considered by some as a principal cluster, but total population numbers and density are lower than Europe's. In other aspects as, for example, rate of urbanization and employment, it closely resembles the European pattern. In Africa, along the Guinea Coast, the Nile

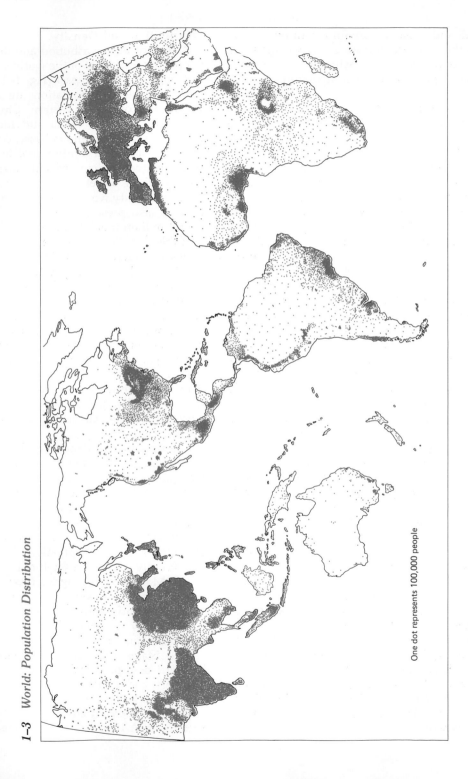

1–3 *World: Population Distribution*

One dot represents 100,000 people

River, and in the eastern highlands, high densities occur, but the total number of people involved in each cluster is relatively small. Similarly, around major urban centers of Latin America, and in the old Aztec, Mayan, and Inca realms, small but locally high-population densities are common. In Asia, Java, the Malaya Peninsula, and in parts of the Middle East, other pockets of high density are found.

Most of the rest of the world's land surface, 80 percent, is more sparsely inhabited. Many of these areas have serious environment problems—cold, dryness, and rugged terrain—which have not been compensated for by mankind. Other areas of sparse population such as some of the humid tropics of South America and Africa are not so easy to explain, especially since similar environments in Asia are densely settled. It is tempting to relate population and density to the broad physical patterns described in Chapter 2, and indeed some have done so. Yet a population-physical environment correlation is an oversimplicity. As we have already seen, technology and political organization are also cultural factors to be considered. Additionally, other aspects of culture, such as desired family size and economic organization, play an important role in influencing population numbers.

Demographic Transformation

Overall, the world's population is increasing at a rate of nearly 2 percent yearly, but this growth is by no means uniform. One explanation of the varied growth pattern is the theory of demographic transformation based on four population stages (Fig. 1–4). Stage one postulates an agrarian society where birth and death rates are high, creating a stable or very slowly growing population. Population density is low to moderate. Productivity is limited, and children, even at a young age, can contribute to the family's wealth. Large families are an asset, particularly since life expectancy is low and family security is dependent upon its members. Employment opportunities aside from agriculturally related ones are few, and technological knowledge is stagnant or nearly so.

In stage two the cultural custom of large families persists, and the birth rate remains high. The death rate, however, drops dramatically because of better sanitation, medical treatment, and greater productivity. Productivity may be increased in the agricultural sector, but more important is the advent of alternative economic activity resulting from industrialization. With industry come urbanization and labor specialization. The principal aspect of stage two is rapid population increase.

In stage three the processes of industrialization and urbanization continue so that society is primarily urban. Large families are no longer desirable. Children in an urban setting are generally an economic liability rather than an asset. The birth rate begins to decline, but the death rate also continues to decline. The population growth rate is reduced, but still the population grows.

Stage four finds the population growth rate stable

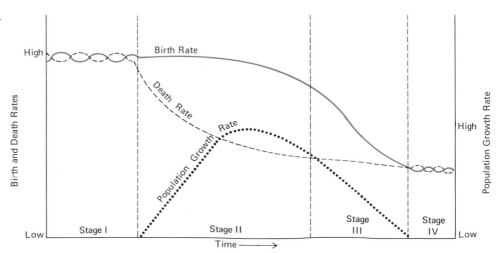

1–4 The Theory of Demographic Transformation

1–5 *World: Annual Rate of Population Growth*

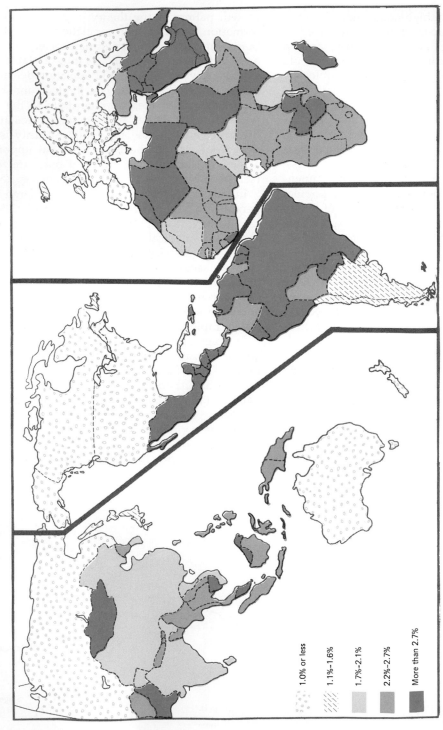

1.0% or less

1.1%–1.6%

1.7%–2.1%

2.2%–2.7%

More than 2.7%

SOURCE: *Population Data Sheet* (Washington, D.C.: Population Reference Bureau, 1977).

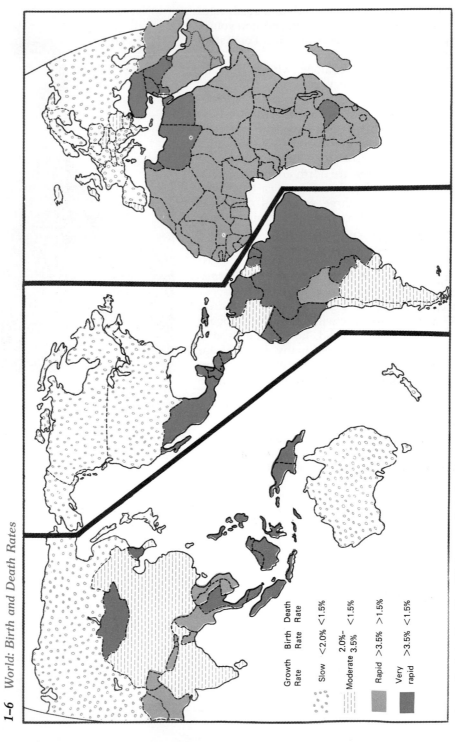

1–6 *World: Birth and Death Rates*

Growth Rate	Birth Rate	Death Rate
Slow	<2.0%	<1.5%
Moderate	2.0%– 3.5%	<1.5%
Rapid	>3.5%	>1.5%
Very rapid	>3.5%	<1.5%

SOURCE: *Population Data Sheet* (Washington, D.C.: Population Reference Bureau, 1977).

or increasing only very slowly. Birth and death rates are low. The population is now urbanized, and voluntary birth control is in general practice. Population density may be quite high.

If birth and death rates along with population growth are plotted for the world, no examples of stage one of demographic transformation are observed (Figs. 1–5 and 1–6). Africa, the Middle East, South Asia, and Southeast Asia have high birth and death rates, but births exceed deaths by a substantial margin causing significant population growth. Examples of stage two areas are north and central Latin America, and parts of Southeast Asia, East Asia and the Middle East. Stage three examples consist of southern Latin America, part of South Asia, and part of East Asia. Anglo-America, Europe, the Soviet Union, and Japan are characterized by stage four conditions.

We must remember that demographic transformation is but a theory based on analysis of the European experience. It cannot be assumed that other areas with a different culture base will follow the European example. If the theory is valid, however, a drop in population growth throughout many parts of the world will occur as stages three and four are reached. If and when population growth will be reduced or stabilized is, of course, unknown.

Malthusian Theory

Another theory that has received widespread attention and has numerous advocates is the Malthusian Theory, or as some state, the Malthusian Doctrine.

Thomas Malthus (1766–1834), an English clergyman, wrote his famous Essay on the Principles of Population in 1798 in which he wrote that population growth would outpace the means of supporting the population unless cultural constraints could check growth in some way. (The Bettmann Archive)

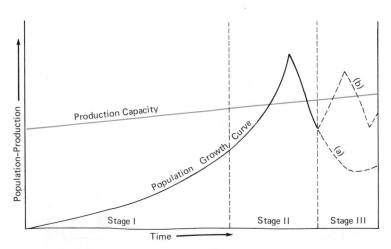

1–7 *Malthusian Theory*

Thomas Malthus, an Englishman, first presented his theory in an essay in 1798. The premises upon which this theory is based are (1) man tends to reproduce prolifically, that is, geometrically: 2, 4, 8, 16, 32; and (2) the capacity to produce food and fiber expands more slowly, that is, arithmetically: 2, 3, 4, 5, 6. If population growth is not checked by society, nature will reduce surplus populations by war, disease, famine, and vice.

If we plot Malthus' idea diagrammatically, three stages of the population-production relation are apparent (Fig. 1–7). In stage one human needs are less than productive capacity. By stage two productive capacity and mankind's needs are about equal. In stage three population has grown to the point where these needs can no longer be met. When stage three occurs, wide fluctuations in population numbers result from kill-off of the population. Once stage three is attained, we cannot be sure what follows. One idea is that stage three perpetuates itself with alternative periods of growth and die-off. Another idea is that the die-off in stage three is so great that stage one is reproduced.

Marxist Philosophy of Population

Karl Marx viewed the population question differently. Marxist philosophy is based partly on the premise that population growth is desirable in a socialist state. According to Marx, economic and social benefits can only be enhanced by an increase in the labor force. Overpopulation and population pressure are blamed on imperfections of the capitalistic economic system, and not on population numbers. Communist nations today still espouse Marx's view from an ideological standpoint, but many from a pragmatic point of view actively encourage family planning measures. In the 1974 United Nations Bucharest World Population Conference, the difference in point of view between Western nations and the Communist Bloc was brought into sharp focus. The Western delegates emphasized the need to limit population growth so that the world's resources would not be overly taxed. The Communist delegates, however, countered that emphasis should be on improvement of the resource base rather than on limiting population growth.

Marxist theory that increasing population can lead to greater prosperity is directly in conflict with Malthusian thought. Malthus painted a dark future

Karl Marx (1818–1883), the German founder of modern communistic and socialistic doctrine, believed that wealth is created by increasing the labor force. Marxist philosophy is based partly on the idea that population growth is desirable.

for mankind, unless population controls are instituted. Marxists believe population growth can lead to great productivity if resources are adequately developed and distributed. Both theories suggest that the population-resource (productive capacity) equation is central to the development process. The idea of the demographic transformation likewise relates population growth to resources, but resources as measured by economic activity.

The Resource Concept

Some Definitions

The resource concept contains three interacting components: resources, obstacles, and inert elements.

Resources are anything that can be used to satisfy a need or desire. They are a means to an end. If a person needs to walk from A to B, his legs are a resource. Obstacles are anything that inhibits the attainment of a need or desire; they are the opposite of resources. If a river or hill must be crossed when walking from A to B, the extra effort inhibits the attainment of the goal (B). These obstacles may be minor ones, since one can wade the river and climb the low hills. If, however, a car were driven, the river and hills would be formidable barriers. Inert elements in our surroundings neither help nor hinder the attainment of needs or desires. In the example, neither the hiker's doctorate in astrophysics nor the arable land (land that can be cultivated) across which he passes helps or hinders his walking ability.

Let us take another case, that of a farmer. The farmer uses several resources: seeds, tools, fertilizer, pesticides, soil, water, sunlight, warm weather, and his skills and knowledge. All of these and other resources are needed to produce a crop. Similarly, minerals located in the hills may be the basic resource of a thriving mining industry.

These simple examples illustrate several points. First, resources are not only material objects. The farmer's knowledge of when and how to plant, cultivate, and harvest crops is absolutely necessary. The mining of a mineral deposit requires skill and organization. Other examples of nonmaterial resources include such diverse things as inventiveness, good government, useful education, cooperation, and adequate social order.

Second, material resources are not just natural resources. The farmer's and miner's tools are culturally derived. Even the farmer's seed and plants are culturally modified. In fact, most material items in the surroundings of an urban dweller are man-made; for example, books, chairs, cars, bridges, and buildings.

Third, as you may have already noted, elements in the environment at the same time may be resources, obstacles, and inert elements. The river is a barrier to the walker, inert to the miner, and a resource to the farmer from which he obtains water to irrigate his fields. The hills are inert to the farmer, a resource to the miner because they contain useful minerals, and an obstacle to the hiker. Finally, resources are not static or finite. The farmer's tools were improved with the development of draft animals

and later the tractor. His seed has been improved by selective breeding. The miner's pick may be discarded for power-driven equipment. Resources are created.

Resources and Culture

Just as the hiker, miner, and farmer use different resources, so diverse cultures have different resources. Each culture group has developed a set of customs, mores, laws, and organizations that effectively structure the lives and attitudes of its members. These cultural controls are evident in how resources are viewed. Those few groups who still live by hunting, gathering, and fishing view their surroundings from the standpoint of nature's productivity of useful plants and animals. They are not concerned about soil quality, growing season, and precipitation amount and distribution, even though these factors bear on natural production. These hunters and gatherers probably do not recognize differences in clay deposits, some of which other groups use as raw materials for brick and pottery manufacture. Hunters and gatherers do view as resources obsidian and other rock deposits from which they can fashion various cutting and hunting weapons; other cultures that have metal instruments would consider these deposits of little significance. Similarly, various plants used for making mats and containers are also resources, but to those of "Western" culture these same plants may be weeds.

Perhaps more important is the role that culture plays in directing economic activity. Resources are basically an economic idea. Some cultures, such as the Bedouin of the Middle East and several Asian cultures, are less materialistic than ours. To them the accumulation of wealth is not a prime goal, and consequently economic organization is structured to provide little more than basic needs. In other cultures, social, political, and economic organization is such that large portions of the population have little or no opportunity to develop resources. Cultural controls inhibit the application of technology and limit individual opportunity. Companies (and formerly nations) that do employ advanced levels of technology are sometimes permitted to exploit these heretofore unused resources primarily for marketing in other parts of the world. Numerous examples readily come to mind, such as petroleum exploitation in the Middle East and Venezuela; copper and tin

Multinational corporations are often permitted to exploit resources in economically poor countries. These companies, possessing capital and technology not locally available, can expand the host nation's store of products destined for export. Petroleum exploration and drilling are a costly activity, and many nations have encouraged multinational companies to develop their oil reserves. (Socony Mobil Photo Library)

mining in Zaire and Malaysia, respectively; and rubber and banana plantations in Indonesia and Central America. The multinational company is to a large degree the product of cultural differences.

In nearly all cultures certain economic activities are socially more acceptable than others. These attitudes tend to direct individuals to use different sets of resources. The true Bedouin of the Arabian Desert is a nomadic herder because in his culture that is the highest occupation one can hold. That large quantities of petroleum underlie the land is of minor import. Neither is he overawed by the possiblity of becoming an oasis agriculturist, even though water and good soil may be available, and the ways of a farmer make greater security of life possible. Among the Indians in the highlands of southern Mexico and Guatemala, the most desirous occupation is as a subsistence maize (corn) farmer. Although the farmer may work at other jobs for which more money is earned, it is because of necessity, not desire. In the United States, urban-oriented, white-collar employment is the goal of most Americans. Many students come to college to prepare themselves for these jobs because of parental and peer pressure, a form of culture control. Our population is so large and diverse, however, that a wide range of economic activities occur. Nevertheless, there are some resources used in other societies that are little used in ours. The wetland areas of the United States are essentially population voids. Those of the lower Mississippi are large and could support intensive cultivations similar to that of southern China and Southeast Asia. Neither do we make much use of wild plants and animals. Finally, there are numerous small patches of cultivable land situated in remote areas that are essentially unused. In India, for example, these places would be made productive.

Resource Development

One of the major fallacies surrounding the resource concept is that resources are static or fixed in amount. Nothing is further from the truth. Resources are created by mankind. The Agricultural Revolution and the Industrial Revolution are excellent examples of multiple-resource innovation. A simple stick becomes a resource when used to poke, pry, and shake trees and bushes for food. When someone discovers that the stick can also be used as a planting tool, its function as a resource is increased, and the stick takes on more value. In a like manner, the rich coal deposits of western Pennsylvania were not resources prior to the coming of the Europeans, because the Indians made no use of coal. In more recent times the widespread use of aluminum has been made possible by generation of cheap electrical power, radioactive elements by their discovery and manufacture, and the use of the southern pine trees for paper and pulp by the development of the Kraft process.

It is not difficult to see that nonmaterial resources can be constantly created. New ideas and better organization have no limit. We can also see that flow resources—material resources that can be replenished, such as trees, crops, animals, and rivers—are used repeatedly and improved upon. What may be more difficult to see is that fund resources, often called nonrenewable resources, are also created. Examples of fund resources are the soil and nonreusable minerals such as coal and petroleum.

Technological improvement can result in the creation of fund resources in two ways. First, a need may be found for a formerly unused mineral or element. Until recently uranium had no use but with the evolution of atomic power has become an important resource. Many rare earths were inert until high-temperature technology created a need for them. It was only in the early part of the twentieth century that techniques for extracting nitrogen from the atmosphere were perfected, giving rise to an important segment of the fertilizer industry. Going back in time, the early phases of the Industrial Revolution led to the use of aluminum, numerous ferroalloys, and natural gas.

Second, perhaps most important in the last 100 years has been the addition to the fund-resource base of low-quality mineral deposits. Prior to the advent of cheap mass transport and processing equipment, only the richest mineral deposits could be mined economically. These deposits represent only a small portion of the total amount of any mineral in the earth's crust (Fig. 1–8). Minerals found in low-quality deposits were part of the inert mass of the crust. Once technology became available to mine these deposits, they became resources. There are numerous examples of low-quality minerals now being mined. Most of the world's copper is produced from ores containing less than 3 percent copper. The rich iron ores of the famous Mesabi Range in northeastern Minnesota have been largely worked out, but taconite, a low-quality ore previously thought worthless, has become the basis of continued exploitation. Other examples of expanded mineral-fund resources include vast quantities of petroleum that could not be obtained without numerous advances in drilling technology, deep underground coal mining with improved methods of ventilation and power equipment to bring the coal to the surface, and the development of processes that remove or compensate for harmful impurities. In the latter case, coal high in sulfur, arsenic, or phosphorus was formerly unusable or of limited value.

Technology has also expanded the agricultural

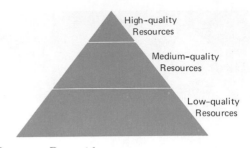

1–8 *Resource Pyramid*

The pyramid concept applies to all types of resources: land, minerals, fish, people, crops.

resource base in several ways, only one of which is examined here. The amount of arable land (land that can be cultivated profitably) has been increased through the development of improved seed and mechanization of cropping. By seed selection and crossing (hybridization), new varieties of crops have been developed which are more drought tolerant or can mature in a shorter period of time than formerly. Growing corn in the United States is now possible farther west in the Great Plains and farther north than was feasible fifty years ago. Wheat is now grown in areas of but 15 inches (38 centimeters) of annual rainfall. Mechanization of agriculture permits the farmer to cultivate what was formerly considered nonarable land. By using power-driven equipment, the farmer can cultivate several hundred acres; and although each acre may yield a small return, the total area farmed provides an adequate income. In the United States and Canada, the expansion of arable land by means of mechanization is well developed. In other areas, such as China, similar quality land is little used despite great need.

Resources and Demand

Another way in which resources can be expanded is by increase in demand. If the demand for a particular item grows, creating higher prices, new resources come into being. There are vast quantities of petroleum in oil shales and tar sands that have not been exploited because cheaper oil is available from other sources. If petroleum prices continue to increase, these very large sources of oil will become profitable to extract. There are many other examples in which demand has led to the creation of new resources.

Loss of Resources

Although demand can create resources, a decrease in demand also reduces resources. Steel-cutting instruments have largely replaced the need for stone and obsidian tools. Synthetic rubber has decreased the need for natural rubber. Artificial heating and the fact that we live and work mainly indoors have decreased our need for heavy woolen clothing. Productivity increases in a crop without a significant change in demand may cause marginal producers to withdraw from farming. Similarly, the development of a new product (analine dye for indigo or the cannon and bomb in place of the catapult) or discovery of a new mineral deposit may lead to the disuse of other products or mines.

Resources can be lost through use. Each drop of oil or gasoline burned and each morsel of wheat or corn eaten represent a decrease in resources. Likewise, some resources are lost if not used. A tree allowed to decay represents a lost resource in one sense but may be an aesthetic resource in another. Crops unharvested may help fertilize the soil and feed wild animals but are also a resource lost.

Unfortunately, resources also are destroyed by improper use. Poor farming techniques lead to erosion and soil depletion. Factories and cities may discharge harmful chemicals and sewage into rivers, destroying water life and ruining water quality for those downstream. War destroys not only people—the most important resource—but also buildings, bridges, and other resources.

Technocratic Theory

A number of authorities have observed the great technological advances made by mankind during the ongoing Industrial Revolution and have derived what is often called the technocratic theory. This optimistic theory is directly counter to that formulated by Malthus (compare Figs. 1–7 and 1–9). The technocrats accept the same population-growth curve as the Malthusians but assume an increase in productive capacity (resources) at a rate greater than population growth. To support their contention, the technocrats point to the expansion of the resource base during the past 400 years and the increase in the standard of living of Western cultures. Furthermore, the technocrats believe that technology will continue to expand, supporting still greater populations. The technocrats believe that science may

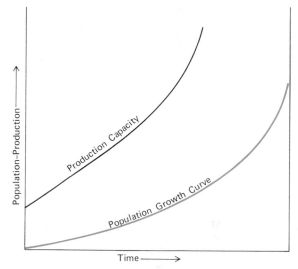

1–9 *Technocratic Theory*

develop to the point that matter can be broken down and reconstituted in any form desired.

The present-day advocates of Malthus (neo-Malthusians) admit that the Industrial Revolution has postponed the day of disaster (Fig. 1–10). They maintain, however, that the expansion of productive capacity cannot continue indefinitely; when population growth surpasses capacity, the prediction of Malthus will be fulfilled. The neo-Malthusians suggest that in

1–10 *Neo-Malthusian Theory*

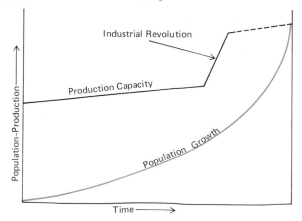

areas such as India, Bangladesh, China, and parts of Africa and Latin America, the prediction of Malthus may soon be upon us. We are left with two points of view, one optimistic and the other pessimistic.

Prospects

Whether a Malthusian or a technocrat, the prospects are for continued population growth for the near future. By the year 2000, 6.5 billion people are expected to be living on the earth. A few daring forecasters estimate a world population of 51 billion by the year 2100; however, long-range predictions must be viewed with great caution. Predictions of resource development are not easy to come by because there is no way to substantiate what resources will be created. The Food and Agriculture Organization of the United Nations believes that technology now exists to feed and clothe the world's population "adequately" for many decades to come. The rub to this statement is that the technology may not be available where most needed. As noted in the next chapters, physical and cultural systems create patterns over the world which directly affect resources, development, and population growth.

Further Readings

ADAMS, ROBERT M., "The Origin of Cities," *Scientific American*, 203 (1960), 153–168. One of a series of articles on mankind in the September issue. Adams examines the role of agriculture in the development of the early cities of Mesopotamia. Other articles in this issue deal with the origin of tools, speech, society, the Agricultural and Scientific Revolutions, and population growth.

BOSERUP, ESTER, *The Conditions of Agricultural Growth: The Economics of Agrarian Change under Population Change* (Chicago: Aldine, 1965), 124 pp. Boserup is an anti-Malthusian who contends that technology produces results from increasing population.

CARR-SAUNDERS, A. M., *World Population: Past Growth and Present Trends* (Oxford: Clarendon Press, 1936), 336 pp. Although some forty years old, this book is a classic examination of demographic characteristics from about 1650 to 1930.

CLARK, W. A. V., and BOYCE, RONALD R., "Brain Power as a Resource for Industry," *Professional Geographer*, 14 (1962), 14–16. This short article is an excellent example of the technocratic point of view.

CUFF, DAVID J., "The Economic Dimension of Demographic Transition," *Journal of Geography*, 72 (1973), 11–16. A brief discussion showing the relationship of demographic transformation stages to economic activity.

Department of Economic Affairs, *Demographic Yearbook* (New York: United Nations, 1948 to present). An annual compendium of useful population statistics including data on total population, population growth rates, urban–rural ratio, births, deaths, life tables, and population movement.

FARMER, RICHARD N., LONG, JOHN D., and STOLNITZ, GEORGE J. (eds.), *World Population: The View Ahead* (Bloomington: Bureau of Business Research, Indiana University, 1967), 310 pp. An authoritative examination of the relationship of population with technology, economic capacity, urbanization, birth and death controls and ethical issues. Also included is a long-range population projection.

HARRIS, DAVID R., "New Light on Plant Domestication and the Origins of Agriculture: A Review," *Geographical Review*, 57 (1967), 90–107. A well-documented review of the literature on early agriculture.

HUNKER, HENRY L. (ed.), *Erich W. Zimmermann's Introduction to World Resources* (New York: Harper & Row, 1964), 220 pp. This short paperback book well illustrates the resource concept.

MEADOWS, DONELLA H., et. al., *The Limits of Growth: A Report for the Club of Rome's Project on the Predicament of Mankind* (New York: Universe Books, 1972), 205 pp. A sobering neo-Malthusian discussion of the race between population growth and technology and pollution.

SPOONER, BRIAN (ed.), *Population Growth: Anthropological Implications* (Cambridge, Mass.: MIT Press, 1972), 425 pp. An analytical review of population growth in various parts of the world as related to agriculture, urbanization, and society in the light of Ester Boserup's theory of agricultural development. This book is for the advanced reader only.

TREWARTHA, GLENN T., *A Geography of Populations: World Patterns* (New York: Wiley, 1969), 186 pp. A basic demographic study with an emphasis on regional patterns. See also Clarke, John I., *Population Geography and the Developing Countries* (New York: Pergamon Press, 1971), 282 pp.

WEINBERG, ALVIN M., "Some Views of the Energy Crisis," *American Scientist*, 61 (1973), 59–60; Cook, C. Sharp, "Energy: Planning for the Future," *American Scientist*, 61 (1973), 61–65; Roberts, Ralph, "Energy Sources and Conversion Techniques," *American Scientist*, 61 (1973), 66–75; and Cheney, Eric S., "U.S. Energy Resources: Limits and Future Outlook," *American Scientist,* 62 (1974), 14–22. These four articles explore the prospects for energy in the immediate and near long-term future. Cheney presents a pessimistic outlook, whereas Weinberg is much more optimistic and challenges some of the projections of the Club of Rome. Cook takes a middle ground and concentrates on conservation, pollution, and efficiency. Roberts presents a future outlook for energy with new power resources.

WOYTINSKY, W. S., and WOYTINSKY, E. S., *World Population and Production: Trends and Outlook* (New York: Twentieth Century Fund, 1953). This 1268-page work contains one of the most extensive collections of demographic and economic statistical data to about 1950, accompanied by a serious discussion of demographic characteristics and economic activity within a historical perspective.

The Physical World as Part of Mankind's Environment

2

cludes a variety of landforms and the loose, overlying material in which soils are formed. Here mankind builds settlements, practices agriculture, gathers timber, minerals, and other useful materials. The great extent of the hydrosphere, water bodies that cover 71 percent of the earth's surface, requires that more intensive economic activities be restricted to the remaining land surface. The hydrosphere is, however, a major component of our environment; it is a source of food, a medium for transportation, and above all, it provides the major moisture source for the atmosphere. The thin gaseous layer that envelops the earth is the atmosphere. Day-to-day variations in the air temperature, humidity, and air movements we call weather. Perhaps more important for our understanding of various areas is climatic character, or atmospheric conditions expressed as averages which occur over months and years. The biosphere includes all living organisms. People have become the most dominant of the biotic forces but included with them are animals and plants.

The Relationship of Man and His Environment

THERE ARE great differences in economic well-being, and, therefore, also biologic well-being. Some parts of the world have high levels of living whereas in other parts there is a constant struggle for enough food. Three principal factors affect the level of living in an area: the physical environment; the political, economic, and social systems of society; and the number of people and their rate of increase. These factors are operating in a framework in which material demands and expectations are rising rapidly, and so also the need for earth resources. In Chapter 1, we briefly examined population; in Chapter 3, we will consider mankind's cultural systems. The focus of this chapter is the physical environment. Not all elements of the physical environment are discussed, but rather some of those of direct importance to economic activity.

The physical environment in which we live consists of several spheres. The lithosphere is the solid portion of our planet. This surface, or crustal area, in-

Professor Sauer has provided us the following:

Every human population, at all times, has needed to evaluate the economic potential of its inhabited area, to organize its life about its natural environment in terms of the skills available to it and the values which it accepted. In the cultural *mise en valeur* (exploitation) of the environment, a deformation of the pristine, or prehuman, landscape has been initiated that has increased with the length of occupation, growth in population, and addition of skills. Where ever men live, they have operated to alter the aspect of the earth, both animate and inanimate, be it to their boon or bane.[1]

Sauer made several important points. Society has an overriding economic concern; survival is based on some form of production. Mankind is continuously concerned with a utilitarian relationship with the environment. The nature of this relationship, however, depends upon the skills a society accumulates

[1]Carl O. Sauer, "The Agency of Man on the Earth," Philip L. Wagner and Marvin W. Mikesell (eds.), *Readings in Cultural Geography* (Chicago: University of Chicago Press, 1962), p. 539.

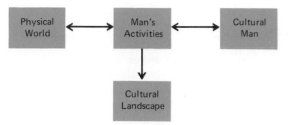

2–1 *Interaction Between Culture and Physical World*
The interaction between cultural man and the physical world, through sets of activities, creates a cultural landscape.

and the value system that motivates it. For example, the way we in the United States make use of a desert environment differs substantially from that of the Bushman of Africa.

Environmental modification begins with its immediate cultural use (Fig. 2–1). Our recent water and air pollution and power problems, as important as they are, sometimes encourage us to think that landscape modification and environmental problems are a recent phenomenon. This attitude is fallacious. Traditional societies do not live and have not lived for thousands of years in a pristine environment, that is, an original unmodified physical environment. The Middle East is recognized as an early center of sedentary agriculture, village life, and higher civilization. The Middle Eastern inhabitants were not without culturally induced environmental problems. Increased salinity of soils, a result of irrigation practices, ultimately rendered many agricultural lands useless and contributed to the decline of some Mesopotamian civilizations.[2] Landscape modification is not new; it is as old as mankind. Most, if not all, landscapes are, then, not "natural" but "cultural," shaped and formed as societies occupy and use the surface of the earth.

Environmental Elements

The numerous components of the physical environment—rocks, soils, landforms, air, vegetation,

[2]Thorkild Jacobsen and Robert M. Adams, "Salt and Silt in Ancient Mesopotamian Agriculture," *Science*, 128 (1958), 1251–1258.

animal life, minerals, and water—are interrelated. Climate is partially responsible for variations in vegetative patterns, part of the soil-forming process, and the shaping of landforms. Organic matter from vegetation affects soil development. The environmental elements are all part of a large ecosystem that often exists in a delicate balance. As important as recognizing the relationship of natural processes is the recognition that mankind interferes with natural processes. Mankind is the most active agent for environmental change on the surface of the earth.

Climate

Climate is of direct importance in our effort to produce food and industrial crops. Agricultural production in great quantities is basic to our civilization. Each of the plants used has specific requirements that favor growth (Table 2–1). Some crops require substantial amounts of moisture (rice); others are more drought tolerant (wheat). Even though wheat yields are generally better with adequate moisture, wheat can be grown in semiarid areas. Coffee requires a year-round growing season, yet does best in the tropics when grown at elevations that provide cooler temperatures. Mankind has modified the character of many plants (cultigens); yet climatic differences ensure great geographic variation in crop cultivation.

The most important climatic elements are precipitation and temperature. Precipitation is moisture removed from the atmosphere and dropped on the surface of the earth. Rainfall is most common, but snow, sleet, and hail are solid forms and are also important. Moisture added to the soil by melting snow can be of value in later seasons. Sleet and hail normally are quite localized in occurrence but can cause damage and significant economic losses.

Precipitation occurs only when a substantial mass of air rises. Ascending air decreases in temperature, and cooler air has a reduced moisture-holding capacity. If enough cooling takes place, air becomes saturated; then condensation occurs, and precipitation may result. There are three principal situations under which the essential lifting of air needed to activitate this process takes place.

Convection is the vertical movement of columns of air. If such air moves high enough, precipitation may be heavy. A great amount of rain may fall in a short time, but the occurrence is often intermittent

Table 2–1   *Physical Requirements for Selected Crops*

	Growing Season (Days) and Temperature Conditions	Annual Precipitation and Distribution (inches)	Other Environmental Conditions
Wheat	90–110, cool to warm	10″–30″, abundant during growth, drier when maturing	rolling to level land, has some drought tolerance
Rice	120+, above 70°	40″ minimum, even on summer maximum	level land facilitates irrigation, poorly drained soils useful, poor drought tolerance
Millets and sorghums	90, 70°	20″+, concentration needed during growth, drier when maturing	rolling to level land, has some drought tolerance
Corn	140+, 75°, warm days and nights	30″–45″, concentration needed during growth, with warm temperatures	rolling to level land, very adaptable, but not drought tolerant
Soybeans	100+, 75°, warm days and nights	30″–45″, concentration during vegetative growth, dry in maturity	rolling to level land, very adaptable, similar to corn, slightly more drought tolerant
Sugar cane	365, warm to hot	40″+, abundant and even for 11 months	rolling to level, photoperiod critical (length of day)
Cotton	180–200, 70°+, warm nights	30″–60″, summer concentration, dry in maturity	rolling to level, requires sunshine and regular moisture

and localized. Runoff is also high; precipitation is not always as meaningful for plant growth under these conditions as the total amount tempts us to believe. Convectional activity is most effective as a source of rainfall in the equatorial portion of the tropics and in the subtropical and middle latitudes during summer. In Figure 2–2, these areas appear as regions of moderate-to-high rainfall.

A front is formed when air masses with quite different temperature and humidity characteristics meet. In a frontal system cooler and drier air remains near the earth's surface as warmer and moister air is forced to slide up and over the colder air. The ascending air is cooled, and precipitation results. Frontal precipitation is most common in middle latitudes, particularly during winter and spring when greatly contrasting air masses are more likely.

The third situation favoring precipitation is when air moves over mountains or hills. This orographic condition may produce large amounts of precipitation, particularly when prevailing wind patterns cause frequent air movement over highlands. A review of Figure 2–2, especially the higher middle-latitude

west coasts of North America and South America, will reveal heavy annual rainfall on the windward side of mountainous zones and meager rainfall on the sheltered or leeward side.

The precipitation map shows the rainy areas of the world to be the tropics and the middle latitudes; in the latter areas, it is particularly heavy on the western sides of continents where exposure to prevailing winds, frontal systems, and the orographic effects aid moisture accmulation. The eastern sides of middle-latitude continents are also rainy, particularly if exposure to major water bodies (moisture sources) are favorable. For example, notice eastern United States and Canada on Figure 2–2. Continental interiors, such as Asia's or North America's, or areas on the leeward side of mountains experience moisture deficiency. Some subtropical areas—northern Africa, Australia, northern Chile and Peru, northwest Mexico, and the southwestern United States—experience meager rainfall. Diverging winds (winds moving in opposite directions) reduce the likelihood of convection, and cold waters offshore accentuate the aridity.

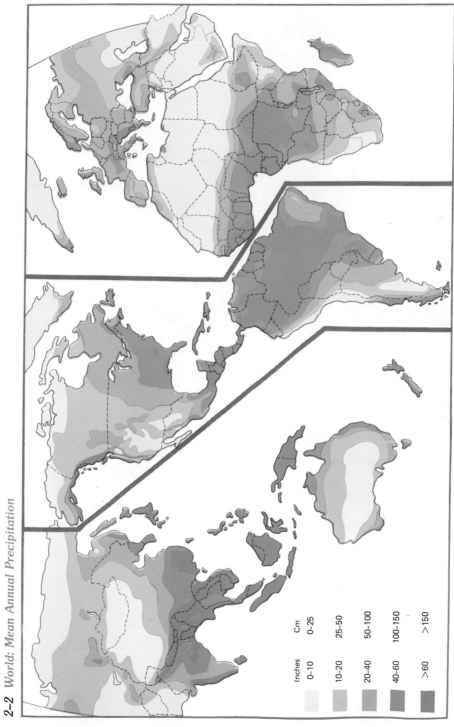

2–2 *World: Mean Annual Precipitation*

Inches	Cm
0–10	0–25
10–20	25–50
20–40	50–100
40–60	100–150
>60	>150

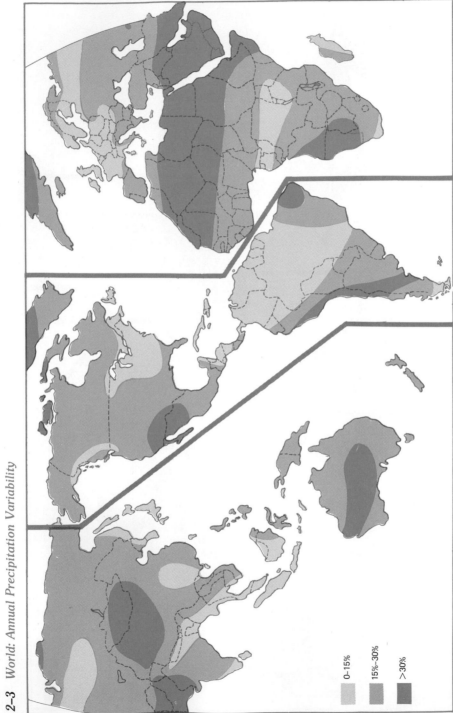

2-3   *World: Annual Precipitation Variability*

0–15%

15%–30%

>30%

The map of average annual precipitation has utility in analyzing the distribution of mankind's activities. Areas of meager rainfall normally are not heavily populated. Exceptions require an exotic water supply (the Nile Valley and southern California). It is also apparent that some areas of very high rainfall have sparse populations. Excessive rainfall and high temperatures contribute to other environmental conditions difficult to overcome, namely infertile soils in tropical regions.

Seasonal rainfall patterns are as important as the yearly amount. Most equatorial areas receive a significant amount of rainfall during all seasons. The adjacent tropical wet and dry regions have a distinct rainy season contrasting with an excessively dry season; the advantage of a year-long growing season is therefore partly negated. The dry summer subtropical climates also experience distinct seasonal variations, except that summers are dry, and winters have the maximum rainfall. In trying to understand land use in many areas of the world, seasonal rainfall patterns are a major consideration.

Variability of precipitation is the actual amount received in a given year expressed as a percentage of annual average. A comparison of Figures 2–2 and 2–3 shows that greatest variability is experienced in areas of minimal rainfall. Unless a special source of water is available, settlement in dry areas is limited. Transitional areas between humid and dry regions (steppes), however, frequently are important settlement zones in which production of staple foods (grains) is significant. Under normal conditions the United States Great Plains, the Black Earth Belt of the Soviet Union, the North China Plain, and the southern edge of the Sahara are significant food-producing regions. Unfortunately, with high variability a recurring phenomenon, each of these and other similar areas have repeatedly experienced human suffering related to drought problems. The most recent example is the Sahel (southern Sahara), where recently thousands have died of starvation. Drought and famine reemphasize the danger of dependence on such areas.

Annual amount, seasonality, and variability of rainfall reveal much about the utility of an environment; however, the actual moisture available also depends upon temperature conditions. High temperatures mean great potential for evaporation and plant transpiration, a high evapotranspiration rate.

An unfavorable relationship between precipitation and evapotranspiration means moisture deficiency for plant growth. Fifteen inches (38 centimeters) of rainfall may be adequate for plant growth in higher middle latitudes but cause moisture deficiency in tropical and subtropical areas.

Temperature and Plant Growth

Perhaps the most important aspect of temperature is the length of the frost-free period, which reflects the influence of latitude on climate (Fig. 2–4). The modifying influence of water on temperature is evident at maritime locations; growing seasons are longer than would normally be expected at those latitudes. Comparing northwest Europe with similar latitudes in the Soviet Union and in eastern North America illustrates this point.

Even winter temperatures are important to plant growth. Many middle-latitude fruit trees require a specific dormancy period with temperatures below a certain level. Cold temperatures (chill hours) are necessary to assure vernalization, or activation of a new flowering cycle.

Photoperiod (the length of day or active period of photosynthesis) and daily temperature during the growing season also affect plant growth. Some plants require a long daily period of photosynthesis to flower (barley), yet others such as soybeans or rice are favored by shorter day length. Each plant variety has a specific daily low and high temperature within which plant growth proceeds, but outside of which it ceases. These temperatures, and the temperature at which growth proceeds most rapidly, are called cardinal temperatures. Cardinal temperatures for most plants are known. The numerous varieties of wheat, for example, have cardinal temperatures that range for the low between 32° and 41°F (0° and 5°C), and for the maximum between 87° and 98°F. (31° and 37°C.). Actually optimum temperatures for the growth of most varieties of wheat are between 77° and 88°F. (25° and 31°C.). Although less than optimum requirements do not exclude the cultivation of a particular crop, departures from the optimum reduce the efficiency with which a crop is produced. Efficiency, translated into cost per unit of production, is a major concern in commercial agriculture.

Climatic Classification

Figure 2–5 presents the global distribution of climates. Climatic classification is based upon tem-

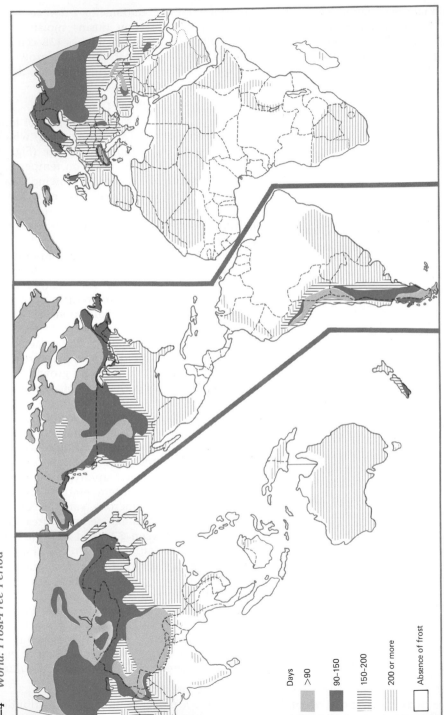

2-4 *World: Frost-Free Period*

Days
>90
90–150
150–200
200 or more
Absence of frost

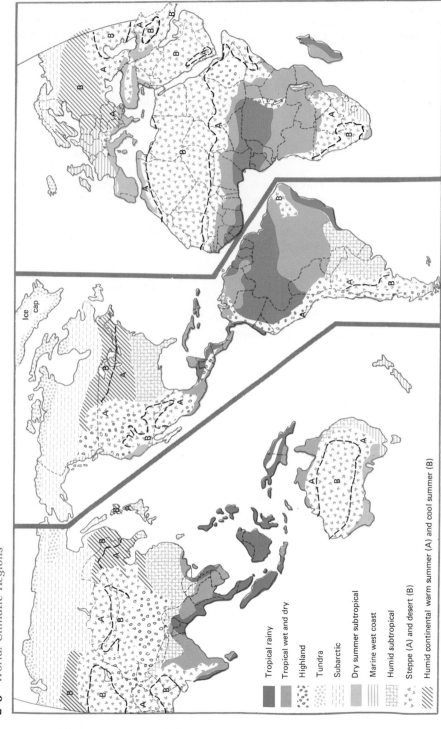

2-5 *World: Climatic Regions*

Tropical rainy

Tropical wet and dry

Highland

Tundra

Subarctic

Dry summer subtropical

Marine west coast

Humid subtropical

Steppe (A) and desert (B)

Humid continental warm summer (A) and cool summer (B)

perature and precipitation conditions (Table 2–2). Although conditions may vary substantially within a climatic region, a generalized classification system is useful to compare other environmental distributions and for area-to-area comparisons. A review of the climatic map reveals a close relationship between lat-

Table 2–2 *Characteristics of World Climate Types*

Climate	Location (Latitude and continental position if distinctive)	Temperature	Precipitation
Tropical rainy	Equatorial	warm, range* less than 5°; no winter	>60″, no distinct dry season
Tropical wet and dry	5°–20°	warm, range of 5°–15°, no winter	25″–60″, summer rainy low sun period dry
Steppe	Subtropics and middle latitudes sheltered and interior positions	hot and cold seasons, dependent on latitude	normally less than 20″ and much less in middle latitudes
Desert	Subtropics and middle latitudes sheltered and interior positions	hot and cold seasons, dependent on latitude	normally less than 10″ and much less in middle latitudes
Dry summer subtropical	30°–40°, western and subtropical portions of continents	warm to hot summers, mild but distinct winters	20″–30″, dry summer, maximum precipitation in winter (normally 10 percent)
Humid subtropical	20°–40°, eastern and southeastern subtropical portion of continents	hot summers; mild but distinct winters	30″–65″, rainy throughout the year occasional winter dry (Asia)
Marine west coast	40°–60°, west coast of middle latitude continent	mild summers and mild winters	moisture throughout the year, some winter maximum tendency, total amount highly variable—20″–100″
Humid continental (long summer)	35°–45°, continental interior and east coast, northern hemisphere only	warm to hot summers, cold winters	20″–45″, summer concentration, no distinct dry season
Humid continental (short summer)	35°–45°, continental interior and east coast, northern hemisphere only	short, mild summers, severe winters	20″–45″, summer concentration, no distinct dry season
Subarctic	46°–70°, northern hemisphere only	short, mild summers, long, severe winters	20″–45″, summer concentration, no distinct dry season
Tundra	60° and poleward	frost anytime, short growing season for limited vegetation	limited moisture (5″–10″), except at exposed marine locations
Ice cap	Polar areas	constant winter	limited precipitation, but surface accumulation
Undifferentiated highland (See Figure 2–5.)			

*range: difference between temperature of warmest and coldest months.

itude, continental position, and climates. The global climatic pattern, therefore, is not difficult to comprehend. Compare, for example, the locational similarity of humid subtropical climates in the United States, in China, and in Argentina.

A fundamental distinction in the classification is that between humid and dry climates. Areas are classified as desert or steppe if the potential for evaporation exceeds actual precipitation. A single specific precipitation limit cannot be used to separate the dry and humid climates; in the mid-latitudes, however, deserts normally have less than 10 inches (25 centimeters) and steppes or semiarid regions less than 20 inches (51 centimeters) of precipitation. Higher latitudes have lower limits, and lower latitudes a higher limit before the dry classification is applied. Twenty-five inches (64 centimeters) of precipitation in higher middle latitudes may provide a humid climate and forest growth, but the same amount in tropical areas may result in a semiarid and treeless environment.

The highland climates shown on Figure 2–5 are a group with variable temperature and rainfall conditions. Along with the other controls of climate—such as latitude, marine exposure, prevailing winds, and atmospheric pressure systems—elevation is a major factor. Precisely what conditions may be expected depends upon elevation and position within mountains. Many mountainous areas are sparsely settled and have limited concern for us; however, in Latin America, East Africa, and Indonesia, highland settlement is important. The variations of highland climate, and therefore vegetation and soils, are treated in the chapters dealing with those regions.

Vegetation

Vegetation patterns exhibit a close association with other environmental phenomena, particularly climate. This relationship is readily evident in the major distinction between grasslands (herbaceous plants) and forests (woody plants). Grassland vegetation on Figure 2–6 is associated with areas of moisture deficiency (steppes and deserts) or with areas of extreme seasonal variation in precipitation, but mankind has extended grasslands in numerous areas with the use of fire and extensive grazing, and so they are found associated with quite varied environmental realms. Primitive societies commonly use fire as a tool to clear land for agriculture or to improve grass cover for domesticated or wild animals. The very cold areas

(tundra) are also treeless but because of a short or nonexistent growing season. Forest growth usually requires a minimum rainfall between 15 and 20 inches (38 and 76 centimeters), depending on evapotranspiration rates, and is associated with humid climates.

The phrase "natural vegetation" should be used carefully. Climatic climax vegetation, as described and classified in Table 2–3, is that expected in an area if vegetation succession is allowed to proceed over a long-time period without human interference. Landscape modification has proceeded for so long that few areas of natural vegetation remain. Most vegetative cover is really part of a cultural landscape, and reflects human activity.

Mankind has cut down wild forms, introduced new plants considered more useful, and inadvertently added undesirable nuisance plants. Examples in the United States, in respective order, are (1) vast forests removed for agriculture, (2) African grasses introduced for use as forage crops, and (3) kudzu, a vine of Japanese origin formerly thought to have utility as forage and as a preventive for erosion in the southeastern United States. Vegetation has also been used as an indicator of climatic or soil conditions, though not always accurately. Nevertheless, the perceptions of environmental utility developed from previous experience and observation have led to clearing some areas and avoiding others.

Vegetation first provided mankind a direct source of food, and a food supply indirectly through its support of animal life. Wild plants were certainly the most significant biotic element with which humans dealt during the earlier millennia. Grass-covered plains facilitated movement and interaction with others. Dense forests functioned as barriers, isolating culture groups, and provided refuge for those trying to remain separate. For example, the Sahara was not the major barrier inhibiting Subsaharan African cultural exchange with the Arab world. In fact interaction did occur. The grassland zone south of the Sahara is a cultural transition zone with features of both African and Arab cultures. The equatorial rainforest was a more effective barrier, not only as a forest that made travel difficult, but also because of other environmental elements. The tsetse fly, for example, inhibited the adaptation of an entire life style—cattle culture.

Vast forested areas in middle-latitude Asia, North America, and Europe have been removed over the

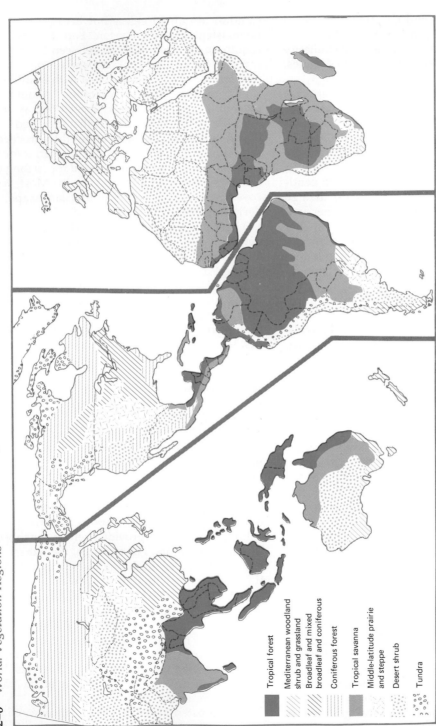

2-6 *World: Vegetation Regions*

Tropical forest

Mediterranean woodland
shrub and grassland

Broadleaf and mixed
broadleaf and coniferous

Coniferous forest

Tropical savanna

Middle-latitude prairie
and steppe

Desert shrub

Tundra

Table 2–3 *Characteristics of World Vegetation Types*

Vegetation	Features
Tropical rainforest	broadleaf, evergreen, stands of diverse species, storied forest, mahogany, teak, rosewood
Tropical forest	mixed deciduous and semideciduous, thorn, shrubs, and grasses dependent on precipitation
Mediterranian woodland, shrub and grassland	evergreen shrub and forest, maquis, chaparral, olive and cork oak
Broadleaf and mixed broadleaf coniferous forest	oak, hickory, chestnut, birch, beech, varieties of pine, mixed or homogenous stands, pines in less favored soil or temperature environments, little remains in heavily settled middle latitudes
Coniferous forest	needleleaf, highly adaptable to variable environments, taiga of high latitude, spruce, fir, pine varieties
Tropical savanna	open grassland ranging to woodland with grassland undergrowth, depending on moisture availability
Middle-latitude prairie and steppe	tall to short grasses varying with moisture available
Desert shrub	deciduous shrubs, bunch grasses
Tundra	grasses, sedges, lichens, mosses

evergreen—no distinctive dormant season when leaves are all dropped.
deciduous—distinctive dormancy with loss of foliage.
broadleaf—exemplified by oak, elm, and maple in middle latitude, mahogany in tropics.
needleleaf—exemplified by varieties of conifers or pines.

centuries (broadleaf and mixed broadleaf and coniferous forests on Figure 2–6). Increases in population and changes in agricultural technology eventually placed great demands on environments with favorable temperature, soil, and moisture conditions. Crop agriculture could only expand at the expense of forest. Forests were not only a barrier to travel but also necessarily had to be cleared and removed so that field agriculture could expand. The vast forested areas that remain, the taiga and tropical rainforests, do so less from the protective concern of mankind than the fact that they are located in relatively less accessible places or contain other environmental features unfavorable for permanent settlement. Even now, concern is expressed over the rapid expansion of Brazilians into the Amazon Basin forests (tropical rainforest) for settlement purposes. Rapid population growth and the need for space may stimulate more clearing.

Human attitudes toward vegetation have begun to change dramatically in recent years. The significance of vegetation in numerous aspects of life is now recognized, including its relationship to other components of our environment (soils and air). Forest vegetation becomes more critical as population increases and consumes ever greater amounts of lumber and paper. Sustained yield forestry (harvesting only at the annual growth rate) of this resource is becoming more common as people recognize that wasteful usage can only result in greater resource problems.

Forests have many uses other than for wood. In the more affluent countries they are prized recreation areas. They are useful as a preventive for erosion, which in addition to destruction of land contributes to siltation problems in rivers, streams, and reservoirs. By reducing water runoff after precipitation, forests also contribute to flood prevention.

Soils

In addition to moisture and sun energy, mineral nutrients are needed for plant growth. Nutrients are derived from minerals in the earth and from organic material added to the soil by vegetation (humus). Underlying parent material, slope of land, climate, vegetative cover, microorganisms, and time all contribute to soil formation, but there are three special

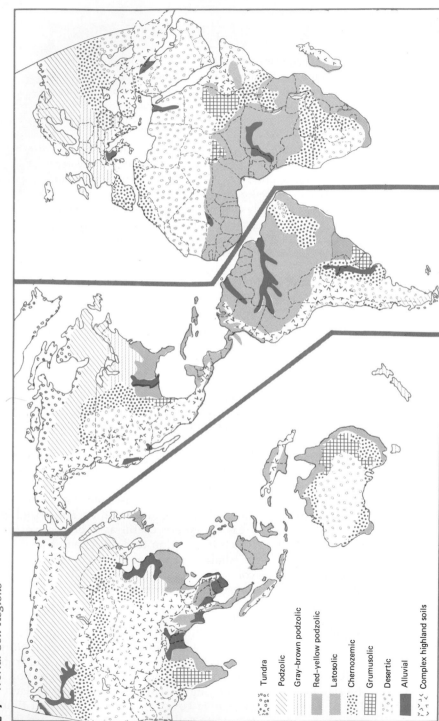

2–7 *World: Soil Regions*

Tundra

Podzolic

Gray–brown podzolic

Red–yellow podzolic

Latosolic

Chernozemic

Grumusolic

Desertic

Alluvial

Complex highland soils

processes important to soil genesis which greatly affect the supply of nutrients and therefore soil fertility: laterization, podzolization, and calcification.

Laterization is a process by which infertile soils are formed in warm humid regions of the tropics. Plentiful rainfall is quite effective in dissolving soluble minerals from the soil (leaching). The soluble minerals removed—calcium, phosphorous, potash, and nitrogen—are among the most important of plant nutrients. Insoluble compounds of aluminum and iron remain, but these elements alone mean infertile soils. The high temperatures and moist conditions inhibit the accumulation of organic material which can function as a source of nutrients. Decomposition of organic material proceeds rapidly and throughout the year, aided by the warm and moist conditions and the microorganisms of the soil. The entire process is most effective in areas of high rainfall and temperature, or the tropical and subtropical humid regions of Figure 2–7. Many of the soils of these regions show a marked decrease in fertility after only a few years' use. Much remains to be learned about the use of tropical environments, and especially their soils.

Podzolization occurs in high latitudes, or cold humid climates, where seasonal temperature variations are distinctive. Leaching is less effective in the normal manner because of either less rainfall or frozen ground during part of the year. Poorly decomposed organic material, however, such as from the pine needles of coniferous trees, combined with water forms weak acid solutions that remove both the soluble minerals and aluminum and iron. What remains are large amounts of silica low in fertility. Figure 2–7 shows podzolic soil variants in humid regions from the subtropics to high latitudes, but podzolization is less effective and gives way to laterization as the tropics are approached.

Calcification occurs on the drier margins of humid regions and in dry areas. Reduced leaching in response to limited precipitation leads to greater humus and calcium carbonate accumulations. Some soils formed under these conditions are very fertile (chernozemic on Figure 2–7) but paradoxically often remain less productive because of the very moisture deficiency that favored the soil-forming process.

Residual soils are those formed in place. They reflect local environmental conditions and are strongly influenced by the foregoing processes. Some soils, however, are transported and have features little related to local environmental conditions. They include alluvium (water deposited) and loess (wind deposited); along with those formed from some volcanic materials, these are among the exceptionally fertile soils of the world. When associated with excessively dry areas, however, their use may require great expenditures of labor or capital to provide the irrigation water necessary to realize their production potential. In humid areas these soils are often the most preferred for cultivation.

Mankind has modified soils in many ways. The most common problems are soil depletion and erosion. Declining fertility resulting from prolonged use can sometimes be overcome by adding chemical or organic fertilizers. Lime is added to reduce acidity, a soil condition detrimental to many plants. Terraces are built to prevent erosion or to aid in the distribution of irrigation water. Crops that place a serious drain on soil nutrients may be rotated with less demanding plants or leguminous plants that have the capability of adding rather than removing nitrogen. Farmers in Iowa and Illinois have placed drainage tile in the ground to remove excessive moisture from poorly drained areas. Other drainage techniques have been used in Asia and Europe to bring into use soils that otherwise would be unproductive. Primitive societies shift from one plot of ground to another as soil fertility decreases. Leaving formerly cultivated land out of production for some years (fallowing) allows soil restoration by natural processes. Efforts to reduce soil destruction or to improve fertility necessitate production costs that must be met by expenditures of labor or capital. Nevertheless, the long-term needs of society require that we accept such costs as essential to environmental maintenance. People, however, are not always willing or able to bear such costs, particularly when short-term profits or immediate survival are the concerns. An exploitative approach leads to maximum returns with minimal inputs over a short period, but associated land-use practices often result in rapid destruction of one of our most basic resources.

An American experience illustrates the environmental damage and human hardship that result from less than cautious resource use. The years 1910 to 1914 are often referred to as the "golden years" of American agriculture because of the high prices received by farmers for agricultural commodities. The

Table 2–4   *World Soil Types*

	Soil Groups	7th Approximation	Location	Features
Soils of Humid Regions	Tundra	Inceptisols	High Latitudes	often poorly drained permafrost
	Podzolic	Spodsols and histosols	High Latitudes	severely leached, acidic, infertile, poorly decomposed organic material, requires lime and fertilizer for use
	Gray-brown podzolic	Alfisols	Middle Latitudes	slightly acidic to basic, good humus accumulation, useful if well managed and fertilized
	Red-yellow podzolic	Utisols	Subtropical Latitudes	severe leaching, low in nutrients, less than moderate fertility, requires careful management
	Latosolic	Oxisols	Tropical Latitudes	severe leaching, low in nutrients, less than moderate fertility, requires careful management
Dry Land Soils	Chernozemic	Mollisols	Middle Latitudes	good organic content, good nutrient supply, highly fertile soils
	Grumusolic	Vertisols	Tropical and Subtropical Latitudes	moderate humus and nutrient supply, difficult to manage because of wet/dry climatic regions
	Desertic	Aridisols		low organic content, but high nutrient content except for nitrogen, fertile soils, possible salt accumulation problem when irrigated
Other	Alluvial	Entisols	Variable	water deposited, good nutrient content, fertile
	Bog	Histosols	Variable	poorly drained, poorly decayed organic material, difficult to use
	Mountain		Variable	highly varied types

(arrow labeled "warmer" spanning the Humid Regions rows; arrow labeled "drier" spanning the Dry Land Soils rows)

SOURCE: *Soil Conservation Service, Soil Classification: A Comprehensive System, 7th Approximation* (Washington, D.C.: Government Printing Office, 1960).

favorable prices continued through World War I and into the 1920s. The high prices and above-average rainfall encouraged farmers on the western margin of the Great Plains to convert grassland into wheat fields. Then followed several years of severe drought coinciding with the depression of the 1930s. The unprotected cultivated lands suffered great damage by wind erosion. The economic hardship wrought upon the farmers led to widespread migration, land abandonment, bank failure, and even town disappearance.[3]

Traditional methods of soil classification have made use of genetic (origin) similarities. Table 2–4 provides such a classification of major soil groups that is useful for comparison with other components of the environment, particularly climate and vegetation. Terminology is included from a more recently

[3]James A. Michener in his novel *Centennial* (New York: Random House, 1974), 909 pp., vividly describes the occupation of the western Great Plains and portrays the problems faced by farmers in attempting to use the land; and John Steinbeck's novel *Grapes of Wrath* (New York: Viking Press, 1939), 619 pp., is focused on the depression years of the 1930s and what happened to the farmers of the western Great Plains.

2–8 *Generalized Associations of Climate, Vegetation, and Soils*

This diagram illustrates the generalized associations of climate, vegetation, and soils. Boundaries are transition zones and correspondence is not perfect, but the general scheme is useful for identification of environmental realms. Regions are labeled in order of climate, dominant vegetation, and soils.

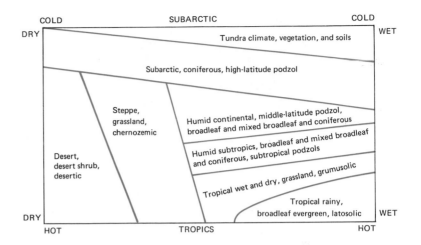

developed means of classifying soils on the basis of similarity in soil properties.

As previously mentioned, the residual soils, which are most common, share a generally close association with climate and vegetation. Climate is intricately related to the processes of vegetation evolution and soil formation. Figure 2–8 is a diagram illustrating this correspondence. Although the associations are somewhat more complex than is suggested by the diagram, the diagram provides an impression of various environmental realms available for human use. For example, using Figures 2–5, 2–6, and 2–7, compare the climate, vegetation, and soils of the southeastern United States with those of China.

Landforms

The surface of the earth is normally classified into four landform categories: (1) plains, having little slope and local relief (variations in elevation); (2) plateaus, level land found at high elevations; (3) hills, having moderate to steep slopes and moderate local relief; and (4) mountains, exhibiting steep slopes and great local relief (Fig. 2–9).

Plains are the most used landforms for settlement and production when other environmental features are favorable. Large areas of land with low slope and little relief have great utility for agriculture. For thousands of years mankind has been an agrarian being producing food by crop cultivation and requiring large amounts of space. Ease of movement over plains, particularly with grassland vegetation, has

facilitated exchange with other societies. Even rivers flowing through plains have had great utility for transportation purposes.

The great densities of population occur in areas where intensive agriculture is practiced, or regions where industrial and commercial activity is concentrated. Certainly the former high-density agricultural populations are associated with plains. The exceptions are frequently more apparent than real (compare Fig. 2–9 and Fig. 1–3). For example, the large population of mountainous Japan is extremely concentrated on coastal and river plains which account for only a small proportion of the total land area. Those plains that do exhibit limited settlement and utilization are less desirable for climatic reasons. Here we include plains in polar or desert areas. Even in mountainous regions, however, the small basin and valley become the focus of settlement. The features of plains that contribute to their utility in peacetime can be handicaps in wartime; with few natural barriers to afford protection they are easily traveled and overrun. Mechanized warfare in World War II in Europe illustrates the ease with which an army can move rapidly over a plains area.

Hills and mountains have a quite different meaning as a habitat for mankind. Some population groups have invaded already settled lands. If forceful enough, they may cause part of the preexisting population, if not assimilated and acculturated, to retreat to less favorable territory such as mountains or hill lands. These rugged areas are difficult to penetrate and

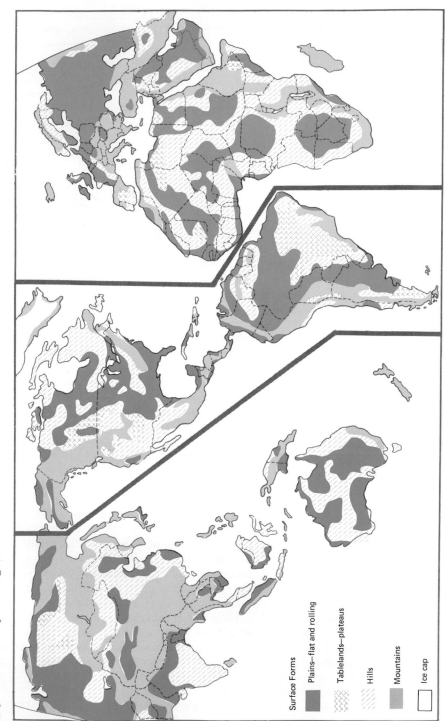

2-9   *World: Landform Regions*

Surface Forms

Plains—flat and rolling

Tablelands—plateaus

Hills

Mountains

Ice cap

Hills and mountains provide a difficult habitat for people. In the mountains of Nepal intricate terracing systems requiring great effort are necessary for intensive crop production. (A.I.D.)

provide a measure of security. The isolation may lead to a distinctive culture. The occupation of less favorable areas may also contribute to economic disparity. Major differences between highland and lowland inhabitants are not unusual. In the United States a distinction can be made between the upland South and the lower South. Conflict between mountain and lowland cultures has been part of the regional history in numerous parts of the world, for example, Southeast Asia. Even today, the achievement of unity in states that incorporate such contrasting environments and cultures is not easy. Separatist movements are expressions of these differences, and isolated highland areas provide excellent bases for guerrilla-type military operations. The Kurds of Iraq and Iran are mountain-dwelling people who, because of their separatist inclination, have been a problem to both governments. Communist and indigenous insurrection is facilitated in Burma where the Karens of the Shan plateau are alienated from the majority of Burmese in the lowland river plains.

In a mobile and affluent society mountains and hill lands are valued for recreational purposes. Distinctive cultural attributes then add to the economic potential for recreational development. Rivers in mountainous or hilly areas may have excessive current or numerous shoals and rapids that detract from their use as transportation routes, but such features also function as focal points at which land travelers

cross with ease, or facilitate mechanical or hydroelectric power development.

Minerals

The politically and economically powerful countries of today are those that have built an industrial structure by processing large quantities of minerals. These countries require huge quantities of fossil fuels (coal, petroleum, and natural gas). Any country striving for inclusion among the modern industrialized nations must either have such resources or acquire them elsewhere. Analysis of Figures 2–10 and 2–11 aids in the identification of those areas that have resource advantages for development.

Figure 2–10 presents the regional pattern of minerals presently exploited. The northern hemisphere concentration is obvious. Certainly the distributions reflect not only the location of minerals but also the use made of them. The mineral-producing regions are suggestive, generally, of areas with significant production and potential. Numerous areas may benefit in the future from resources yet to be discovered and developed if the inclination and wherewithal arise.

Modern power sources are coal, petroleum, and natural gas. These sources provide the vast quantities of power needed. Recently developed nuclear energy as yet contributes only minute quantities of power. Water is another source, but hydroelectric

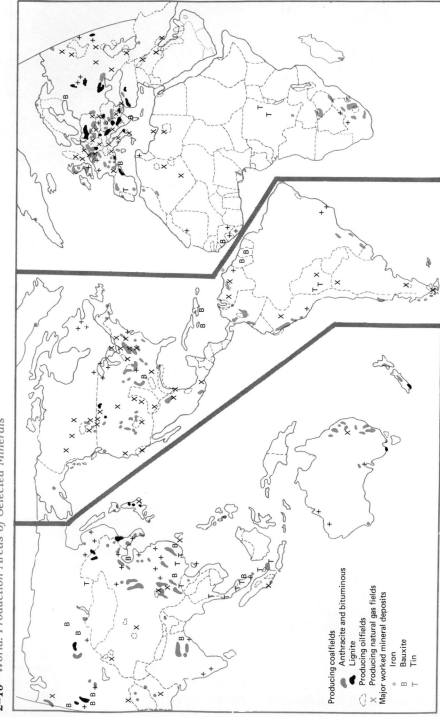

2–10 *World: Production Areas of Selected Minerals*

Producing coalfields
  Anthracite and bituminous
  Lignite
Producing oilfields
Producing natural gas fields
✕ Major worked mineral deposits
+   Iron
B   Bauxite
T   Tin

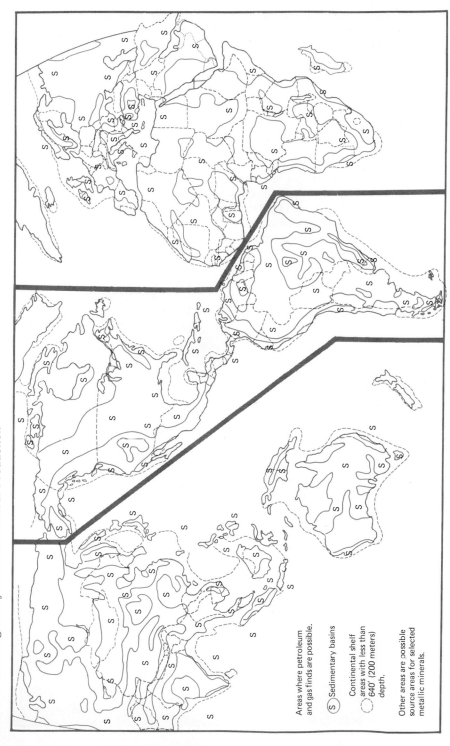

2-11 *World: Regions of Potential Mineral Production*

Areas where petroleum
and gas finds are possible.

Ⓢ Sedimentary basins

◯ Continental shelf
areas with less than
640' (200 meters)
depth.

Other areas are possible
source areas for selected
metallic minerals.

power, although it may be of localized importance, normally provides only a small proportion of a nation's total power needs.

Petroleum and natural gas are of organic origin (hydrocarbons formed from marine organisms) and are associated with subsurface rock structure of marine origin (sedimentary basins). A comparison of Figures 2–10 and 2–11 illustrates the association of current-producing regions with sedimentary structure and suggests where future discovery and production are possible.

The use of petroleum and natural gas has increased rapidly since World War II in all industrialized areas but particularly in the United States. This trend has given those less developed countries who share significantly in the world's oil wealth a special im-

portance in international politics, far greater than their size or military strength could otherwise provide.

Coal is also found in association with sedimentary structure, and in fact is a sedimentary rock of plant origin. Coal is available in huge quantities, especially in the United States, in the Soviet Union, and in China. When available, however, petroleum has been used to meet expansion needs and as a substitute for coal. Petroleum and natural gas are cleaner and contribute less to the air pollution problem. Whether the increasing cost of petroleum will generate a reverse trend—that is, a move back to coal for power—is not yet clear.

Iron, aluminum, and copper are the most important metallic minerals used in material production. Other

Coal is one of the United States' most abundant energy sources. Openpit mining using mechanized equipment provides low cost means of extraction yet creates problems of pollution and environmental degradation. (A.I.D.)

metallic minerals such as chromium, copper, zinc, lead, gold, and silver are found in the crystalline rock areas shown in Figure 2–11. Nonmetallic minerals such as nitrogen, calcium, potash, and phosphate are used for fertilizer. Salt, building stones, lime, sulfur, and sand are other commonly used nonmetallic minerals.

It is possible that discoveries of significant minerals in other parts of the world will occur, outside the current-producing regions, and greatly affect our evaluation of the industrial potential of such areas. At present a crude balance of power exists between the United States and the Soviet Union. Their power and position in world society are based upon the effective use of industrial and power resources. Europe is a close third party but somewhat more vulnerable because of limited petroleum supplies. Japan is clearly vulnerable because of the great dependence of its industrial structure on imported resources. But could the centers of power one day shift, based upon newly developed resources for industrial production? What is the potential of a country such as Brazil?

Summary

We have reviewed a select few of the many components of our physical environment. This emphasis is on the highly varied character of the physical environment and those aspects that have inhibited or supported mankind's activities in numerous different ways. Two considerations are of utmost importance. First, a basic dilemma is provided by the need to resist environmental deterioration at the same time that increasing population and expectations are placing greater demands on environment. The problem will not be easily solved. Nonuse is not a real solution, but neither is environmental destruction. Somehow, what must be found is an intelligent use of environment for supplying the needs of the larger society. This accomplishment will necessarily be at a cost and sacrifice for at least some members of society. This intelligent use, however, will be elusive and complicated by the varied nature of environment, of which there is yet so much to learn. Second, the meaning of these environments cannot be separated from the culture of the user society, and societies differ. The motivations, attitudes, accumulated technology, and ways of organizing economically and politically—all cultural considerations—greatly affect the use of these environments.

Further Readings

Introductory textbooks dealing with the entire spectrum of the physical environment include the following: Hidore, John J., *Physical Geography: Earth Systems* (Glenview, Ill.: Scott, Foresman, 1974), 418 pp.; Strahler, Alan N., and Strahler, Arthur H., *Elements of Physical Geography* (New York: Wiley, 1976), 469 pp.; Trewartha, Glenn T., Robinson, Arthur H., and Hammond, Edwin H., *Fundamentals of Physical Geography,* 2nd ed. (New York: McGraw-Hill, 1968), 384 pp.

Focus on selected components of the environment are provided by the following: Billings, W. D., *Plants, Man, and the Ecosystem,* 2nd ed. (Belmont, Calif.: Wadsworth, 1970), 160 pp.; Cruickshank, James G., *Soil Geography* (Newton Abbott: David and Charles, 1972), 256 pp.; De Laubenfels, David J., *A Geography of Plants and Animals* (Dubuque, Iowa: William C. Brown, 1970), 133 pp.; Eyre, S. R., *Vegetation and Soils* (Chicago: Aldine, 1963), 324 pp.; Hidore, John J., *The Geography of the Atmosphere* (Dubuque, Iowa: William C. Brown, 1972), 136 pp.; and Steila, Donald, *The Geography of Soils* (Englewood Cliffs, N.J.: Prentice-Hall, 1976), 222 pp.

The Association of American Geographers (Washington, D.C.) has published a series of resource papers designed as supplements for undergraduate college geography. The papers provide general topical coverage and numerous references on the subject of concern. Following are several focusing on the physical environment and related topics: Bryson, Reid A., and Kutzback, John E., *Air Pollution,* No. 2 (1968), 42 pp.; Dury, George H., *Perspectives on Geomorphic Processes,* No. 3 (1969), 56 pp.; Harman, J. R., *Tropospheric Waves, Jet Streams, and United States Weather Patterns,* No. 11 (1971), 37 pp.; Price, Larry W., *The Periglacial Environment, Permafrost, and Man,* No. 14 (1972), 88 pp.; and Tuan, Yi-Fu, *Man and Nature,* No. 10 (1971), 49 pp.

Detwyler, Thomas R. (ed.), *Man's Impact on Environment* (New York: McGraw-Hill, 1971), 731 pp.; Mathews, W. H., Smith, F. E., and Goldberg, E. D., *Man's Impact on Terrestrial and Oceanic Ecosystems* (Cambridge, Mass.: MIT Press, 1971), 540 pp.; Thomas, W. L., Jr. (ed.), *Man's Role in Changing the Face of the Earth* (Chicago: University of Chicago Press, 1956), 1193 pp.; and Wagner, Philip L., *The Human Use of the Earth* (New York: Free Press, 1960), 270 pp., are all volumes on people as users and modifiers of the earth environment.

Following are articles that focus either on a specific environmental realm or on problems associated with their utilization: Chang, Jen-Hu, "The Agricultural Potential of the Humid Tropics," *Geographical Review,* 58 (1968), 333–361; Dubois, René J., "Humanizing the Earth," *Science,* 179 (1973), 769–772; Ebert, Charles H. V., "Irrigation and Salt Problems in Remmark, South Australia," *Geographical Review,* 61 (1971), 355–369; Gomez-Pampa, A., Vazques-Yanes, and Guevara, S., "The Tropical Rainforest: A Nonrenewable Resource," *Science,* 177 (1972), 762–765; Grigg, David, "Ecological Problems in Agricultural Development," *The Harsh Lands,* (London: Macmillan, 1970), pp. 157–282; Jansen, Daniel, "Tropical Agroecosystems," *Science,* 182 (1973), 1212–1219; Meyer, Norman, "National Parks in Savannah Africa," *Science,* 178 (1972), 1255–1263; Wells, Philip V., "Postglacial Vegetational History of the Great Plains," *Science,* 167 (1970), 1574–1582; Storrie, Margaret C., and Jackson, C. I., "Canadian Environments," *Geographical Review,* 62 (1972), 308–332; Tosi, Joseph A., Jr., and Voertman, Robert F., "Some Environmental Factors in the Economic Development of the Tropics," *Economic Geography,* 40 (1964), 189–205; and Vankat, John L., "Fire and Man in Sequoia National Park," *Annals, Association of American Geographers,* 67 (1977), 17–27.

Macinko, George, "Man and the Environment: A Sampling of the Literature," *Geographical Review,* 63 (1973), 378–390, is a review of recent literature dealing with population-environment studies.

Mankind and Culture

3

CULTURAL ATTRIBUTES, a reflection of mankind's past activities, achievements, and conditions, strongly affect contemporary actions. Culture is part of the operational environment, just as is the physical world. Culture is, therefore, an important consideration in our effort to understand the condition of mankind in various regions of the world.

Mankind is unique in that learned behavior is accumulated and transferred to successive generations. These behavior patterns have utility in assuring sustenance and preservation of social order. With this accumulation of learned behavior (culture), mankind makes decisions and executes life styles. Some behavior is based upon earlier experiences of society (inherited culture), and other actions are based on the experiences of societies with which we have contact (diffused culture). The entire set of elements that identify a group's life style comprises the culture of a society (Table 3–1).

Another way of viewing culture is as a hierarchy of traits, complexes, and realms, A *trait* is the way a society deals with a single artifact, for example, how people plant seeds. A *complex* is a group of traits employed together in a more general activity, such as agricultural production. A *culture realm* is an area in which numerous activities or culture complexes are adhered to by most of the population. Not all of the larger world regions that we shall examine exhibit easily definable culture realms. Some are transition zones, in which numerous distinct cultures have met and clashed. Cultural pluralism is frequently evident even within political units, adding to the difficulty of achieving national unity.

A *culture hearth* is a source area where a culture complex has become so well established and advanced that its attributes are passed to future generations within and outside the immediate hearth area. It is no longer accepted that mankind's progress stems from achievements spread from a single hearth. More likely, several different hearths contributed to culture and became the basis for advancement over much larger areas.

Those areas that have served as major culture hearths have advanced because of a cumulation of their own achievements, and also from cultural exchange with other areas. In fact, some of the most significant of our culture hearths have functioned as crossroad areas, places across which many people have traveled, met, and exchanged goods and ideas. Contact, for whatever reason, leads to cultural diffusion (the spreading of ideas and artifacts). So, Buddhism was spread from its source area in India along the inner Asia trade routes and finally reached China in the fifth century A.D. During the sixteenth and following centuries, as Europeans from Iberia settled various portions of Latin America, systems of land tenure and organization were transferred from the homeland to the New World. Knowledge of these systems is central to understanding contemporary Latin America.

Societies have advanced at very uneven rates. For example, the Bantu cultures of western Africa advanced more rapidly than those of the Bushmen or Hottentots of southern Africa. Furthermore, two different societies may progress culturally but along different paths. The culture complex utilized for food production in China during the nineteenth century had traits different from those of Europe during the same period. This cultural distinctiveness need

Table 3–1   *Selected Elements of Culture*

Philosophy and Ideology	(beliefs, concepts, and
Religions	attitudes of individuals
Political	and groups)
Social	
Economic	
Skills and Tools	(acquired abilities for
Domesticated plants and	manipulating the human
animals	and natural
Use of metals	environment)
Use of inanimate power	
Means of movement	
Medical practices	
Social Organizations	(institutions)
Political	
Economic	
Social	
Educational	
Religious	
Communications	(skills by which cultural
Language	expressions are
Graphic arts	transferred)
Literature	
Music	
Drama	

not be construed as a difference in level of achievement but rather as a difference in kind. The cumulative nature of cultural evolution, its dynamic quality, the unevenness with which it occurs, and the differing orientation—all contribute to fundamental and intriguing variations among the 4 billion people of our contemporary world.

The Growth of Culture

Most of mankind's major accomplishments have occurred during the last few thousand years, a short time span when measured against the 2 million years or so that hominids (the family of humanlike beings) have occupied the earth. When the long Paleolithic, or Old Stone Age, period ended approximately 10,000 years ago, people had spread over most of the habitable portions of the earth, although low-population densities were prevalent. Cultural accomplishments

included the manufacture of stone tools, the use of fire (perhaps the most significant achievement), and the construction of shelters. The Neolithic, or New Stone Age, period (which followed the Paleolithic) marks the beginning of one of mankind's great revolutions (Fig. 3–1). It should not be thought of as an abrupt occurrence but rather as a period of several thousand years during which mankind's relationship to the environment changed slowly but fundamentally as new technologies were developed. People shifted from direct use of the environment by hunting and gathering to systems of agriculture.

Early Primary Culture Hearths

The Middle East contained one of the world's foremost culture hearths (the Fertile Crescent), or more accurately, several hearths in proximity (Fig. 3–2). The earliest plant domestication apparently took

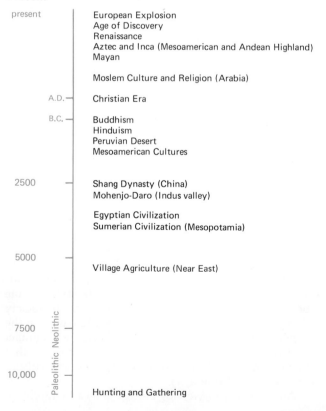

3–1   *Time and Culture from the Paleolithic to the Present*

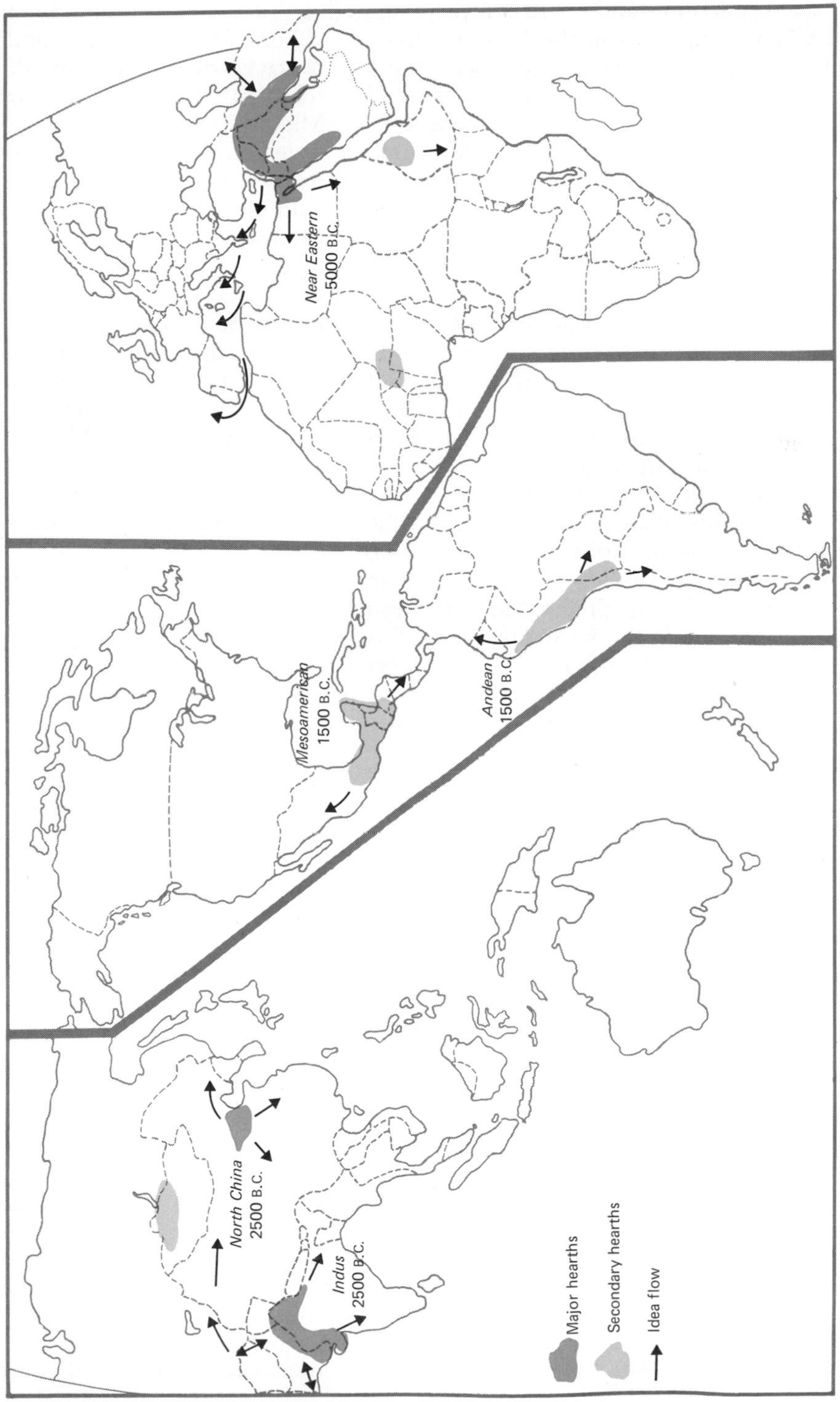

3–2 World: Early Culture Hearths

Near Eastern
5000 B.C.

Mesoamerican
1500 B.C.

Andean
1500 B.C.

North China
2500 B.C.

Indus
2500 B.C.

Major hearths

Secondary hearths

Idea flow

place in adjacent hill lands, and by 10,000 years ago agricultural villages appeared in the Mesopotamian lowlands. From the fourth millennium B.C., there followed flourishing civilizations, city-states, and empires. The participants of these civilizations codified laws, made use of metals, developed the functional use of the wheel, established mathematics, and contributed several of the world's great religions (Zoroastrianism, Judaism, Christianity, and Islam). These are only a few of their major achievements.

The crossroads nature of any location is important. The Middle East was one such location; the diffusion of culture from this hearth has been prolific. Early contact by way of inner Asian trade routes brought small grains to China. Grain and livestock, which may have come from the Indus Valley, were diffused across northern and eastern Africa. The diffusion of the grid-street pattern used for urban settlement, first identified in the Indus Valley, later spread to the Middle East, then to portions of the Mediterranean, and ultimately to the New World.[1] Numerous culture traits spread westward through the Mediterranean civilizations. Diffusion was the means by which this great culture hearth served as a major source area for Western civilization.

A second of the world's great culture hearths developed in the Indus Valley. A mature civilization existed there by 2500 B.C. The exchange of ideas and materials with Mesopotamia began early and continued over a long period of time (3000–1000 B.C.). Much of this exchange of culture was by way of ancient Persia (Iran). The two hearths may even have attracted migrant peoples from the same areas. The Indus Valley experienced repeated invasions and migrations of people from northwestern Asia and central Asia. Each new invasion meant an additional infusion of racial and cultural traits. The Indus and the adjacent Ganges valley served as a source area for the Indian culture realm. Significant contributions were made through literature, architecture, the working of metals, and city planning. Philosophy evolved which later contributed to Hinduism (the caste system), and the idea of living the sinless and good life may have been transferred to the Middle East, where it was incorporated into Christianity.

The middle Hwang Ho Valley and its tributaries

[1] Dan Stanislawski, "The Origin and Spread of the Grid-Pattern Town," *Geographical Review*, 36 (1946), 105–120.

served as a hearth area for the evolution of Chinese culture. China, however, undoubtedly benefited from contacts with other areas. Wheat and oxen, a part of the north China agricultural complex, may have Middle Eastern origins. Rice, pigs, poultry, and the water buffalo have Southeast Asian origins. Nevertheless, the long period of cultural achievement, with distinct early Neolithic ties, and the peculiarly Chinese traits maintained even after contact with other areas, clearly suggest that this area is a hearth from which an original culture emanated. Crop domestication, village settlement, distinctive architecture, incipient manufacturing, and metalworking were in evidence at the time of the Shang Dynasty (1700 B.C.). Chinese culture spread into south China, and northeast into Manchuria, Korea, and eventually Japan.

Each of the early primary culture hearths appears to have accomplishments acquired independently, yet each was also a recipient of ideas and commodities from other areas. Each hearth was able to maintain a distinct identity that was transferred to succeeding generations as population increased, or to invading or conquered people eventually assimilated and acculturated. China is a good example. The cultural revolution, which has continued into recent times, spread from original hearths to other areas, and culture realms have evolved that have clearly different traits and complexes. The Middle East, the Indian Subcontinent, and East Asia are distinctive culture realms that can trace their origins to the primary hearths.

Secondary Culture Hearths

Two areas in the Americas served as culture hearths for major civilizations. Northern Central America and southern Mexico (Mayan civilization) and central Mexico (Aztec civilization) supported large sedentary populations. This large Mesoamerican region had civilizations with cities, political-religious hierarchies, numeric systems, and domesticated crops (maize, beans, and cotton). The Middle Andean area (Andes of Peru and Bolivia) contained the second American culture hearth. The Inca civilization advanced more slowly than its Middle American counterpart but by the sixteenth century had developed irrigation, worked metals, domesticated the white potato, established complex political and social systems, set up a transport network, and controlled an empire.

The high civilizations of the Americas reached their heights relatively late and were disrupted by European conquest. These culture hearths are considered secondary because the civilizations that evolved from them never became the source from which lasting culture realms developed. For example, the derivation of the Latin American culture realm as we know it presently is clearly of indigenous American and European origin.

Areas in West Africa, the Ethiopian highlands, and central Asia probably functioned in a similar manner. The Bantu language family seems to have its source area in West Africa. The Ethiopian highlands were a secondary domestication center for various grains (wheat, millets, sorghums). Central Asia was also a domestication center for grains and certainly a trade route along which ideas and commodities were exchanged between major civilizations.

Europe as a Culture Hearth

Europe is an additional primary culture hearth but must be viewed in a different perspective from other earlier hearths. First, it flowered much later than the civilizations of the Middle East, southern and eastern Asia, and Latin America. Second, the Industrial Revolution associated with Europe has affected virtually all parts of the earth. Many of our contemporary global problems have origins in the confrontation between modern European culture and the traditional systems employed elsewhere. This confrontation does not necessarily imply a deliberate attempt to erase traditional ways of life; but where elements of the European system have been imposed or willingly adopted, the adjustments required for accepting the new ways have frequently meant difficult transitions.

Europe has a Mediterranean and Middle East heritage. Technology such as the wheel and plow agriculture, social concepts, and religions are of Middle Eastern origin: they spread through the Minoan (based on Cyprus), Greek, and Roman civilizations and into northern Europe.

When Middle Eastern civilizations were ascending, Europe was a peripheral area inhabited by "barbarians" not yet receiving or accepting the innovations of the Neolithic period. Rome had a civilizing influence on Western Europe through the introduction of order, roads, sedentary agriculture, and establishment of towns as market centers. Later barbarian invasions disrupted the Roman Empire, and several

centuries of limited progress followed (the Dark Ages, the period from the fall of the Roman Empire in A.D. 467 to the rediscovery of America in 1492). Nevertheless, a long period of slow agricultural change stimulated the Renaissance and indirectly the Industrial Revolution.

By the thirteenth century, agricultural and manufactural trade was common along trade routes that crossed Europe. The Crusades (during the late eleventh to fourteenth centuries) led to new contacts and stimulated thought and interest in learning. Spurred by internal competition and the new interest in science, exploration, and trade, the Europeans extended their influence and culture worldwide during the fifteenth to twentieth centuries. They explored, traded, conquered, and claimed new territories in the name of their homelands. Modern European states emerged with the capability of extending their power over other areas. Figure 3–3 shows the vast extent of areas directly influenced by Europe. European people resettled in newly "discovered" lands of the Americas, southern Africa, Australia, and New Zealand.

What took place has been aptly described as a "European explosion." Professors Spencer and Thomas have argued that the European explosion has stimulated a process of cultural convergence.[2] Their thesis is that isolated cultures are largely relics of the past, and assimilation or extinction are their not so different options; that methods of organizing politically, producing food, using inanimate sources of power, and consumption are becoming more similar. Racial and cultural blends in progress are almost certain to continue. The technologic aspects of European culture are those most readily accepted. Languages, religions, and attitudes adjust more slowly, however, and frequently become the sources of friction from which conflict arises.

This argument does not assert that contributions of traditional cultures will be discarded, or that they will totally disappear. Rather, large and viable culture complexes will be modified as cultural blends are created. Japan provides a good example of where European traits have been accepted in modified form. Also, arguing for convergence does not assert that regional disparities in economic, social, and biological well-being will cease. Indeed, herein comes the con-

---

[2]J. E. Spencer and W. L. Thomas, *Introducing Cultural Geography* (New York: Wiley, 1973), pp. 185–205.

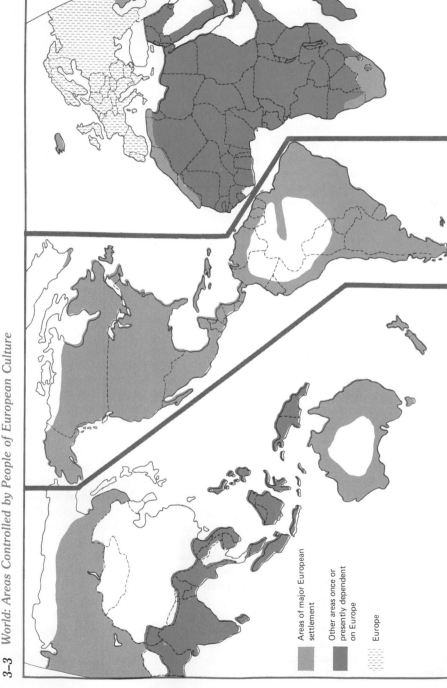

3-3   *World: Areas Controlled by People of European Culture*

Areas of major European settlement

Other areas once or presently dependent on Europe

Europe

verse argument. Portions of the world provide evidence that the old axiom "the rich get richer and the poor get poorer" has basis in fact. Acceptance by a people of the European way of doing things (economic pursuits) means that particular areas will have unique and great value because of location, material resources, and human resources; other areas will not be so well endowed. Even in the United States, regional disparity has long been a fact of life and a continuing problem (the Appalachians). Some look at the world and see rich and superindustrial powers continuing to widen the gap between haves and have nots. Among the underdeveloped nations, they recognize a few countries progressing but most falling further behind as their population continues to grow at unprecedented rates. Most certainly regional disparity will remain a problem. There are conditions inherent in modern political and economic systems (and in human nature) that preclude simple solutions to these adjustment problems.

Special Elements of Culture

Language

Language is a set of meanings given to various sounds used in common by a number of people; it is the basic means of cultural transfer from one generation to the next. The relative isolation in which various societies evolved led to the use of a great number of languages, frequently with common origins but not mutually intelligible. The analysis of language patterns areally is useful for reconstructing early cultural diffusion (Fig. 3–4).

Language differences may function as a barrier to the exchange of ideas, acceptance of common goals, or the achievement of national unity and allegiance. Members of most societies are not bilingual and do not speak the language of neighboring societies if different from their own. Sometimes the linguistic differences are overcome through a lingua franca, a language used over a wide area for commercial or political purposes by people with different speech. The use of Swahili in eastern Africa, English in India, and Urdu in Pakistan are examples. In these three areas many different mother tongues are represented.

The political leadership in many countries that acquired independence after World War II has found

achievement of national unity and stability difficult. Internal problems and conflict frequently stem, in part, from cultural differences; linguistic variation is often one of these differences. An example is provided by East and West Pakistan. From 1947 to 1972 these two regions functioned as one political unit. The political leadership sought to promote a common culture and common goals.[3] A single national language was deemed necessary, where in fact several mother tongues were in use. Bengali is the language of Bangladesh (formerly East Pakistan) derived centuries ago from Sanskrit. It provided the East a cultural element around which unity could be achieved. West Pakistan contained a number of languages: Baluchi, Pashto, Punjabi, Sindhi, and Urdu (a lingua franca). The problem resulting from this language diversity was just one of many that eventually led to the establishment of Bangladesh as a separate and independent state. Lack of a common language within a country is a contributor to plural societies; plural societies frequently mean political instability or hindrances to the development process.

Religion

Religions have their origin in a concern for comprehension and security, a concern that seems as old as mankind. Animism (the worship of natural objects believed to have souls or spirits) is among the earliest religious forms, and still exists in some isolated culture groups. It includes rituals and sacrifices to appease or pacify spirits but is simple in that complex organization is usually absent. Most such religions were probably quite localized; certainly those that have survived are held by isolated and small culture groups. The more important contemporary religions are codified, organized in hierarchical fashion, and institutionalized to assure transfer of basic principles and beliefs to following generations and other people.

Modern religions are classified in several ways. Ethnic religions are those that can, in origin, be tied to a particular area and group of people. Examples of ethnic religions are Shintoism (Japan), Judaism, and Hinduism (India). Universalizing religions are those considered by adherents as appropriate, indeed desirable, for all mankind. Buddhism, Christianity, and Islam all have ethnic origins but have become

[3]Glenn V. Stephenson, "Pakistan: Discontiguity and the Majority Problem," *Geographical Review,* 58 (1968), 195–213.

3–4 *World: Major Languages*

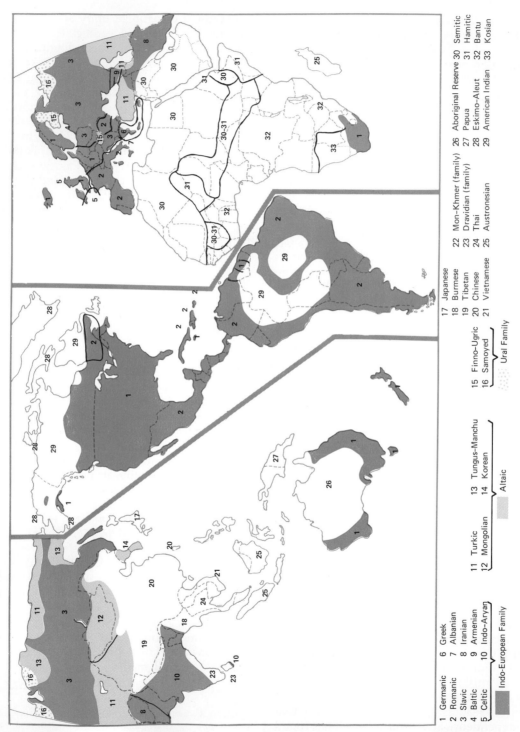

1 Germanic	6 Greek	11 Turkic	13 Tungus-Manchu	17 Japanese
2 Romanic	7 Albanian	12 Mongolian	14 Korean	18 Burmese
3 Slavic	8 Iranian			19 Tibetan
4 Baltic	9 Armenian			20 Chinese
5 Celtic	10 Indo-Aryan			21 Vietnamese

15 Finno-Ugric
16 Samoyed

22 Mon-Khmer (family)   26 Aboriginal Reserve   30 Semitic
23 Dravidian (family)   27 Papua   31 Hamitic
24 Thai   28 Eskimo-Aleut   32 Bantu
25 Austronesian   29 American Indian   33 Kosian

Indo-European Family

Altaic

Ural Family

Table 3–2 *Major Religions of the World*

Religion	Adherents* (millions)	Precepts
Hinduism	450	Ethnic—many gods, reincarnation, ordered universe, caste system with major social and economic implications
Buddhism	225(?)	Originator Guatama—spread from India to other Asian regions, erase suffering by seeking truth, encourages monasticism, inhibited sciences, numerous duties
Confucianism	500(?)	Originator Confucius—ethical precepts for living (the now), emphasis on family structure, close state ties, education through Confucianism, merges easily with Buddhist philosophy
Taoism	55	Originator Lao-tse (China)—commune with nature, seek harmony, but not through rigidity of government, numerous dieties
Shinto	50	Japan—numerous dieties, strong state ties and educative role until after World War II
Christianity Roman Catholic	580	Monotheistic, Middle Eastern origins—supernational organization under papal authority, outgrowth of major schism in Christian Church culminating in 1054. After reformation (fifteenth century) remains strong in Southern Europe
Eastern Orthodox	130	Outgrowth of schism culminating in 1054—evolves national structures, often close state ties historically, Eastern Europe
Protestant	260	Outgrowth of sixteenth century schisms of Roman Catholic Church and reformation—strong emphasis on private judgment rather than papal authority, middle-class material philosophy and protestant philosophy compatible, strong in Northwest Europe
Judaism	14	Middle East origin—dispersion, strongly ethnic, Zionist movement and state concept, monotheistic
Islam	500	Middle East origin (Arabia)—Mohammed, monotheistic (Allah); Koran as inspirational source, proselytizing religion; fostered learning and scholarship during Europe's Dark Ages, inhibited learning and reason after twelfth century

*Numbers of adherents are approximations. Consistent data on religious affiliation are difficult to obtain.

universal in nature over the centuries. Proselytizing is often considered a responsibility of adherents. Missionaries as well as traders, migrants, and the military have been the means by which religions spread from one people to another. Universalizing religions become such as they lose their association with a single ethnic group.

The impact of religious ideology on civilization is great (Table 3–2). Sacred structures contribute to the morphology of rural and urban landscapes. The distribution of citrus in southern Europe is directly related to certain Jewish observances. Food taboos account for the absence of swine in the agricultural system of Jewish and Moslem people in the Middle East. The Hindu taboo on meat eating, a response to respect for life, has the opposite effect: an over-abundance of cattle requiring space and feed but return little in the way of material benefits. In the United States the economic impact of religion is seen in taxation policies (organized churches are usually exempt from taxation), work taboos on specified days (Blue Laws), institutional ownership of land and resources, and attitudes toward materialism and work (the Protestant ethic).

One implied function of religion is the promotion of cultural norms. The results may be positive in that societies benefit from the stability that cohesiveness and unity of purpose bring. Unfortunately conflict also frequently arises from differences that have religious underpinnings; such differences may evoke intolerance, suppression of minorities, or incompatibility among different peoples. Examples include warfare between Catholics and Protestants in northern Ireland, the partition of the Indian subcontinent in 1947 as a response to Hindu-Moslem differences, and the Crusades of the Middle Ages in which Christians attempted to wrest control of the Holy Land from Islamic rule.

Figure 3–5 shows the world distribution of dominant religions. Reviewing this map with that of culture hearths suggests some of the world's major patterns of diffusion.

Political Ideology

Political ideologies also have major implications for societal unity, stability, and the use of land and resources. Political ideologies, however, need not be common to the majority of a population. People may be either apathetic or powerless to resist the imposition of a system of rule and decision making. Small groups (oligarchies) or in some instances individuals (dictators) have been guided by a particular philosophy and by imposition have made their thinking basic to the functioning of society.

Most modern governments assume some responsibility for the well-being of the people included in the state. Some socialist governments assume complete control over the allocation of resources, the investment of capital, and even the use of labor. Individual decision making consequently is limited. In other societies governments assume a responsibility for providing an environment in which individuals or people in corporate fashion may own and decide on the use of resources. National and economic development programs reflect differences in political systems. Socialistic governments take an approach different from that of capitalistic countries. These differences are treated in the regional chapters on Anglo-America, the Soviet Union, India, and China.

Social and Political Organization

Political organizations are control mechanisms that may spring from the common bonds of a few families or of millions of people (Table 3–3). The promotion of specified behavior patterns and institutions is deemed necessary to stimulate common bonds. Responsibilities of political organizations include resolution of conflict between and within societies and control over the distribution, allocation, and use of resources. The ability to execute these functions in a satisfactory manner depends not only on those in authority and the political system used but also upon the unity, support, or passivity of the people.

The Band and Tribe

The band is the simplest and formerly most common political organization that has existed throughout the entire period of mankind's cultural experience. A band usually consists of no more than a few dozen people occupying a loosely defined territory within which game, fish, insects, and vegetable matter are gathered. The community functions on a cooperative basis; no individual can for long direct group activity

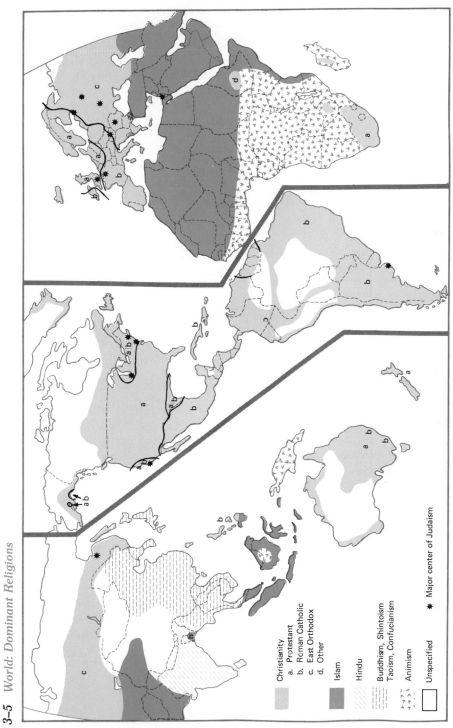

3–5 *World: Dominant Religions*

Christianity
a. Protestant
b. Roman Catholic
c. East Orthodox
d. Other

Islam

Hindu

Buddhism, Shintoism
Taoism, Confucianism

Animism

Unspecified

∗ Major center of Judaism

Table 3–3 *Sociopolitical Organization*

Traditional Classes	Fried System	Features
Band	Egalitarian	Weak territoriality, voluntarism as basis for social order, sharing economy. Survive in remote areas. Equal access to resources.
Tribe	Rank	Vague to rigid territoriality, control by "elders" through kinship and lineage, reciprocal and redistributive economy. Survives in less modernized areas. Equal access to resources.
State	State	Precise territorial definition, social control by coercion, variable economy with tendency to commercial, public and private property, unequal access to resources. Kinship allegiance replaced by nationalism. Stratified society. Worldwide occurrence.

Note: The Fried system is not perfectly coincident with the traditional classes. See references at end of Chapter 3.
SOURCE: Morton H. Fried, *The Evolution of Political Society* (New York: Random House, 1967).

without voluntary cooperation. There are neither exclusive rights to resources and territory, nor is power concentrated in the hands of a specified number of persons. Groups of this sort are aptly called sharing societies.

Probably no more than 100,000 such people remain scattered over remote and isolated parts of the earth. Contact between these static societies and more advanced groups leads to extinction of the former. Of the few remaining groups the Bushman of the Kalahari, the Motilone of Venezuela, and the Jivaro of Ecuador are examples.

Increases in population density and advances in technology (agricultural production) require more specific allocation of resources and space. A focus of authority and ranking of position evolve in which community structure is based upon kinship; elders or heads of extended families lead and direct tribal activity. Hundreds, or even thousands of people, may be included in a kinship-structured tribe. Boundaries may be either vaguely or specifically defined depending on external pressures, trespass by neighbors, or colonial imposition. The Ibo of Nigeria are examples of an ethnically related group, which, prior to European penetration of Africa, were subdivided into hundreds of patrilineal clans or lineages, each with a specific territory. The Ibo are but one of many tribal groups that Nigeria must unify in its effort to build a modern state.

States

As population, size of territory needed, and economic specialization increase, a more formal organization becomes necessary. The focus of power becomes more distinct; authority is channeled through a larger hierarchy, and society becomes more stratified. The importance of kinship as a medium for organization disappears. The state emerges as a specifically defined territory, occupied by a people with a distinct bond to territory (sense of territory). It is organized politically, if not economically, and controls the entire territory and people. Thus the state includes territory, people, and an organizational system. It might be said that the territory is the state and the people component elements, whereas, in lesser forms of political organization, the society was the state and the territory of secondary importance to identity.

City-states were among the earliest of states, but expansion based upon conquest led to the formation of empires. Preindustrial states emerged in the Mesopotamian, Indus, and Nile valleys, in ancient China, and likely in Subsaharan Africa. The Inca and Aztec civilizations of the Americas provide more recent examples. The difficulties in controlling diverse peoples with varied language, religion, and economy were frequently beyond the capabilities of early governments. Traditional transport systems, of course, further handicapped control. Disruption and breakup of such states were frequent.

The evolution from tribal organization to modern states was a long and slow process in Europe. Finally, the emergence of England, France, Norway, and Russia, among others, during the later Middle Ages, signified that differences in language, religion, and feudal organization were overcome, and the nation-state emerged to become the model for much of the world.

The modern state is a product of the Industrial Revolution. The increasing complexity of the industrial society leads to greater levels of organization for maintaining order, assuring communication, providing protection, and promoting the common culture. Increased interaction and interdependence of individuals also lead to allegiances to the higher organization rather than to kin or individuals. Nationalism is one result of industrialization and urbanization of society. Intense loyalty long directed to a state organization occupying a specified territory leads to the formation of a nation-state.

Nation-States

The nation-state exists when the state and its territory are occupied by a people with a high degree of cohesiveness, common goals, and culture that are transferred from generation to generation by a common language, and that maintain a loyalty to the larger political organization. An emotional attachment is experienced that is not hindered by linguistic, ethnic, racial, or religious distinction. Such conditions are most difficult to achieve. Although the nation-state may be recognized as desirable from the standpoint of achieving stability, few states exist that do not exhibit some traits that detract from complete national unity or allegiance. Unity is more easily achieved when a degree of cultural homogeneity exists.

Problems of State Development

The existence of the modern state should not be taken to mean that states evolved along a singly defined path. Indeed, many areas of the world exhibit strong tribal tendencies and even resemble remnant band societies within areas exhibiting the boundaries and organizational structure of the modern state (Fig. 3–6). Political leaders strive to enhance the stability and visibility of the modern state in the world community of nations, but internally the confrontation between tribal loyalties and the modern state provides one of the basic problems of the developing world. Tribalism in conflict with modern state structure is not a sign of inherent backwardness but rather the outgrowth of imposition of state organization by colonial powers on an existing tribal political structure.

Perhaps Africa provides the most appropriate examples. Upon acquiring independence, the new states became rapidly aware, sometimes by violent experience, of the difficulty of forging stable political units from a dual heritage. On the one hand were Europeanization, modernization, and territorial definition from the colonial era, and on the other, the indigenous heritage of a great variety of tribal groups with poorly defined territorial boundaries, linguistic differences, and variations in cultural achievement and acculturation under the Europeans. The civil strife and secessionist attempts within Zaire during

3–6 *Contemporary Political Systems*

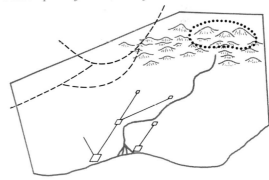

☐   Cities and Towns

──────   Modern state boundary—precisely defined.

─ ─ ─ ─ ─   Tribal boundary—vaguely defined.

•••••••••   Band territory—indefinitely bounded.

Tribal and band identity weakens and disappears with modern state development and strengthening.

the 1960s and the more recent attempt at Ibo secession in Nigeria (Biafra) should not have surprised us. These newly independent states contain diverse ethnic and language groups and people with variable economic achievement. Such differences add greatly to the difficulty of achieving unity and stability. Although Europeans may have fallaciously thought of Nigeria as a nation, in fact it consisted of several "national" groups.

We must also guard against the notion that instability or divisiveness are characteristic only of recently independent and developing nations. Racial antagonism has been a problem—for example, in Malaysia, Uganda, Peru, and the United States. A strong culture group extending control over a weaker people rarely makes social, economic, or political equalization a priority. It may impose lower order or position in a stratified society using physical traits as a method of identification. Sometimes it is formally aided by agencies and laws of the state (apartheid in South Africa). Reverse discrimination has been charged in instances where former subservient populations now rule newly independent states. This reversal of roles has been the stimulus for migration of Asians from east Africa in recent years.

An implicit suggestion that uniformity of religion or language ensures the unity necessary for political stability should be avoided. Religion was the reason for being for Pakistan from 1947 to 1972. Other social and economic differences, combined with physical fragmentation of the state territory, ultimately led to disintegration and formation of Bangladesh.

The lessons of history may suggest and encourage the formation of states with cultural homogeneity; yet, were it always so, many states would be so small that functioning in the modern economic framework might be impossible. The variety of resources necessary for modern economic systems would fall far short of the desirable level. It is most apparent that a continuing problem will be the achievement of political stability and economic progress in plural societies.

Multinational Alliances

The formation of alliances among states for the promotion of common goals is not a recent phenomenon. The military alliance has been most common, usually with the objective of providing mutual aid and protection in the event of military attack from a specified source. A contemporary example is the North Atlantic Treaty Organization (NATO), founded in 1949.

The motivation for the European Economic Community (Common Market) was a concern for the economic growth of Western Europe. Currently nine full members (six originally) have as their goal economic integration; the intent is that specific programs and policies assure economic development and prosperity for all member nations. The assumption is that one large economic unit can be more effective than numerous small countries. Member countries have high levels of development and collectively control resources that are complementary. Paradoxically, they are nations of diverse cultural background and have frequently engaged in warfare among themselves. The Common Market represents a significant step toward unification. Countries of Europe that formerly followed competitive and independent approaches now try interdependence and cooperation as a means for achieving progress. An important question is whether national interest of individual countries can best be met through the framework of large economic units or individually. The Common Market has served as a model for other multinational groups such as the Latin American Free Trade Association (LAFTA) and the Central American Common Market (CACM).

Economic Activity

Economic activities are those in which mankind engages to acquire food and other wants. They are the most basic of all the activities and are found wherever there are people.

Primary activity is the direct harvesting of earth resources. Fishing off the coast of Peru, pumping oil from wells in Libya, extracting iron ore from mines in Minnesota, and growing corn in Iowa are all examples of primary production. The commodities that result from these activities acquire a value from the effort required in production and from consumer demand. Processing these commodities is classified as secondary activity. A primary commodity such as cotton is processed into fabric or one of numerous other products. Fabric is cut and assembled as apparel. Textile and apparel manufacturing are two secondary activities in which an item is increased in value by the changing of its form to enhance its usefulness (form utility). Economic activities in which a service is performed are classed as tertiary. Wholesaling and retailing are tertiary activities in which primary and secondary production is made available to the consumer. Other tertiary activities include governmental, banking, educational, medical and legal services, journalism, and the arts.

Transportation is a special kind of economic activity. Transportation linkages are provided in numerous ways; human transport, pack animals, automobile, rail, air, water, and pipeline are all currently used. The cost of transportation is a measure of distance; if cost is too high, then movement of goods is not feasible. If resources and goods cannot be effectively exchanged, then self-sufficiency must be the basis upon which needs are met. An efficient transportation system is vital for a modern society. The nature of the transportation system in an area reveals much about the economic organization that exists.

Each economic activity, including transportation, involves the creation of wealth. The value of primary commodities increases as they are transported to locations where they are needed (place utility). The value added by improving place utility is further increased by processing (form utility). Each move or additional manufacturing stage adds more value to the items involved. Value, however, does not accrue on an equitable basis with each step in the process. The primary levels of production are not great wealth producers; for some individuals, yes, but normally not for the society at large. Most of the final value of a product is accrued from secondary, tertiary, and transportation activities.

Textile and apparel manufacturing occurs throughout the world. In some areas cloth and clothing are made in the home using few tools and often as an auxiliary activity. Elsewhere these products are made in modern factories emphasizing power equipment and labor specialization. (United Nations)

Economic Organization

Numerous factors contribute to the specific manner in which economic activity is organized and accomplished. First, many facets of our physical world affect our decisions regarding economic activity. Second, the environment may be used in various ways, depending on technology available. Third, even when potential for land use is analyzed relative to technology, a spate of other economic, social, and political conditions influence our decisions.

Location relative to existing markets, or degree of accessibility, may encourage or preclude some forms of production. Long-established methods of land tenure may maintain an uneven allocation of basic resources such as land. Governments aid in economic production by assuring proper infrastructure (credit systems, roads, education), or hinder by suppressing attempts to make necessary adjustments in the economic system. Traditions and attitudes have worked as both positive and negative forces in the functioning of economic systems.

Two extremes are recognized: the traditional and the modern commercial economic systems (Figs. 3–7 and 3–8). The most traditional of all economic systems is the subsistence economy in which a family or small band engages in both the production and limited processing required for local consumption. Group members function as producers, processors, and consumers of their own commodities. Self-sufficiency and sharing are the distinctive features. Present-day true subsistence societies are few but are exemplified by the Tasaday of the Philippines, the Bushman of Africa, and the Campa of Peru. Very few people are so isolated that they do not engage in some kind of exchange, even if only occasionally.

When we try to classify an entire country with respect to the nature of its economic system, it becomes impossible to identify one as purely subsistence. Even in a country that is obviously less developed, some people engage in subsistence activity, some produce and exchange on a reciprocal or barter basis, and others are clearly in an exchange or money economy. It is appropriate, therefore, to think of countries as existing along a continuum between the extremes of traditional subsistence and modern commercial. Although it is most difficult to locate a country precisely along such a continuum, there are features that are indicative of the level of modernization achieved (Fig. 3–7).

Importance of Primary Production

The majority of the labor force in a traditionally oriented country is engaged in primary production,

TRADITIONAL	← - - - - - - - - - →		MODERN
Subsistence	*Reciprocal*	*Peasant*	*Exchange*
Commodity Sharing	Barter	Minor exchange for capital	Full commercial
No urban foci			Major urban development
Simple technology			Complex technology
Animate power (muscle)			Inanimate power
Localized economy			Regional specialization
	Production Systems		
Gathering Nomadic herding Primitive agriculture Intensive subsistence agriculture			Commercial agriculture Commercial fishing Commercial grazing Commercial forestry Manufacturing and commerce

3–7 *Economic Organization*

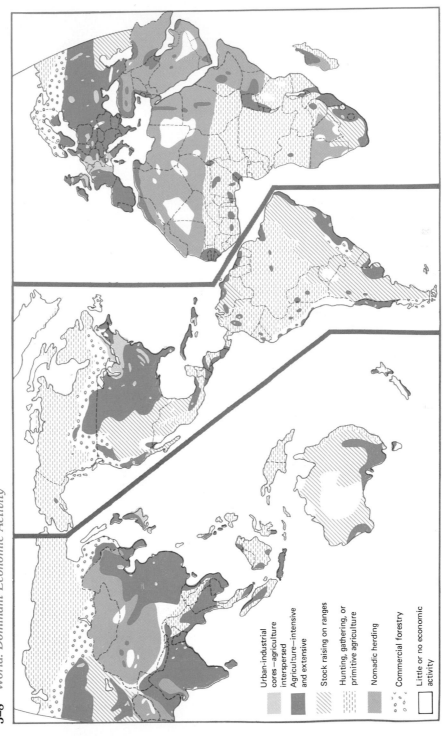

3-8 *World: Dominant Economic Activity*

Urban-industrial
cores—agriculture
interspersed

Agriculture—intensive
and extensive

Stock raising on ranges

Hunting, gathering, or
primitive agriculture

Nomadic herding

Commercial forestry

Little or no economic
activity

Of all primary activities, agriculture is by far the most important. Where people and draft animals provide the power needed for farming, as much as eighty percent of the labor force may be in primary production. Where power-driven machines are used, only a small part of the labor force is needed for cropping. (United Nations)

as much as 80 percent in some countries. Any society that is restricted largely to the primary activities is limited in the wealth produced from its resources. Per capita gross national product for traditional economies is low. There are exceptions of course; Kuwait, because of oil, is a good example. The modernized economy has a high proportion of its labor force in the secondary and tertiary activities. The primary sector may engage less than 10 percent of the labor force when extreme levels of industrialization and tertiary activities are reached. The concentration in processing and services means greater national income and has implications for the internal distribution and activity of population.

Production Inputs

The production inputs in traditional society are largely labor applied to land—agriculture. The modernized society substitutes capital for labor, evident in the automation of industry and mechanization of agriculture. Implicit in the shift to capital intensive production is a shift to inanimate power sources and high per capita consumption of fuel minerals. Therefore, per capita energy consumption is indicative of the extent of modernization.

Division of Labor and Regional Specialization

The division of labor reaches a high level in modern countries. It is evident at the local levels where members of a community perform a variety of tasks. Division of labor with increased modernization is also evident at the regional level. Regional specialization exists when several areas produce goods for which they have a particular advantage, or which are not produced elsewhere, and then contribute their specialized production to the larger economic system. Regional specialization occurs at both the national and international levels.

Regions engaged in specialized production make use of the principle of comparative advantage. Comparative advantage of an area may stem from climate, soils, labor, power, capital and enterprise, transportation, institutional advantages, or some combination of the above. Whatever the advantage, it means an area gains by specialized production and trading for other commodities needed. Occasionally areas produce goods for which they have no distinct advantage. This type of production is possible when other areas that could provide the same goods more

efficiently choose not to do so because of alternative possibilities that return even higher income.

Urban Corelands

High levels of urbanization are also characteristic of modernized societies. Economies and savings are realized when the secondary and tertiary activities that dominate modern society are agglomerated; hence the basis for cities. Examination of the United States and Canada, the European countries, the Soviet Union, and Japan on Figure 3–8 reveals urban-industrial corelands. Numerous cities in proximity, high-population density, a high proportion of national manufacturing, and high standards of living characterize such areas. Corelands also function as national educational, financial, political, and cultural centers. A distinctive feature is the high level of commercial and social interraction among the urban centers that are a part of the coreland. Less modernized countries often have only one major city that functions as the coreland. The absence of a multicity coreland is indicative of the more traditional economic system.

Urban-industrial corelands have peripheral economic regions (extended hinterlands) that do not exhibit features of a coreland but that do have clear functional ties to the core. The economic hinterland may be located within the country in question or may include other regions of the world.

Trade Relationships

The trade relationship between corelands and their peripheral counterparts reveals their differing character. Corelands are suppliers of manufactured goods. Internal exchange and per capita trade with other corelands of the world are high. Some manufactured goods flow to less industrialized hinterlands, but the volume is far less. The hinterlands are suppliers of raw materials, power supplies, and food products for corelands that often cannot supply all of their high per capita needs. This exchange is reminiscent of the colonial relationship, but the latter terminology should possibly be avoided. Figure 3–9 illustrates the common trade relationship found between rural areas specializing in primary production and urban regions specializing in secondary and tertiary activities. It is a distinction that exists, regardless of whether exchange is domestic or international.

A problem, however, lies in the fact that such

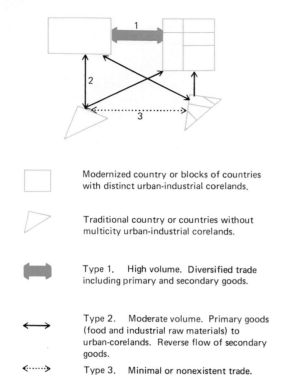

Modernized country or blocks of countries with distinct urban-industrial corelands.

Traditional country or countries without multicity urban-industrial corelands.

Type 1. High volume. Diversified trade including primary and secondary goods.

Type 2. Moderate volume. Primary goods (food and industrial raw materials) to urban-corelands. Reverse flow of secondary goods.

Type 3. Minimal or nonexistent trade.

Petroleum accounts for an increasingly large proportion of international trade. Petroleum flow is largely from traditional countries to urban-industrial corelands.

3–9 International Trade: Flow Types (excluding mineral fuels)

relationships often favor urban-industrial cores. The value of primary goods has generally not increased as rapidly as that of secondary goods. The likelihood of regional income disparity is great. If the exchange is between countries, it may cause balance-of-payment problems for less developed regions. The country that is a supplier of primary goods to industrialized regions outside its own territory also faces competition—price fluctuations or political pressures over which it can exert little control. Furthermore these countries often have few customers or depend heavily on two or three commodities for most of their exchange revenues. This situation contributes further to the possibility of economic problems should the trade be disrupted.

The more traditional countries often suffer the problem of dual economies. One part of the economy concentrates on commercial production of primary goods, much of which may flow to an external coreland. Other areas within the country with a large proportion of the population may remain traditional in economic structure. Economic dualism is commonly exhibited in distinct regional differences and contributes to internal social and political disunity.

The intent of the foregoing discussion is not to suggest that all is well with modernized societies and point to the woes of those countries less progressive. Although the latter evidence poverty, health problems, dual economies, and regional disparities, the problems of modernized societies are also formidable. The high level of per capita consumption requires immense quantities of material and power resources, such that it is almost frightening to think of modern society as a model for the remainder of the world. Space, quality living, suitable air and water are demanded at a very time when life style places increasing pressure on our total environment.

Further Readings

Grossman, Larry, "Man–Environment Relationships in Anthropology and Geography," *Annals, Association of American Geographers,* 67 (1977), 17–27. The author, with lengthy and useful references, provides a good philosophical discussion of the man–environment relationship as approached by geographers since 1900.

Lee, Richard B., and Devore, Irven (ed.), *Man the Hunter* (Chicago: Aldine, 1968), 415 pp. Contains numerous articles dealing with social, economic, political, and demographic aspects of "hunter societies."

Samuelsson, Kurt, *Religion and Economic Action* (New York: Basic Books, 1961), 157 pp. Provides an interesting discussion of the controversial thesis that Protestantism, particularly Calvinism, provided the conditions necessary for the growth of capitalism.

Soja, Edward, *The Political Organization of Space* (Washington, D.C.: Association of American Geographers, Resource Paper No. 8, 1971), 54 pp. Contains a useful review of systems of political organization and their territorial implications. An interesting discussion of the evolution of political systems is found in Fried, Morton H., *The Evolution of Political Society* (New York: Random House, 1967), 270 pp. See Stephenson, Glenn V., "Cultural Regionalism and the Unitary State Idea in Belgium," *Geographical Review,* 62 (1972), 501–523, for an example of the role of cultural variations in state organization; see also Burghardt, Andrew F., "The Bases of Territorial Claims," *Geographical Review,* 63 (1973), 225–245, and McColl, Robert W., "The Insurgent State: Territorial Bases of Revolution," *Annals, Association of American Geographers,* 59 (1969), 613–631.

Sopher, David E., *Geography of Religions* (Englewood Cliffs, N.J.: Prentice-Hall, 1967), 118 pp. Provides an introduction to religion as a component of culture influencing the relationship between mankind and environment. See also McCasland, S. Vernon, Cairns, Grace E., and Yu, David C., *Religions of the World* (New York: Random House, 1969), 760 pp.; and Schneider, Louis (ed.), *Religion, Culture, and Society* (New York: Wiley, 1964), 663 pp.

Spencer, J. E., and Thomas, W. L., *Introducing Cultural Geography* (New York: Wiley, 1973), 409 pp. The authors present an introduction to cultural geography written from a historical perspective.

Rich and Poor Nations: An Overview

4

For about 40 percent of the world's population, life's basic necessities are easily obtained. For this group the worry of obtaining sufficient food, clothing, and shelter is of secondary importance. The remaining 60 percent suffer from inadequate food, clothing, and shelter. Many must struggle constantly for survival. For some, life consists of an empty stomach, a piece of cloth or a cheap set of clothing, and a small dirt-floored hut for a shelter. Others are a bit more secure, but food supply is barely sufficient, often of poor quality, and limited in variety. Clothing is adequate to protect the body but offers few frills. Housing does little more than provide shelter from rain and cold.

Within all nations there are differences in economic well-being. In the United States, for example, there are many in the urban slums and poor rural areas (Appalachia) who live in poverty. Conversely, in India, where most of the people are poor, there are some who live affluently. The bulk of the population in any country, however, falls into either the moderately wealthy or poverty-stricken class. We can, therefore, consider nations either rich or poor. Both rich and poor are relative terms. By rich is meant freedom from worry of obtaining sufficient food and adequate clothing and housing.

When we are speaking of rich and poor nations, our reference is to per capita levels of well-being, not to the nations as a whole. The total accumulation of wealth in India or China, for example, is much greater than that of a small nation in Europe. Yet the average Indian or Chinese is not nearly so prosperous as his European counterpart. We are concerned primarily with individual economic well-being rather than the aggregate wealth of a nation.

The Widening Gap

Recognition of a rich and a poor world came forcefully to many in the late 1940s and early 1950s. Designation of advanced or developed and underdeveloped, undeveloped, backward, emerging, less developed, and developing countries became common in the popular press. Nations such as the United States, the Soviet Union, and many Western European countries initiated foreign-aid programs to less wealthy neutral and friendly countries. Much of the United Nations' work has been devoted to assisting the material progress of poor nations. In fact, the United Nations designated the 1960s as the "Development Decade," a period to emphasize and assist the modernization of the poor world. In recent years the World Bank and other regional development banks, financed largely by the rich nations, have partly supplanted individual national programs. The effects of all these actions, however, are not readily apparent.

Trends in Per Capita GNP

Over the past twenty years the gap between rich and poor nations has increased. The prospects are for a greater disparity. Statistical data are not available to show the widening gap fully, but enough information is extant to illustrate some general dimensions and trends. Per capita gross national product (GNP) is considered one of the best indicators of economic

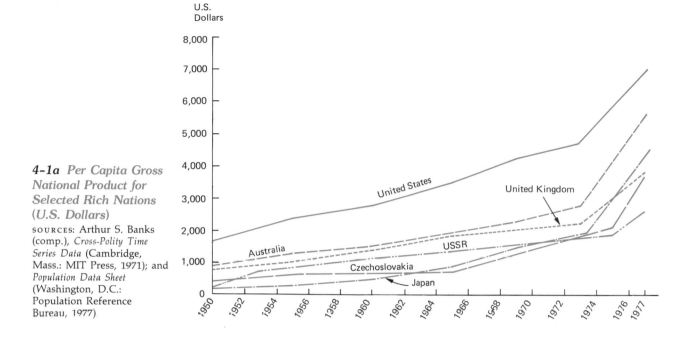

4-1a *Per Capita Gross National Product for Selected Rich Nations (U.S. Dollars)*

sources: Arthur S. Banks (comp.), *Cross-Polity Time Series Data* (Cambridge, Mass.: MIT Press, 1971); and *Population Data Sheet* (Washington, D.C.: Population Reference Bureau, 1977)

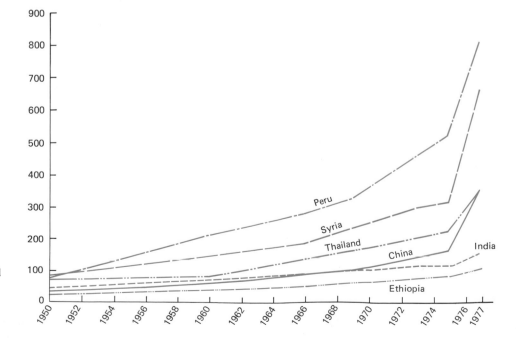

4-1b *Per Capita Gross National Product for Selected Poor Nations (U.S. Dollars)*

sources: Arthur S. Banks (comp.), *Cross-Polity Time Series Data* (Cambridge, Mass.: MIT Press, 1971); and *Population Data Sheet* (Washington, D.C.: Population Reference Bureau, 1977).

well-being.[1] Since the end of World War II, nearly every nation has shown an increase in per capita GNP, although in many cases, because of inflation, the gain does not result in greater purchasing power. In fact, in some nations effective buying power has decreased. Figures 4–1a and 4–1b graphically demonstrate the increasing disparity between rich and poor nations. Note that the vertical scale (dollar values) is different in these figures. If a poor nation's per capita GNP were plotted on the rich nations' chart, it would be difficult to see any increase in income.

On both figures the rate of change in per capita GNP can be interpolated. Ethiopia, representative of many African nations, has increased its per capita GNP an average of 5 percent annually, equivalent to $2.22 yearly. India is only slightly better off. Thailand, Syria, Peru, and, recently China have shown a substantially greater per capita GNP and a greater increase in rate of change. These increases must be viewed carefully. Peru appears to be making very rapid progress with an average rate of change about 15 percent yearly. In absolute terms, however, per capita GNP has increased only a bit more than $26 annually.

The rich nations far outdistance the poor nations in absolute per capita increases, although rates of change vary considerably for both rich and poor nations. Recovery from World War II damage was not completed by 1950, so the high growth rate for nations such as Japan, the Soviet Union, and Czechoslovakia is partly a result of economic recovery rather than new advances. In the same period the United Kingdom, Australia, and the United States approximated average yearly increases of 9 to 10 percent. In absolute increases the rich nations' values are most startling. Over the period shown, the average yearly increase in per capita GNP was about $199 for the United States, $175 for Australia, $122 for Czechoslovakia, $160 for Japan, $90 for the Soviet Union, and $110 for the United Kingdom.

Trends in Agricultural Production

Part of the reason for a widening gap between rich and poor nations lies in the different levels of agricul-

tural productivity. Figures 4–2a and 4–2b chart the changes in per capita agricultural and food production by means of index numbers.[2] These numbers are based on the assignment of a 100 value for the 1952–1956 period; all other values are in relation to the base years' period. For the poor nations, per capita production has increased only slightly, and most of the increase was during the first part of the period covered. In fact, since about 1960 no real change in total agricultural production has occurred. The lack of change in the poor nations does not mean that agricultural production has remained static; rather, it signifies that production has kept pace with population growth. This is no mean feat when one realizes that the population growth rate in the poor nations is two to three times that of most rich nations (Fig. 1–5). Per capita production in the rich nations has increased substantially over the period covered, and indications are that the trend will continue.

One reason why the rich nations are experiencing greatest per capita farm production is that yields per unit of land for most crops have improved dramatically. Improved seed, more extensive use of fertilizers and pesticides, and better management techniques have facilitated increased yields. In the poor nations, use of these technological innovations has not been so widespread. The trend in yields per land unit is demonstrated in Figures 4–3a, 4–3b, and 4–3c in which yields of the world's three principal cereals are presented. These crops are chosen, in preference to others, because they represent an important element in the diet of most of the world's people and are cultivated widely. A three-year moving average is used to reduce the effects of weather and to show more clearly the general trend in yields.[3] Figure 4–3a shows that wheat yields for most rich nations are higher and increasing more rapidly than for poor nations. Wheat in the Soviet Union, in Australia, and in the United States is grown mainly in dry lands, areas not well suited for high yields. Mechanization,

[1]Gross national product is the total value of all goods and services produced and provided during a single year. GNP values are subject to some error, particularly in nations where a significant part of the population is engaged in subsistence activities. These activities are often undervalued or not reported.

[2]The index numbers reflect the level of total agricultural or food production for the years cited compared to the base-period. If the index number is 106, production has increased 6 percent over the base period.

[3]A moving average is a smoothing method designed to show general trends, not comparisons on a year-to-year basis. The yield value for each year is obtained by summing the yields for a consecutive three-year period, dividing the sum by 3, and assigning the result to the middle year.

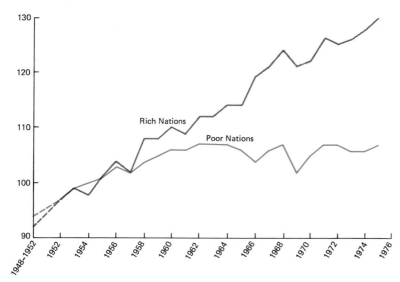

4–2a *Per Capita Total Agricultural Production Index Numbers: Rich and Poor Nations (1952–1956 = 100)*
SOURCE: Food and Agricultural Organization, *Production Yearbook* (Rome: United Nations, 1975).

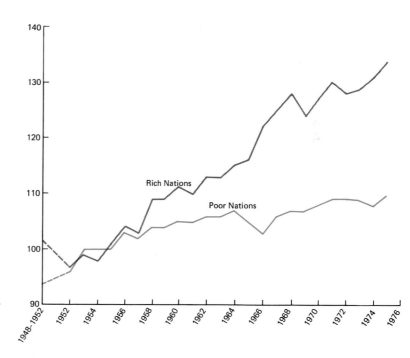

4–2b *Per Capita Food Production Index Numbers: Rich and Poor Nations (1952–1956 = 100)*
SOURCE: Food and Agricultural Organization, *Production Yearbook* (Rome: United Nations, 1975).

however, permits the farmer to cultivate a large area so that production is feasible. In France and Hungary, wheat is raised under climatic conditions favorable to high yields. In the poor nations production is not mechanized, and wheat is generally grown in more humid areas than those of the United States or Soviet Union.

A more striking contrast in yields per land unit between rich and poor nations is observed with maize (corn) and paddy rice. Yields for these crops have

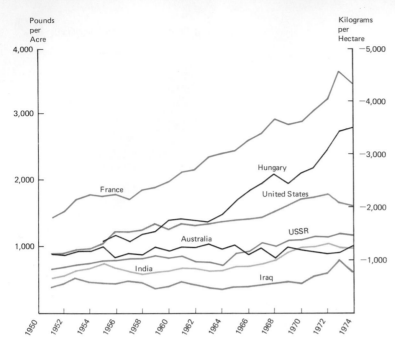

4–3a *Yields of Wheat for Selected Nations (3-Year Moving Average)*

SOURCE: Food and Agricultural Organization, *Production Yearbook* (Rome: United Nations, 1975).

4–3b *Yields of Maize (Corn) for Selected Nations (3-Year Moving Average)*

SOURCE: Food and Agricultural Organization, *Production Yearbook* (Rome: United Nations, 1975).

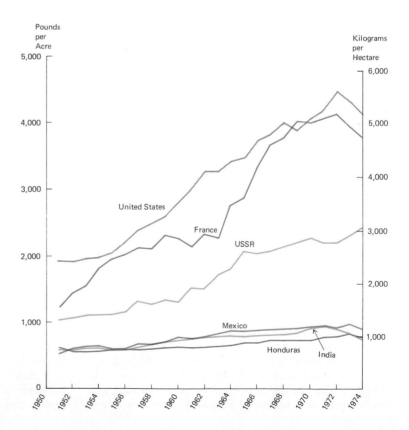

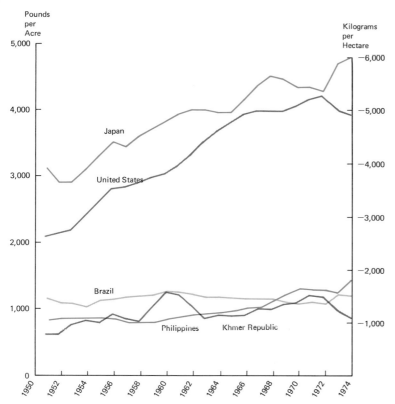

4–3c *Yields of Paddy Rice for Selected Nations (3-Year Moving Average)*
SOURCE: Food and Agricultural Organization, *Production Yearbook* (Rome: United Nations, 1975).

Improved high-yielding grains, often called "miracle grains," have been produced in research stations such as is shown here. Rice plants have been genetically modified to increase yields by two or more times that of the traditional local varieties. (United Nations)

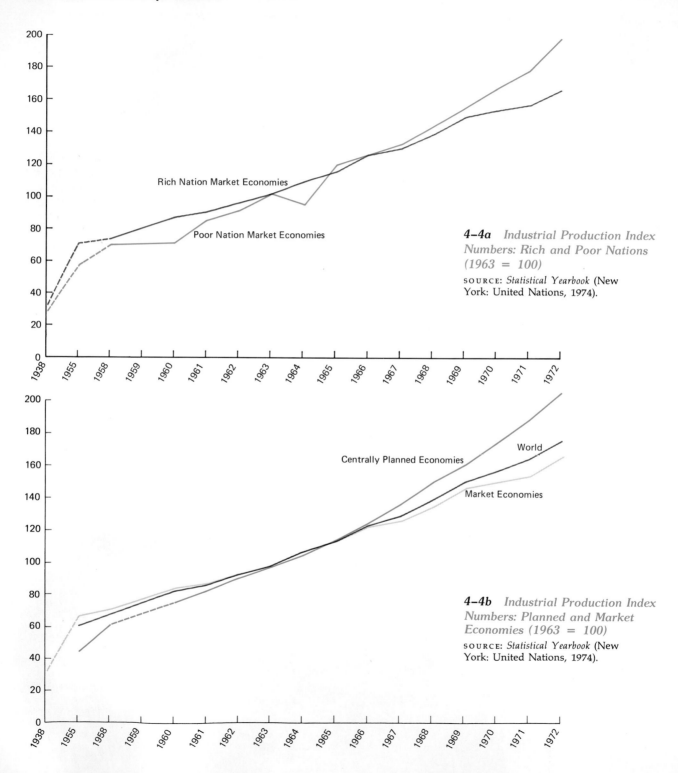

4–4a *Industrial Production Index Numbers: Rich and Poor Nations (1963 = 100)*

SOURCE: *Statistical Yearbook* (New York: United Nations, 1974).

4–4b *Industrial Production Index Numbers: Planned and Market Economies (1963 = 100)*

SOURCE: *Statistical Yearbook* (New York: United Nations, 1974).

not changed much in the poor nations since 1951. Both Mexico and the Philippines are centers where much research has been carried out on maize and rice, respectively. Only in the Philippines and to a lesser degree in Mexico, however, is there an indication that cultivation of these new plants has had any impact. Even so, yields in the rich nations are not only several times greater but increasing at a more rapid rate.

Trends in Industrial Production

Another indicator of the widening gap between rich and poor nations is the varied levels of industrial production. Unfortunately data on manufacturing are less prevalent than on agriculture. With the data available, along with some interpretation of world manufacturing patterns, some significant findings are revealed. Most poor nations have a limited industrial sector; the contribution that industry makes to the GNP is normally small. In the rich world, however, manufacturing is very important and characteristically provides a large part of the GNP. These points are made simply to demonstrate that industry is a much more significant economic activity in rich nations than in poor nations. With that in mind, Figure 4–4a tells us two things. First is that in relative terms poor and rich nations have for many years grown in industrial production at about the same rate; perhaps in recent years the growth rate has been higher in the poor nations, but not enough time has elapsed to discern a true trend. Second, even if poor countries are growing in industrial production at a faster rate than rich countries, it is certain that the absolute increase in production is much higher in the rich nations. This point is easily seen when we remember the industrial base in rich nations is much greater than in poor nations; thus an equal rate of increase of index numbers means a greater absolute value.

Lastly, it should be noted that Figure 4–4a shows only "market economies," that is, capitalistic and other noncentrally planned economies (non-Communist). The reason for excluding centrally planned economies is simply that data for many Communist nations are unavailable. This exclusion, however, does not affect the validity of the chart, for, as seen in Figure 4–4b, industrial production index numbers vary little between centrally planned and market economies. That market economies have a

lower index value simply is indicative that most poor nations are market economy nations; data for China are not included in the figures.

Measurement of Rich and Poor Nations

Most authorities on economic development differentiate rich and poor nations by one or more of three measures. They are per capita GNP, per capita consumption of inanimate energy, and percent of labor force in primary activities. Of these measures per capita GNP is most widely employed. Although this measure seems the best indicator of economic well-being, some caution is warranted. Per capita data are obtained by dividing total national values, in this case wealth generated, by the nation's population. Per capita data are thus average values. In many poor nations, average values do not apply to large segments of the population, because wealth is concentrated in the hands of a small upper class. Per capita figures are therefore generally higher than is actually the case for the majority of the population. In the rich nations, per capita values are more meaningful. Second, it should be recognized that GNP values, even under the best of accounting procedures, are only estimated. Caution is required in comparing nations; minor variations should be discounted. Finally, many poor nations—for example, Ethiopia—have per capita GNP values so low that it is inconceivable that the vast majority could physically survive, but they do. We must assume therefore that actual per capita wealth generated must be higher than reported or of a nature much different from that of the rich nations.[4]

Per Capita GNP

We have already compared, from a historical perspective, per capita GNP between selected rich

---

[4]Gross national product is often undervalued in poor regions because part of the economy is not commercialized. For example, in rich areas proceeds from the barber trade are counted as contributing to the GNP. In poor areas where barbering is done at home, no contribution to the GNP is acknowledged. In general, transactions in which money is not exchanged are not counted toward determining GNP, even though the work performed may be the same as one in which money was paid.

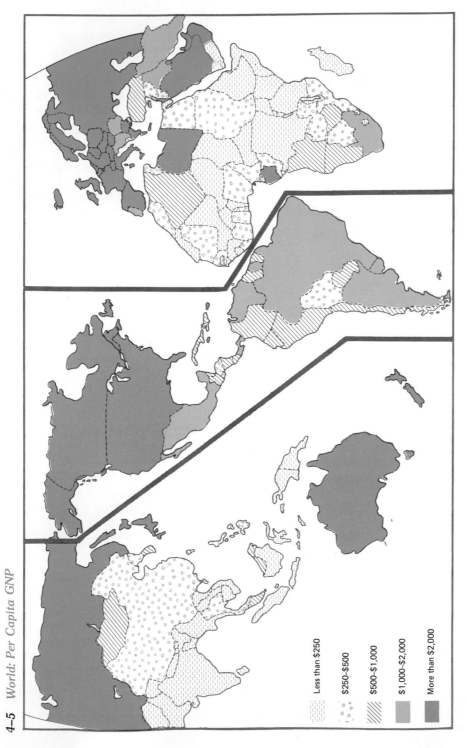

4–5 *World: Per Capita GNP*

Less than $250

$250–$500

$500–$1,000

$1,000–$2,000

More than $2,000

SOURCE: *Population Data Sheet* (Washington, D.C.: Population Reference Bureau, 1977).

and poor nations (Figs. 4–1a and 4–1b). It now remains to view the present distributional GNP pattern on a world scale (Fig. 4–5). This map clearly shows a striking regionalization. Areas of high per capita GNP ($2,000 or more) include the United States and Canada; Europe and the Soviet Union, except for Portugal, Albania, Romania, and Yugoslavia; Japan, Australia, and New Zealand along the western margin of the Pacific Ocean; Puerto Rico and Venezuela in Latin America; and some of the oil-rich areas of northern Africa and the Middle East. In the last-mentioned group are Saudi Arabia, Libya, Qatar, and the United Arab Emirates. Israel also is part of the richest group nations. Few nations are in the $1,000 to $2,000 range. They include several Latin American nations; a few oil-producing countries in the Middle East; South Africa; and Portugal, Romania, and Yugoslavia in Europe. Nations with $500 to $1,000 per capita GNP mainly include parts of Latin America and north Africa and the Middle East. Countries with less than $500 per capita GNP encompass a large part of the world's population. Those between $250 and $500 include Bolivia in Latin America; parts of Subsaharan Africa; China, Philippines, and Thailand. Nations with per capita GNPs of less than $250 include much of central and eastern Africa; southern and southeastern Asia; and Haiti in Latin America. We can conclude that most areas of the world fall readily into a relatively rich or poor category.

Per Capita Inanimate Energy Consumption

One of the Industrial Revolution's characteristics is a shift from animate power, man and beast, to inanimate energy, mineral fuels and water. The degree to which a nation is able to supply inanimate energy from internal and imported sources is an important indicator of the application of modern technology and consequently of productivity. Just as per capita GNP is a measure of productivity in terms of value, per capita inanimate energy use is a measure of production in terms of power expended. There is a close relationship between energy consumption and degree of economic activity. Low per capita energy consumption is associated with subsistence and other forms of nonmechanized agricultural economies, whereas high per capita energy use is linked with industrialized societies. Intermediate levels of energy characterize regions that have aspects of both the industrialized urban areas and the more traditional nonmechanized rural scene.

The distributional pattern of per capita inanimate energy is similar to that of per capita GNP (Fig. 4–6). Per capita energy consumption of more than 1,500 gallons (5,678 liters) of crude oil is limited to three small Middle East nations, the United States, and Canada.[5] In the last two nations, the automobile is ubiquitous and accounts for a substantial share of power used. Per capita energy users in the 750- to 1,500-gallon (2,839–5,678 liters) range include a belt of nations in Eurasia from France, Iceland, and the United Kingdom on the west through the Soviet Union on the east. Australia and Japan are also high per capita energy users. Only a few nations are in the 250- to 750-gallon (946–2,839 liters) per capita category. These are Argentina, Chile, Surinam, and Venezuela in Latin America; South Africa; Israel; Ireland; and part of southern Europe. Nations using between 125 and 250 gallons (473 and 946 liters) per person include Portugal and Albania in Europe; Iran, Iraq, and Saudi Arabia in the Middle East; Gabon and Libya in Africa; South Korea and Taiwan in eastern Asia; and Cuba, Mexico, and Colombia in Latin America. Finally, areas of per capita energy consumption of less than 125 gallons (473 liters) comprise nearly all of the southeastern quarter of Asia; the northern and southern extremities of the Middle East; most of Africa; and most of Central America, and tropical South America.

Percent of Labor Force in Primary Activities

As we have seen so far, there is a relationship between economic activity on the one hand and wealth generation and energy use on the other. In general, nations with a large part of their labor force in primary activities are less able to produce income and use relatively small amounts of power on a per capita basis. Conversely, nations that have a strong secondary and tertiary component in the labor force usually have a greater per capita GNP and energy consumption. The idea for using percent of labor force as an indicator is based on these relationships. Furthermore, any nation with a labor force dominantly in primary production has a limited opportunity for

[5]One gallon of crude oil is roughly equivalent to one gallon of gasoline.

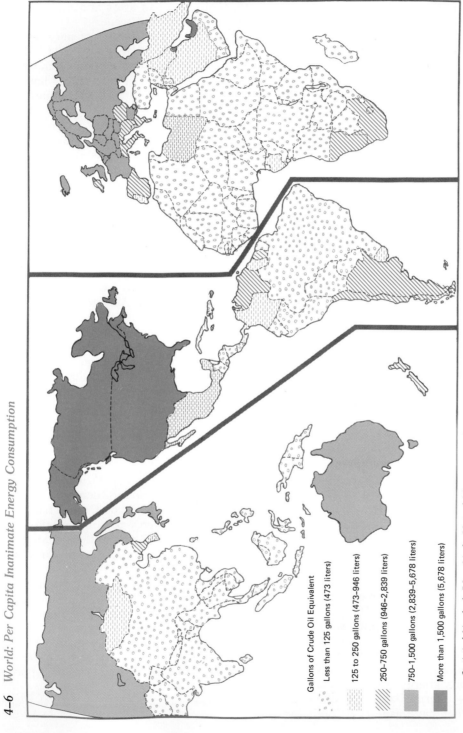

4-6   *World: Per Capita Inanimate Energy Consumption*

Gallons of Crude Oil Equivalent

Less than 125 gallons (473 liters)

125 to 250 gallons (473–946 liters)

250–750 gallons (946–2,839 liters)

750–1,500 gallons (2,839–5,678 liters)

More than 1,500 gallons (5,678 liters)

SOURCE: *Statistical Yearbook* (New York: United Nations, 1975).

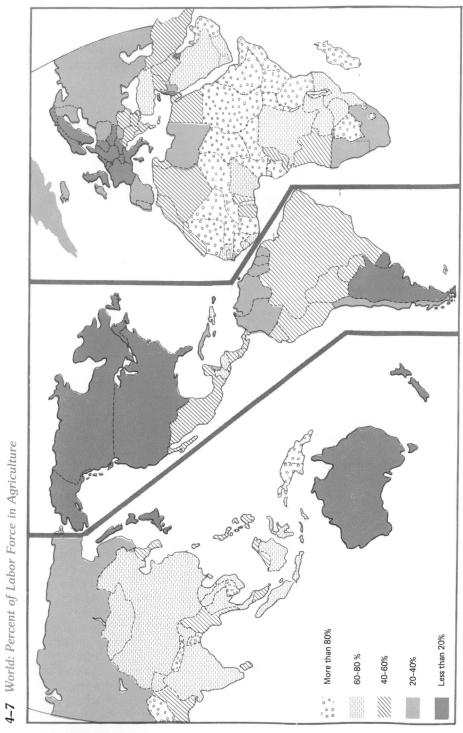

4-7 *World: Percent of Labor Force in Agriculture*

More than 80%

60-80 %

40-60%

20-40%

Less than 20%

SOURCE: Food and Agricultural Organization, *Production Yearbook* (Rome: United Nations, 1977).

labor specialization. In theory at least, labor specialization and production diversity are basic to economic growth. Certainly the prospects for high levels of individual production are diminished if the worker must not only grow his crops but also process, transport, and market them and be a "jack-of-all-trades" in providing housing, tools, and clothing.

Of all the primary activities, agriculture is by far the most important. Roughly 98 percent of the primary labor force in the world is in agriculture with the remaining 2 percent about equally divided among hunting and fishing and mining. A map of the labor force in agriculture, for which data are most accurate, is a good representation of primary occupation dominance. The broad pattern of the agricultural labor force is similar to the distributions of per capita GNP and energy consumption (Fig. 4–7). Areas with less than 20 percent of the labor force in agriculture include Anglo–America, Australia–New Zealand, northwest Europe, Japan, Israel, Kuwait, and southeast South America. Areas with 20 to 40 percent include parts of Europe, the Soviet Union, South Africa, Venezuela, Chile, Cuba, and Colombia. Much of northern and middle Latin America, southern Europe, northern Africa, and part of the Middle East have between 40 and 60 percent of their labor force in agriculture. In addition, Sri Lanka, Burma, Malaysia, Pakistan, the Philippines, Taiwan, Korea, and Mongolia fall in the same class. Most of southeastern Asia has 60 to 80 percent in agriculture; other areas include parts of Subsaharan Africa, Turkey, and northern Central America. Areas with more than 80 percent in agriculture include much of Africa, Bangladesh, Afghanistan, and the Khmer Republic.

Relationships Among Per Capita GNP, Energy Consumption, and Labor Force

In the preceding sections, reference was made to the similarity in the worldwide distributions of per capita GNP, energy use, and percent of labor force in agriculture. These similarities or relationships are apparent by simply comparing Figures 4–5, 4–6, and 4–7. For example, the United States and several other nations stand out as rich nations in all three aspects, whereas India and adjacent areas are classed as poor nations by these same measures. Many other examples are easily found. Although visual comparisons are useful and valid, another simple technique

of illustrating these relationships is the scatter diagram. In many respects scatter diagrams show these relationships more clearly. Figure 4–8 is a scatter diagram of the correlation between per capita GNP and per capita energy consumption. Basically the diagram shows that GNP and energy use are positively correlated. That is, as one variable (GNP) increases, the second variable (energy) also increases. There are, however, a few exceptions. Switzerland and Libya are examples of variants to the general pattern. A scatter diagram of GNP or energy and percent of labor force in agriculture would show an inverse correlation. That is, as GNP or energy increases, percent of labor in agriculture decreases. This statement can be checked by constructing a scatter diagram using data provided in the Appendix.

Other Measures

Although GNP, energy, and agricultural labor force are most often used to determine rich and poor nations, other measures are occasionally used. Two additional measures worthy of recognition are life expectancy and food supply. The life expectancy measure would seem to be the ultimate indicator of development. After all, some cultures are less materialistic than others, so the standard measures mask some important cultural attributes of society. All cultures, however, place a value on preservation of life, although in some, life is not so highly treasured as in others. Furthermore, life expectancy is a measure of the end result of economic activity, how well the system works to provide life support. A nation whose inhabitants can expect an average life span of but forty years has failed in its most important function. Figure 4–9 shows the distribution of life expectancy at birth and correlates generally with the previous maps we have examined. In assessing this map, remember that data on life expectancy are not as reliable as some other measures.

If we accept life expectancy as an ultimate measure of a rich or poor nation, then food supply is a measure of life quality. Two measures of food supply are fundamental. The number of calories available is an indicator of dietary quantity, and protein supply is a measure of dietary quality. Adequate quantity is considered at least 2,400 available calories per person daily. Adequate protein supply is attained if at least 60 grams of protein are available per person daily. These two factors are combined in Figure 4–10. Only

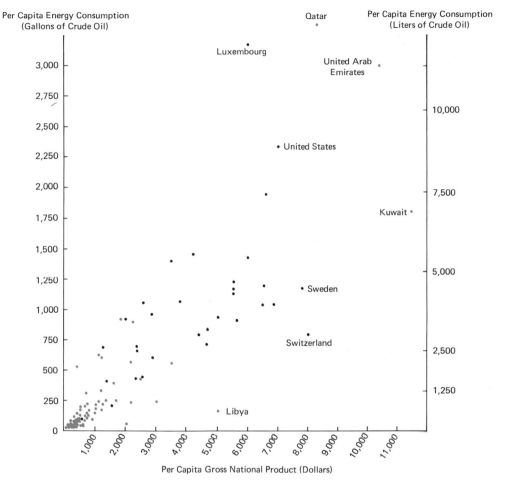

4–8 *Scatter Diagram: Relationship of Per Capita GNP to Per Capita Energy Consumption*

NOTES:
1. Many poor nations are not shown, since they cluster heavily in the lower left corner of the diagram.
2. Nations that are included in the rich regions are indicated by black dots; those in poor regions by colored dots. (See Fig. 4–11.)

category one provides both adequate calories and protein. Category one areas include Anglo-America, all of Europe except Albania, the Soviet Union, Australia–New Zealand, northeast and southern Africa, Japan, Taiwan, and South Korea, and much of Latin America. Category two—sufficient calories but inadequate protein—characterizes Albania and parts of Southeast Asia, Africa, and Latin America. Category three—inadequate calories but sufficient protein—is more widespread and is associated generally with livestock-raising areas in the east African

highlands and along the southern fringe of the Sahara Desert. Category four—inadequate calories and protein—characterizes much of southeastern Asia, tropical Africa, much of the Middle East, and the areas of Latin America with a high Indian population.

Major World Regions

We can now combine the various measures of rich and poor nations into a single map and identify some broad regional patterns. This map is, of course, highly generalized and lumps some rich nations

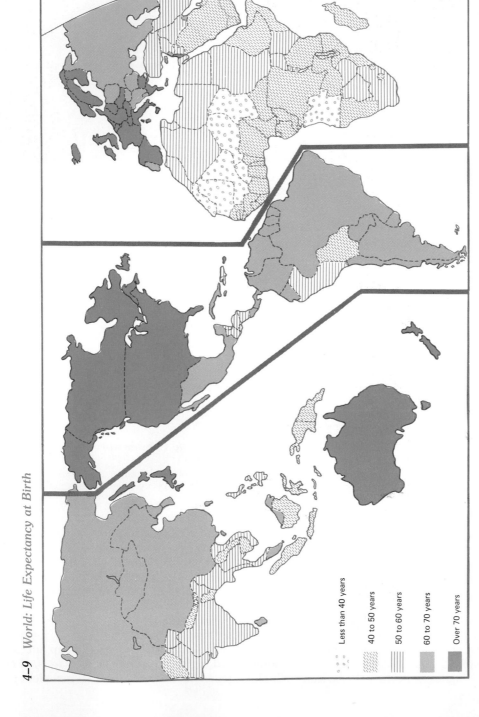

4-9   *World: Life Expectancy at Birth*

Less than 40 years

40 to 50 years

50 to 60 years

60 to 70 years

Over 70 years

SOURCE: *Population Data Sheet* (Washington, D.C.: Population Reference Bureau, 1975).

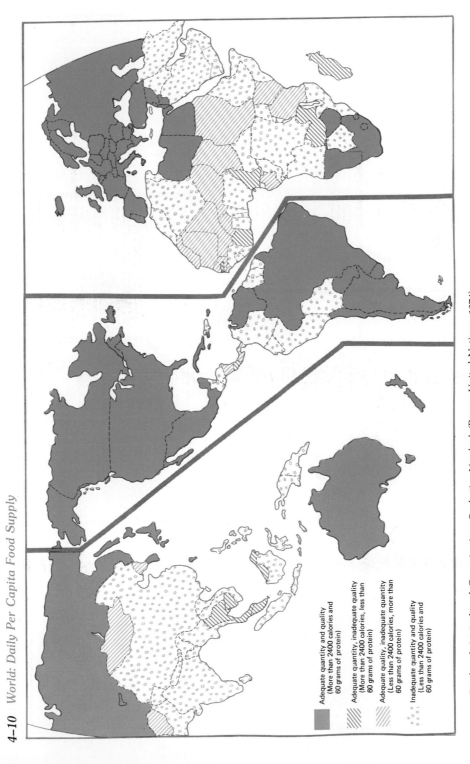

4–10 *World: Daily Per Capita Food Supply*

Adequate quantity and quality
(More than 2400 calories and
60 grams of protein)

Adequate quantity, inadequate quality
(More than 2400 calories, less than
60 grams of protein)

Adequate quality, inadequate quantity
(Less than 2400 calories, more than
60 grams of protein)

Inadequate quantity and quality
(Less than 2400 calories and
60 grams of protein)

SOURCE: Food and Agricultural Organization, *Production Yearbook* (Rome: United Nations, 1974).

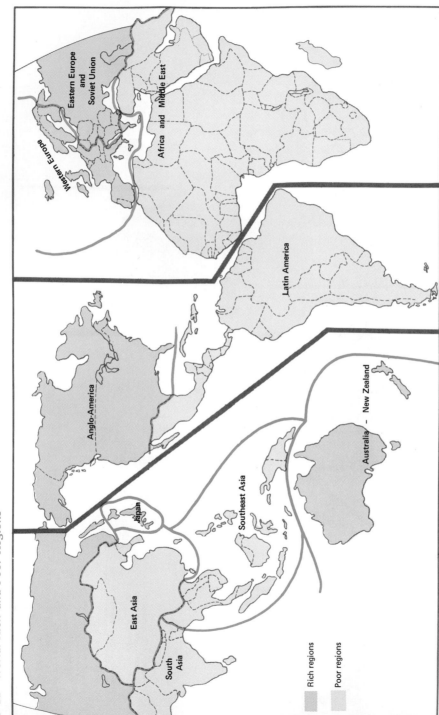

4-11 World: Rich and Poor Regions

with poor nations. In a like manner, some poor nations are included in the rich regions. Albania and Portugal are examples of the latter case. In generalizing, such discrepancies cannot be avoided. Some nations are classified as rich by all measures, and many are equally classed as poor. In these cases there is no problem in categorization. A number of nations, however, fall among the rich in some categories but among the poor in others. All along the continuum from very poor to very rich are nations in various stages of development, and to place them into one of two classes, rich or poor, involves a subjective judgment.

The map (Fig. 4–11) shows a strong correlation between rich and population areas predominantly of European origin. Anglo-America by its very name implies European roots. Latin America, a poor region, also has a European connotation, but this name is really a misnomer because many parts of the region are inhabited by peoples of Indian background. Much of Mexico, northern Central America, and Andean South America have a large and viable Indian culture component. Where people of European tradition do form the dominant population (Argentina, Chile, and Venezuela), the measurement indicators tend toward a rich index. Like Anglo-America, Australia's and New Zealand's populations are European. Of all the rich regions only Japan is an enigma, for it is truly non-European. The poor regions—Latin America, Africa and the Middle East, and South, Southeast, and East Asia—have no common cultural background. Diversity of culture is, in fact, the rule.

Some Additional Characteristics of Rich Nations

We have already noted several characteristics of rich nations: high per capita GNP and energy use; a small part of the labor force in primary activities and a consequent emphasis on secondary and tertiary occupations; and a longer life expectancy and generally a better and more abundant food supply. In addition, a review of Figure 1–5 shows rich nations have a low rate of population growth. To a large

degree, these nations have progressed through a demographic transformation.

Economic Characteristics

The most basic economic characteristic of the rich world is the widespread use of technology. The fruits of the agricultural and industrial revolutions are widely applied, and new techniques are quickly diffused and adopted. One need only look back a few years and compare them with the present. These new technologies mean new resources, a larger base with the opportunity for still greater wealth generation. The use of high technological levels leads to increased labor productivity and the creation of an improved infrastructure. Infrastructure means support facilities such as roads, communications, energy and water supply, sewage disposal, credit institutions, and even schools, housing, and medical services. These support facilities are necessary for accelerated economic activity and are a prerequisite for specialization of production. Productivity is enhanced by labor and regional specialization.

The heavy dependence of rich nations on minerals further differentiates them from poor nations. Not only are we in an iron and steel age but also in a fossil fuel age, a cement age, a copper and aluminum age—the list is almost endless. To an increasing degree, rich nations are importing these minerals from poor nations; the recent oil crisis is a case in point and illustrates the increasing interdependence of nations. This interdependence takes the form of trade in which the rich export manufactured goods, and in some cases food, and import raw materials. This trade arrangement has led to trade surpluses for most rich nations, since their exports have increased substantially in price, and to trade deficits for many poor nations, since the price of raw materials has increased more slowly; petroleum is a recent exception. Moreover, the rich nations gain further revenues by banking and investment in poor nations. Removal of this wealth from the poor nations has encouraged some poor nations to nationalize or exert greater control over foreign investments.

High productivity, a favorable trade balance, and new technologies result in elevated personal and corporate incomes. Individuals only need to spend a part of their income for food and shelter. Other income is used for services, products, and savings. The more money spent, the more demand for eco-

Rich nations are very dependent on minerals. Many of the minerals needed come from poor nations. Shown here is an open-pit tin mine in Malaysia owned and operated by a British company and using local laborers. (United Nations)

nomic growth, and money saved is usually invested in industries and businesses. For entrepreneurs (organizers who gather together labor, financing, and all those things necessary to construct an effective economic activity), the rich nations offer numerous advantages and are, in turn, benefited by the entrepreneurs' organizational skills.

Cultural Characteristics

None of these economic characteristics can stand alone. They are all intertwined. Moreover, they are, in fact, but an outgrowth of cultural characteristics. It is probably no happenstance that all rich nations except Japan have a strong European heritage. Mankind's attitudes and value systems are reflected in economic performance. The followers of the Judeo-Christian ethic consider the desire to work an important attribute and the accumulation of material things the representation of work performance. Success and peer respect are measured largely by wealth. Japan does not have a strong Judeo-Christian tradition, but Japan's cultural ethic toward work and success is similar to the European example.

Another cultural attribute of rich nations is the importance given to education. Literacy rates among nations are not comparable because the measure of literacy varies greatly. What is clear, however, is that the measure of literacy in rich nations is more rigid, and yet the literate percent of the population is above 80 percent. Education is also enhanced by communications infrastructure, and new ideas are quickly spread. Finally, education is geared partly toward economic advancement; diverse disciplines such as engineering, geology, economics, agronomy, and chemistry have obvious direct applicability for resource development.

An educated population is an extremely important resource. In addition, an educated and urban population, as are most populations in the rich world, more readily accepts change. In fact, sociological literature demonstrates such a population will accept change more readily than any other group. Acceptance of change means that technology is more easily adopted, new products and services are received, and the population is more mobile.

Some Additional Characteristics of Poor Nations

Population Characteristics

We have already learned that poor nations are characterized by low per capita GNP and energy use; a high proportion of the labor force in primary pursuits; a relatively short life span; and a diet often deficient in either quantity or quality, or both. A brief study of the population growth-rate map (Fig. 1–5) shows that poor nations have a high rate of growth, resulting from a continued high birth rate and a declining death rate. The age structure of a poor nation's

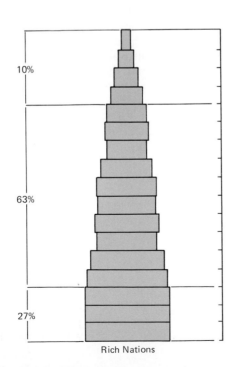

4–12 Population Pyramids: Rich and Poor Nations

SOURCE: *Trends in Developing Countries* (Washington, D.C.: World Bank, 1975).

10%

63%

27%

Rich Nations

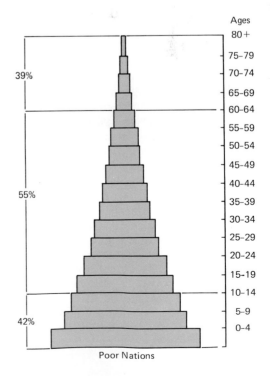

Ages
80+
75–79
70–74
65–69
60–64
55–59
50–54
45–49
40–44
35–39
30–34
25–29
20–24
15–19
10–14
5–9
0–4

39%

55%

42%

Poor Nations

population is consequently different from that of a rich nation. Figure 4–12 graphically illustrates this difference. For the poor nation, a large segment of the population is outside the most productive age group. Fully 40 percent of the population is less than fifteen years old. In a sense per capita comparisons are unfair because all people are counted equally, yet in poor nations a smaller percent of the population is made up of mature laborers.

A youthful population also effectively reduces the ability of women to engage in economic activities because they must tend the young. Moreover, women's status in the diverse cultures of most poor nations excludes them from many occupations. In the rural sector, women may work with men in the fields during times of peak labor requirements, but their major role is in the home where they may engage in some craft industry such as weaving for household use and for sale or barter. In the cities, where life is more cosmopolitan, women find employment as domestics, secretaries, and more recently in industry. For many illiterate women, however, these two last-mentioned opportunities are not available.

Cultural Characteristics

Literacy in poor nations is generally low. Most rural inhabitants cannot read or write or do so at minimal levels. In the cities literacy is more prevalent, but, still, large numbers of the urban poor lack effective command of the written word. The inability to read excludes a large mass of the population from learning new technologies and from engaging in more remunerative occupations. Despite some major campaigns to improve literacy, many nations have made little headway in educating their population. The dearth of teachers and a large and growing number of school-age children are formidable obstacles. Further, child labor is prevalent; many parents lack appreciation for education and need their children to work in the fields or at odd jobs in the city.

Most cultures in the poor world are conservative and, therefore, resist change. Herein lies a fundamental paradox. These cultures by and large wish to preserve their customs and mores, yet at the same time want to partake of the material benefits that Western society enjoys. Economic development leads to cultural change, the destruction of old traditions, and the acquisition of new behavior patterns. The

family, often extended, sometimes loses part of its cohesiveness; village life becomes subservient to the larger urban centers; labor and regional specialization lead to commercialization of the economy; and loyalties to the local community give way to national allegiance. These changes and others are almost inevitable if economic development takes place. Most poor nations are now somewhere along the continuum of economic development, and the disruption caused by the conflict between old and new ways of life is characteristic. This conflict accentuates the usually preexisting cultural pluralism of the poor world's nations.

Cultural pluralism occurs where two or more ethnic groups exist within a single nation. Each group has its own set of institutions, language, religion, life style, and goals. In some places these ethnic groups live apart, but in others they are intermingled. In both cases, however, if the differences among the groups are great or if mutual antagonism exists, joint action for development is difficult. In fact, opposing goals may lead to inaction, and attempts to alter the status quo cause conflict. Numerous examples of cultural pluralism are found within the poor world. The rural black of various parts of Africa clings to the traditional ways of life, but many blacks in the city have adopted the Western economic system and cultural goals. In the various parts of southeastern Asia, national unity is weak because of village orientation and the diversity of ethnic groups. In Latin America, cultural pluralism is evident in areas of large Indian populations. Cultural pluralism does not preclude economic development but does create an additional obstacle. Belgium is one example of a rich nation composed of two ethnic groups (Flemish and Walloon). In Eastern Europe, the diversity of ethnic groups has led to war and disruption. With the domination of the Soviet Union in the area, these disruptions have been suppressed.

Economic Characteristics

Economically the poor nations share many similar traits. Most have a dual economy, one part organized for domestic consumption and another for export trade. All nations, of course, produce for both internal and external markets, but in rich nations the farmer or factory worker produces the same items for both markets and rarely knows whether or not they are consumed locally. In the poor world, however, the producer for the local market provisions only the domestic market, whereas a different producer provides for the export trade. The domestic producer generally uses antiquated techniques and often is subsistence-oriented; that is, he sells only his surplus production. Export producers often have access to modern technological innovations and have the means to apply them. They are fully within the market economy and must remain competitive with rivals. The domestic producer does not have the same incentives or the same ability.

Poor nations rely mainly on flow (renewable) resources, principally those related to agriculture. This orientation does not necessarily mean that the people live in ecological balance with their surroundings. In many regions, increasing population pressures, coupled with static technology, have led to exploitation of agricultural resources to the point where soil depletion and erosion have become serious problems —in Andean South America and the Middle East, for example. Yet in parts of China where dense populations have lived for hundreds of years, agricultural resources have actually been improved. Fund resources—for example, minerals—are important export commodities in many countries but enter the domestic market usually as finished goods imported from rich nations. Finished goods and food products are the main imports of poor nations, and raw materials and agricultural products are the principal exports.

Poverty is widespread. The vast majority of the population is within the lower class, and a small upper class controls the economic and political life. In most countries the middle class is embryonic but in others it is growing as urbanization progresses. Limited income for the large bulk of the population means little savings. Accumulated wealth that does exist is in the hands of the upper class and often not available for investment in new activities. Limited income results from several factors: low levels of productivity, a limited infrastructure, a lack of applied technology, and a societal and economic structure not designed to facilitate economic development but rather to maintain existing conditions. High levels of unemployment and widespread underemployment also contribute to low incomes because there are so few alternative occupations available to most of the labor force.

Some Development Theories

Control Theories

A number of theories have been advanced to explain why some nations are rich and others poor. None are totally accepted, and some have strong opposition. One of the earliest theories is environmental determinism. This theory has been discredited but continues to have advocates. In essence, the theory is based on the premise that the physical environment within which a society lives controls or channels what people do. Certain climatic regions, those in the mid-latitudes, are said to be more stimulating for economic activity than either the polar or tropical climates. A review of the maps measuring economic development does show a propensity for rich nations in mid-latitude locations and poor nations in the tropics.

Another "control" type theory is cultural determinism. According to this theory, currently quite popular, a person's range of action is determined largely by culture. We can certainly see that parental guidance and peer pressure are strong influences and are reinforced by custom, mores, and laws. Should the culture place emphasis on a work-ethic, most members of that society will work. Conversely, if the possession of worldly goods is not an attribute of a culture, economic well-being, as measured, will probably be low. Some previously cited examples of groups that place emphasis on economic activities that are not particularly productive include the maize culture of Mayan Indians, Bedouin nomadic herding, and the adherence to village life, local self-sufficiency, and ethnic unity of numerous groups in southeastern Asia.

Colonialism and Trade

Colonialism and more recently the effects of foreign trade are related explanations that have numerous adherents. Colonialism in the traditional sense of the word is largely a thing of the past. In the years immediately following World War II, most colonial nations gave up, or were forced to give up, their possessions. Only a few vestiges of colonial rule remain. Mercantilism was the philosophy by which most mother countries traded with their colonies. Under mercantilism, the colony supplied raw materials and foodstuffs needed by the mother country. The mother country in turn used its colonies as a market for finished products and other surpluses. To assure that colonies did not compete with the mother country, products the mother country had in abundance could not be produced in the colonies. Moreover, the colonies could trade only with the mother country. This arrangement obviously was to the advantage of the mother country and severely limited the economic options of the colonies. Today trade among nations is controlled by tariff structures. Most rich nations have a tariff or duty on manufactured goods and low or no duties on raw materials and foodstuffs they need. The effect of this tariff structure is to perpetuate the trade patterns that existed in the colonial period. Most poor nations also have a similar tariff pattern. They use tariffs to protect their new manufacturing activities. Interestingly, the United States' trade with Japan is similar to that of a poor nation with a rich nation, for the United States supplies foodstuffs and raw materials to Japan and in return receives manufactured goods.

Circular Causation

Another economic theory is circular causation which results in either a downward or upward spiral effect. In the downward case a farm family, for example, barely produces enough to feed themselves; they have little or no savings. Should a minor crop failure reduce the harvest, they have nothing to fall back on. They eat less, can work less, produce less, and so on. The upward spiral can also be illustrated by the farm family. Again the family produces only enough to feed themselves. By good fortune, they obtain some additional capital to buy fertilizer which increases crop yields; so they eat better, work harder to produce more, and sell the surplus, therefore increasing the family's capacity to buy more fertilizer and improved seed. Although our example is a single family, the theory is equally applicable to groups and nations. The circular causation idea reinforces the old adage that the rich get richer and the poor get poorer.

Rostow's Stages

Finally, Walter Rostow, by comparing historical economic data, has advanced the idea of five stages of economic growth. In stage one, a traditional society exists; most workers are in agriculture, have limited savings, and use age-old productive methods. Indeed, all of the characteristics of a truly poor society are exhibited. In stage two, the "pre-conditions for take-off" are established. This stage may be initiated internally by an awakening of the population to a desire for a higher standard of living or by external forces that intrude into the region. In either case, production increases, perhaps only slightly, but fundamental changes in attitude occur, and individual and national goals are altered. Stage three is the "take-off." New technologies and capital are applied to increase production greatly. Manufacturing and tertiary activities become increasingly important and lead to migration from the rural environment to the bustling urban agglomerations. Infrastructural facilities are improved and expanded. During this stage, political power is transferred from the landed aristocracy to an urban-based power structure. Stage four is the "drive to maturity," a continuation of the processes begun in stage three. Urbanization progresses, and manufacturing and service activities become increasingly important. The rural sector loses much of its population, but those who remain produce large quantities with mechanized equipment and modern technology. Stage five, the final stage, is "high mass consumption." Personal incomes are high and abundant goods and services are readily available. Individuals no longer worry about securing the basic necessities of life and, should they choose, can devote more of their energies to noneconomic pursuits. Rostow also identified the take-off dates for some nations (Table 4–1).

Summary

Whether or not any of the development theories are accepted, they provide a useful yardstick against which we can measure and examine the development process in various parts of the world. Certainly a wide range of rich and poor nations span the globe. Some poor nations are undergoing economic development, but the gap between rich and poor is widening. This fact is evident by viewing the historical trends in some of the standard measures of economic well-being. Despite considerable diversity, all rich nations share certain characteristics, and the poor nations have their

Table 4–1 *Rostow's Take-Off Dates for Selected Nations*

Nation	Take-Off Date
United Kingdom	1783–1802
France	1830–1860
United States	1843–1860
Germany	1850–1873
Japan	1878–1900
Soviet Union	1890–1914
Argentina	1935–
Turkey	1937–

SOURCE: Walter Rostow, *The Stages of Economic Growth: A Non-Communist Manifesto,* 2nd ed. (Cambridge: Cambridge University Press, 1971).

particular common features. Although most measures of rich and poor nations are economic, per capita GNP and energy consumption and percent of labor force in primary activities, characteristics of each group also contain a number of cultural features. We must remember that the economy is but one aspect of culture, and to assign purely economic theory to explain the development process or lack of development may be misleading. Unfortunately our knowledge has not progressed to the point where we can definitely attribute any theory to explain why some nations have developed extensively but others have made little progress. It is hoped that by analyzing individual countries more closely, more insights will become apparent.

Further Readings

A number of books have been published on the aid given to poor nations by rich countries. A sampling includes the following: Mikesell, Raymond F., *The Economics of Foreign Aid* (Chicago: Aldine, 1968), 300 pp.; Hayter, Teresa, *French Aid* (London: Overseas Development Institute, 1966), 230 pp.; Holbik, Karel, and Myers, Henry A., *West German Foreign Aid, 1956–1966: Its Economic and Political Aspects* (Boston: Boston University Press, 1968), 158 pp.; and Goldman, Marshall I., *Soviet Foreign Aid* (New York: Praeger, 1967), 265 pp.

DALTON, GEORGE (ed.), *Economic Development and Social Change* (Garden City, N.Y.: Natural History Press, 1971), 664 pp. This volume traces, by case study, the impact of development from colonial times to the present. The role of economic development as an agent of cultural change has led to the establishment of a journal entitled *Economic Development and Cultural Change* that contains numerous articles worthy of examination.

Department of Economic and Social Affairs, *Statistical Yearbook* (New York: United Nations, 1948 to present). An annual set of statistical data covering a wide range of topics. Particularly useful are data on economic activities including agriculture, forestry, fishing, mining, manufacturing, energy use, trade, transport and communications, consumption of selected items, national accounts, and several other useful sets of statistics.

Food and Agriculture Organization, *Production Yearbook* (Rome: United Nations, 1946 to present). An annual compendium of agricultural statistics including area harvested, yields, and total production by country. Additional data are available on prices, pesticide and fertilizer consumption, and farm machinery.

GAUTHIER, HOWARD L., "Geography, Transportation, and Regional Development," *Economic Geography,* 46 (1970), 612–619. Gauthier makes a case for a transport infrastructure in the development process.

GINSBURG, NORTON, *Atlas of Economic Development* (Chicago: University of Chicago Press, 1961), 119 pp. This atlas presents some forty-eight different measures of development on world maps including those used in this book. The commentary for each map is brief and to the point.

GINSBURG, NORTON, "From Colonialism to National Development: Geographical Perspective on Patterns and Policies," *Annals, Association of American Geographers,* 63 (1973), 1–21. A broad view of the problems of development and the potential role of geography in assisting the development process.

HUKE, ROBERT E., "San Bartolome and the Green Revolution," *Economic Geography,* 50 (1974), 47–58. A study of the diffusion of new high-yielding rice in one area of the Philippines.

HUNTER, ROBERT E., and RIELLY, JOHN E. (eds.), *Development Today: A New Look at U.S. Relations with the Poor Countries* (New York: Praeger, 1972), 286 pp. A collection of short articles covering a range of topics from the role of multinational corporation in economic development to the politics of foreign assistance legislation in the United States Congress.

ROSTOW, WALT W., *The Stages of Economic Growth: A Non-Communist Manifesto,* 2nd ed. (Cambridge: Cambridge University Press, 1971), 253 pp.; and Rostow, Walt W. (ed.), *The Economics of Take-off into Sustained Growth* (New York: St. Martins' Press, 1963), 482 pp. In the former citation Rostow presents his now famous stages of growth. In the latter book Rostow's stages are examined by a group of leading economists.

UCHENDU, VICTOR C., "The Impact of Changing Agricultural Technology on African Land Tenure," *Journal of Developing Areas*, 4 (1970), 477–486. Uchendu explores the idea that the traditional multiple-tenure interests enjoyed by many on the same piece of land are gradually reduced as increasing levels of technology are applied.

United Nations Educational, Scientific, and Cultural Organization, *Statistical Yearbook* (Louvain: UNESCO, 1962 to present). The data presented in this yearbook are not so inclusive nor accurate as other United Nations annuals. The data on education in terms of enrollments at various levels, research personnel, library, and book production are particularly useful.

Anglo-America

James S. Fisher

PART TWO

Anglo-America: The Natural Resource Base

5

ANGLO-AMERICA CONSISTS of two vast countries encompassing more than 14 percent of the land area of the world: Canada with 3.85 million square miles (9.98 million square kilometers) and the slightly smaller United States with 3.68 million square miles (9.52 million square kilometers). Only the Soviet Union is larger than both, and China is but slightly larger than the United States. Reference to size is not simply a literary form of chest-beating but in fact implies a great variety of resources that can be marshaled for the support of the population. The 240 million people, most of whom reside in the eastern United States and adjacent area of Canada, form one of the world's most densely settled regions (Fig. 1–3).

Anglo-Americans have achieved a high level of development in an economic and technical sense. Most people earn their living by participating in secondary and tertiary economic activities, the basis of urbanization. Nearly 75 percent of the population is urban. The population's high material level of living means high per capita consumption. Anglo-America, by virtue of its size, population, economic, social, and political achievements, and attendant problems, is an excellent example of the development process.

In many parts of the poor world, however, the development process operates in a cultural milieu different from that of Anglo-America. Differences in culture often cause differences in results, even though similar technologies are used. Environmental maintenance, resource consumption and supply, regional economic disparity, and unequal sharing of material benefits among various population groups are problems faced not only by the inhabitants of Anglo-America but also are problems and pitfalls of all nations, rich and poor, that pursue similar goals and levels of achievement.

Anglo-America: Its Resource Strengths and Problems

The native and immigrant populations who developed the United States and Canada had the advantage of an immensely rich environment. The initially small population often used this environment as though it contained an endless supply of resources. The result has been costly. The two countries, particularly the United States, now with much larger populations, consume vast quantities of resources and by all indications will continue to do so. We should, therefore, analyze the region's physical endowment, giving some attention to its role in development and its current meaning. Although it would be foolish to argue that Anglo-American achievement was based on a rich and bountiful environment alone, it would be equally unrealistic not to recognize the role of land, water, minerals, and numerous other environmental features that have provided great advantages for the complex and interacting processes incorporated in the development experience.

Land Surface Regions of Anglo-America
Anglo-America is composed of nine surface regions (Figs. 5–1 and 5–2). The surface forms provide highly varied environments and have aided or inhibited

agriculture, functioned both as a barrier and a route-way for movement, and in other ways contributed to the varied settling and use of Anglo-American space by its population.

The Coastal Plain

The Coastal Plain borders Anglo-America from Long Island, New York, to Texas. Traces of this plain are evident on Cape Cod, and the islands of Nan-

5–1 *Anglo-America: General Locations*

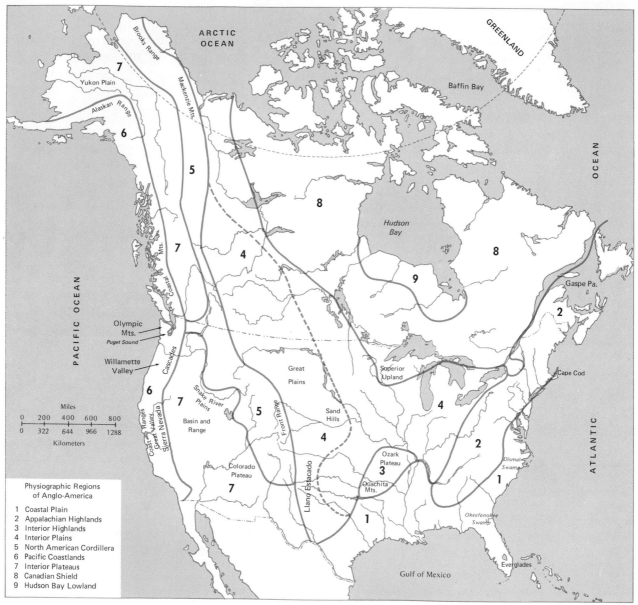

5–2 Anglo-America: Land Surface Regions

tucket and Martha's Vineyard off the New England coast. The inner edge of the plain is near sea level at its narrow northern extremity in New Jersey and about 1,000 feet (305 meters) above sea level at San Antonio and Austin, Texas, on the western border. The widest extent occurs in the Mississippi Valley where the plains extend inland to southern Illinois. The plain is of recent origin and composed of sedimentary materials deposited when overlain by the sea. Although the sedimentary layers are no longer horizontal, and some areas are even hilly, there are no major surface features that inhibited movement or initial settling by Europeans.

Some portions of the plain were given less attention as habitats for early settlement because of limited accessibility to the coast, infertile soils, or poor drainage. For example, the Pine Barrens of Georgia were considered infertile and were sparsely settled during the early nineteenth century when plantation systems flourished on the Georgia coast and farther inland. Poorly drained areas are typified by the Dismal Swamp, the Okefenokee, and the Everglades. On the other hand, the Black Prairies of Texas and Alabama, along with the alluvial Mississippi Valley, provide some of the best soil resources of the South.

The Appalachian Highlands

The Appalachian Highlands cover a vast area extending from Alabama to Newfoundland. Although of related geologic history, the highlands contain distinct topographic regions: the Piedmont, the Blue Ridge Mountains, the Ridge and Valley, the Appalachian Plateau, and the New England section (Fig. 5–3).

The zone of contact between the Piedmont and the Coastal Plain is called the Fall Line. Rivers flowing from the Piedmont onto the Coastal Plain at this contact zone have cut through soft sedimentary materials to form stream beds on the harder rock beneath. The resulting shoals and rapids associated with the Fall Line were focal points for converging Indian trails crossing rivers and later formed the head of navigation for settlers using rivers for transportation. A line of cities associated with the Fall Line includes Columbus, Macon, Augusta, Columbia, Raleigh, Richmond, Washington, Baltimore, and Philadelphia.

The Piedmont and the Blue Ridge exhibit sharp landform differences. The Piedmont is a rolling upland plain that forms the eastern margin of the Appalachians from Pennsylvania southward. The Blue Ridge Mountains are distinguished by their relative height. Some peaks attain elevations in excess of 6,000 feet (1,829 meters) in the wider southern portion in Georgia, North Carolina, and Tennessee, but farther north in Virginia the Blue Ridge are much

narrower and lower. The economic and social distinctions between the Blue Ridge and the Piedmont are almost as sharp as the topographic differences. The former remains more representative of "poverty Appalachia," whereas the latter, the Piedmont, is one of the most rapidly growing industrial regions in the United States.

The Ridge and Valley portion of the Appalachians contains a landscape of parallel ridges and valleys that extend from Alabama to New York. The area experienced severe folding and faulting of the earth's crust. The original layers of horizontal sedimentary rock, when deformed, exposed strata of varying resistance to erosion. The Great Valley is most distinctive and extends from Alabama to northern New York. The Coosa in Alabama, the Valley of East Tennessee, the Shenandoah, the Cumberland, and the Hudson Valley are some of the local names applied to the Great Valley taken from the numerous rivers that drain the valley. Other valleys are neither as continuous nor as wide. Like the Great Valley, however, their limestone floors have contributed excellent agricultural soils and served as routeways for early settlers seeking land in the interior. Modern transportation systems follow these valleys.

The westernmost portion of the Appalachian Highlands, also a region of sedimentary rock layers, is not so severely folded and faulted. Despite the name Appalachian Plateau, some parts have been severely eroded into hill lands and mountains (West Virginia), yet other areas such as the Cumberland in Tennessee retain a distinctive plateau character.

The role of the southern Appalachian Highlands in the evolution of the United States social and economic geography has been peculiar. It lies close to all of the early European settlements, yet many parts have remained relatively isolated and even lagged culturally. The physiographic character of the area has variously hindered population movement, influenced the direction taken by settlers, contributed to relative isolation, and functioned as a main roadway (the Piedmont) for initial settlement across the South.

Presently, some of the nation's finest resources exist within the Appalachian poverty area. Some of the best recreational and timberland in eastern United States and Canada is included. The most important of America's coalfields is found in the Appalachian

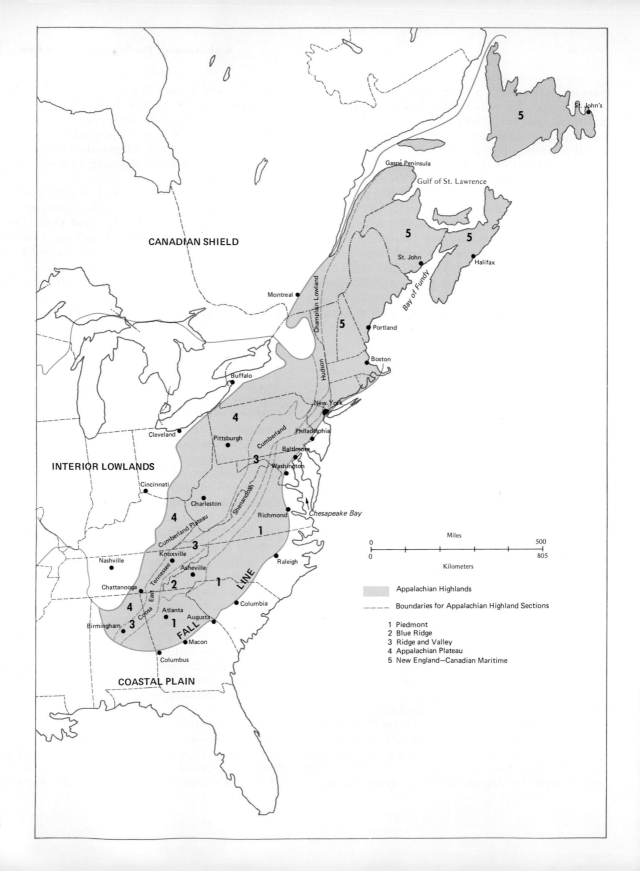

CANADIAN SHIELD

Gaspé Peninsula

Gulf of St. Lawrence

5 St. John's

5

5 Halifax

St. John

5

Montreal

Champlain Lowland

Portland

Hudson

Boston

Bay of Fundy

Buffalo

4

New York

Cleveland

Pittsburgh

Cumberland

Philadelphia

3 Baltimore

Washington

INTERIOR LOWLANDS

Cincinnati

Charleston

Shenandoah

Richmond

Chesapeake Bay

4 Cumberland Plateau

1

3

Knoxville

Tennessee

Nashville

Asheville

Raleigh

East

Chattanooga

2 **1** LINE

Columbia

4 Coosa Atlanta

1 FALL

Birmingham

3 Augusta

Macon

Columbus

COASTAL PLAIN

Miles
0   500
0   805
Kilometers

Appalachian Highlands

Boundaries for Appalachian Highland Sections

1 Piedmont
2 Blue Ridge
3 Ridge and Valley
4 Appalachian Plateau
5 New England—Canadian Maritime

Plateau, yet the plateau is one of the most severe economic problem regions in the United States.

The New England section of the Appalachians consists mainly of eroded and glaciated uplands that provide an uneven topography. The highest elevations are between 3,000 and 6,000 feet (914 and 1,829 meters) in the Green, White, and Taconic mountains in the west. Elevations decrease eastward to sea level. The Appalachian Highlands continue northeastward into the Gaspé Peninsula of Newfoundland.

The Interior Highlands

The Interior Highlands have a geologic history, structure, and physiography similar to that of the Appalachian Plateau and the Ridge and Valley regions. The Arkansas River Valley separates the northern Ozark Plateau from the Ouachita Mountains to the south. A similarity of culture and problems further relates the Southern Appalachians and Interior Highlands.

Interior Plains

One of the largest continuous plains areas of the world, the Interior Plains, extends from central Tennessee and Kentucky in the southeast and Texas and Oklahoma in the southwest northward into the MacKenzie Valley of northern Canada. The plain consists mostly of nearly horizontal or slightly folded sedimentary rocks and contains no major barriers to movement. Because much of the area north of the Ohio and Mississippi rivers has experienced continental glaciation, glacial landscape features such as moraines, till plains, former glacial lake beds, and disarranged drainage are common.

Much of the Interior Plains is at low elevations. The western portion (Great Plains), however, is an exception. The eastern edge of the Great Plains is 2,000 feet (610 meters) above sea level. Elevation gradually increases to the west which reaches 4,000 to 5,000 feet (1,219 to 1,524 meters) at the border of the Rocky Mountain system. Some portions of the Great Plains have almost imperceptible changes in elevation (the Llano Estacado of Texas); others are rolling, such as the grass-covered sand dunes of western Nebraska (Sand Hills).

From a topographic standpoint, the vast Interior Plains provide one of the more favorable regions of the world for agriculture. Within the vast area, however, are also great variations in climate which limit its agricultural use. Some portions are excessively dry, and others have growing seasons that are very short and cool.

The North American Cordillera

In recent geologic time, great mountain-building processes disturbed part of western Anglo-America, forming the North American Cordillera. The results are high elevations and a mixture of landforms. Portions of the Rocky Mountains, such as the Front Ranges of Colorado, consist of intrusions of materials from deep within the earth that disturbed and uplifted the sedimentary layers of rock lying nearer the surface. The Canadian Rockies are formed of folded and faulted sedimentary materials. North and northwest of the Rockies are the MacKenzie, the Richardson, and the Brooks ranges which, with the Rockies, form a mountain system that extends from Mexico through Alaska. The great heights and youthful nature provide some of the most spectacular scenery in Anglo-America, much of it relatively remote and high and protected from any great onslaught of settlement. Within the larger system, however, are many valleys and basins where small settlements have been established.

The Pacific Ranges

The Pacific Ranges are a system of mountains and valleys that extend from southern California northward along the Canadian coast and westward to the Alaskan peninsula. The inner part of this region is formed by the Sierra Nevada of California, the Cascades of Washington and Oregon, the Coastal Mountains of Canada, and the Alaska Range. Ice-carved valleys and volcanic peaks accentuate their beauty but also evidence their height and significance as a barrier. The Columbia, the Fraser, the Stikine, and the Skeena are the only major rivers with origins in the interior which traverse the entire system. The western portion is formed of yet another chain of mountains with strong linear features (parallel moun-

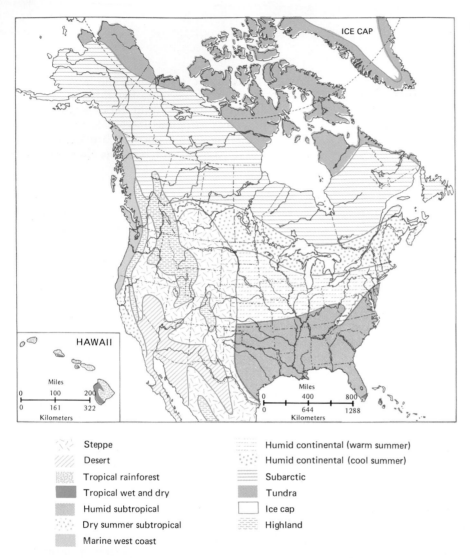

ICE CAP

HAWAII

Miles
0   100   200
0   161   322
Kilometers

Miles
0   400   800
0   644   1288
Kilometers

≈≈ Steppe	⋯ Humid continental (warm summer)
⫽ Desert	⋰ Humid continental (cool summer)
▦ Tropical rainforest	☰ Subarctic
▓ Tropical wet and dry	▒ Tundra
▨ Humid subtropical	□ Ice cap
⋮ Dry summer subtropical	▒ Highland
▨ Marine west coast	

tains and valleys), the Coast Ranges of California. The linear character becomes less distinct in Oregon and culminates in the Olympic Mountains of Washington. Between the Coast Ranges and the Sierra Nevada and Cascades lie several lowland areas. The Great Valley of California, a fertile alluvial trough, is one of the most productive agricultural regions of the United States. The Cascades and the Coast Ranges enclose the productive Willamette Valley and the Puget Sound Lowland. The Coastal Mountains are discontinuous in the form of islands off the Canadian coast, and the lowlands are submerged, reappearing only in Alaska as basins south of the Alaska and Aleutian ranges.

The Interior Plateaus

Between the Pacific Ranges and the Rocky Mountains lie the Interior Plateaus. The essential character of this region is that of a level upland; elevations are normally in excess of 3,000 feet (914 meters), but there are notable exceptions in the Yukon Plain of Alaska and Death Valley of California (the latter is

282 feet [86 meters] below sea level). The Basin and Range segment of Nevada, California, and Utah is characterized by faulted mountain ranges half buried in alluvial debris.

The Colorado Plateau (Arizona and Colorado) illustrates the previous nature and geologic history of the area. Many layers of sedimentary rock now at 9,000 to 11,000 feet (2,743 to 3,753 meters) above sea level are exposed by the Colorado River, which has cut the Grand Canyon and provided spectacular evidence of the great depth of underlying sedimentary rock. In eastern Washington, Oregon, and Idaho, formerly active volcanoes deposited thick layers of lava, some of which virtually covered mountains. The Snake River Plain has cut a canyon through the lava which nearly rivals the Grand Canyon in splendor. In Canada and Alaska, the more intermontane area consists of alluvial basins, plateaus, and mountains.

The Canadian Shield and Hudson Bay Lowland

In northern and northeastern Canada lies the great Canadian Shield, which nearly encircles Hudson Bay and extends southward to the United States, forming the Superior Upland of Wisconsin, Michigan, and Minnesota. The Canadian Shield was scoured by ice during continental glaciation and has a relatively smooth surface with little local relief. Unfortunately other products of glaciation are absent or thin soils, stony surfaces, and disarranged drainage with swampland. Although utility for agriculture is limited, the Canadian Shield has provided a wealth of furs, timber, and minerals. The Hudson Bay Lowland is a sedimentary region along the southern margin of the bay where the rocks of the Canadian Shield are beneath the surface.

Climatic Regions

Anglo-America exhibits the types of climate expected on a large continent extending from subtropical to polar latitudes. Extreme southern Florida, however, exhibits a small area of tropical wet and dry. The eastern portion of the continent shows a strong latitudinal influence in the sequence of humid subtropical, humid continental (warm and cool summer varieties), subarctic, and polar climates (Fig. 5–4). The most important differences among these climates are the shorter and cooler summers and longer and more severe winters as one proceeds northward from

the Gulf of Mexico. The interior west, because of remoteness from major moisture sources and location leeward of major highlands, is dominated by dry climates (desert and steppe). Along the coast from California to Alaska is a subtropical to high latitude west coast sequence of climates (dry summer subtropical, marine west coast, subarctic, and polar climates). Comparison of this sequence with that of Western Europe (Fig. 2–5) shows a basic similarity but also a significant difference: the highlands are more effective in Anglo-America in preventing the extension of these humid climates great distances to the interior. The great diversity of climates in Anglo-America is an expression of variations in precipitation and temperature regimes.

Precipitation Patterns

Two humid regions are evident in Figure 5–5. The larger eastern region extends from the Texas Gulf Coast to Hudson Bay and eastward to the Atlantic. The general pattern within the region is one of decreasing precipitation with increasing distance from coastal areas, except in the southern Appalachians where higher precipitation occurs because of an orographic effect. Although the southern Appalachians receive somewhat more rainfall than adjacent areas, they are not effective as a barrier to the moisture-laden winds of either the Gulf of Mexico or the Atlantic. Maritime air masses from these water bodies are the moisture source for areas far to the interior.

A second humid region covers a smaller area along the west coast from California (the Sierra Nevada) to southern Alaska and the Aleutian Islands. The great variations in precipitation are a reflection of elevation and exposure to rain-bearing winds. Locations at lower elevations in sheltered positions receive much less rainfall. Seattle, Washington, and Vancouver, British Columbia, illustrate this difference (Fig. 5–6).

A region of low precipitation (less than 20 inches [51 centimeters] per year) extends from the American southwest to the Northwest Territories of Canada. In the far north it spreads from western Alaska to Greenland, though we normally do not think of the extreme north as dry because of low evapotranspiration rates. The limited precipitation is most important in human terms in the southern half of this large region. The agricultural potential of the immense Interior Plains is considerably reduced because of the large area (Great Plains) in which moisture is deficient. Even the marginal area (steppe), which can be used for the production of drought-resistant crops, suffers recurrent problems because of the high variability of precipitation experienced in such areas. The vast steppe and desert area is the result of interior location and the barrier effect of highlands.

The interaction of continental air masses from the Canadian north (cool and dry) with the maritime air masses of the Gulf and Atlantic (warm and moist) provide much of the day-to-day weather variation in the eastern part of North America, particularly in winter and spring. The interaction of these air masses in storm systems is the source of much frontal precipitation, and the flow of maritime air contributes to convectional activity, especially in summer. The vast interior, however, receives much less of this precipitation simply because of its great distance from the Gulf and Atlantic. The west coast receives moisture from air masses moving eastward from the north Pacific. Although copious amounts of rainfall may be experienced along the coast, the effect of the highlands is pronounced. Areas immediately east of

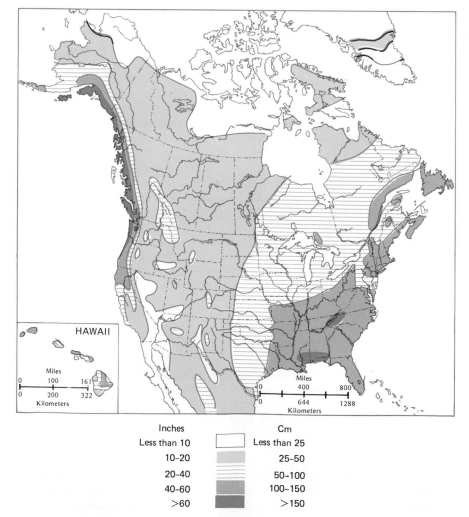

5–5 *Anglo America: Annual Precipitation*

Inches		Cm
Less than 10		Less than 25
10–20		25–50
20–40		50–100
40–60		100–150
>60		>150

HAWAII

Miles
0   100   161
0   200   322
Kilometers

Miles
0   400   800
0   644   1288
Kilometers

5–6 *Anglo-America: Mean Monthly Precipitation for Selected Cities. Annual precipitation is given in inches and centimeters: 33 inches (84 centimeters)*

SOURCES: Weather Bureau, *Monthly Summaries of Climatological Data* (Washington, D.C.: Government Printing Office, 1970–1975); and *Average Climatic Water Balance Data of the Continents*, Part VI (Centerton, N.J.: C. W. Thornthwaite Associates, 1964).

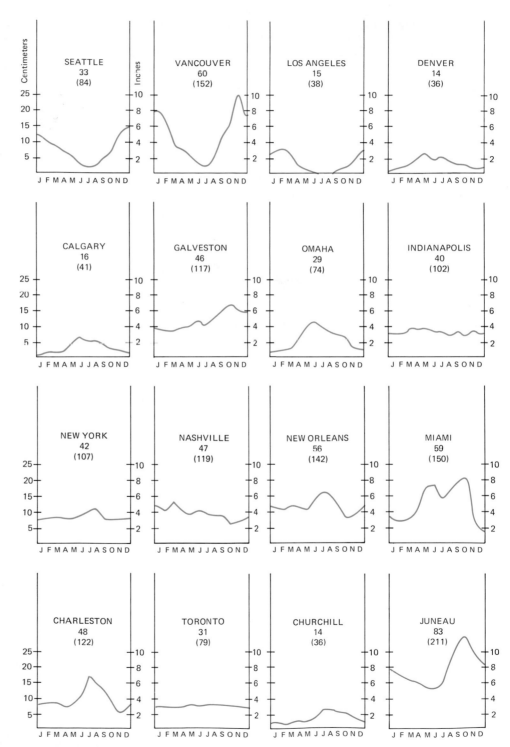

the Coastal Ranges, the Cascades, and Sierra Nevada receive such small amounts of precipitation as to render them moisture deficient.

Water Resources

The distribution of precipitation is of major importance, not only because of the direct importance of water for agricultural activities, but also because of the critical relationship between water and other functions of an urban-industrial society. The fact that most of the Anglo-American population is found in areas where precipitation is most plentiful does not preclude water problems, for these are precisely the areas in which water consumption is high. One of the important problems in the northeastern United States involves adequate and proper development of water resources for municipal, industrial, and recreational use. The planning and use of water resources are complicated by numerous state and municipal governments which, as water users, are frequently competitive. Regional planning on a broad scale may be one means of adequate provision of water resources in the future.

The water problems of the American southwest are different. A high proportion of the much more limited water resources is used for irrigation and therefore is competitive with urban-industrial water demand. The utility of irrigation water for increasing agricultural productivity and income, as well as accelerated urban-industrial needs, is illustrated by a long history of legal disputes between California and Arizona and the United States and Mexico for rights to Colorado River water.

Water is possibly the most basic natural resource of any society. Only small quantities are necessary for direct human consumption, but great quantities are necessary for the support of vegetation and the production of agricultural commodities. Most such water is derived from precipitation directly, but some is withdrawn from streams and wells for irrigation. As a society proceeds from an "underdeveloped" to a "highly developed" condition, the transformation of the economic system greatly reshapes the demand for water. Water needs increase with population growth, both for direct consumption (domestic use) and expanding agriculture. Even more significant is the growth of water needs associated with industrial

expansion. Industrial societies use immense quantities of water as solvents, as a waste carrier, and as a coolant. Pulp and paper mills, petrochemical industries, petroleum refineries, and steel mills are large users of water. More than 60,000 gallons (227,118 liters) of water may be used in the production of a ton of steel, but not all is consumed. Industrial water may be used and returned to surface water, at a high temperature, or unclean, and with a great impact as a pollutant. Many industrial plants recycle their water for reuse or cleanse used water before discharge back into streams and lakes.

In recent years in the United States, less than 10 percent of the water removed from surface sources was for domestic use. More than 50 percent was for industrial use. The remainder, approximately 40 percent, was used as irrigation water. Most such water is surface water drawn from streams, rivers, or lakes as opposed to groundwater that is removed by wells. Two-thirds of the irrigation water and three-fifths of municipal water are from surface sources.

Temperature

The amount of energy available for the conversion of nutrients and water into vegetable matter shows as much regional contrast as precipitation (Fig. 5–7). Much of southern Anglo-America has a growing season in excess of 200 days, which allows the production of a great number of "subtropical" crops (cotton, peanuts, citrus, and even some sugar cane). The length of the growing season decreases with higher latitudes. Much of Canada and Alaska are not suitable for agricultural production. Growing season, however, is not the only detrimental environmental feature for Canadian agriculture. The highlands of the west and thin soils and poorly drained glaciated areas of the east further detract from the utility of the Canadian environment for agriculture.

The temperature regimes of the United States and Canada are largely attributable to three climatic influences: latitude, altitude, and marine. Foremost is latitude, evidenced in the east-west orientation of isotherms on Figure 5–7. Plant growth seasons become increasingly shorter with increased latitude. The effect of higher altitude in the western portion of the continent is also evident where more severe minimum temperature regions exist in highland

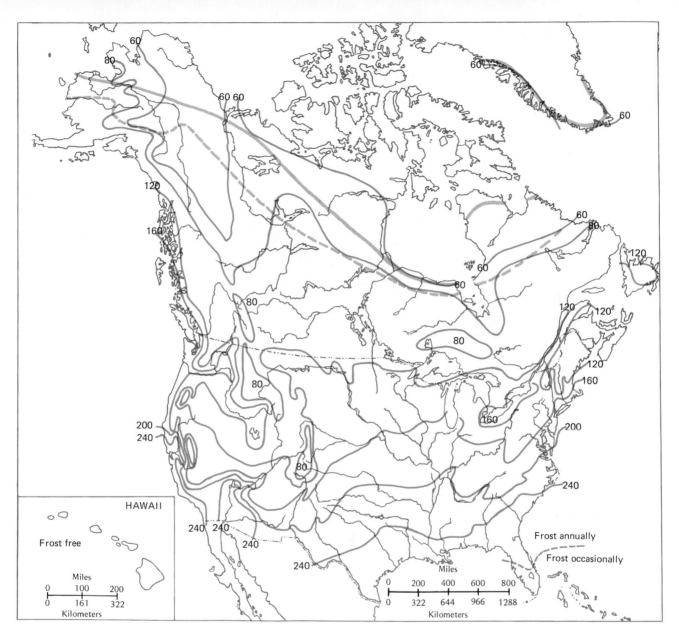

Growing Season	Frost-Free Days
Long	More than 240
	200–240
Moderate	160–200
	120–160
Short	80–120
Minimal	60–80
Frost anytime	Less than 60

PERMAFROST —— Continuous
------ Discontinuous

HAWAII

Frost free

Miles
0   100   200
0   161   322
Kilometers

Frost annually

Frost occasionally

Miles
0   200   400   600   800
0   322   644   966   1288
Kilometers

*5–7 Anglo America:
Frost-Free Period*

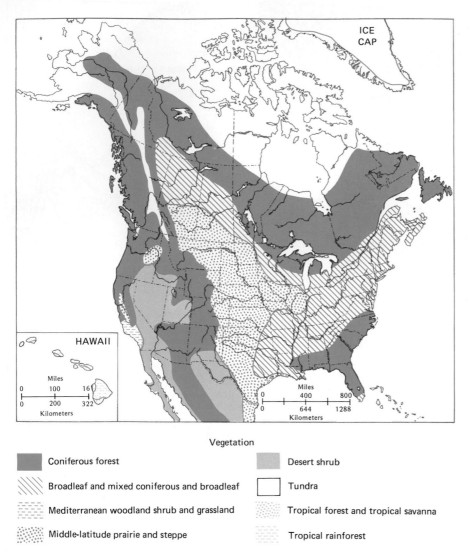

5–8 *Anglo-America: Natural Vegetation*

Vegetation

Coniferous forest	Desert shrub
Broadleaf and mixed coniferous and broadleaf	Tundra
Mediterranean woodland shrub and grassland	Tropical forest and tropical savanna
Middle-latitude prairie and steppe	Tropical rainforest

areas. The marine influence is reflected in the long growing seasons associated with coastal zones as contrasted with interior locations.

Vegetation Patterns

A narrow band of treeless tundra extends across far northern America from Alaska to Greenland (Fig. 5–8). In Alaska and northwest Canada, the tundra extends southward in highland areas. South of the tundra is a vast coniferous forest extending from Newfoundland to Alaska. This boreal forest, along with the similar taiga of the Soviet Union, is one of the largest expanses of forest remaining in the world. The dominant species of this forest are spruce, fir, and pine. Extensions of the boreal forest into the United States occur only in the highlands of the Far West, in the Lake Superior area, and in New England.

Despite Canadian emphasis on forestry, two-thirds of the boreal forest remains unused and, to some degree, protected because of poor accessibility. The southern margins of the forest and those areas adjacent to waterways have been most intensively exploited. The boreal forest gives way to deciduous forests through a broad transition zone of white and yellow birch, poplars, and maples (broadleaf species) in the humid eastern half of the continent.

The original deciduous forest roughly coincided with the northeastern quadrant of the United States. It consisted of oak, elms, hickory, beech, maples, and, in less fertile areas, pine. The deciduous forest is perhaps the most modified vegetation region in Anglo-America. Its very existence suggested to early European settlers advantages in climatic and soil resources. Utilization of such resources required removal of trees. Remnants of this forest remain only on nonfarm land, in areas where agriculture has ceased, or on farm woodlots, and even then not in original form. This great deciduous forest extended much farther south than is generally realized. Much of the Piedmont consisted of oak-hickory forest, but generations of farmers removed the original forest, produced cotton and corn, and ultimately abandoned the land. Today varieties of pine stand on the Piedmont where there were once immense hardwoods. Indeed, pine has become a common denominator in the landscape of much of the lower South. The forests that have been reestablished by natural processes or planting are indeed part of the American cultural landscape.

It is fortunate that all forest resources did not occupy land considered desirable for agriculture. The forest resources that remain provide the basis for a major part of Canadian industrial structure (lumbering and pulp and paper manufacturing). Similar industries are of primary importance in the United States Southeast, the Pacific Northwest, and secondarily in the north central and New England states.

The grasslands of interior Anglo-America result from limited moisture. The progression is from tall grass prairies on the eastern margin of the Great Plains, transitional between humid and dry climates, to short grasses on the western margin. Actually, the prairie extended eastward beyond the Great Plains into Iowa and Illinois at the time of European settling. Frequent "oak openings," forests interrupted by expanses of prairie, characterized the transition zone. Many prairie openings were initially encountered by early settlers even in the far eastern portion of the continent; old fields left by Indian communities were readily usable for agriculture. The prairies of the Great Plains also may have been extended eastward by the repeated use of fire by Indians.

Although grassland areas are indicative of moisture deficiency, their use for agriculture is not precluded. The excellent soils commonly associated with prairies have led to agricultural development. Unfortunately, although limited precipitation is a factor in the processes of fertile soil formation, the same characteristic contributes to the hazards of recurring drought. In areas where supplemental irrigation for farming is not available, extensive wheat farming and grazing are common, though crop yields and animal-carrying capacity are low. Even with ranching, resource destruction can occur. Overgrazing has frequently damaged natural grasslands, which are then replaced by woody shrubs such as sagebrush or mesquite.

Soils

The spatial correspondence of soil groups with climatic and vegetative patterns is evident from a comparison of Figures 5–5, 5–8, and 5–9. The associations are not unexpected, since climate and vegetation are intricately involved in soil-forming processes. Elimination of areas subject to severe temperature, precipitation, or topographic constraints allows focus on the gray-brown and red-yellow podzolics, chernozemic soils, and alluvial soils as those most significant to food and fiber production. One of the most favored large regions of the world for "modern agriculture" is the area from central Ohio westward, across Indiana, Illinois, and Iowa. The fertility of the gray-brown podzols, merging with chernozemic soils to the west, the favorable topography, the moderate growing season, and reliable rainfall combine to make this one of the world's most productive agricultural areas.

The chernozemic soils of the United States and Canada (Great Plains) are perhaps even more fertile as measured by nutrient content but are less productive because of limited precipitation and recurring drought. Utilization of these soils requires farming systems that incorporate drought-resistant crops (wheat) or grazing systems based on pasturage.

The red-yellow podzolics have experienced severe leaching and require substantial inputs of fertilizer to maintain productivity. Many problems of southern agriculture, however, are not inherent in the soils but the result of farming methods. Row crop production

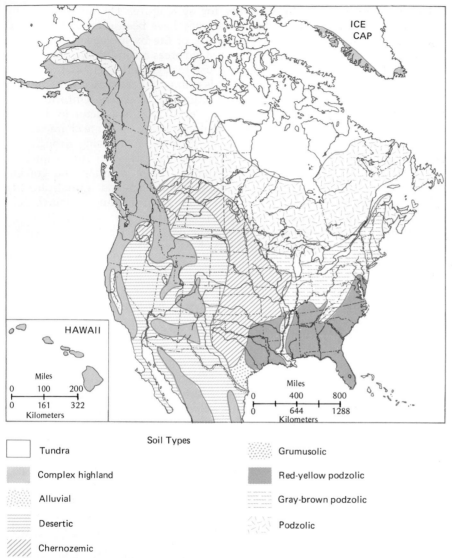

Soil Types

☐ Tundra		⣿ Grumusolic	
Complex highland		Red-yellow podzolic	
Alluvial		Gray-brown podzolic	
Desertic		Podzolic	
Chernozemic			

for many years on the hilly Piedmont accelerated severe erosion problems that eventually contributed to the abandonment of crop agriculture in many areas. There are also exceptionally favorable soils in the South. The Black Soil Belts of Texas and Alabama, the limestone valleys of the Appalachians, and the alluvial Mississippi Valley contain some of the best soil resources of the United States.

Early use of Anglo-American soil resources was often destructive for several reasons. Settlers occupying such vast areas were imbued with the notion of almost unlimited space and expansion possibilities. Indeed, the forest had to be cleared for agriculture, and clearing became a basic task of early agricultural systems. Farmers often engaged in a system where labor was expended on new land rather than in maintaining old land. In older areas of settlement, east of the Appalachians, much land had cycled in and out of agricultural use by the time of the Civil War. Elsewhere, large areas showed the destructive effects of soil erosion by water and wind by the 1930s. Destruction in a variety of environments resulted from overgrazing, cropping of land subject to drought, water and wind erosion, and improper row cropping. Hardly any major agricultural region of the United States or Canada was immune from the problem.

Since the 1930s major conservation efforts have been implemented. The efforts of the Soil Conservation Service in the United States and comparable agencies in Canada have promoted both removal of submarginal land from agriculture and have improved agricultural methods. It is important to recognize, however, that one key reason for success in improving the use of soil resources is often overlooked. During most of the period since the 1930s, the Anglo-American agricultural problems have been related to surplus production and low prices. From the standpoint of land needed for food production, population pressure was low. Thus programs could be implemented which removed land from production or reduced the intensity of production and yet were not detrimental to the larger economic system. Few areas of the world have been in such a position. The Soviet Union, by contrast, has sought to increase land area in agricultural production, sometimes using land comparable to the marginal land of the Great Plains. The questions now are: What of the future as pressure for domestic production increases, and how do other countries solve land-use problems where population pressure may be much greater and destructive practices are already widespread?

Resources for Industrial Growth and Development

The high material level of living that most Anglo-Americans enjoy is based upon an expanding use of minerals and fossil fuels in the production of consumer and capital goods. The consumption of these items will continue at current high levels, and in many instances may increase at rates faster than

Table 5–1 *United States: Power Consumption by Source*

Power Source	Percent Contributed
Coal	17.9
Petroleum	42.5
Natural gas	34.6
Water	4.4
Nuclear	.6
	100.0

SOURCE: U.S. Bureau of the Census, *Statistical Abstract of the United States* (Washington, D.C.: Government Printing Office, 1974).

current population growth, as they have in the past.

The world power position of the United States has been based to a great extent on the utilization of mineral wealth. The degree to which the United States will be able to maintain this position in world society will depend partly on the continued availability of mineral fuels and crucial metallic minerals. Canada, though less densely settled, has an economy strongly based upon natural resources (forests, oil, hydroelectric power, and a variety of metals).

Power Supply

The nonindustrial society depends primarily upon animate sources of power (muscle). Societies proceeding through industrial development become increasingly dependent upon inanimate power sources for both consumer and industrial uses. The United States' experience illustrates the importance of an adequate power base for development. It can be argued that power can be bought, but in the last decade only at increasingly high prices and with potential detrimental effects on balance of payments. Domestic availability of power resources is an index to a nation's potential and vulnerability.

As it was in Europe and the Soviet Union, coal was the power source for American industrial expansion. Even prior to World War II, however, the contribution of coal to the energy supply was decreasing relative to the contribution of petroleum and natural gas (Table 5–1). The increased fuel demanded by an expanding automotive industry did not provide expanding markets for coal. In addition, substitution of petroleum and gas as power sources for heating and industrial use inhibited expansion of the coal in-

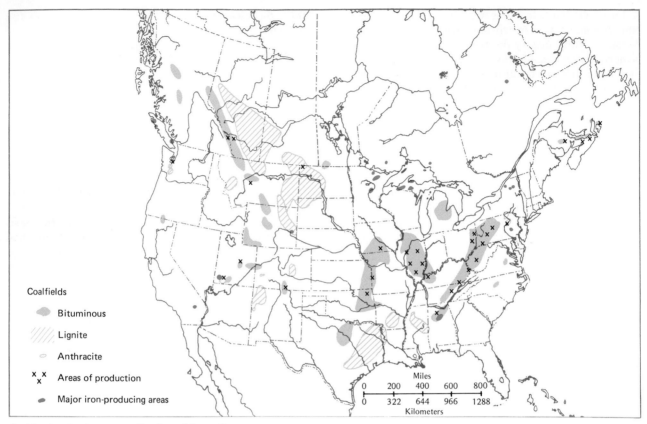

Coalfields

- Bituminous
- Lignite
- Anthracite
- x x x Areas of production
- Major iron-producing areas

Miles
0 200 400 600 800
0 322 644 966 1288
Kilometers

5–10 Anglo-America: Coal and Iron Ore

dustry. The absolute production of coal has alternately decreased and increased.

Attendant depressed economic conditions in the coal production areas (Pennsylvania, West Virginia, and Kentucky) were less from the lack of expansion than the result of streamlining mining technology and mechanization. While the coal-producing industry realized economic benefits from automation, the need for miners decreased from more than 500,000 prior to World War II to approximately 145,000 in the late 1960s, most of the decline occurring during the 1950s. The severely affected areas were eastern Kentucky, West Virginia, and Pennsylvania, major contributors to "poverty Appalachia."

The major coal-producing states are West Virginia, Kentucky, Pennsylvania, and Illinois (Fig. 5–10). Large quantities of coal are also available in western states, but production is limited by small local need and the great distances from the major eastern markets. Tremendous quantities of bituminous coal remain, enough to supply United States energy needs for several hundred years—a comforting thought, perhaps, but not so simply implemented. The main use for coal currently is for fuel in the generation of electrical power (about 60 percent of the market) and the remainder mainly as an industrial fuel. Only small quantities are exported. Its use for home or transportation purposes would require costly conversion efforts and the application of new and expensive technology.

Canada also contains large reserves of coal; however, their location greatly reduces utility. Two Maritime Provinces—New Brunswick and Nova Scotia—and two Prairie Provinces—Alberta and

Saskatchewan—contain most of the reserves. The great area of need for coal as a power resource is in the urban-industrial regions of Ontario and Quebec. Because of the great distances involved, Canadians find it more practical to import Appalachian coal. Another result of the unfavorable location of Canadian coal is that Canada has placed a correspondingly greater emphasis on both petroleum (cheaper to transport) and water as power sources.

Environmental concerns increasingly affect the feasibility of using various resources. Coal is a prime example. The supply of low sulfur coal will depend on economic means for removing sulfurous pollutants. The increasing restrictions and regulations, for air-quality control, on industry and individual use of fuel and the concern for avoidance of the scarred landscapes once left behind by the mining industry are factors affecting any effort at expanded use of coal as a power source.

Petroleum and natural gas have become increasingly important sources of power in recent years. Combined, they account for more than 70 percent of United States energy consumption. The largest single use of petroleum is as automotive fuel (55 percent); heating needs provide approximately 20 percent of the petroleum market. Ten percent is used as an industrial raw material, such as for road oil, lubricants, and raw material for petrochemical industries (ammonia, carbon black, synthetic rubber, plastics, and synthetic fibers).

Petroleum production in the United States began in 1859 in Titusville, Pennsylvania. The expansion of petroleum use was particularly rapid with the advent of automotive transportation. For many years its rate of use increased more rapidly than the increasing demand for energy because of substitution for coal, dieselization of trains, and use as a heating fuel. It once accounted for nearly one-half of the nation's energy requirements, but its proportional share has declined slightly as natural gas has, in turn, become a substitute for oil as a heating fuel.

The future of oil and gas in the energy picture of the United States and Canada is difficult to assess. First, the United States is both a major producer and consumer. Although coal production and consumption have been nearly equal until recent years, the same cannot be said of petroleum. The United States has produced approximately 17 percent of the world's petroleum in recent years but has also accounted for nearly one-third of the world demand, almost 5 billion barrels annually. There is little doubt that demand will increase. Domestic demand and availability of oil depend upon many related factors such as substitution, technology, economics, government, and international demand.

Proven reserves, a commonly used term, has limited meaning, since it refers to oil known to be available by actual drilling and removable at a given cost and technology. Actual reserves of oil are much greater. The proven reserves to production ratio have been approximately 10 to 1 in recent years. This ratio does not mean that the United States will run out of petroleum in ten years. Present estimates indicate that more than 500 billion barrels of oil remain; at present recovery rates, 250 billion barrels are therefore available. The recovery rate will likely increase, and continued new information on the Alaskan reserves will likely require major upward adjustment of these figures. Thus, depending upon the consumption level, enough oil is beneath the surface to last for some decades. Nevertheless, oil imports account for more than 21 percent of the oil consumed in the United States. The economics of oil have encouraged major companies to use imported oil, and thus the dependence upon foreign oil has expanded in recent years. Venezuela, Canada, and the Middle East are the major suppliers. The degree of United States' dependence upon foreign oil will depend upon both price, which has increased greatly in recent years, and availability, as other growth areas seek larger quantities of oil. Newly developing countries may place greater dependence upon oil as fuel than did the United States or Europe during their industrialization process.

The present major oil- and natural gas-producing regions of the United States are in Texas, Louisiana, California, Oklahoma, and Wyoming (Fig. 5–11). Alaska's north slope will in all likelihood catapult that state as a ranking source of petroleum for domestic use.

In recent years the use of natural gas as fuel has increased more rapidly than other fuels, even petroleum, largely because of low cost. Low cost will undoubtedly maintain a high demand and encourage continued use. In recent years, as the rate of energy use increased at a rate of 3.1 percent, natural gas use increased at 6.9 percent. Natural gas availability, however, is more limited than petroleum, so the

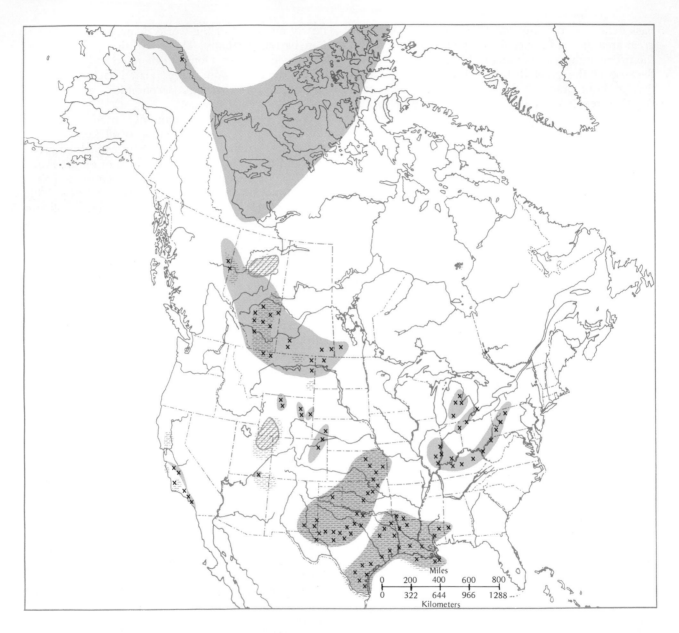

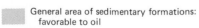
General area of sedimentary formations: favorable to oil

x x x
x x x   Producing oil fields

Natural gas supply areas

Oil shale or tar sands

––––––   Canadian energy line
Refineries west of line supplied by domestic crude
Refineries east of line supplied by imported crude

5–11   Anglo-America: Petroleum and Natural Gas

rapidly increasing demand may not be easily met.

Canada, though not one of the world's oil giants, does produce significant quantities of oil and natural gas in the Prairie Provinces of Alberta and Saskatchewan. Again, the locational aspect of the Canadian oil and gas industry is most significant. A line south from Pembroke to Brookville, Ontario, serves as an "energy line." Canadian refineries east of the line depend largely upon imported oil, mainly from Venezuela. Refineries west of the line utilize Canadian domestic oil and gas, serve localized markets, and export to the United States. Only in recent years have oil exports exceeded imports.

Oil Shale and Tar Sands

The solid organic materials associated with other minerals in shale formations of Utah, Colorado, and Wyoming (the Green River Formation) represent one of the world's largest deposits of hydrocarbons; the energy potential is immense. The contribution that these resources will make to energy supply by the year 2000 will probably be very limited, but advances in technology or price of petroleum could change this assessment dramatically. Technology for oil production from shale rock exists, but efficiency of production is far from competitive with other forms of power resources. Moreover, were it necessary and economically feasible to use oil shale, immense environmental problems must be overcome. Vast quantities of rock must be processed, and restoration policies minimizing environmental destruction are a significant factor in making any decision regarding large-scale production.

Canadian tar sands along the Athabasca River in Alberta also contain great quantities of oil. Like the United States oil shale, their significance will depend upon improved technology and cost considerations.

Waterpower

Mechanical waterpower (waterwheel) use was of major importance in early phases of industrial growth (eighteenth and nineteenth centuries). Mechanical waterpower, however, meant that only small quantities of power could be harnessed at any point; industry necessarily was dispersed. Since then, technology has developed which allows the movement of water through turbines for the purpose of generating electrical power and thereby amassing large quantities of power at a point. Waterpower is now used to provide 75 percent of Canada's electric power needs and 20 percent of the United States electrical power. Nevertheless, waterpower as a contributor to all power consumed continues to rank far below coal, oil, and gas. Despite the likelihood of continued expansion of waterpower development, the total share of power so provided, electrical and total, will likely continue to decline.

Present developed waterpower is probably about 25 percent of the total available in Anglo-America. Development is proportionately higher in several Anglo-American regions. The West Coast (Columbia River Basin), the Tennessee River Valley and southern Piedmont, and the St. Lawrence Valley are all areas where waterpower generation is proportionately higher. Canada, with its limited quantity of coal, has placed especially great emphasis on the generation of electrical power with water. It has contributed directly to the massive Canadian aluminum industry, an industry for which Canada has no raw material and only a limited market.

Nuclear Power

The extent of uranium or thorium use in the future for energy purposes is difficult to predict. The likely continued advances in nuclear technology, the cost of alternative fuels, dependence upon and cost of foreign oil, and environmental concerns will all stimulate or inhibit the development of this energy resource. Whether or not it will become a major energy source cannot be said. At present, however, the United States' supplies of uranium may not be adequate to supply large long-term demand. Nuclear technology is in its infancy and continually advancing. The likelihood of improving the efficiency of production is substantial (Fig. 5–12).

Metal Resources

The United States illustrates how iron ore, complemented by coal, underlies modern industrial structure. The United States is both the major producer (17 percent of the world total) and the major consumer (25 percent) of iron ore. Despite recent substitutions of aluminum and even plastics, iron remains the metal consumed in greatest quantity.

The Lake Superior District (Fig. 5–10) has been the major source of United States ore for many years. The development of the Soo Canal in 1855, on the St. Mary's River between Lakes Huron and Superior,

5–12 *United States Nuclear Power Plants, Units in Operation, 1974*

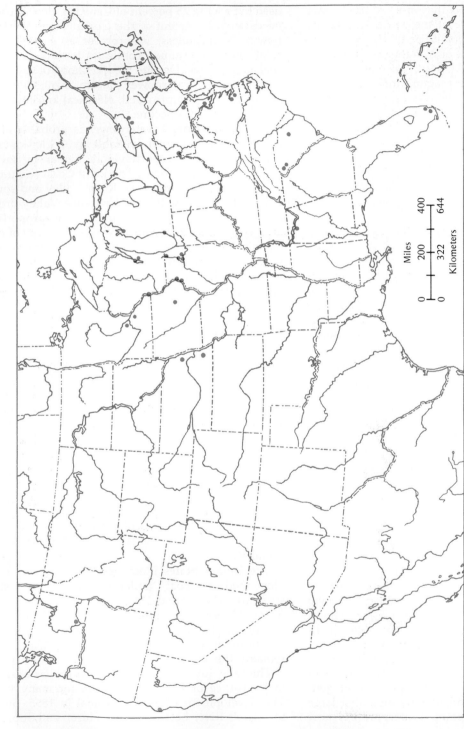

SOURCE: Atomic Energy Commission, *Nuclear Power Growth, 1974–2000* (Washington, D.C.: Office of Planning and Analysis, 1974).

facilitated the use of Lake Superior ores. The great Mesabi Range, the single most important ore deposit in the United States, began supplying ore in 1890. Despite increased independence upon foreign ores, more than one-half of the iron consumed in the United States continues to come from the Lake Superior area. Other significant domestic sources include the Adirondack Mountains (New York) and the Birmingham, Alabama, area. The Birmingham-Gadsden area of Alabama is somewhat unusual in that coal, iron ore, and limestone (for flux) are all found in proximity, thereby providing low assembly costs on the raw materials required for steel production. Many scattered iron deposits are found throughout the West, but most important are those near Cedar City, Utah. Distance from the major United States markets, however, dictates use in the smaller steel centers of the West (California and Utah).

It became apparent during the 1940s that high-grade ores (with an iron content of 60 percent) were becoming less readily available. Dependence upon foreign ores accelerated during the 1950s and 1960s. About one-half of the imported ore normally comes from Canada which, considering its population size, must be considered one of the mineral-rich nations. Canadian ore is available in the Lake Superior District at Steep Rock, Ontario, and major deposits have been developed in the past decade in Labrador. The building of a railroad connecting the Labrador District (Schefferville, Wabush Lake, and Gagnon) with Sept-Isles at the mouth of the St. Lawrence River, plus the development of the St. Lawrence Seaway, have made it possible for Canadian ore to move cheaply to numerous Great Lakes ports at which steel industries are located. Furthermore, the development of high-grade ores in Venezuela and large ore carriers encouraged use of Latin American sources. Venezuela, Chile, and Peru are additional major suppliers of iron ore. The development and subsequent improvement of the St. Lawrence Seaway and the construction of large ore carriers have facilitated the use of higher-grade foreign resources. This trend corresponds to the principle of the use of best resources first. That is, other considerations aside, it is less costly to extract iron from ore with a high iron content than from low-grade ore. In another context one might use good soils before poor soils, meaning more profit or lower cost for the consumer. In the case of iron ore, however, water transport is low cost for long-haul bulk shipment of raw material. Improvement of internal waterways simply increased accessibility to high-quality (cheaper to use) foreign ore.

Some authorities predict increased United States dependence upon high-grade foreign sources of ore, whereas others suggest that though dependence will continue, it may actually decrease in the future. Technologic advancements now allow the use of low-grade ores such as taconite, a very hard rock with low iron content, which is concentrated into ore pellets (with more than 60 percent iron content). The taconite industry has expanded rapidly in Minnesota and Michigan, partially in response to favorable tax concessions granted the industry. Steel manufacturers find some advantages in beneficiated ores, so taconite may one day be as economical to use as foreign ores.

Aluminum

During the past three decades, aluminum has become an extremely useful and sought-after metal. It is used extensively in the transportation and construction industries. Although aluminum is a common earth element, its occurrence in form that allows use for metal manufacture is limited. The United States consumes nearly 37 percent of the world's aluminum but produces only 15 percent of its own needs, mostly in Arkansas. Canada is the world's third ranking aluminum producer and number one exporter. The Canadian industry is based totally on the processing of imported ores and the use of substantial local hydroelectric power. The major consumers of aluminum are industrialized countries such as the United States, yet high proportions of the world's reserve exist in underdeveloped countries. The major Western world sources and those for the United States and Canada are in Jamaica, in Surinam, and in Guyana. Greater independence will depend upon continued improvements in technology that allow use of lower-grade domestic ores.

Conclusions

The United States and Canada are large countries with great quantities of basic resources. Anglo-America, however, also has a large population that consumes materials at high per capita rates. The great resource base and ability to use it have contributed both to Anglo-America's high material level of living and the primate power position in world society. Maintaining this position requires continued high resource consumption. To assume a major change in the basic nature of Anglo-America's social and economic system is unreal.

The varied climatic and soils regions of Anglo-America allow diversified agricultural production of food and industrial raw materials. Not only are domestic needs met, but large quantities of agricultural commodities (wheat, soybeans, cotton) are normally available for export. Surplus grains from both the United States and Canada are frequently needed to alleviate hunger in other parts of the world. Agricultural commodity export significantly affects the balance-of-trade position of both countries. The American agricultural economy was for several decades plagued by surplus production, now alleviated by lesser numbers of farmers and increased worldwide demand for food.

Anglo-America also is well endowed with industrial resources. Many decades of high-level consumption, however, and higher demand levels put increasingly strenuous pressures on domestic resources. The result has been an increase in the dependence upon foreign materials and fuel. Even if programs aimed at energy saving are successful, it is probable that they will only slow the rate at which consumption increases.

United States dependence upon foreign sources for basic needs has been to a substantial degree economic; for example, it has been cheaper to use foreign iron ore than domestic sources. Technology or politics may change this situation but, if not, dependence on foreign areas may increase. Underdeveloped areas, however, are increasingly demonstrating a desire to control the price, production, and processing of their own resources (Jamaican bauxite), even when developed nations are the ultimate consumers. Such trends will mean higher prices for foreign resources with attendant impacts on the domestic economy. It could also spur or provide incentive for domestic exploration and technologic advances for domestic lower-grade material use.

The questions relating to Canada are somewhat different. Canada is coal deficient, except in the Maritime Provinces, but can with relative ease and low cost obtain coal for its industrial needs from the United States. In other respects Canada is mineral rich. The concerns relate to what degree of dependence should the Canadian economy place upon primary production, of which petroleum and metals are a significant part. Furthermore, to what extent should the much smaller Canadian population allow its neighboring industrial giant to draw upon its petroleum, gas, copper, iron ore, nickel, and other metals as well as forest resources?

Another question is posed by the progress of the presently labeled underdeveloped countries. They may, presuming progress on their part, seek a greater share of the world's resources. Indeed, in the interest of social and economic progress elsewhere, the United States, which now consumes at such a high level, may find it necessary to strive for greater domestic independence, and leave resources for other growth areas.

Further Readings

General texts useful for an overview of Anglo-America include the following: Paterson, J. H., *North America: A Regional Geography,* 5th ed. (New York: Oxford University Press, 1975), 384 pp.; White, C. Langdon, Foscue, Edwin J., and McKnight, Tom L., *Regional Geography of Anglo-America,* 4th ed. (Englewood Cliffs, N.J.: Prentice-Hall, 1974), 617 pp.; and the *Oxford Regional Economic Atlas of the United States and Canada* (Oxford: Clarendon Press, 1967), 128 pp.

Additional general geography works on the United States include the following: Estall, Robert, *A Modern Geography of the United States* (Baltimore: Penguin Books, 1972), 401 pp.; Alexander, Lewis M., *The Northeastern United States* (New York: Van Nostrand Reinhold, 1967), 122 pp.; Hart, John F., *The Southeastern United States* (New York: Van Nostrand Reinhold, 1967), 106 pp.; and Akin, Wallace E., *The North Central United States* (New York: Van Nostrand Reinhold, 1968), 160 pp.

For additional general works that focus specifically on Canada, see the following: Gentilcore, R. Louis (ed.), *Canada's Changing Geography* (Scarborough, Ontario: Prentice-Hall of Canada, 1967), 224 pp.; Warkentin, John (ed.), *Canada: A Geographical Interpretation* (Toronto: Methuen, 1968), 608 pp.; Wilson, George W., and others, *Canada: An Appraisal of Its Needs and Resources* (New York: Twentieth Century Fund, 1965), 453 pp.; and Nicholson, Norman L., *Canada in the American Community* (New York: Van Nostrand Reinhold, 1963), 128 pp.

Data of great variety are found in the United States Department of Commerce, U.S. Bureau of the Census, *Statistical Abstract of the United States* (Washington, D.C.: Government Printing Office, 1974); and *Canada Yearbook* (Ottawa: Dominion Bureau of Statistics), an annual publication of statistics with narratives on the resources, history, institutions, and social and economic condition of Canada.

Information on the resource position of Anglo-America is included in the following: Landsberg, Hans H., *Natural Resources for U.S. Growth* (Baltimore: Johns Hopkins Press, 1964), 260 pp.; Committee on Resources and Man, National Academy of Sciences, *Resources and Man* (San Francisco: W. H. Freeman, 1969), 259 pp.; and Patton, Donald J., *The United States and World Resources* (New York: Van Nostrand Reinhold, 1968), 128 pp.; the United States Department of the Interior, Bureau of Mines, *Mineral Facts and Problems* (Washington, D.C.: Government Printing Office, Annual), provides a great deal of information on the production, processing, and use of specific minerals. See also Thomas, Trevor M., "World Energy Resources: Survey and Review," *Geographical Review,* 63 (1973), 246–258; Clements, Donald W., "Recent Trends in the Geography of Coal," *Annals, Association of American Geographers,* 67 (1977), 109–125.

For readings dealing with aspects of the regional environment, see the following: Beard, Daniel P., "United States Environmental Legislation and Energy Resources: A Review," *Geographical Review,* 65 (1975), 229–244; Hare, F. Kenneth, and Ritchie, J. C., "The Boreal Bioclimates," *Geographical Review,* 62 (1972), 333–365; Kollmorgen, Walter, and Kollmorgen, Johanna, "Landscape Meterology in the Plains Area," *Annals, Association of American Geographers,* 63 (1973), 424–441; Storrie, Margaret C., and Jackson, C. I., "Canadian Environments," *Geographical Review,* 62 (1972), 309–332; Lewis, G. Malcolm, "William Gilpin and the Concept of the Great Plains Region," *Annals, Association of American Geographers,* 56 (1966), 33–51; and Thornbury, William D., *Regional Geomorphology of the United States* (New York: Wiley, 1965), 609 pp. A popular account of the origin and character of the American physical landscape is found in Farb, Peter, *Face of North America* (New York: Harper & Row, 1963), 316 pp.

Anglo-America: Early Differences, Experiences, and Technologic Changes

6

THE ANGLO-AMERICAN ENVIRONMENTS, as we have seen, are vast and varied. These environments have been used, and often misused, by numerous people with different cultural heritages and experiences. Furthermore, the New World experience, a relatively short one when measured against Old World cultures, did not have beginnings in a single location. At least four early European settlement foci are identifiable which served as source areas for cultural "imprints" that have lasted to the present; possibly a fifth should be recognized (Fig. 6–1).

The result of European immigrants settling a new land over a few centuries is not simply the product of people and land; also critical is the time when settling occurred. The time was one of new ideas and rapid technologic advance. The purpose of this chapter, therefore, is to stress certain historic aspects of this experience, with the intent of demonstrating the importance of time, attitudes, demographic conditions, and technologic change in the development process of two countries.

Early Settlement

Early exploration of Anglo-America was carried on by the Spanish, the Portuguese, and the French. The Spanish were the first to establish a permanent colony (St. Augustine, Florida, in 1565), and with the exception of the British rule for the twenty years from 1763 to 1783, they controlled Florida until 1819. Despite the effort and long control of Florida, however, the Spanish never were successful in establishing a viable society from which settling and diffusion of culture could proceed inland. The great role of the Spanish people and culture in Latin America contrasts, therefore, with only a peripheral role in Anglo-America. It was northwest Europe rather than southwest Europe (Iberia) that provided the dominant cultural influence for Anglo-America.

The first permanent settlements from which distinctive American culture traits evolved were the English Jamestown Colony, established in 1607, and the French settlement at Quebec, established in 1608. Soon following was the Plymouth Colony (English, 1620) and later settlement in New York (Dutch, 1625). Germans and Scandinavians also made their appearance, but it was above all others the English cultural imprint that was the most profound and lasting on the new continent. The English came in greater numbers and, over all, exercised the greatest control in the development process. English dominance should not preclude recognition of the value of native American, African, Asian, or other European peoples (the French in Quebec) to American culture.

Transfer of European Ideas

Initial settling coincided with a time when expansionism was a motivating force in Europe. The colonies were a product of that mood. Social, political, and economic conditions in Europe were favorable to migration and colonization. European ideas regarding greater equality before the law were transferred to the colonies, where along with the mutual effort required for survival, they promoted egalitarian notions. Commercialism, long an acceptable activity and source of profit in Europe, found fertile ground in the colonies. That companies sponsoring settlements sought a profit as private enterprises was indicative of the heritage immediately transferred to this part of the New World.

Early Economic Orientation

The European colonists remained close to the Atlantic coast for nearly two centuries. Initial sponsors such as the London Company (Jamestown) were motivated by a desire for profit; hope even existed for mineral wealth. But the apparent lack of gold or silver meant that settlement into the interior would not immediately be established as they were in Latin America early in the sixteenth century. The search quickly focused on an alternative, a crop that could provide income. Tobacco, cotton, rice, and indigo became early income producers, as were fish and timber.

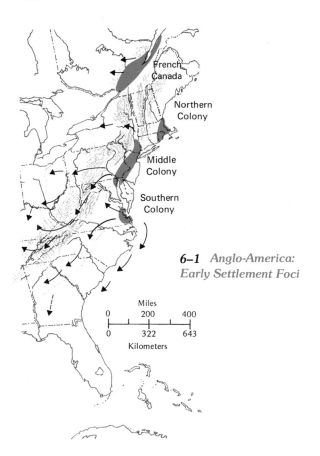

6–1 *Anglo-America: Early Settlement Foci*

Miles

0   200   400

0   322   643

Kilometers

Transport technology encouraged tidewater or coastal orientation for the production of such crops. Transport was slow and expensive, relative to modern forms. Human or animal and wagon transport over land was far more expensive than water transport. Even in the late eighteenth century it was cheaper to move a ton of iron ore across the Atlantic to England than overland 100 miles. The result, for those commercially motivated, was a decided emphasis on tidewater or coastal location to facilitate trade, migration, or intercolony movement. Thus, early plantations were often restricted to accessible waterways, such as the James River in Virginia, with subsistence farmers occupying inland areas more remote from water. The Virginia and North Carolina Piedmont was one such inland area; and later it became a source area for people who established the subsistence mountain culture of the Appalachians. In New England, the emphasis on commercial fishing and lumbering provided a strong coastal orientation. If great mineral wealth had been found in the interior, the entire settling process might have been different. Expensive land transport might have been overcome by the high value per unit of weight of the mineral commodity.

French Interests

Furs, particularly in the case of the French, provide an interesting exception and illustration of the way in which a high-value product overcomes high transport cost. The fur trade drew small numbers of people inland. Here was a high-value product that could be moved great distances overland or by water. Fur trade, however, usually involved trade and interaction with Indians, so it required little in the way of interior settlement other than for protection and manning trading posts. By 1700 the territory claimed as New France was far greater than that occupied. Settled New France consisted of the thinly populated areas around the Bay of Fundy, Acadia (Nova Scotia), and along the St. Lawrence (Quebec to Montreal). French people, in contrast to French government, had little interest in colonial settlement, perhaps because commercial agriculture was a less likely success in the more severe northern environment. Fishing contributed to a strong peripheral focus (Prince Edward Island-Isle St. Jean). Exploration and fur trade, however, led to vast claims that were later difficult to hold without permanent settlement and occupation. The end result was that New France became part of the British Empire in 1763.

More than a century was required to occupy the initial settlement core areas effectively. During the early settling phase, distinctive differences among the colonies began to emerge which ultimately contributed to significant regional differences within Anglo-America.

New England

A strong sense of commerce was established early in New England. Agricultural efforts were necessary to sustain the populace, but there was no special crop that could provide great wealth or form the basis for trade as tobacco did in the southern colonies. Instead, wealth was accumulated by fishing, trade, and forestry. The white pine forest provided useful lumber for shipbuilding and trade. The codfish on the offshore banks were another resource that could be traded. These resources, plus the wealth generated by their exchange, became a source of capital and established commercialism early in the northeast. By the late eighteenth century, capital was available for incipient industrial growth, and non-agricultural pursuits were already a tradition. Water-power potential in mechanical form (waterwheel) was substantial, and the ocean-shipping capacity for movement of raw materials and manufactured goods existed. Shortly after independence, New England's incipient industry emerged as a competitor with Europe. Along with this development came the idea of tariff protection in some form for domestic industry.

The Southern Colonies

The people of the southern colonies pursued a different path almost from initial settling. Tobacco became a commercial crop almost immediately, and later, as settlements were extended south at various coastal points, indigo, rice, and cotton were added. Many inland settlers, hampered by lack of access to water, were essentially subsistence farmers. Where commercialism was feasible, the plantation—a distinct agrarian system in a spatial, social, and operational sense—began to evolve. "Plantation" originally meant no more than a clearing in the woods, but it soon implied a system that required more than family labor inputs. Indentured and slave labor led to a dis-

tinction between labor and management, particularly when applied at large scale. Thus emerged the commercial plantation, larger than "family-size," with division between labor and management, social distinctions, and attendant forms of organization and layout of buildings.

The source of wealth for commercially oriented people was agriculture. The markets for their agricultural products were in Europe, and the southern producer wanted no tariff system that might inhibit the movement of his source of wealth. Basic sectional differences appeared early, one of which was the attitude regarding tariff policies. Another was the socially stratified society that evolved along with the economic system based upon slave labor.

The plantation was established from tidewater Virginia to Maryland and southward at various coastal points to Georgia. Inland from the tidewater and coastal agricultural colonies, and beyond easy water routes, were smaller free labor farms (yeoman farmers), particularly on the North Carolina and Virginia Piedmont. Later, particularly after 1800 when an improved cotton gin became available, both the yeoman farmer and the plantation culture spread throughout the lower South. Generally the plantation system prevailed in the choice areas for agricultural settlement. Plantation culture thereby came to dominate the social, economic, and political life of the lower South.

The Middle Colonies

New York, Pennsylvania, and portions of New Jersey and Maryland were a distinct early settlement core that contrasted with New England and the southern colonies. Settling occurred by a greater variety of people; English, Dutch, Germans, and Swedes were early participants. Neither the cash crops of the South nor the lumbering and fishing activities of New England were as significant as a source of income. Nevertheless, the area became an important American source region for both people and ideas. The middle colonies provided the settlers moving southward into the Appalachians, and westward down the Ohio Valley, and on into the Midwest. They used the Indian grain maize (corn) to fatten hogs and cattle, a system that spread into the American Midwest. They made tools, guns, wagons, and worked the iron deposits discovered in eastern

Pennsylvania. Although one might expect the middle colonies to have been transitional between the southern and northern colonies, they were not.

The Lower St. Lawrence

Although English culture has dominated Canadian evolution, it has not been exclusively so. The United States has in fact functioned as a source area for Canada. Canada's most distinctive settlement focus, however, was French, located along the lower St. Lawrence between Montreal and Quebec. The French spread themselves over a vast area as fur trappers, traders, and missionaries. Their numbers, however, were so small that claims over much of Canada and the Mississippi Valley could not be sustained. It was only in the lower St. Lawrence where Frenchmen settled as farmers that a lasting French cultural imprint was made. The French evolved as a traditional rural populace with a distinct culture. Now as a more urban element, they give a distinctive French character to an entire Canadian province. French Quebec, in contrast to the English provinces, provides a dichotomy to Canada as a country that requires careful and intelligent handling to avoid political disunity.

The Southern Appalachians

The southern Appalachians were not an original settlement area. Nevertheless, the Appalachians from Virginia and West Virginia southward (including the Blue Ridge and the Ridge and Valley) functioned as a secondary settlement area from which distinctive cultural traits were eventually diffused. The region acquired its settlers during the early eighteenth century. The migrants were descendants of the earlier Scots-Irish, German, and English settlers in the middle colonies. They moved southward along the Piedmont, and the Appalachian Valley, and westward into the plateaus of eastern Tennessee, Kentucky, and western Virginia. These migrants were intelligent, resourceful, and daring. Indeed they were settling the upland South even before the wave of settlers moved across the Midwest or lower South. These settlers carved out small subsistence farms and operated as slaveless yeoman farmers and thereby contributed to some of the distinctive traits that contrast the southern Appalachians with portions of the lower South. The southern Appalachians have remained an area of small farms and an almost totally white population.

In the more isolated portions of the Appalachians, a distinct culture evolved, not from a spirit of progressiveness, but rather as a culture of archaism resulting from isolation and an inability to change with the remainder of the country. As transport systems improved, the remote Appalachian coves and valleys remained unaffected, except when transport was necessary to remove a special resource such as lumber or coal. The distinctiveness of the mountain culture can be seen in Elizabethan speech and music, the use of distinctive suffixes attached to place names (cove, gap, and hollow), and mountaineer attitudes. Low education levels and poverty are widespread contemporary problems.

Labeling the southern Appalachians as a culture center implies that it functioned as a source area for other regions. It has done so for both the Interior Highlands of Arkansas and Missouri and the hill country of central Texas.[1] By the mid-nineteenth century, population levels in the Appalachians were such that out-migration was necessary. The surplus population found refuge in hill lands farther to the interior which provided some similarities in environment and the isolation for a culture that these people were not eager to change.

Americans and Canadians: Their Demographic Experience

The seventeenth- and eighteenth-century Anglo-American population grew slowly, even in New England and the Chesapeake Bay area where English interests and activity were most intensive. At the time of American Independence, 167 years after Jamestown was established, the colonies contained

[1]Terry G. Jordan, "The Imprint of the Upper and Lower South on Mid-Nineteenth Century Texas," *Annals, Association of American Geographers,* 57 (1967), 667–690.

only about 3 million people. Nearly 100 years passed before Canada contained a similar population. The early size differential between the United States and Canadian populations was established and maintained largely because of differences in net migration. The population difference, as we shall see later, has major implications for Canadian self-identity.

United States Population Growth

The rapid population growth experienced by the United States between 1800 and the present was a response to high birth rates, declining mortality, and immigration. The high United States birth rate declined by 1900 (3.2 percent) and continued declining to a low of 1.8 percent in the 1930s when young people married at an older age than previously. The severe economic depression of the 1930s and World War II had profound but opposite effects on the demographic experience of both the United States and Canada. The restraint on family size in the 1930s came to an end after World War II, possibly in response to wartime delays in family growth and the economic prosperity in the years following the war. Birth rates again rose and reached a new high of 2.7 percent, bringing a new era of relatively rapid population growth, or the "baby boom." More recently the birth rates have declined to a new low of 1.5 percent. The baby boom has been replaced by what some refer to as a "birth dearth" (Table 6–1).

The declining growth rate over the past century has not been in exact correspondence with changes in birth rates. Mortality rates declined and also are a factor in the demographic experience. The decline of infant mortality and the extension of life expectancy beyond seventy years mean greater numbers of people alive at any given time. Immigration has been another important factor in the growth experience of the United States. Approximately 44 million people have immigrated to the United States since 1820. The immigrants enlarged the population as they came but also increased the population base from which future growth was derived. The greatest numbers came from Germany, Italy, Great Britain, Ireland, the Soviet Union, and Canada. The flow was particularly great from 1880 to 1910; in the last of these decades more than 8 million immigrants entered the United States, accounting for 40 percent of United States population growth. Recent legal immigration has been at the rate of about 400,000 persons annually and accounts for about 16 percent of U.S. growth during the 1960s and 1970s. The amount of illegal immigration is not precisely known. In 1975, United States governmental officials estimated that 6 to 8 million illegal aliens were living in the nation.

Considering the differential between birth and death rates and immigration, the general trend has been declining growth rates during the last century. The rate has declined from more than 3 percent during the early nineteenth century to less than 1 percent at present. The current low rates of growth do not mean small population increments, however, for now the population base is large. Even a 1 percent rate of growth means an addition of more than 2 million people per year.

The predictions were for high rates of growth during the 1960s and the 1970s. Those born during the baby boom were expected to provide another period of relatively rapid growth as they reached childbearing age; however, as the proportion of the population in the childbearing age increased, the fertility rate decreased (rate of childbearing among women aged fifteen to forty-four). Whether or not the lower rates will prevail depends upon whether the young women are merely delaying childbirth or actually will have fewer children in the years ahead.

Canadian Population Growth

The Canadian demographic experience has been generally similar to that of the United States insofar as birth and mortality rates are concerned; however, a substantial difference in immigration has led to a Canadian population vastly smaller in size. A large proportion of Canada is not suitable for settlement and has attracted far fewer immigrants than the United States. Also contributing to a smaller population is the fact that several million people have migrated from Canada to the United States since 1867.

From 1867 (population about 3.5 million) to 1900, Canada grew mainly by natural increase. But even its natural increase was limited somewhat because of the low fertility rate, a response to youthful migration to the United States. Net migration was negative during most of this early period. After 1900, however, there was a large influx of immigrants from Europe, raising fertility rates and slowing the decline of birth rates. Like the United States, Canada experienced a low birth rate during the depression and a sharp rise after World War II, despite urbanization and industrialization. The Canadian baby boom was also followed by a birth rate decline in the late 1960s and early 1970s.

Immigration to Canada has exceeded 9 million, but Canada has not been able to retain this number. Emigration is estimated to have been over 6 million. Immigrants to Canada have often become return emigres or later continued on to the United States. Positive net migration has aided Canadian population growth in only two periods: the first three decades of this century and during the post–World War II years. The decade of the 1950s was the decade of greatest Canadian population growth, a response to high birth rates applied to a larger population base and positive net migration.

Low Population Growth and the Future

Despite the fact that both the United States and Canadian populations are still increasing, the birth rates of these two countries have decreased to a historic low. The dramatic decrease in the birth rate is similar to what is being experienced in approximately thirty other highly developed nations. This minority of developed countries is atypical; most other nations are experiencing high birth rates and lowering death rates. Therefore, while some industrialized nations are rapidly approaching zero population growth, other less developed countries are experiencing quite the opposite. Furthermore, some are not receptive to the view that their population growth is too rapid.

Zero population growth for the United States, if it is achieved, probably will not occur for at least another fifty years. The United States, like any other country rapidly reducing its birth rates, will likely have a total population fluctuating around some base level or growing at a modest rate. The present youthful population, even with lowered fertility rates, has the potential to create another small "baby boom" simply because of the large number of people involved.

Figure 6–2 shows a contemporary population pyramid and one as it might appear in A.D. 2000. Obviously the reduced birth rates have major implications for the age structure of the population. The now numerous youthful group will age and be replaced by a smaller youth group. The changing age structure has numerous economic and social implications, but people disagree on whether or not these changes will prove troublesome. Will the proportionately smaller youthful population mean possible labor shortages, especially for particular industries? Will some industries experience dramatic market declines? Will educational institutions experience an oversupply of facilities and personnel? Will the facilities required and the cost of caring for a disproportionately large aged population place a disturbingly high tax and social security burden on the economically active population? Or will this new age structure allow an immense improvement and solid attack

	Canada		United States		
	1901	1975	1900	1975	
Population	5,400,000	23,500,000	76,100,000	216,700,000	
Birth rate in percent	3.5	1.6	3.2	1.5	
Death rate in percent	1.5	.7	1.7	.9	
Immigrants per 1000 population	9	9.5[a]	8	1.8[a]	
Growth rate	3.4	.8	2.3	.6	

Table 6–1 Anglo-America: Demographic Features

[a]Value for 1975.

SOURCES: *Population Data Sheet* (Washington, D.C.: Population Reference Bureau, 1977); U.S. Bureau of the Census, *Statistical Abstract of the United States* (Washington, D.C.: Government Printing Office, 1974); and *Canada Yearbook* (Ottawa: Dominion Bureau of Statistics, 1973).

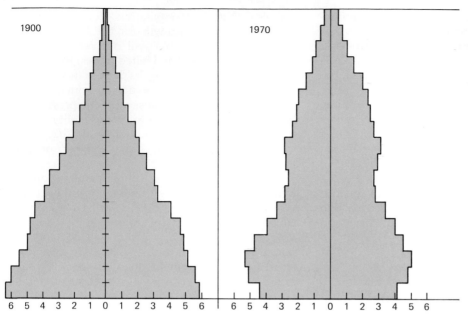

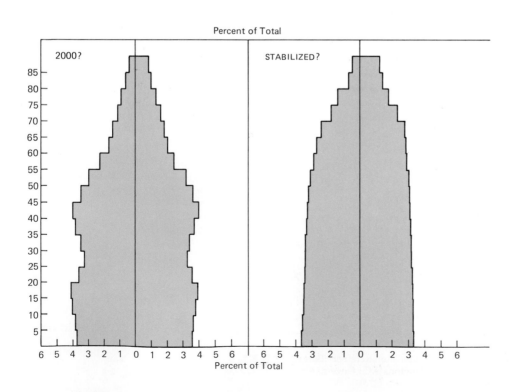

6–2 *United States: Population Pyramids, 1900 to 2000*
SOURCE: Adapted from Charles F. Westoff, "The Population of the Developed Countries," *Scientific American*, 231 (1974).

on some of the economic and social problems currently facing the United States?

Another consideration resulting from growth differences between developed and underdeveloped areas is the proportion of total population in under- developed areas. The population of developed countries is already a "minority" group and will become increasingly so. The smaller developed population, however, finds other countries looking increasingly to them for supplies of food and aid.

6–3 Anglo-America: Population Distribution

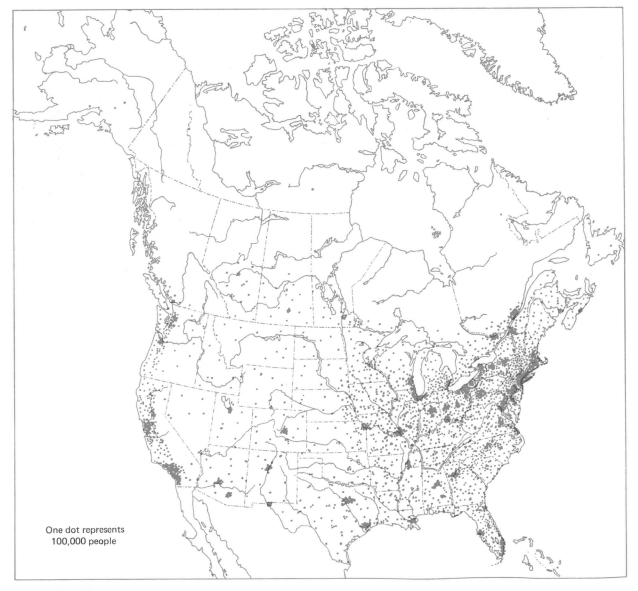

One dot represents 100,000 people

Population Distribution

Most of the United States population is located east of the Mississippi River (Fig. 6–3). Within this area it is even more concentrated in the northeastern quadrant bounded by the Mississippi and Ohio, the Atlantic, and the Great Lakes. Population densities are somewhat lower in the South, except in industrialized growth areas such as the Piedmont. Most of the west coast population is concentrated in several distinctive lowland areas which include the Los Angeles Basin, the Great Valley of California, the valleys of the Coastal Ranges in the vicinity of San Francisco–Oakland, and the Willamette Valley and Puget Sound Lowland. The remainder of the western United States is sparsely populated, particularly west of the 100th meridian. Exceptions are the higher population densities found at the oasis type locations exemplified by Phoenix, Arizona, and Salt Lake City, Utah.

Most of the Canadian population is within 200 miles (322 kilometers) of the United States border. The Prairie Provinces provide the only significant exception. If it were not for the sparse population immediately north of Lake Superior, the distribution might be described as a long east-west ribbon, north of which lies most of the vast Canadian space, only sparsely inhabited. More than 60 percent of Canada's population is located on the Ontario Peninsula and in the St. Lawrence Valley of southern Quebec (Fig. 6–3). The Maritime Provinces contain approximately 10 percent of the population and the Prairie Provinces about 17 percent. The already populated urban-industrial areas of Quebec and Ontario have been experiencing positive net migration, suggesting the continued concentration of population common in industrial societies.

Population Redistribution

Population redistribution began even before the initial settling of the more habitable parts of the United States and Canada was complete. The shift from an agrarian to industrial society began early in the nineteenth century. Industrial growth, particularly during the latter half of the nineteenth century, was the basis for major urban growth which has continued during this century but spurred more by the expansion of tertiary activities than by industrial growth.

The northeastern quarter of the United States has

Table 6–2 *United States: Population by Residence and Race, 1970*

Residence	Race			Average Annual Change 1960–1970
	THOUSANDS		PERCENT	PERCENT
Black				
Standard Metropolitan Statistical Areas	16,771		74.3	2.8
Central cities		13,140	58.2	2.8
Outside central cities		3,630	16.1	2.8
Nonmetropolitan Areas	5,810		25.7	− 0.5
TOTAL	22,581		100.0	1.9
White				
Standard Metropolitan Statistical Area	120,579		69.8	1.4
Central cities		49,430	27.8	− 0.1
Outside central cities		71,148	40.0	2.5
Nonmetropolitan Areas	57,170		32.2	0.6
TOTAL	177,749		100.0	1.1

SOURCE: U.S. Bureau of the Census, *Statistical Abstract of the United States* (Washington, D.C.: Government Printing Office, 1974).

Table 6–3   *Population of the Standard Metropolitan Statistical Areas of the United States*

	Total in 1970	Population in 1970 (thousands)	Percent of Total	Population Change 1960–1970 by 1970 Boundaries	
				Thousands	Percent
All SMSAs	243	139,419		19,824	
SMSAs by Size					
More than 3,000,000	6	37,710	27.0	4,002	20.2
1,000,000–3,000,000	27	42,946	30.8	7,584	38.2
500,000–1,000,000	32	21,936	15.7	3,348	16.9
250,000–500,000	60	19,761	14.2	2,769	14.0
100,000–250,000	92	14,973	10.7	1,892	9.5
Less than 100,000	26	2,091	1.5	229	1.2

SOURCE: U.S. Bureau of the Census, *Statistical Abstract of the United States* (Washington, D.C.: Government Printing Office, 1974).

led the country in urban growth until recent years. More than 50 percent of the northern population was urban in 1900, and now exceeds 80 percent in some states. Even those people classified as rural are such only in residence, since most are urban workers (Table 6–2). The west coast and the Southwest have also become highly urbanized. The South and the Great Plains (the Canadian segment included) have urbanized more slowly, as one would expect in an area with an agrarian orientation. In recent decades, however, urban growth in the South has been substantial, as the region shifted from an agrarian to an industrial economy. In fact, the urban nature of the South is somewhat obscured by urbanization measured by residence. The recency of industrialization, or shift from agricultural to industrial employment, has been made by many people who have been able to maintain rural residence. Therefore, a state such as North Carolina, which appears to lag in urbanization, has a level of industrialization which, when measured by employment, is above the national average. The rural nonfarm people are a growing segment of the population.

The United States and Canadian populations are continuing to concentrate in urban areas. The most rapid population growth during the recent decades is in the larger SMSAs (Standard Metropolitan Statistical Areas). Table 6–3 shows the greatest absolute and percentage growth to be in the SMSAs between 1 and 3 million in size. Furthermore, those SMSAs

growing most rapidly are in the southern half of the nation. The bulk of the population growth in Arizona is in just two cities, metropolitan Tucson and Phoenix.

The rapidly increasing metropolitan population should not obscure another significant movement, that of population leaving the larger central cities and moving to the suburbs, thereby adding even more to the land requirements of urban areas. Metropolitan area growth is a response to population growth and population shifts within metropolitan areas, particularly since World War II.

Population migration from rural areas has long been a source of population for urban centers. It is now, however, the urban centers themselves that are major sources of population growth. Three-fourths of the population growth of the 1960s was the result of population increase within the cities themselves, the remainder from rural migration and immigration. Rural areas will become even less important relatively as population sources areas. The President's Commission on Population estimated that of a growth of 81 million in metropolitan areas between 1972 and 2000 (based on a two-child family), 4 million will be from rural areas, 10 million from immigration, and the remainder from natural increase within cities.[2] A problem, however, is the fact that rural migrants with

[2]Report of the Commission on Population Growth and the American Future, *Population and the American Future* (Washington, D.C.: Government Printing Office, 1972), p. 29.

Percent
100
90–99.9
75–89.9
Under 75

6–4 *United States: 1970 Population as a Percent of Maximum Population, by Counties*

SOURCE: U.S. Bureau of the Census (Washington, D.C.: U.S. Department of Commerce, 1970).

124

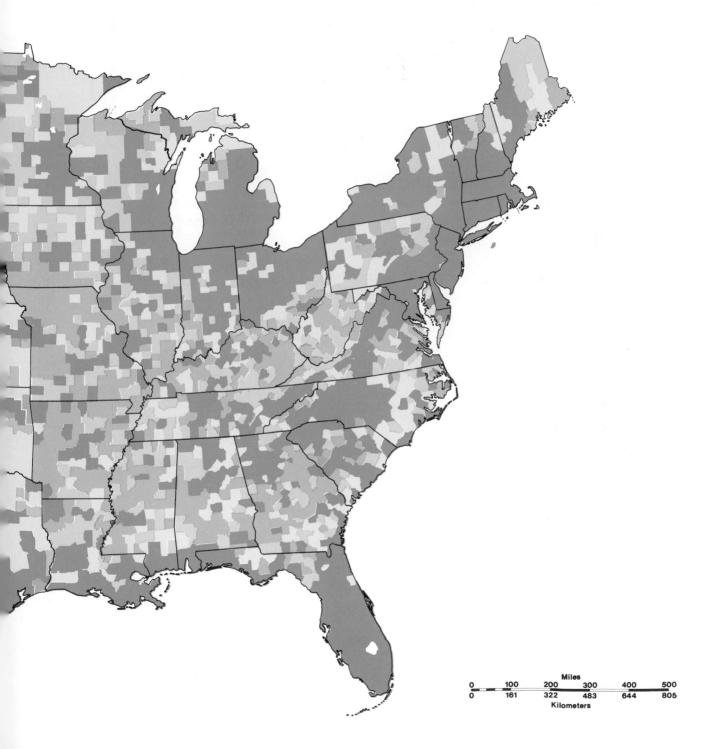

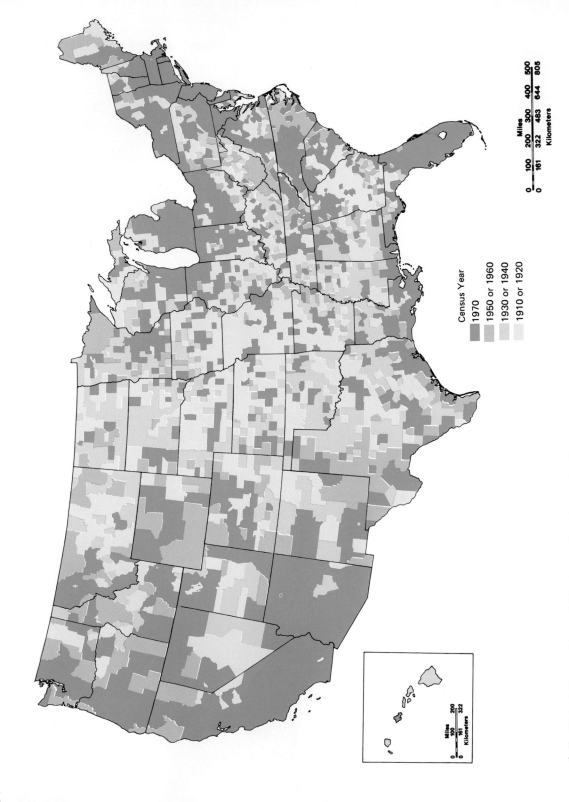

6-5 *United States: Year of Maximum Population by Counties*

SOURCE: U.S. Bureau of the Census (Washington, D.C.: U.S. Department of Commerce, 1970).

Census Year

1970
1950 or 1960
1930 or 1940
1910 or 1920

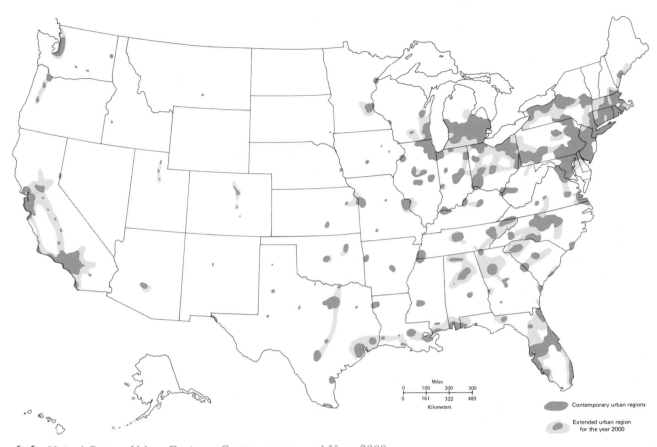

6-6 *United States: Urban Regions, Contemporary and Year 2000*
SOURCES: Commission on Population Growth and the American Future, *Population and the American Future*
(Washington, D.C.: Government Printing Office, 1972); and *Standard Metropolitan Statistical Area*
(Washington, D.C.: U.S. Bureau of the Census, 1974).

limited skills from economically depressed areas may contribute a disproportionate share of the economic and social problems of large cities.

The series of population maps (Figs. 6-4 and 6-5) should be compared carefully. Figure 6-4 is suggestive not only of the major rural areas that have lost population but it also indicates that population decline began long ago in many areas, in fact often within a few decades after initial settling. Areas that have experienced population decline include much of the Coastal Plain from Virginia to Texas, the middle and lower Mississippi Valley, and the Great Plains from central Texas to the Canadian border, and por-

tions of the mountain west. It is difficult to generalize about the Appalachian Region. This region is often considered a large area with severe economic conditions and declining population. Indeed, it has been a source area for many migrants. The significant areas of population loss during the 1960s included the plateau and hill portions of northern and eastern Tennessee, eastern Kentucky, West Virginia, Pennsylvania, and the hill lands of southeastern Ohio. Only a narrow string of Blue Ridge counties lost population in the eastern Appalachians. Obviously much of the area gained in population, particularly the Piedmont and the Ridge and Valley areas of eastern Tennessee, northwest Georgia, and northeast Alabama. The Appalachian Region is not uniform in its demographic experience.

Continued reference to a westward movement of population may be somewhat misleading. Certainly the national center of population has moved westward from the northeast, and the west coast of the United States has experienced great population growth during the twentieth century. Many areas in the West, however, have been sparsely populated and experienced population loss and decline (mining centers). More accurately, the traditional agrarian areas have been regions of population decline. Some attained their peaks in population by 1900. Meanwhile, urban areas gained in population. The areas peripheral to the Great Lakes in Canada and the United States constitute the older urban industrial regions that have long experienced population growth from natural increase, immigration, and rural to urban migration. The Piedmont, the southern Ridge and Valley area, the Gulf Coast intermittently from southern Florida to Texas, and the Far West and Southwest are urban growth areas. It is more accurate to suggest population as moving in several directions but mainly to expanding urban regions (Fig. 6–6). Those areas remaining disproportionately rural, such as parts of the Coastal Plain, the Great Plains, the Midwest, and Appalachians, lose population.

Technologic Change and Anglo-American Development

The nineteenth-century development experience of Anglo-America is rooted in the Industrial Revolution. The commercialization of agriculture was stimulated by market expansion coinciding with the revolution. Raw materials needed for growing industry required improved and expanding transportation. With these needs came numerous technologic inventions and improvements that were both contributors to and products of the Industrial Revolution. For example, an improved ginning technology was necessary to reduce the laborious removal of seeds from cotton. When improved ginning was achieved (Whitney, 1783), much of the lower South was viewed

as suitable for settling and cotton culture. The technologic change thereby contributed to the rapid nineteenth-century settlement of the lower South. The Colt revolver gave Europeans a superiority over Indians which probably hastened their loss of control over plains territory. Barbed wire allowed the separation of cattle and crops and isolation of cattle for breeding purposes. Barbed wire and the windmill (for pumping water) gave a new value to the Great Plains as an agricultural region and were necessary before the cattle kingdoms would give way to stock raising and cropping. Countless other technologic changes occurred which significantly aided in resource use and changed the value of particular places. Perhaps none, however, have been as important to resource use and life style as those changes occurring in nineteenth- and twentieth-century transportation.

Anglo-American Transport Systems

Transportation has had a major impact on population flow, distribution, and evolving settlement patterns from initial settling to the present. It provides focus on particular places; but as technology changes, the significance of such places is altered. Also, the development of a transport system with the capability of moving great quantities of materials at relatively low cost has been essential to the utilization of resources for industrialization in both the United States and Canada. Frequent changes in transport technology have lowered the cost and increased the feasibility of moving goods and also frequently altered the significance of various places. Transportation, therefore, has played a crucial role in the development experience of Anglo-America.

Overland and water transportation, the latter in the form of rafts, barges, and sailing vessels, were the major forms of transport available from initial settling until the 1830s. The great cost advantage of water transport had a major effect on early settlement, both for those people inclined toward agriculture and those with a commercial bent. Early cities were the result of trade, and transportation advantages provided by water were crucial to their progress.

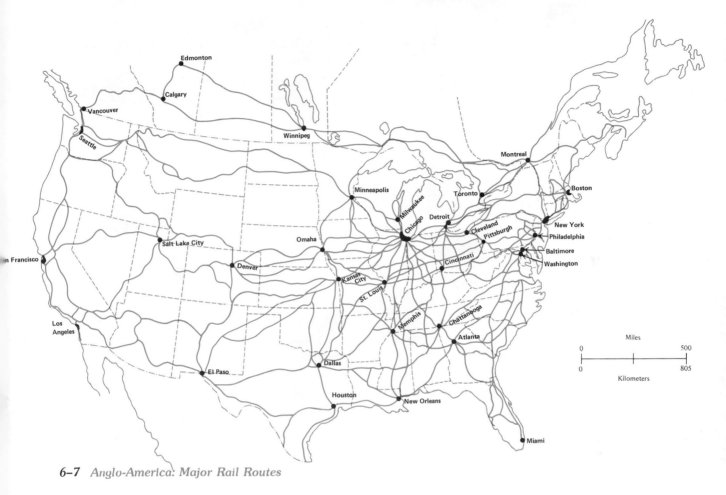

6–7 *Anglo-America: Major Rail Routes*

By 1800, Boston, New York, and Philadelphia were of similar size and as Atlantic ports functioned in a similar way. Numerous other coastal and tidewater ports existed; but these three cities exemplify the larger of the early urban centers, nearly all of which were Atlantic ports. These cities functioned for hinterlands that were part of the larger European exchange system.

By the early nineteenth century water transport had increased in importance, particularly inland. With the development of steam power, the utility of the Ohio-Mississippi and other river systems appreciably changed. A canal-building era was stimulated which was ushered in by the Erie Canal (1825). It was a success, an exception to what was more typically a financial disaster. The cost and time required for shipping from Buffalo to New York via the Erie

Canal and Hudson Valley were reduced from $100 per ton and twenty days to $5 per ton and five days. The canal expanded, by way of the Hudson and Mohawk valleys, the hinterland of New York, and in no small part explains New York City's growth as a national focus. Transportation improvement in the form of the Erie Canal allowed the marketing in the East of efficiently produced agricultural commodities from beyond the Appalachians. Furthermore, later connecting canals on the Great Lakes complemented the Erie Canal and facilitated the movement of grains, forest products, and minerals. The general effect of improved transportation was to extend commercial activities far to the interior, drawing upon new and rich resources. The value of New England as an agricultural resource area, which was not rich to begin with, declined. Production became concentrated on

Improved forms of transportation have aided urban growth and industrialization. Toronto, on the edge of Lake Ontario, has access to both land and water transport systems. (United Nations/Toronto Transit System)

some specialized commodities. The vast new hinterland centered on New York, giving that city a distinct lead over other eastern cities, particularly Boston. It is not merely a historical incident but a classic example of the impact of transportation change, one that has been and will be repeated often in newly developing areas.

The railroad era began even before the short canal boom ceased. By mid-nineteenth century, new focal points were identified. The railroad, though occasionally competitive with canals, generally complemented water transportation and greatly increased the significance of the Great Lakes as an interior waterway. Rail networks emerged with focus on selected coastal ports; New York grew much faster than Philadelphia or Boston. The networks also converged on other "interior ports," such as Chicago and St. Louis.

The rails connected farm, forest, and mineral resource areas with ports and cities. Continued improvements in rail technology made long-haul transport feasible (and the concentration of raw materials at a few selected points). The rails aided in the opening of agricultural land in the West and the tapping of copper, lead, and zinc in the Far West. Grains could be moved to and from ports. Coal was hauled to Great Lakes ports, and iron ore from Great Lakes ports to inland cities (Pittsburgh). By the late 1860s a major turning point had been reached in urban-industrial growth which would not have been possible without improved transportation. Transport technology provided special focus on selected cities, most often on water (Fig. 6–7). Massive urban and industrial growth proceeded hand in hand for the next half century and established major or national routeways between these selected cities. A common result of such transport technology was selective growth. Some port facilities declined in importance as others, fewer in number, became the center of attention.

Urban growth from 1860 to 1920 was stimulated mainly by industrialization, a massive expansion never before seen anywhere in the world. Except for Los Angeles, the leading industrial centers of 1920 were identifiable as important commercial centers by 1860. During these years the railroads and waterways, as transport complements, enhanced the initial advantage of specific centers that grew to a position of national eminence.

Since 1920 tertiary growth has surpassed industrial expansion as a stimulus for urban growth. The railroad was partly replaced by the car and plane. Railroads have retained importance as long-haul freight carriers but lost their passenger and short-haul

6-8 Anglo-America: National Highway Systems

——— National Interstate Highways of the U.S.
꜀꜀꜀꜀꜀꜀ Incomplete segments of the Interstate System
——— Limited access highways of Canada
——— Portions of the trans-Canada Highway System (nonlimited access)

| 0 | 100 | 200 | 300 | 400 Miles |
| 0 | 161 | 322 | 483 | 644 Kilometers |

Construction of expressways has greatly affected movement and residential patterns. The movement of people to the suburbs is one example of the impact of expressways. With suburbanization have come perimeter shopping centers and relocation of some manufacturing and warehousing. (FHWA)

carrier functions. Both automotive and air transport have captured passenger traffic, and the auto-highway system is extremely important for freight movement. The national routeways and flows established during the railroad era, however, were essentially maintained during the automotive and air era, and so the growth of established large centers continued.

The settled portions of the United States and Canada contain an intensive network of paved roads.

In the United States these range from paved county roads to the limited access Interstate Highway System (Fig. 6–8). The present highway system will have a "selecting" effect in localized areas but will not re-route national flows. Rather, the system will aid in the integration of existing economic regions.

In recent decades internal waterways have received renewed attention. The completion of improvements on the St. Lawrence Seaway in 1959 extended the

Great Lakes system from the west end of Lake Superior (Duluth-Superior) and southern end of Lake Michigan (Chicago) to the Atlantic via the St. Lawrence and also the New York State Barge Canal (the former Erie Canal). Thus, not only are the Great Lakes important as an internal waterway, but they now also function as an international water route connecting cities such as Cleveland, Detroit, and Chicago with Europe. Canals also connect Chicago, the nation's most important inland urban-industrial concentration, with the Mississippi system. The second major internal waterway is the Mississippi with its extensions along numerous tributaries (Fig. 6–9). The waterways are used for massive movements of grain, coal, iron ore, and petrochemicals, bulk commodities. Waterways in effect surround the highly urbanized and industrialized northeastern United States and penetrate more than 2,000 miles (3,208 kilometers) along the southern border of Canada, providing major transport advantages to the continental portions of each country.

Further Readings

For perspectives on the historical geography of various regions in Anglo-America, see the following: Brown, Ralph H., *Historical Geography of the United States* (New York: Harcourt Brace Jovanovich, 1940), 596 pp. and by the same author, *A Likeness of the Eastern Seaboard: A Mirror for Americans* (New York: American Geographical Society, 1943), 312 pp.; Mitchell, Robert D., "The Shenandoah

6–9 *Anglo-America: Inland Waterways*

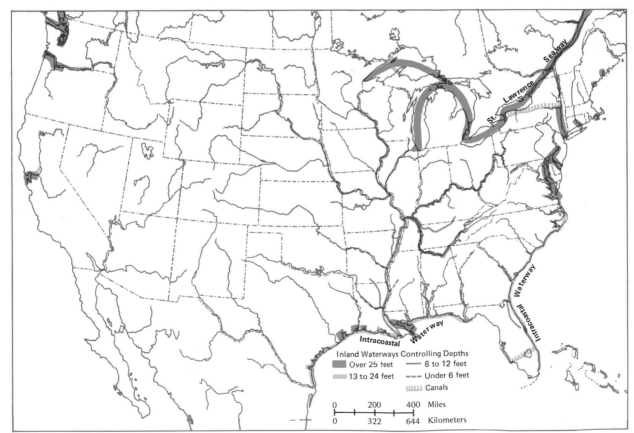

Valley Frontier," *Annals, Association of American Geographers,* 62 (1972), 461–486; Meinig, Donald W., "The Mormon Culture Region: Strategies and Patterns in the Geography of the American West, 1847–1964," *Annals, Association of American Geographers,* 55 (1965), 191–220; DeVorsey, Louis, Jr., *The Indian Boundary of the Southern Colonies, 1763–1775* (Chapel Hill: University of North Carolina Press, 1966), 267 pp.; pages 74–82 of Spencer, Joseph E., and Horvath, Ronald J., "How Does an Agricultural Region Originate?," *Annals, Association of American Geographers,* 53 (1963), 74–92; Kerr, D. G. G. (ed.), *A Historical Atlas of Canada* (Toronto: Thomas Nelson and Sons, 1960), 120 pp.; Webb, Walter Prescott, *The Great Plains* (New York: Grosset & Dunlap, 1931), 525 pp.; Meinig, Donald W., *Imperial Texas: An Interpretive Essay in Cultural Geography* (Austin: University of Texas Press, 1969), 145 pp.; and Clark, Andrew H., *Acadia: The Geography of Early Nova Scotia* (Madison: University of Wisconsin Press, 1968), 450 pp.

Jordan, Terry G., "The Imprint of the Upper and Lower South on Mid-Nineteenth-Century Texas," *Annals, Association of American Geographers,* 57 (1967), 667–690, and his "The Texan Appalachia," *Annals, Association of American Geographers,* 60 (1970), 409–427, are interesting papers dealing with the diffusion of culture traits across the South; Zelinsky, Wilbur, *The Cultural Geography of the United States* (Englewood Cliffs, N.J.: Prentice-Hall, 1973), 164 pp., is a short but excellent and challenging treatment of the cultural geography of the United States. The Report of the Commission on Population Growth and the American Future, *Population and the American Future* (Washington, D.C.: Government Printing Office, 1972), 186 pp., provides a useful discussion of the United States demographic experience and the implications for the future. For Canada, see Kalbach, Warren E., *The Impact of Immigration on Canadian population* (Ottawa: Dominion Bureau of Statistics, 1970), 405 pp.

Anglo-America: Economic Growth and Transformation

7

COMMERCIAL ECONOMIC activity began in Anglo-America shortly after European settling. Commercial activity was in fact a motivation for sponsors of colonies. Subsistence agriculture, however, was essential for the provision of food for local use. Furthermore, accessibility to markets was so limited when located away from water that only subsistence activities were possible. Wherever transportation was suitable, usually meaning accessibility by water, commercialism rapidly became the norm. Even in early Anglo-American economic history, the driving of live animals from frontier farms eastward to markets was not uncommon. Agricultural commodities, lumber, furs, and fish were produced and gathered for exchange. Primary production and tertiary activities (trade) were important long before settling was complete or before manufacturing became significant. Of all activities, agriculture became the mainstay and remained so for more than two centuries.

Shortly after independence, people who had accumulated wealth from commerce began to recognize the financial possibilities of manufacturing. The growth of population, domestic markets, and transportation stimulated manufacturing. Much of the industrial expansion of the middle and late nineteenth century was based on domestic needs and potential, whereas earlier commerce had been more externally oriented.

A complete change in economic emphasis occurred during the century after 1850. Table 7–1 illustrates these changes as evidenced in employment by economic sector. Few Anglo-Americans today are farmers, only 3.7 percent in the United States. Slightly less than 23 percent are employed in manufacturing, and the remainder, instead of being producers of commodities, are engaged in the distribution of goods and provision of services. Since the turn of the century and particularly after World War II, the tertiary sector has rapidly become the dominant source of employment and the basis of continued urbanization. The employment changes illustrate the transformation from agrarian to highly developed urban-industrial societies. The decline of employment in primary activities (mainly agriculture) does not mean their disappearance but rather a complete change in the manner in which they are accomplished. For example, mechanization and automation have greatly reduced the need for labor in two such differing activities as agriculture and coal mining.

Anglo-American Agriculture

Five characteristics of United States and Canadian agriculture are especially important. First, the agriculture of both countries evolved in large and contrasting environments. Second, there existed a strong commercial emphasis during early settlement wherever accessibility was such that markets could be acquired for specialized products. Third, agriculture in Anglo-America has been dynamic, both in the application of improved and changing methods and in the distribution of production. Regional shifts and adjustments are constantly in the evolution of American agriculture, shifts that are the product of changing methods, transportation, and economic conditions.

Economic Sector	United States (1975)		Canada (1974)	
	PERCENT		PERCENT	
Primary	4.0		7.8	
Agriculture		2.6		5.2
Other primary		1.5		2.6
Secondary	28.6		28.7	
Manufacturing		22.7		22.2
Construction		5.9		6.5
Tertiary	66.3		63.5	
	100.0		100.0	
Total employment in thousands	84,783		9,137	

Table 7–1 *Anglo-America: Employment by Economic Sectors*

SOURCE: U.S. Bureau of the Census, *Statistical Abstract of the United States* (Washington, D.C.: Government Printing Office, 1976); and *Canada Yearbook* (Ottawa: Dominion Bureau of Statistics, 1976).

Fourth, the Anglo-American agricultural regions are immensely productive. The dynamism and productivity, however, have been paradoxically both a source of problems and benefits. Fifth, although the two prosperous countries contain highly productive agricultural systems, the periodic changes in crops and methods have reduced the competitive position of many farmers. Serious economic and social problems have resulted.

Availability of Agricultural Land

Anglo-American agriculture evolved in a large and rich environment. Of the total land area of the United States, approximately one-fifth is classified as cropland, not all of which is cultivated in any given year (Fig. 7–1). Land in crops actually has decreased by 45 million acres (22 million hectares) in recent years. Pasture and range land, also part of the food-producing resource base, make up about 39 percent of the land area. The slight decline in this land-use category, 6 percent in the last decade, results from reclassification of land rather than actual change in use. Nearly 5 acres per person (2 hectares) exist for agricultural production of food and industrial raw materials. Obviously the land resources useful for agriculture are substantial relative to population. Few other countries have such a favorable population to agricultural land ratio.

Canada has only 4.4 percent of its land in cropland and pasture. The Canadian population, however, is substantially smaller than that of the United States, and so the ratio of people to agricultural land is much the same (4.58 acres, or slightly less than 2 hectares, per person). The export of primary commodities, which include farm products (wheat in particular), is important to the Canadian economy, despite the small proportion of total land area used for agriculture.

Both the United States and Canada have substantial quantities of land available for agriculture when total population is considered. The figures cited are based upon land actually used for production in recent years and do not reflect additional land that might be used should it become necessary. Both countries have in fact experienced a decline of agricultural land use in marginal areas that are not essential for food supply, areas that were settled during earlier expansion periods when agriculture was the dominant economic activity. Much additional land not presently used for agriculture could be converted into cropland. For example, in the lower Mississippi Valley and the Southeast, 50 million acres (20.2 million hectares) were considered convertible from woodland and pasture in 1970. Under more favorable economic conditions (higher prices and lower production costs), the figure could reach 100 million acres (40.5 million hectares). Even such conditions, however, would not necessarily assure use of those re-

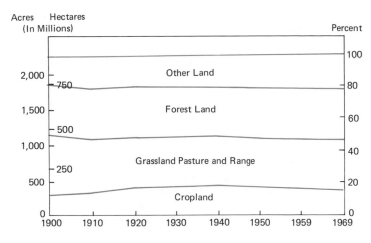

7–1 *United States: Rural Land Use, 1900 to 1969*

SOURCE: Adapted from *Our Land and Water Resources* (Washington, D.C.: Department of Agriculture, 1974).

sources. Farmers with the equipment and capital to expand production may be in short supply.

Agricultural Regions

The largest expanse of highly productive land is that part of the Interior Plains referred to as the "Corn Belt" (Fig. 7–2). This region, extending from western Ohio to Nebraska, is a large expanse of moderately rolling to flat plains, highly suited to mechanized agriculture. The outstanding climatic conditions are the reliable rainfall (30 to 40 inches, 76–102 centimeters, annually) and a moderately long growing season of 160 to 180 days. The gray-brown podzolic and chernozemic soils are among the best of Anglo-America. This favorable combination of features provides a resource base that can be used for the production of a great variety of grains, forages, and vegetables. Individual farmers, however, tend to specialize.

The single most important crop throughout the region is corn, sometimes produced as a "feed grain" and marketed commercially. In other instances, it is fed to animals to "finish" cattle and hogs. Farms producing corn for commercial markets commonly include other crops such as soybeans or even wheat. Other production systems include dairying and vegetable production. Farmers can choose from several production systems that yield a high return on their investment. Stated differently, the Midwestern farmers have an absolute advantage over the other areas in the production of a number of commodities. The advantage accrues from the excellent resource base of their area. Most often, however, farmers select "commercial grain" farming or "grain and livestock finishing," systems that give much of the region its agricultural character.

The favorable position of the Corn Belt region becomes even more apparent when location relative to transportation routes and urban-industrial districts is considered. Various parts of the region lie between or adjacent to the inland waterways of the United States and Canada and have been laced with a dense network of railroads. The transport system facilitates easy marketing of efficiently produced commodities in nearby eastern markets and worldwide.

Surrounding this superlative agricultural region are several others. The northeast, where Anglo-American agriculture had its beginnings, does not provide such large expanses of favorable terrain and fertile soils. Rather, productive land is the exception: the Connecticut Valley, the Hudson Valley, and limestone soils of southeast Pennsylvania (Lancaster County). Topography is commonly hilly to mountainous and covered with thin and frequently infertile soils. Nevertheless, the northeast is a significant agricultural area. As better interior lands became accessible, the northeastern farmers found it necessary to "retreat" to the better land, to specialize in the production of items needed in nearby cities, and to make good use of their one great advantage—proximity to growing northeastern markets. Thus evolved the vegetable and poultry production of the Delmarva Peninsula and New Jersey, and the dairying of New England, New York, and Pennsylvania.

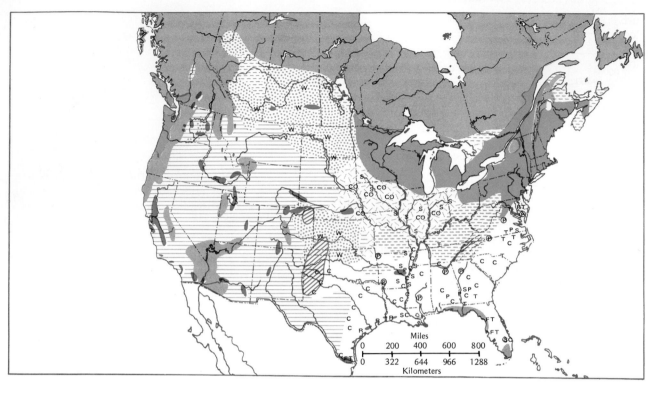

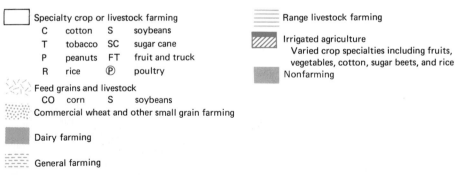

7-2 *Anglo-America: Agricultural Regions*

The commodities are of high value but require great capital inputs, have high transport costs, and sometimes are highly perishable. Nearness to markets is an important advantage.

The farmers on the periphery of, or sometimes literally within, huge urban agglomerations derive operational characteristics and benefit from this urban proximity but also often are in conflict with expanding urbanization. Closeness to urban centers can mean high taxes, pressure to relocate, or complaints from new urban dwellers concerning odors and noise. Although agriculture may be profitable in such a setting, it is also commonly unstable in that farmers and farmland frequently are lost to rapid urban growth. Urbanization has resulted in significant loss of farmland. The common presumption, however, that urban growth is the primary reason for loss of farmland over the past several decades is incorrect. Far more land has been idled or shifted to less intensive use (grazing or forestry) because of lack of profitability. Clearing, drainage, or expanded irrigation usually adds more land to agricultural use than urbanization removes.

Southern United States, from southern Maryland and Virginia to Texas, is a large area with variable resources for agriculture. The humid subtropical climate means adequate rainfall, 30 to 60 inches (76 to 152 centimeters), and long growing seasons (more than 200 days). Evapotranspiration rates, however, are high, and irregularities in rainfall are frequent enough that drought damage to crops is common. Agricultural productivity is often affected by slope, drainage, and soils. Large parts of hill land South, the Appalachians and Ozarks, have slopes too steep for sustained use as cropland. Where farming does occur, it is often quasi-subsistence. Other areas, such as the Piedmont, have been used for agriculture during favorable economic periods, such as when cotton prices were high, but the land has deteriorated with continued row crop production (crops such as cotton, corn, or tobacco planted in widely spaced rows may accelerate erosion). Many coastal areas of swamp and sandy "pine barrens" also are undesirable for intensive agriculture. Within the southern region, however, are subregions which, because of topography and soils and the subtropical environments, provide some exceptionally good agricultural resource areas. Included are the alluvial Mississippi Valley, the Nashville and Bluegrass basins, the black soil belts of Alabama and Texas, and the limestone valleys of Appalachia.

The term "Cotton Belt" as an appellation descriptive of agricultural activity dominating a large portion of the South is inappropriate. The South is a diverse farming region with several subregions. Highly diversified agriculture is based upon numerous specialties in different areas. Cotton, which was so important across much of the lower South, is now produced mainly in the western sector and the far western states of California and Arizona. In the East, the Mississippi Valley (Tennessee, Arkansas, Mississippi, and Louisiana), the coastal plain of Texas, and Lower Rio Grande remain the areas of specialized cotton production. In many other parts of the South, cotton has disappeared or has become of secondary importance. The diversity of southern agriculture is illustrated by tobacco in southern Maryland, Virginia, North and South Carolina, Kentucky, Tennessee and Georgia; poultry in North Carolina, northern Georgia and Alabama, central Mississippi, and northwest Arkansas; peanuts in Virginia, Alabama, and Georgia; rice in Louisiana, Texas, and Arkansas; sugar cane in Louisiana and Florida; citrus in Florida and Texas; soybeans in the Mississippi Valley; and beef cattle throughout the South.

The United States Great Plains and the Prairie Provinces (Alberta, Saskatchewan, and Manitoba) of Canada contain an environment in which farmers face recurrent drought, and even "normal" rainfall may mean water deficiencies. Settlement and production practices drawn from experiences in humid regions have had to undergo substantial adjustment in seeking a stable agricultural economy in these dry lands.

Most characteristic of the wetter parts of the plains states and provinces is wheat farming, sometimes diversified to include livestock raising (basically cattle). The drier margins are given over almost exclusively to cattle raising. Areas such as the Sand Hills of Nebraska, stabilized dunes with poor moisture retention, and the rolling Flint Hills of Kansas are examples of specialized ranching areas. Only where irrigation water is available is the pattern of dry farming and ranching eliminated. Irrigation agriculture often results in intensive cropping systems; it allows the high plains of Texas to function as one of the major cotton-producing regions of the United States. Eastern Colorado produces irrigated sugar beets and corn with nearby livestock finishing farms similar to those of Illinois and Iowa. The plains are food surplus areas, and much of the production is transported elsewhere to supply domestic and international demands. It is the products of this region in Canada that are most competitive with United States farm output in the international marketplace.

Along the northern edge of the Corn Belt is a dairy and general farming region that extends from Minnesota, across the Great Lakes states and Canada's Ontario peninsula to New England and Nova Scotia. The humid and cool summer of the region is favorable for forage crops and feed grains but gives way northward to harsh environments with short growing

Irrigated cotton farming on the high plains of Texas is highly mechanized and productive. Such a farming system minimizes the use of labor and maximizes capital inputs in the form of fossil fuels, fertilizers, and machinery. (U.S.D.A. photo)

The Salinas Valley, long famous for its lettuce production, exemplifies the fertility and productivity of California agriculture. The flat valley bottoms with alluvial soils are rich agricultural lands within the Coastal Ranges. (U.S.D.A. photo)

seasons and infertile, thin soils. The Canadian section is narrow in latitudinal extent and represents a transition to the nonagricultural northern lands.

While the productivity of the plains states is hampered by dry conditions and distance from eastern markets, areas farther west are even more handicapped. Many of the western areas are nonagricultural because of rugged terrain, aridity, or inaccessibility, and are used extensively as grazing land. The exceptions to this generalization are extremely important. Aside from the Pacific Northwest (northern California, western Oregon, and Washington), agriculture, when carried on at all, is intensive at oasis locations. The valleys east of the Cascades, the Snake River plains, the Salt Lake oasis, the Imperial Valley, the Los Angeles Basin, and the Great Valley of California are some of the areas with highly productive agriculture. Farmers in these areas make use of fertile soils and level terrain and take advantage of national transport systems that allow marketing in the East; but they are agricultural systems utterly dependent upon irrigation. These conditions have made California the most diversified and most productive agricultural area in either the United States or Canada. The Pacific Northwest has a humid climate and is therefore less dependent upon irrigation. The Puget Sound Lowland and Willamette Valley contain a productive dairy industry.

Commercialized Agriculture

The commercial emphasis in Anglo-American agriculture began early. Financial backers for early settlements sought a profit. Tobacco was the first successful venture, but rice, indigo, and cotton followed as colonies were established southward along the Atlantic coast. The frontier farmer, or the farmers who settled more remote areas such as the Appalachian valleys, were generally subsistence oriented, or produced in addition to provisional foodstuffs items such as livestock which could be driven to markets. Grains produced on the frontier could be marketed by conversion into a high-value product such as whiskey.

The expansion of commercial agriculture to the interior was stimulated by expanding markets and the transportation improvements that allowed cheap and efficient movement of commodities. The urbanization and industrialization of Europe were a stimulus for expansion of commercial agriculture in various parts of the world, including Anglo-America, during the nineteenth century. The distribution of public land to settlers through various land acts and the building of railroads in both the United States and Canada were not first of all to assure a plentiful supply of agricultural products in the East but rather to settle and occupy territory. These motivations were the product of an expansionist period. Railroad expansion and its improvement, however, allowed the settlers to participate in a commercial system.

The spread of wheat culture westward illustrates the role of transportation. Wheat for commercial use was first produced along the eastern seaboard from New England to Virginia. The opening of the Erie Canal allowed the use of lands south of Lake Erie (better wheatlands) and in the Ontario Peninsula. The initial railroad era linked the interior to Chicago with rails. Wheat production spread to Illinois, later to Wisconsin, and then with the transcontinental rail lines into the Great Plains. With expanding transportation, commercial agriculture spread inland but usually left the more intensive activities yielding higher economic returns near the markets.

The agricultural history of California similarly illustrates the importance of transportation. Mexican ranches producing cattle for hides and tallow were replaced by American wheat farms when railroads appeared, and with later improvements such as refrigeration, the farmers shifted to more intensive fruit and vegetable production.

The effects of transportation improvement and expanding settlement meant that land with greatly varying capabilities became accessible for many kinds of production. The principle of comparative advantage, therefore, is allowed to operate, leading to widespread regional specialization. The principle of comparative advantage means that some locations have a definite advantage over other areas in the production of one or several items. If adequate transportation exists to allow commercial exchange, regional specialization will result. Farmers having an advantage in some form of production will choose a system providing them the greatest returns and concentrate on one or two production systems. Areas with no advantage then may concentrate on the production of items that other areas ignore. Areas, therefore, specialize in limited types of production, either because of advantage or by default, and through exchange contribute their production to the larger economic system. Expressed differently, the high degree of regional specialization is possible because an efficient transport system allows the use of resources best suited to particular kinds of production.

Continued Adjustment of Agriculture

Anglo-American agriculture is often described as dynamic, capable of continued change. It might be more appropriate to say that it has had to be dynamic. As one of the world's major agricultural regions, Anglo-America is young; one would expect change in an area that undergoes rapid settlement for agricultural purposes. Settlement and development imply changing spatial patterns. For example, the "corn and livestock" region of the Midwest is based upon a system of grain production and cattle fattening that originated in the East when lean cattle driven from frontier areas to market areas were fattened before slaughter. The system spread westward along the Ohio Valley and into the Midwest, an excellent area for the production of feed grains. Similarly, rice and cotton production has not been static but in fact has undergone major spatial shifts. The coastal zone of South Carolina and Georgia was a major rice-producing region until the latter half of the nineteenth century. The disruption resulting from the Civil War, hurricanes, and coincidence of these problems with competition from newer lands in Louisiana, Arkansas, and California led to the disappearance of rice on the southeastern coast. Likewise, cotton production has concentrated on the better lands of the Mississippi Valley and westward. Government production controls and acreage reductions, eroded land, and farmers unable to compete have led to cotton acreage declines in older production areas of the southeast.

Land distribution policies in both the United States and Canada were implemented in response to a felt need to occupy and settle space, not simply because of a need for food. In both countries some lands were settled which were poor and unnecessary for agriculture. Farmers settled Appalachian hilly land where the steep slopes were easily eroded. Others occupied steppe margins (the western edge of the Great Plains) where drought meant disaster. The early nineteenth-century retreat of agriculture in the New England uplands is an event that has been repeated in many areas. Stated otherwise, we have withdrawn some of our agricultural efforts to concentrate on some of our better lands. The peak in agricultural cropland was attained sometime early in the twentieth century (Fig. 7–1). Numerous areas have experienced cropland decrease, whereas other choice areas increased cropland acreage. Land clearing, for example, for cropland expansion continues in the Mississippi delta (Fig. 7–3). Another adjustment by farmers is to increase intensity of production (increased effort) on better land while reducing efforts on poorer land.

During the nineteenth century a substantial immigrant and native-born population was available as a labor force. Frequently agricultural laborers preferred to seek their own land, not impossible in an atmosphere of expansionism, free land, and national desires to occupy territory. Eventually urban-industrial growth attracted labor, which, with the availability of land for oneself, was to make farm labor scarce. Labor-saving devices, such as mechanical reapers (nineteenth century), were significant technological advances even before the power revolution in agriculture. The growing industrial capacity of the two countries allowed the supply of implements.

Mechanization of agriculture means more than simply the substitution of capital for labor, which in itself often requires difficult adjustments. Mechanization frequently requires operational reorganization, the need for larger capital expenditures, and greater scale of production. The benefits of mechanization or other technologic improvements rarely help all farmers but usually only those who have the wherewithal (land or money) to make the required adjustments. From time to time many farmers have found it impossible to make the required changes and have not survived as farmers (Fig. 7–4). For those farmers who can find alternative employment, cessation of farming is not necessarily bad. A shift of inefficient and low-income farmers into other activities can be beneficial for the larger economy. The problem, however, is that not all farmers have been able to find satisfactory alternatives. Large numbers of poor tenant farmers and farm laborers have migrated to cities, ill-prepared for urban life, jobs, stresses, and the urban discrimination that may exist for both poor blacks and whites.

Agricultural Productivity

Anglo-American agriculture is extremely productive. Implicit in the high degree of regional speciali-

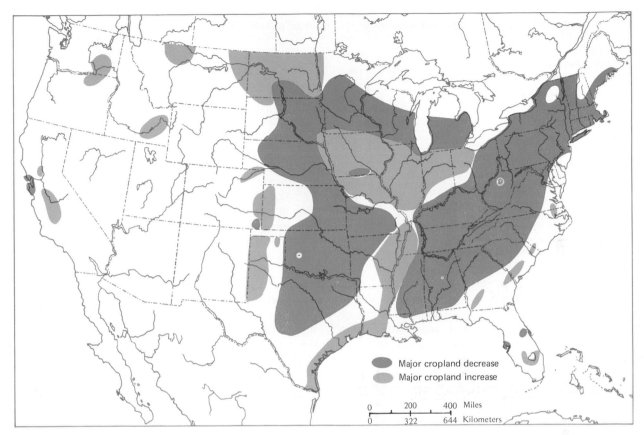

7–3 *United States: Areas of Major Cropland Increase or Decrease, 1944 to 1969*

7–4 *United States: Farm Population as a Percent of Total Population*

SOURCE: U.S. Bureau of the Census, *Statistical Abstract of the United States* (Washington, D.C.: Government Printing Office, 1974).

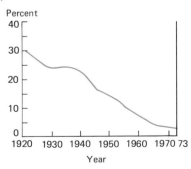

zation is the production of many crops using the best resources for a particular crop. By world standards, yields per land unit for many commodities rank high (Fig. 4–3). Furthermore, productivity has increased greatly during the past three decades and includes major increases per unit of land and per person.

Two examples illustrate increased productivity. Corn yields averaged fewer than 30 bushels per acre (1,878 kilograms per hectare) in the 1930s. The current average is nearly 80 bushels (5,009 kilograms) with yields of 150 bushels per acre (9,391 kilograms per hectare) not uncommon. The value of agricultural land in the Midwest has consequently increased greatly. Land that provided one-fourth to one-half bale of cotton in the 1930s (at 500 pounds [1,102 kilograms] per bale) now produces a bale or more,

	1935–1939 Average	1970–1974 Average
Corn for grain		
Bushels per acre	26.1	84
Hours required per 100 bushels	108.0	6
Cotton		
Pounds per acre	226.0	470
Hours required per 500 lb. bale	209.0	25
Persons Supplied per Farm Worker	1940	1972
At home	10.3	42.0
Abroad	.4	10.4

Table 7–2   Agricultural Productivity in the United States

SOURCE: U.S. Bureau of the Census, *Statistical Abstract of the United States* (Washington, D.C.: Government Printing Office, 1974, 1976).

and the better cotton lands of the Mississippi Valley and Texas may yield 2 to 3 bales. The productivity of the individual farmer also has increased greatly (Table 7–2). Prior to World War II the hours required to produce 100 bushels (2,534 kilograms) of corn or a bale of cotton were 108 and 209 hours, respectively: these figures have been reduced to 9 and 39 hours and require less land as well as labor. Cotton, once planted on 45 million acres (18 million hectares) annually, now is harvested from less than 10 million acres (4 million hectares), but total output remains about the same.

In summary, fewer farmers are producing far more commodities on less land. The increased production has several causes. Mechanization is one. Although mechanization may aid in increasing yields, this is not always the outcome. A more important result is the increase in output per person who, with the aid of machinery, can do the work of many. Viewed differently, one person can farm a large area.

Another cause of the great increase in productivity is improved seed and plant varieties, more effective fertilizers, new pesticides, and chemicals for disease and weed control. Indeed, even now, rapid progress is being made in the development of "hybrid wheats," which will probably mean a great increase in yield and production similar to that which occurred with corn over a number of decades. Hybrid wheat may cost more to produce, and possibly only the more adaptable farmers will make the needed changes.

Paradox: Productive Agriculture and Poor Farmers

As is frequently true in a time of rapid change and innovation, many adjustments are required by farmers who wish to participate in improved production methods. Some are capable of doing so; others are not. Many farm laborers no longer needed because of mechanization or farm reorganization have migrated to cities, particularly from southern areas where the landless farm population was large. Other farmers remain, however, who are marginal producers with low incomes. Such farmers contribute little to the national agricultural economy, and in fact most agricultural output occurs on a small number of farms that are generally large and capital intensive. It cannot be presumed that the marginal farmers will necessarily contribute more if food demand increases. Their shortage of capital and land, as individuals, may prevent their participation even if needed.

Anglo-American Manufacturing

Manufacturing activities are basic to the modern Anglo-American economies. The development of secondary activities accounts for much of the spatial structure of the United States and Canada. Although employment in manufacturing is not growing as fast as it once was, absolute employment has continued to rise, and nearly 20 million people (23 percent of the labor force) are now employed in manufacturing activities in the United States. The more than 2,000,000 so employed in Canada represent about 22 percent of the Canadian labor force. Those employed in manufacturing and the great numbers required in the distribution of goods and needed services are the basis for massive Anglo-American urbanization.

The proportion of the total labor force employed in manufacturing has decreased somewhat as relatively faster growth has occurred in tertiary activities, a characteristic of developed societies. Manufacturing, however, remains basic to the stability of the American economy. Approximately 30 percent of the United States income is directly derived from manufacturing, but even more important is the great number of employment opportunities provided in the tertiary realm as goods are distributed to consumers and services provided for manufacturers and others.

Evolution of American Manufacturing

Like other economic activities in Anglo-America, manufacturing has expanded and changed rapidly in terms of productivity and structural patterns and has undergone distinct spatial adjustments.

The evolution of manufacturing activities was first hindered by a colonial policy that favored mercantilism rather than independent manufacturing. With independence, industry had an opportunity for expansion in an entirely new setting. Population expansion, resource development, great innovations in transportation, new forms of power supply, and an immense confidence and sense of destiny were combined during the nineteenth century to promote unparalleled industrial expansion.

Industry developed first in southeastern New England and the middle Atlantic area from New York to Philadelphia and Baltimore, two of the early American settlement cores. Prior to the railroad era, access to water meant great transportation advantages. Industrial power was available as mechanical waterpower. The humid northeast contained many streams suitable for waterpower development. New England had become an important agricultural and settlement area out of necessity, even though soils and slopes were basically poor for agriculture. Great wealth had been accumulated from other pursuits such as lumbering, fishing, and ocean shipping.

People seeking financial backing for new industrial activity could find entrepreneurs with capital and a venturesome spirit. Capital availability was particularly important in stimulating the early growth of the textile industry in New England, which dates from the 1790s. The surplus rural populations of New England, suffering from the economic competition of newly opened land in the interior, also facilitated industrial growth during the nineteenth century. Such people readily stopped farming for work in the textile mills and leather and shoe factories that dominated New England industry by 1900.

Shipbuilding, food processing, papermaking and printing, and ironworking were other early industries of New England and the middle colonies. Ironworking became an important industry in eastern Pennsylvania, though the iron industry was scattered and localized in numerous communities. Prior to the steel age, iron ore and charcoal (from hardwood trees) were the basic components for ironworking. With high transport cost on bulky and heavy goods, many small iron deposits were used in conjunction with local hardwood forests. New technology was to change that during the mid-nineteenth century.

Digressing for a moment, recall in Chapter 4 Rostow's "Stages of Economic Development." An early or "traditional stage" never really existed in the Americas for many people of European descent. Settling and development began by a people with experience from areas in Europe where "preconditions" for take-off existed. The United States and Canada were created by British and other European elements already in transition. The traditional society (feudal) was decaying, and the worth of the individual was emerging; democratic ideals were taking root. Capitalism was developing as those with excess wealth diverted funds into new economic enterprises: trade, transport, power development, and manufacturing. Thus, those people building Anglo-America brought the United States to the "take-off" stage by the 1840s and Canada by the 1890s. In both cases independence was necessary to allow the pursuit of industrial growth.

The take-off began an era of great industrial expansion, aided by railroads that made possible the movement of materials over great distances and gave new areas and places an entirely new locational significance. The railroads themselves became a major market for steel (available from the 1850s).

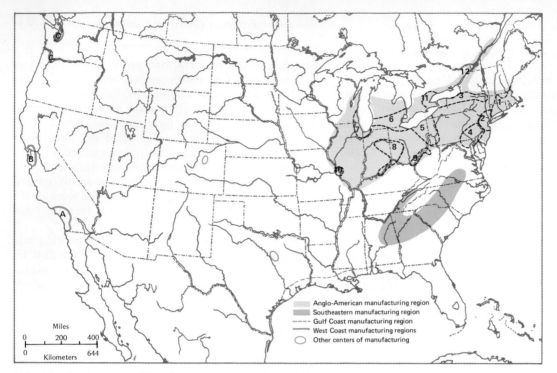

7–5 *Anglo-America: Manufac-turing Regions and Districts*

Anglo-American Manufacturing Region

1 New England District
    Electrical machinery
    Machinery
    Fabricated metals
    Textiles
    Electronic products
    Apparel

2 Greater New York District
    Apparel
    Printing and publishing
    Machinery
    Food processing
    Fabricated metals
    Chemicals

3 Central New York District
    Electrical machinery
    Chemicals
    Optical machinery
    Iron and steel

4 Mid-Atlantic District
    Apparel
    Iron and steel
    Chemicals
    Food processing
    Machinery

5 Pittsburgh-Cleveland District
    Iron and steel
    Machinery
    Electrical equipment
    Rubber
    Machine tools

6 Southeast Michigan District
    Automobiles
    Iron and steel

7 Lake Michigan District
    Iron and steel
    Fabricated metals
    Machinery

Printing and publishing
Electrical machinery

8 Southwest Ohio-Eastern Indiana
   District
    Iron and steel
    Fabricated metals
    Machinery
    Electrical machinery
    Paper manufacturing

9 Great Kanawha and Middle Ohio
   Valley District
    Chemicals
    Primary metal
    Glass

10 St. Louis District
    Transportation equipment
    Iron and steel
    Fabricated metals
    Food processing

11 Ontario Peninsula District
    (Canada)
    Iron and steel
    Machinery
    Chemicals
    Food processing

12 St. Lawrence Valley District
    (Canada)
    Pulp and paper
    Primary metals (aluminum)
    Textiles
    Apparel

Southeastern Manufacturing Region
    Textiles
    Apparel
    Transportation equipment
    Furniture
    Food processing
    Lumber
    Primary metals

Gulf Coast Manufacturing Region
    Petroleum refining
    Chemicals
    Primary metals (aluminum)

West Coast Manufacturing Regions

A Los Angeles–San Diego District
    Aircraft
    Electrical equipment
    Automobile assembly
    Apparel
    Petroleum refining

B San Francisco District
    Food processing
    Shipbuilding
    Machinery

C Pacific Northwest (Portland-
   Seattle) District
    Aircraft
    Lumber products
    Food processing

Other Centers of Manufacturing
Kansas City
    Food Processing
    Automobile assembly

Minneapolis–St. Paul
    Food processing
    Machinery
    Fabricated metals

Dallas–Fort Worth
    Transportation equipment
    Food processing

Denver-Pueblo
    Food processing
    Chemicals
    Iron and steel

By 1865 the United States was experiencing its drive to maturity, achieved according to Professor Rostow by 1900. During this time the improvements in railroads (steel rails), continued immigration, and an extremely favorable population-resource balance aided growth. From 1865 to World War I, the immense growth of industry was directly related to urbanization; it was the primary basis of city growth. During this formative period (1840–1900), the northern states were to experience the bulk of the industrial growth, but the southern lag is not accurately explained as simply "southern indifference" to industrialization. As stated previously, there existed distinctive differences between northern and southern colonies from the days of early settlement. Southern colonies early developed a commercial agricultural system, financially and socially rewarding for a select group.

In contrast, New England, partially because of a less rewarding agricultural resource base, concentrated on commercial shipping and fishing and along with the middle Atlantic area developed traditions in nonagricultural as well as agricultural pursuits. When railroads became the essence of transportation systems and the dominant market for steel, specific northern locations took on a new locational meaning and value. For example, New York became the primary focus of movement from the interior eastward by way of the Mohawk Valley and the Hudson River. The new and growing steel industries required great quantities of coal and iron ore. As Appalachian bituminous coalfields and Mesabi iron ore increased in importance, the locations with utility were those between the coal and iron ore: Pittsburgh, at the junction of the Monongahela and Allegeny rivers, and Cleveland, Erie, and Chicago among others, on the Great Lakes. Rails complemented the lakes by moving coal westward to meet iron ore moving eastward by water. The areas between the Appalachians and the lakes, and along the periphery of the lakes, gained importance for the assembly of materials for production. Furthermore, as the process of growth proceeded, the agricultural goods of the rich interior moved eastward to market which stimulated domestic industry.

The South, removed from the new national routeways for most materials, continued its agricultural production for external markets. Although perhaps overstated, it was like a colonial appendage. Manufacturing did exist in the antebellum and postbellum South but never achieved the rate of growth or dominance of manufacturing in the North. Furthermore, after the Civil War, when the weakness of southern industrial strength was evident, other conditions made a reversal of the traditional economy even more difficult. The South had embraced an agrarian philosophy into which many of its leaders virtually retreated after the Civil War.

The southern agrarian system was more than an economic system; it was also a social system with specific political overtones. Those who opposed a concerted effort to build a new South (urban-industrial) recognized the social and political features of an economic revolution: the relationship among population elements would change as the rigid social structure was broken in an urban setting. The retreat into agrarianism was aided by a narrow political structure that gave disproportionate power to the planter interests. Even the education systems reflected this dominating attitude; the leadership neglected its responsibility for promoting widespread public education.

Thus by the 1930s when the economic development process is said to have reached maturity, great regional variations existed. Part of the United States had evolved as a major urban-industrial region, whose internal character consisted of numerous specialized urban-industrial districts with intervening agricultural regions. The industrial coreland was an area of relatively high urbanization, high industrial output, high income, immense internal exchange, and interaction. The South was a region of low urbanization, limited industrial growth, and poverty for great numbers of white and black in overpopulated rural areas.

Manufacturing Within the United States Coreland

The American manufacturing region is a large area that consists of numerous urban-industrial districts with which certain types of industrial specialties can be associated (Fig. 7–5). Southern New England is the most industrialized area in the United States measured by employment. The district's prominence is based upon the nineteenth-century growth of the textile and leather-working industries. During this century, however, the region has suffered some severe economic problems. First, an area that is dependent upon one or a few products runs the risk of severe economic consequences should competition in the form of more efficient producers arise. Second, locational significance changes; New England and the meaning of its location changed as it achieved industrial maturity. New England has a high labor cost in a high tax area with a power supply based upon imported resources. In addition, population increases farther west have meant that New England's location relative to national markets has changed adversely. Eastern New England is in some respects now peripheral to the core of national markets. The area's dominant industries (textiles and shoes) are failing to grow nationally and experiencing a regional shift to the South. New England has tried to emphasize high-value products such as electronic equipment, electrical machinery, firearms, machinery, and tools that can withstand high transport charges, power, and labor cost. New England, however, illustrates that the industrial structure of a region may change but not always as a conscious choice.

Metropolitan New York contains the largest manufacturing complex in the United States. Its location at the mouth of the Hudson, its function as the major port for the rich interior, and its own huge population have combined to generate and support nearly 11 percent of the United States' manufacturing. The tendency is toward diversified manufacturing that includes printing, publishing, machinery, food processing, metals fabricating, and petroleum refining. A heavy concentration of garment manufacturing and the lack of primary metals processing are other features characterizing Greater New York's industry. The functions of New York and Montreal are notably parallel.

There are three manufacturing districts in which steel industries are characteristic. Inertia, immense capital investments, and linkages with other industries assure considerable locational stability for such industries. The first district is the area enclosed by lines extended from Baltimore to Philadelphia, Bethlehem, and Harrisburg, Pennsylvania. Massive steel-producing capacity exists near all of these cities and supports shipbuilding (along the Delaware River and Chesapeake Bay), and many other machinery industries. The steel industry has expanded here because of proximity to large eastern markets (other manufacturers) and accessibility to external waterways. Waterway accessibility is a factor of growing importance as dependence on foreign sources of iron has increased. In addition, those cities on waterways have become major petroleum refining centers and petrochemical manufacturers.

The second major district is a large triangle with points at Pittsburgh, Toledo, Ohio, and Erie, Pennsylvania. This district is the oldest steel-producing center of the United States. Initial advantages were derived from location between Appalachian coalfields and the Great Lakes, by way of which iron ore came from the Superior Ranges, especially in northern Minnesota. The locational significance of Pittsburgh and its steel-producing suburbs has changed. South American iron ore is moved to eastern coastal works, and Canadian ores come by way of the St. Lawrence and now move even farther inland along the St. Lawrence Seaway (Great Lakes). The eastern district (Baltimore-Philadelphia) is nearer eastern markets and foreign ores; Detroit and Chicago are more easily reached by Canadian ore and closer to Midwestern markets.

The southeastern Lake Michigan (Gary, Chicago, and Milwaukee) area has a vast array of machinery manufacturing that is supplied by the massive steel industry at the southern end of Lake Michigan. The steel industries built in Chicago, Illinois, and Gary, Indiana, have benefited from a superb location. Ore moving on the Great Lakes meets coal from Illinois, Kentucky, and West Virginia. Chicago had become a major transportation center by the late nineteenth century, where rails met and complemented water transport, making the southern Lake Michigan area an excellent location to assemble materials and distribute manufactured products. The St. Lawrence Seaway has simply given renewed importance to the location, for now Chicago and other inland cities can function as a mid-continent port from which ships can sail an almost great circle route to Europe. Southern Michigan and adjacent areas in Indiana and Ohio are distinguished by the emphasis on automotive production, both parts and assembly. These industries are linked not only to the Detroit steel industry but also to steel manufacturers in the Chicago area and along the shores of Lake Erie (Toledo, Lorain, and Cleveland). The automotive industry serves as a huge market for major steel-producing districts on either side.

Figure 7–5 illustrates the additional districts within the American manufacturing region. This large region of urban-industrial districts contains more than 60 percent of the United States and Canadian manufacturing capacity and the majority of the Anglo-American market. The existing complementary transportation system of water, roads, and rails, which has been strengthened for manufacturing districts adjacent to the Seaway, provides great advantages for both assembling materials and distributing finished products.

The problems that may jeopardize the vitality of the American manufacturing region may be less directly those affecting the functioning of industries and more directly those ephemeral concerns of residential quality, social conflict, air and water pollution, urban water supply, and the governing and integration, or lack of it, of numerous but contiguous political units.

The Southern Economic Revolution

It is difficult to bracket with precise dates the southern economic revolution, partly because it remains in progress. The beginnings date from the 1880s and are first perceptible in the attitudes of "New South" advocates who believed industrialization was necessary. There were few adherents, however, to the "New South" philosophy. In fact, by 1900 the old agrarian values were as dominant as ever.

The first major manufacturing activity to become distinctly identified with the South was the textile industry. The textile industry had evolved as the dominant force in New England's nineteenth-century industrial growth; but by the early twentieth century the industrial maturity of the New England area was reflected in high wage rates, unionization, costly fringe benefit programs, high power costs (imported coal), and obsolete equipment and buildings. The response was a regional shift in the textile industry. New England plants closed, and new plants were established in the South. This regional shift from New England to the South involved an industry that has not been a growth industry for the nation as a whole; it has thus been a regional shift benefiting one region at the expense of another. Firms locating within the South found advantages in the quantity of labor available at relatively low cost; the agrarian South had a surplus of landless rural inhabitants willing to accept alternative employment. Other advantages accrued from better location with respect to materials used (cotton), and lower cost power and taxes. By 1930, more than one-half of the United States textile industry was located in the South. Presently, more than 90 percent of cotton textiles, 75 percent of synthetic fibers, and 40 percent of woolen textiles are of southern manufacture.

A question in addition to "why the South" is "why particular southern regions." The textile industry located in a zone adjacent to the Blue Ridge, the Piedmont, and some in the Ridge and Valley. Particularly heavy concentrations developed in North and South Carolina. The Piedmont exemplifies the reasons for this internal location pattern. The Piedmont experienced greater changes in agriculture than nearly any other part of the South during this century. The eroded land, boll weevil devastation of cotton, large rural population, and government controls in agriculture after 1930 encouraged change. Furthermore, within the South, Piedmont location frequently meant greater availability of power; it was being developed at hydroelectric stations flanking the mountains, first on the Piedmont and later in the Tennessee Valley Authority area. Thus, while specific reasons engendered a regional shift, New England to the South, spatial variations in labor availability and power further affected industrial concentration within the South.

The rate of southern industrial growth has increased, particularly since World War II, but with distinct differences. In addition to the labor-oriented textile and apparel industries, material-oriented pulp and paper, food processing, and forest industries have grown rapidly. In the Gulf southwest petroleum refining and petrochemical industries have contributed much to Gulf Coast industrial expansion. The economic transformation has now reached a stage where the South itself provides a significant regional market that generates further industrial growth (multiplier effect). This regional market orientation is exemplified by the automotive industry; auto assembly operations exist in Louisville, Atlanta, and Dallas, which serve regional markets. The transformation from an agrarian to urban life style means higher incomes and new consumption patterns that greatly increase the market importance of a formerly rural population.

Southern Manufacturing Regions

A distinctive manufacturing region coincides with much of the southern Piedmont and adjacent areas of Alabama. This region, from Danville, Virginia, to Birmingham, Alabama, is characterized by light industry (textiles, apparel, food processing, and furniture). The chief attraction has undoubtedly been the availability of suitable labor at costs below industry wage scales elsewhere. As an industrial region, the Piedmont has traits quite unlike the core districts of the American manufacturing region. It is comprised of industries that for the most part are located in small cities, towns, and not infrequently rural areas. The labor supply is commonly rural in residence. Atlanta and Birmingham are two major exceptions. Not all southern industry is labor oriented. Hydroelectric power supplies in the Appalachians have undoubtedly aided Piedmont industrial growth. Coal from the Appalachians serves as the major fuel source for electric power generation throughout the South.

Southern material resources have also been the basis for substantial industrial growth, particularly along the Gulf Coast. As a manufacturing region, the Gulf Coast also is distinguishable from the American manufacturing region in that density of manufacturing centers is low, that is, a series of distinctly separate industrial nodes extending from Corpus Christi, Texas, to Mobile, Alabama. The material base for much of the region's industry and recent growth includes petroleum, natural gas, salt, sulfur, agricultural products, and coal as a source of power. Coastal location facilitates the exchange of goods for both domestic and international trade. Industries include petroleum refining, petrochemicals and other chemicals; alumina smelting and aluminum refining based upon Jamaican and Guyana ores; processing of sugar and rice; and steel manufacture based upon local and imported ores. The Gulf, with industries based upon material resources, has benefited and can benefit from high growth industries (petrochemicals) capable of generating many linked industries.

The Birmingham-Gadsden, Alabama, steel industry began with the unique circumstance of coal, iron ore, and limestone all available in proximity. The area has become the major steel center of the South. The Coastal Plain is also not without industry. Pulp and paper industries and pine plywood industries have expanded greatly since World War II, mainly because of advantages in rapid forest growth. Atlanta and Fort Worth are both noted as centers of aircraft manufacture but are atypical of southern industrial centers. They are cities whose growth is more the result of their functions as regional centers rather than manufacturing.

Manufacturing Growth on the West Coast

Approximately 10 percent of the United States manufacturing is located on the Pacific Coast. The largest single concentration is in the greater Los Angeles area, where aircraft, defense industries, food processing, petrochemicals, and apparel are dominant.

The productivity of California's agriculture and commercial fishing stimulated the food-processing industries that became the state's first major and dominating industry until the 1940s. World War II generated the defense industries, aircraft and shipbuilding, and the former has continued to be a major employer in the Los Angeles area in postwar decades. In addition, automobiles, electronic parts, apparel, and petrochemicals have achieved importance in California's industrial structure. Much of California's industrial growth is not based on local material resources or access to eastern markets but on a rapidly growing local market. California, particularly Southern California, has received a large number of migrants from other parts of the United States. Industries, especially those in which material and power costs are not high, have followed the migrants.

Manufacturing in the Pacific Northwest includes food processing (dairy, fruit, vegetable, and fish products), forest products industries, primary metals processing (aluminum), and aircraft factories. The emphasis, however, is on the processing of local primary resources and use of hydroelectric power from the Columbia River system. The region's great distance from eastern markets and smaller local markets has inhibited growth.

Canadian Industrial Growth

Secondary activities (manufacturing) are as important to the Canadian economy as to the United States. The manufacturing economy is in fact closely integrated with the United States economy, as indicated by trade flows and the high level of United States investment in Canadian industry.

The take-off period in Canadian economic development began in the 1890s and coincided with an immense boom in economic growth. The relatively later start (United States take-off began in the 1840s) is probably attributable to several factors: (1) the harsh physical environment attracted fewer immigrants; (2) since political independence was later, there was no concerted effort to industrialize until late in the nineteenth century; (3) economic ties with the United States were limited until "maturity" was approached in the larger economy to the south, which then stimulated Canadian development.

Nineteenth-century commercial activities were based upon supplying local needs and staples for Europe, particularly Britain. After 1840, however, as Britain moved toward "free trade," Canada was not necessarily in an advantageous position.

By the late nineteenth century, the United States was achieving maturity and was on the verge of a drive for high mass consumption. It consisted of a large relatively wealthy market close to Canadian population and resources. Canada was resource rich, and the boom occurring in the United States could benefit from the resources of Canada and stimulate Canadian growth. The shift from a mercantile relationship with Britain to an industrial relationship with the United States began. At the same time a belated extension of railroads into the Prairie Provinces stimulated migration and interprovincial trade. Wheat was hauled eastward, encouraging city growth at rail and lake terminals. Supplies and lumber moved west for building the settlements on the prairies where eventually 80 percent of Canadian agricultural land was to exist.

Canadian industry and exports have been tied closely to the production and processing of staples: first, fishing and furs; later, wheat, forest products, and metals. A maturing of the Canadian economy since World War II has reduced this primary commodity dominance, particularly in Ontario, but it nevertheless remains an identifying feature of the Canadian economy. Wheat, primary metals (raw or partially processed), forest products, and tourism are still the major export earners.

The early twentieth-century growth was a response to the United States market and capital. Canadians feared that their location relative to the United States would make them simply a supplier of materials for the United States, an economic colony. The tariff became a protective device to encourage manufacturing in Canada, and then exportation, thereby assuring primary production and secondary processing in Canada. Some Canadians have argued that the tariff policy has meant higher prices for commodities they consume and, therefore, a lowered level of living. Tariffs, however, have forced the use of resources, human and material, by encouraging manufacturing at home, as is evident in the immense investment of capital by United States firms. Yet another dilemma is now over the great role United States companies play in the Canadian economy. To a large extent there is a high degree of United States–Canadian economic integration.

Economic integration is a logical outgrowth of modernization. Industrial states consume great quantities and varieties of materials which, even in the case of large countries, cannot always be found within their own territory. Trade is the exchange of complementary resources and can stimulate economic growth. If large countries such as Canada and the United States find it mutually beneficial to trade, and thus to some extent to function as one economy, it also can be advantageous for many small countries with less resource variety.

The Distribution of Canadian Manufacturing

The St. Lawrence Valley and the Ontario Peninsula form the industrial heart of Canada and are contiguous with the United States industrial core. It may be thought of as the northern edge of the Anglo-American industrial core, specializing in the production and processing of materials from Canadian mines, forests, and farms.

Montreal contains approximately 13 percent of Canadian industry. In this sense it parallels New York. Both produce a variety of consumer items intended for local and national markets (food processing, apparel, publishing). Both function as significant ports for international trade. Outside Montreal, the industrial structure is more specialized.

The immense hydroelectric potential of Quebec is a major power source for industry along the St. Lawrence and has been the basis for Canada's importance in aluminum production. The Saguenay and St. Maurice rivers (both tributaries of the St. Lawrence) have provided power for aluminum refining and smelting: Alma, Arvida, Shawinigan Falls, and Beauharnois. Bauxite is brought by water from Jamaica and Guyana. The aluminum produced is more than Canada consumes, and the country is the world's leading exporter. The aluminum industry illustrates Canadian industry as a processor and supplier for other nations, using national resources. In this case the national resource is power from the St. Lawrence Valley and tributary streams. Other metals-processing industries located near production centers are copper and lead at both Flin Flon and Noranda, nickel at Sudbury, and magnesium at Haley.

The St. Lawrence Valley and its tributary valleys are also Canada's major areas of wood pulp and paper manufacturing. Canada is the world's leading supplier of newsprint, most of it sent to the United States and Europe.

Outside Montreal and vicinity, the most intense concentration of industry is found in the "golden horseshoe," extending from Toronto to Hamilton at the western end of Lake Ontario. This region produces most of Canada's steel (Hamilton) and a great variety of other industrial goods such as auto parts, assembled autos, electrical machinery, and agricultural implements. It is also one of the most rapidly growing industrial districts in Canada. Two factors

have contributed to the past growth and present advantages of the region: (1) over 60 percent of Canada's market is found along the southern edge of Ontario and Quebec, and market orientation of industry appears to be strengthening; and (2) Canadian industry has evolved behind a protective tariff. The tariff was initially important when foreign capital, especially United States and British capital, was invested in industries processing Canadian resources for foreign use. It has become even more important as Canadian industry has matured, as industries have sought to serve expanding Canadian markets. Foreign companies have found it necessary to locate in Canada to avoid tariffs, but in doing so they have tended to locate in larger Canadian industrial centers (Windsor, Hamilton, Toronto, and Montreal) and close to the city containing the parent United States firm. For example, many Detroit firms with subsidiary operations in Canada have located immediately across the river in Windsor.[1] This condition is suggestive of the degree of Canadian and United States economic integration and the effect of tariff structures. Had the tariffs not been used, Canadian industry probably would be even more oriented to

[1] D. Michael Ray, "The Location of United States Manufacturing Subsidiaries in Canada," *Economic Geography*, 47 (July 1971), 389–400.

Much of Canada's forest resource is located in sparsely settled areas. Here selective clean cutting is being practiced in the mountains of British Columbia. Canada's forests provide lumber and newsprint for export to the United States and Europe. (British Columbia Forest Service photo)

primary materials processing. The economic integration is further suggested by the fact that each is the other's most important trading partner, a situation not unlike that which is found among Western Europe's Common Market members.

Industrialization and Urbanization

The industrial growth of the United States and Canada during the late nineteenth and early twentieth centuries stimulated employment in tertiary activities. New industrial jobs have a "multiplier effect" by generating employment opportunity in the service activities needed to support the new workers. Employment expands in wholesaling, retailing, education, government, the professions, and a host of other urban-oriented activities. Thus, secondary and tertiary activities expand, and the cumulative growth may generate new industry and other economic activities because a larger market exists. Economies accrue to secondary and tertiary activities when they are located in towns and cities. The growth of these activities, therefore, has been the basis for tremendous urban expansion. Manufacturing activities were probably the principal stimulus for urban growth until the 1920s, but during the past fifty years tertiary growth has become the major stimulus.

The total rural population in 1900, when industrialization was well under way, was 46 million or 60 percent of the United States population. By 1970, this rural population had grown to 54 million but only 25 percent of the total United States population. Rural to urban migration and the natural increase of population in cities expanded the urban population to nearly 160 million people in the United States alone and another 15 million in Canada. Furthermore, the vast majority of the rural are nonfarm people who live in rural areas and small towns but work in urban areas. They send their children to school and shop in urban areas; they are urban in most ways except residence.

Present and Future Regions

From a population standpoint, both the United States and Canada consist of a number of urban re-

gions, many contiguous, which continue to intensify and increase. Figure 6–6 shows the present urban regions of the United States and the urban regions projected for A.D. 2000. Urban development is so great in some areas as to justify the use of the term "megalopolis." Professor Gottmann used this term to refer to the massive urban region extending from northern Virginia to southern New Hampshire.[2] The core of this region is formed by the cities of Washington, Baltimore, Philadelphia, New York, and Boston but includes a host of nearby cities that combined form a massive urban region with more than 40 million people. Megalopolitan areas include numerous cities, adjacent or in such proximity that the boundary among urban centers is almost indistinguishable. The interaction between these high-density centers becomes so great that single-city identification is almost impossible, yet administratively this is how they continue to function. Within the region are contained massive industrial complexes with much of the nation's manufacturing capacity and the nation's most intense market concentration.

Gottmann's designation of "mainstreet and crossroads of the nation" implies far more functional significance for the region than just manufacturing.[3] Indeed, it is a center of political and corporate management and decision making. The ports—particularly New York, but including Boston, Philadelphia, Baltimore, and others—are the focus of international and national connectivity, as well as the major terminus of inland rail and auto transport routes. Furthermore, the nation's most prestigious financial and educational centers are located within the region. This massive urban region indeed functions as "downtown U.S.A."

Even agriculture within and adjacent to a megalopolis has a distinctive character. Farmers occupy high-value land because of urban land-use potential, not its inherent food-producing capabilities. A high output of dairy products, vegetables, poultry, and other specialty items are produced for the adjacent markets. Although production costs are high, transport cost to market is low. Farmland, however, like recreation land, faces tremendous pressures from urban en-

---

[2]Jean Gottmann, *Megalopolis: The Urbanized Northeastern Seaboard of the United States* (Cambridge, Mass.: MIT Press, 1961), 810 pp.

[3]Ibid., pp. 7–9.

croachment. Proximity to the urban region means advantages to the farm and a high utility value for potential recreation land but also problems. High taxes and inflated offers from developers lead to reduced numbers of highly productive farms.

The eastern megalopolis may extend from southern Virginia to southern Maine by the year 2000. The projection for the year 2000, however, shows a Great Lakes megalopolitan region connecting with the eastern megalopolis. The Great Lakes megalopolitan region is already well advanced and by the year 2000 may be a major part of a massive urban-industrial coreland extending from Illinois-Wisconsin to the Atlantic shore. The Chicago-Milwaukee urban region forms the western end and extends across Michigan and Indiana to connect with Cleveland and Detroit. Extensions continue eastward along several "settlement chains." One extends from Detroit across the Ontario Peninsula to Buffalo, New York, and Toronto, and in lesser fashion along Lake Ontario and the St. Lawrence Valley to Montreal. Another extends from Cleveland to Pittsburgh and Washington-Baltimore. A third extension may exist from Cleveland to Buffalo and connect with the eastern megalopolis by way of the Mohawk Valley cities. A review of a map showing major urban settlements and transportation routes will aid in the identification of additional extensions from the megalopolitan regions. A series of less intense, but nevertheless distinct, urban regions will extend along the Piedmont and into northeast Georgia and northern Alabama. Florida, the Gulf Coast, and California contain other extended urban regions.

Urban Problems

There exist common problems with which all of these urban regions must deal. One major problem is administrative. The actual city, in a functional sense, remains subdivided into many independent political units. Governments exist and try to provide services and maintain jurisdiction over the inner city, while numerous suburbs attempt the same for themselves. Frequently the administrative functions overlap those provided by county governments. Consolidation of services and government remains a major issue in urban regions. Further complicating planning and problem solving in urban areas is the fact that cities and suburbs differ. Economic, social, and racial segregation means communities with unequal ability to generate funds for the provision of services. The problem becomes particularly acute in central cities which experience an in-migration of blacks and poor and out-migration of whites and relatively more prosperous people. Consolidated urban political regions, or political boundaries coincident with "real cities," would mean a more equitable tax base. Not surprisingly, however, this idea is often opposed.

Population numbers are but one aspect of urban growth; the other is land area absorbed. The physical extent of land absorbed into urban confines has been substantial. Farmers who view themselves as having been there first, with some frustration find a new life style, system of land use, and tax scale encroaching upon their domain. Although some "suffer" the consequences of developers' prices with tongue in cheek, others genuinely resent being squeezed from some of the nation's best agricultural land. Urban growth often has occurred at such a rapid pace that planning has failed to prevent unsightly development, inadequate services for residents and businesses, and acute traffic congestion.

Further Readings

The June 1972 issue of the *Annals, Association of American Geographers* is devoted entirely to the regional geography of the United States. The several articles individually or as a whole provide an excellent supplementary source of information. Included are the following: Jackson, J. B., "Metamorphosis"; Meinig, D. W., "American Wests: Preface to a Geographical Introduction"; Vance, James E., Jr., "California and the Search for the Ideal"; Durrenberger, Robert, "The Colorado Plateau"; Mather, E. Cotton, "The American Great Plains"; Hart, John F., "The Middle West"; Prunty, Merle C., and Aiken, Charles S., "The Demise of the Piedmont Cotton Region"; Louis, George K., "Population Change in Northern New England"; Lewis, Pierce F., "Small Town in Pennsylvania"; Borchert, John R., "America's Changing Metropolitan Regions." An earlier complementary work is Borchert, John R., "The Dust Bowl in the 1970's," *Annals, Association of American Geographers,* 61 (1971), 1–22; Hart, John F., *The Look of the Land* (Englewood Cliffs, N.J.: Prentice-Hall, 1975), 210 pp., provides a good synthesis of the origins and character of the American rural landscape. Also see Hollon, Eugene W., *The Great*

American Desert, Then and Now (New York: Oxford University Press, 1966), 284 pp.; Miller, Elbert F., "Economic and Social Changes in the Columbia Basin, Washington," *Land Economics*, 41 (1965), 335–346; Preston, Richard E., "Urban Development in Southern California between 1940 and 1965," *Tijdschrift voor Economische en Social Geografie*, 58 (1967), 237–254; Thomas, William L., Jr. (ed.), "Man, Time and Space in Southern California," *Annals, Association of American Geographers*, 49, Supplement (1959), 1–120.

Useful papers that deal with American agriculture include the following: Gregor, Howard F., "The Large Industrialized American Crop Farm: A Mid-Latitude Plantation Variant," *Geographical Review*, 60 (1970), 151–175; Hart, John F., "Loss and Abandonment of Cleared Farmland in the Eastern United States," *Annals, Association of American Geographers*, 58 (1968), 417–440; Kiefer, Wayne E., "An Agricultural Settlement Complex in Indiana," *Annals, Association of American Geographers*, 62 (1972), 487–506; Prunty, Merle C., "The Renaissance of the Southern Plantation," *Geographical Review*, 45 (1955), 459–491; Hewes, Leslie, "Causes of Wheat Failure in the Dry Farming Regions, Central Great Plains, 1939–1957," *Economic Geography*, 41 (1965), 313–330; Kollmorgen, Walter M., and Simonett, David L., "Grazing Operations in the Flint Hills-Bluestem Pastures of Chase County, Kansas," *Annals, Association of American Geographers*, 55 (1965), 260–290, Lewthwaite, Gordon R., "Wisconsin and Waikato: A Comparison of Dairy Farming in the United States and New Zealand," *Annals, Association of American Geographers*, 54 (1964), 59–87.

Two volumes that treat various aspects of governmental agricultural policy are as follows: Paarlberg, Don, *American Farm Policy* (New York: Wiley, 1964), 375 pp., and Hildreth, R. J. (ed.), *Readings in Agricultural Policy* (Lincoln: University of Nebraska Press, 1968), 463 pp. For a shorter discussion of government programs in relation to development, see Brunn, Stanley D., *Geography and Politics in America* (New York: Harper & Row, 1974), specifically pp. 318–362; Beale, C. L., "The Negro in American Agriculture," in J. P. Davis (ed.), *The American Negro Reference Book* (Englewood Cliffs, N.J.: Prentice-Hall, 1966), 969 pp., provides not only an excellent review of the problems of the Negro farmer but also a thoughtful analysis of problems faced by all American farmers since World War II. See also Higbee, Edward C., *Farms and Farmers in an Urban Age* (New York: Twentieth Century Fund, 1963), 183 pp.

For readings that focus on the process of American metropolitan and industrial growth, see the following: Borchert, John R., "American Metropolitan Evolution," *Geographical Review*, 57 (1967), 301–332; Borchert, John R., "America's Changing Metropolitan Regions," *Annals, Association of American Geographers*, 62 (1972), 352–373; Pred, Allan R., *The Spatial Dynamics of U.S. Urban Industrial Growth, 1800–1914: Interpretation and Theoretical Essays* (Cambridge, Mass.: MIT Press, 1966), 225 pp.; Pred, Allan, "Industrialization, Initial Advantages, and American Metropolitan Growth," *Geographical Review*, 55 (1965), 158–185. A somewhat older but classic description of America's first megalopolitan region is Jean Gottmann's *Megalopolis, the Urbanized Northeastern Seaboard of the United States* (Cambridge, Mass.: MIT Press, 1961), 810 pp. See also Higbee, Edward C., *A Question of Priorities: New Strategies for Our Urbanized World* (New York: Morrow, 1970), 214 pp.; Easterbrook, W. T., and Aitkin, Hugh G. J., *Canadian Economic History* (Toronto: Macmillan of Canada, 1963), 606 pp.

Suggested articles on specific manufacturing topics include the following: Ray, D. Michael, "The Location of United States Manufacturing Subsidiaries in Canada," *Economic Geography*, 47 (1971), 389–400; Pred, Allan, "Manufacturing in the American Mercantile City: 1800–1840," *Annals, Association of American Geographers*, 56 (1966), 307–328; Pred, Allan, "The Concentration of High-Value-Added Manufacturing," *Economic Geography*, 41 (1965), 108–132; Lonsdale, Richard E., and Browning, Clyde, "Rural-Urban Locational Preferences of Southern Manufacturers," *Annals, Association of American Geographers*, 61 (1971), 255–268; Pred, Allan, "Toward a Typology of Manufacturing Flows," *Geographical Review*, 54 (1964), 65–84; Weiss, Leonard, *Case Studies in American Industry* (New York: Wiley, 1967), 361 pp., provides basic economic background and focus on electric power production and steel production. For the steel industry, see also the following: Logan, M. I., "The North American Iron and Steel Industries: Three Case Studies on the Location of Complex Manufacturing Industires," in Rutherford, J., Logan, M. I., and Missen, G. J., *New Viewpoints in Economic Geography* (Sydney: Martindale Press, 1966), pp. 314–331; Estall, R. C., *New England: A Study in Industrial Adjustment* (London: G. Bell and Sons, 1966), 296 pp.; and by the same author, "Changing Industrial Patterns of New England," *Economic Geography*, 39 (1963), 189–216.

Anglo-America: Problems in a Developed Realm

8

Rᴵᴄʜ ᴀɴᴅ ᴅᴇᴠᴇʟᴏᴘᴇᴅ countries such as Canada and the United States are not without significant and sometimes pressing problems. The kinds of difficulties that these countries must overcome are partially an outgrowth of the development process itself. The problem of supply and use of enormous quantities of resources have already been discussed. Unbalanced economic growth, the effective integration of various regions into a national economy, and the social, economic, and political situation of minority groups are other fundamental problems with which Canada and the United States must deal.

Income Disparity and Regional Problems

It is no surprise that there are large numbers of people who have not been able to acquire the ma-

terial benefits of the average Anglo-American (Table 8–1). How many poor exist in America depends, of course, on one's definition of "poverty," but possibly between 12 and 15 percent of the population should be included. The United States Bureau of the Census estimates that some 33 percent of the nearly 24 million black Americans are in poverty. This group represented approximately 32 percent of all poor in the United States. Although a higher proportion of all blacks live in poverty than whites, the absolute number of poor white are about double that of the blacks. The causes of poverty are numerous and often work in concert. Even a cursory examination of poverty and its spatial dimensions illustrates that the broader aspect of poverty is not simply explained.

The United States' poor are almost equally divided between metropolitan and nonmetropolitan areas. The poor in metropolitan areas are often a smaller proportion of the total metropolitan population than the poor in nonmetropolitan areas, but they are more intensely concentrated in ghetto communities. The nonmetropolitan poor are dispersed areally, which adds to the problem of employment and provision of services. The greatest concentrations of poor are found in the states of New York, Illinois, and Texas, and secondarily in other northern and southern states.

The nonmetropolitan component of poor people live in small towns and rural areas. They are less concentrated and, therefore, are less visible in some respects. The South, in particular, contains large numbers of poor people in rural areas. Numerous poor are found throughout the southern Coastal Plain from Virginia to Texas and in the upland South, particularly the Appalachian Plateau and Ozarks. Even among the rural population, however, poverty is not limited to the South, though it is most widespread there. Smaller areas in peripheral New England, the upper Great Lakes (Michigan, Wisconsin, and Minnesota), the northern Great Plains, and the Southwest also have been identified as poverty regions.

It is unwise to assign a simple, single cause or solution to a problem that exhibits such variation in social, economic, and physical setting. Certainly racial biases and cultural attitudes have contributed immense barriers to blacks, Mexican-Americans, American Indians, and Appalachian whites both in rural

Poverty in metropolitan areas is concentrated in ghettos. These ghettos are often occupied by distinct ethnic groups as exemplified by this Puerto Rican scene in New York City. (United Nations/John Rabaton)

Table 8–1
Poverty in the
United States

	Money Income of Families					
	(Percent below $5,000 – 1976)					
Race & Region			Race & Region			
White	10.2		Black	28.4		
Northeast	8.8		Northeast	25.9		
North central	8.9		North central	23.2		
South	11.0		South	33.0		
West	10.6		West	19.6		
		All families	12.0			
White	10.2		Black	28.4		
In metropolitan areas	8.2		In metropolitan areas	25.5		
Central cities		10.5	Central cities			26.8
Outside central cities		6.9	Outside central cities			21.0
Outside metro areas	14.0		Outside metro areas	38.0		
Nonfarm		13.2	Nonfarm			38.4
Farm		18.1	Farm			54.4

SOURCE: U.S. Bureau of the Census, *Statistical Abstract of the United States* (Washington, D.C.: Government Printing Office, 1976).

and metropolitan areas. These biases not only affect employment opportunity directly but have also contributed to unequal education and training. In portions of the Appalachians, agriculture has been the basic activity since initial settling. As agriculture evolved into a commercial and profitable enterprise elsewhere, making use of capital and modern technology, isolated areas with small farms and poor land (slope and soil) have been unable to adjust to modern ways of doing things. They have suffered simultaneously from isolation and lack of access to modern markets. The populace slips into poverty as it fails to modernize—a major problem in a dynamic and changing society.

Another contributing factor appears as an imbalance in the supply and demand for labor with particular qualities and skills, a problem that also has a spatial implication. Large parts of the nineteenth- and twentieth-century South experienced relatively high-population density in rural areas where agriculture remained labor intensive; agricultural labor rarely realized more than a low-to-modest income. Other portions of the country experienced rapid urban-industrial growth and the development of labor skills that provided the participants higher incomes. The result was distinct regional income variations and greater proportions of the total southern population in the poverty categories. Such regional income discrepancies contribute to migration, but unfortunately the migrants are inadequately prepared for participation in those sectors of urban life that are materially rewarding. Coupled with limited resources, social bias, and the frequent massive concentration of the poor, progress becomes most difficult.

Changes in the significance of location, the need for specific kinds of resources, or the decline in particular kinds of economic activities also contribute to unemployment and poverty. New England experienced a decline in the dominant textile industry. Miners from Pennsylvania to Kentucky saw jobs dissipate as the demand for coal failed to increase, while the mines were automated. Farmers in the South witnessed a declining and changing agricultural scene that had major implications for laborers, tenants, and farm owners. The results are often pockets of economic problems if not widespread poverty. Migration may concentrate poor people in cities, but often many who remain as residual farm people, miners, or unwanted factory laborers are unable to adjust to a changing society and also exist in poverty. The solution to poverty is usually far more complicated than merely changing location.

The numerous causes contributing to poverty may operate independently or in concert; more often it is the latter. Moreover, these causes are of varying importance from one region to another. Although we cannot study all the poverty regions with the thoroughness needed for ample understanding, it is beneficial to examine one area and the approach taken to solve its problems.

Appalachia

Appalachia brings to mind for some a picture of beautiful mountain forests, rushing streams, isolated coves, upper valleys with picturesque and quaint farmsteads, hillside orchards, and prosperous farms on valley floors favored with soils formed on limestone bedrock. Others may envision slow mountain folk, limited in formal education, living a life of isolation and backwardness by choice. More recently, surely because of the publicity and attention being given the Appalachian region as a problem area, a different image is emerging. It includes an image of poor whites existing in isolation, commonly in poverty, and with little hope for betterment.

The problems of Appalachia are attributed to the long period of isolation and an inability to participate in the modernization and commercialization of agriculture. The prosperity of the coal-mining era was short-lived and created more wealth elsewhere. After 1940, mining rapidly automated and concentrated on fewer mines, leaving a residue of unemployed people and scarred landscapes. Lumbering operations that removed the wealth and beauty of an area in a single generation can be viewed in much the same way. Both the physical and human ecology of the region have been severely affected. Soil erosion, the scars of strip mining, floods worsened by poor agricultural techniques, and the removal of forest vegetation from watersheds were major problems before the early twentieth century. Human poverty, low expectation, violence, and human stagnation have been a way of life for generations.

Appalachia includes uplands that extend from the Maritimes of Canada to central Georgia and Alabama and includes several distinct physiographic regions.

8–1   United States: Appalachian Region

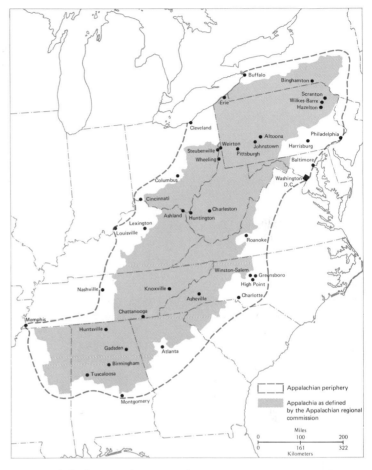

The Appalachians, with severe economic and social problems, are usually thought of in a more restricted fashion as the plateau, Ridge and Valley, and mountain areas south and west of the Mohawk and Hudson valleys. But even here, the hardcore problem area is not spatially coincident with a physiographic region. Figure 8–1 identifies the Appalachian corridor or problem area as defined by the Appalachian Regional Commission. The Appalachian Regional Commission administers the massive funds allocated under the Appalachian Regional Development Act of 1965. The essence of the program is a concerted effort to eliminate poverty. The greatest emphasis thus far has been on highway development. It is believed that part of the reason for Appalachia's problems has been isolation, the lack of accessibility to the area's interior. Early subsistence farmers in the more remote areas failed to make the adjustment to commercial systems, possibly because of isolation from markets as well as poor resources. The lumber and coal resources were harvested here, but the lucrative markets were elsewhere. Coal and lumber stimulated transportation development only to the degree necessary to haul the resources outside the region.

Initial advantages can propel a process of circular causation and growth, but it may well be that disadvantages can do the same in reverse fashion (Chapter 4). Isolation, poverty, low education levels, and limited incentive can feed off each other and contribute to further problems. Those who believe that transportation improvements will stimulate economic growth argue that industry will locate near highways, and the use of recreational potential and tourism can be promoted.

Table 8–2 *Population Change in the Appalachian Region, 1960–70*

SMSA	1970 Population	1960–70 Population Change	Natural Increase	Net Migration	Net Migration Rate[a]
ARC-Appalachian Cities					
Altoona, Pa.	135,356	− 1,914	7,107	− 9,021	− 6.6
Asheville, N.C.	145,056	14,982	11,854	3,128	2.4
Birmingham, Ala.	739,274	18,067	70,440	− 52,373	− 7.3
Charleston, W. Va.	229,515	− 23,410	24,657	− 48,067	−19.0
Chattanooga, Tenn.–Ga.	304,927	21,758	31,658	− 9,900	− 3.5
Erie, Pa.	263,654	12,972	26,435	− 13,463	− 5.4
Gadsden, Ala,	94,114	− 2,836	9,158	− 11,994	−12.4
Huntington–Ashland, W. Va.–Ky.–O.	253,743	− 1,037	23,577	− 24,614	− 9.7
Huntsville, Ala.	228,239	74,378	36,635	37,743	24.5
Johnstown, Pa.	262,822	− 17,911	15,159	− 33,070	−11.8
Knoxville, Tenn.	400,337	32,257	39,480	− 7,223	− 2.0
Pittsburgh, Pa.	2,401,245	− 4,190	163,037	−167,227	− 7.0
Scranton, Pa.	234,107	− 424	3,862	− 4,286	− 1.8
Steubenville–Weirton, O.–W. Va.	165,627	− 2,129	12,838	− 14,967	− 8.9
Tuscaloosa, Ala.	116,029	6,982	13,778	− 6,796	− 6.2
Wheeling, W. Va.	182,712	− 7,630	8,076	− 15,706	− 8.3
Wilkes-Barre–Hazleton, Pa.	342,301	− 4,671	7,318	− 11,989	− 3.5
Peripheral Cities					
Atlanta, Ga.	1,390,164	372,976	172,927	200,049	19.7
Baltimore, Md.	2,070,670	266,925	214,549	52,376	2.9
Binghamton, N.Y.	302,672	19,072	31,727	− 12,655	− 4.5
Buffalo, N.Y.	1,349,211	42,254	125,912	− 83,658	− 6.4
Charlotte, N.C.	409,370	92,589	50,777	41,812	13.2
Cincinnati, O.	1,384,851	116,372	152,532	− 36,160	− 2.9
Cleveland, O.	2,064,194	154,711	199,681	− 44,970	− 2.4
Columbus, O.	916,228	161,304	111,003	50,301	6.7
Harrisburg, Pa.	410,626	38,973	31,524	7,449	2.0
Lexington, Ky.	174,323	42,417	19,967	22,450	17.0
Louisville, Ky.–Ind.	826,553	101,414	86,566	14,848	2.0
Memphis, Tenn.–Ark.	770,120	95,537	104,166	− 8,629	− 1.3
Montgomery, Ala.	201,235	1,591	23,702	− 22,111	−11.1
Nashville, Tenn.	541,108	77,480	55,104	22,376	4.8
Philadelphia, Pa.	4,817,914	475,017	430,144	44,873	1.0
Roanoke, Va.	181,436	22,633	14,111	8,522	5.4
Washington, D.C.	2,861,123	797,033	380,353	416,680	20.2
Winston-Salem, N.C.	603,895	83,646	69,090	14,556	2.8

[a] 1960–70 net migration as percent of 1960 population.

SOURCE: U.S. Department of Commerce, Bureau of the Census, *General Demographic Trends for Metropolitan Areas, 1960 to 1970* (Washington, D.C.: Government Printing Office, 1971).

Others argue differently.[1] Economic growth potential may be greater for cities that have good interaction potential with other cities or population regions. Thus those centers with the greatest potential for growth and economic improvement are not the interior cities of Appalachia but rather the cities peripheral to or fringing on the region (Table 8–2). Furthermore, small and isolated Appalachian communities contain a small labor supply, part of it in sparsely populated rural environs. It is easy for rural areas to become overindustrialized relative to labor supply, even where low wage-labor intensive industry is involved.

It is evident from Figure 8–1 and Table 8–2 that the greater number of growth centers are peripheral to the Appalachian corridor. The northern edge, in Pennsylvania and New York, may be thought of as relatively prosperous. The same applies to the peripheral Piedmont extending from Virginia to Alabama. It is possible that highways may simply aid some of the corridor people in leaving for the more advantageous periphery without really changing conditions within the corridor.

Tennessee Valley Authority

The attention given Appalachia, and through other regional commissions, the Ozarks, the Four Corners (the area adjacent to where New Mexico, Arizona, Utah, and Colorado meet), the Great Lakes, New England, and the Atlantic coast, is not the first effort at solving regional problems by a comprehensive approach. The Tennessee Valley Authority (TVA) is often cited as the first example of comprehensive regional planning in the United States on a large scale (Fig. 8–2). The inception of TVA dates from the depression of the 1930s. The area was agrarian, poor, and flood-weary. The TVA was assigned the responsibility of planning and actual development. Attention was focused on flood control of the Tennessee River and its tributaries, improvement of navigation on the river to Knoxville, and hydroelectric power generation. The area that became the basic unit for planning was the Tennessee River's drainage basin. In addition to the major objective,

related projects were to enhance economic and social conditions by improving agriculture and education and developing the area's recreation potential.

Objective evaluation of the TVA program has always been difficult because its beginning was a controversial political issue. Nevertheless, there are some obvious successes. TVA has aided flood control, the transportation improvement on the Tennessee River has been valuable, and the hydroelectric power has been a marketable product useful for individual farm families and industries. The demand for power, however, now greatly exceeds that generated by hydroelectric facilities. A large proportion of TVA power is sold beyond the borders of TVA and is generated by using Appalachian coal.

That a portion of the TVA region is included in the area designated as the Appalachian problem region also suggests that all problems were not solved. The difficult social and economic condition of many people, particularly in the eastern area, has not been substantially improved. The fact is that economic development potential is variable. It becomes evident when we evaluate the resources of particular areas. The earth surface varies in soil, slope, and resources contained beneath the surface. Population concentrations, transportation routes, and human preferences provide focus on some areas and indifference for others. It may be that some areas simply should not be expected to provide "equal economic opportunity" for significant numbers of people.

Black America

It is not unusual for political units to contain a number of subgroups distinguishable by race, ethnic and linguistic difference, or economic achievement. Such divisions can provide major obstacles to the achievement of a unified political organization and social and economic satisfaction. The difficulty of integrating such groups into a larger society stems not only from outward cultural differences but also from the seemingly selfish nature of mankind. The United States illustrates this problem explicitly in the black-white relationship that has been so difficult to resolve.

---

[1]Carl W. Hale and Joe Walters, "Appalachian Regional Development and the Distribution of Highway Benefits," *Growth and Change*, 5 (1974), 3–11.

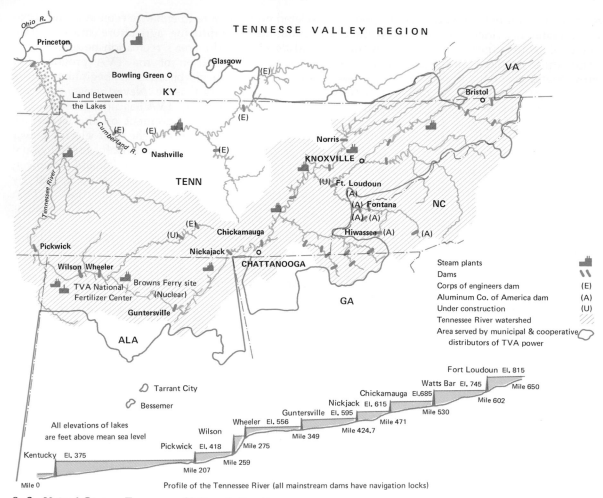

8-2 *United States: Tennessee Valley Authority*

SOURCE: State of Tennessee, *Statistical Abstract of Tennessee, 1969.*

In the United States, the initial patterns of black residence and the black-white relationship were an outgrowth of the diffusion of the plantation system across the lower South. The early English settlers at Jamestown quickly sought a source of income from commercial crops because mineral wealth in the form of gold and silver was not found. Despite experimentation with several crops, tobacco became the favored commercial crop within a few years; it was a readily accepted commodity in Europe.

Tobacco culture required an intensive labor effort for land clearing, planting, weeding, harvesting, and curing. Tobacco culture depleted the soil after a few years of continuous growth; therefore, the colonial system required almost continuous clearing of new farmland, which added to the labor requirements. Early efforts in the use of indentured labor were not highly successful. Once the indentured servants fulfilled their obligations in years to a sponsor, the availability of land provided incentive for independent effort. The slave system provided a cheap, controlled, and stabilized labor force and grew rapidly, particularly during the eighteenth century. The slave labor system was heavily adopted in areas where northwestern Europeans established commercial production using the large estate system.

The result was the southern plantation, an entire landed social and economic system, engaged in commercial production on relatively large holdings and using slave labor. The original center for the system was tidewater Virginia and adjacent portions of Maryland and North Carolina. As settlement proceeded southward, slavery accompanied commercial crops such as rice and indigo to coastal cities of Wilmington, Charleston, and Savannah. Charleston eventually became a major port of entry for slaves. Oglethorpe's early Georgia colony (Savannah, 1733), though it began without slaves, found the "one thing needful" to be a cheap and controllable labor force. In the hill lands, and northward into the middle Atlantic and New England colonies, slavery remained of minor importance and was eventually declared illegal. Economic emphasis there was often on less labor-intensive efforts.

The plantation, once established, became the basic system used in those areas of the lower South considered best from the standpoint of agricultural production. The diffusion of the plantation system went hand in hand with the diffusion of slave labor and established the initial distributional pattern of black residence across the South. Just as it is erroneous to believe that plantations were dominant across the entire South, so too is it incorrect to assert that the blacks were equally distributed across the South. Selected areas such as the outer Piedmont, the inner Coastal Plain, a narrow coastal zone of islands and river banks in Georgia and South Carolina, the Black Soil Belt of Alabama, the Tennessee Valley of northern Alabama, the Mississippi Valley (Arkansas, Tennessee, Mississippi, Louisiana, and Missouri), and by 1860, portions of Texas, were areas of dominantly black population. The Nashville Basin and Bluegrass Basin were other areas of some plantation occupancy and, therefore, black residence. That distribution left much of the South at an opposite extreme. The southern Appalachians, including the inner edge of the Piedmont, the Interior Highlands (Ozarks), portions of the Gulf Coast, and lower Texas were either without blacks or contained few in proportion to the white yeoman farmers. A small percentage, possibly one-seventh, who were freedmen lived outside the South or in southern urban areas where slightly less rigid social pressures allowed their existence as free black artisans.

The distribution of blacks did not change immediately after the Civil War. Although no longer a slave, the freedman had not changed in other respects. They were, after all, agricultural laborers with little training other than in farming, with no land and no capital. In the aftermath of the Civil War, they did not migrate in large numbers but instead entered into a system of tenancy with the white landowners in the same locations previously important for black residence. With a few exceptions, the rural black population of today's South identifies the former plantation regions.

By 1900, a large rural black population lived as a landless tenant labor force. By World War I, the great century of white immigration from Europe (1814–1914) was over. The black could find employment outside the South to replace the greatly diminished flow of European immigrants. Although the black migration slowed somewhat during the 1930s when economic conditions in urban areas were also bad, the migration has continued to the present (Table 8–3). From the rural South to southern and northern urban centers, the massive black redistribution has given rise to a black population that is now southern rural and urban, and northern urban (Fig. 8–3). The rural black is truly rare in northern communities.

The migration has implications that go far beyond population redistribution. Many blacks have clearly improved their economic and social position in urban areas, but it was a group progress not won easily and not without failure for many individuals. The migrants have often been the better educated and more motivated persons, meaning an economic and social loss for the area of origin, but paradoxically they have also lacked the skills needed in urban areas. Furthermore, the racial and social bias added to the difficulty of obtaining suitable housing, a proper education, and access to economic opportunity. The result for many has been existence in a ghetto with its distinctive structure and seeming hopelessness for the future.

Year	Total	Northeast	North Central	South	West
		Number (in millions)			
1970	22,580	4,344	4,572	11,970	1,695
1960	18,860	3,028	3,446	11,312	1,074
1950	15,042	2,018	2,228	10,225	571
1940	12,886	1,370	1,420	9,950	171
1930	11,891	1,147	1,262	9,362	120
1920	10,463	679	793	8,912	79
1910	9,828	484	543	8,749	51
1900	8,834	385	496	7,923	30
		Percent			
1970	100	19.2	20.2	53.0	7.5
1960	100	16.1	18.3	60.0	5.7
1950	100	13.4	14.8	68.0	3.8
1940	100	10.6	11.0	77.0	1.3
1930	100	9.6	10.6	78.7	1.0
1920	100	6.5	7.6	85.2	.8
1910	100	4.9	5.5	89.0	.5
1900	100	4.4	5.6	89.7	.3

Table 8–3 *Black Population of the Conterminous United States*

SOURCE: U.S. Bureau of the Census, *Census of Population* (Washington, D.C.: Government Printing Office, 1910–1970).

The spatial pattern of residence for urban blacks stems from their economic weakness and social position. Hence we derive the American phenomenon of highly concentrated black neighborhoods in the older residential portions of cities, often vacated as economically progressing whites flee to city peripheries and suburban communities. The process has progressed so far that numerous cities have become more black than white. The strength of black numbers is already evident in the increasing numbers of black city mayors, but noticeably central city mayors, not suburban.

The 1954 Supreme Court decision is spotlighted as a landmark in black-white American history. The process that led to a new social and economic position for American blacks, however, began much earlier. The late nineteenth-century South contained advocates of a "New South," envisioned as urban, industrial, and economically strong. Opponents of the new doctrine envisioned a social and cultural disaster that would be born of an urban-industrial society. It would mean a racially mixed labor force, a new black consciousness, and education systems that would undermine the status of an agrarian

Southern society. The opponents of the New South were correct in their fears, for out of urbanization new economic gains and opportunity for some, and better education, even if slowly attained, has arisen a new black consciousness. The new status, hopes, and demands for a participating role by blacks could come only with a break from the old agrarian system. Migration was symptomatic of that break whether to northern or southern cities and was stimulated by numerous other economic, social, and political forces.

Canadian Identity and Unity

French Canada

A major concern of Canadian political and community leaders has been that of national unity and identity. Canada's political organization, over 100 years ago, as a federation was at the insistence of French descendants that any system for union preserve French identity and influence. Thus was pro-

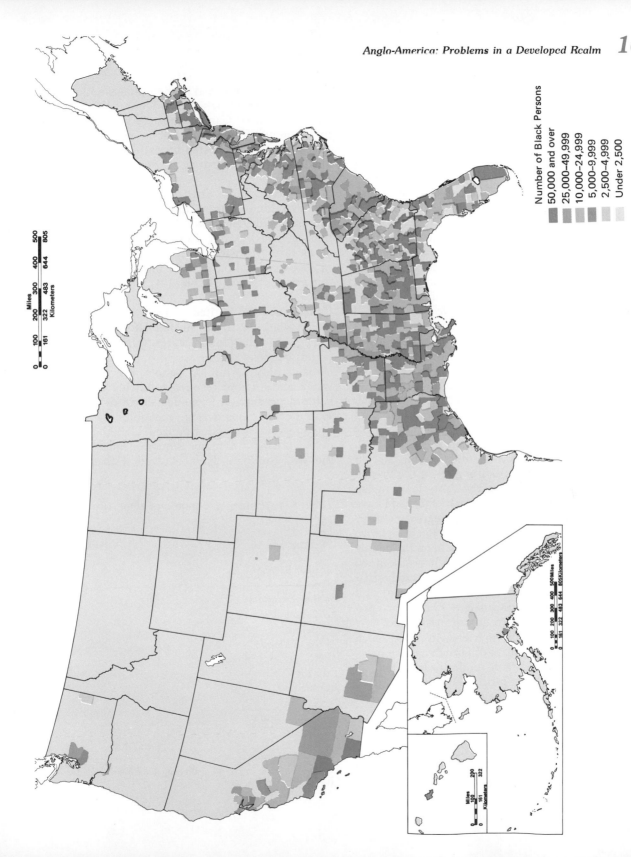

8–3 *United States: Number of Blacks, by Counties, 1970*

SOURCE: U.S. Bureau of the Census (Washington, D.C.: Government Printing Office, 1970).

Number of Black Persons

50,000 and over
25,000–49,999
10,000–24,999
5,000–9,999
2,500–4,999
Under 2,500

moted a confederation of colonies with already distinctive cultural differences—French and English Canada.

The French settled the lower St. Lawrence and later, Acadia, the area around the Bay of Fundy. Great numbers of French settlers did not follow, however, and furs and fishing remained major economic activities. By 1763 the British had overcome French control in Anglo-America. Despite vast holdings by France used for extracting furs, it was only the lower St. Lawrence that was to remain French in culture. The British allowed the French agriculturists to remain. The approaching United States Revolution, however, caused "loyalists" from the British colonies farther south (New England and the middle Atlantic) to move to the loyal colonies of Nova Scotia and Newfoundland, in numbers great enough to justify a new colony (New Brunswick). Others settled farther inland in Ontario. The rapid extension of people of British descent further weakened the position of the French descendants, except along the lower St. Lawrence where the French remained firmly established.

The French have maintained their identity and seem intent on continuing so. Their distinctiveness is not only linguistic, though that in itself is enough to promote a separate cultural identity, but also in religion. The Roman Catholic religion of the French Canadians is an abrupt contrast to Protestant English Canada. In other respects, the French of Canada have been stereotyped to a misleading degree even by other Canadians. It is a misconception to characterize them as quaint, rural, agrarian, unchanging people with high birth rates. Quebec, which is about 80 percent French, is 75 percent urban, part of industrial Canada, second only to Ontario in income, and clearly integrated into the Canadian core in an economic sense. It once was more agrarian than now and experienced unusually high growth rates, which engendered some of the fears of French culture overtaking English culture. With modernization and an urban life style, however, birth rates have decreased sharply below the national average. Continued immigration will further reduce the French proportion of Canada.

The French, though 70 percent urban, do not yet experience an occupational structure similar to that of English Canadians. Although they have a social hierarchy and political power, occupationally they remain overrepresented in the primary and unskilled areas. The French insist on identity—their language remains—and so Canada must forge a national unit in a bilingual framework, not an easy task. Violence and separation suggested by extreme French nationalists are not the attitude of the majority of French Canadians.

The French and English may be thought of as charter ethnic groups. Immigration patterns of the present century, however, have seen significant numbers of Poles, Dutch, Germans, and Italians, sometimes to the consternation of the French who often oppose immigration because of a diminution of their numerical strength (Table 8–4).

Canada and the United States

Canadian identity is further affected by Canada's proximity to the United States. Canada is one of the largest countries in the world but in all its vast area contains only 23 million people. It is rich in forest, water, and mineral resources, but its value for agricultural production is modest because of harsh environments. Considering available resources and proximity to the industrialized United States, it is not surprising that major economic linkages between the two countries have evolved. Canada and the

Table 8–4 *Principal Canadian Ethnic Groups*

Ethnic Origin	Percent of Total Population	
British Isles	44.6	
French	28.7	
Other European	23.0	
German		6.1
Italian		3.4
Netherlands		2.0
Ukrainian		2.7
Asiatics	1.3	
Other	2.4	
Eskimo		.1
Native Indian		1.4
Negro		.2
	Total 100.0	

SOURCE: *Canada Yearbook* (Ottawa: Dominion Bureau of Statistics, 1973).

United States have become each other's most important trading partners. This economic relationship is further evident in the high proportion (nearly 40 percent) of Canadian industries that are controlled by United States parent firms. The trade relationship has, to a large degree, been based on the removal and processing of Canada's special material resources. Canada's concern over a "colonial" relationship generated a long-standing tariff policy that has had the effect of encouraging United States firms hoping to market in Canada to establish industry within Canada. Paradoxically the success of the policy in turn generates the concern over foreign influence and control. Economic prosperity in Canada is significantly tied to United States prosperity, but this fact can only contribute to the problem of Canadian identity. Political efforts to redirect trade relationships toward other areas may be in direct conflict with the normal economic relationships expected between two large and well-endowed countries in proximity.

Canada and the United States in Retrospect

Canada and the United States are two large countries in area and resources. They have achieved a high level of living for the majority of their populace and a powerful position in the world. The motivations of these two nations are such that they will certainly attempt to maintain their positions. Out of their development experience has come a technology that can be of great benefit to the remainder of the world, though it has not always been employed as such. In fact, resources of underdeveloped areas have frequently been used to promote the welfare and expansion of United States and Canadian economic activity. As other countries develop and consume more material resources, the question of how much resource material the United States and Canada can gather from elsewhere, or how much it should, becomes difficult to answer and one of the important future issues.

The processes of immigration and settlement have treated the two areas differently. The United States'

greater population has in fact aided in the accumulation of greater wealth through the processing and use of its resources. The United States, however, is now consuming at a scale whereby the cost and availability of basic goods may well encourage or require limited growth or even stabilization. Quite differently, it is possible that for a less populous Canada, greater economic independence and internal industrial growth may only occur with domestic population growth and expansion. The value of population for developing and using resources is even now reflected in Canadian immigration policies. Although we may view stabilization as necessary for some large and developed countries, for others substantial economic development may occur in the context of population growth and increasing demand.

Internal problems for both countries include the need for integration of minority groups into the larger society. Indeed, as the United States progressed, entire regions as well as minority groups lagged in the acquisition of wealth and position within the system. Effective means are needed for integrating the poverty pockets into the larger economy.

American society will be required to make other adjustments. North Americans are becoming an even more urban society. The immense urban formations, "megalopolis," may require new approaches to government and planning. These are such highly integrated urban systems that many needs such as water, transportation, revenue, and recreation facilities must be planned in a regional framework that extends far beyond the traditional political city.

Further Readings

General works dealing with poverty are as follows: Hansen, Niles M., *Rural Poverty and the Urban Crisis: A Strategy for Regional Development* (Bloomington: Indiana University Press, 1970), 352 pp.; Morrill, Richard L., and Wohlenberg, Ernest H., *The Geography of Poverty in the United States* (New York: McGraw-Hill, 1971), 148 pp.; and Harp, John (ed.), *Poverty in Canada* (Scarborough: Prentice-Hall of Canada, 1971), 357 pp.

Caudill, Harry M., *Night Comes to the Cumberlands* (Boston: Little, Brown, 1963), 394 pp., has written an excellent work providing historic and contemporary insights into the people and problems of a portion of poverty Appalachia.

In addition, see Ford, Thomas R. (ed.), *The Southern Appalachian Region: A Survey* (Lexington: University of Kentucky Press, 1962), 308 pp.; Deasy, George F., and Griess, Phyllis R., "Effects of a Declining Mining Economy on the Pennsylvania Anthracite Region," *Annals, Association of American Geographers*, 55 (1965), 239–259; Estall, R. C., "Appalachian State: West Virginia as a Case Study in the Regional Development Problem," *Geography*, 53 (1968), 1–24; and Gauthier, Howard L., "The Appalachian Development Highway System: Development for Whom?," *Economic Geography*, 49 (1973), 103–108.

A special issue of *Economic Geography*, 48 (1972), 134, is entitled "Contributions to an Understanding of Black America," and includes the following: Morrill, Richard L., "Geographical Perspectives on the History of Black America"; Adams, John S., "The Geography of Riots and Civil Disorders in the 1960's"; Rose, Harold M., "The Spatial Development of Black Residential Subsystems"; Brown, William H., Sr., "Access to Housing: The Role of the Real Estate Industry"; Deskins, Donald R., Jr., "Race, Residence, and Workplace in Detroit, 1880 to 1965"; Jenkins, Michael A., and Shepherd, John W., "Decentralizing High School Administration in Detroit: An Evaluation of Alternative Strategies of Political Control"; and Craig, William, "Recreational Activity Patterns in a Small Negro Urban Community: The Role of the Cultural Base." Other works that deal with the problems of the black minority include the following: Davis, George A., and Donaldson, O. Fred, *Blacks in the United States: A Geographic Perspective* (Boston: Houghton Mifflin, 1975), 270 pp., and Rose, Harold M., "The Development of an Urban Subsystem: The Case of the Negro Ghetto," *Annals, Association of American Geographers*, 60 (1970), 1–17.

On the issue of French Canada, see Corbett, Edward M., *Quebec Confronts Canada* (Baltimore: Johns Hopkins Press, 1967), 336 pp.

Western Europe

Louis De Vorsey, Jr.

PART THREE

Western Europe: A Varied Home for Mankind

9

WESTERN EUROPE IS the home of almost 350 million of the earth's most productive and prosperous people. The twenty-four political units they inhabit cover an area that is a good deal smaller than the United States or Canada (Figs. 9–1 and 9–2). In fact, Western Europe's area of 1,387,550 square miles (3,593,755 square kilometers) makes up just 3 percent of the earth's total land surface. As the map showing its area and latitudinal position reveals, many of Western Europe's most populous and advanced countries are at the same latitude as is much of Canada's thinly populated north. How is it that so many people have managed to create such productive and comfortable life styles in this relatively small and northerly region? Clearly there is no simple answer to such a question. It is nonetheless worth asking, and even a partial answer will give us a better understanding and appreciation of this important region of the developed world.

Western Europeans often appear closer to being the masters of their physical environment than the people of any other great culture region. As a result of their accomplishments, modern Europeans feel the effects of many elements of their physical environment in more indirect and subtle ways than the people in much of the rest of the world. This chapter focuses on only the most significant of the physical elements making up the human habitat of Western Europe. Western Europe's location in the world, configuration, climate, major landforms, and biophysical resources are discussed in an effort to shed some light on the course of Western Europe's emergence as a pole of productivity, wealth, and influence in the modern world.

Western Europe: The Center of the Land Hemisphere

The countries making up Western Europe occupy a strategic and central location on the earth. By moving far into space, astronauts and cosmonauts have been able to gain a unique view of our planet. In one glance they have seen and photographed a hemisphere at a time. If a photograph were taken from a space station directly above the small ✕ in Western Germany on the map shown in Figure 9–3, it would reveal the most important hemisphere of all, the Land Hemisphere. This hemisphere is the half of the globe that contains about 90 percent of all the earth's inhabited land area and about 94 percent of the world's total population and economic production.

As a result of their central location, Western Europeans enjoy relatively easy contact with almost the entire habitable world and its resources. Of course, the significance of any particular location means little until people begin to focus flows of resources, products, and information on it. So the significance of Western Europe's centrality in the world became meaningful and important only after the Europeans mastered the arts and technology of distant oceanic navigation in the fifteenth and sixteenth centuries. Then the newly charted sea-lanes to the New World, India, Africa, and East Asia began to function as networks of circulation for people, materials, and ideas with Western Europe their center. The colonial empires of many Western European

9–1 *Western Europe: Area and Position Relative to Anglo-America*

ICELAND
Reykjavik

● Cities with 1,000,000 or more inhabitants
○ Cities with less than 1,000,000 inhabitants

FINLAND

NORWAY   SWEDEN
Helsinki

Oslo   Stockholm

Göteborg

DENMARK
Copenhagen

Edinburgh
UNITED
Glasgow
Newcastle
KINGDOM
Lubeck
Liverpool   Leeds   Hamburg
NETHERLANDS   Berlin
Dublin   Manchester
Bremen
IRELAND   Birmingham   WEST
Amsterdam   GERMANY
London   BELGIUM   Rotterdam
Brussels   Cologne
Bonn   Frankfurt
LUXEMBOURG   Mannheim
Paris   Karlsruhe   Stuttgart
Munich   AUSTRIA
Zurich
FRANCE   SWITZERLAND

Lyon   Milan
Turin   Venice
Genoa   Florence
Nice
SPAIN   Marseille
ITALY
PORTUGAL   Madrid   Barcelona   Rome
Lisbon   Naples
GREECE

Granada   Athens

Miles
0   100   200
0   161   322
Kilometers

9–2   *Western Europe: Nations and Principal Cities*

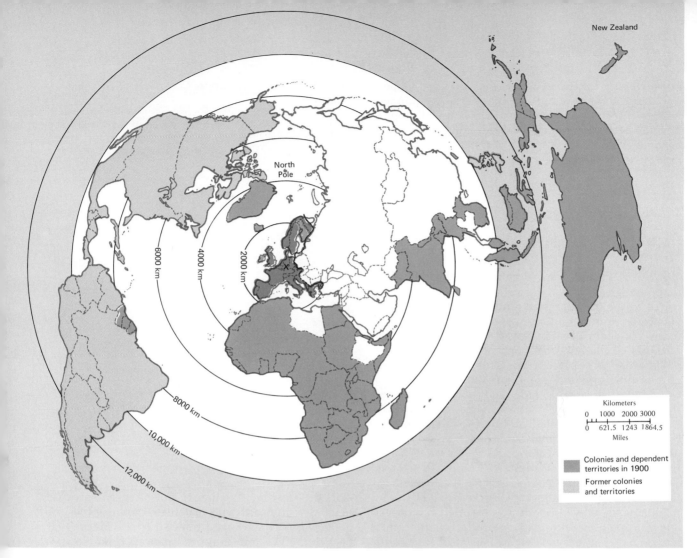

New Zealand

North
Pole

6000 km
4000 km
2000 km

8000 km

10,000 km

12,000 km

Kilometers

| 0 | 1000 | 2000 | 3000 |

| 0 | 621.5 | 1243 | 1864.5 |

Miles

Colonies and dependent
territories in 1900

Former colonies
and territories

9–3   Western Europe: Center of the Land Hemisphere and Colonial Possessions

nations are partly a reflection of Europe's central location. Later, the advent of railroad technology opened the vast continental interiors and enlarged the world circulation network that could be focused on Western Europe. In the modern era, long-range aviation and high speed oceanic travel further emphasize the tremendous significance of Western Europe's centrality. It is not surprising to find that the Western Europeans have consistently been in the forefront as pioneers of commercial aviation. Count Ferdinand von Zeppelin of Germany organized the world's first commercial airline in 1910, employing his famous airships, or Zeppelins as the lighter-than-air craft were known. In the first three years

of operation, over 35,000 paying passengers were transported in the Zeppelins. By 1919, the first regular international airmail service was established between London and Paris. More recent years saw the British Overseas Airways begin the first regularly scheduled jet airliner service in 1952. Nor is it surprising to find the British and French combined to develop the world's first supersonic commercial airliner, the controversial Concorde.

A recent decision by the Soviet Union has opened that nation's Arctic Sea route to foreign ships. As ships begin to ply this new route, the distance between the ports of northwestern Europe and eastern Asia will be considerably shortened. Soviet atomic-

173

powered icebreaker technology may eventually open the Arctic to year-round use by specially equipped vessels. It seems safe to predict that Western Europe's position at the center of the Land Hemisphere will become even more important as transportation technology improves in the years ahead.

Western Europe's Maritime Orientation

Western Europe, viewed at the continental rather than the global scale, forms an irregular Atlantic fringe to the vast Eurasian landmass. At its western extremity, Eurasia intermingles with the sea as a maze of peninsulas and island groups. Western Europe's sheltered coasts and many harbors provided an almost perfect setting for the development of maritime-oriented economies as the Europeans extended their world trade and political linkages through the past five centuries. Western Europe's irregular and fragmented outline provides a higher ratio of coastline to total land area than any other major cultural area of the world. Only a small handful of the countries of Western Europe lack direct access to the 70 percent of our planet covered by the oceans. Even landlocked Switzerland, however, maintains an active ocean trade via the Rhine River and Europort in Rotterdam. Norway, on the other hand, enjoys an easy access to the sea thanks to its harbor-studded 1,650 miles (2,655 kilometers) of coastline (Table 9-1). The Norwegians have long looked to the sea for their livelihood. As a result of their effort and ingenuity, this country, with a total population only slightly larger than that of Alabama, boasts the world's third largest merchant marine fleet. The seas beyond Norway's fringing *skerries* (low offshore islands) have long proved a more inviting field for investment and enterprise than have the *fjelds* (wind-swept rocky hills) and mountains of the harsh interior.

Norway is an extreme case, but much the same thing can be said about a great deal of Western Europe. No part of Western Europe is far from the sea and its challenges. A glance at a map or globe reveals that numerous seas, gulfs, and bays—such as the North, Bothnian, Baltic, Irish, Mediterranean,

Table 9-1   *Western Europe: Coastlines of Selected Countries*

	Estimated Length of Coastline	
	Miles	Kilometers
Belgium	34	55
Denmark	686	1,104
Finland	735	1,183
France	1,373	2,210
West Germany	308	496
Greece	1,645	2,647
Iceland	1,080	1,738
Ireland	663	1,067
Italy	2,451	3,944
Malta	50	80
Monaco	3	5
Netherlands	198	319
Norway	1,650	2,655
Portugal	743	1,196
Spain	2,038	3,280
Sweden	1,359	2,187
United Kingdom	2,790	4,490
United States, Atlantic coastline only	1,612	2,594

Adriatic, Biscay, Ligurian, and Aegean—provide Western Europeans with matchless opportunities for ocean-borne contacts and trade. It is no wonder that flags of the Western European nations from Norway to Greece are commonplace in ports the world over.

The "Continental Architecture" of Western Europe

The term "continental architecture" is a bit unusual. It includes the many landform areas such as the plains, uplands, and mountains that form the physical framework or "skeleton" on which the Western Europeans have built their landscapes. At first glance the physical map of Europe seems complicated and confusing (Fig. 9-4). A bit of study, however, soon reveals certain broad landform similarities over extensive areas. Once recognized, these broad physiographic regions can be very helpful in understanding the diverse human and economic patterns that make up modern Western Europe.

Western Europe occupies a portion of each of the four great physiographic subdivisions of Europe. These are from north to south (1) the Northwestern Highlands; (2) the Great European Plain; (3) the Central Uplands; and (4) the Alpine Mountain systems of southern Europe. A glance at Figure 9–4 reveals that these divisions follow a general west-to-east trend across Europe. This orientation is in sharp contrast to Anglo-America, where the major physiographic divisions are aligned from north to south. The west-east orientation in Europe has had a profound influence in the development of the region through time. For one thing, it has allowed for the relatively easy penetration of marine influences into much of Western Europe. A lofty range of north-south trending mountains, such as the Rockies, across the breadth of Europe would have severely limited the penetration of these influences. It is safe to state that Western Europe would have developed in a far different way had such been the case. One might even hazard the suggestion that had the Alps and Northwestern Highlands formed a continuous barrier from the Mediterranean to the Arctic, Western Europe would be in the underdeveloped rather than the developed world. Such an alignment in the northerly latitudes of Western Europe would most certainly create a far more severe climate than is now found. As a result, the range of agricultural opportunities would be greatly limited and transportation extremely difficult. Space does not allow an exhaustive discussion of how Western Europe might have developed under such hypothetical conditions. Such speculations are worthwhile, however, because they assist in demonstrating the significance of basic geographic patterns to human economic development.

The Northwestern Highlands

The Northwestern Highlands include practically all of the northern countries of Sweden, Norway, and Iceland as well as a large portion of Finland, the British Isles, and the Brittany Peninsula in northwestern France. Generally speaking, the area is underlain by very hard and geologically ancient rocks. In a few areas like Iceland, volcanism is still actively building masses of igneous rock. Rugged, hilly uplands and wind-swept plateau surfaces,

however, dominate the landscapes over most of the Northwestern Highlands. Isolation and ruggedness, plus a northerly latitudinal position and exposure to the gales and winds of frequent Atlantic storms, create many serious problems for the people living there.

Several thousand years ago, the climate of this part of the earth differed significantly from the present. Annual snowfalls accumulated on highland surfaces and failed to melt in the cool summers.

The Rhine river is a major transport artery of Western Europe and is used by Switzerland, France, West Germany, and the Low Countries. Along the Rhine are roads, railways, and manufacturing plants. (United Nations)

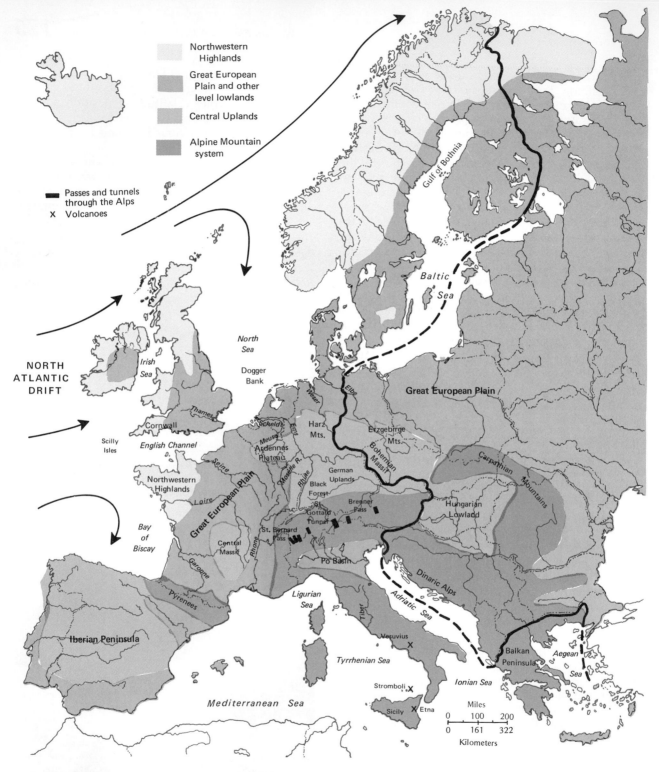

Legend:

Northwestern Highlands

Great European Plain and other level lowlands

Central Uplands

Alpine Mountain system

Passes and tunnels through the Alps

X Volcanoes

NORTH ATLANTIC DRIFT

Iceland

Gulf of Bothnia

Baltic Sea

Great European Plain

North Sea

Dogger Bank

Irish Sea

Thames

Cornwall

Scilly Isles

English Channel

Northwestern Highlands

Seine

Loire

Bay of Biscay

Garonne

Pyrenees

Iberian Peninsula

Mediterranean Sea

Central Massif

Rhone

Scheldt

Meuse

Moselle R.

Rhine

Ardennes Plateau

Harz Mts.

Erzgebirge Mts.

Bohemian Massif

German Uplands

Black Forest

St. Gottard Tunnel

St. Bernard Pass

Brenner Pass

Po Basin

Ligurian Sea

Tiber

Vesuvius X

Tyrrhenian Sea

Stromboli X

Sicily X Etna

Adriatic Sea

Dinaric Alps

Ionian Sea

Balkan Peninsula

Hungarian Lowland

Carpathian Mountains

Aegean Sea

Weser

Elbe

Miles
0   100   200
0   161   322
Kilometers

9-4 *Western Europe: Physiographic Regions and Physical Features*

Gradually the great thicknesses of the accumulating snowfields led to the formation of massive glaciers of continental scale. These ponderous masses flowed from higher areas toward the sea under the incredible weight of ice accumulated over many centuries. In their slow but relentless progress, the glacial lobes ground and gouged the land over which they flowed. As a result, many areas in the Northwestern Highlands are practically devoid of soils and exist as vast stretches of barren rock waste that continue to defy man's attempts to put them to productive uses (Fig. 9–5). These landscapes go a long way toward explaining why Norwegians, Icelanders, and other northern Europeans have traditionally turned to the sea and its resource potential in their efforts to find a rich and rewarding life style. Certainly the location of some of the world's most productive fishing grounds just off these coasts has encouraged their residents to choose fishing rather than farming as a way of life. Here the mixing of warm and cold ocean currents in conjunction with broad areas of relatively shallow waters over smooth bottoms called "banks" provide almost ideal conditions for lucrative commerical fishing.

Other visible relics of the glacial heritage of the Northwestern Highlands are the vast areas of marshland and countless lakes that dot the landscapes of much of Finland, Sweden, and Ireland. The many lakes help provide the region with reservoirs for its chief local energy source, hydroelectricity. The fossil fuels, petroleum and coal, are almost entirely absent from the geologically ancient Northwestern Highlands. Consequently, hydroelectric power installations have played a very significant role in the industrial development there (Fig. 9–6). Norway and Sweden are among the world's leading countries in terms of hydroelectricity produced.

Still another reminder of the glacial heritage is the deep coastal fiords that characterize northwestern Europe's coastal zone. These glacially excavated flooded valleys allow the sea to penetrate deeply into the land. In Norway and Scotland, the stark beauty of the fiords has become the basis for a lucrative tourist trade.

The agricultural potential of the Northwestern Highlands region is severely limited by ruggedness of topography, thin and infertile soils, remoteness, and the excessively cloudy wet cool climate. Grazing sheep and cattle frequently typify agricultural effort

here. In more sheltered inland locations, the verdant grass cover is replaced by hardy coniferous forests. The vast brooding expanses of trees represent the westernmost extent of the great taiga forest belt that girdles subarctic Eurasia from the Sea of Okhotsk to the North Sea. The trees are cropped to provide the raw material for a host of forest-related industries such as paper-pulp manufacturing, lumber, and construction materials. Large segments of the national economies of Sweden, Finland, and Norway are based on forest-related industries. In fact, Sweden and Finland consistently rank as Europe's first and second most important exporters of both lumber and wood pulp.

In some protected valleys and coastal lowlands where better soils and a more moderate climate exist, some crop farming is carried on. Compared to other regions of Europe, however, the Northwestern Highlands is not an area of great agricultural promise or productivity. Even tolerant crops like oats and barley are largely limited to the southern and eastern fringes of the Northwestern Highlands.

Metallic minerals are found in some of the ancient rock masses of the highlands and frequently support important mining operations. By far the best known of these mining centers is north of the Arctic Circle near the Swedish cities of Kiruna and Gallivare. Here, in the land of the still semi-nomadic Lapps of the far north, ultramodern cities have been built to support the mining of one of the world's great deposits of very high-grade iron ore. Such cities in the interior of the Northwestern Highlands are the exception rather than the rule. This region is one of Europe's most thinly settled areas. As has been shown, there is little in the habitat of the Northwestern Highlands to encourage people to settle here in large numbers. It is largely a land of glaciers, rugged uplands, quiet crystal lakes, rapid-flowing rivers, and dark brooding forests that yield their wealth grudgingly to a small but hardy population.

The Great European Plain

The Great European Plain stretches from the Ural Mountains on the east to the Pyrenees Mountains on the west as a broad, undulating lowland south of the Northwestern Highlands. In Western Europe it narrows considerably and is interrupted in places by relatively shallow bodies of water such as the Baltic Sea, North Sea, and English Channel. Gen-

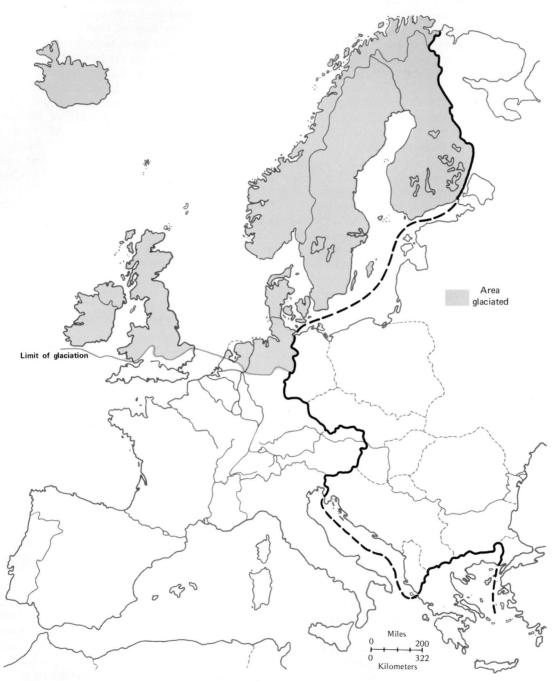

Area
glaciated

Limit of glaciation

Miles
0   200
0   322
Kilometers

9-5 *Western Europe: Extent of Continental Glaciation*

Coalfields
Tidal power
Geothermal

HYDROPOWER DOMINANT

Oil

Oil

Natural gas

Ruhr

HYDROPOWER DOMINANT

Miles
0   200
0   322
Kilometers

9–6   Western Europe: Energy Sources

erally, the Great European Plain is underlain by geologically more youthful rocks including many sedimentary layers. Many of these sedimentary rocks are relatively soft and are weathered more easily than the harder crystalline rocks of the highlands adjacent to the Great European Plain. Erosion has proceeded more rapidly, and the land presents a "softer" and more gently rolling surface even where local folding and warping have thrust up ranges of hills and escarpments to break the plain. Although usually of only moderate elevation, these ranges are of considerable local importance and lend a great degree of variation to the landscapes of the region. It is, on the whole, a region that has presented mankind with a stimulating degree of challenge rather than with imponderable barriers. Even more important, the plain is richly endowed with a wide range of resources, the fundamental building blocks of modern industrial societies during the past two or three centuries. It is not surprising, therefore, to discover that the Great European Plain in Western Europe is one of the earth's most populous and highly developed regions.

During the glacial periods, part of this region was covered by the great continental ice sheets that slowly flowed out of the Northwestern Highlands. Here on the plains the glaciers were at their outer limits and deposited great quantities of earth materials rather than scouring and removing them as they did farther north. As a result, the topography, drainage, and soils in glaciated areas are more varied and complex than those located beyond the maximum reach of glacial effects. These differences contribute to the agricultural variety that marks much of the Great European Plain today.

As the map of physiographic provinces shows (Fig. 9-4), the Great European Plain on the mainland of Western Europe is bounded by highlands to the south and seas to the north. The great rivers flow across the region in a northerly direction to meet the sea. These rivers—such as the Garonne, Loire, and Seine in France; the Schelde and Meuse in Belgium; the mightly Rhine of Germany and the Netherlands; and the Ems, Weser, and Elbe in Germany— have provided the people of Western Europe with natural routeways for travel and trade. The Great European Plain itself has served as Europe's major east-west routeway. It is not surprising to find that the highways, railways, rivers, and canals of the Great European Plain are integrated into one of the world's most dense and efficient transportation networks. Ease of circulation was a basic precondition for the development of this region, development that has allowed its great population to achieve one of the world's highest levels of life.

Many of the world's greatest cities and industrial centers have grown here as Western Europeans have made increasingly extensive use of their habitat and reached out to form linkages with the rest of the world. Perhaps even more fundamental to the growth of these great industrial centers and cities than the excellent transportation potential of the plain have been the vast coal deposits below ground (Fig. 9-6). Coal was found by primitive Western Europeans at and near the surface where rivers and streams had eroded into the carboniferous strata. As a fuel resource, coal remained of purely local significance for a very long time. It was not until the middle decades of the eighteenth century that coal became a factor of major importance to large numbers of Western Europeans. During this period the steam engine was perfected, and coke was utilized in British blast furnaces to process iron in large quantities. These and a host of other technological breakthroughs formed the foundation for the fundamental changes in European life which we now think of as the Industrial Revolution.

The agricultural potential of the Great European Plain in Western Europe is immense. In Denmark and the Netherlands, for example, almost 75 percent of the total land area is used for some form of agricultural activity. The range of crops is also broad, including most of the temperate climate foods and fibers known today. Contributing to the agricultural potential is the mild, moist marine climate that prevails. Rainfall, though not heavy, is evenly distributed throughout all seasons of the year, and temperatures are moderate in summer and winter. On the whole, the climate of this portion of Western Europe is usually considered stimulating and useful to mankind.

From almost every point of view, the Great European Plain area of Western Europe is a region richly endowed to provide the bases for the development of modern technologically advanced societies. In such a habitat, alert, inventive, and energetic populations have been able to create some of the earth's richest and most varied life styles.

The Central Uplands

The Central Uplands stretch as a belt of hilly and rugged plateau surfaces across the central part of Europe generally south of the Great European Plain. They are composed of geologically ancient rocks and resemble portions of the Appalachian Mountains of the United States. Like the Appalachians, they are often rounded and of moderate elevation and frequently heavily forested. To the east they are represented by the ranges of Czechoslovakia which form the Bohemian Massif. Westward, the Central Uplands are represented by the broad hilly region between the Black Forest and Harz districts of central and southern Germany. The German Uplands continue to the Ardennes Plateau of Belgium on the north and the Central Massif of France on the south. Although the Central Uplands are of only modest elevation, they have a tradition of remoteness and remain heavily forested and thinly settled.

Many of Europe's great rivers cross the Central Uplands in deep valleys which make movement across the region difficult. Most routeways and important settlements are located in these valleys and form the chief focus for life in the Central Uplands. Compared to the Great European Plain, the Central Uplands are not as productive or densely populated. They are, however, vastly more productive and populous than either of the other two physiographic divisions, the Northwestern Highlands and the Alpine Mountains.

The Central Uplands of Europe are important for their extensive and varied mineral deposits. Mining and metalworking had an early start in this region. In general, these early industries were based on the ores, waterpower, and abundant charcoal available locally. With improved technology and the increasing use of coal for fuel, these industries shifted to the region's margins near the great coalfields. Coal was found at the surface in river valleys where the valleys passed from the Central Uplands to the Great European Plain. Typical of this coal-producing zone is the Ruhr Valley of Western Germany. The Ruhr ranks among the world's greatest mining and industrial regions today.

The Central Uplands are somewhat less well suited for agricultural enterprises than the Great European Plain. Much of the terrain is rugged and steeply sloping. Such areas are poorly suited for most agricultural use and are left in forests such as the Ardennes of wartime fame. The higher elevations of the Central Uplands have cooler temperatures and more abundant precipitation than on the lower European Plain. Cool wet conditions favor the growing of grass and other fodder crops that support large herds of grazing animals. Certain more well-suited valley areas are important agricultural regions. Densely populated industrial regions lying in and around the margins of the Central Uplands are ready markets for agricultural products.

The Central Uplands is a region of great diversity and contrast. Its margin, along with the adjacent Great European Plain, supports one of the earth's largest and most productive population groups. It might well be said that the heart or core of Western European culture lies where these two extensive and productive physiographic regions meet.

The Alpine Region

This physiographic division, the Alpine region, stretches across the southern flank of Europe from Portugal and Spain on the west to the Greek Peninsula on the east. High mountains, rugged plateaus, and a predominance of steeply sloping land characterize the region and provide its distinctive character. Three major peninsulas—the Iberian, Italian, and Balkan—are included in the Alpine region as are a number of Mediterranean island groups.

Geologically the lofty folded Alpine mountains are relatively youthful. In many respects they are similar to the Rocky Mountains of the United States. Both systems of ranges were probably formed about 70 million years ago as a result of great pressures in the earth's crust. The pressures that formed the Alpine system were generated far to the south and caused the rock layers to fold against the more ancient and resistant rock masses of the Central Uplands. Like the Rockies, the Alps have many mountains more than 10,000 feet (3,048 meters) above sea level, and some are over 15,000 feet (4,572 meters). Breaks in the Alpine system are few but extremely important. These breaks, or "passes," in the mountains have served as traditional focal points for routes connecting northern Europe with the Mediterranean Basin. Certain of these Alpine passes such as the Brenner, St. Bernard, and St. Gotthard have been among the most important routes of human movement in all history. Today, modern motor and railway tunnels speed travelers through the passes in comfort at all seasons. Petroleum pipelines, conveying oil from the ports on the Mediterranean Sea to the industrial regions of central Europe, are built through the same historic Alpine passes that saw the march of Caesar's armies.

The modern inhabitant of the Alpine region is frequently reminded of the geological youthfulness of the area. Earthquakes, a symptom of geologic activity, are common. In recent years the city of Skopje in Yugoslavia was nearly destroyed, as were Lisbon in 1755 and Messina in 1908. Volcanic activity too is a sign of geologic unrest and youthfulness. Etna, Vesuvius, and Stromboli are among the better known volcanoes found in southern Europe. In the province of Tuscany in modern Italy, volcanic hot springs and steam geysers are tapped to provide geothermal energy for industrial uses. Some are utilized for producing electricity, and others serve as sources of chemicals such as boron.

Industrial activity in southern Europe is concentrated in centers such as Milan, Turin, Barcelona, Marseille, Genoa, and Athens. These centers are located on coastal plains or enclosed lowland basins which are included in the Alpine region. Turin and Milan are closely associated with the populous and productive Po River Basin of northern Italy. Coal and petroleum are largely absent from this part of Europe. As an alternative source of energy, the rapidly flowing Alpine rivers, sometimes called the "white coal" of the Alps, have been extensively used to produce hydroelectricity. Tiny Switzerland is notable for having almost 4 percent of the world's total hydroelectric power capacity.

High elevations, thin soils, and steep slopes severely limit agriculture over large areas of the Alpine region. In other areas, however, agriculture flourishes with many subtropical crops. Vineyards, fruit orchards, and olive groves lend a distinctive character to the agricultural landscapes found along the southern flank of Europe. The sunny conditions typical of the dry summer subtropical climate permit these and many other crops to flourish. The six leading crops of Spain are typical of the Alpine region as a whole. They are wheat, olives, barley, oats, grapes, and fruit. The proportion of people engaged in agriculture is far higher in the countries of this southern region of Western Europe than in the three regions to the north. For example, only about 4 percent of the labor force in the United Kingdom is employed in agriculture. In Spain, on the other hand, over 40 percent of the labor force is engaged in farming.

The Alpine region is quite different from the three other physiographic divisions making up Western Europe. It is a region of tremendous challenge, frequent disappointment, and sometimes even disaster to its inhabitants. It is not surprising that thousands of southern Europeans leave their homelands to seek better economic opportunities in the more highly developed industrial north every year. They follow the same lure of a better life which led millions of them to the United States and southern South America in the period 1860 to 1920.

The Climate of Western Europe

The Alps can be viewed as a "climatic divide" in Western Europe. To the north most of the heavily inhabited area of Western Europe enjoys an extremely temperate and moist marine climate. South of the Alps is the dramatically different dry summer subtropical climate. Even an orbiting astronaut far out in space would visually notice the contrast between the two regions. To the north, the marine climate produces a lush green landscape of verdant forests, fields, and farms. Ireland's nickname, "Emerald Isle," emphasizes this characteristic regional greenness. To the south of the gleaming glaciers and snowfields that crown the lofty Alps, the landscape changes rapidly to one where the browns and yellows of parched Mediterranean fields and purple gray mountains predominate.

When the ancient Romans left their Mediterranean homeland to conquer most of Western Europe, they found a strange alien and hostile world. It is easy to sympathize with Tacitus, who, in about A.D. 100, wrote: "The climate in Britain is disgusting from the frequency of rain and fog." He did find, however, that "the cold is never severe." Although the marine climate of the north and the dry summer subtropical climate of the south are different, they are both permissive insofar as mankind is concerned. Countless millions of vigorous and productive humans have flourished in both climatic regions in the past and continue to flourish there at the present time. It would be unwise to say, as some have, that one of the climates is superior to the other. More realistic is the recognition that the two climatic regimes are very different and so provide differing challenges and opportunities for the people living in them.

The Marine Climate

As its name indicates, this climate owes its essential characteristics of moderate temperatures and abundant supplies of moisture to the Atlantic Ocean which lies to the west of Europe. Water heats and cools more slowly than land and so acts as a moderating influence on climate. Europe falls in the earth's prevailing westerly wind belt and is affected by conditions originating over the Atlantic. The surface waters of the Atlantic to the west of Europe are warmer than one might expect, for they receive a great drift of warm tropical water pushed in a clockwise direction by the prevailing winds of the northern hemisphere. Along the coast of the United States from Florida to Maine, this warm water forms the

9–7 *Western Europe: Average Monthly Temperature and Precipitation at Selected Stations*

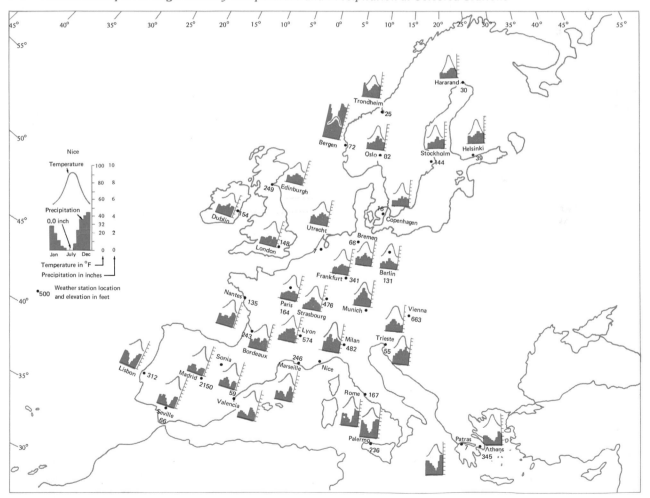

SOURCE: Department of Agriculture, *Agricultural Geography of Europe and the Middle East* (Washington, D. C.: Government Printing Office, 1948).

Gulf Stream. In the Atlantic off Nova Scotia and Newfoundland, the Gulf Stream becomes diffused into a broad area of relatively warm water drifting to the east. This relatively warm surface water is called the North Atlantic Drift. Thanks to the Prevailing Westerly winds and North Atlantic Drift, northwestern Europe has mild moist winters and cool moist summers. These conditions result in a lush green landscape. The climagraph for Dublin typifies the cool moist conditions that prevail over much of Western Europe north of the Alps (Fig. 9–7).

The Dry Summer Subtropical Climate

The southern margins of Europe share a climate common to the vast Mediterranean Basin. It is characterized by clear, dry, hot summers and moderately moist, mild winters. Dry summer subtropical climate is found in central and southern California as well as in central Chile, southernmost Africa, and southern Australia. Another name for this type climate is Mediterranean.

The Mediterranean lies close to the thirtieth parallel of north latitude which marks the approximate center of the belt of relatively high pressure known as the Subtropical High. In summer, this belt of desert-making high pressure shifts a few degrees to the north to cover the Mediterranean Basin. When it comes, it brings the clear sunny skies and dry air that are common to the Sahara, the earth's largest desert. In winter, the Subtropical High pressure belt shifts a few degrees to the south, and the Mediterranean Basin is influenced by the marine air masses and cyclonic disturbances of the Prevailing Westerly wind belt. These marine air masses and storms bring cloudy skies, cooler temperatures, and moisture to provide the winter rains common to the region. The climagraphs for Nice and Athens clearly show the pattern of hot dry conditions that characterize Mediterranean summers (Fig. 9–7). Notice particularly how dry the three summer months—June, July, and August—are in Nice. This seasonal drought is far more important to plant life than is the fact that Nice receives a total amount of precipitation slightly greater than that of Dublin in an average year.

The result of this wet-dry climatic rhythm is seen in almost every aspect of the landscape in the southern part of the Alpine region of southern Europe. The summer drought is extremely hard on many plants common to the rest of Europe. The natural vegetation of the Mediterranean Basin is made up of plant species that resist excessive evaporation and loss of moisture. Some have thickened stems or bark, thorns, waxy coatings, small leaves, or hairy fibers. Succulent, water-storing plants such as cactus also do very well here. The term "xerophytic" is used to describe these drought-resistant plant types.

Human life too is geared to the alternation of wet and dry seasons. Farmers plant and tend crops during the winter and spring and harvest them in early summer. Fruit and other deep-rooted tree crops such as the olive are well adapted and form an important part of the Mediterranean agricultural scene. The olive is so representative of the region that its limits of cultivation are taken by some as the limits of the dry summer subtropical climatic region. Wheat probably originated in the area of the eastern Mediterranean and remains the chief grain crop of the region. It thrives in the moisture of the Mediterranean winter for its germination and growth, and the aridity of the summer for its maturation and harvest.

The reliably sunny summer weather of Mediterranean Europe has become one of the region's greatest modern economic assets. It is the basis of a flourishing tourist industry that provides a large and profitable income. Hundreds of thousands of prosperous Europeans from the northern industrial centers of Western Europe enjoy their yearly vacation holidays in the sun and blue waters of the Mediterranean coast.

Taken as a whole, Western Europe can be seen to have a varied climatic background that has proved favorable for human development. It is, as we have seen, a temperate area of reasonably humid, seasonal climate. It lacks any regions that could be accurately described as deserts as well as any regions where excessive heat and humidity would handicap human performance. Although far from perfect and often subject to rapid changes in daily weather, Western Europe is climatically well suited for human activities.

Conclusions

In this chapter several of the more significant aspects of Western Europe's location and physical geography have been discussed to show how they are related to the region's rich and varied life styles.

On the whole, the view has been continental rather than regional or local in scale. This view has been necessary but not entirely satisfactory, since the meaning of these patterns and elements to the individual living in Western Europe has not been dealt with directly. To understand Western Europe and its position in our unfolding modern world, some attempt at insight into the personal and individual level is imperative. The following passage, written by one of Western Europe's great humanists, Salvador de Madariaga, is particularly well suited. In his "Introduction" to the splendid book *Europe from the Air*, Madariaga observes:

. . . Europe is a continent of modest dimensions. There are no boundless open spaces, no mountain ranges towering into the sky, no rivers resembling inlets of the sea, no icy cold and no torrid heat. Everything is moderate—not too hot nor too cold. . . . The physical shape of our continent is exceedingly complicated, mountain ranges, inland seas, and the configuration of its Atlantic coastline divide and subdivide it into numerous areas—rather like a large building with several wings. It is an important point that these different 'rooms' in Europe are separated from one another by obstacles which are just big enough to make the division clear, but not big enough for complete isolation. This circumstance may well be at the root of the main features of the European character. For in the 'rooms' all that is best in the European tradition has accumulated over the centuries. Here lies the origin of the strongly pronounced local characteristics, whereas the comparatively easy traffic between the 'rooms' has at the same time made possible a certain intermingling, a duologue of the blood. And it is probably this duologue, to this tension between characteristics, that we owe the wealth of intellect and willpower which marks the European. . . . Such is Europe. A landscape of quality, not of quantity, rich in nuances and tensions, where humanity has achieved clear definition not only in the individual but also in the nations. . . .[1]

Further Readings

CHANDLER, T. J., *The Climate of London* (London: Hutchinson, 1965), 240 pp. A fascinating discussion of the effect of large cities on their local climate.

[1]Emil Egli and Hans Richard Muller (translated from the German by E. Osers), *Europe from the Air* (London: George G. Harrap, 1959), pp. 10–11.

CHURCH, R. J. HARRISON, and others, *An Advanced Geography of Northern and Western Europe* (London: Hulton Educational Publications, 1967), 480 pp. Contains excellent discussions on the physical environments found in northwestern continental Europe.

EGLI, EMIL, and MULLER, HANS RICHARD (translated from the German by E. Osers), *Europe from the Air* (London: George G. Harrap, 1959), 223 pp. A magnificent collection of vertical and oblique photographs focusing on the physical and human landscapes of Europe.

GOTTMANN, JEAN, *A Geography of Europe* (New York: Holt, Rinehart and Winston, 1969), 866 pp. Probably still the best single-volume work of its type. A literate and humanistic view of Europe by an internationally acclaimed geographer.

JORDAN, TERRY G., *The European Culture Area: A Systematic Geography* (New York: Harper & Row, 1973), 381 pp. An innovative recent treatment of the geography of Europe from a topical or systematic point of view. Well worth reading in part or whole.

KENDREW, W. G., *The Climates of the Continents* (New York: Oxford University Press, 1942), 327 pp. Although written over fifty years ago, this classic by Kendrew is still well worth reading.

LOBECK, ARMIN K., *Physiographic Diagram of Europe* (New York: Geographical Press, Columbia University, 1923), 8 pp. This excellent physiographic diagram is an invaluable aid to understanding Europe's varied physical environment. It is accompanied with a lengthy textual explanation.

MEAD, W. R., and SMEDS, HELMER, *Winter in Finland* (London: Hugh Evelyn, 1967), 160 pp. Two extremely gifted European geographers present a readable and well-illustrated monograph on the struggle with winter faced by the people of the northernmost independent country in the world.

O'DELL, ANDREW C., *The Scandinavian World* (London: Longmans, Green, 1957), 541 pp. A detailed and well-illustrated treatment of the northern reaches of Western Europe.

SHACKLETON, MARGARET R., *Europe: A Regional Geography* (London: Longmans, Green, 1951), 525 pp. A very solid treatment of physical geography in a regional framework.

STEERS, J. A. (ed.), *Field Studies in the British Isles* (London: Thomas Nelson, 1964), 525 pp. This book contains thirty-three essays written by leaders of the field study tours that were a part of the 1964 meeting of the International Geographical Union Congress held in London. The essays represent a comprehensive survey of the major regions of contemporary Britain and Ireland and are rich in detail concerning the physical background to development.

Western Europe: A Blend of Old and New

10

One thing that has never failed to impress Anglo-American visitors in Europe is the way European life and landscapes include elements of the past in the patterns of the present. Standing near the ancient Tower of London, built by William the Conqueror, one can see ultramodern, high-speed hydrofoils or hovercraft skimming along the Thames. A few steps farther on and one can purchase a colorful London tabloid newspaper (boasting the world's largest daily circulation) on a newsstand located against the ruins of a Roman temple. In the countryside, the visitor can drive along roads that have been altered but little since Roman engineers built them. Traces of relic field patterns established by ancient Celtic farmers lie adjacent to modern highly mechanized farms that are veritable "food factories." Everywhere in Western Europe the landscape bears eloquent testimony to the fact that the present is but the past flowing into the future. Western Europeans are heirs to a long and rich heritage of cultural as well as economic development.

Through this long period of cultural growth, the people of Western Europe have created a rich and varied cultural landscape. This cultural landscape, so fascinating to the tourist of today, can be studied like a document to reveal many things about the people who shaped it and gave it its distinctive character. The cultural landscape is composed of patterns of geographic features such as buildings, towns, highways, canals, forests, farms, factories, and mines. These patterns forming the cultural landscape often owe their origin and present locations to conditions and forces existing in the distant past. If we are to "read" the cultural landscape of today intelligently, we must know something about the major historic forces that have shaped it. A selective review of those historic forces and processes that help us to understand present-day Europe is undertaken in this chapter.

Prehistoric People in Western Europe

Western Europe was inhabited for many thousands of years before the beginning of the last great ice age. The people depended on hunting, fishing, and gathering for their existence. They possessed only the crudest stone tools and weapons and moved about seeking game. Such primitive people utilized the resources of the Western European habitat as they found them. They did very little to alter the habitat. They lived in protected caves or under ledges and built no lasting structures. They used the forests and swamps as hunting and gathering grounds. It can be seen that this way of life did little or nothing to improve the habitat, and so the total number of people that could be supported was small. Small total numbers and few tools severely limited the people's ability to alter their habitat through the period of the Old Stone Age in Western Europe. Perhaps the most important tool available to these earliest Europeans was fire. The deliberate and accidental use of fire by Stone Age people appears to have resulted in certain long-term influences in the Western European landscape. For example, the frequent use of fires set to trap or drive game in the course of hunting resulted in subtle changes in the natural vegetation.

Introduction of Agriculture

Following the melting of the continental ice sheets, Neolithic, or New Stone Age, people began to move into the area of Western Europe from Asia by way of the Mediterranean Basin. These newcomers brought an important new way of life with them. Essential to this new way of life were domesticated animals and crops. It was in the Mediterranean that the ancient Greeks began to develop the origins of Western European culture. Ancient Greece is often considered the "culture hearth" from which an essentially European way of life spread.

As the Neolithic herdsmen and farmers spread into Western Europe, they first sought lands that were naturally clear of thick forest. They found large tracts of clear grassy land suitable for grazing and primitive farming in many places. Frequently these were areas of porous glacial sands or fine wind-deposited soils called *loess*. Porous limestones, called "chalk" by Europeans, also seem to have formed lighter soils that were naturally grassy. From these natural clearings in the forest cover of Western Europe, the primitive farm communities began to expand.

The Neolithic farmers and herdsmen utilized the resources of the Western European habitat far more efficiently than the hunting and gathering peoples who had preceded them. They were able to store surplus foodstuffs for periods of shortage. They had time to build protective homes and establish settlements for mutual protection. Life became more secure and regular. Certain individuals could take time off from food producing to develop other skills and arts. The more secure and regular life led to increases in the populations of Neolithic settlements. This increasing population resulted in increasing pressure on the resources of the habitat. The naturally clear and easily cultivated lands were eventually all taken, and Neolithic Europeans were faced with a challenge.

These early Europeans reacted to the challenge with an aggressiveness that continues to be an important European characteristic. They began to clear the land of the dark brooding forests with primitive but effective methods. In many respects forest clear-ing in Neolithic Europe resembled forest clearing in pioneer America. Trees were cut and burned where they fell. The ashes of the burned wood helped fertilize the soil. Stumps were left to rot, and the primitive farmer cultivated around them.

The lasting result of this process of forest clearing has been considered as "Perhaps the greatest single factor in the evolution of the European landscape."[1] Western Europe was thus profoundly and permanently altered, for the once terrifying forests slowly gave way under the stone axes and fires of the Neolithic farmers. The habitat was profoundly modified by these people to provide for their needs. Recent scientific discoveries based on the study of ancient pollen grains indicate that vast areas of unforested heath and moorland in Europe today are the heritage of Neolithic forest clearing.

Use of Metals

As the Neolithic farmers grew in number and enjoyed a more reliable food supply, they developed new skills. One of the most important of these involved the smelting of ores and fashioning of metal tools and weapons. Metalworking, like herding and farming, spread into Europe from Asia by way of the Mediterranean about 6,000 years ago. At first, easily worked copper and gold were used to produce ornaments and jewelry. These metals were found in many of the highland areas. By adding a small amount of tin to copper, primitive metalworkers made an alloy called bronze. Bronze was much tougher than copper and could be sharpened. Bronze became an important commodity in Western Europe. Bronze hoes and axes began to replace those of stone in the hands of the prehistoric farmers and woodsmen. Many sorts of implements and weapons were made of the useful new metal, and the people were equipped to play an even greater role in changing their landscape.

The use of metals had many significant influences in the early geographic patterns of Western Europe. Mining and smelting became important activities. They usually took place close together where deposits of workable ores were found. Ore-rich areas were found in the highlands which had been earlier

---

[1]H. C. Darby, "The Clearing of the Woodland in Europe," in William L. Thomas, Jr. (ed.), *Man's Role in Changing the Face of the Earth* (Chicago: University of Chicago Press, 1956), p. 183.

avoided by the Neolithic farmers in their quest for productive cropland. The early European metal-workers sought out new elements of the Western European environment. They exploited new and remote areas far from the agriculturally productive plains. In their quest for rich ore deposits and abundant supplies of wood to use in smelting them, people spread over the face of highland Europe. Such remote areas as the Erzgebirge Mountains, the Pyrenees, Cornwall, and the Scilly Isles became important mining centers.

Development of Trade

Trade was stimulated as a result of these developments. No longer did everyone merely subsist on what animals they raised or crops they grew. Groups of people began specializing in activities such as mining, smelting, and smithing. They traded their skills and products for the grains of the farmer and leather of the herdsman. Commerce and trade led to the development of routes and better methods of transportation. Trackways began to evolve as a network connecting the many small scattered centers of population and production. Products of the farm, forge, and pasture moved along these routes. Perhaps even more important than the trade in goods was the spread of people and ideas along these primitive trackways. These shared experiences and ideas were the first foundation stones of Western European culture.

Western Europe developed very slowly with iron replacing bronze as the chief toolmaking material by about 500 B.C. Settled villages began to appear as the centers of stable agricultural communities. Western European life was still chiefly focused on subsistence farming groups clustered in the plains and valley regions best suited for agriculture. Here and there more isolated groups worked rich ore deposits in the forested hills. A primitive web of trackways tied the farmers and miners together and permitted a meager commerce based on barter to emerge. Fishermen, herdsmen, and workers in clay, wood, and other materials also contributed to the continent's infant economic pulsebeat. Total numbers of people were small, however, and untamed forests still dominated the landscape. In all of Britain, for example, the population at about the time of Christ has been estimated at less than 500,000.

The Heritage and Gifts of Rome

Civilization spread from Asia to Western Europe by way of the Mediterranean Basin. The process of spreading the new ideas and techniques such as agriculture, pottery making, and metalworking took long periods of time. Whereas people in Italy and Greece were well advanced in the arts of civilization at an early date, the people of Western Europe north of the Alps lagged far behind.

Rome began in the hilly country of central Italy near the Tiber River. The traditional date for the founding of the city of Rome is 753 B.C., but we can be sure that people were living and working there for a very long time before this date. Their productive farms yielded surpluses of foodstuffs that permitted specialized craftsmen to develop trade with other groups. Gradually the strength and wealth of the Roman farmers allowed them to extend their influence and control over the surrounding region and thus ensure their security against attacks by enemies. Within 500 years, they controlled most of the Italian Peninsula and were ready to expand overseas to challenge the powerful Phoenicians for control of the Mediterranean Basin. Roman victories in a long series of conflicts called the Punic Wars made the Mediterranean "Mare Nostrum," or "Our Sea." In other words, Rome was in control of the great centers of Western civilization, and the Mediterranean Sea was a Roman lake. From the Mediterranean the Romans extended their empire across the Alps to take all of Europe west of the Rhine River. Even distant Britain was conquered and made an integral part of the Roman Empire for 400 years (Fig. 10–1).

Rome ruled over most of Western Europe for more than five centuries. During this period Roman culture had a great influence on the people and landscape of the region. Even those areas to the east of the Rhine that were not directly under Roman rule were influenced. Roman techniques and ideas were adopted readily as the less advanced Germanic people living beyond the Rhine realized their value and importance. It is safe to say that only a few remote corners of Western Europe escaped some impact of Roman culture.

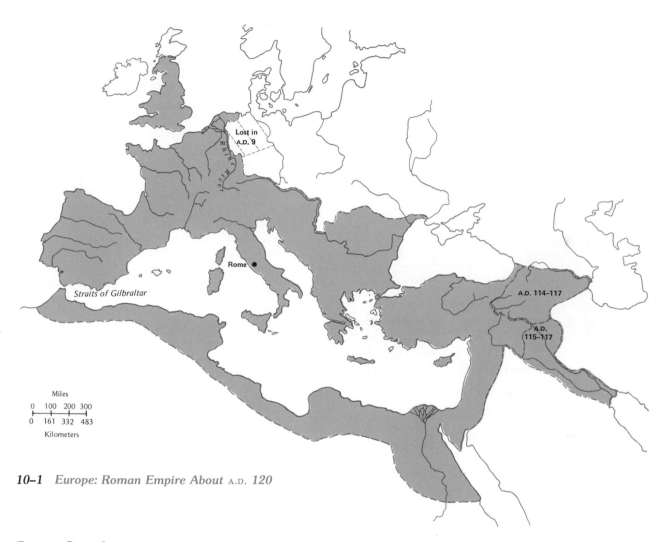

Lost in
A.D. 9

Rhine River

Rome ●

Straits of Gilbraltar

A.D. 114–117

A.D.
115–117

Miles
0   100   200   300
0   161   332   483
Kilometers

10–1   *Europe: Roman Empire About* A.D. *120*

Roman Contributions

Roman culture was complex and involved almost all phases of human life. Rather than reviewing its total impact, we will concentrate on a few of the major gifts that Rome brought to Western Europe. The Romans really acted as cultural "middlemen" in bridging the mountain barriers that separated the primitive northerners from the highly civilized southerners of Mediterranean Europe. The Romans carried new crops, tools, materials, animal breeds, inventions, and ideas across the Alpine passes and on the sea routes that passed through the Straits of Gibraltar to the Atlantic coasts of Western Europe.

Among their other accomplishments, the Romans were great engineers. They used their skills abundantly in Western Europe to increase the productivity of this part of their empire. Highways were constructed to allow for the efficient movement of armies as well as the free flow of trade between one region of the empire and another. Impressive stone bridges were built to span rivers, and swamps were filled and drained to permit roads to cross. Western Europe enjoyed a system of land transportation under the Romans which was not equaled again until the nineteenth century. The easier circulation provided by the Roman roads helped knit areas with similar cli-

Rome, the seat of a culture that has affected all parts of the world, reveals its heritage in old buildings and its vibrance in new streets and cars. (Italian Government Travel Office)

matic and physical conditions into smoothly functioning economic regions. Fertile plains areas began to specialize in the growing of grain crops like wheat, while sunny slopes were devoted to grape cultivation for wine making. Like the Roman roads that helped bring them about, many of these specialized agricultural regions remain as parts of the present-day scene in Europe.

In the change-over from subsistence to specialized

or commercial farming, the Romans introduced many new agricultural tools and techniques. Better plows and implements, the use of manure, field rotation, and double-cropping schemes enabled the Western Europeans to produce more food than ever before. Some of the surpluses were sold through the new money economy of the Roman Empire. Trade and commerce increased with the increasing agricultural productivity.

To serve the requirements of Western Europe's quickening economic life, urban centers were needed. Towns served as the focal points for life in the Roman Empire. Until the coming of the Romans, Western Europe had no real towns or service centers. Many of the great cities of present-day Western Europe began as Roman military posts or trading centers (Fig. 10–2). London, now one of the world's three largest cities, was nothing but a fishing camp along the marshy banks of the Thames when the Romans conquered Britain. Paris, regarded as one of the world's most cosmopolitan cities, was a rude island village in the Seine River until Caesar sent one of his lieutenants to secure it as a bridge site.

The Romans applied more advanced methods and techniques to the primitive metalworking industry of Western Europe. They introduced greatly improved bellows to provide much higher temperatures in iron smelting. Steel, the valuable metal on which so much of modern life depends, was produced in Western Europe for the first time under the Romans. They introduced improved windmills and waterwheels to the Western Europeans and used them extensively in making the habitat more productive. Building with bricks and cement and architectural styles were also introduced into the landscape.

Spatial Integration

Perhaps the most important gift the Romans gave the Western Europeans was the gift of spatial integration. Under the Roman Empire, the Western Europeans enjoyed the fruits of civilized life for the first time. These benefits that made life more enjoyable for all came as the result of the integrated use of the environment and its resources by people. In unity there was strength, and with this strength Western Europeans made a giant step ahead in their effort to tame nature. The legal codes, money economy, and administrative government of the Roman Empire were the greatest gifts brought by these Mediter-

10–2   *Western Europe: Modern Cities Based on Roman Towns*

10–3 *Europe: Roads of the Roman Empire*

SOURCE: Modified from O. A. W. and M. S. Dilke, "All Roads Lead to Rome," *Geographical Magazine*, 46 (1974).

Miles
0   200
0   322
Kilometers

ranean people as they conquered, for these gifts allowed Western Europe to be integrated into a productive whole (Fig. 10–3).

The Barbarians, Christianity, and the Middle Ages

Eventually the Roman Empire lost its ability to control and direct the affairs of Western Europe. Contributing to the decline of Roman power was the growing strength of the Western European groups or "barbarians" who lived beyond the limits of the empire. They too had benefited from the techniques and ideas that made the empire so productive. Eventually they were able to surpass their Roman teachers in the arts of war and invade the city of Rome itself. In the year A.D. 476 a barbarian general named Odoacer sacked Rome. This date is usually taken as the end of the Roman Empire in Western Europe. From that point onward, the center of the Roman Empire shifted to Constantinople (Istanbul) in the East.

Under the barbarians the integrated spatial and administrative patterns of Roman Europe crumbled. The universal law and order of Rome was replaced by local rulers who owed their position and power to their ability in the destructive art of war. Trade and commerce declined; roads, bridges, and canals fell into disuse. Piracy and banditry flourished and made travel dangerous. As a result, the towns lost their vigor as cultural centers, and a decline in learning, science, and technology set in. This period in European history is known as the Dark Age and lasted from about A.D. 500 to 1000. The economic and administrative integration that had allowed Western Europe to flourish under the Romans was replaced by economic and governmental disintegration in the Dark Ages. In this period many formerly productive areas deteriorated and were allowed to revert to forests. Productive farmland created by drainage and irrigation under the Romans was lost not to be reclaimed again for a thousand years. The landscape of Europe reflected the change from Roman integration to barbarian disintegration in a multitude of ways as towns, villages, and productive farm lands fell into disuse and abandonment.

The Rise of Christendom

While the economic pattern of Western Europe was being devastated by the change from integration to disintegration, another force was gaining strength in the minds and hearts of the people. This force was Christianity. Christianity was to become the next great integrating influence in the historical development of Western Europe. Christianity was only one of many religions practiced in the Roman Empire. It grew slowly at first but became a great force in the empire and was finally adopted as the official state religion in A.D. 392. Christianity was spread through the empire and to the barbarians beyond by the zealous efforts of such early Christian missionaries as St. Patrick and St. Boniface.

Christianity was a great unifying force. Essentially it involved the spiritual and intellectual areas of Western European life. All Christians believed in one God and followed the leadership of a centralized clergy. The old universal languages, Greek and Latin, were kept alive and allowed educated Christians to communicate in spite of local language differences. In many respects the Christian Church represented the most important thread of continuous cultural development throughout the 1,000 years that followed the collapse of the Roman Empire in Western Europe. Early Christian centers in Western Europe, shown in Figure 10–4, frequently served as places from which new ideas and techniques diffused to influence people throughout the countryside.

With the passage of time, the church began to play a more active role in the economic life of Western Europe. It had grown wealthy from the collection of "tithes" from worshippers. Most of this wealth was invested in land. Also, many devout kings and nobles gave the church gifts of land and treasures as tokens of appreciation and loyalty. As time passed, church leaders became great landholders. Also, many monasteries were founded. The archbishops and bishops, along with the parish priests and monks, were in a position to influence the way in which vast stretches of the Western European habitat were developed in these centuries. What they did was significant and often lasting.

The monasteries became important cultural centers. The devoted monks who inhabited them kept

10–4 *Western Europe: Early Christian Centers of Diffusion*

of economic life. The monks experimented with improved crops and animals as well as cultivation techniques. Their religious ties with the Vatican in Rome provided excellent channels for the spread of better agricultural methods. The agriculturally productive monasteries became impressive elements of the landscape and served as examples that the people often copied.

Manual labor was included along with obedience, chastity, and poverty in the vows the monks made. Clearing forests, breaking new land, draining marshland, and building irrigation systems were as much a part of monastic life as were meditation and the artistic embellishment of manuscript pages of Biblical texts. Other crafts and skills, such as pottery making and brewing, were also carried on by the monks. They invented new processes while keeping alive the old arts for future generations.

The clearest evidence of the importance of Christianity and its contribution to the growth of Western European culture can be seen in the landscape of the present day. This evidence is in the form of the churches, monasteries, and great cathedrals that still dominate much of the Western European landscape. The splendor of a church in a town or country district today often gives a clear indication of its wealth and population centuries ago. Many of these churches and cathedrals still carry on important religious and cultural functions more than a thousand years after being established. They are among the most impressive works of mankind and still generate a feeling of awe and reverence.

The Renaissance Discoveries

The culture of Western Europe was enriched very slowly through the ten centuries of the Middle Ages. The Crusades brought wider contacts and an interest in distant places. During the thirteenth and fourteenth centuries increased trade and commerce in the Mediterranean area brought great wealth to Italian city-states such as Venice, Florence, and Genoa. The merchants and bankers of these Mediterranean city-states acted as middlemen in the increasing trade between Western Europe and the growing Moslem culture to the east. New wealth stimulated an urban way of life based on commerce and craftsmanship.

The church played an active role in the economic life of Western Europe. The monasteries were important cultural centers from which some great universities evolved. The monasteries were also centers of crop and livestock improvement and served as "agricultural experiment stations." (French Government Tourist Office)

alive the ancient knowledge and established schools and libraries. Some of these schools later became great universities. They provided shelter and aid for travelers as well as for the sick and helpless in society. The monasteries also became important centers

This revival in urban activities began to spread over the rest of Western Europe during the period known as the "Renaissance." The word *Renaissance* means "rebirth" in French. It describes what was going on in Western Europe during the period from about 1400 or 1450 until about 1700.

Foundations of Modern Science

The Renaissance is often described as the period when Western Europeans rediscovered themselves and the world. Much of the learning of the ancient Greeks, Babylonians, and Egyptians had been considered pagan and therefore suppressed by the Christian Church through the Middle Ages. These ancient learnings were rediscovered by humanist scholars and again stimulated mankind's minds. They helped to provide individuals like Copernicus, Galileo, Bacon, and Descartes with fundamental challenges in scientific thought. These men, together with others such as Kepler, Newton, Harvey, and Boyle, laid the foundation for modern science. The scientific method was adopted in the search for solutions to human problems. Conclusions were based on controlled observation and experiment rather than on the words of Biblical authority or superstition.

The discoveries made in theoretical science attracted the attention of practical people who began to apply the new theories to everyday problems. For example, once Boyle's "laws" regarding the behavior of gases had been propounded, the perfection of the steam engine was only a matter of time. The new science of the Renaissance ultimately equipped the Western Europeans with the machines and technology of the Industrial Revolution.

Geographical Discoveries

Along with the discoveries being made in the sciences and arts were new geographical discoveries. The trade monopolies enjoyed by the Italian city-states had caused prices of imported goods from Asia to become excessively high in Western Europe. The writings of the ancient geographers taught that the earth was spherical. Renaissance people began to think seriously of the possibility of finding a new route to Asia via the ocean to the west.

Portuguese living on the westernmost fringe of Western Europe were among the first to undertake planned voyages of geographical exploration. They felt that India and the "Spice Islands" could be reached by sailing around Africa. The voyages of the early Portuguese navigators were only possible because of advances in the art of navigation that had slowly developed in the Mediterranean area. The compass was available to tell direction, and the astrolabe was used to determine latitude or position. Sailing charts called "Portolani" allowed the explorers to record their discoveries more accurately and permitted others to follow their routes more easily.

The Spanish, who shared the Iberian Peninsula with the Portuguese, were anxious to join in the race of discovering going on in the fifteenth century. Their efforts to expel the Moslems from Spain, however, took all of their energy. At last in 1492, Spain reduced the last Moorish stronghold of Granada. With the Moslems expelled from Western Europe, Spain immediately turned to thoughts of new trade routes and new lands in the west. Christopher Columbus received royal support in the form of ships and men to sail west to India. Columbus didn't find India; he found instead two new continents: North and South America. It took Western Europeans several years to realize that Columbus had in effect discovered a "New World," and that the earth was much larger than most people had believed possible.[2]

Colonial Expansion

In the two centuries that followed the discovery of the Americas, Western Europeans began to spread over the world to explore and trade. The Atlantic Ocean and western margins of Europe became the chief focal points in the newly developing world trade patterns. Trade was increasing in value and volume as the Western Europeans began for the first time to benefit from their position at the center of the Land Hemisphere. Western Europeans were now in a position to utilize the resources of almost the whole world to satisfy their needs and desires. At first they attempted to get these resources through trade. Grad-

[2]There is an abundance of evidence to indicate that Western Europeans had voyaged to the New World centuries before Columbus. Chief among them were Viking navigators from Scandinavia who had established settlements in Iceland and Greenland before A.D. 1000. Very little ever came from these early tentative contacts, however, and not until Columbus' voyages and reports did the New World begin to influence the lives of Western Europeans in important ways and on a large scale.

ually the Western Europeans found that they were better equipped with weapons and better organized than many of the less materially advanced culture groups they encountered.

Taking advantage of their technological superiority over the native peoples of America, Africa, and Asia, the Portuguese and Spanish began to establish colonies. In due course, many of the other Western European powers followed this example. The spread of colonial empires brought the spread of Western European culture over much of the earth. In mercantilistic theory the colonies were created and maintained for the profit and benefit of the mother country. In practice many potential benefits such as improved medical care, better education, and higher levels of living eventually came to some of the native people of the colonies.

By 1700 the people of Western Europe felt the wind of change blow through almost every aspect of their culture. The Renaissance had shaken Western Europe into an awareness of the wonders of science and technology as well as of the immensity and variety of the world beyond the oceans. It was still just an awareness, though. In many respects life in Western Europe was little changed since the end of the Roman period. The mass of people were still illiterate, abjectly poor, and often hungry. Large estates dominated the countryside, with the feudal landowners enjoying great power and prestige. The clergy, landowners, and military leaders controlled the political life of the times with little concern for the betterment of the common people. The towns were growing in significance, but the urban part of society was only a small fraction of the whole. Merchants, lawyers, bankers, doctors, and other professionals were accepted as necessary evils but not admired by the majority of people. The people of Western Europe still followed a traditional agrarian way of life. Population had increased slowly over the centuries but was still concentrated in the easily farmed lowlands and valleys of the Great European Plain and adjacent highlands. The modern map of Western Europe with its huge cities and sprawling industrial regions of mines and factories was not even dreamed of two and a half centuries ago. One more set of important historic forces, the Industrial Revolution, was necessary to alter profoundly the character and face of Western Europe.

The Industrial Revolution

The series of technological advances and social changes that are usually called the Industrial Revolution began in England about two hundred years ago. In the course of a few generations, these new technical and social forces changed the face of Western Europe profoundly, more profoundly in fact than it had been changed by the countless generations living there since the Stone Age. In the words of one writer:

It was a revolution which has completely changed the face of modern Europe and of the New World, for it introduced a new race of men—the men who work with machinery instead of with their hands, who cluster together in cities instead of spreading over the land in villages and hamlets; the men who trade with those of other nations as readily as with those of their own town; the men whose workshops are moved by the great forces of nature instead of the human hand, and whose market is no longer the city of the country, but the world itself.[3]

It might be added that we today are living in the "Machine Age" which is the outgrowth of the Industrial Revolution. The computers, atomic power, and supersonic travel of today are further refinements in the accelerated drive to control and exploit the habitat which began in Western Europe over two hundred years ago.

The Industrial Revolution grew out of Renaissance discoveries in science and geography. Western European countries began to profit from trade with the newly discovered lands beyond the seas. Technologically less advanced people such as the Indians of North and South America were eager to trade precious furs and metals for even commonplace European manufactured goods. Things like knives, hatchets, glass, guns, and woolen cloth were the basis of what has come to be called the Commercial Revolution of the sixteenth and seventeenth centuries. Great wealth began to flow into Western Europe, and the tempo of life quickened. Ancient guilds, which had controlled production since the Middle Ages, began to break down. Individual craftsmen and artisans of the guilds could not meet the increasing demands for manufactured goods of all sorts.

[3]J. Salwyn Schapiro, *Modern and Contemporary European History* (Boston: Houghton Mifflin, 1953), pp. 11–12.

The Cottage System

The domestic or cottage system of production was developed to meet these growing demands. Under this system an entrepreneur or middleman would distribute raw materials to be manufactured into finished goods in the cottages or homes of workers. Pay was usually on the basis of piecework. The cottage households, including women, children, and sometimes servants, all pitched in to get the job done. No great visible changes in the landscape accompanied this system, since most of the cottagers continued to farm for a portion of their livelihood. They were supplementing their subsistence way of life and perhaps earning cash for the first time. The hours worked under this system were probably long, but at least they were spent in the familiar surroundings of the countryside. The tools and processes were simple and traditional. Spinning wheels, looms, anvils, and the cobbler's bench were often items of household furniture.

In time, the cottage industries became more specialized. In late eighteenth-century Britain, for example, houses were built with large window areas to provide ample light for the spinning and weaving done in them. Several devices were invented to increase the productivity of workers in the cloth industry. The demand for cotton, woolen, and silk cloth was becoming tremendous in the eighteenth century. The "flying shuttle" was a greatly improved hand loom invented by John Kay in 1733. As the workers began to weave cloth more rapidly, they required greater quantities of thread. This need was met by James Hargreaves in 1765 when he perfected his improved spinning machine known as the "spinning jenny." Both of these inventions were operated by the workers' hands and were really only tools that increased individual productivity.

The Factory System

The rapidly expanding markets opening in an increasingly prosperous Europe and in the new overseas colonies could not be satisfied by the traditional cottage system of production. A new method of production was needed that was not entirely dependent on human muscle and dexterity. Enterprising people in England began to develop the factory system of production. This system brought large numbers of disciplined workers together and provided them with some inanimate source of power. This combination was essential if the demand for manufactured cloth and other goods was to be satisfied. In the first part of the eighteenth century, factories began to appear. Perhaps the first true factory built in England was the silk mill built at Darby in about 1720. It was a five-story brick building with 300 workmen and machinery driven by the water of the Derwent River. Although animals walking on treadmills provided the power in a few factories, most were driven by waterwheels. Waterpower linked to ingenious new machines was the chief source of power during the early phases of the Industrial Revolution.

This source of power allowed the factory system to develop, and production increased. Countless factories were built along the banks of rapidly flowing streams and rivers. Frequently such locations were remote and far from Western European centers of population and ports. Raw materials such as wool, cotton, and iron ore were bulky, and the cost of transporting them to remote new factory sites was expensive. These limitations were serious handicaps to the early industrialists. Waterpower lacked flexibility; seasonal floods and droughts often made power unavailable for several months each year. Severe freezes also stopped the wheels of industry. Great strides were made by the pioneer industrialists using waterpower, however, and many important inventions were perfected. It was an improvement but not a final answer to the problem of how to convert the raw materials of the Western European habitat into manufactured goods to satisfy a growing world market.

New Power Sources

The answer came in 1769 when James Watt made a major improvement on the crude steam-driven pumps that were being used to pump water out of British mines. Thomas Newcomen had first devised these steam-driven pumps in 1704. They were crude, heavy, and extremely wasteful of fuel. Watt's improvements added flexibility and made steam power available for a countless variety of uses. His efficient and powerful steam engine made the factory system independent of remote waterpower sites. A flexible source of tremendous power was now available to allow humans to exploit their habitat in a way never before dreamed possible.

Steam was produced by heating water in a boiler. To create that heat a cheap fuel was required. As we saw earlier, much of Western Europe was underlain

Scottish
Lowlands

Newcastle

Midlands

Wales

Ruhr

Sambre-
Meuse

Saar

Silesia

Miles
0   200
0   322

10–5 *Europe: Early Coal-Based Industrial Centers*

by thick seams of coal. In many areas these coal seams were near the surface. Along certain streams and coasts, currents and waves exposed the coal layers at the surface. The Londoners of Tudor times called coal "sea coal" because it was dug out of coastal seams in the north of England where waves had exposed it. The coal also arrived by sea because London had no deposits of coal close at hand. Although coal had been used in many ways before, it was not a particularly important fuel. It became tremendously important when steam engines were perfected. The steam engines allowed mankind to convert this mineral resource of the earth into the key of productivity—controlled mechanical power.

The steam engine, combined with Western Europe's vast coal supply, provided the power necessary for the Industrial Revolution to take place. But power is of little value if it is not harnessed and put to useful work. To do this work, complex machines were required. To build durable machines, durable materials are required. The best and most widely used tool and machine-building material is iron. Iron ore, like coal, is found in abundance in Western Europe.

Iron Ore Processing

Iron is never found in its pure form in nature. It is found in chemical combinations with other elements in many different rocks. When these rocks contain reasonably large amounts of iron economically feasible to mine, they are called ores. To separate the metallic iron from the other substances forming the ores requires a great amount of heat. Traditionally this heat was provided by charcoal. Rich ores were mixed with charcoal and limestone in crude blast furnaces or large forges.

In the early eighteenth century, iron making had not improved much since the days of the Romans. It was a slow process almost totally dependent on waterpower and charcoal. In the eighteenth century, just when the demand for cheap iron was increasing, England's production was decreasing. Many centuries of iron making had depleted much of Britain's forest cover. The use of wood in constructing merchant and naval ships had also brought about critical shortages. Wood and charcoal were used for hundreds of other purposes from cooking to leather tanning.

Many people had experimented with coal in iron making. Coal was plentiful and produced great heat, but it didn't make good iron. The coal added to the impurity problem that charcoal minimized in iron making. Early in the eighteenth century an Englishman, Abraham Darby, succeeded in smelting iron with coke. Darby was fortunate because his ironworks at Coalbrookdale in Shropshire were close to a very pure type of bituminous coal found near the surface. It made a superior coke and contributed to the success of Darby's experiments. Coke is formed by baking coal to drive out most of its impurities.

By the 1740s, Darby's techniques were becoming widely known, and coal began to replace charcoal as the most important fuel resource. Iron began to be used in a great many new ways. It soon replaced wood, copper, brass, and tin in many tools and machines used in the rapidly growing factories of the late eighteenth century. Further improvements in the iron manufacturing industry allowed the output of this vital metal to increase as the demand created by other phases of industrial growth increased.

By 1760, John Smeaton perfected the "steam blast," which utilized steam power to replace the old water-driven bellows needed to provide a blast of air in the blast furnace. Another leap forward came with Henry Cort's invention of "puddling," which produced cheap steel from pig iron. By the end of the eighteenth century, the iron and steel industry was freed from a dependence on woodland for fuel and on running streams for power. It, like most other industries, began to move toward the new source of power, the coalfields. A pattern of distribution that still characterizes much of Western Europe's industrial and urban life was beginning to form (Fig. 10–5).

During the late eighteenth and early nineteenth centuries, the European countries located in the iron-and-coal-rich regions emerged as the world's first great industrial powers. They reached out from their central position to seek raw materials for their factories and markets for their products in the far corners of the world.

Further Readings

BEAUTIER, ROBERT-HENRY, *The Economic Development of Medieval Europe* (Harcourt Brace Jovanovich, 1971), 350 pp. A richly illustrated review of a fascinating and often misunderstood period of European development.

BERESFORD, M. W., and ST. JOSEPH, J. K., *Medieval England: An Aerial Survey* (Cambridge: Cambridge University Press, 1958), 275 pp. This volume contains a splendid collection of aerial photographs and essays showing how much of the medieval landscape can be observed and analyzed through aerial photography techniques of data gathering.

BOLAND, CHARLES M., *They All Discovered America* (New York: Pocket Books, 1963), 430 pp. An inexpensive but interesting paperback that presents most of the better known accounts of pre-Columbian contacts between the Old and New Worlds.

DARBY, H. C., (ed.), *A New Historical Geography of England* (Cambridge: Cambridge University Press, 1973), 750 pp. From the coming of the Anglo-Saxons to the time of World War I forms the span of this definitive volume by a group of historical geographers.

EAST, W. GORDON, *An Historical Geography of Europe* (London: Methuen, 1956), 450 pp. A widely read and highly regarded book that tends to be selective in the topics treated. Contains an excellent bibliography.

FOX, EDWARD W. (ed.), *Atlas of European History* (New York: Oxford University Press, 1957), 65 pp. A handy reference for most aspects of European historical geography.

HAYS, DENYS, *Europe: The Emergence of an Idea* (Edinburgh: Edinburgh University Press, 1957), 132 pp. An interesting investigation of how the idea of Europe as a continent developed.

HERRMANN, PAUL, *Conquest by Man* (New York: Harper & Row, 1954), 450 pp. A fascinating history of travel including several accounts of pre-Columbian contacts between the Old and New Worlds. The book was a Book-of-the-Month Club selection.

HOSKINS, W. G., *The Making of the English Landscape* (London: Hodder and Stoughton, 1960), 239 pp. Hoskins deals with the rich historical evolution of the English landscape as a product of human activity. An excellent approach skillfully handled.

LAMBERT, AUDREY M., *The Making of the Dutch Landscape: An Historical Geography of the Netherlands* (London and New York: Seminar Press, 1971), 400 pp. Without doubt this book is the definitive work on this topic. It is clearly written and illustrated with excellent photographs and sketch maps.

PIGGOT, STUART, *Ancient Europe from the Beginning of Agriculture to Classical Antiquity* (Chicago: Aldine, 1970), 340 pp. A clearly written and well-illustrated account of man's material, social, and cultural growth in ancient Europe.

POUNDS, NORMAN J. G., *An Historical Geography of Europe, 450 B.C.–A.D. 1330* (Cambridge: Cambridge University Press, 1973), 475 pp. Pounds provides a set of five widely separated historical reconstructions of Europe's past geography; lucidly written by an expert historical geographer.

SCHAPIRO, J. SALWYN, *Modern and Contemporary European History* (Boston: Houghton Mifflin, 1953), 946 pp. A tightly written textbook offering a comprehensive view of Europe from 1815 to the 1950s.

SMITH, C. T., *An Historical Geography of Western Europe before 1800* (London: Longmans, Green, 1967), 600 pp. Clearly written and illustrated, an excellent treatment of a fascinating topic.

VAN DER MEER, F., and MOHRMANN, CHRISTINE, *Atlas of the Early Christian World* (London and Edinburgh: Thomas Nelson, 1958), 200 pp. This volume presents excellent maps with a good balance of interesting text.

WHITE, JAMES L., *The Origins of Modern Europe* (New York: Washington Square Press, 1966), 450 pp. An illustrated paperback, both readable and authoritative.

Western Europe: Landscapes of Development

11

IN CHAPTER 3, Europe was identified as a world-culture hearth, or a place from which fundamental changes in human life have flowed. By comparison with most other world-culture hearths, Europe flowered much later but spawned many profound and far-reaching changes. Also, it is acknowledged that many of the world's most serious contemporary problems result from the tensions generated as an exploding European culture, particularly its technology, confronts traditional non-European societies around the globe. Professor Terry Jordan goes so far as to suggest that

The world is in the process of being Europeanized in numerous, fundamental ways. . . . European culture may one day be world culture, as regional differences fade in an increasing acceptance of the European way of life.[1]

[1]Terry G. Jordan, *The European Culture Area: A Systematic Geography* (New York: Harper & Row, 1973), p. 15.

European Culture

Certainly the evidence of this "Europeanization" process is increasingly visible in the material and tangible aspects of life in even the farthest corners of the inhabited world. The acculturated bulldozer-driving or radar-operating Eskimo of the polar north now frequently has more in common with his American, Canadian, Danish, or Russian co-workers than with his tradition-bound hunting uncles and father. A recent observer at Thule, Greenland, noted:

Within a generation or two the Polar Eskimo hunting culture may die slowly through attrition. Or the outside world of technology may finally engulf it. Recently, a major oil discovery was made on Ellesmere Island, just west of Thule. Two mining companies are planning explorations on Peary Land, north of Thule . . . no longer will Thule be the farthest of lands.[2]

Similar accounts could be written concerning the Bedouins of the oil-rich Old World deserts or the Stone Age Indian tribes of Latin America's rainforests—in fact, about primitive and traditional societies around the world. What is it about the Europeans and their culture that makes them such a potent force in the march of world change at the present time? Many geographers and others have concerned themselves with providing answers to this question. Although none have been entirely successful, their attempts can be helpful in an effort to gain an understanding of Western Europe, one of the most influential of the world's developed regions. In their innovative book *Culture Worlds* of some years ago, Professors R. J. Russell and F. B. Kniffen stressed the aggressiveness and self-confidence that have marked Europeans from a very early period and listed as other outstanding European culture traits:

1. Field agriculture, where each field is owned by someone, and the fields are large by world standards. Also, mechanization and consequent high productivity per farm worker are the rule.
2. Industrialization.
3. Urbanization.
4. A high degree of occupational specialization.[3]

[2]Fred Bruemmer, "The Northernmost People," *Natural History*, 83 (February 1974), 33.
[3]R. J. Russell, F. B. Kniffen, and E. L. Pruitt, *Culture Worlds: Brief Edition* (New York: Macmillan, 1967), p. 5.

11-1 *Europe: Traits of "European-ness" According to Jordan*

Country with 8–12 traits

Country with 4–7 traits

Country with 3 or less traits

Miles
0   200
0   322
Kilometers

SOURCE: Adapted from Terry G. Jordan, *The European Culture Area: A Systematic Geography* (New York: Harper & Row, 1973).

More recently Professor Jordan defined the European culture area as being all Old World areas in which the people (1) have a religious tradition of Christianity, (2) speak one of the numerous related Indo-European languages, and (3) are of Caucasian race. To these three basic traits he added ten more that he found necessary to form a detailed areal definition of present-day Europe. These are:

1. A well-educated population.
2. A healthy population.
3. A well-fed population.
4. Birth and death rates far below world average.
5. An annual average national income per capita far above the world average.
6. A population that is dominantly urban.
7. An industrially oriented economy.
8. A market-oriented agriculture.
9. An excellent transport system.
10. Nations that are old.

It has already become apparent that the first nine of his additional traits of "European-ness" are also key traits of the developed or rich nations of the world. By evaluating the countries of Europe in terms of a modified index of these culture traits, Jordan produced a map of their "European-ness" similar to Figure 11–1. It will be quickly noted that almost all of the countries possessing the highest degree of "European-ness" are located in Western Europe. In fact, only East Germany and Czechoslovakia are found outside Western Europe. Significantly, however, three Western European countries, Portugal, Spain, and Greece, exhibit fewer than three-fourths of the traits. These countries are some of the least developed and poorest in Western Europe. The relationship that appears to exist is clearly one where "European-ness" in a country's culture is positively associated with technological development and widespread wealth for its citizens.

Cultural Landscapes and Change

As shown in Chapter 10, the culture of the Europeans developed through a period of several thousand years. In this evolution, the peoples of Europe have created a rich and varied cultural landscape that documents their development. Today's Western European landscapes are the creation of the many millions of people who have lived, worked, and died here since the dawn of human history in a rich and varied physical environment. As the distinguished geographer Jean Gottmann once observed, "Sweat has flowed freely to bring European landscapes to their present shape."[4] A study of the main patterns formed by these landscapes can reveal much about the processes that have enabled the European people to develop one of the world's highest material and cultural standards of living. Even more important, such a study can also add greatly to an understanding of the types of change that many of the poor countries of the world may experience as they attempt to develop their economies and societies along essentially European lines. Undertaking a complete review of Western Europe's complex cultural landscapes would be a massive task far beyond the scope of this or any other single book. For this reason a highly selective approach is required; only a small number of the major patterns formed by mankind's countless imprints on the face of Western Europe can possibly be undertaken. In the remainder of this chapter, Western Europe's patterns of political fragmentation, population distribution, industrialization, urbanization, and agriculture are described and discussed.

Western European Political Fragmentation

One of the things that baffled early European explorers and colonizers was the lack of a sense of property and exclusiveness they encountered among the American Indians and other native peoples. As Edward Soja has pointed out,

Conventional Western (European) perspectives on spatial organization are powerfully shaped by the concept of property, in which pieces of territory are viewed as "commodities" capable of being bought, sold, or exchanged at the market place. Space is viewed as being subdivided into compartments whose boundaries are "objectively"

[4]Jean Gottmann, *A Geography of Europe* (New York: Holt, Rinehart and Winston, 1969), p. 58.

determined through the mathematical and astronomically based techniques of surveying and cartography.[5]

Indeed, the fragmented look of the political map of Europe is in itself ample proof of this thesis.

As discussed in Chapter 9, the total area of Western Europe is only about one-third that of Canada. Yet it is divided into eighteen fully autonomous and independent countries (see Appendix). Additionally there are the six semi-independent "micro states": Andorra, Liechtenstein, Malta, Monaco, San Marino, and Vatican City. It is no wonder that Western Europe appears to be politically fragmented, its modest area being divided as it is into twenty-four separate political systems. As interesting as the fact of this fragmentation is in its own right, our concern is with Western Europe's high state of development and relative prosperity in today's world. We should, therefore, ask, Has this pattern of political fragmentation played any role in Western Europe's emergence as a developed region? This question, like many others posed or implied in these discussions, can only be partially answered. Even partial answers are of value, however, in that they suggest modes of thought and lines of inquiry for further exploration and research.

In many ways the individual countries have served as cultural cradles for the various national groups living in Western Europe. Until about a century ago when steam power was put on wheels to create the first rail transportation systems, movement was difficult. The great mass of people spent their lives almost within walking distance of the places where they were born. Even modest ranges of hills helped to form natural regions or compartments where distinctive patterns of living, working, and speaking developed in relative isolation. In the period before the Industrial Revolution, slow improvements in transportation and communications had gradually brought many of these neighborhoodlike regions together to form larger groupings.

Core Areas

The larger groupings adopted patterns of living that were typical in particularly influential regions called core areas. For example, the English and French

nations each developed from core areas located in productive river basins. The characteristically English way of life originated in the basin of the Thames River in southeastern England and gradually spread over much of the British Isles. Similarly, the French pattern of life and culture first developed in the fertile region drained by the Seine River and its tributaries. As Professor Norman J. G. Pounds wrote:

A core-area must have considerable advantages in order to perform [its] role. Simply put, it must have within itself the elements of viability. It must be able to defend itself against encroachment and conquest from neighboring core-areas, and it must have been capable at an early date of generating a surplus income above the subsistence level, necessary to equip armies and to play the role in contemporary power politics that territorial expansion necessarily predicates.[6]

Such core areas were frequently bound together by the ties of a common language or religion. Even today, language and religious beliefs provide some of the strongest foundations on which states are formed. The linguistic map of Europe, showing where the various languages are spoken, has frequently been used to establish political boundaries following wars (Fig. 3–4). A look at the maps in a history book or historical atlas will show that the political boundaries of Western Europe have been changed frequently in the past as certain groups became strong and expanded at the expense of their neighbors.

Nation-State Evolution

The formation of nation-states as we know them today is a relatively recent development in human organization. In the period following the withdrawal of the Roman Empire from Western Europe, the major language groups began to evolve along paths that eventually led to the present political and language patterns. Certain states, like Portugal and Spain, unified early. They led the Western Europeans to a position of world influence through ocean exploration. Others, such as Germany, remained unconsolidated and divided until only a hundred years ago. Some, such as France, grew through the efforts of strong kings. Still others grew by the voluntary

[5]Edward W. Soja, *The Political Organization of Space* (Washington, D.C.: Association of American Geographers, Commission on College Geography, 1971), p. 9.

[6]Norman J. G. Pounds and Sue S. Ball, "Core-Areas and the Development of the European States System," *Annals, Association of American Geographers,* 54 (1964), 24.

association of small regions, such as the Swiss Cantons or counties, which joined in a confederation.

Whatever their origin, the many countries of Western Europe represent the will and ambitions of the people living in them. The political fragmentation that has characterized Western Europe in the past several centuries has had a profound influence on the lives and activities of the millions of people living in the culture area. These countries govern and administer the people and resources of given portions of the culture area in different and often contrasting ways. This difference has resulted in an exceptional degree of variety in the ways that the local resources have been developed.

For example, many Western European governments have traditionally desired to be self-sufficient as a guarantee against hardship in case a war or other emergency should cut them off from foreign sources. As a result of this desire for national self-sufficiency, or autarky, many of Western Europe's farmers receive aid called subsidies from their governments for growing certain crops. These subsidies have a great influence in determining just what crops are grown and just which areas are farmed. If the principle of self-sufficiency were completely done away with by all the countries in Western Europe, the pattern of agriculture would doubtless change.

National Diversity

The great degree of national variety and difference in Western Europe is another of the things that never fails to impress visitors from the United States. Although all of the countries share many things in common, they are still distinctively different from one another. This national and cultural diversity is one of the most distinguishing features of Western Europe. As one authority put it, "European man is not a standardized concept but one of diversity."

Nearly one-tenth of the world's population lives in Western Europe. Until the present, Western Europe was split into a large number of individual and frequently competing countries. This large number of countries is an inheritance from the preindustrial past and has frequently made it difficult for the resources of Western Europe to be exploited efficiently for the maximum benefit to all. Since World War II, however, the leaders and people of Western Europe have increasingly recognized the advantages of greater unity. This recognition has been particularly true in such areas as economic development and military defense. If Western Europe as a whole can begin to work as a single economic unit, it may easily emerge as the world's third great force along with the United States and the Soviet Union. If and when this union comes about, the pattern of political fragmentation that has characterized Western Europe for so long may change drastically. Many serious observers feel that a true "United States of Europe" is already evolving. Many others are equally convinced that the roots of European nationalism reach too deep, and that such a development is more chimerical than real.

The Trend Toward Western European Unity

In a memorable speech delivered at Zurich University in 1946, Britain's flamboyant wartime leader, Winston Churchill, called for the creation of a "European Family . . . with a structure under which it can dwell in peace, in safety and in freedom." He went on to say, "We must build a kind of United States of Europe. In this way only will hundreds of millions of toilers be able to regain the simple joys and hopes which make life worth living."[7] His was only one of countless voices that urged the unification of Europe in the dark days that followed the destructive havoc of World War II. A sense of common misfortune resulting from the nationalistic rivalries and economic competition that had led to World War II developed into a resolve that Europe must never again experience the indiscriminate horror of modern warfare. Along with this resolve grew a conviction on the part of many European leaders that the traditional national frameworks of the European states were too narrow to permit the kind of economic growth that would be required to guarantee widespread prosperity in the second half of the twentieth century. Europe, many felt, must revive in an economic framework of a sufficiently large scale to allow it to compete

[7]S. Patijn, *Landmarks in European Unity* (Leyden: A. W. Sijthoff, 1970), p. 29.

with the world's "superpowers," the United States and the Soviet Union.

Of course, rhetoric and the internationalist schemes voiced by idealists are one thing, and the realities of national political life and government quite another. Although individuals and groups throughout Western Europe began to think more seriously than ever before about the idea of economic and political unification, the old pattern of political fragmentation was strongly rooted and resisted change.

In the months immediately following the resumption of peace, two organizations emerged to implement the flow of Marshall Plan economic and technologic aid from the United States. These were the Organization for European Economic Cooperation (OEEC) and the European Payments Union (EPU). Eighteen European nations joined the OEEC to share in its benefits. Even the traditional neutrals—Switzerland, Austria, Sweden, and Ireland—took part. Absent were all of the Communist countries of the east, and Finland, which was still under Soviet domination through the payment of staggering war indemnities. In 1948 the General Agreement on Tariffs and Trade (GATT) lowered tariff walls and went a long way toward eliminating the threat of one of Western Europe's old problems, tariff wars. The North Atlantic Treaty Organization (NATO), a major military alliance stretching from the United States and Canada to Greece and Turkey, grew out of agreements concluded in the same year.

There could be no doubt that the peoples and governments of postwar Western Europe were actively entering into a number of cooperative intergovernmental schemes for their common welfare. Throughout this period, the International Committee of the Movements for European Unity played an active role in directing thought and discussion at all levels. In their efforts the committee clearly and forcefully worked toward something more effective than simple international cooperation. Rather, they strove for the creation of a truly supranational union or federation of the individual states of Europe.

Perhaps the most important force of all to encourage this development was the highly successful European Coal and Steel Community (ECSC), which included France, West Germany, and the three small Benelux states—Belgium, the Netherlands, and Luxembourg. Among other goals, the ECSC was organized to ensure that the full potential of the Ruhr coal supplies in West Germany and the Lorraine iron ore deposits in France be achieved as Western Europe's vital steel industry was rebuilt. Industrial and economic interdependence replaced traditional rivalry. The economies of modernized, large efficient mills and mines brought benefits to the workers in both of these important industrial regions as production and profits soared.

It was no coincidence that the members of ECSC became the founders of the most important current force for Western European unity, the European Economic Community (EEC), or "Common Market" as it is more commonly known. The long-standing tradition of a customs union between the Benelux countries, Belgium and Luxembourg since 1921 (the Netherlands joined in 1947), was coupled to the obvious success of the ECSC in the 1950s. In 1955 the governments of Belgium, Luxembourg, and the Netherlands urged the member states of the ECSC to take a new step on the road toward European integration. They stated that they "consider that the establishment of a united Europe must be sought through the development of common institutions, the progressive fusion of national economies, the creation of a large common market and the progressive harmonization of social policies."[8] Meetings that followed in Messina in 1955 and in Venice in 1956 led to general agreement on the idea of the Common Market. It was ratified in the Treaty of Rome signed in March 1957, and the European Economic Community or Common Market became a major fact of European life. Many observers have taken the view that the Treaty of Rome ranks with the Constitution of the United States in terms of its historic significance to the world. Although it has already had profound and far-reaching impacts, it is perhaps still too soon to say just how important the Common Market will be in either European or world terms.

In 1958 the European Common Market began its active life in an effort to remove gradually the difficulties that existed among the economic policies of its six members and bring about prosperity and harmonious development through balanced expansion. Basic to the accomplishment of these goals were the following necessary but potentially challenging

[8]Ibid., p. 93.

Agriculture
(>12%)
Industry
(>45%)
Services
(>43%)

(Above EEC average)

Miles
0   200
0   322
Kilometers

11-2 *Western Europe: Dominant Sectors of Employment in the Common Market*

SOURCE: *Basic Statistics of the Community* (Luxembourg:
Statistical Office of the European Communities, 1975).

measures that continue to be vital issues in Western European political life:

1. Removal of customs duties and import and export quotas among members.
2. Abolition within the community of obstacles to the free movement of persons, services, and capital.
3. Inauguration of common agricultural and transport policies.
4. Establishment of a system insuring competition.
5. Adoption of procedures for coordination of domestic policies and for remedying any balance-of-payments problems.
6. Removal of differences in national laws necessary for operation of the Common Market.
7. Creation of a European social fund to educate and train displaced workers and to raise their standard of living.
8. Establishment of a European investment bank to facilitate economic expansion.
9. Association of dependent overseas territories with the community.
10. Establishment of a common tariff and commercial policy for states outside the community.[9]

Needless to say, these goals were easier to list than they were to accomplish in the event-filled decade of the 1960s. Bitter partisan political battles and even riots took place as the Common Market countries moved from traditional and strongly nationalistic policies toward the ideal of economic and political unity. As the 1970s opened, the economic success of the Common Market was clear to all.

The glowing accomplishments of the Common Market and the passing of France's anti-British president Charles De Gaulle in 1970 paved the way for the enlargement of the community. Norway, Denmark, Ireland, and the United Kingdom all began to seek membership actively in 1972. In a referendum held in September of that year, however, the Norwegians voted against the move. As a result, only the United Kingdom, Ireland, and Denmark completed membership negotiations and became fully fledged partners in the Common Market with Belgium, France, the Federal Republic of Germany, Italy, Luxembourg, and the Netherlands on January 1, 1973. The Western European scene is clearly destined to be dominated by these nine wealthy and vigorously

developing states in the years ahead. Figures 11–2, 11–3, and 11–4 provide some background for an understanding of the problems and potentials that lie before the Common Market in the mid-1970s. Clearly, development, wealth, and opportunity are not evenly spread throughout the market.

It should not be assumed that the Common Market is entirely inward looking. Such is clearly not the case. Greece and Turkey are formally associated with the community with a long-term view of full membership when their economies are strong enough to allow them to compete on an even footing with the present members. Recent political tensions arising from the Greek-Turkish competition in Cyprus, however, may delay their joining. In 1970 and 1972, Malta and Cyprus, respectively, also signed association agreements. Trade agreements are in effect between the community and Israel, Lebanon, Egypt, Spain, and Yugoslavia. Negotiations toward the same goal are in progress with Jordan, and regular economic and diplomatic contacts are maintained with the countries of Latin America. Eighteen African countries, all formerly colonies, enjoy full access to the market. They also receive economic aid from the special European Development Fund. A recent agreement has also been reached which outlines a uniform economic policy to be followed as Common Market members increase their contacts with the Communist countries of Eastern Europe.

Population Patterns of Western Europe

Although less familiar than the political pattern, Western Europe's pattern of population distribution is even more important to our understanding of this highly developed region. Just where people choose to live and center their activities is the sum total of the complex interworking of a wide range of factors, both human and physical. It might even be argued that if one could know where and why the people of any area lived, one would be a long way toward achieving a full understanding of the area's physical geography and cultural history, as well as of the people's way of life.

In Chapter 9, mention was made of the significance of environmental conditions as factors in the pattern

---

[9]Based on a similar list in J. Warren Nystrom and George W. Hoffman, *The Common Market*, 2nd edition (New York: D. Van Nostrand, 1976), p. 96.

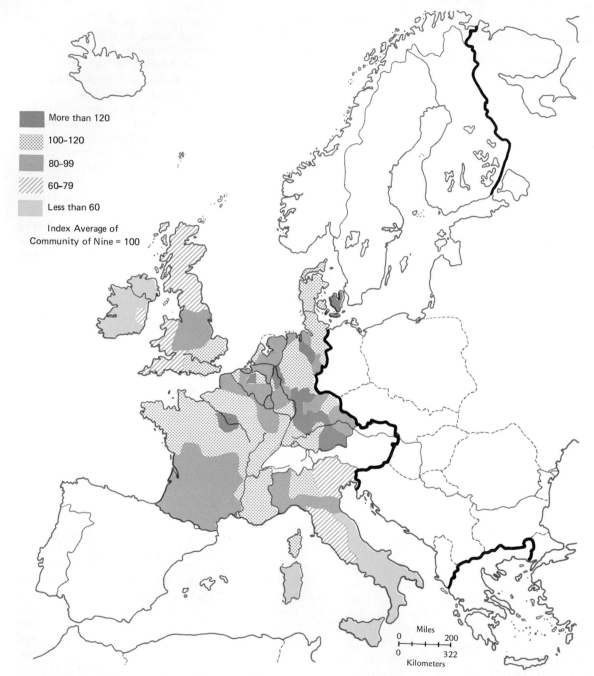

Legend:
- More than 120
- 100–120
- 80–99
- 60–79
- Less than 60

Index Average of
Community of Nine = 100

Miles
0   200
0   322
Kilometers

11–3 *Western Europe: Gross Domestic Product per
Head of Household in the Common Market*
SOURCE: *Basic Statistics of the Community* (Luxembourg:
Statistical Office of the European Communities, 1975).

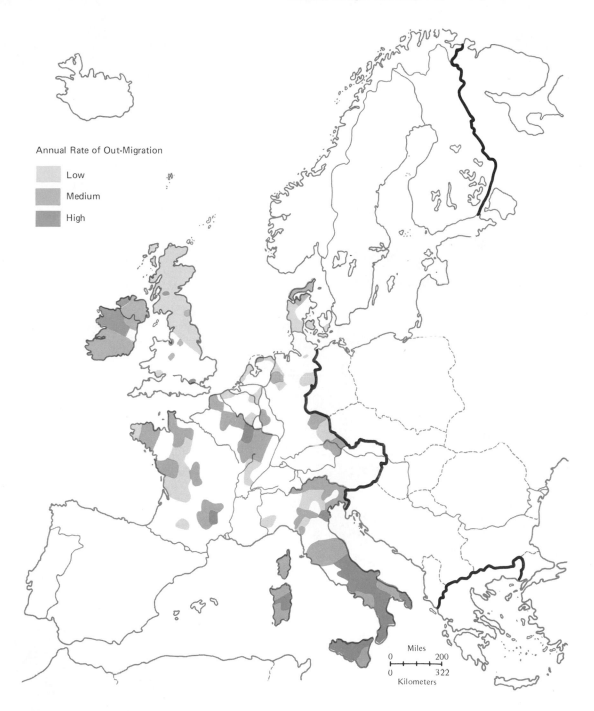

11-4 *Western Europe: Net Out-Migration in the Common Market*

One dot represents 100,000 people

Berlin

Miles
0   100   200
0   161   322
Kilometers

11–5   *Western Europe: Population Distribution*

of population. For example, the rather harsh environments of the Northwestern Highlands have not attracted the vast numbers of people who have chosen to live on the richly endowed Great European Plain. It should be kept in mind, however, that modern Western Europeans are not mere passive pawns moving in response to their environments. On the contrary, as pointed out in the same chapter, the Western Europeans are probably closer to being the masters of their physical environment than are the people of any other great culture region of the world. Still, it is true that most human decisions of a spatial nature reflect to some extent the physical environment.

Population Distribution

Although by world standards Western Europe is densely populated, its 350 million inhabitants are unevenly distributed within their homelands. As the map of population distribution (Fig. 11–5) shows, northern Europe and the Alps have sparse populations. On the other hand, the Great European Plain, particularly near coal sites, has a very high density. Belgium and the Netherlands, two of Western Europe's most densely settled countries, lie here. The Po Valley and adjacent northern and central Italy as well as coastal areas of the south are also densely settled. So too are the areas flanking the Rhine River corridor from Switzerland northward. Other densely populated areas include the central lowlands of Scotland, portions of central England, and around London Basin, the Rhône-Saône corridor in southern France, and coastal Portugal and Spain. These areas of dense productive population help to form the economic and political core of Western Europe and thus are deserving of our major attention.

Population Change Through Time

It is significant that population distribution has changed through time. For example, at about the time of the beginning of the Christian era almost two thousand years ago, conditions were very different. At that time the densest population areas in Western Europe were located along the Mediterranean Sea. It was here—in southern Spain, France, Italy, and Greece, where the great classical empires of Greece and Rome developed flourishing agrarian economies and an urban focus to life—that dense populations were possible. Most of Europe north of the Alps was

thinly peopled. Table 11–1 gives a summary of population change regionally as it took place through time from A.D. 1 to the present.

Table 11–1 also reveals some interesting population changes that took place in Western Europe in past periods. Notice the decrease in total population during the first eight centuries of the Christian era. Following the decline of the Roman Empire and its well-integrated economic system, Europe suffered a long period of economic collapse, famines, epidemics, and invasions of barbarians from the north and Huns from the east. The toll in human life was enormous. In the 700s, the great Moselm empire of the Middle East and North Africa conquered the Iberian Peninsula and threatened to overrun Western Europe. It is no wonder that the term "Dark Ages" is used to identify this period of Western European history.

Gradually population began to grow again as relative stability returned to Europe in the period of the Middle Ages. By 1340, it had reached 57 million, double what it was at the beginning of the era. Contrast this 1,340-year doubling time for Europe's population with today when it is estimated that it will take abut 125 years to double. In 1348, disaster struck in the form of the Black Death, as the bubonic plague was called. Between 1348 and 1350, it is estimated that one-quarter of all Europeans died from the plague. From 1348 to 1379, England's total population dropped from an estimated 5.7 million to 2 million, more than 50 percent in the course of one lifetime. The Hundred Years' War added to the toll of lives in the period following the first outbreaks or plague. Wars should not be overlooked as a factor in contributing to Western Europe's high death rates. The Barbarian invasions and Hundred Years' War contributed to net decline in total numbers. In the period 1618–1648, the Thirty Years' War once again made war ravages a factor in overall population growth. This war struck particularly hard in central Europe, where Germany and Bohemia probably lost one-third of their population as a direct result of the war.

The Demographic Transformation

With the Peace of Westphalia in 1648, Europe once again achieved relative peace and stability. Families were large, with an average of from six to eight children sharing in the labor associated with the almost universal agricultural way of life. The Black Death

Table 11–1 Western Europe: Estimated Population, 1 to 1977

Year (A.D.)	Population (millions)	Iberia, Italy, Greece		France, Benelux		British Isles		North Europe Scandinavia, Finland, Iceland		Central Europe Germany, Austria, Switzerland	
		(Millions)	(%)	(Millions)	(%)	(Millions)	(%)	(Millions)	(%)	(Millions)	(%)
1	27.0	16.5	61.2	6.6	24.4	0.3	1.1	0.3	1.1	3.3	12.2
350	18.9	10.0	52.9	5.2	27.5	0.3	1.6	0.2	1.1	3.2	16.8
600	13.2	7.2	54.5	3.1	23.5	0.7	5.3	0.2	1.5	2.0	15.2
800	22.0	11.6	52.7	4.9	22.3	1.2	5.4	0.3	1.4	4.0	18.2
1000	25.9	14.1	54.4	6.1	23.6	1.5	5.8	0.4	1.5	3.8	14.7
1200	37.7	17.6	46.7	9.8	26.0	2.9	7.7	0.5	1.3	6.9	18.3
1340	57.3	21.0	36.6	18.9	33.0	5.6	9.8	0.6	1.0	11.2	19.5
1400	32.2										
1500	43.1	15.1	35.0	16.2	37.6	3.9	9.0	0.6	1.4	7.3	16.9
1650	76.0	26.0	34.2	30.0	39.5	7.0	9.2	2.0	2.6	11.0	14.5
1700	80.3	26.0	32.4	27.2	33.9	7.9	9.8	4.5	5.6	14.7	18.3
1750	96.6	30.8	31.9	32.2	33.3	9.8	10.2	5.6	5.8	18.2	18.8
1820	132.3	42.0	31.7	35.7	27.0	21.0	15.9	6.3	4.8	27.3	20.6
1900	235.1	70.5	30.0	54.9	23.3	39.2	16.7	11.7	5.0	58.8	25.0
1930	280.0	80.0	28.6	60.0	21.4	50.0	17.8	15.0	5.4	75.0	26.8
1950	324.5	99.0	30.5	71.5	22.0	60.5	18.6	16.5	5.1	77.0	23.7
1977	345.3	111.3	32.2	77.6	22.5	59.2	17.1	22.3	6.5	74.9	21.7

SOURCE: Terry G. Jordan, The European Culture Area: A Systematic Geography (New York: Harper & Row, 1973).

also disappeared as a scourge during this period. From 1650 to 1750, Western Europe's population grew by 20 million to place the total at approximately 100 million on the eve of the Industrial Revolution. The year 1750 is a momentous date for another reason, since it marks the point at which Western Europe began the demographic transformation (see Chapter 1). Figure 11–6 is based on the actual population statistics for England and Wales over the past 275 years. Notice that these data have been arranged in a manner similar to that shown in Figure 1–4, the model of the demographic transformation.

Stage one is characterized by rather static population growth brought about by the very high death rates that counterbalanced the high birth rates of the still traditional agrarian way of life in early eighteenth-century England and Wales. At one point in the late 1730s, the birth rate was about 3.9 percent and the death rate was 3.9 percent. As a result, there was no natural increase in the population in that year. By 1750, a clear trend of continuing high birth

rates accompanied by sharply dropping death rates shows that England and Wales had entered stage two of their demographic transformation with the consequent accelerating growth of total numbers. A huge surplus of births over deaths could spell disaster to a traditional agricultural society, but in Western Europe it coincided with the Industrial Revolution and the opening of overseas empires and other immigration opportunities on a world scale. Rather than disaster, it provided the substance of Europe's most important export of all times—people. Western European explorers and colonizers followed by immigrants carried their culture to the far corners of the world and set in motion processes of change which are still going on.

By 1880, the population of England and Wales had grown to approximately 26 million, and a noticeable decline in birth rates occurred. The urban-industrial way of life had come of age, and attitudes toward family size had begun to change. Children were no longer viewed as economic assets, as child

11–6 Demographic Transformation of England and Wales

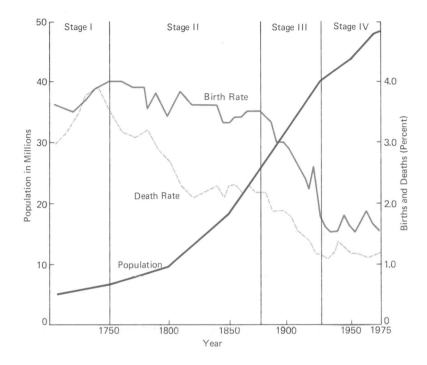

labor laws were enacted and formal education became widespread. More women joined the work force and sought freedom from the burdens of numerous pregnancies and large families. In a word, the fundamental change of attitude toward family size was beginning to affect the decisions of Western Europeans in the industrializing countries. The result was a shift toward smaller families with a consequent sharp drop in birth rates. At the same time, growing affluence and improved medical knowledge and public health care brought about an equally sharp decline in death rates. Population growth was considerably slowed, however, and the fruits of affluence were being shared by the mass of the population in a way never before experienced in the history of mankind.

At the present time, the countries of Western Europe are in Stage Four of the demographic transformation with low birth and death rates, with the result that only a very few are increasing in population by as much as 1 percent or more. Many economists and planners feel that a modern industrialized society should have no serious problem in coping with population increases of this size.

These trends, which began in England and Wales,

diffused across Europe from west to east until, by the 1930s, the bulk of Europe was clearly in Stage Four of the demographic transformation. As this process was taking place, equally important changes were occurring which influenced the spatial elements of Western Europe's population pattern. In particular, the patterns of industrialization, urbanization, and agriculture were causing a rapid areal change in which people were increasing, which is only vaguely suggested by the data in Table 11–1. As the table shows, there was a regional shift from southern Europe to the British Isles and central Europe. The causes for this shift and many other areal population changes can be found in the discussions that follow.

The Pattern of Industrialization

A major portion of Chapter 10 was devoted to a description of the Industrial Revolution of the eighteenth century. As was shown, the industrial activity of the second half of that century was characterized

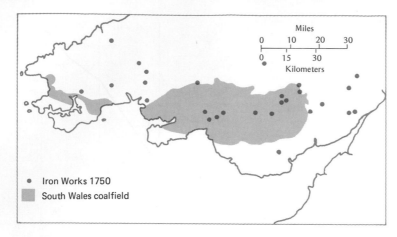

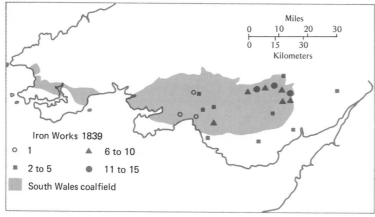

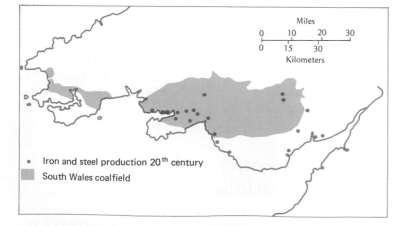

11–7 Southern Wales: Industrial Coalescence, 1750 to Present

SOURCE: Adapted from E. G. Bowen, Wales: *A Study in Geography and History* (Cardiff: University of Wales Press, 1960).

by the emergence of coal as the most significant single factor in the development of large-scale industrial enterprises. As older, waterpowered factories became outmoded, their sometimes poorly located river valley sites were abandoned in favor of growing industrial complexes located near rich coalfields and better served by transportation facilities. As the American industrial geographer E. Willard Miller observed, "The Industrial Revolution was largely a revolution in energy consumption and coal predominated as the source of energy for mechanical power until after World War I."[10] This pattern of industrial coalescence in the vicinity of easily worked coal deposits began in Britain and spread to the rest of Western Europe in the nineteenth and early twentieth centuries.

Locational Shifts in Industry

Figure 11-7 shows the spatial shift that took place in the iron and steel industry of south Wales from 1750 to the present day. Here an impressively large charcoal iron industry was flourishing in the mid-eighteenth century with no significant relationship to the coal seams that underlay the region. The charcoal-iron industry of 1750 was so large, in fact, that the wood supply was growing short, and the industry was beginning to drift to the rugged and still forested valleys of western Wales and the interior. The map for 1750 in Figure 11-7 shows this dispersed pattern quite clearly. Abraham Darby's successful experiments with coke as a blast furnace fuel were studied by other ironmasters and led to the widespread adoption of the new technique. By the end of the eighteenth century, scarcely any charcoal was being used in the blast furnaces of south Wales. In this period the occurrence of coal, iron ore, and limestone in proximity to one another became the crucial factor in the location of the iron industry.

The 1839 map shows a distinct clustering of ironworks on the northeastern outcrop of the coalfield near easily mined deposits of iron ore and limestone. In the modern period the Welsh iron and steel industry is increasingly dependent on imported iron ore and steel scrap for its operations. This shift in raw material supply has resulted in a noticeably peripheral pattern, with the steelworks being located near harbors or with good access to them. These kinds of locational shifts did not take place without having severe impacts on both people and the environment. Derelict buildings, mine head works, spoil heaps, and serious subsidence often blight the landscape in the older industrial regions of Western Europe.

[10]E. Willard Miller, *A Geography of Manufacturing* (Englewood Cliffs, N.J.: Prentice-Hall, 1962), p. 130.

The Complexity of Locational Factors

Of course, the total areal pattern of Western European industrialization is infinitely more complex than this one interesting example can suggest. A whole range of factors operates in determining the locations of industrial enterprises. Such considerations as the source and character of raw materials required, power requirements of processes involved, labor and skills, markets, and transportation needs vary from industry to industry and so operate differently in influencing locational decision making. Also important are such factors as governmental policy, capital requirements, water needs, and climatic conditions. Clearly it is impossible to evaluate the industrial map of Western Europe fully in these complex terms in this short chapter. Suffice it to note that there is a strong tendency for industries to remain concentrated in those areas indicated in Figure 11-8. It is in these areas that several of the industrial locational factors of significance are found in proximity, and it is in these areas that most new factories are being built at the present time.

In the British Isles, France, the Benelux countries, and Germany, there is often a close areal relationship between coalfields and industrial activities. These nations were among the earliest countries to develop strong industrial sectors. In fact, three of these—Britain, France, and Germany—accounted for more than 70 percent of Europe's total industrial output at the outbreak of World War I. As a consequence of their early development in a period when coal was the chief industrial energy source, the coalfields were the focus of much major industrial development. In Britain and in France, notable exceptions to this pattern were found in the case of their primate cities: London and Paris. Here the huge and affluent urban populations formed markets that attracted a wide range of market-oriented industrial activities. The Renault and British Ford automotive works in Paris and in London, respectively, are typical of these industries.

Recent Industrial Development

In more recently industrialized countries, coal has played a much less significant locational role. Italy and Sweden are among the best examples of industrialization in the absence of local coal supplies. In both countries hydroelectricity, a more recently perfected energy source, has played a strong role in

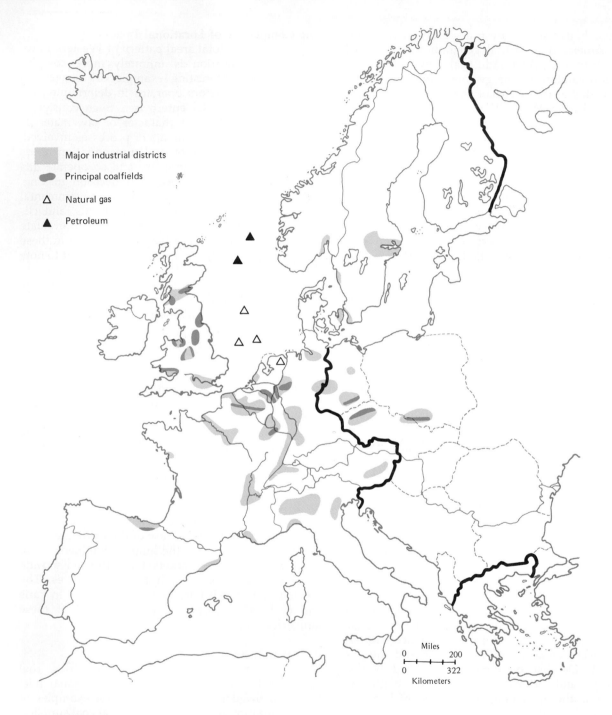

Major industrial districts

Principal coalfields

△ Natural gas

▲ Petroleum

11-8 Western Europe: Industrial Regions

industrial location. In Italy about three-fourths of all manufacturing takes place in the northeast in the Po Valley. Within this region no single factor accounts for the industrial pattern. Most industrial materials must be brought into the area from outside, thus giving a distinct advantage to cities that are well served by transportation facilities. Often these are very old places, like Turin and Milan, with rich cultural traditions of an earlier age. Also, hydro-electric power distributed cheaply over wide areas lends a high degree of flexibility to industrial location choice. Similar conditions prevail in central Sweden around Stockholm on the Baltic and Goteborg on the Atlantic. In both this lake-studded northern area and Italy's northeastern plain, a diversity of light industrial activities blend in the urban and rural landscape with a greater degree of visual harmony than is usually found in the older and more concentrated coalfield-oriented industrial belts of Britain, France, and Germany.

Tariffs and Boundaries as Locational Factors

Another set of locational factors has been traditionally of great importance to Western European industrialization but now seems to be declining in significance. They are tariff policies and the location of international boundaries, aspects of political fragmentation discussed earlier. High tariff walls, which were a widespread feature in Western Europe prior to World War II, encouraged the development of a broad range of industries within each European country. Often these industries had relatively inefficient plants that could not have competed with larger foreign producers. They were literally kept alive through the protection of high tariffs that forced the price of imported goods to levels higher than would otherwise have been necessary. Likewise, a factory located near an international political boundary often suffers a distinct disadvantage, because the boundary acts as a barrier to the distribution of products when import duties and tariffs are imposed. The effect is a great addition to the cost of the factory's products at the boundary and consequent lowering of their competitive position in the market of the neighboring country. Also, fears of invasion or war damage in times of tension made border zone locales less attractive in the past.

These problems are greatly minimized in present-day Western Europe. The nine member nations of the Common Market now form one huge economic unit, thanks to the elimination of barriers to the movement of raw materials, labor, capital, and finished goods.

The Pattern of Agriculture

In Western Europe, cultural factors and physical environmental conditions have interacted to produce an extremely varied pattern of agricultural activity. Despite the region's high state of industrial and

Table 11–2   Western Europe: Pattern of Land Utilization

Country	Agricultural Area (percent of total area)	Use of Agricultural Area	
		Cultivated (percent)	Permanent Meadow and Pasture (percent)
Belgium	51.4	53.2	46.8
Denmark	69.1	89.3	10.7
France	59.1	57.6	42.4
Germany, West	54.0	60.2	39.8
Ireland	68.9	25.3	74.7
Italy	58.0	70.0	30.0
Luxembourg	51.3	47.4	52.6
Netherlands	57.3	40.0	60.0
United Kingdom	76.6	38.3	61.7

SOURCE: *Basic Statistics of the Community* (Luxembourg: Statistical Office of the European Communities, 1975).

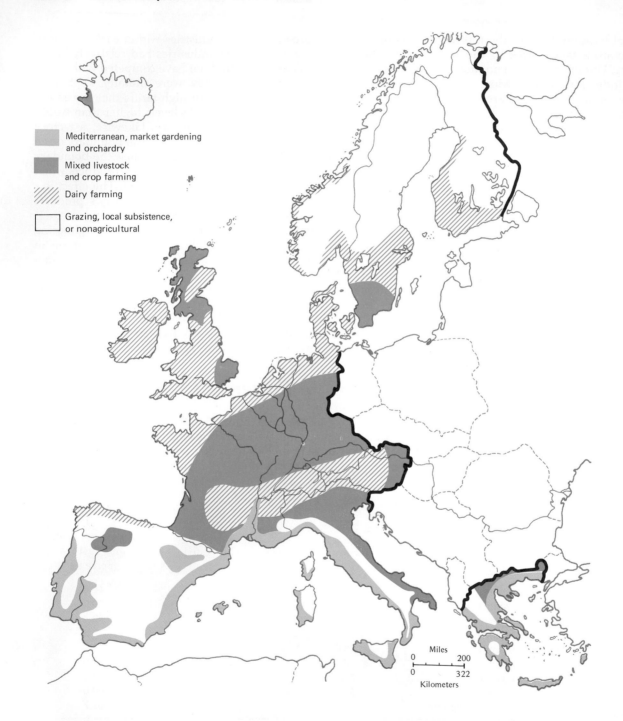

Mediterranean, market gardening and orchardry

Mixed livestock and crop farming

Dairy farming

Grazing, local subsistence, or nonagricultural

Miles
0   200
0   322
Kilometers

11–9 Western Europe: Agricultural Regions

Table 11–3 *Western Europe: Labor Force by Main Employment Sectors (percent)*

Country	Agriculture	Industry	Services	Unemployed
Belgium	3.7	41.2	55.1	2.4
Denmark	9.6	32.3	58.1	3.6
France	12.0	39.2	49.2	2.3
Germany, West	7.3	47.6	45.1	2.2
Greece	34.2	25.7	40.2	2.9
Ireland	24.3	31.1	44.6	5.8
Italy	16.6	44.1	39.3	2.9
Luxembourg	6.6	49.0	44.4	0.0
Netherlands	6.6	35.5	57.9	3.0
Norway	11.4	33.9	54.7	1.5
Portugal	28.8	33.8	37.4	—
Spain	26.5	38.0	35.5	1.2
Switzerland	7.1	46.3	46.6	0.0
United Kingdom	2.8	42.3	54.9	2.1

SOURCE: *Basic Statistics of the Community* (Luxembourg: Statistical Office of the European Communities, 1975).

associated development, agriculture remains the dominant form of land use. Table 11–2 illustrates this fact clearly in the case of the nine Common Market countries. It can be seen even in the small and heavily urbanized and industrialized countries of Belgium and Luxembourg that agricultural land exceeds one-half of the total area.

When the whole of Western Europe is considered, three fairly distinctive types of agricultural systems can be identified. In each case a large area of the region is dominated and characterized by the particular system involved. The region dominated by the Mediterranean system of market gardening and orchardry is shown on the map of Western European Agricultural Regions (Fig. 11–9) as forming a fairly continuous ribbon along the Mediterranean coast. To the north is a belt coinciding with the Alps and adjacent highlands where dairy farming dominates the scene. Across the bulk of the Central Uplands and much of the Great European Plain, the area is dominated by a system of mixed live-stock and crop farming. Along the northern flank of Western Europe, dairy farming again becomes dominant. A few very rugged or remote areas such as the interiors of the Iberian, Italian, and Scandinavian Peninsulas remain outside the mainstream of the modernized market-oriented agricultural life of Western Europe.

Mediterranean Market Gardening and Orchardry

Agriculture has an ancient heritage in the Mediterranean Basin where, since very early times, wheat has grown through the moist, mild winter season to provide a harvest in the spring. Several drought-resisting deeply rooted vines and trees, particularly the grape and olive, also characterized classical Mediterranean landscapes. Although many vestiges of this agricultural tradition still remain, the Mediterranean market gardening and orchardry region is increasingly integrated into the modern commercial tempo of Western European agriculture. In many respects agricultural activities here closely resemble those in California where a similar climatic background prevails. As the name of the region suggests, most of the activity is extremely labor intensive. The resulting high proportion of the total labor forces of countries like Spain, Greece, Portugal, and Italy that are engaged in agricultural pursuits is shown on Table 11–3.

Recent developments in the Mediterranean market gardening and orchardry region result from accelerating specialization and commercialization. These developments are in response to the huge market found in the industrialized and urbanized countries to the north. Improved transportation facilities make it possible for Mediterranean growers

A vineyard in the Rhône Valley of France forms part of the Mediterranean Market Gardening and Orchardry region of southern Europe. The French Mediterranean is noted for its mer de vignes (sea of vines). (French Government Tourist Office)

to speed highly perishable fruits and early vegetables to the affluent markets of Germany, France, Britain, Benelux, and Scandinavia. A notable trend has been for whole districts to specialize in one or a few crops. Within the French Mediterranean belt, for example, vast areas have been turned into a huge *mer de vignes,* or "sea of vines," so dominant are the vineyards. Along the Mediterranean belt in coastal Spain, irrigated areas called *huertas* present a similar scene, with citrus trees in place of vines. Olives too dominate extensive areas, lending a distinctive gray-green appearance to the landscape. Early tomatoes are cultivated extensively for the northern European market. Perishable melons, apricots, and table grapes flow to the same market. Tobacco, another labor-intensive crop, is also an important source of income, particularly in Italy and Greece.

Cattle have never been a very significant element in the Mediterranean agricultural system. Goats and sheep have traditionally provided the milk for the region's famous cheeses. In some mountainous areas, large flocks are still maintained by small numbers of herders who follow a seasonal herding rhythm called transhumance. Transhumance involves the grazing of animals in high mountain pastures during the dry Mediterranean summer and back in the lower areas during the moist winter when grass flourishes.

The future of Mediterranean agriculture appears bright, since the farmers enjoy a considerable advantage over other subtropical producers in the world. They are near to the rich and growing markets of urbanized northern Europe. Also, local industrialization promises to increase steadily and provide growing markets within the region. Problems will occur as modernization and mechanization proceed to eliminate the need for the large labor force still engaged in agriculture. Planners hope the surplus labor can be usefully employed in encouraging local industrial growth.

Dairy Farming

As the map of Western European Agriculture (Fig. 11–9) indicates, much of the region to the north of the Mediterranean is devoted to dairy farming. The dairy belt stretches from the British Isles and Brittany Peninsula and along the shores of the North and Baltic Seas to Finland. A second large area of dairying has developed in the Alps of Switzerland and Austria.

It is significant that much of this area is occupied

by peoples of Germanic origin who traditionally showed dietary preference for dairy products derived from cows. This preference is not found in the Mediterranean lands where olive oil replaces the northerner's butter in most national cuisines. Additionally, these are some of the cloudiest and coolest portions of Western Europe. Many field crops, such as wheat, often fail to mature or suffer fungus attack under these conditions, whereas hay and a number of other fodder crops flourish. In the rugged Alpine terrain, the steep, grass-clad slopes can be grazed or cropped for hay with little or no danger of soil erosion. Row crops, however, would create serious erosional problems. Other extensive, almost level, areas of mountain pasture are at too high an elevation for cultivation of food crops. They do, however, provide luxuriant grazing for cattle during the summer months.

The long tradition of dairy farming in northern Europe is clearly evidenced in the fact that almost every major dairy cattle breed originated here. The breeds carry names that indicate their origin. The large Brown Swiss was originally bred in Schwyz Canton in central Switzerland. The popular Jersey and Guernsey breeds originated on the small British Channel islands near the coast of France. The Holstein originated in the area known as Schleswig-Holstein between Denmark and Germany. These breeds and others less well known were carried with the colonists who traveled overseas in more recent centuries. Herds of these distinctly northern European animals are a familiar sight in New Zealand and Australia as well as in the United States and Canada.

Within the dairy farming areas of Western Europe, there are many important islands of intensive market gardening. These areas are usually closely associated with large urban concentrations. The demand for garden-fresh vegetables in large cities often results in a circular zone of market gardening or, as it is termed in the United States, truck gardening, on their outskirts. In some areas like the Netherlands flowers replace vegetables.

The nature and intensity of dairying vary somewhat from region to region in northern Europe. Denmark, for example, has developed a strong specialization in butter production. Over 16 percent of all butter that is involved in international trade

Dairy farming is an important activity in northern Western Europe and in the highlands of Switzerland and Austria. Shown here is a Danish dairy barn with typical cleanliness and building design; Danish dairy products are famous for their quality throughout Western Europe. (Royal Danish Ministry for Foreign Affairs)

originates in Denmark. The skim milk, which is a by-product of butter production, is used to fatten pigs. The pigs in turn are utilized in the production of world-famous Danish ham and bacon. In the Netherlands much grazing takes place on permanent man-made polder pastures where high water tables preclude the cultivation of many crops. Here specialization in the production of cheese and condensed and powdered milk allows the Dutch to export much of their output to markets throughout the world. In the United Kingdom much fluid milk is marketed locally for human consumption.

The proximity of Western Europe's most densely settled urban and industrial areas seems an ideal insurance that dairy farming will continue to prevail as a vital agricultural activity. Changes will continue, however, as small-farm operators diminish as the mechanization and specialization of the industry continue. Dairy farming has already ceased to be a traditional way of life over much of northern Europe, so serious social problems do not appear to be in view as new technology continues to be applied to dairying.

Mixed Livestock and Crop Farming

The intensive crop specialization of the Mediterranean market gardening and orchardry region is separated from the animal-dominated dairy farming belts to the north by a broad zone where a mixed form of farming characterizes the landscape. Here environmental conditions facilitate a wide range of agricultural activities. As a result of those environmental features and the many cultural traditions found over such a broad region, an extremely mixed and varied agricultural landscape is found. Diversity in this area is further heightened by political fragmentation that has discouraged the evolution of vast regional belts of agricultural specialization such as those found in the United States, Canada, Australia, and the Soviet Union.

Essentially the mixed livestock- and crop-farming region is based on the much earlier and basically subsistence farming systems that evolved here during the medieval period. Crops came to dominate the scene on fertile, easily tilled soils like the loess lands stretching across northern France to central Germany. Livestock often assumed significance in valley areas of heavy clay soils or on cool, moist uplands. Swine were particularly important among Germanic groups,

who grazed them in herds in heavily forested areas.

To maintain soil fertility in the absence of chemical fertilizers, crop rotation schemes played an important role. So too did animal manures. In many respects the medieval farming scene was one that would have delighted the modern ecology-conscious conservationist. As with many traditional subsistence agricultural systems, a balance had been struck and mankind seemed in harmony with nature. That was true at least until population numbers increased beyond the carrying capacity of the system. When the carrying capacity was surpassed, catastrophe struck in the form of famine and plague, frequent visitors to the medieval scene.

As population increased in postmedieval Europe, this traditional way of agriculture began to change. New crops such as the white potato and maize (corn) were introduced from the New World. Each found an important niche in Western Europe's agricultural system. The potato was ideally suited to the cool, moist conditions of the north, and maize to the sunnier conditions of the mid-south. Table 11–4 shows the productions of each of these New World imports in a number of Western European countries. Both

Table 11–4   *Western Europe: Production of New World Crops*

Country	Production (1,000 of tons)	
	POTATOES	MAIZE
Belgium Luxembourg	1,459	22
Denmark	745	0
France	7,905	9,299
Germany, West	14,630	577
Greece	683	573
Ireland	1,277	0
Italy	3,052	4,801
Netherlands	5,706	7
Norway	671	0
Portugal	1,110	518
Spain	5,161	2,006
Sweden	1,109	0
Switzerland	970	99
United Kingdom	6,911	6

SOURCE: *Basic Statistics of the Community* (Luxembourg: Statistical Office of the European Communities, 1975).

Table 11–5 *Western Europe: Percentage of Farms Under 25 Acres (10 Hectares) in Size*

Country	1960	1975
Belgium	75	53
Denmark	47	32
France	56	36
Germany, West	72	55
Greece	96	N/A
Ireland	49	38
Italy	89	86
Luxembourg	40	32
Netherlands	54	47
Spain	79	N/A
Switzerland	77	N/A
United Kingdom	N/A	27

N/A = Not Available.

SOURCE: *Basic Statistics of the Community* (Luxembourg: Statistical Office of the European Communities, 1975.)

potatoes and maize form important livestock feedstuffs and further emphasize the mixed character of the agriculture. In addition to new crops, new techniques and crop rotation schemes were introduced as Western Europe's agricultural sector was caught up in the process of economic development that began in the late eighteenth century.

Table 11–5 illustrates one of the most striking changes currently occurring in Western European mixed agriculture. This change is the rapidly declining importance of small farms. Small farms can only be expected to produce a reasonable income in today's terms if converted to an intensive activity such as market gardening or "factory farming" with some type of livestock such as chickens or hogs. Notice how dramatically this traditional small-family-operated type of farm declined in the ten-year period from 1960 to 1970 in highly developed countries like France, Belgium, and Germany where mixed livestock and crop farming are widespread. It is clear that many problems of adjustment lie ahead for the Western Europeans, as agriculture ceases to be a way of life for many and becomes a technologically complex form of business enterprise. We can be sure that many political questions will hinge on how well or how poorly individual governments cope with the manifold problems encountered as Western Europe's pattern of agriculture continues to change.

The Pattern of Urbanization

Urbanized societies dominate the Western European scene today. Such societies, in which the majority of people live in towns and cities, result from a new step in humanity's social evolution. This step, which reflects a fundamental change in economic development along essentially European or Western lines, is causing many problems in countries around the globe. Modern cities form large and dense agglomerations involving their populations in a degree of human contact and social complexity never before witnessed. Few people seem to comprehend fully either the newness of such great urbanization or the speed with which this process has been taking place. In the words of an eminent sociologist, Kingsley Davis, "Before 1850 no society could be described as predominantly urbanized, and by 1900 only one— Great Britain— could be so regarded. Today, only 65 years later, all industrial nations are highly urbanized, and in the world as a whole the process of urbanization is accelerating rapidly."[11]

The Spread of Early Cities

One should not, however, jump to the conclusion that cities as such are new to Western Europe. They are not, of course. What is new is the overwhelming influence of the city form and function that has become the norm of modern Western European societies. The city of the classical age was first introduced into Europe by the ancient Greeks and Phoenicians as they spread their commercially oriented civilizations around the Mediterranean Basin. The Romans carried the urban form of life deep into Western Europe as they founded new cities in their conquered provinces to the west, east, and north. The many cities that functioned as the administrative and commercial centers of the Roman Empire are shown on Figure 10–2. This pattern suggests how widely the idea of urban life was diffused in Western Europe by the third century of the present era.

The Decline of Roman Urban Life

With the breakup and decline of the Roman Empire in Europe, a period of urban decline was experienced.

[11]Kingsley Davis, "The Urbanization of the Human Population," *Scientific American,* 213 (September 1965), 41.

The early cities had depended heavily on commercial exchange that only flourished under a climate of political stability ensured by Rome's military superiority. When the Pax Romana disappeared from Western Europe, commerce declined sharply, and the subsistence societies that arose in the wake of Roman withdrawal had little need for cities. Some commercial towns reverted completely to the status of agricultural villages, while others were abandoned and forgotten. A few with considerable religious or political significance were able to flourish in a modest way even through the Dark Ages, but by and large the urban way of life ceased to be a dominant feature of the Western European scene until the Middle Ages when commercialism once again became important.

Medieval Revival of Cities

During the Middle Ages, many of the moribund Roman towns were revived and once more became the centers of urban commercial activity. Also many new towns were established both within and without the limits of the old empire. Often the new towns grew up around a fortified preurban core where a feudal lord or ecclesiastical authority assured a degree of security and protection. As they grew, the towns sought and gained political autonomy that was spelled out in formal charters from the ruling prince of the area. Many authorities identify the three essential attributes of the medieval European city as the charter, a town wall, and a marketplace. Self-government required self-protection, hence the need for city walls, which are still a feature in many West European cityscapes. As the walled cities grew in population, the confined space within the wall created patterns of extreme crowding and close placement of multistoried buildings with little open space for streets. These conditions continue into the present in many of Western Europe's towns causing tremendous problems as their affluent residents adopt the automobile in increasing numbers. In the West German cities of Bremen and Lubeck, for example, 84 and 91 percent of the total street mileage is less than 23 feet (7 meters) in width. A study of 141 West German cities indicated that well over three-quarters of the total urban street mileage is too narrow for safe two-way traffic in any reasonable amount. A number of Western European nations are already in the position of having one passenger car for every four or five citizens. The pressure to create parking lots and wider streets is growing in cities across the region. In some areas such as Florence and Rome in

Western Europe's love affair with the automobile has created serious problems of traffic congestion, especially in the old urban centers. (United Nations)

Italy, and Athens in Greece, air pollution, primarily resulting from auto exhaust, is threatening priceless works of art and sculpture as well as ancient marble temples. Coping with the automobile is one of the major problems facing most Western European city authorities at the present time. Pollution and congestion are the unwanted by-products of Western Europe's current love affair with the automobile.

Industrialization and Contemporary Urbanization

Clearly it was industrialization that brought the present pattern of urbanization to the people of Western Europe on such a massive scale. In about the year 1800, nearly 10 percent of the people in England and Wales were living in cities of 100,000 or more. This proportion doubled to 20 percent in the next forty years and doubled once more in another sixty years. By the year 1900, as we have seen, the British were an urbanized society, the first in the modern world. Western Europe as a whole has a population that is slowly increasing at the present time. Its urbanized population, on the other hand, is growing more rapidly. One commonly used index or urbanization is the percentage of a country's total population that lives in large urban areas of over 100,000 population. The pattern that emerges when these data are mapped, as on Figure 11–10, shows that only a handful of the most peripheral Western European countries have fewer than 30 percent of their people living in large metropolitan centers. If, however, different criteria were used, such as how many people live in towns and cities of over 5,000 or over 20,000 in population (often used by various national census bureaus), a much higher percentage of urban living emerges. Notice on Figure 11–11, which portrays urban percentages as they are defined by the national government involved, that only Portugal claims less than 50 percent of the people as urban dwellers.

Suburbanization

It is important to note that much of modern European urbanization is really what is better described as suburbanization. Modern trends of urban living in Europe, as in other areas of the developed world, bring about decreasing population densities in urban centers. As a result, ever-increasing amounts of rural and agricultural land are being converted to urban uses. In England and Wales, for example, only about 5 percent of the total land area was classified as urbanized in 1900. As the society became increasingly urbanized, land was converted from rural to urban use at the rate of about 1 percent of the total area per decade. As a result, fully 11 percent of England and Wales was classified as urban in 1972. This slow but inexorable process is continuing to change Britain's famed "green and pleasant land" to an increasingly urbanized, built-over, and paved landscape. To most Britons urbanization is a disconcerting prospect, to say the least. In West Germany and the Netherlands, the prospect is even worse. The rate of conversion from rural to urban landscape is 1.6 percent each decade in Germany and 2.8 percent in the Netherlands. By the year 2000, a generation away, one-quarter of the Netherlands will be urbanized, while the totals for Britain and West Germany will be about 15 and 18 percent, respectively.

The causes for this progressive urbanizing of the landscape result from changing urban space standards. The densely populated central city is a thing of the past as low-density housing on the outskirts becomes a dominant feature of the Western European urban pattern. Sprawling suburbs so common in Anglo-America are now increasingly obvious features as more and more agricultural land disappears before the bulldozer's onslaught. In France, urbanization is officially considered "the main problem in environment planning" nationally.[12] During the period from 1954 through 1962, France's total population rose 8.1 percent, but its urban population rose by 13.8 percent. The graph, Figure 11–12, shows the breakdown of French population in the years 1851, 1946, 1962, and as it is projected for the year 2000. As the graph shows, Paris has been a major focus for urbanization in France since World War II. In their national planning to cope with the problems of increasing urbanization of the Paris region, French officials have adopted a scheme of encouraging growth in selected peripheral cities that are designated as "Regional Metropolises." These are shown on Figure 11–13. Great emphasis will be put into equipping these cities with high-level facilities in the areas of research, higher education, medical care, government, culture, and communications in an effort to counterbalance the lure of Paris, the primate city. As now scheduled, these cities should be functioning as full-scale re-

---

[12]Service de Presse et d'Information, *France, Town and Country Environment Planning* (New York: Ambassade de France, 1965), p. 14.

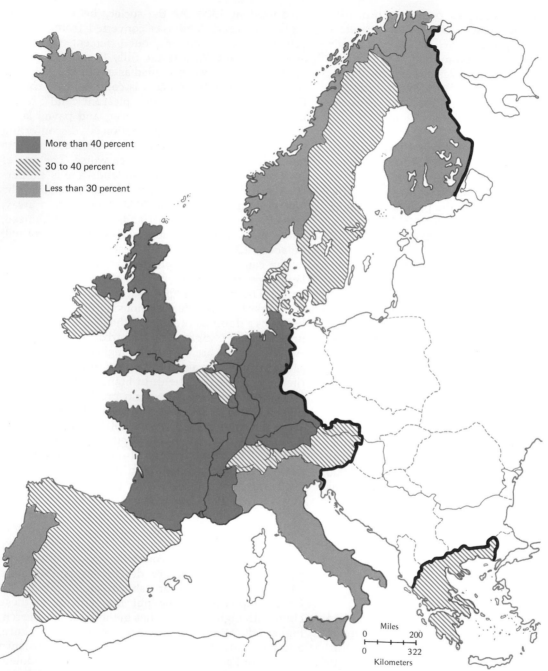

11–10   *Western Europe: Percent of Population in Urban
Areas Greater Than 100,000*

SOURCE: *Basic Statistics of the Community* (Luxembourg:
Statistical Office of the European Communities, 1975).

Legend:

More than 40 percent

30 to 40 percent

Less than 30 percent

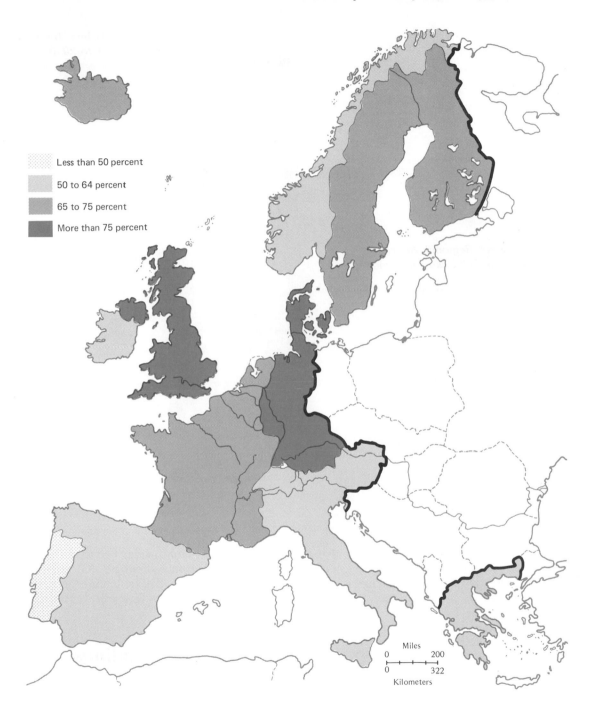

Less than 50 percent

50 to 64 percent

65 to 75 percent

More than 75 percent

11–11 *Western Europe: Percent of Population Classed as Urban by Each National Government*

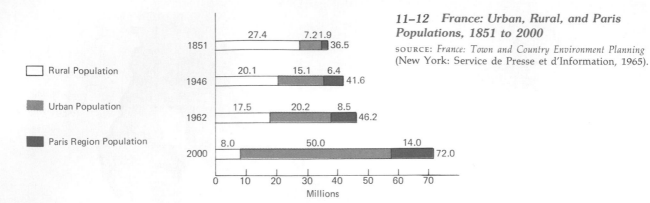

11–12 *France: Urban, Rural, and Paris Populations, 1851 to 2000*

SOURCE: *France: Town and Country Environment Planning* (New York: Service de Presse et d'Information, 1965).

□ Rural Population

▨ Urban Population

■ Paris Region Population

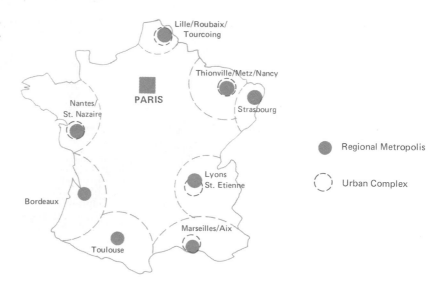

11–13 *France: Regional Metropolises and Their Spheres of Influence*

SOURCE: *France: Town and Country Environment Planning* (New York: Service de Presse et d'Information, 1965).

gional metropolises by 1985. It is felt that planning on such a comprehensive scale is necessary to preserve an acceptable life style as France looks forward to the prospect of her population increasing to over 62 million by the year 2000.

In this review of the pattern of urbanization, we have been limited to only a few facets of both the process and the problems that is poses for the countries of Western Europe. Needless to say, a whole spectrum of changes and potential problems confront the people of Western Europe as they continue to alter their life styles to fit in with the tempo and demands of urbanized life in the last quarter of this century. Traditional values and modes of operation

will be under increasing pressure. Some will be abandoned, some modified, and some will be salvaged and reinforced. The manner in which these changes manifest themselves will doubtlessly be varied.

Further Readings

ASHTON, T. S., *The Industrial Revolution, 1760–1830* (London, New York, and Toronto: Oxford University Press, 1948), 120 pp. A popular and very readable review of the Industrial Revolution, especially as it occurred in Britain.

BAILY, RICHARD, *The European Community in the World* (Lon-

Paris, primate city of France and former Roman center, has experienced increasing urbanization problems. The French government hopes to encourage future growth in selected regional centers. (French Government Tourist Office)

don: Hutchinson, 1973), 200 pp. An excellent world view of the European Community as a fact of modern international affairs.

CLAYTON, K. M., and KORMOSS, I. B. F. (eds.), *Oxford Regional Economic Atlas Western Europe* (London: Oxford University Press, 1971), 100 pp. An excellent set of clear meaningful maps illustrating most facets of the Western European scene.

CLOUGH, S. B., MOODIE, T., and MOODIE, C., *Economic History of Europe: Twentieth Century* (New York: Walker, 1968), 380 pp. A readable collection of extracts from the key documents and speeches outlining European national life from World War I to the decade of the 1960s.

CLOUT, HUGH D., *Agriculture* (London: Macmillan, 1971), 62 pp. This concise paperback in the "Studies in Contemporary Europe" series is a fresh and readable approach to an old and important European way of life now rapidly changing.

CLOUT, HUGH D., *Rural Geography: An Introductory Survey* (New York: Pergamon Press, 1972), 200 pp. Richly illustrated with an abundance of clear sketch maps, this readable study is devoted almost entirely to the European scene.

LOBEL, M. D. (ed.), *Historic Towns: Maps and Plans of Towns and Cities in the British Isles, with Historical Commentaries, from Earliest Times to 1800* (London and Oxford: Lovell Johns-Cook, Hammond & Kell, 1969), 150 pp.

NYSTROM, J. WARREN, and HOFFMAN, GEORGE W., *The Common Market,* 2nd edition (New York: D. Von Nostrand, 1976), 148 pp. An interesting but dated paperback tracing the evolution of the Common Market and its characteristics.

231

O'Donnell, James, and Scott, Dermott (eds.), *European Community Directory & Diary* (Dublin: Institute of Public Administration, 1974), 232 pp. An extremely valuable directory and reference on the European Community.

Patijn, S., *Landmarks in European Unity: 22 Texts on European Integration* (Leyden: A. W. Sijthoff, 1970), 220 pp. A collection of the key speeches and documents relating to European economic and political integration since 1946.

Pounds, Norman J. G., *The Geography of Iron and Steel* (London: Hutchinson University Library, 1959), 200 pp. An excellenet review of the world distribution of iron and steelmaking with a strong emphasis on Western Europe.

Pounds, Norman J. G., *The Ruhr: A Study in Historical and Economic Geography* (London: Faber and Faber, 1952), 280 pp. An excellent geographic study of one of Western Europe's most important industrial regions.

Rostow, W. W., *The Process of Economic Growth* (Oxford: Clarendon Press, 1960), 300 pp. In Professor Rostow's words, this is an historian's book about economic theory. Rostow's model of economic development has been very influential in shaping national policies through much of the world.

Weil, Gordon L., *A Handbook on the European Economic Community* (New York: Praeger, 1965), 480 pp. As its title indicates, this is a very good place to start in any effort to learn about the nature and workings of the European Community or Common Market.

Western Europe: Its Place in Today's World

12

Throughout most of the three preceding chapters, Western Europe has been viewed in general terms. At no point have any of the individual countries of the region been discussed as distinctive entities. A review of the current realities of Western Europe's economic and political life suggests that the region's political units fall into four categories. These are shown on the map of "Western Europe; National Groupings" included here as Figure 12–1. Shown are (1) six tiny micro states: Andorra, Liechtenstein, Malta, Monaco, San Marino, and the Vatican City; (2) the nine full members of the European Economy Community or "Common Market": Belgium, Denmark, France, Federal Republic of Germany, Ireland, Italy, Luxemberg, Netherlands, and United Kingdom; (3) the four neutral states: Austria, Finland, Sweden, and Switzerland; and (4) the five peripheral states: Greece, Iceland, Norway, Portugal, and Spain. It can further be seen that the current individual patterns of national development in Western European countries are strongly influenced by membership in one or another of these groupings. The bulk of this chapter is devoted to the largest and most important of these groupings, the Common Market nations.

The Common Market Nine

As the map (Fig. 12–1) clearly shows, the nine member states of the European Economic Community or Common Market stretch across the breadth of Europe and Italy on the south to the British Isles on the north. Together they form the world's largest market. Its total population is approximately 260 million, greater than that of either the United States or the Soviet Union. The Common Market's combined monetary reserves are almost five times greater than those of the United States and more than nine times larger than the USSR's. In terms of world trade, the Common Market is easily the world's largest importer and exporter, although total United States production remains significantly larger. To help move their prodigious volume of foreign trade, the Common Market countries support a merchant fleet that is more than three times larger than that of the United States. In terms of steel, the most essential industrial output, the Common Market countries also outproduce both the United States and the USSR by impressive amounts. So too does the Common Market lead in the production of several important agricultural outputs such as pigs, wines, milk, butter, and cheese. If final proof of the productive and economic capacity of the Common Market is needed, it might be found in the statistics of motor vehicles produced. The Common Market produces almost 2,000,000 more passenger cars than does the United States. Perhaps the data presented in Table 12–1 will provide a good basis for an appreciation of the tremendous significance of the Common Market in today's world of growing economic competition and contrasts.

Five countries of the Market are monarchies: Britain, the Netherlands, Belgium, Denmark, and Luxembourg. Although the traditions and trappings of royalty remain vigorous and alive, their real political power is almost nonexistent. In all cases the nine members have national governments that are

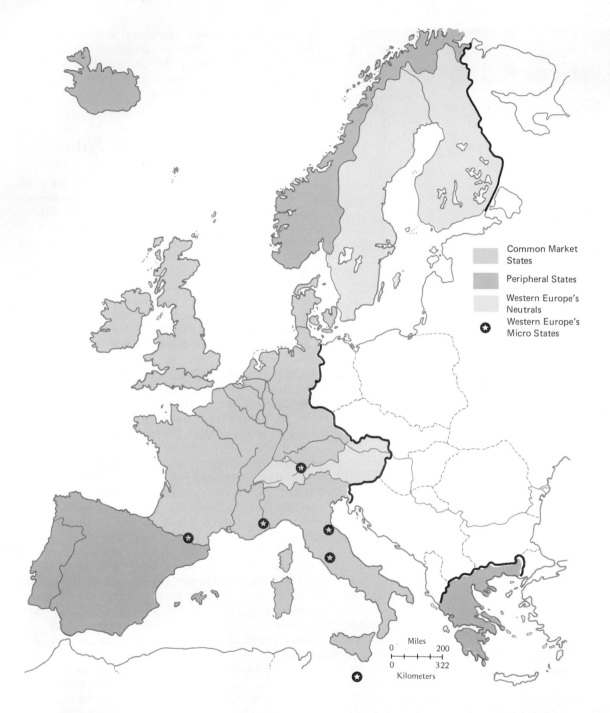

Common Market States

Peripheral States

Western Europe's Neutrals

Western Europe's Micro States

12-1 Western Europe: National Groupings

Table 12–1 *Developed World Comparisons*

	Common Market	USA	USSR	Japan	World
Total population (millions)	260	217	259	114	4,083
Gross national product (millions of dollars)	1,276,516	1,529,902	678,580	509,332	6,246,990
Average annual rate of GNP growth per capita (percent)	4.1	3.7	2.6	8.6	—
Passenger cars in use (thousands)	58,877	104,269.7	1,300	15,852.2	245,920
Annual per capita energy consumption (gallons of oil)	1,312	2,320	1,056	762	—
Steel production (thousands of tons)	155,587	132,196	136,229	117,131	704,800
Total imports (millions of dollars)	171,698	55,296	16,890	30,657	411,920
Total exports (millions of dollars)	167,931	57,053	17,170	29,549	395,440
Merchant fleet (thousands of tons)	69,088	14,587	19,236	39,740	342,162
Foreign aid to developing countries (millions of dollars)	6,142	8,450	204	2,957	—

SOURCES: *Population Data Sheet* (Washington, D.C.: Population Reference Bureau, 1977); *Statistical Yearbook* (New York: United Nations, 1974); *Basic Statistics of the Community* (Luxembourg: Statistical Office of the European Communities, 1975).

democratic and representative in the fullest modern sense of these terms. Since all Common Market states are members of the Council of Europe, it is possible for individual citizens to bring actions against their own national governments under the Council's Human Rights Convention for safeguarding individual rights. All except Ireland are members of the North Atlantic Treaty Organization and thus closely allied with the United States and Canada in common defense schemes. France, it should be noted, has withdrawn its military forces from direct NATO command but remains a member in other respects.

Since World War II, individual mobility and mass communications have allowed young Europeans to have much more in common than ever before. Eurovision, an international TV broadcasting organization, makes possible the transmission of important events to audiences throughout Western Europe

simultaneously. Most of the elements of modern "Pop" culture are universalities crossing national and linguistic boundaries with ease in modern Europe. People living within the Common Market need no passports to visit other countries within the organization. Border-crossing formalities and requirements have been greatly lessened throughout Western Europe making it simple for hordes of tourists to move easily and provide the income for the increasingly important tourist industry. Most Common Market governments have joined in plans and programs aimed at large-scale youth exchanges deliberately designed to assist in breaking down the old psychological barriers created by the intense nationalism of the past. Historians, geographers, and other Common Market academics have been working on projects aimed at producing school textbooks with European rather than nationalistic

Table 12–2 The Federal Republic of Germany: Selected Economic Characteristics

Total Area in Thousands of Square Miles (square kilometers)	Percent of Total Area in Farms	Percent of Farmland Cultivated and in Temporary Pasture	Total Labor Force (in millions)	Percent of Labor Force		
				PRIMARY ACTIVITIES	SECONDARY ACTIVITIES	TERTIARY AND UNKNOWN
96 (249)	54	60	26	7.3	47.6	45.1

SOURCE: *Basic Statistics of the Community* (Luxembourg: Statistical Office of the European Communities, 1975).

stress and interpretation. European schools in Brussels and Luxembourg have worked to develop truly European curricula and outlooks. These schools were first opened to provide for the needs of the children of the staff members of the growing Common Market administrative hierarchy. Their rapid growth and commendable results have led to the basic idea being adopted in many other cities. To encourage pupils to approach their studies on a European rather than a national basis, a large part of their time is spent in reading and speaking other European languages. With the entry of the United Kingdom and Ireland into the Common Market, hopes for a Channel Tunnel have risen if only as a response to the expected increase in the volume of goods flowing between the British Isles and the Continent.

With these observations concerning the Common Market Nine in mind, we now undertake a brief review of each member country. As has been mentioned more than once in preceding sections, the roots of nationalism and individuality run deep in Western Europe. It would be foolhardy ever to forget this fact while attempting to understand Western Europe's role in either today's or tomorrow's world.

The Federal Republic of Germany

The Federal Republic of Germany, better known as West Germany, is a highly developed, prosperous, industrialized nation, ranking among the world's leading economic powers (Table 12–2). These facts are little short of remarkable when it is recalled that West Germany is only a portion of the greater German Empire which threatened to conquer all of Europe in World War II. Through the efforts of the allied powers, chiefly the Soviet Union, France,

Britain, and the United States, Hitler's plans for European and world dominance were broken after a brutal total war that devastated much of the Continent. Germany surrendered unconditionally to the allies on May 8, 1945. As a part of the postwar policy, Germany was divided into occupation zones administered by the military authorities of the four chief allies. Originally it had been agreed that for economic purposes Germany would be treated as a single unit. This plan for an economically unified Germany failed as the ever-increasing postwar differences between the Soviet Union and the Western allies made themselves felt. In 1948 the Soviets withdrew from the four-power governing organizations of Germany and Berlin, which had also been divided. Berlin became cut off from the area under allied control, since it was located deep within the Soviet-controlled zone.

As the differences between the Soviet Union and the United States and its allies crystallized into what became known as the "Cold War" of the late 1940s, 1950s, and into the 1960s, two separate and distinctive political units were formed: West and East Germany (more properly the Federal Republic of Germany and the Democratic Republic of Germany). Winston Churchill struck a prophetic note in a telegram he sent to President Truman in the summer of 1945 by declaring that "An iron curtain is drawn down upon their [the Russian] front. We do not know what is going on behind. There seems little doubt that the whole of the regions east of the line Lübeck-Trieste-Corfu will soon be completely in their hands."[1] The

---

[1] Quoted in Norman J. G. Pounds, *Divided Germany and Berlin* (New York: Van Nostrand Reinhold, 1962), p. 5.

Table 12–3 Western Europe: Common Market Country Dependence on Foreign Energy Sources (Percent of total supply)

	West Germany	France	Italy	Netherlands	Belgium	Luxembourg	United Kingdom	Ireland	Denmark
1960	9.6	41.0	58.1	50.9	32.6	99.8			
1961	12.8	43.7	62.8	49.6	35.8	99.4			
1962	17.7	44.3	68.1	58.6	40.9	99.5			
1963	22.4	52.6	70.2	63.2	49.4	99.7	26.3	71.9	98.0
1964	27.2	54.1	71.5	65.8	54.7	99.7			
1965	32.6	55.0	73.5	62.8	59.3	99.3			
1969	42.8	63.8	78.2	47.7	73.5	99.3	43.8	73.0	99.6
1970	47.9	71.0	81.8	42.3	81.8	99.2	45.1	76.1	100.0
1971	50.6	73.1	81.8	26.2	84.6	99.5	49.5	80.5	100.0
1972	53.2	75.0	81.1	15.6	82.7	99.5	49.7	77.8	99.6
1973	54.8	78.0	83.0	6.3	86.4	99.6	48.2	80.7	99.6
1974	51.1	82.3	83.1	9.1	91.2	99.4	50.2	82.2	99.5

SOURCE: *Basic Statistics of the Community* (Luxembourg: Statistical Office of the European Communities, 1975).

term "Iron Curtain" was soon to become a household word along with "Cold War" in Anglo-America as well as Europe. It has only been in the 1970s that the tensions that so seriously divided Europe and the world for three decades have begun to disappear. Today the foreign policy of West Germany is marked by efforts to improve relations between Eastern and Western Europe and reach closer accords with the Communist regime in East Germany.

On the economic side, West Germany made a recovery following World War II that can only be termed miraculous. There have been periods of business recession, but high levels of employment and a stable currency have been maintained more successfully than in most developed countries. The average rate of growth of West Germany's gross national product during the first ten years of the Common Market was an impressive 6.6 percent. In 1970, while the rate was declining in other countries, it was maintained at 5.5 percent. Contributing to

West Germany's economic good health has been the steady productivity of the coal industry. In 1971, German mines produced a total of 117 million tons from sixty-seven operating collieries. A recent agreement reached between the German federal government and the state of North Rhine Westphalia with the private mine owners to ensure the future of the industry in the Ruhr will stabilize that region's annual output at 80 million tons. West Germany like all members of the Common Market is a net importer of energy supplies. Also, as Table 12–3 indicates, West Germany's dependence on foreign energy sources, largely petroleum, is increasing dramatically.

Imports of oil have risen substantially each year. Natural gas is also being used in growing quantities. The gas is increasingly being imported from the nearby Netherlands. Agreements with the Soviet Union have been reached to purchase large amounts of natural gas from that source. West Germany's large energy requirement is met as shown in Table 12–4.

Table 12–4 The Federal Republic of Germany: Fuel Sources as Percent of Total Energy Production

Oil	Solid Fuel	Natural Gas	Nuclear	Tidal, Hydro, and Geothermal
54.9	36.6	7.3	.6	.6

SOURCE: *Basic Statistics of the Community* (Luxembourg: Statistical Office of the European Communities, 1975).

West Germany's major industries are steel and iron, chemicals, machinery, electrical equipment, and automobiles. The industrial sector of the German economy requires approximately one-third of the total foreign oil imports; transportation about one-quarter; and residential-commercial use approximately one-third; the remainder is consumed as a nonenergy raw material. It can be seen that West Germany's growing reliance on imported oil represents the most serious threat to its economy's continued development. Like countries throughout much of the rich world, West Germany's dependence on energy imports has emerged as the single most serious problem area in the mid- and late-1970s.

West Berlin

Countless Germans living in both West and East Germany still regard Berlin as the rightful national capital and believe that it will ultimately be restored to its former position when Germany is finally reunited. The East German Communist regime maintains its governmental functions in the eastern sector of the city, and so Berlin functions as the capital of the Democratic Republic of Germany.

Following World War II, Berlin was not included in any of the occupation zones established by the victorious allied occupation armies. The city was made a separate area with a four-power control similar to that established for the whole of Germany. When the Soviets withdrew from that four-power organization in 1948, they also withdrew from the joint administration of Berlin. As a result of their subsequent policies, Berlin became a divided city, with West Berlin an enclave or island of West German territory deep inside the frequently hostile East German state. West Berlin presently has a population of 2.1 million and covers an area of 186 square miles (482 square kilometers). It shares a 28-mile (45 kilometers) boundary with East Berlin along which the famous Berlin Wall was constructed by the Communist regime to keep East Germans from moving to the economically more attractive West. As can be imagined, West Berlin's viability as a metropolis and as a Western enclave is very tenuous, since its nearest hinterland or reliable source of supply and markets is West Germany, 110 miles (177 kilometers) away.

From 1945 to 1961, as West Germany's economy was restored and began to boom, almost 3 million East German refugees escaped to the West via West Berlin. On August 13, 1961, the world was shocked when the Communists sealed off the boundary and began to build a concrete block and barbed wire wall along it. Berlin was a major focal point of Cold War tensions between the United States and USSR during the 1950s and 1960s to the extent that it sometimes seemed that war between the two superpowers was almost inevitable. Today conditions appear to be normalizing between East and West Germany, but West Berlin's long-term future remains unclear.

France

Almost four-fifths the size of Texas, France is the largest Western European state and fulcrum of the Common Market (Table 12–5). The country combines great natural wealth with a central location and is endowed with a wide variety of terrain, two-thirds of which is nearly level or gently rolling. Without doubt, France has exercised the strongest influence on the Common Market to date. Somewhat unique among the major industrial countries of Western Europe,

Table 12–5   France: Selected Economic Characteristics

Total Area in Thousands of Square Miles (square kilometers)	Percent of Total Area in Farms	Percent of Farmland Cultivated and in Temporary Pasture	Total Labor Force (in millions)	Percent of Labor Force		
				PRIMARY ACTIVITIES	SECONDARY ACTIVITIES	TERTIARY AND UNKNOWN
210 (544)	59	58	22	12.0	39.2	49.2

SOURCE: *Basic Statistics of the Community* (Luxembourg: Statistical Office of the European Communities, 1975).

Table 12–6 France: Fuel Sources as Percent of Total Energy Production

	Oil	Solid Fuel	Natural Gas	Nuclear	Tidal, Hydro, and Geothermal
	66.4	22.0	6.8	1.5	3.3

SOURCE: *Basic Statistics of the Community* (Luxembourg: Statistical Office of the European Communities, 1975).

France boasts an important agricultural sector with substantial indigenous resources of primary raw materials, a diversified and modern industrial plant, and a capable labor force. Government ownership of industry is important but far from total with about 30 percent of all industrial output being produced by nationalized enterprises.

France's gross national product of over $200 billion ranks fifth in tthe world after the United States, the Soviet Union, Japan, and the Federal Republic of Germany. Per capita GNP equals $3,620. French exports total well over $20 billion with agricultural products contributing over one-fourth of the amount. The principal products of France's agricultural sector are grains, sugar beets, wine grapes, dairy products, livestock and meat, and fruits and vegetables. About 14 percent of the total labor force is engaged in agricultural pursuits. As these figures demonstrate, they contribute much to France's economic well-being. The principal deficiencies in agricultural production are fats and oils, hard wheats, long-grain rice, citrus and other tropical fruits, and natural textile fibers such as cotton and wool. Imports must be relied upon to satisfy these needs.

France is amply provided with raw materials such as iron ore, soft coal, bauxite, and uranium. Hydroelectric power sources are also well developed. As a result, France stands as one of Western Europe's major suppliers of metals and minerals. On the world scene, France emerges as one of the leading producers of iron ore, coal, and bauxite. In addition, it possesses large deposits of antimony, magnesium, pyrites, tungsten, and some radioactive minerals. Self-sufficiency is enjoyed in salt, potash, fluorspar, and sulfur.

Recent years have seen a declining production of the low-iron-content French ores in favor of high-grade imports from Africa. Similarly French coal production has been lagging because of the attractiveness of competitive fuel oil and cheaper coal imports. Serious problems of readjustment have been created in several traditional French mining regions.

France, like West Germany, has very limited domestic reserves of petroleum and must rely heavily on imports of this vital modern energy source. As Table 12–3 shows, France is increasingly dependent on foreign energy sources. Approximately three-fourths of its energy requirements must be met by imports. Of these, foreign oil makes up a disturbingly large share and represents an alarming area of concern for French governmental and industrial leaders (Table 12–6).

Historically one of the world's leading manufacturing countries, France is active in all major branches of industrial activity. The aluminum and chemical industries rank among the largest in the world as do the mechanical and electrical sectors. The French are very dependent on the automobile in their daily lives. In meeting this need, the French automobile industry produces over 2,000,000 vehicles yearly. The electronic, telecommunication, and aerospace industries have all contributed to the present high level of technological development that characterizes much of French industry. Military use of atomic energy, plus certain civilian applications, have achieved an impressive level of sophistication excelled only by the United States and the United Kingdom. France has joined with the United Kingdom in producing the first supersonic commercial jet air transport, the Concorde. France launched its first earth satellite in 1965, and has pursued an active space research program that includes an equatorial launch site in French Guiana.

Italy

Modern Italy is the heir of more than 2,000 years of Roman and Renaissance civilization (Table 12–7). In this role Italian leaders frequently remind their Common Market colleagues that the idea of European unification is as old as Rome, and that their Roman forebears laid the foundations for the lasting Euro-

Table 12–7   *Italy: Selected Economic Characteristics*

Total Area in Thousands of Square Miles (square kilometers)	Percent of Total Area in Farms	Percent of Farmland Cultivated and in Temporary Pasture	Total Labor Force (in millions)	Percent of Labor Force		
				PRIMARY ACTIVITIES	SECONDARY ACTIVITIES	TERTIARY AND UNKNOWN
116 (301)	58	70	19	16.6	44.1	39.3

SOURCE: *Basic Statistics of the Community* (Luxembourg: Statistical Office of the European Communities, 1975).

The Mezzogiorno section of Italy, a region of poverty, suffers from a lack of water and from thin stony soils. Major efforts have been exerted to improve the economy of the region. (United Nations)

pean union of Christendom. Although these facts cannot be denied, it is also true that the modern state of Italy is but a century old. The entire Italian Peninsula came under a single government under King Victor Emmanuel II of the House of Savoy only in 1870. From 1870 to 1922, Italy was a constitutional monarchy with a parliament elected with limited suffrage. In 1922, Benito Mussolini came to power at the head of a Fascist government dedicated to the founding of a corporate state. World War II saw Italy allied with Germany and participating in the conflict against the Allies. Following the invasion of Sicily, the Italians sided with the Allies against Germany. The Germans were finally driven out of Italy in April 1945, and in 1946 a plebiscite led to the establishment of the present republican form of government.

Compared with many other Western European countries, Italy is poorly endowed by nature. Much of the country is unsuited for farming because of mountainous terrain or unfavorably dry climate. There are no significant deposits of coal or iron ore. Most other important industrial mineral deposits are widely dispersed and generally of poor or indifferent character. Natural gas deposits, found in the Po Basin in the post–World War II period, constitute the country's most important mineral resource. The gas is being rapidly depleted and does not constitute a long-term energy supply. As a result of these natural deficiencies, most raw materials for manufacture must be imported. Other handicaps affecting Italian development have been the low level of agricultural productivity as well as in certain industrial sectors, and the need to upgrade the skill levels of much of the labor force.

12–2 Italy: The North and South

gorized as hilly; and the remainder is mountainous. Unfortunately these mountains are not high enough to ensure year-round snow accumulations. As a result the rivers are dry for long periods of the year, flowing only during the moist Mediterranean winter season. Ironically winter is often a period of flooding, water-logged fields, and serious soil erosion. Irrigation schemes are consequently very difficult and exceedingly expensive undertakings in the Mezzogiorno. Such undertakings are essential, however, if the region is to share in the benefits of development which characterize most of the rest of Italy and the Common Market. As stated in a report prepared for the International Bank for Reconstruction and Development:

> The problem of water supply, no less than the land reclamation program, is one of the key problems of Southern Italy which is still far from being solved; the task is to improve an indispensable service, the insufficiency of which stands seriously in the way of any attempt at promoting the economic and social development of large areas in Southern Italy.[2]

Although still lagging, the Mezzogiorno has begun to move ahead in terms of economic development. The effects of massive investments in regional agricultural improvement schemes and industrial development are beginning to be felt. It will, however, take a long time before the Italian south begins to equal the north in terms of productivity and prosperity for its workers. As the map of Out-Migration (Fig. 11–4) indicates, people still leave the Mezzogiorno at a disturbing rate in search of a better life.

The same map suggests that conditions are considerably different in northern Italy, as indeed they are. Here Italy's essentially private enterprise economy has been flourishing on a broad front. The

The south of Italy, a region known as the *Mezzogiorno* (Land of the Midday Sun), comprises one of the Common Market's major regions of under-development (Fig. 12–2). It covers an area the size of Pennsylvania and Delaware and includes the Italian Peninsula south of Rome and the islands Sicily and Sardinia. Here rugged landscape, poor soils, and Mediterranean climate have had their most profound impact on the nature of the Italian economy. Only 15 percent of this vast area is considered level enough for cultivation; another 50 percent is cate-

[2]Quoted in Sergio Barzanti, *The Underdeveloped Areas Within the Common Market* (Princeton, N.J.: Princeton University Press, 1965), p. 28.

Table 12–8 Italy: Fuel Sources as Percent of Total Energy Production	Oil	Solid Fuel	Natural Gas	Nuclear	Tidal, Hydro, and Geothermal
	79.7	6.6	9.6	.7	3.4

SOURCE: *Basic Statistics of the Community* (Luxembourg: Statistical Office of the European Communities, 1975).

Table 12–9 *Belgium: Selected Economic Characteristics*

Total Area in Thousands of Square Miles (square kilometers)	Percent of Total Area in Farms	Percent of Farmland Cultivated and in Temporary Pasture	Total Labor Force (in millions)	Percent of Labor Force		
				PRIMARY ACTIVITIES	SECONDARY ACTIVITIES	TERTIARY AND UNKNOWN
12 (31)	51	53	4	3.7	41.2	55.1

SOURCE: *Basic Statistics of the Community* (Luxembourg: Statistical Office of the European Communities, 1975).

country's total gross national product is approximately $108 billion giving a per capita GNP of about $1,960. Italy's economic growth rate was outstanding during the period 1954–1963, when it averaged in excess of 6 percent. This rate was exceeded only by Japan and West Germany among the industrial nations of the non-Communist world. High levels of investment, particularly in industrial equipment and construction, plus low labor costs, sparked this high-growth rate. In 1971 serious problems in the form of a severe economic recession beset the economy. Even with the relative stagnation of the early 1970s, Italy ranks as the world's seventh most important industrial power.

As Table 12–3 indicates, Italy is even more dependent on foreign energy sources than France. What is even more disturbing for the Italians is the overwhelmingly important role of oil in meeting their energy requirements (Table 12–8). Solving the chronic regional disparities that exist between the north and south of Italy will not be easy as Italy, along with the rest of the Common Market, moves forward into an era of sharply rising energy costs.

Belgium

The kingdom of Belgium is geographically and culturally at the crossroads of Western Europe (Table 12–9). During the past 2,000 years there has been an almost constant flow of different cultures through this narrow segment of the Great European Plain. Today the Belgians are sharply divided along ethnolinguistic lines with French-speaking Walloons occupying the southern half of the country and the Flemish, speaking a Dutch dialect, the north. The Flemish-Walloon tensions flared into violence in the 1960s and seemed to threaten the continued existence of a unified Belgium. Street fighting was severe in the capital, Brussels, which has a French-speaking majority but is located in the Flemish portion of the country. In effect, Belgium is a state formed of two distinct national groups. Many observers feel that its continued existence will rely on how successfully the central government can manage to allow each group to achieve the almost complete autonomy they seem to demand.

Belgium achieved its independence in 1830, when a constitutional monarchy was established following a popular uprising. Despite a policy of scrupulous neutrality prior to the two world wars, Belgium was attacked and occupied by German armies in both 1914 and 1940. These traumatic events, plus a reaction to bellicose postwar Soviet behavior in Europe, have caused Belgium to become one of the most vigorous advocates of European integration and collective security. Belgium is a member of NATO, the European Common Market, the Organization for Economic Cooperation and Development (OECD), and the Benelux Customs Union. Perhaps the cultural schism that threatens Belgian internal unity makes the people even more conscious than their neighbors of the great need for unity on the larger European scale.

Belgium emerged from World War II with an industrial base much less damaged than its neighbors. The immediate postwar era saw rapid reconstruction, trade liberalization, and high economic growth rates. The pace began to slacken, however, and it was the establishment of the Common Market in 1958 that brought about a new surge in the Belgian economy. Confidence in the opportunities provided to Belgian

Table 12–10 *Belgium: Fuel Sources as Percent of Total Energy Production*

Oil	Solid Fuel	Natural Gas	Nuclear	Tidal, Hydro, and Geothermal
63.7	25.0	11.3	0	0

SOURCE: *Basic Statistics of the Community* (Luxembourg: Statistical Office of the European Communities, 1975).

producers in the huge new market led to bold new investments in manufacturing plants and equipment. The growth rate rose through the late 1960s to reach 6 percent annually in 1970. The gross national product is approximately $32 billion, which yields a per capita GNP of about $3,210.

Agriculture plays a relatively minor role in the Belgian economy. In recent years agriculture has been responsible for slightly less than 4 percent of the GNP, employing roughly the same percentage of the labor force. Livestock and poultry raising are the dominant agricultural activities, while the traditional crops, sugar beets, potatoes, wheat, and barley still form an important part of the rural scene.

At the hub of a major West European crossroads with a dense concentration of industry and population, Belgium has played an impressive role in international trade. Its industrial development is essentially based on coal, metal fabrication, textiles, chemicals, and iron and steel production. The country's industries contribute $16 billion in exports and comprise about 50 percent of the total GNP. Belgium does not, however, possess any significant stores of natural resources and must import most raw materials, as well as fuel, machinery, transport equipment, and about one-fourth of its food requirements. The country's development and prosperity are, as a result, largely the product of a highly skilled labor force

and managerial expertise. The cornerstone of Belgian industry remains the iron and steel and metal fabricating sectors, which supply approximately 40 percent of the exports by value.

As can be seen on Table 12–3, Belgium's reliance on imported energy sources increased dramatically during the decade of the 1960s. In 1971 the Belgians were even more dependent on foreign energy than were the Italians.

As can also be seen from the information presented in Tables 12–3 and 12–10, Belgium's future development, like that of the other petroleum-deficient Western European countries, is clearly dependent on ensuring a reliable future supply of energy at reasonable cost, since the skilled labor and industrial plant are practically valueless without it.

The Netherlands

With approximately 860 people per square mile (332 per square kilometer), the Netherlands ranks as Western Europe's most densely populated country (Table 12–11). It may come as a surprise to many to learn that the Netherlands is even more densely populated than Japan. Less surprising is the fact that the Dutch people have been energetically adding to their low-lying territory through land reclamation and drainage schemes for centuries. Dike building

Table 12–11 *The Netherlands: Selected Economic Characteristics*

Total Area in Thousands of Square Miles (square kilometers)	Percent of Total Area in Farms	Percent of Farmland Cultivated and in Temporary Pasture	Total Labor Force (in millions)	Percent of Labor Force		
				PRIMARY ACTIVITIES	SECONDARY ACTIVITIES	TERTIARY AND UNKNOWN
16 (41)	57	40	5	6.6	35.5	57.9

SOURCE: *Basic Statistics of the Community* (Luxembourg: Statistical Office of the European Communities, 1975).

appears to have begun about A.D. 1000 in the extensive tidal marshes of the southern Netherlands. The dikes protected lands that lay above the low-tide level but were subject to periodic flooding during high-tidal conditions. Sluice gates allowed surplus water to drain out of ditches at low tide but were closed to keep out the incoming tidal flow. The protected lands that resulted were known as "polders." It was not until the perfection of the windmill for pumping and later steam-driven pumps that the Dutch were able to begin creating agriculturally productive polders in areas that lay below sea level. Later, freshwater lakes were drained by pumping to provide additional land. In more recent years the Zuider Zee, once a prominent feature on European maps, has been diked and drained to create a huge new land area and the freshwater Lake Ijssel.

In 1953 the Dutch suffered a severe disaster when storm winds of up to 100 miles (161 kilometers) per hour drove the waters of the Rhine-Mass-Schelde across farmlands and towns killing an estimated 1,800 persons and causing another 100,000 to be evacuated (Fig. 12–3). So great was the catastrophe that the flood damage was conservatively estimated at over one-half billion dollars. To prevent a recurrence of such a flood in the future, the Dutch have undertaken the immense "Delta Project," which consists of four huge sea dikes to close most of the mouths of the Rhine-Mass-Schelde distributary system. The New Waterway will remain open and become the main channel for Rhine River shipping. This ambitious project is nearly completed.

It is not without good cause that many observers of the Netherlands speak of the "Dutch Miracle."

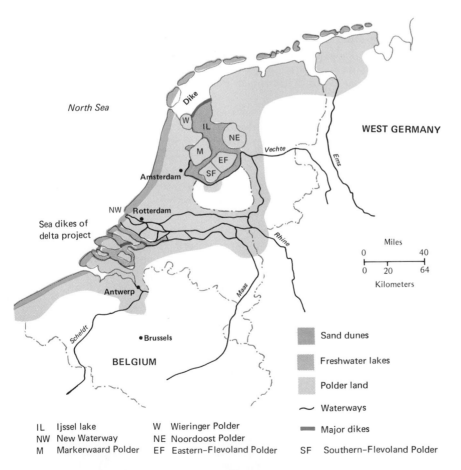

12–3 *The Netherlands: Land Reclamation*

IL	Ijssel lake	W	Wieringer Polder
NW	New Waterway	NE	Noordoost Polder
M	Markerwaard Polder	EF	Eastern-Flevoland Polder
		SF	Southern-Flevoland Polder

They are properly impressed by the fact that the Dutch have literally created almost 40 percent of their country by diking and draining polders for over a thousand years. Here more than anywhere else in Europe is it appropriate to speak of a "human landscape." Productive agricultural enterprises flourish on much of the reclaimed land, and in many areas new towns and industrial areas have been developed.

Like neighboring Belgium, the Netherlands was neutral during both world wars. In May 1940, however, German troops overran the country, and Queen Wilhelmina fled to London where a government-in-exile was established. Five years later the German army in the Netherlands capitulated, and the occupation ended. In 1949, Indonesia gained its independence from the Netherlands. At the present time the Netherland Antilles, Aruba, Curacao, Bonaire, Saba, Saint Eustatius, and a portion of Saint Maarten are considered integral parts of the Netherlands but are becoming increasingly autonomous. Surinam, on South America's north coast, is now semi-independent. Many tourists and oil refining facilities provide an important source of income for the Antilles close to the Venezuelan coast.

The Dutch economy is characterized by private enterprise, but the central government plays a strong role. A limited base of natural resources, primarily gas, oil, and coal, plus the strategic location have been developed into an economy unusually dependent on foreign trade. Industry is modernized and characteristically competitive, backed by a diligent and highly skilled labor force and a vigorous and adaptable business community.

Although almost completely modernized and characterized by very high crop yields, the role of agriculture in the Dutch economy is small. As an industry, agriculture relies heavily on imports, particularly in the area of livestock feed. The livestock industry accounts for about two-thirds of the total farm output.

The Netherlands ranks as the Common Market's most efficient dairy and livestock producer. Expansion in this sector has taken place as a result of the Market's Common Agricultural Policy (CAP) which favors efficient producers. About 6.5 percent of the Dutch GNP is derived from agriculture.

In 1950, the Netherlands' industrial production totaled slightly more than $2 billion and contributed about 40 percent of the GNP. By 1971, it had risen to $17 billion and contributed 49 percent of the GNP. Today, well over one-half of the GNP is derived from manufacturing. Part of this rapid industrial development was made possible by the discovery in the mid-1950s of the largest proven natural gas deposit in Western Europe at Groningen in the northern part of the Netherlands. A modern hydrocarbon-based chemical industry has developed to utilize this important new natural resource and has contributed substantially to the economic development of the Netherlands in recent years. It is significant to note that Dutch dependence on foreign energy sources declined sharply during the period 1960–1974 (Table 12–3), whereas all of her Common Market partners were becoming increasingly dependent. As Table 12–12 shows, natural gas plays a very significant role in meeting the energy needs of the Netherlands at the present time.

As a charter member of the Benelux Economic Union and the Common Market and boasting the world's largest port, the Netherlands has emerged as one of the most important entrepôt countries of Europe if not of the globe. Rotterdam became the world's most important port in 1962 when it surpassed New York in terms of tonnage handled. It serves as a major transit port for the transshipment of raw materials entering the Rhine corridor and of finished products and coking coal entering world trade from central Western Europe. An impressive industrial complex is growing in and around Rotterdam, including the refining of imported petroleum.

Table 12–12 The Netherlands: Fuel Sources as Percent of Total Energy Production

Oil	Solid Fuel	Natural Gas	Nuclear	Tidal, Hydro, and Geothermal
60.4	5.6	33.8	0.2	0

SOURCE: *Basic Statistics of the Community* (Luxembourg: Statistical Office of the European Communities, 1975).

Table 12–13   *Luxembourg: Selected Economic Characteristics*

Total Area in Thousands of Square Miles (square kilometers)	Percent of Total Area in Farms	Percent of Farmland Cultivated and in Temporary Pasture	Total Labor Force (in millions)	Percent of Labor Force		
				PRIMARY ACTIVITIES	SECONDARY ACTIVITIES	TERTIARY AND UNKNOWN
1 (3)	51	47	.2	6.6	49.0	44.4

SOURCE: *Basic Statistics of the Community* (Luxembourg: Statistical Office of the European Communities, 1975).

Luxembourg

Closely linked with Belgium in an economic union for over a half century, tiny Luxembourg bears many similarities to its larger neighbor (Table 12–13). Its neutrality also violated by Germany in the two world wars, Luxembourg is a staunch charter member of NATO. The government is a constitutional monarchy with executive power in the hands of the heriditary Grand Duke and cabinet. Internally, Luxembourg is divided into 126 communes, each administered by an elected council in a system closely patterned on that of Belgium.

A high level of industrialization provides the inhabitants of this Rhode-Island-sized country with one of the highest per capita GNPs in the Common Market. The overall economic growth rate is approximately 3.5 percent.

Despite large-scale attempts to diversify, Luxembourg's industrial scene is dominated by iron and steel. These basic enterprises account for about one-half the country's total industrial production. Steel output is about 5.4 million tons or about 1 percent of the total world production, a real feat for a country of only 350,000 people. Needless to say, its per capita production of over 16 tons is the highest of any country in the world. Almost one-half of the iron ore smelted is produced from domestic mines, while practically all of the remainder comes from nearby French workings. All coal consumed is imported, primarily from West Germany. As Table 12–3 indicates, Luxembourg relies on imports for nearly all its energy supply.

A growing number of light industries have been encouraged in an effort to diversify the country's industrial base. Combined, these newer industries contribute only about one-half as much to the GNP as do the traditional giants, iron and steel. Rubber, chemicals, and fertilizer are among the most important of the diverse industries. As Table 12–13 shows, almost 50 percent of Luxembourg's total work force is employed in industrial activities.

Agriculture absorbs another 10 percent of the labor force, mostly small-scale livestock raising and mixed farming. The vineyards of the Moselle Valley provide the raw material for a flourishing wine industry. Lusembourg's excellent dry white wines are exported widely. Many of the more than 40 percent of workers who are shown in Table 12–13 as being in tertiary activities are engaged in Luxembourg's flourishing financial activities. There are over sixty large banks in this small country, making it one of the world's important financial centers. Trading in multinational corporate securities is also an important feature of Luxembourg's active commercial and financial sector. Favorable tax treatment by the government has encouraged many large holding companies to establish their headquarters here.

Denmark

Long an agriculturally dominated economy, Denmark shifted its emphasis to industry about twenty-five years ago (Table 12–14). As a result, industrial exports outstripped agricultural sales for the first time in 1964. In recent years industrial products have accounted for about 65 percent of its total export trade. Over 37 percent of Denmark's workers are engaged in industrial pursuits. Manufacturing contributes about 39 percent of the total GNP.

As Table 12–3 suggests, Denmark is very poorly endowed with fuel resources. Some low-grade iron

Table 12–14   *Denmark: Selected Economic Characteristics*

Total Area in Thousands of Square Miles (square kilometers)	Percent of Total Area in Farms	Percent of Farmland Cultivated and in Temporary Pasture	Total Labor Force (in millions)	Percent of Labor Force		
				PRIMARY ACTIVITIES	SECONDARY ACTIVITIES	TERTIARY AND UNKNOWN
17 (43)	69	89	2	9.6	32.3	58.1

SOURCE: *Basic Statistics of the Community* (Luxembourg: Statistical Office of the European Communities, 1975).

ore is found in the south, but on the whole Denmark depends on imports for its principal industrial raw materials. Major import items include industrial raw materials, fuels, machinery, and equipment. Consumer goods account for less than one-fifth of the import total.

One of the reasons for the successful development of the Danish economy in the postwar period has been its adaptability, an adaptability that has resulted in a wide range of specialization. The Danes boast of being able to "one-up" their trading partners by selling such things as fine smoking pipes in England, whiskey in Scotland, and chewing gum in the United States. Beer is sold widely throughout the world from Europe's two largest breweries which are located in the country. Clearly Denmark's one really important domestic "raw material" is its skilled labor. In recognition of this attribute, great importance is attached to education and training, and large sums are invested in scientific research.

The most extensive use of Denmark's 16,619 square miles (43,044 square kilometers) of territory

remains agricultural. By far the largest portion of the agricultural land is cultivated for animal foodstuffs. Ninety percent of farm gross income derives from animal production, chiefly butter, cheese, bacon, beef, veal, poultry, and eggs. The principal export market for these products is divided mainly between neighboring West Germany and the United Kingdom.

Ireland

The Irish Republic occupies about five-sixths of the island of Ireland (Table 12–15). The northern portion of the island is occupied by the six counties making up Northern Ireland, or Ulster, which is an integral part of the United Kingdom. Unlike predominantly Protestant Northern Ireland, the Irish Republic is about 94 percent Roman Catholic. Recent years have seen Northern Ireland embroiled in a bitter civil dispute that often erupts in the form of bombings and street warfare. The current struggle stemmed from the complaints of Northern Ireland's Catholic minority concerning their civil rights. By mid-1971 the tensions between the Protestant and

Table 12–15   *Ireland: Selected Economic Characteristics*

Total Area in Thousands of Square Miles (square kilometers)	Percent of Total Area in Farms	Percent of Farmland Cultivated and in Temporary Pasture	Total Labor Force (in millions)	Percent of Labor Force		
				PRIMARY ACTIVITIES	SECONDARY ACTIVITIES	TERTIARY AND UNKNOWN
27 (70)	69	25	1	24.3	31.1	44.6

SOURCE: *Basic Statistics of the Community* (Luxembourg: Statistical Office of the European Communities, 1975).

Catholic communities had reached a crisis condition with increasing violence requiring the British Army to take several drastic steps to restore order. The basic problem still lingers, and bitter civil warfare can break out at any time. Needless to say, the government and people of the Republic of Ireland are vitally concerned about these activities so near their northern border. English is the common language, but Irish Gaelic or Erse, currently encouraged by the government, is spoken in some places. Gaelic is an official language in Ireland along with English. The curious combination is very obvious to the tourist or visitor from abroad these days, since highway directional signs are printed in English and Gaelic.

Ireland is notably lacking in most industrial natural resources. Only some small deposits of zinc, lead, and copper are mined in commercial quantities. What industrial development Ireland has had until very recently was oriented primarily toward the domestic market. Exceptions to this general pattern were those industries specializing in foods and beverages, woolen textiles, automobile assembly, and copper mining. At the present time, efforts are being exerted toward accelerating industrial development. To this end, Ireland's entry into the Common Market is seen as a great boon. One of the additional impediments to industrialization in Ireland has been the absence of a pool of skilled workers.

As Table 12–15 indicates, Ireland has a far greater percentage of its labor force engaged in primary activities (basically agriculture) than any other Common Market country. Not surprisingly, agriculture remains a major factor in national development schemes. Agricultural output has failed to keep pace with industrial growth in recent years. Exports, however, are still largely agricultural in origin. Cattle and dairy products form the most important exports and go mainly to the nearby United Kingdom. Crops of importance include turnips, potatoes, hay, sugar beets, barley, and beets, all of which do well in the island's cool moist climate.

Ireland's monetary and banking system too is closely linked to the larger British system. The Irish pound maintains a one-to-one relationship with its British equivalent. As a rule, British coins and currency are freely accepted in Irish shops. The reverse is usually true only in Northern Ireland and not through the island of Great Britain. Although eco-

nomically very closely linked with the United Kingdom, Ireland scrupulously adheres to a policy of political independence from its large and affluent neighbor. In recent years Irish troops have served with distinction in a number of United Nations peacekeeping operations around the globe. Ireland remains outside NATO, however, and in 1969 the Prime Minister stated that Ireland's international policy was neutrality like that of Sweden, Austria, and Switzerland. This view was reaffirmed when Ireland joined the Common Market in 1973.

The United Kingdom[3]

There can be little doubt that Britain has the oldest industrialized economy in the world (Table 12–16). In the eighteenth century, island-based Britain was able to gain mastery of the seas and establish an empire scattered on all the continents. The nineteenth century saw this empire enlarged and reinforced to make the British supreme as a world power. Changing world conditions punctuated by two costly world wars severely weakened Britain's position in the twentieth century. The old empire gave way to the Commonwealth (a loose political and economic association between the United Kingdom and many of the former colonies) as political independence was achieved by the colonies in the interwar and post–World II periods.

When the Common Market was formed, there was considerable concern on the part of several Western European countries that did not join. It was feared that a division would develop between the countries that were in the Common Market and those that remained outside. Britain, in particular, was faced with special problems because of its long-standing connections with the Commonwealth. Others were hesitant to join and possibly compromise their status as neutrals. The idea of a looser sort of trading relationship became popular with a number of political leaders in the non-Market countries during the late 1950s.

The outcome of a convention of these non-Market states held in Stockholm in 1959 was the European Free Trade Association (EFTA), which came into

---

[3]The terms "United Kingdom" and "Britain" are used synonymously to refer to England, Wales, Scotland, and Northern Ireland, and to The United Kingdom of Great Britain and Northern Ireland.

Table 12–16   *United Kingdom: Selected Economic Characteristics*

Total Area in Thousands of Square Miles (square kilometers)	Percent of Total Area in Farms	Percent of Farmland Cultivated and in Temporary Pasture	Total Labor Force (in millions)	Percent of Labor Force		
				PRIMARY ACTIVITIES	SECONDARY ACTIVITIES	TERTIARY AND UNKNOWN
94 (244)	77	38	25	2.8	42.3	54.9

SOURCE: *Basic Statistics of the Community* (Luxembourg: Statistical Office of the European Communities, 1975).

being on May 3, 1960. In the words of the convention, EFTA's objectives were:

1. To promote in the area of the association and in each member state a sustained expansion of economic activity, full employment, increased productivity and the rational use of resources, financial stability and continuous improvement in living standards.
2. To secure that trade between member states takes place in conditions of fair competition.
3. To avoid significant disparity between member states in the conditions of supply of raw materials produced within the area of the association.
4. To contribute to the harmonious development and expansion of world trade and to the progressive removal of barriers to it.

Seven countries—Austria, Denmark, Norway, Portugal, Sweden, Switzerland, and the United Kingdom—formed Western Europe's EFTA bloc or "Outer Seven" during the decade of the 1960s. Britain, with more than one-half of EFTA's total population and nearly 60 percent of its total production, clearly dominated the organization. The EFTA countries enjoyed substantial economic growth, although at a slower rate than that which was being enjoyed by the six Common Market countries.

By the mid-1960s, however, it was becoming clear that Britain's brightest economic prospects lay in the direction of the Common Market rather than the Commonwealth and EFTA. The information presented in Table 12–17 helps explain how this opinion gained wide support and backing. As can be seen, Commonwealth trade was declining while trade with Western Europe was increasing. The importance of Western Europe in Britain's future economic outlook became obvious. It was emphasized by the realization that the rate of increase in trade with the Common Market was leveling off. Once the organization was crystallized, a formidable barrier to British imports would result. For instance, German cars sold in France, and French cars sold in Germany, would not be subject to any import duty. British cars sold to Common Market countries, on the other hand, would be subject to the common external tariff of the community. Also, it was obvious that Com-

Table 12–17   *Destination of United Kingdom Exports (percent)*

Area	1958	1960	1965	1966
Commonwealth	38	37	28	26
EEC	14	15	19	19
EFTA	11	12	14	16
Other W. Europe	2	2	4	4
W. Europe Total	27	29	37	38
U.S.A.	9	10	10	13

SOURCE: *Basic Statistics of the Community* (Luxembourg: Statistical Office of the European Communities, 1975).

12-4 *United Kingdom: North Sea Oil and Gas Areas*

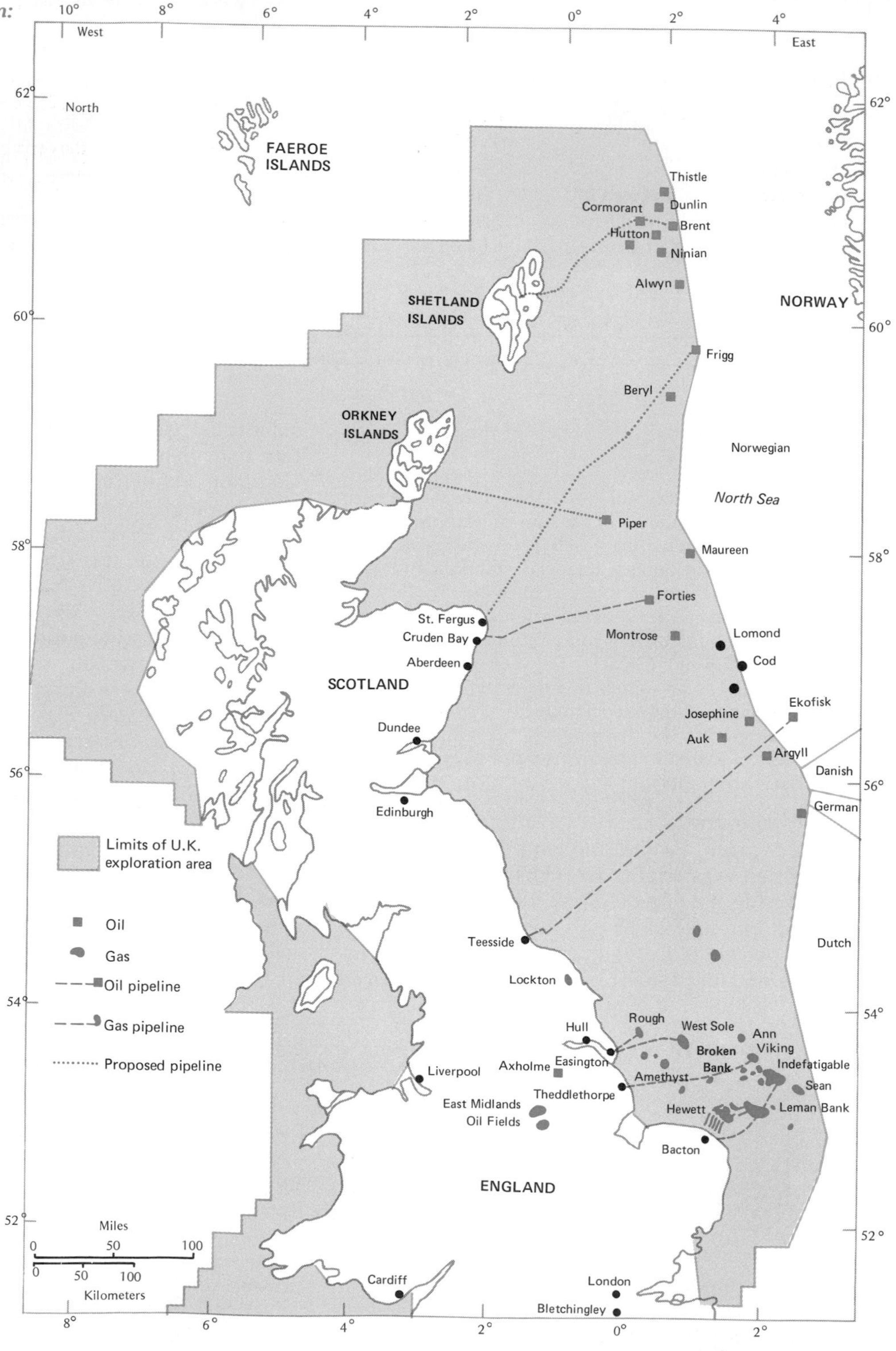

North

FAEROE ISLANDS

SHETLAND ISLANDS

ORKNEY ISLANDS

NORWAY

Norwegian

North Sea

Thistle
Cormorant
Dunlin
Hutton
Brent
Ninian
Alwyn

Frigg

Beryl

Piper

Maureen

St. Fergus
Cruden Bay
Aberdeen

SCOTLAND

Montrose
Forties
Lomond
Cod

Josephine
Auk
Ekofisk
Argyll

Dundee

Danish

Edinburgh

German

Limits of U.K. exploration area

Oil

Gas

Oil pipeline

Gas pipeline

Proposed pipeline

Teesside

Lockton

Dutch

Hull
Rough
West Sole
Ann
Viking
Broken Bank
Indefatigable
Amethyst
Sean
Axholme
Easington
Theddlethorpe
Hewett
Leman Bank
East Midlands Oil Fields
Bacton

Liverpool

ENGLAND

Miles
0   50   100
0   50   100
Kilometers

Cardiff

London

Bletchingley

monwealth countries were intent on rapidly developing their own industrial sectors and could no longer be considered as ready markets for British goods. Indeed, in certain areas such as textiles, Commonwealth competition was creating serious problems for British producers in the world market. In 1973, Britain, together with Denmark and Ireland, joined the Common Market.

Leaders in both Britain and the Common Market enthusiastically hailed its entry as a new renaissance. They saw it as a step that could immensely increase the security and stability of the Market and the prosperity and quality of life for the peoples of Western Europe. Only time will reveal how accurate these hopes and aspirations will be. The present nine Common Market nations are certainly a more impressive economic unit than was the original six, thanks largely to Britain's large population and productive capacity.

Some serious thoughts concerning the advisability of membership began to develop during the early 1970s as a period of rapid inflation gripped Britain. To clear the air once and for all, the Labor government held a national referendum in 1975 on the question of Britain's continuation as a member of the Common Market. The vote was resoundingly in favor, and it now seems certain that Britain's future development will be undertaken as a member of this powerful economic bloc.

The most notable feature of Britain's economy is the importance of industry, services, and trade. It ranks fourth in world trade behind the United States, West Germany, and Japan. Britain takes 10.4 percent of the world's exports of primary products and contributes 10.1 percent of the world's exports of manufactured goods which account for about 85 percent of its total exports. In its share of invisible world trade (such as financial services, civil aviation, travel, and overseas investment), Britain ranks second behind the United States.

Coal, the paramount energy source for the early industrialization of Britain, has been declining in importance during the present century, and an alarming reliance on imported petroleum has developed (Table 12–3). Presently, however, hopes are running high that Britain will be able to meet its petroleum needs with domestic production in the 1980s. These hopes are based on the huge deposits of petroleum and natural gas that were first discovered during the mid-1960s under the bed of the North Sea (Fig. 12–4). In 1972, natural gas from the North Sea accounted for 88 percent of all gas available in Britain. More than one-half of all gas sold in Britain is for industrial and commercial uses with the remainder going to domestic consumers. In 1973, the North Sea wells produced an average of 3,000 million cubic feet (84.9 cubic meters) a day.

Between the start of exploration in 1964 and April 1974, 592 exploration or appraisal wells and 198 production wells were drilled in the North Sea. Some twenty-five huge and expensive deep-sea drilling rigs operated during 1973, and the number climbed to forty in 1974. The expense of these rigs is staggering. In the British Petroleum Company's Forties Field, a single drilling platform costs $50 million. Newer generation rigs will be considerably

The recent discovery of petroleum and natural gas in the North Sea has led to extensive exploration. The cost of sea drilling and the roughness of the North Sea make drilling a costly enterprise. (A.I.D.)

more expensive. Weather conditions in the North Sea leave much to be desired. Extreme wave heights to 95 feet (29 meters) are encountered as are wind velocities of 80 mph (129 kilometers) with gusts reaching 120 mph (193 kilometers) in particularly stormy weather. Rig "down time," a considerable expense factor, is consequently far greater in the North Sea than in most other offshore oil-producing areas of the world.

Still, reliable sources of energy are absolutely imperative if Britain and other Western European countries are to maintain and improve their standards of living in the future. The stakes, astronomic costs of exploration and production, must be invested now as a guarantee that future economic development will be possible in Western Europe. The recent "energy crunch" brought about by the Arab petroleum-producing nations' boycott in 1973–1974 made this fact painfully obvious. In the case of Britain, the future outlook for energy production seems bright. In 1975, the first offshore petroleum from the Argyll field came on stream and moved Britain into the ranks of the producing nations. All indications now point to British petroleum production reaching nearly 5 million barrels (7.9 trillion liters) a day in 1984. This level of production will put Britain in the "Big League" of oil producers, outstripping Venezuela, Indonesia, Nigeria, and every Middle East country except Saudi Arabia and Iran. Although it is hazardous to speculate too far into the future, it does seem that an energy renaissance is in store for the world's senior developed country. It will be interesting to see how this renaissance is translated to the day-to-day lives of the people of Britain and Western Europe and the world at large.

Further Readings

ALLEN, KEVIN, and MACLENNAN, M. C., *Regional Problems and Policies in Italy and France* (London: Allen & Unwin, 1970), 352 pp. A well-researched volume devoted to an examination of regional development problems in two key Common Market countries.

BARZANTI, SERGIO, *The Underdeveloped Areas within the Common Market* (Princeton, N.J.: Princeton University Press, 1965), 437 pp. An excellent study of regional underdevelopment in one of the world's most developed areas.

BOAL, F. W., "Social Space in the Belfast Urban Area," in JONES, EMRYS (ed.), *Readings in Social Geography* (London: Oxford University Press, 1975), pp. 149–167. This paper is a fascinating analysis of Belfast, Northern Ireland, in terms of the interaction taking place between the residents of its socioeconomic regions. Roman Catholic-Protestant contrasts are graphically delineated.

British Membership of the European Community (London: Her Majesty's Stationary Office, 1973), 87 pp. An excellent review of the background and significance of Britain's entry into the European Community. A valuable chronology of events from 1945 to 1972 is included.

BUSTEED, M. A., *Northern Ireland* (London: Oxford University Press, 1974), 48 pp. A concise geographically oriented study of one of Western Europe's most troubled areas. An excellent background for understanding the complex social and economic problems found here.

CAROL, HANS, "Stages of Technology and Their Impact upon the Physical Environment: A Basic Problem in Cultural Geography," *Canadian Geographer,* 8 (1964), 1–7. An interesting view in which the author establishes five stages of technology. It represents a stimulating geographic approach to the nature of economic development.

CLOUT, HUGH D., *The Geography of Post-War France: A Social and Economic Approach* (Oxford: Pergamon Press, 1972), 165 pp. An excellent systematic and regional synthesis on Western Europe's pivotal power in the current period.

CLOUT, HUGH D., *The Massif Central* (London: Oxford University Press, 1973), 48 pp. A well-illustrated and highly readable study of one of France's most underdeveloped regions.

Information and Documentation Centre for the Geography of the Netherlands, *A Compact Geography of the Netherlands* (The Hague/Utrecht: Ministry of Foreign Affairs, 1974), 40 pp. An outstanding effort, the best publication of this kind produced by any national government.

KOHSTAMM, M., and HAGER, W., *A Nation Writ Large?* (London: Macmillan, 1973), 275 pp. An excellent review of the major foreign policy problems facing the movement toward Western European unity.

POUNDS, NORMAN J. G., *The Economic Pattern of Modern Germany* (London: John Murray, 1963), 133 pp. A short, readable study of Western Europe's most powerful economic force. A bit dated in certain aspects but still worth reading.

ROSTOW, W. W., *The Stages of Economic Growth* (Cambridge: Cambridge University Press, 1971), 200 pp. ROSTOW presents an important counterview to many Marxist arguments concerning economic growth. His book contains much valuable historical statistical material on Western European development.

Eastern Europe and the Soviet Union

Roger L. Thiede

PART FOUR

Eastern Europe: The Land Between

13

EASTERN EUROPE lies between the Soviet Union and Western Europe (Fig. 13–1). Eight countries make up the region: East Germany, Poland, Czechoslovakia, Hungary, Yugoslavia, Albania, Romania, and Bulgaria. These countries have a total population of 132 million people living in an area approximately 14 percent the size of the United States. The individual countries of Eastern Europe vary in population and area. Albania is the smallest, both in population (2.5 million) and area (11,100 square miles or 28,749 square kilometers). Poland is the largest with 35 million people and an area of 120,725 square miles (312,678 square kilometers).

Political Identity

The Eastern Europe that emerged from the destruction of World War II was a region whose course of political, economic, and social development was abruptly altered. Soviet-trained Communists stepped into positions of leadership. Leftist groups, often with support from the Soviet Army, established "People's Republics" or "People's Democracies." Soviet-style institutions were imposed in all eight Eastern European countries within three years after the end of the war. The Communist party became the sole political party. Land, industries, banks, and all but the smallest commercial establishments were nationalized. The established churches came under close governmental supervision and control as the Marxist-Leninist-Stalinist philosophy became the basic code of life.

A primary objective of these new governments was economic reconstruction and development. Emphasis was placed on secondary activities. The ideological goal was to create a Communist society. Theoretically this new utopia promised no exploitation and no government. Before communism could be achieved, however, the Soviet Marxist dogma specified a period of "socialism." During this stage of development, remnants of the former society would be eliminated; the state would own all the means of production, and each person would be required to "give according to his ability," but only "take according to his work." The dictatorship of the Communist parties of Eastern Europe was to direct these new command (i.e., centrally controlled and planned) economies under Moscow's guidance.

Eastern Europe is a distinct unit on the basis of the Soviet-inspired Communist governments initially established in Yugoslavia, Albania, Romania, Bulgaria, Hungary, Czechoslovakia, Poland, and the German Democratic Republic (East Germany). Not since the early nineteenth century when Russian troops entered Paris had the Russians been so prominent in European affairs, but this time the successor to the Russian Empire, the Union of Soviet Socialist Republics, was in a position for more effective and enduring control. The tsars of old never enjoyed such an influence over so large a piece of Europe.

It was not long, however, before cracks appeared in this Soviet-dominated region. The Communist nationalist leader of Yugoslavia, Marshall Tito, broke away from the Stalinist Soviet Union in 1948 and embarked on his own variant of socialism. Tight control was held on the remainder of Eastern Europe until Stalin died in 1953. After Stalin's death, the process of modification and revision of many of his

13–1 *Eastern Europe: National Political Units and Cities*

practices in the Soviet Union led to Albania's withdrawal from the Soviet Bloc. Condemning Soviet Union "revisionism," Albania became the chief European ally of China, the USSR's principal antagonist within the Communist world.

During the post-Stalinist years, even the six countries that remained as satellites were able to exert more distinctive and independent ways. The Soviet Union recognized that there were "different paths" to the building of communism. The acceptance of this doctrine was in part the realization that the Soviet way was not necessarily the way for the Eastern European countries. Signs of unrest, however, had become evident in some of these countries in the 1950s. Hungary's 1956 bloody uprising that was put down by the Soviet Army was the most dramatic demonstration of the underlying tensions in the region.

Eventually countries like Hungary and Czechoslovakia were able to introduce economic reforms that the Soviet Union considered too "Western" for its own tastes. Poland has ceased entirely the collectivization of agriculture. The Catholic Church is able to speak to many millions of Poles with authority. Romania has exerted an independence that would have seemed impossible in the early postwar years. It has refused to participate in maneuvers of the Warsaw Pact, the military alliance of the Soviet Union with its six satellite Eastern European countries. Romania has balked at economic plans put forth by the Council for Mutual Economic Assistance (CMEA). CMEA is the organization designed to promote increased economic cooperation and even integration of the Soviet, Polish, East German, Bulgarian, Czechoslovakian, Hungarian, Romanian, and the Asiatic Mongolian economies.

Moscow, however, still considers Eastern Europe vital to its interests. Soviet troops are stationed in East Germany, Poland, Czechoslovakia, and Hungary. The Soviet Union is willing to accept some expressions of independence from its fraternal People's Democracies. Kremlin leaders, however, are unwilling to sit idly when they consider a challenge has been made to their control of its satellites. Proof of this attitude is seen in the Soviet invasion of Czechoslovakia in 1968.

Soviet domination often obscures the diversity that is contained within Eastern Europe. The considerable cultural, historic, economic, and environmental varieties are explored in this chapter.

Topography

The relief map of Eastern Europe (Fig. 13-2) shows the major contrast between the north and the central and southern sections. The low-lying Great European Plain increasingly widens as it crosses the north. Most of East Germany and Poland are included in this plain drained by three major rivers, the Elbe, Oder, and Vistula. Central Eastern Europe is a region of hills and low mountains separated by fertile plains. Most notable of these relief features are the low, nearly circular, mountain ranges of the Bohemian Massif, whose crests form the boundary of eastern Czechoslovakia. Enclosed within the Massif is the Bohemian Plain wherein lies Prague, the capital of Czechoslovakia.

Farther south is the higher mountain system of the Alps. The Carpathian Mountains are an eastward extension of this system. These mountains form part of the boundary between Poland and Czechoslovakia and continue in a southerly orientation into central

Eastern Europe's major river, the Danube, flows through Hungary's capital and principal city, Budapest. The river is a major transport artery. Budapest is actually two cities, Buda on the left and Pest on the right. In the foreground is the Liberation Monument that commemorates the end of Nazi occupation in 1945. (Eastfoto)

*13–2 Eastern Europe: Land Surface
Regions*

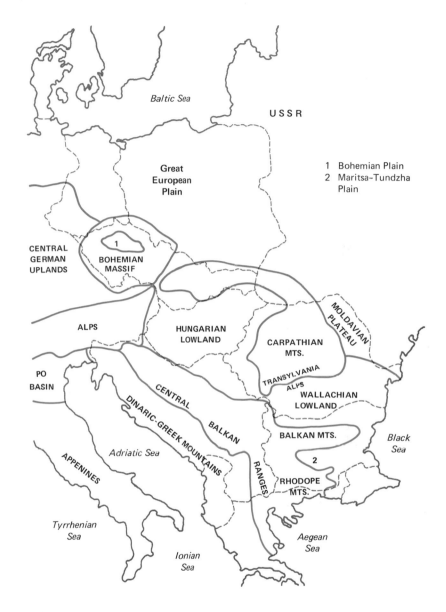

Romania. The Transylvanian Alps in Romania and
the Balkan Mountains in Bulgaria are continuations
of the Carpathians. Between the Transylvanian Alps
and the Balkan Mountains lies the Wallachian Lowland of Romania and Bulgaria. To the west and south
of the Balkan Mountains are the Rhodope Mountains
in southern Bulgaria and the Dinaric-Greek Mountains and the central Balkan Ranges of Yugoslavia
and Albania.

The large Hungarian Lowland is almost completely
enclosed by these mountain systems. This lowland
area occupies most of Hungary and neighboring
parts of Yugoslavia, Romania, and Czechoslovakia.
The Hungarian Lowland is drained by Eastern
Europe's major river, the Danube. From its origin in
southern West Germany, the Danube flows through
Austria and across the break in the Alps and Carpathian Mountains into Hungary. Along the Yugoslav-

Romanian border, the Danube has cut the Iron Gate, a deep gorge through the Balkan mountains, and passes through the Wallachian Lowland to the Black Sea.

It is essential to recall that the topography of an area is not an end in itself but a factor that may aid in understanding more significant features of human geography. For example, not only does the complexity of the mountain ranges of the Balkan Peninsula contribute to the diversity of ethnic groups found there but also complicates the construction of a well-integrated transportation network over much of Yugoslavia, Albania, and Bulgaria, whereas in the north the easily traversed Great European Plain more readily exposes the Poles to conquerors from both east and west. Furthermore, relief accentuates the difficulties of creating intraregional transport linkages within Eastern Europe as a whole.

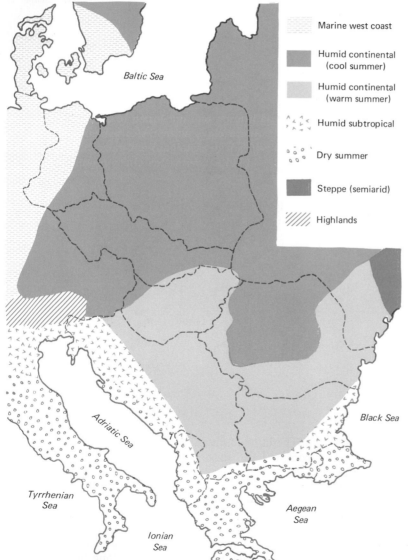

13–3 *Eastern Europe: Climatic Regions*

Legend:
- Marine west coast
- Humid continental (cool summer)
- Humid continental (warm summer)
- Humid subtropical
- Dry summer
- Steppe (semiarid)
- Highlands

Baltic Sea

Adriatic Sea

Tyrrhenian Sea

Ionian Sea

Aegean Sea

Black Sea

Climate and Soils

Climatically, Eastern Europe may be viewed as a bridge connecting the marine climates of northwestern Europe and the severe continental climates to the east (Fig. 13–3). The continental influence is felt most strongly in eastern Poland, which has the coldest winters as well as the longest periods of snow cover (excepting high-mountain altitudes). Western East Germany lies within the marine west coast climate zone. Here summers are cool, and winters are moderate. Precipitation throughout northern Eastern Europe is adequate. Annual range of rainfall is generally between 20 and 30 inches (51 and 76 centimeters) a year, with higher amounts received on the mountain slopes.

The original natural vegetation of mixed forests has given way to agriculture in northern Eastern Europe. The podzolic soils of the north blend into the gray-brown podzolic soils toward the south (Czechoslovakia, southern East Germany, and southern Poland). The fertility of these soils, particularly the podzolic, is not naturally high. They require fertilization to maintain their productivity.

In central Eastern Europe is the warm summer humid continental climate. The length of the growing season is longer than to the north. Like northern Eastern Europe, precipitation is generally adequate, falling year round with a summer maximum. The smallest amounts of rainfall received in Eastern Europe are found on the east side of the Carpathians in eastern Romania.

Together with the original deciduous forest vegetation that covered much of the warm summer humid continental climate, central Eastern Europe includes two large natural grassland areas: the Hungarian and Wallachian plains. The predominant soil types in these two areas are the fertile chernozemic soils. Needless to say, these two areas, especially the Hungarian Lowland, are among the most agriculturally productive areas of Europe.

Climate, soils, and vegetation patterns are more complex in southern Eastern Europe. On the western and southeastern margins of the Balkan Peninsula are mild winter climates. In the interior of western Yugoslavia the climate is humid subtropical with winter temperatures above freezing, and with plentiful rainfall. Actually, in the Dinaric Alps, precipitation reaches its maximum in Eastern Europe with over 60 inches (152 centimeters) a year. Along the western coast of Yugoslavia, over most of Albania, and in southern Bulgaria, a dry summer subtropical climate exists with moderate moist winters and hot dry summers. Soils in the mountainous areas are generally thin and unproductive, but in the river valleys there are fertile alluvial soils. The natural vegetation is principally coniferous and deciduous forests in the humid subtropical zone with xerophytic plants in the dry summer subtropical areas.

Cultural Diversity

Territorial Instability

The expression "Shatter Belt" frequently is applied to Eastern Europe. This term describes the instability of the region politically: its inability to resist the greater military and political power of its past and present neighbors, particularly Germany, Austria, Russia, and the Ottoman Empire. Consequently, throughout much of modern European history, the boundaries of Eastern European countries have frequently changed. Figures 13–4a, b, c, and d indicate some of this instability with the political boundaries of 1815, 1914, 1919, and 1945.

Eastern Europe in 1815 was, with the exception of the tiny kingdom of Montenegro and some small German states in the north, under the domination of four great powers. Prussia and Russia controlled the north, Austria the center, and the Ottoman Empire the south. During the following century the most notable changes occurred in the Balkan Peninsula, where Turkish power rapidly declined. There, small independent national states appeared. In the north the German Empire was formed (1871) consolidating the lands of Prussia and other German states. Russia slightly increased its territory at the expense of Turkey and Austria. In 1869, Austria became part of the dual monarchy of Austria-Hungary upon granting Hungary a coequal role.

The defeat of Germany and Austro-Hungary in World War I signaled a further upheaval in the political boundaries of the region. The Austro-Hungarian Empire was dissolved, creating in its place the three independent states of Austria, Czechos-

13–4a *Eastern Europe: Political Units in 1815*

13–4b *Eastern Europe: Political Units in 1914*

3–4c *Eastern Europe: Political Units in 1919*

13–4d *Eastern Europe: Political Units in 1945*

lovakia, and Hungary. Other territories formerly held by the empire became parts of new or expanded independent countries. A new Poland was re-created from Prussian, Russian, and Austrian lands. Romania gained lands from Hungary west of the Carpathians. The new kingdom of Yugoslavia was formed uniting Serbia with other territories of the south Slavs. In 1919, the expression of national independence in Eastern Europe reached heights hitherto unknown in modern times. The powers that formerly had dictated Eastern Europe's fate were either in abeyance or had been removed from the scene. Russia was in chaos as an aftermath of the 1917 Communist Revolution. Not only did Russia lose territory to Romania and Poland, but in addition its Baltic lands were divided into the independent states of Estonia, Latvia, and Lithuania.

This new experience of independence which swept across Eastern Europe was to last fewer than twenty years. Nazi Germany sought revenge for the defeat of Germany in World War I and attempted to gain control over all Europe. First with the threat of force, then by invasion, Germany began its temporary domination of Eastern Europe. Initially the Soviet Union was an ally of Germany and annexed parts of Romania and Poland as well as the Baltic states. Hitler's uncontrolled appetite for land and power, however, soon ended this uneasy alliance. The Germans invaded the Soviet Union. Germany's inability to withstand the force of the Russians in the east and the United States and Britain in the west led to its near annihilation. The boundaries were again redrawn after World War II.

The Soviet Union emerged as the unquestionable dominant force in Eastern Europe in 1945, and turned the region into a vast buffer zone between it and the West. Moscow oversaw the establishment of regimes in all of the region modeled upon its own system and initially in accordance with its own objectives. The major boundary changes that occurred, and remain to this day, were at the expense of Germany and to the advantage of the Soviet Union. The Polish boundary was shifted westward, and Poland was compensated for land lost to the Soviet Union by the acquisition of German territory. The Oder and Neisse rivers became the new boundary between Polish and German lands. The German territory of East Prussia was divided between the Soviet Union and Poland. The Soviets turned their occupation zone in Germany

into the German Democratic Republic (East Germany). Moscow acquired the very eastern tip of Czechoslovakia, thereby giving it a strategic foothold in the Hungarian Lowland. The Baltic States were reincorporated into the Soviet Union, and that part of Romania called Bessarabia was annexed.

Postwar Population Change

During the immediate post–World War II period there were massive migrations of people within Europe, in part associated with boundary changes. About 2 million Poles migrated out of former Polish territory acquired by the Soviet Union. Considerably fewer Hungarian Slovaks were sent to Czechoslovakia, Czechoslovakian and Romanian Hungarians to Hungary, and Bulgarian Turks to Turkey. Thousands of Italians fled territories gained by Yugoslavia. The largest of all these migrations, however, was expulsion or fleeing of Germans. Over 12 million Germans left Eastern Europe. Most of them were from former German territories acquired by Poland and the Soviet Union. About 3 million Germans, however, left Czechoslovakia, and smaller numbers were removed from Hungary, Romania, and Yugoslavia. Today, small minorities of Germans are found in all Eastern European countries except Bulgaria and Albania.

The most tragic ethnic change in Eastern Europe resulted from the Nazi liquidation of the Jews, principally in Poland and Czechoslovakia. Polish estimates place the number of Polish Jews exterminated at more than 1 million. The Nazis also considered the gypsies an inferior "race" and began their liquidation.

Varieties of Language

Eastern Europe's vulnerability to its stronger neighbors was and is promoted by its variegated cultural character. The absence of cohesiveness among the peoples of Eastern Europe furthered the region's division and subjugation. One measure of the cultural diversity is the multiplicity of languages spoken in the region (Fig. 13–5 and Table 13–1).

All of the peoples of Eastern Europe except the Hungarians and small Turkish minorities speak a language belonging to the Indo-European family of languages. Hungarian belongs to the Uralian language family and Turkish to the Altaic family. Both Hungarian and Turkish languages use the Roman alphabet; however, they are very different from the Indo-European tongues in structure and vocabulary.

13–5 *Eastern Europe: Language Patterns*

Indo-European Family of Languages

Slavic Group

Western Slavic languages
- Polish P
- Czech C
- Slovak S
- Lusatian L

South Slavic languages
- Slovene
- Croatian
- Serbian
- Macedonian
- Bulgarian B

Eastern Slavic languages
- Russian Ru
- Belorussian Be
- Ukrainian U

Germanic Group
- German G
- Danish
- Swedish

Romance Group
- Romanian R
- Italian I
- Vlach V

Illyrian Group
- Albanian A

Hellenic Group
- Greek Gk

Baltic
- Latvian
- Lithuanian
- Gypsy Gy

Uralian Family

Ugric Group
- Hungarian H

Altaic Family

Turkish Group
- Turkish T

Symbol indicates where a significant minority of each language is found.

The majority of the Eastern Europeans speak Slavic languages, which are usually divided into three subgroups: the eastern, western, and southern Slavic tongues. Eastern Slavic languages (Russian, Ukrainian, and White Russian) are spoken in the USSR, whereas the western and southern Slavic languages are found in Eastern Europe. Polish, Czech, and Slovak, which prevail in northern Eastern Europe, are the major Slavic languages of the western sub-group. Southern Slavic languages include Serbo-Croatian, Slovene, Macedonian, Montenegrian, and Bulgarian and are spoken in the Balkans.

Non-Slavic, Indo-European languages of Eastern Europe are German, a Germanic language; Romanian, a Romance language related to Italian and French but with strong Slavic influences; Albanian, probably the oldest of the Indo-European tongues spoken in Europe; and the language of the Gypsies.

Table 13–1 *Eastern Europe: Relationship of Language, Alphabet, and Religion*

Language	Alphabet	Dominant Religious Tradition
Indo-European Family		
Slavic Group		
Polish	Roman	Roman Catholic
Czech	Roman	Roman Catholic
Slovak	Roman	Roman Catholic
Slovene		
(Croatian)	Roman	Roman Catholic
Serbo-Croatian		
(Serbian)	Cyrillic	Eastern Orthodox
Macedonian	Cyrillic	Eastern Orthodox
Bulgarian	Cyrillic	Eastern Orthodox
Russian	Cyrillic	Eastern Orthodox
Ukrainian	Cyrillic	Eastern Orthodox
Belorussian	Cyrillic	Eastern Orthodox
Germanic Group		
German	Roman	Protestant
Romance Group		
Romanian	Roman	Eastern Orthodox
Illyrian Group		
Albanian	Roman	Moslem
Hellenic Group		
Greek	Greek	Eastern Orthodox
Altaic Family		
Turkish Group		
Turkish	Roman	Moslem
Uralian Family		
Ugric Group		
Hungarian	Roman	Roman Catholic

The overwhelming majority of the German-, Polish-, Hungarian-, Romanian-, Bulgarian-, and Albanian-speaking peoples are found in their respective states. Czechoslovakia consists largely of the Czech speakers in the western part of the country and those using Slovak in the east. Yugoslavia is a multilingual state with Serbo-Croatian, Montenegrian, Slovene, and Macedonian all official languages.

With the exception of Czechoslovakia and Yugoslavia, the present Eastern European states are more homogeneous than ever before in respect to their linguistic character. Small minorities of diverse language groups are, however, found in all countries. The Lusatians in East Germany speak a western Slavic language of the same name. Minorities of Polish- and Hungarian-speaking peoples live in Czechoslovakia. Hungarian is also spoken in parts of Romania and Yugoslavia. The Turkish language is used by small groups in Bulgaria and Yugoslavia. Other minority languages include Italian, Albanian,

13-6   Eastern Europe: Dominant Religions

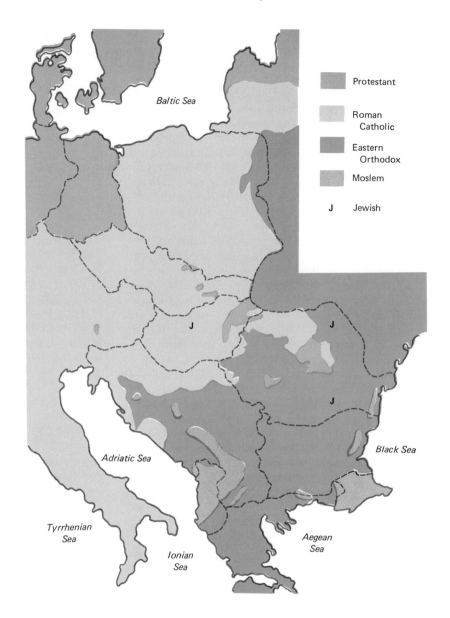

and Romanian in Yugoslavia, the Gypsy language that is much dispersed across the Balkans, and widely scattered pockets of German.

Pattern of Religious Heritage

Greater complexity is added to the cultural diversity map when traditional religious affiliations are considered (Fig. 13–6 and Table 13–1). Although the Marxist governments have imposed severe restric-tions on the church throughout Eastern Europe, religion, particularly Roman Catholicism, continues to exist even if it does not thrive. Significantly the religious tradition has been an important element in the formation of a national character and the antago-nisms that developed among cultural groups.

Not only are the three branches of Christianity (Roman Catholicism, Protestantism, and Eastern Orthodoxy) found in Eastern Europe, but there are also adherents to Islam and small Jewish minorities. Numerically Eastern Europeans are predominantly Roman Catholic. The majority of the Poles, Czechs, Slovaks, Hungarians, and Croats are or were Roman Catholic. Protestant minorities are found in Czechoslovakia, Hungary, Romania, and Yugoslavia. The East German population nominally is overwhelmingly Protestant (Lutheran).

The second largest religious group in the region is the Eastern Orthodox Christians. Orthodoxy is predominant among the Serbs, Macedonians, Bulgars, and Romanians. Whereas the Roman Catholic religion was brought to Eastern Europe by missionaries from Rome, Orthodoxy was disseminated from Constantinople (Istanbul), the seat of eastern Christianity. Missionaries brought the alphabet and consequently a written language for the people they converted. Today, the language of the Serbs, Macedonians, and Bulgars is written in the Cyrillic alphabet as are the eastern Slavic languages. The Romanians, although traditionally Orthodox in faith, use a Roman or Latin alphabet derived from their pre-Christian Latin heritage.

On the other hand, the Roman Catholic cultural areas of Eastern Europe and the Protestant areas (once Roman Catholic) use the Roman alphabet. Consequently the Serbs and the Croats speak the same language (Serbo-Croatian). The Serbs, however, principally of Orthodox heritage, use the Cyrillic alphabet, whereas the Catholic Croats use the Roman alphabet. The animosity between the Serbs and the Croats has at times been among the most intense of Eastern Europe. Furthermore, the possibility exists that a change in the tightly controlled government of Yugoslavia, as well as of those in several other countries of Eastern Europe, could reawaken old cultural enmities.

Population

Density and Distribution

Population density in Eastern Europe is lower than in the more industrial and urbanized Western Europe (Table 13–2). East Germany is the most densely populated country of the region with about 400 people per square mile (154 per square kilometer).

Table 13–2 *Eastern Europe: Population Densities*

	Pop./Sq. Mile	Pop./Sq. Km.
Albania	227	87
Bulgaria	205	79
Czechoslovakia	306	117
East Germany	398	154
Hungary	297	115
Poland	286	110
Romania	235	91
Yugoslavia	220	85

The lowest population density (205 per square mile or 79 per square kilometer) is in Bulgaria.

The distribution of the population throughout the region is highly variable (Fig. 13–7). The largest area of high population density is found in the north. This high-density zone forms a band stretching across southern East Germany, northwestern Czechoslovakia, and the most southern part of Poland. A smaller high-density zone also surrounds Budapest, Hungary. Moderate population densities include much of the remaining Great European Plain, the Wallachian Lowland, the Hungarian Lowland, and adjoining areas in southern Czechoslovakia, in northeastern Yugoslavia, and in central Romania. Lower population densities prevail in the mountain systems covering much of Yugoslavia, Albania, and Bulgaria as well as in the Carpathians.

Population Growth

With the exception of Albania, the countries of Eastern Europe fit into the last stage of the demographic transformation: low birth rates, low death rates, and low rates of natural increase (Appendix). Annual growth rates are in all cases (excepting Albania) near 1 percent or less. East Germany's population is actually declining at a rate of −0.4 percent. Romania has a 1.1 percent annual growth rate; Yugoslavia and Poland, a 1 percent increase. Albania's high birth rate (3.3 percent) and low death rate (0.8 percent) result in an annual rate of population growth of 2.5 percent. This figure is comparable to the high rates of natural increase common in poor nations.

Much of World War II was fought on the territory of Eastern Europe. As a result of this conflict (the

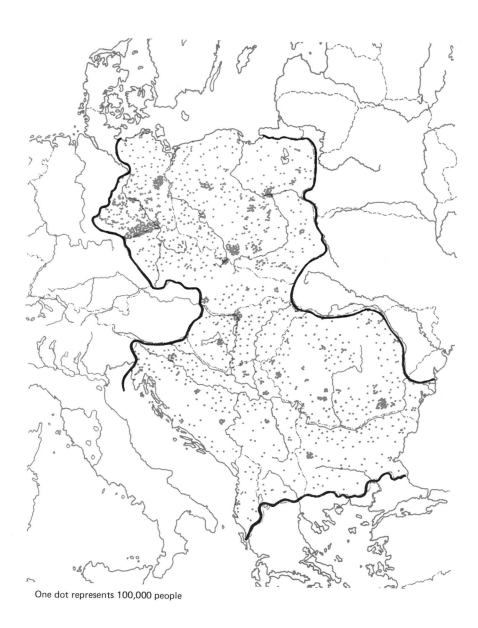

One dot represents 100,000 people

13–7 *Eastern Europe: Population Distribution*

military and civilian losses and the migration of peoples), Eastern Europe's total postwar population was 10 percent less than before the war. The "People's Democracies" that ascended to power in Eastern Europe ascribed to the basic Marxist philosophy of population growth: namely that growth is desirable for the development of the economy, and that population problems only exist because of the unequal distribution of wealth. The population of Eastern Europe grew at generally high rates in the postwar years with the notable exception of East Germany. Prior to the construction of the Berlin Wall, many East Germans emigrated to the West.

Policies such as subsidies to families and the availability of day-care centers affirmed the Marxist pronatal attitude. Other governmental policies, however, helped reduce the birth rates and the growth of population. The role of the woman was changed in the new order. Women were needed in the growing labor force and gained rights to increased education and training. Liberalized birth control laws were enacted which greatly increased the availability of abortions. These factors, along with the family's need for the woman's income and the shortage of adequate housing, helped reduce the birth rate. By the mid-1960s, the governments of East Germany, Czechoslovakia, Bulgaria, and Romania feared an inadequate future labor supply. They adopted policies to encourage an increase in the birth rate. These measures included restricted availability of abortions, increased subsidies for families, and extended maternity leaves. These countries, however, as Eastern Europe on the whole, continue to experience general declining growth rates.

Urban Population

Compared to Western Europe, Eastern Europe has a smaller proportion of its population living in cities. In Eastern Europe, four of the eight states have less than one-half of their population classified as urban. These relatively low proportions reflect the greater predominance of agriculture as a way of life in the region and the smaller share that industrial and service activities play in the economy.

There is, however, considerable contrast in the level of urbanization. East Germany is one of the most urbanized states in Europe with 74 percent of its people living in cities. At the other end of the scale is Albania, the least urbanized of Europe's states

with only 34 percent of its population living in cities. The level of urbanization in the other four Balkan countries ranges from 39 percent for Yugoslavia, 41 percent for Romania, and 49 percent and 56 percent for Hungary and Bulgaria, respectively. Slightly more than one-half of the peoples of Czechoslavakia (56 percent) and Poland (53 percent) reside in cities. This proportion compares favorably with the level of the urban population of the lesser urbanized Western European states of Austria, Ireland (both 52 percent), and Norway (45 percent).

Most cities in Eastern Europe are very old. Some, such as Budapest and Belgrade, date back to the time of the Roman Empire or even before. During the medieval period, there was an increase in town construction. The medieval town became increasingly important as the center of political and church affairs and as the focal point of growing commercial activities. Much of the most impressive architecture (churches, town halls, palaces, and merchant buildings in such cities as Prague, Krakow, and Budapest) dates from this period. World War II, however, destroyed many of these original buildings.

Compared to the rapid construction of new cities and the expansion of old cities with the growth of industry in Western Europe in the nineteenth century, Eastern Europe's urban building was much less noticeable. The Balkan states were virtually untouched by industrialization. In parts of northern Eastern Europe, urban growth did accelerate during the nineteenth century. Industrial districts developed on the edges of large cities, such as Warsaw, Budapest, Berlin, and Prague. As a result of industrial expansion, urban growth also took place in such regions as Silesia (then divided between Germany and Poland), the western part of Czechoslovakia, the Lodz district of Poland, and the Saxony area of what is now East Germany.

Although many of the larger cities of Eastern Europe grew between the two world wars as a result of the migration of the surplus rural population, urban development as a whole slackened except in those regions that formerly were part of Germany.

The Communist regimes of Eastern Europe place a high priority on industrial growth. As a consequence, post–World War II urban development has proceeded briskly. For example, shortly after the war in Poland, only about one-third of the Polish population was urban compared to about 53 percent now. Romania's

An East German iron and steel plant. Construction on the plant and the accompanying new town began in 1950. Iron ore is imported from the USSR, and coking coal is brought from Poland. A barge canal, highway, and railroad service the plant and growing town. (Eastfoto)

urban population has increased from 28 percent in 1945 to 41 percent, and Albania's from 20 percent to 34 percent. In addition to the reconstruction and expansion of existing cities, several new towns have been built, which underscores the emphasis on industrialization. Some of these new towns are Eisenhüttenstadt (iron and steel) in East Germany, Nowa Huta (iron and steel) in Poland, Victoria (chemicals) in Romania, and Havirov (mining) in Czechoslovakia.

Another indicator of Eastern Europe's level of urban development is the relatively low percentage of its population living in large cities. There are only five cities in Eastern Europe with a population greater than 1 million (Budapest, Bucharest, Warsaw, East Berlin, and Prague) and eight more with populations between 500,000 and 1,000,000 (Sofia, Belgrade, Lodz, Leipzig, Krakow, Zagreb, Wroclaw, and Dresden) (Table 13–3). The largest city of each country is the

Table 13–3 *Eastern Europe: Cities with More Than 500,000 Inhabitants*

Cities	Population
Budapest, Hungary	2,033,500
Bucharest, Romania	1,507,300
Warsaw, Poland	1,377,100
Berlin (East), East Germany	1,090,100
Prague, Czechoslovakia	1,083,700
Sofia, Bulgaria	919,000
Lodz, Poland	777,800
Belgrade, Yugoslavia	746,100
Krakow, Poland	651,300
Leipzig, East Germany	579,200
Zagreb, Yugoslavia	566,200
Wroclaw, Poland	557,200
Dresden, East Germany	505,400

SOURCE: *Demographic Yearbook* (New York: United Nations, 1974).

national capital, which is also the major economic and cultural focus of the state. With the exceptions of Belgrade and Warsaw, the capitals of Eastern Europe are good examples of primate cities (cities that are disproportionately large in population and important in the economic, political, and cultural life of the country). East Berlin is about two times larger than Dresden, the nation's second city. Prague and Sofia are more than three times the size of their countries' second cities. Tirana is about four times larger than Albania's second urban settlement, while Bucharest has more than seven times the population of Cluj (Romania's second largest city). Few cities in the world hold such a dominant position in their country as Budapest. Budapest contains 19 percent of Hungary's population and is more than eleven times the size of the second-ranked city.

Levels of Economic Development

The highest priority has been placed on the industrialization of Eastern Europe by the Communist governments. This emphasis has meant a primary concentration on heavy industry, the manufacturing of producer goods, and the development of the mineral resource and energy base of the state. Prior to World War II only Germany and Czechoslovakia had a well-developed industrial sector in their economy. Today, despite the considerable progress not only in industrial production but agriculture as well, Eastern Europe still has the least industrialized and poorest economic landscape of Europe.

The levels of economic development and the levels of living are varied. This diversity is most evident when the five criteria of development discussed in Chapter 4 are applied to the region, particularly the three standard indicators (per capita GNP, per capita consumption of energy, and percent of the labor force in agriculture) (see Appendix). The two supplementary measures (daily food supply and life expectancy) are more uniform; however, the low level of calorie intake of Albania (2,390 calories per person per day) raises doubts about the inclusion of Albania among the rich nations.

The three standard measures suggest more convincingly that the term "developed" needs qualification for some of the countries of the region. East Germany, Czechoslovakia, Hungary, and Poland can be included readily among the rich states. Yugoslavia, Romania, Bulgaria, and most notably Albania, show greater contrasts to the richer states of Europe. Their per capita GNPs vary from $2040 (Bulgaria) to $600 (Albania). Per capita energy consumption is relatively low and the share of the labor force in agriculture relatively high compared to their richer neighbors to the west. On the other hand, the data suggest that these last four countries cannot be unquestionably classified as poor. Perhaps they are transitional, neither rich nor poor. In the case of Albania, however, convincing arguments can be presented for its inclusion among the ranks of the poor nations of the world.

Agriculture

The historical legacies, physical environments, and local reactions to Soviet policies have all contributed to creating a varied agricultural landscape. At the end of World War II, much of Eastern Europe suffered from chronic rural overpopulation. The existence of large estates and the shortage of land for the peasants in the best agricultural areas in the central and northern sections were a major agricultural problem. In the Balkans, however, where Turkish influence was felt and where the terrain was rugged, agriculture was at a subsistence level on small, highly fragmented farms.

Collectivization

Numerous attempts at land reform were made throughout Eastern Europe prior to World War II. None could compare, however, with the drastic changes made following the establishment of the postwar socialist regimes. The land was nationalized; and where large estates existed, the land was redistributed among the peasant population. The next step was the collectivization of agriculture. Land was consolidated into either large-scale, Soviet-style state farms or collective (communal) farms. The state farms are funded and directed by the central governments. The operation of the collective farms is directly the responsibility of the farmers within the framework of state directives.

Throughout much of the Balkans small subsistence-level farms persist. In the mountainous sections of Yugoslavia, traditional labor-intensive agriculture remains a dominant economic activity. (Sovfoto)

The objectives sought through collectivization reflected the same points used to justify the collectivization of land in the Soviet Union. A purer Marxist-Leninist form of agricultural organization would make it easier to control this important sector of the economy and thereby facilitate the enactment of programs of industrial development. Furthermore, it was anticipated that the availability of foodstuffs for the growing number of industrial workers would be increased. Equally important was the recognition that agriculture could be more productive with large-scale units, which allow greater application of mechanical equipment and specialization of labor.

The results of the collectivization are varied. There was large-scale resistance throughout the region. In Poland and Yugoslavia, the collectivization process has been unsuccessful. Presently only 14 percent of the agricultural land of both Poland and Yugoslavia is held by collective and state farms. In the other countries, collectivization has proved much more

A state farm in Yugoslavia emphasizing dairying. Collectivization of agriculture in Eastern Europe has met with varying degrees of success. In Poland and Yugoslavia collective and state farms control only a small part of the agricultural lands. In contrast, the other nations have the bulk of their farmland in the socialist sector. (Eastfoto)

durable and successful. All of Albania's farms are in the socialist sector (state and collective farms). In Hungary, Romania, East Germany, Czechoslovakia, and Bulgaria, the socialist sector controls 84 to 92 percent of the agricultural land.

General State of Agriculture

On the whole, agriculture plays a larger role in the economy of Eastern Europe than in Western Europe. One indicator of agriculture's relative position is provided by the data showing the percent of the labor force of each country employed in agricultural pursuits.[1] (Appendix). As expected, the lowest proportion of farm workers is encountered in the most industrialized and wealthiest states of East Germany and Czechoslovakia. Agriculture is more mechanized in these two countries than in the rest of the region. Everywhere, however, agricultural equipment is scarce. Other limiting factors have been a shortage of labor and the chronic problem of lack of incentives. Levels of productivity in both East Germany and Czechoslovakia are below the standards of Western Europe, and neither is self-sufficient in foodstuffs.

Compared to a quarter of a century ago, the proportion of agricultural workers in Hungary has been cut in half. Today, less than one-quarter of Hungary's labor force is so employed. Hungary has long been important as an agricultural producer and today remains a net exporter of farm goods. Of all the countries with a predominantly socialized agricultural organization, Hungary has gone the furthest in allowing collectives to make their own production decisions. In Hungary, however, the growth of agricultural production has suffered from poor financial incentives for the farmer.

Of the six CMEA countries, only Poland is dominated by private farm holdings. These private farmers, however, are closely regulated by the central government. Furthermore, the state is theoretically committed to eventual collectivization of all agricultural lands. The production of foodstuffs in Poland is insufficient to meet domestic needs, and Poland

[1]The use of percent of labor force employed in agriculture overemphasizes, however, the role agriculture plays in the economy. In terms of dollar value added to the GNP, Eastern European farmers are only one-third to one-half as productive as their fellow industrial workers.

must import food. The situation became especially severe in 1970 when food riots broke out because of shortages and high prices. To encourage agricultural production, the government embarked on a policy to raise agricultural prices. The problem of introducing greater efficiency through increased mechanization in Polish farms is exacerbated by the small size of the private farms. These farms average only 12 acres (5 hectares). The government has established agricultural circles to encourage cooperation among the private farmers. These circles have not been as successful as was hoped. In spite of these problems, Poland has been the most successful of the Eastern European states in increasing the productivity of its land, which is low compared to Western Europe.

The four Balkan states—Yugoslavia, Albania, Romania, and Bulgaria—are more agricultural than the northern Eastern European countries. In Albania, more than six out of every ten workers are engaged in agriculture. For Yugoslavia and Romania, this proportion is about one-half of the total work force. Romania and Bulgaria are net exporters of foodstuffs. Yugoslavia is generally self-sufficient in its food needs; however, during years of poor harvest, some food must be imported. Compared to the levels of pre–World War II agricultural production, the Balkan states have made substantial gains. Despite these improvements, their per capita and per area productivity remain among the lowest in Europe.

Land Use

Although there is considerable variation in the quality of the agricultural resources of Eastern Europe, much of the region suffers from a marginal resource base. A major unfavorable factor is the quality of the soils. Most of East Germany and Poland have relatively infertile, sandy, podzolic soils. In the mountainous and hillier sections of the region, thin and relatively unproductive soils are encountered. The best soils are the fairly rich chernozemic soils of the Hungarian Lowland, the Wallachian Lowland, and the Moldavian Plateau in Romania. In eastern Romania and in other scattered sections, insufficient precipitation frequently brings ruin to the crops. On the other hand, several areas suffer from too much groundwater and require drainage. The growing season is for the most part adequate, although in much of East Germany and Poland, the

Table 13–4 Eastern Europe: Land Use in Percent of Total Land Area

	Agricultural Land			
	CROPLAND	PASTURE AND MEADOW	FORESTED	OTHER
Albania	19.3	23.9	43.2	13.5
Bulgaria	40.7	13.5	33.7	12.1
Czechoslovakia	41.7	13.7	37.7	6.9
East Germany	44.6	13.5	27.2	14.6
Hungary	60.0	13.7	15.9	10.4
Poland	48.9	13.5	27.4	10.2
Romania	44.2	18.6	26.6	10.5
Yugoslavia	32.0	24.8	34.6	8.6

SOURCE: Food and Agricultural Organization, *Production Yearbook* (Rome: United Nations, 1973).

relatively cool summers together with poor soils limit cultivation to hardy crops.

The share of the total land area used for agricultural purposes (growing crops and raising livestock) ranges from 43 percent (Albania) to 74 percent (Hungary) (Table 13-4). The amount of land in crops varies even more greatly. Only 19.3 percent of Albania is cropped. Hungary, in contrast, has 60 percent of its land area suitable for cultivation. Hungary's proportion of cultivated land is the highest in Europe after Denmark (61 percent). Only 43 percent of Eastern Europe as a whole is cultivated.

Types of Agricultural Production

The cooler northern section (East Germany, Poland, and Czechoslovakia) is a major grain-growing region with rye the dominant grain. The area planted to rye, however, has steadily declined, largely in favor of wheat. The better soils, such as the productive loess soils in southern Germany and Poland, are devoted to wheat production. There has also been an increased emphasis in industrial crops (sugar beets, oils, and fiber plants). Also in the north the important staple, potatoes, is widespread throughout the area. Vegetable and fruit growing are prominent in the vicinity of the urban areas.

The availability of adequate meat supplies has been a chronic problem in the northern agriculture zone. The area planted to fodder crops has increased generally throughout the region but is still insufficient to meet the needs of the meat and dairy industries.

Cattle and hogs are the principal livestock in this part of Eastern Europe. The number of sheep has gradually declined in East Germany and Poland but has increased in recent years in Czechoslovakia at the expense of the cattle herds.

In the southern part of Eastern Europe, the principal grain-growing region is the Hungarian and Wallachian Lowlands. Wheat and maize are the most important grains here and in other parts of Bulgaria, Yugoslavia, and Albania. Emphasis has been placed on the cultivation of industrial crops, most notably sugar beets. Bulgaria is Eastern Europe's major producer of cotton, tobacco, and vegetables. Bulgaria, Romania, and Hungary all export vegetables, both fresh and processed. They are sent principally to the Soviet Union and northern Eastern Europe. The dry summer subtropical climate zone of Yugoslavia and Albania produces a variety of specialized crops such as citrus, olives, and vegetables. Vineyards, primarily for wine grapes, are a significant feature of the dry summer zone of Yugoslavia. Vineyards are also important in the Hungarian Lowland, in Romania, and in Bulgaria. Livestock production has encountered the same difficulties in this part of Eastern Europe as in the north, most notably an inadequate feed base. Both Yugoslavia and Romania, however, export livestock products for a source of western hard currencies. The raising of sheep both for meat and wool, partially reflective of Islamic influence, is an important phase of the livestock industry in the four Balkan states.

13–8 *Eastern Europe: Mineral Resources and Industrial Districts*

Industry

Resource Base

Eastern Europe is not well endowed with industrial resources (Fig. 13–8). The region in its entirety is deficient in energy resources, iron ore, and other minerals. There is, furthermore, a disparity in the distribution of the few resources that do exist. The only hard coal deposit of substantial size is the Silesia-Moravian field. Most of this deposit lies in Poland, but a small part extends into Czechoslovakia. The production of hard coal in Poland exceeds that of all the other countries in Eastern Europe. Poland's total output of hard coal is 28 percent of the output of the United States. A coalfield of secondary importance is found in western Czechoslovakia, and small scattered deposits are found elsewhere. Although Eastern Germany is deficient in high-quality coal and must rely principally on the Polish deposits, it has large reserves of low-grade coals. Subbituminous coals are widely scattered throughout Eastern Europe. With the exception of the East German lignite fields, these fields are small. The principal use of low-grade fuels is for the production of electricity and for domestic heating.

There is also a shortage of liquid fuels in Eastern Europe. Romania's Ploesti Field has the oldest commercial oil well in the world. Prior to the discoveries in the North Sea, it contained Europe's principal petroleum deposits. The total reserves of Romania, the Ploesti Field, the newer Bacau Field in the eastern part of the country, and other scattered small deposits are small. They are calculated to last only a few more years at the present rates of production. Yugoslavia's reserves of petroleum and natural gas are presently adequate for its needs, and production is increasing. Consumption of these fuels, however, is rapidly growing, and there are serious doubts that Yugoslavia will be able to meet its domestic requirements by the mid-1980s. Albania presently is a small exporter of petroleum products. All the remaining Eastern European countries have little or no oil and gas reserves. They are dependent on imported petroleum and gas principally from the Soviet Union. Liquid fuel consumption is increasing throughout Eastern Europe. The dependency on foreign sources, particularly the Soviet Union, will inevitably become greater.

Hydroelectric potential is limited in Eastern Europe except for the Balkan states. Presently there is little developed hydroelectric power, and most plants are quite small. Potential development of hydroelectricity is greatest in Yugoslavia. Extensive construction is underway to more than double 1970's output by 1985. The largest of these projects is the joint Yugoslavian and Romanian dam and power plant at the Iron Gate, the deep gorge where the Danube cuts through the Balkan Mountains. This station, completed in the mid-1970s, was designed to be the world's third largest generator of electricity when operating at capacity.

Uranium as fuel for nuclear power stations is found in several Eastern European countries, most notably Czechoslovakia, East Germany, Romania, Yugoslavia, and Hungary. Generation of electricity by nuclear power is still in its infancy, but plans are under way for increasing output.

Ferrous and nonferrous metals are also in short supply in Eastern Europe. Most critical is the limited

Romania's Ploesti oil field contains continental Europe's most extensive petroleum deposits. Although the field is old, exploration and drilling continue, since present reserves are calculated to last only a few more years. (Eastfoto)

amount of iron ore. Even Poland must rely on the USSR for iron ore in order to keep operating its iron and steel industry, the largest in Eastern Europe. Yugoslavia is in the best overall position. Although most of its metallic deposits are relatively small and production is limited, it possesses a wide variety of raw materials. Included in its metallic inventory are iron ore, copper, lead, zinc, chromium, and bauxite. Among the most important metals in other Eastern European states are bauxite in Hungary, chromium in Albania, and copper and tin in East Germany. Most notable of the region's nonmetallic minerals are Poland's sulfur and East Germany's potash deposits. In both countries these minerals provide the raw materials for chemical industries.

Industrial Growth

The greatest strides made in the economic development in Eastern Europe during the postwar period have been in manufacturing. The greatest gains have been posted by the least industrialized countries, particularly Bulgaria. Although the gap between the richer and poorer industrial states has been narrowed, East Germany and Czechoslovakia remain the most industrialized countries.

In the early postwar years, emphasis was on increasing production. In the CMEA countries and Albania, tight centralized management of the economy often resulted in inefficiency, waste of raw materials, and low levels of labor productivity. Since the mid-1960s, increasing concern has been placed on production efficiency. Governments have become most cost conscious and have attempted, with the notable exception of Albania, to reduce central direction and increase the profits of industry. As a consequence, growth rates have generally been lowered in comparison to the early years of development.

Yugoslavia has gone the farthest in reducing centralized control over industry. It has attempted to establish a Western market-type economy in place of a Soviet-style economy which arbitrarily sets prices. Plant managers and workers have a greater decision-making role. The hope is that these reforms will reduce waste and make Yugoslavian industry more competitive with the industries of Western Europe. The Soviet Union has condemned Yugoslavia for this action as being nonsocialistic.

The Location of Industrial Activity

A large industrialized region extending from Western Europe cuts across the northern part of Eastern Europe. The western parts of Czechoslovakia, Poland, and all of East Germany are part of this area of concentrated industrial activity. The principal industrial areas of Eastern Europe found within this part of the region are the Silesian-Moravian district of Poland and Czechoslovakia, the Bohemian Basin of Czechoslovakia, and the Saxony District of East Germany (Fig. 13–8).

The Silesian-Moravian district was formerly Germany's second heavy industrial region. Following the war, Poland acquired the largest part of Silesia; a small section was given to Czechoslovakia. Silesia is Poland's major heavy industrial district based on local coal resources. The coal is actually of poor coking quality and must be supplemented with imported coke from the Soviet Union. Several industrial cities in the district form Poland's largest conurbation. In addition to coal mining and the production of iron and steel, manufacturing includes agricultural machinery, machine tools, and various chemicals. The Czech section of Silesia is centered in the city of Ostrava, the largest iron and steel center of the district.

The Bohemian Basin in western Czechoslovakia is also an iron- and steel-producing district. Coal mines are worked in this area; however, the deposits are small. Czechoslovakia's capital city, Prague, is located within the basin and is Czechoslovakia's largest city and center of industrial production. The capital city's industry is quite diversified and includes important chemical, food, and machinery industries.

East Germany's Saxony District contains the country's greatest concentration of industrial production. There is some iron and steel production, but the region is highly diversified and famous for its manufacture of chemicals, textiles, and engineering industries. The Saxony District is a highly urbanized district. Both Leipzig and Dresden, East Germany's second and third largest cities, are found within the district.

Manufacturing outside the industrial region is chiefly in the large cities. For example, cities like Berlin, Budapest, Warsaw, and Belgrade are major focal points of diverse industries. Other centers of production are generally smaller and often associated with the location of a raw material (chemical industries at Ploesti). Several new industrial centers have been established in Eastern Europe in order to develop the more backward parts of a state such as southern Yugoslavia or Slovakia in Czechoslovakia.

Foreign Trade

Foreign trade among the six Eastern European CMEA members is largely with the Soviet Union and other CMEA states. Romania has the lowest proportion of trade with these Communist countries—about 48 percent. Nearly 72 percent of Bulgaria's trade is with the Soviet Union and the CMEA states. The amount of trade carried on with non-Communist countries varies from about one-half (Romania) to one-quarter (Bulgaria). Both countries have substantially increased their amount of western trade in recent years.

Of all the imports of CMEA, fuels and raw materials are most prominent. These commodities are especially important for East Germany, Hungary, Czechoslovakia, and Poland. Machinery and equipment of various types are the major imports for Bulgaria and Romania. The major supplier of raw materials and machinery is the Soviet Union. Romania exports petroleum; Poland and Czechoslovakia export coal. East Germany, Czechoslovakia, and Poland are major suppliers of machines to both Eastern Europe and the Soviet Union. Foodstuffs of various sorts are principal export items from Bulgaria, Romania, and Hungary. East Germany and Czechoslovakia are Eastern Europe's major food importers.

In contrast to the CMEA countries, Yugoslavia's trade has been predominantly with non-Communist states (about 58 percent), in particular with West Germany and Italy. Yugoslavia's trade with the Soviet Union has been substantial and has increased somewhat in recent years. Albania, on the other hand, conducts about 90 percent of its trade with the Communist world, about 50 percent with China alone.

The Uncertain Future

Eastern Europe has traveled a long road since those early post–World War II years when it was enclosed behind the "Iron Curtain." Yugoslavia has gone the furthest in establishing relationships with the West. Westerners may travel to and in Yugoslavia with relative ease. Yugoslavians are permitted to migrate to Western Europe to hold temporary jobs. More so than elsewhere, the Yugoslav Communist party has relaxed its grip on the economy and society; yet the party is still the sole legal political force. Much of Yugoslavia's future depends on the ability of its political leaders to hold the diverse elements of the country together after the death of the popular Marshall Tito. Latent internal ethnic hatreds and boundary disputes with Albania and Bulgaria could threaten the security and economic progress of Yugoslavia.

Albania, on the other hand, remains tightly within the grasp of a conservative Stalinist-type Communist party. Although the government has recently opened the door to a trickle of outsiders, Albania remains the most closed society of Europe. The course of future events in Albania depends to a great degree on Soviet-Chinese relationships and China's willingness to continue its substantial support of that country.

The six CMEA countries are closely tied to the Soviet Union, although they increasingly are taking steps at some variance with those of the USSR. The attitude toward the Soviet Union may vary, though, from that of Bulgaria, which has proposed becoming part of the Soviet Union, to Romania's nationalistic defiance of the Soviet Union. Romania's strong nationalistic stance, however, does not threaten the Soviet Union with the development of Western democratic tendencies.

Economic realities also promise a continuing close relationship with the Soviet Union. Dependence is still on the Soviet Union for raw materials, capital goods, and credit essential for economic growth. The decrease in the rate of economic development and the rising expectations of the citizenry of Eastern Europe have made these countries increasingly interested in greater contacts with the West. The CMEA countries are desirous of acquiring technology and capital from the industrialized West. Furthermore, increased markets for their goods in the West could stimulate production at home. Much of the progress of increased East-West relations depends on détente between the United States and the Soviet

Union. Warmer relations between the Soviet Union and the United States would most likely result in greater Western contacts with the European CMEA countries. Particularly meaningful in this respect are the recent increased relations between East Germany (one of Moscow's most loyal allies) with Western states, notably West Germany. The internal stability of the Eastern European government reflects the economic welfare of its people. This welfare relates to the ability and desire of the governments to bring about changes necessary for greater economic efficiency. The changes themselves still must have the tacit approval of the Soviet Union. The nature of the modifications of the economic and political life of Eastern Europe in the end will reflect changes within the Soviet system itself.

Further Readings

BENES, V. L., and POUNDS, N. J. G., *Poland* (New York: Praeger, 1970), 416 pp. This volume is mostly concerned with modern political history. Two chapters, however, present an excellent examination of the Polish lands, their resources, and economic development.

BROMKE, A., and RAKOWSKA-HARMSTONE, TERESA (eds.), *The Communist States in Disarray, 1965–71* (Minneapolis: University of Minnesota Press, 1972), 363 pp. This volume is a collection of articles focusing on recent political events in Eastern Europe. There is one chapter on each of the eight Eastern European countries.

CAMPBELL, ROBERT W., *The Soviet-Type Economies: Performance and Evolution,* 3rd ed. (Boston: Houghton Mifflin, 1974), 259 pp. An essential book for anyone wanting to better understand the complexities of the economic systems of Eastern European countries and how they compare with one another and with those of the Soviet Union and other Communist states.

POUNDS, N. J. G., *Eastern Europe* (Chicago: Aldine, 1969), 912 pp., and OSBORNE, R. H., *East-Central Europe: An Introductory Geography* (New York: Praeger, 1967), 384 pp. These two excellent and thorough studies of the region provide a useful background to the region. Less detailed, but nonetheless informative and enjoyable reading, are the appropriate chapters in GOTTMAN, JEAN, *A Geography of Europe,* 4th ed. (New York: Holt, Rinehart and Winston, 1969), 866 pp.

SCHOPFLIN, GEORGE (ed.), *The Soviet Union and Eastern Europe: A Handbook* (New York: Praeger, 1970), 614 pp. An encyclopedic book containing articles on the politics, history, economics, society, and arts of Eastern Europe.

Geographies on the individual countries are few, and many are now quite old. Two recent studies are MATLEY, I. M., *Romania, A Profile* (New York: Praeger, 1970), 292 pp.; and TODOROV, NIKOLAI, and others, *Bulgaria: Historical and Geographical Outline* (Sofia: Foreign Language Press, 1965), 278 pp. The Bulgarians present a view of their own country. An older but useful study is WANKLYN, HARRIET, *Czechoslovakia* (New York: Praeger, 1954), 446 pp. See also SKENDI, STAVRO (ed.), *Albania* (New York: Praeger, 1958), 389 pp., a comprehensive study of the Albanian land and its politics, economy, and culture as of the late 1950s; HOFFMAN, GEORGE W., *The Balkans in Transition* (New York: Van Nostrand Reinhold, 1963), 124 pp.; and POUNDS, N. J. G., *Divided Germany and Berlin* (New York: Van Nostrand Reinhold, 1962), 128 pp.

The USSR: Its Physical Environment and Territorial Growth

14

T HE WORLD'S FIRST Communist government seized control of the largely agrarian Russian Empire in 1917. Today the Union of Soviet Socialist Republics (Soviet Union, or USSR) is the foremost industrial, political, and military power in the communist world and second only to the United States in the world. The Communists inherited the earth's largest state, in areal extent, which occupies over 40 percent of Eurasia. Within its borders were more than 100 ethnic groups of great economic and religious and linguistic cultural diversity. The emerging Communist leadership contained elements of several cultures. The new government, however, was dominated by the Russians, the largest of these ethnic groups.

The Land

The sheer physical magnitude of the USSR, one-sixth of the earth's land surface, is in marked contrast to the size of other European states. Its 8.6 million square miles (22.2 million square kilometers) are forty-one times larger than France, the largest European country. The United States including Alaska is only 43 percent the size of the Soviet Union (Fig. 14–1). One passes through eleven time zones traveling the 6,000 miles (9,656 kilometers) from the Baltic Sea to the Bering Straits. At its maximum north-south extent, the Soviet Union is almost as wide as the United States is long, approximately 2,900 miles (4,667 kilometers).

To many the size of the USSR evokes the image of unlimited raw materials awaiting exploitation. To others it conjures a picture of vast tracts of virgin lands awaiting the settler. Unquestionably the size of the country enhances the potential occurrence of vital natural resources. Indeed, the USSR is extremely well endowed with raw materials requisite for its world-power status. The presence of these natural resources does not, of course, mean that their development is assured. Technology and sufficient capital for their exploitation must be available, and the resources must be economically accessible to their potential consuming regions. Many of the Soviet Union's raw materials are found in regions remote from population concentrations, and their exploitation requires a large capital investment that the government may be unable or unwilling to pay. Furthermore, much of the Soviet land presently is considered inhospitable for settlement and unsuitable for agriculture because it is too cold, too wet, or too dry.

The USSR's large area has enabled the state to survive the attacks of nomadic tribesmen and to outlast the invasions of more modern would-be conquerors, notably Napoleon and Hitler. Areas remote from the battle lines have provided a refuge

14–1 *USSR–USA: Land Area Comparison*

and sites for relocated industries. Agriculture also has benefited from size through the opportunity to expand cultivated land; however, many areas are of only marginal quality because of scanty precipitation and short growing seasons.

The Soviet Union has a boundary of over 37,000 miles (59,544 kilometers), of which about one-third is on land bordering twelve countries. The western boundary historically has been the most critical for the security of the state. The long southern boundary lying in mountainous terrain remote from the core of the state has been less threatened by its weaker neighbors. Today, however, the USSR has deployed from forty to forty-five army divisions along its southern frontier for protection against its major antagonist, China.

Landforms

The landforms of the Soviet Union vary from lowlands below sea level to mountain peaks above 24,000 feet (7,325 kilometers). Several broad landform regions are noted as one crosses the USSR (Fig. 14–2). The vastness of some of these regions often gives the traveler a feeling of monotonous landscape. European Russia, the area west of the Ural Mountains, is dominated by the Great European Plain. The plain covers about one-quarter of the country and stretches from the western border to the Ural Mountains and from the Black Sea to the Arctic Ocean. The plain is primarily a lowland with elevations below 650 feet (198 meters), but interspersed are low-lying hills rising to 750 feet (229 meters) above the plain. The northern part of the plain has undergone glaciation. The glaciers rounded the hills and exposed the mineralized ancient crystalline bedrock of the Fenno-Scandinavian Shield. Glaciation also greatly disturbed the drainage pattern, creating numerous lakes and swampy areas. To the south the glaciers deposited materials forming many morainic features such as the Moscow-Smolensk Ridge.

East of the Volga River the plain abuts the Ural Mountains. These mountains are greatly eroded and low, particularly in the central section, and permit easy movement from west to east. The Urals, traditionally the boundary between Europe and Asia, are today best known for their rich and varied mineral wealth.

Between the Urals and the Yenisey River lies the vast West Siberian Lowland. It extends for 1,000 miles (1,609 kilometers) east to west and 1,200 miles (1,931 kilometers) from the Kazakh Upland in the south to the Arctic Ocean in the north. The elevation of this very level lowland does not exceed 400 feet (122 meters) above sea level. Much of the lowland between the Ob and Irtysn rivers is an extensive bog land, the Vasyugane Swamp. The rivers flow slowly from the south across this flat lowland and drain into the Arctic. Spring thaws first affect the upper courses of the rivers, but downstream the lower portion of the river remains frozen, forming ice dams that cause flooding of a large area.

Southwest of the West Siberian Lowland is the Turan Lowland. The center of this lowland is the Aral Sea, which, like the Caspian, has no outlet to the world's oceans. The water level of both of these bodies of water is dropping, and their ecological balance is undergoing gradual changes. These changes are particularly critical for the valuable fisheries of the Caspian Sea.

East of the Yenisey River rises the Central Siberian Plateau. This upland region lies mostly between 900 and 1,200 feet (274 and 366 meters) above sea level. Rivers have cut into the rock structure of the plateau, forming deep gorges with immense hydroelectric potential. The Lena River and Lena Plateau separate this large massif from the complex mountain system that covers most of eastern Siberia.

Mountainous and rugged areas cover not only much of eastern Siberia but also mark the southern boundaries of the USSR from Europe to the Pacific Ocean. The highest mountains in the Soviet Union are found in the extensive ranges that lie in central Asia. Peak Communism, 24,590 feet (7,495 meters), in the Pamir Mountains is the highest point in the USSR. The Carpathians, the Crimean, and the Caucasus mountains fringe the south end of the Great European Plain. The low-lying Kopet Dag Mountains form part of the border of Iran and the Soviet Union.

Natural Regions

The magnitude of the Soviet landmass, together with its high latitudinal location, are important elements in the severe continental climates that dominate the country. The southernmost part of the country lies approximately at 35° N. latitude, about the same as Memphis, Tennessee (Fig. 14–3). Moscow is more northerly than Edmonton, Canada. Approximately one-tenth of the USSR is within the Arctic

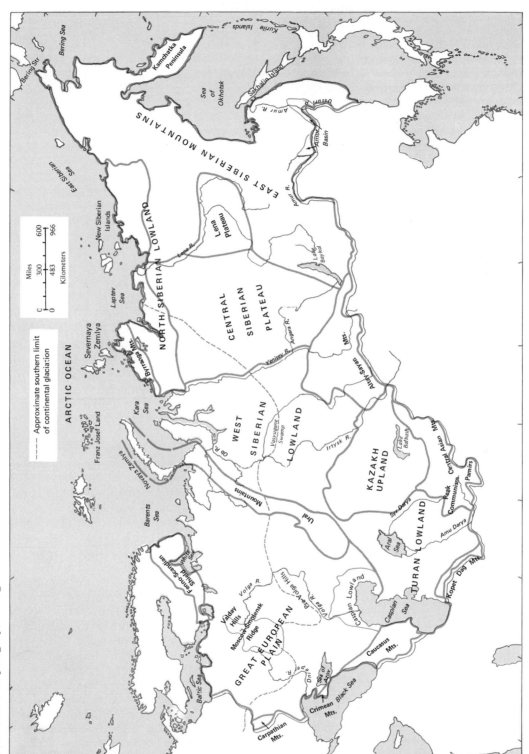

14-2 USSR: Physiographic Regions

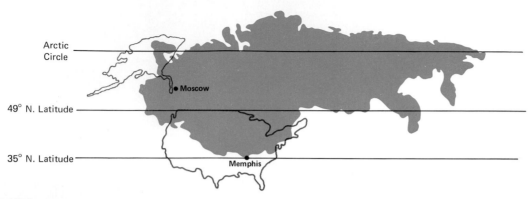

14–3 **USSR–USA:** *Comparison of Latitudinal Position*

Circle, and over 75 percent lies poleward of the 49th parallel, the northern boundary of the United States. Climatic character is evident in vegetation patterns and to a lesser degree the soils of an area (Chapter 2). A useful device for studying the physical environment is the "Natural Region." These regions are essentially vegetation zones with related and generalized climatic and soil characteristics. "Natural Region," however, is a misleading term. Large parts of the Soviet Union have been altered by mankind and are not in a natural or primordial state.

Tundra

Northernmost of the Soviet natural regions is the tundra (Fig. 14–4). This zone, together with tundra conditions that exist elsewhere in the mountains, covers about 13 percent of the Soviet Union. The mean temperature of the tundra's warmest month is below 50°F. (10°C.) but above 32°F. (0°C.). This region's very short growing season results in sparse vegetation characterized by such hardy plants as reindeer moss, lichens, and shrubs. The tundra is treeless because of the limited heat received in the high latitudes, high winds, and the presence of permafrost (permanently frozen ground that restricts root growth). Tundra soils are usually infertile and poorly developed, and many areas are water-logged during the very short cool summers. Needless to say, the tundra with its bleak and cold climate has been little affected by mankind. Widely scattered indigenous tribes, hunters, trappers, and miners are among the few who have penetrated this remote and inhospitable region.

Taiga

Forests cover almost 50 percent of the Soviet territory. These forests are divided into three natural regions: the vast taiga, the mixed forest of European Russia, and the broadleaf forest of the far east. The taiga lies south of the tundra. Except for the higher elevations in the low Ural Mountains, it reaches no farther south than 55° N. latitude in European Russia and western Siberia. Most of eastern Siberia lies within the taiga. The climate of the taiga is primarily subarctic. Lower temperatures have been recorded in the taiga than in the tundra because of the intense development of the winter Siberian high pressure cell and local topographic conditions. The mean January temperature of Verkhoyansk, for example, is − 59°F. (− 49°C.). A record low of − 96°F. (− 71°C.) makes Verkhoyansk the coldest place in the world outside Antarctica. Although summers are short and cool with average July temperatures of 60°F. (15°C.), temperatures may reach 100°F. (37.7°C.); thus the taiga has a great range in temperature from the coldest to the warmest month. The differences between the mean January temperature and the mean July temperature for Verkhoyansk is 112°F. (63°C.). Such extremes, however, are not found throughout the taiga. The western sections have generally warmer winters and cooler summers. In general, the growing season is less than 100 days, and consequently agriculture is very limited.

The dominant soil of the taiga is podzolic. Grayish in color and leached of many nutrients, podzolics are infertile and highly acidic from the decaying needles of the coniferous trees. Areas of permafrost cover virtually all of eastern Siberia except the

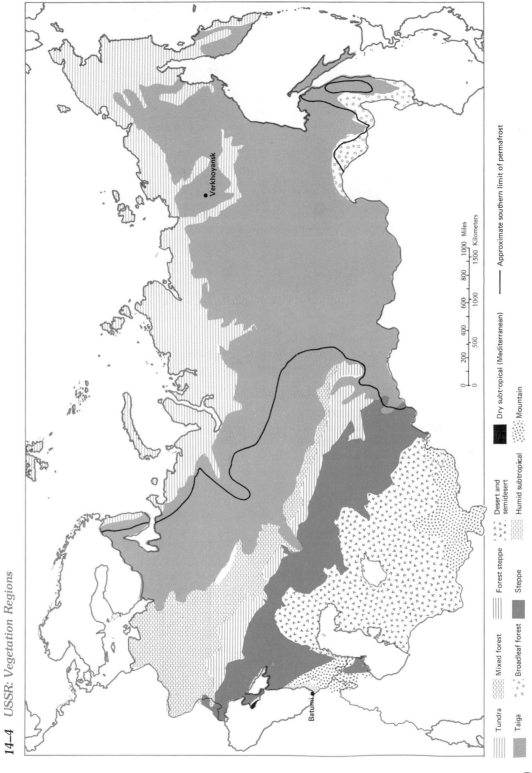

14–4 *USSR: Vegetation Regions*

Verkhoyansk

Batumi

Tundra

Taiga

Broadleaf forest

Mixed forest

Forest steppe

Steppe

Desert and
semidesert

Dry subtropical (Mediterranean)

Humid subtropical

Mountain

—— Approximate southern limit of permafrost

0	200	400	600	800	1000 Miles
0	500	1000		1500 Kilometers	

The taiga in the USSR contains one of the world's greatest forests. The forest is mainly coniferous, and in western Siberia the land is a flat-to-rolling plain. Because of poor soils, permafrost, and a short growing season the region is only sparsely inhabited. Tree growth is very slow, often taking more than 100 years before reaching a size suitable for lumbering. (TASS from Sovfoto)

extreme southeast portions. In western Siberia, only the northern portion of the taiga is affected by these large areas of frozen soils; the European taiga has essentially no tracts of permafrost. Besides retarding root growth, permafrost can contribute to excessive water-logging of the top soil in the summer when water cannot seep through the frozen subsoil. Water-logging may lead to lateral soil movements that can cause havoc to railroad lines, roads, and even buildings if they are not firmly anchored in the frozen subsoil.

Coniferous forests are the most common of the taiga vegetation. Tree growth is slow, and trees rarely exceed 3 feet (about 1 meter) in circumference. Slow growth and remoteness from population centers reduce the economic significance of these forests. Although the Soviet taiga contains one-third of the world's timber, it is not one continuous forest. Extensive swamps and meadows are frequently interspersed among the forest lands. Moderate precipitation and relatively low rates of evaporation in the western taiga lead to excessive moisture. Furthermore, in western Siberia the flatness of the land retards drainage and intensifies the retention of moisture in the ground.

Taiga lands are generally unsuitable for agriculture, which is limited to isolated areas in the vicinity of populated centers. Under Soviet programs nomadic native groups of the region are turning to more

sedentary forms of herding and some agricultural pursuits. The traditional wealth of the taiga—timber, fur-bearing animals, and precious metals—is still significant. The discovery and development of new industrial resources such as the oil and gas fields in northwestern Siberia, however, are growing in importance. Major population concentrations are found in the southern taiga, especially along the Transiberian Railway in eastern Siberia.

Mixed Forest

Between the taiga and the grasslands of southern western Russia lies the triangular-shaped region of the mixed forest. The name "mixed forest" indicates the occurrence of both deciduous broadleaf and coniferous needle-leaf trees. From north to south, the presence of conifers decreases, whereas deciduous trees become more dominant and eventually take over completely. Temperatures and the length of growing season increase toward the south, but precipitation decreases, and the native forest cover becomes smaller as the open meadows increase. This region more than any other in Russia has been culturally modified, for within the mixed forest evolved the modern Russian state. Centuries of settlement have resulted in land clearing for agriculture and the rise of cities. The longer growing season and the gray-brown podzolic soils in the mixed forest are more suitable for agriculture than in the taiga. Hardier type grains, rye and flax, are confined to the northern sections, whereas wheat is more prevalent in the southern part of the region.

Broadleaf Forest

The third forest region is the broadleaf forest of the southern part of far eastern USSR. This region has cold, dry winters and hot, humid summers. As the name describes, the vegetation of this area are broad-leafed deciduous trees of Asiatic origin. Some conifers and open grassy meadows also may be encountered. A varied agriculture is possible here; however, poorly drained soils and deficiency of summer rainfall necessitate land improvement projects in some areas.

Steppes

The steppe zone of the USSR consists of two subregions: the forest steppe, a transitional region between the forested regions, and the treeless grassy true steppe. The forest steppe is characterized by woods separated by extensive grasslands. The trees are largely oaks in the west and birch in the east. In the true steppe, trees grow only in the river valleys, and grasslands stretch as far as the eye can see. The soils of the forest steppe are more leached than in the true steppe, where less effective precipitation results from higher temperatures and higher evaporation rates. Consequently, in the true steppe there is a greater buildup of humus and minerals in the topsoil producing rich chernozemic (from the Russian meaning "black earth") soils. The entire steppe zone is important for agriculture. Conditions for farming are best in the forest steppe, where the soil is fertile and precipitation generally sufficient and reliable. In the true steppe, increasingly arid conditions and variability of precipitation enhance the chances of drought.

Desert

South of the steppe lands, principally in the trans-Volga area, aridity increases until true desert conditions exist. Precipitation in the desert is generally less than 10 inches (25 centimeters) a year, and very hot dry conditions prevail in the summer. Winters are cold with some light snow, particularly in the more northerly sections. The vegetation of the desert consists mostly of clumps of grass and xerophytic plants that can store moisture; there are large expanses of bare earth, rock, and sand. In widely scattered oases and along the few rivers flowing through the desert, a rich plant life is supported on alluvial soils. Large parts of the desert are used for livestock grazing. Crop cultivation is limited to areas watered by streams or by irrigation projects.

The Subtropical South

Two small but important regions are the humid subtropical region along the east coast of the Black Sea and the dry summer subtropical region along the

The southern shore of the Crimean peninsula is an area of dry summer subtropical climate. During Tsarist times many of the Russian nobility built palatial homes along the coast. Today these homes have been turned into rest and vacation hotels. (TASS from Sovfoto)

south coast of the Crimean Peninsula. The Crimean Mountains help protect the narrow coastal region of the peninsula from severe cold winds of the north. The mild climate and moisture from the Black Sea contribute to a varied agriculture, including nuts, fruits, and vineyards. The Crimea is a famous resort area. Old palatial residences of the wealthy, including the tsar's former palace at Livadia, have been turned into rest and vacation hotels for Soviet workers and bureaucrats.

The humid subtropical area is also a favorite vacation spot. This area is renowned for its specialized agriculture, of which tea and citrus fruits are most important crops. Moisture-laden winds from the

Along much of USSR's southern border are mountains. In central Asia these mountains remain the home of pastoralists who graze their sheep in the mountains during the summer and winter them in the valleys. Many of the once-fortressed towns on the mountains have been abandoned and are now in ruins. (TASS from Sovfoto)

sea move inland bringing large amounts of precipitation to this area. Batumi receives about 98 inches (249 centimeters) a year. The soils are quite fertile and support a luxuriant vegetation.

Mountain Areas

The climate, soils, and vegetation of the mountain regions are diverse, reflecting the location of the mountains, their local relief, and most important, altitude. Many valleys, foothills, and mountain meadows of the mountain systems in central Asia and in the Caucasus support relatively large populations on productive agricultural lands.

Territorial Growth of the USSR

The Soviet Union is the end product of some 700 years of expansion that began with the small principality of Muscovy (Moscow) and is now the world's largest state. Even before Muscovy, however, there existed the first Slavic state of Kievan Rus, which was centered on the city of Kiev, now capital of the Ukraine. Kiev was one of the many small city-states formed by eastern Slavs throughout the steppe and forests of the Great European Plain. Its importance rested partly on its location on the commercial Dnieper River route connecting the Baltic and Black Seas. Kiev also served as a fortress to defend the Slavic lands from the incursions of nomads from the south and east. The city rose to preeminence among the Slavic states under the leadership of Norsemen from Scandinavia. They played a leading role in the commercial and military affairs of the Slavs from as early as the ninth century A.D. and were soon assimilated with the Slavs. Originally these Norsemen, or Varangians, were called *Ros* or *Rus*, a term later applied to the Slavic people and their country, Russia (*Rossiya* in Russian). Kievan Russia prospered by commercial ties with the city of Constantinople (Istanbul) and peoples of the north. It grew in military strength and extended its power over the other Slavic states. Finally, it fell to a northern Russian prince in the twelfth century. The Russian lands, which had been given some degree of national unity under Kiev, disintegrated into several princely states. The self-

interest of the princes, accompanied by internecine wars, weakened the states and laid them open to the most powerful of the nomadic invaders from Asia. Batu, the grandson of Genghis Khan, overran and destroyed many of the great Russian cities and forced the submission of the others. The Golden Horde, a great Mongol state, was formed in the thirteenth century and extended from the Urals to the Carpathians.

The Ascent of Muscovy

Mongol domination lasted more than 200 years, during which Russian relations with Europe were disrupted. This "Tatar yoke" left a permanent mark on the unified Russian state that eventually emerged. The presence of a common enemy did not, however, lead to a united national movement under the Russian princes. On the contrary, it perpetuated the feuds with the princes vying one against the other for recognition from the Khan as the Grand Duke, the first of princes. Starting from a small wooden fortress on the banks of the Moscow River, the obscure principality of Moscow rose to preeminence during the years of Tatar domination. Moscow was located in the mixed forest zone of the Great European Plain. This location did have some advantages. Although Moscow was not immune from Mongol attack, it was farther removed from the enemy than the devastated grasslands from which the Slavs migrated. Moscow had room for these new settlers and could grow without immediately infringing on the lands of its more powerful neighbors. It was centrally located within the Russian lands of the north, and its access to water transport was advantageous. Locational features did not, of course, ensure Moscow's eventual supremacy.

The ruthlessness and ambition of its princes, although subservient and obsequious to the Khans of the Golden Horde, played a vital role in the rise of Moscow. The Muscovite prince Ivan I was named Grand Duke of the Russians in the early fourteenth century; he offered the highest tribute (bribe) to the Mongols. This office made Ivan the chief tax collector for the Khan, enabling him to expand his own treasury. The political favor of the Khan and growing wealth of the Muscovite Grand Duke led to the purchase and expansion of territory. His prestige grew among the Russian princes. Furthermore, the Eastern Orthodox Church enhanced Muscovite influence from the early fourteenth century when the

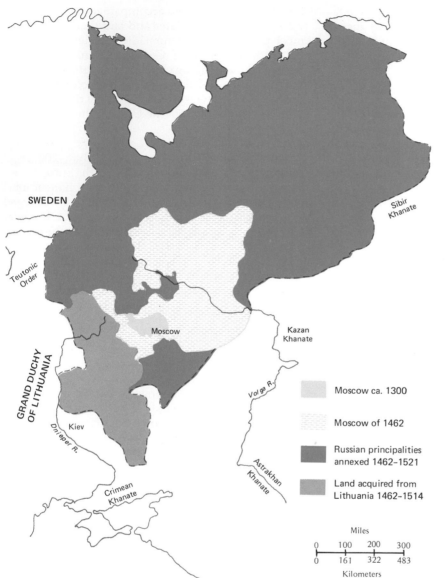

14–5   *Growth of Moscovy, 1300 to 1521*

SWEDEN

Sibir Khanate

Teutonic Order

Kazan Khanate

Moscow

GRAND DUCHY OF LITHUANIA

Volga R.

Kiev

Dnieper R.

Astrakhan Khanate

Crimean Khanate

Moscow ca. 1300

Moscow of 1462

Russian principalities annexed 1462–1521

Land acquired from Lithuania 1462–1514

Miles

0   100   200   300

0   161   322   483

Kilometers

head of the church (the Metropolitan) moved to Moscow.

In the years from 1300 to 1462, Moscow purchased, conquered, or otherwise acquired many Russian principalities. In the process it grew from a small duchy of approximately 500 square miles (1,295 square kilometers) to more than 15,000 square miles (38,850 square kilometers) (Fig. 14–5). During the next seventy years, the size of Muscovy increased by more than two and one-half times. Moscow took from Lithuania a large parcel of land. Competition from the remaining Russian principalities was effectively eliminated, and the Muscovite grand dukes continued to consolidate their power as the absolute rulers of the Russians. Ivan IV, who is referred to as the Dread or Terrible, assumed the title of "Tsar"

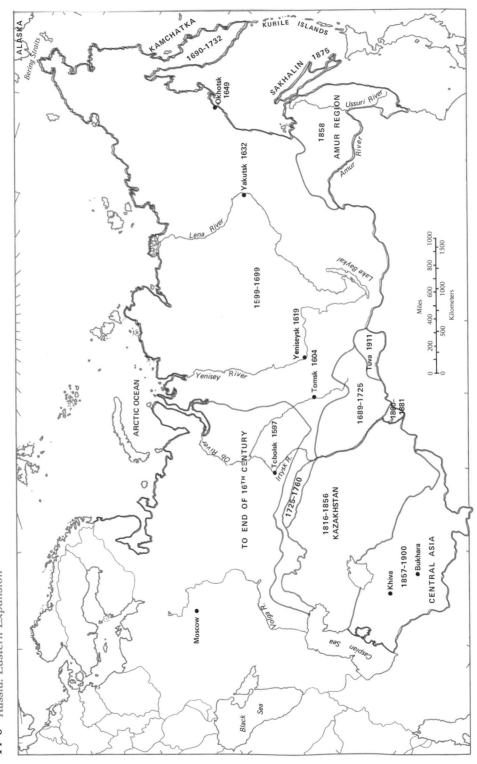

14-6 *Russia: Eastern Expansion*

ALASKA

Bering Straits

KAMCHATKA

1690-1732

KURILE ISLANDS

SAKHALIN 1875

Okhotsk 1649

AMUR REGION

Ussuri River

1858

Amur River

Yakutsk 1632

Lena River

1599-1699

Lake Baykal

Yeniseysk 1619

Tuva 1911

Tomsk 1604

1689-1725

1860-1881

Yenisey River

ARCTIC OCEAN

TO END OF 16TH CENTURY

Tobolsk 1597

Irtysk R.

Ob River

1725-1760

1816-1856

KAZAKHSTAN

Khiva

1857-1900

•Bukhara

CENTRAL ASIA

Volga R.

Moscow

Caspian Sea

Black Sea

Miles

Kilometers

1000

1500

800

600

1000

400

500

200

0

0

(Caesar or Emperor) of all the Russians in 1533, and Moscow's supremacy was complete. The Golden Horde's power had declined rapidly in the fifteenth century. The remnants of this once great empire broke up into the Khanates of Kazan, Astrakhan, Sibir, and Crimea. The lands of the first three of these Khanates were incorporated into Russia in the second half of the sixteenth century under Ivan IV. The Crimean Tatars in the Ukraine were able to withstand Russian expansion for another 200 years. Within a period of 300 years, Moscow rose from the obscurity of a lesser Russian state to master of a large empire. Not only had most of the Russian states been forced into a unified nation, but also numerous other people, especially in the north and east, had come under Moscow's control.

The Eastern Expansion

The demise of the Mongol rule in the east permitted a rapid expansion of the Russian state to the Pacific. The Russians followed the course of the rivers and portages across western Siberia and beyond in search of a treasure of furs and the glory of conquest. At strategic points they established fortified settlements. The speed with which they moved across Asia is evident in the founding dates for the following trading and military posts: Tobolsk, 1597; Tomsk, 1604; Yeniseysk, 1619; Yakutsk, 1632; and Okhotsk, 1649 (Fig. 14–6). Thus, in only 65 years following the death of Ivan the Terrible, the Russians advanced more than 3,000 miles (4,828 kilometers) from western Siberia to the Pacific.

This vast territory was sparsely settled, which contributed to the speed and relative ease of movement. Those indigenous tribes that were encountered were no match for the superior weapons of the Russians. During the first half of the eighteenth century, Russians appeared on the shores of Alaska, and settlement followed. The Russians actually penetrated the North American continent as far south as Fort Ross, California (1812). Their presence, however, was much too tenuous, and the settlement was soon abandoned. Alaska itself was sold to the United States in 1867 for 7 million dollars because it was not deemed worth the expense of retaining.

In the Siberian southeast, the Russians encountered opposition to their expansion and had to bide their time. In the Amur region the Russians met the Chinese who were expanding from the south. The Treaty of Nerchinsk in 1698 divided this land between the Russians and Chinese. It was almost 200 years later that Russia was able to wrest from a weak and disintegrating China the territory north and east of the Amur and Ussuri rivers. Sakhalin Island came under Russian and Japanese control during the mid-nineteenth century. Shortly afterward, Russia acquired complete control of the island. Russia was forced to cede the southern portion of Japan as a consequence of the Russo-Japanese War of 1905.

Russian authority was established over some of the nomads of western and northern central Asia in the 1930s. It was not until the first half of the next century, however, that this territory was annexed by the tsar. The second half of the nineteenth century saw the Russians continue their conquest of Kazakhstan and central Asia, occupying the formerly powerful Khanates of Bukhara (1868) and Khiva (1873). The advance of the Russians into central Asia brought Russia and the British in India into dangerous proximity. Several times in the late nineteenth and early twentieth centuries, Russian and British conflicts nearly erupted into war. The last parcel of Siberian territory to be brought under Russian control during the tsarist years was that of Tuva, on the Mongolian border, in 1911.

European Expansion

In contrast to the eastern expansion, the westward and southern extension of the Russian lands proceeded much more slowly. Powerful states confronted Russia in the west and south. Sweden, Teutonic knights, Lithuania, Poland, Cossacks, and Crimean Tatars, all posed major threats to Moscow's expansion. Political and military supremacy was translated into territorial growth at the loser's expense. The removal of these antagonistic powers from the lands on Russia's margins reduced the threats to the internal security of the state. It also provided some new lands for agricultural expansion. The Russian Eastern Orthodox Christians were also motivated in part by a fervor to restore to their rule coreligionists who had been brought under Roman Catholic or Moslem control. Furthermore, beyond these lands lay the Baltic and Black seas. Access to these seas became increasingly vital to Russia as political, economic, and military relations with Europe intensified from the beginning of the eighteenth century.

Figure 14–7 depicts the Russian (Muscovy) boun-

14–7 *Russia: European Expansion, 1462 to 1914*

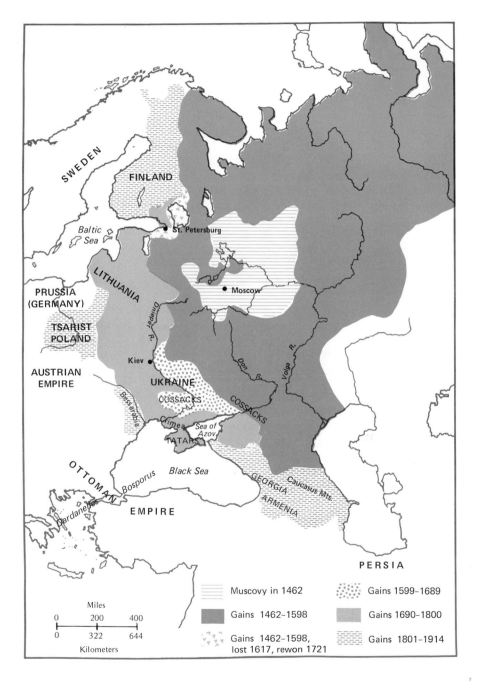

Muscovy in 1462

Gains 1462–1598

Gains 1462–1598, lost 1617, rewon 1721

Gains 1599–1689

Gains 1690–1800

Gains 1801–1914

daries for five dates, from 1462 to the eve of World War I. Prior to the seventeenth century, Moscow's principal acquisitions resulted from the incorporation of competing Russian states and victories over a strong Lithuania. During the seventeenth century, the western Russian boundary changed very little. Russia lost its Baltic Sea outlet to Sweden and gained some ground against Poland and the Turks with their subjects, the Crimean Tatars, in the southwest and south. In the middle of the seventeenth century, Russia was able to take from Poland a parcel of the Ukraine on the right bank of the Dnieper as well as the city of Kiev. At this time, Russia was also able to acquire an oath of allegiance from the Cossacks of the southern steppe. These independent people, however, were able to withstand direct incorporation into Russia for another 100 years.

The 1700s saw impressive Russian gains in Europe. The power of Sweden, Poland, and Turkey declined vis à vis Russia under two strong leaders: Peter the Great, during the early eighteenth century, and Catherine the Great, during the last quarter of that century. Peter was successful in defeating Sweden in several important battles. He reacquired an outlet on the Baltic, where he promptly began to build his new capital, St. Petersburg (Leningrad). This city represented his European outlook and his modernization of old Russia. In the south, the Tatars and Turks were still too strong for Peter's attempts to acquire an outlet to the Azov and Black seas. Not until the reign of Catherine was Turkish influence in the southern steppes finally broken and the Black Sea littoral incorporated into the empire. Russia was closer than ever to the long-standing objective of access to the Mediterranean through the Dardanelles and Bosporus. In the west, Russia, along with Austria and Poland, divided Poland among themselves, and in 1795 that state was wiped off the map.

Early in the nineteenth century, Russia acquired three significant parcels of land. The largest of these was Finland, which Russia annexed in 1809 from Sweden with the encouragement of Napoleon. The Grand Duchy of Warsaw was created by Napoleon from Polish lands held by Prussia and Austria. With the defeat of Napoleon, the bulk of this Duchy was given to Russia (1815). Both Finland and Poland were to be autonomous units within the Russian Empire; however, many of their rights, especially those of Poland, disappeared over the next 100 years. Bessarabia in the southwest was won from an ever-weakening Turkish Ottoman Empire in 1812.

Russian conquest of the Caucasus was accomplished in the nineteenth century. The king of the Christian state of Georgia appealed for the tsar's help against Moslem Turkey in 1801, and that kingdom was annexed. Russian interests in the Caucasus led to conflict not only with Turkey but also with Persia (Iran). Russian successes, however, won territories from these two antagonists during the first half of the 1800s. Some of the fiercest fighting with which the Russians had to contend was with the numerous mountain peoples of this complex ethnographic area. In the 1860s, however, these tribes were somewhat pacified, and the Caucasus was made part of the massive Russian Empire.

Territorial Changes Under the Soviets

On the eve of World War I in 1914, Russia suffered from serious social and economic problems exacerbated by the discontent of many of the ethnic groups within its borders. These problems, together with the severe strain the war placed on the tsarist government, led to its collapse. The short-lived provisional government that replaced the tsar attempted to continue the war effort against the Germans, Austrians, and Turks. Continued losses and internal problems, effectively used by the Communists, however, toppled the government in nine months. The new Communist government sued for peace, and by treaty lost large territories. These losses included Finland, the Baltic states of Lithuania, Estonia, and Latvia, and Poland, Bessarabia, and large areas of White Russia and the Ukraine as well as parts of the Caucasus (Fig. 14–8). The victory of the Allies over Germany the following year led to the creation of the independent states of Finland, Estonia, Latvia, Lithuania, and Poland. Bessarabia and Kars remained parts of Romania and Turkey, respectively. The Bolsheviks were able to reestablish their control over the Ukraine during the Civil War of 1918–1922. They were unsuccessful, however, in their campaigns in the west against Poland. For the next twenty years, the new Union of Soviet Socialist Republics turned inward as the Communist party and Stalin consolidated their power.

The next move for territorial acquisition came in 1939. As Hitler attacked Poland, the USSR invaded Finland, the Baltic states, Bessarabia, and Poland. It annexed outright Estonia, Latvia, and Lithuania as well as Bessarabia and parts of Poland and Finland. The German attack on the USSR shortly afterward forced the Soviets to abandon these lands temporarily. As the Allies forced Germany to its knees with the Russians advancing westward, these territories plus additional land were incorporated into the USSR. The prewar Polish boundary was moved west about

14–8　USSR: Boundary
Changes, 1917 to 1945

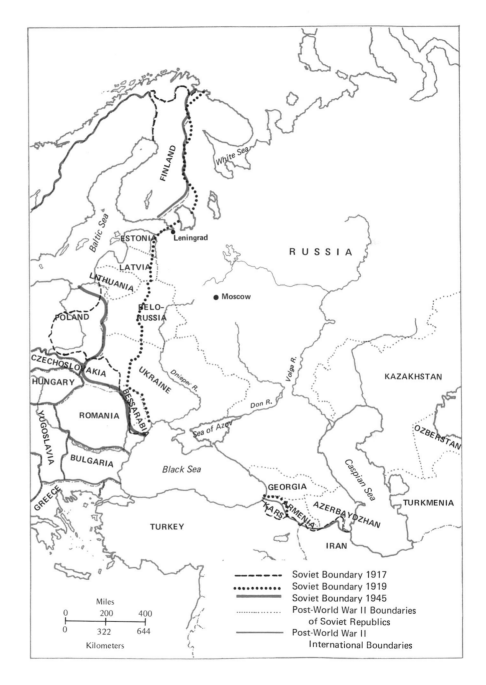

200 miles (322 kilometers), and further territorial acquisitions were made at the expense of Finland, Germany, Czechoslovakia, and Romania. As a reward for a month-long participation in the war against Japan, the Soviet Union acquired control of all of Sakhalin Island and the Kurile Islands in the Far East.

Although the contemporary area of the Soviet Union is slightly less than that of the Russian Empire in 1914, it remains vast in comparison with the other states of the world. The post–World War II Soviet presence in Eastern Europe, moreover, means the Russians exert more control over the earth's surface now than ever before.

Further Readings

BERG, L. S., *Natural Regions of the U.S.S.R.* (New York: Macmillan, 1950), 436 pp. The classic study of Soviet landscape zones by an eminent Soviet geographer.

FLORINSKY, MICHAEL T., *Russia: A History and an Interpretation* (New York: Macmillan, 1957), 1511 pp. A valuable work for the investigation of the complex process of the formation of the Russian state.

KERNER, ROBERT, *The Urge to the Sea* (Berkeley: University of California Press, 1942), 212 pp. This account of Russian expansion places undue stress on the "urge to the sea" or a drive for warm water ports as the dominating theme of Russian expansion; however, it is necessary reading for anyone interested in the growth of the Russian territory.

LONGWORTH, PHILIP, *The Cossacks* (New York: Holt, Rinehart and Winston, 1970), 409 pp. A very readable account of the relations between several Cossack groups and Moscow. One chapter is devoted to the role of the Cossacks in the Siberian expansion.

McNEILL, WILLIAM H., *Europe's Steppe Frontier, 1500–1800* (Chicago: University of Chicago Press, 1964), 252 pp. The conflict among Russia, Turkey, Poland, and Austria in the steppes of Europe is a major theme in this book by a well-known historian.

MELLOR, R. E. H., *Geography of the U.S.S.R.* (New York: St. Martin's Press, 1965), 381 pp. A general geography of the Soviet Union which is especially good for its chapters on the physical environment and the expansion of the Russian state.

MIROV, NICHOLAS T., *Geography of Russia* (New York: Wiley, 1951), 362 pp. An older textbook but one that is very good for a survey of the physical geography of the USSR.

PARKER, W. H., *An Historical Geography of Russia* (Chicago: Aldine, 1968), 416 pp. This book not only provides useful information on the expansion of Russia but also depicts the economic geography of the state at various periods of history.

SUSLOV, S. P., *Physical Geography of Asiatic Russia* (San Francisco: W. H. Freeman, 1961), 594 pp. A detailed investigation of the geology, hydrography, climate, soils, vegetation, and fauna of Siberia and central Asia. Many photographs and maps greatly enhance the value of this book for the appreciation of the landscape zones in this part of the USSR.

The USSR: The Population

15

THE UNION OF SOVIET SOCIALIST REPUBLICS with a population of 259 million people is the third most populous state in the world. Population size and its nature play an important role in Soviet economic development. An understanding of the most significant population characteristics contributes to our appreciation of the economy's spatial distribution as well as contemporary and possible future economic and political problems. The salient population features examined in this chapter are ethnic composition and distribution; population age and sex structure and related manpower problems; population distribution and variations in regional population growth; and urbanization trends.

Soviet Ethnic Groups

In the Soviet census 105 ethnic groups are recognized. These peoples vary widely in their language, history, religious tradition, physical characteristics, and geographic distribution. The presence of different ethnic groups within a sovereign state frequently is a threat to the unity and even the existence of the state. The Soviet Union, however, has maintained a high degree of cohesiveness with apparently little conflict among its diverse ethnic groups.

To minimize potential ethnic conflict, the USSR has adopted a governmental structure that allows a modicum of rights for the ethnic groups, depending in part on each group's size, culture, and economy. The rights afforded include use of their native language in schools, courts, place of business, newspapers and books; maintenance of customs; and for fifty-three of these groups to occupy a politically recognized territory. The laws, customs, and acts of the ethnic groups cannot, however, conflict with the dictates of the central government, which is synonymous with the national Communist party controlled by the dominant ethnic group in the country—the Russians. Stalin defined the nature of the government of the Soviet Union as "national in form, but socialist in content."

Although ethnic groups are granted rights protecting their customs and traditions that do not conflict with the Marxist-Leninist beliefs, a cultural and economic process of integration has continued throughout the Soviet period. Russians have migrated into other ethnic areas and in many now are the major population group. Russian is a required language for students. More than 13 million non-Russians (11.5 percent) speak Russian as their first language, and more than 40 million (35.6 percent) use Russian as a second language.

There is no question that Soviet power has brought benefits to many non-Russians. Illiteracy has been nearly eliminated throughout the country. Life expectancy among even the most remote ethnic groups has been increased. Industries and improved agricultural techniques have been introduced in many economically underdeveloped areas. It is also true, however, that frequently these economic advances have meant that Russians, along with Ukrainians and Belorussians, have moved into non-Russian areas to manage and work in these new enterprises. Overall, the non-Russians have paid a high price for their material and educational improvements. Many have lost the right to continue customs, traditions, and religious practices unacceptable to Moscow. Others,

Table 15–1   *USSR: The Soviet Socialist Republics (SSR)*

		Capital
Union of Soviet Socialist Republics (USSR)	(Soviet Union)	Moscow
1. Russian Soviet Federated Socialist Republic (RSFSR)	(Russia)	Moscow
2. Estonian Soviet Socialist Republic	(Estonia)	Tallin
3. Latvian Soviet Socialist Republic	(Latvia)	Riga
4. Lithuanian Soviet Socialist Republic	(Lithuania)	Vilnius
5. Belorussian Soviet Socialist Republic	(Belorussia)	Minsk
6. Ukrainian Soviet Socialist Republic	(Ukraine)	Kiev
7. Moldavian Soviet Socialist Republic	(Moldavia)	Kishinev
8. Georgian Soviet Socialist Republic	(Georgia)	Tbilisi
9. Armenian Soviet Socialist Republic	(Armenia)	Yerevan
10. Azerbaydzhan Soviet Socialist Republic	(Azerbaydzhan)	Baku
11. Kazakh Soviet Socialist Republic	(Kazakhstan)	Alma Ata
12. Turkmen Soviet Socialist Republic	(Turkmenia)	Ashkhabad
13. Uzbek Soviet Socialist Republic	(Uzbekstan)	Tashkent
14. Kirgiz Soviet Socialist Republic	(Kirgizia)	Frunze
15. Tadzhik Soviet Socialist Republic	(Tadzhikstan)	Dushanbe

most recently the Estonians, Latvians, and Lithuanians, have lost their independence as self-governing sovereign peoples.

Political Territorial Units

Highest of the political territorial units of the USSR is the Soviet Socialist Republic (SSR). According to the Soviet Constitution, the USSR is " . . . a federal state formed on the basis of a voluntary union of equal Soviet Socialist Republics. . . ." The Union, as seen from the history of the formation of the USSR, is anything but voluntary. It was forged by Bolshevik force. Among the many rights the constitution bestows upon the SSR is the right of secession; however, in reality no SSR is allowed to secede, for such a move would indicate "bourgeoisie nationalist tendencies." Rough guidelines set up for the formation of socialist republics include a population of more than 1 million; location on the periphery of the country; and the ethnic group for which the republic is named constitutes a majority of the population. The fifteen republics represent large ethnic groups (Table 15–1 and Fig. 15–1). They do not, however, represent the fifteen largest population groups. The Tatars, Jews, Germans, and Chuvash all rank among the fifteen largest ethnic groups. None of them has their own Soviet Socialist Republic.

Thirty-eight other ethnic groups have an administrative unit of lesser status. The peoples of these units have various rights of language and custom but are subordinate to the laws of the SSR in which they are located as well as to Moscow. Where there are no significant ethnic groups, the territory is divided into administrative oblasts.

Ethnic Composition

Ethnic composition by language groups is shown in Table 15–2. The eastern Slavs belong to the Indo-European language family and are the largest group in the USSR. Seventy-four percent of the nation's population are eastern Slavs, with the Russians alone comprising 53 percent of the total population. There are over 40 million Ukrainians and 9 million Belorussians. The Russians are most prominent in central and northern European Russia, Siberia, and Kazakhstan (Fig. 15–2); however, they are distributed throughout the USSR, where they form important minorities, especially in cities. The Ukrainians and Belorussians are likewise widely dispersed, although the majority are found within the limits of their own SSR.

Other ethnic groups representing the Indo-European family of languages include the Latvians and Lithuanians in the Baltic region, the Armenians in

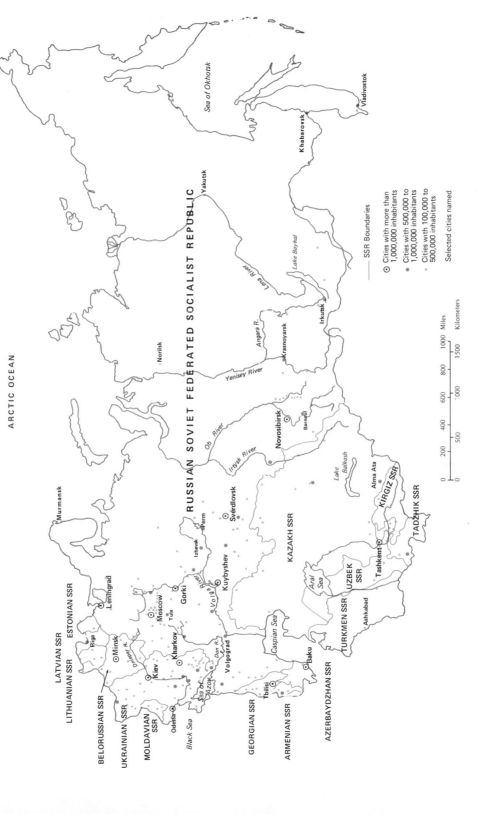

15-1 *USSR: Soviet Socialist Republics and Cities Over 100,000 Population*

ARCTIC OCEAN

RUSSIAN SOVIET FEDERATED SOCIALIST REPUBLIC

Sea of Okhotsk

Vladivostok

Khabarovsk

Yakutsk

Lena River

Lake Baykal

Irkutsk

Norilsk

Angara R.

Krasnoyarsk

Yenisey River

Novosibirsk

Barnaul

Ob River

Irtysh River

Sverdlovsk

Perm

Izhevsk

KAZAKH SSR

Lake Balkhash

Alma Ata

KIRGIZ SSR

TADZHIK SSR

Murmansk

Leningrad

Gorki

Moscow

Tula

Volga River

Kuybyshev

Aral Sea

Tashkent

UZBEK SSR

TURKMEN SSR

Ashkabad

Riga

Minsk

Dnieper R.

Kiev

Kharkov

Don R.

Volgograd

Sea of Azov

Odessa

Black Sea

Caspian Sea

Baku

Tbilisi

LATVIAN SSR

ESTONIAN SSR

LITHUANIAN SSR

BELORUSSIAN SSR

UKRAINIAN SSR

MOLDAVIAN SSR

GEORGIAN SSR

ARMENIAN SSR

AZERBAYDZHAN SSR

— SSR Boundaries

⊙ Cities with more than 1,000,000 inhabitants

● Cities with 500,000 to 1,000,000 inhabitants

· Cities with 100,000 to 500,000 inhabitants

Selected cities named

0   200   400   600   800   1000 Miles

0   500   1000   1500 Kilometers

297

	Population (in Thousands)	Percent of Total Population
Indo-European Family of Languages		
Slavic Group		
Russian	129,015	53.4
Ukrainian	40,753	16.9
Belorussian	9,052	3.8
Total Eastern Slavs	178,820	74.1
Polish	1,167	0.5
Baltic Group		
Lithuanian	2,665	1.1
Latvian	1,430	0.6
Armenian Group		
Armenian	3,559	1.5
Romance Group		
Moldavian	2,698	1.1
Iranian Group		
Tadzhik	2,136	0.9
Germanic Group		
German	1,846	0.8
Altaic Family of Languages		
Turkic Group		
Uzbek	9,195	3.8
Tatar	5,931	2.5
Kazakh	5,299	2.2
Azerbaydzhan	4,380	1.8
Chuvash	1,694	0.7
Turkmen	1,525	0.6
Kirgiz	1,452	0.6
Bashkir	1,240	0.5
Uralian Family of Languages		
Finnic Group		
Mordvinian	1,263	0.5
Estonian	1,007	0.4
Japhetic Family of Languages		
South Caucasian Group		
Georgian	3,245	1.3
Jewish	2,151	0.9
Others	9,017	3.6

Table 15–2 USSR: Major Ethnic Groups (Population greater than 1 million in 1970) Total Population: 241,720,000

SOURCE: USSR Central Statistical Committee, *Itogi Vsesoyuznoy Perepis Naseleniya 1970 Goda* (Moscow: Statistika, 1973).

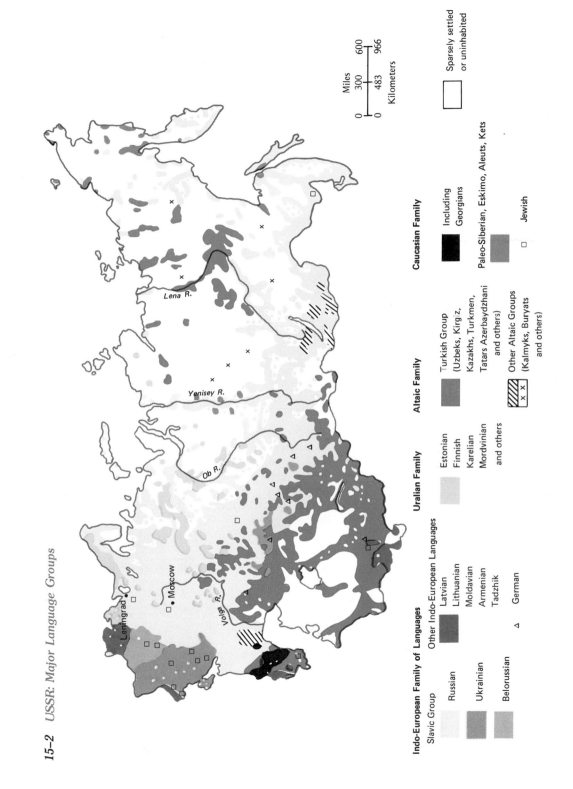

15-2 *USSR: Major Language Groups*

Indo-European Family of Languages

Slavic Group

Russian

Ukrainian

Belorussian

Other Indo-European Languages

Latvian
Lithuanian
Moldavian
Armenian
Tadzhik

△ German

Uralian Family

Estonian
Finnish
Karelian
Mordvinian
and others

Altaic Family

Turkish Group
(Uzbeks, Kirgiz,
Kazakhs, Turkmen,
Tatars Azerbaydzhani
and others)

Other Altaic Groups
(Kalmyks, Buryats
and others)

Caucasian Family

Including
Georgians

Paleo-Siberian, Eskimo, Aleuts, Kets

□ Jewish

Miles
0   300   600
0   483   966
Kilometers

Sparsely settled
or uninhabited

Lena R.

Yenisey R.

Ob R.

Volga R.

Leningrad

Moscow

299

Soviet central Asia was once an important center of Moslem culture. Samarkand is one such center that has flourished under Soviet control. The Moslem heritage is exemplified by a distinct architectural style. (TASS from Sovfoto)

the Caucasus, and the Tadzhiks, an Iranian people, in central Asia. There are also Germans, Moldavians, and Poles. Except for the Moslem Tadzhiks, these groups are traditionally Christian. The Latvians are Lutheran; the Lithuanians Catholic; the Armenians have their own ancient Christian church; and the Belorussians, Ukrainians, Russians, and Moldavians are Eastern Orthodox.

There are over 30 million people speaking Turkic languages. These groups have an Islamic heritage and are found principally in central Asia, the middle Volga, and the Caucasus. The Uralian family of languages is represented by a number of groups found in northern European Russia and in western Siberia. Most prominent of this family are the Estonians with their close relations, the Finns and Karelians.

The Jewish people form the eleventh largest ethnic group in the USSR. They are predominantly an urban population found mostly in European Soviet Union. Moscow's attempt to establish a Soviet homeland for the Jews in the Far East, the Jewish Autonomous Oblast, has not been successful in attracting the Jewish population. Less than 7 percent of the population of the oblast is Jewish. Other groups comprising the nationality patchwork include the numerous Caucasian peoples in the Caucasus, of whom the Georgians are most notable. There are also Mongols, Koreans, Gypsies, and the many small indigenous tribes of Siberia, such as the Eskimos.

Ethnic Demographic Differences

The Slavs, Baltic peoples, and other groups of modern European culture have passed into that stage of the demographic transformation in which birth rates, death rates, and rates of natural growth are low. In contrast, other ethnic groups with a more traditional and agricultural life style are experiencing higher birth rates which, with low death rates, result in faster population growth.

Ethnic groups in central Asia, the Caucasus, and Moldavia have a higher rate of population growth than the Slavs and other European types. In 1959, the Eastern Slavs comprised about 76 percent of the nation's population, whereas in 1970 they made up 74 percent. The Russian population for the same period declined from 55 percent to 53 percent. It is possible that within two to three decades, the Russians will be less than 50 percent of the population. This fact in itself, however, should not lessen their domination over the entire country. It must be remembered that a handful of people representing the top hierarchy of the Communist party effectively controls the entire country. The presence of so many nationalities remains a reality, however, with which the government must contend. Minority resentment against the Russians continues, and the government watches for potential disruptions such as the Lithuanian demonstration in 1972, and the continued efforts of Jews and other ethnic minorities to leave the Soviet Union.[1]

Demographic Characteristics of the Population

From the early part of the twentieth century to the present, the population as a whole has passed from a stage of high birth rates and moderately high death rates to a stage of low birth and death rates. In 1913, the birth rate was 4.5 percent; in 1959, 2.5 percent; and now 1.8 percent. The death rate for this same period went from 2.9 percent to 0.9 percent. The resultant rates of natural increase were 1.6 percent in 1913, 1.7 percent in 1959, and 0.9 percent presently. Life expectancy in European Russia at the end of the nineteenth century was thirty-one years for men and thirty-three years for women. Presently, life expectancy is sixty-five years for men and seventy-four years for women (sixty-seven and seventy-five years, respectively, in the United States).

Age and Sex Structure of the Soviet Population

Figure 15–3 shows the distribution of the Soviet population by age and sex. This population pyramid is a valuable tool to demonstrate some of the Soviet demographic characteristics. One striking characteristic is the great imbalance in the male and female population. For the entire Soviet Union, there are 85.5 males for every 100 females (a population of 95

[1]For example, in September 1974, 3,500 Germans petitioned the West German prime minister in Moscow for his assistance in their efforts to emigrate.

to 99 males per 100 females is considered normal). It will take at least two generations before a normal ratio is attained. The abnormal ratio between the males and females from age forty and higher reflects the destructive effects on men, in particular, of wars, revolution, and collectivization. It is estimated that 14 million lives were lost in World War I and subsequent civil strife, 5 million during collectivization of agriculture in the late 1920s and early 1930s, and as many as 20 million in World War II—an astounding total in a mere thirty years.

The shortage of males has required the use of female labor in many occupations including heavy labor. Agriculture particularly has a high female labor component, since many young men have migrated to the city for higher paying industrial jobs and greater urban amenities. The availability of labor is also affected by variations in the birth rates. During World War II, the birth rate dropped sharply; therefore, severe labor shortages appeared in the late 1950s and early 1960s. To remedy this deficit, the Soviet government adopted a number of policies such as shortening the period of secondary schooling by a year, stepping up the call for volunteer work on Saturdays and Sundays, and requiring students to interrupt their studies to work where manpower deficiencies existed. Also, laws were passed giving women greater equality in Soviet society.

During the 1960s, the number of young workers entering the labor force steadily increased; however, the lower birth rates in the 1960s generated further labor shortages from the late 1970s. The decline in the birth rate in the 1960s resulted primarily because fewer women born during the war years were entering the prime childbearing years. Other factors that affect the birth rate include the high proportion of women in the labor force, greater economic expectations and achievement, especially with both husband and wife working, and the drain of family resources with numerous children. Furthermore, severe shortages in the availability of housing discourage large families. Also, abortions are readily available in the Soviet Union as are various forms of contraception.

There are important regional variations in the size of the work force. Birth rates are generally lower among the more economically advanced groups, the Slavs and Baltic people, than among the groups in central Asia. The former groups, especially the Slavs,

15–3 *USSR: Population Pyramid*

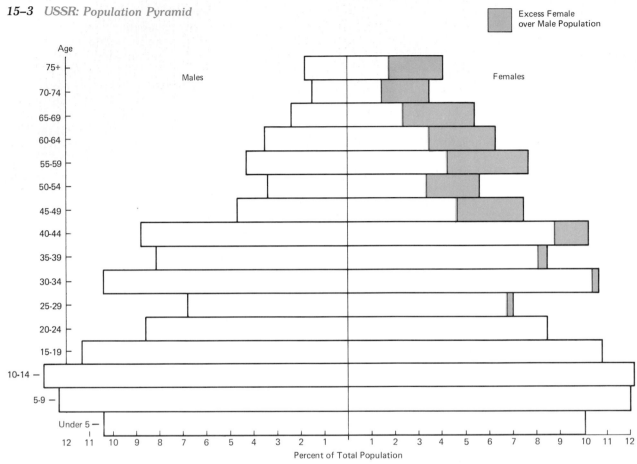

SOURCE: Frederick A. Leedy, "Demographic Trends in the USSR," Joint Economic Committee, Congress of the United States, *Soviet Economic Prospects for the Seventies* (Washington, D.C.: Government Printing Office, 1973).

are much more a part of the national economy than the latter, who for the most part have not been fully integrated into the modern industrialized economy. Ensuring an adequate labor supply requires improvement in overall, individual labor productivity through increased capitalization and job training. Also, a more concerted effort to expand the participation of the people of the less developed regions in the modernized Soviet economy can help overcome labor deficits.

Although the Soviet Union theoretically accepts the Marxist position regarding population growth, discussed in Chapter 1, it does not have a clearly defined pronatal policy. Bachelors and childless couples pay an additional income tax of 6 percent, and awards and monetary stipends are given to mothers with four or more children. Such payments are small and are not a real incentive for large families. On the other hand, the availability of government-provided contraceptive devices and abortions, together with numerous economic factors, encourage small families. The declining birth rates and prospective labor shortages have stimulated governmental discussion of a more forceful pronatal policy. Such a policy, if enacted, might include extended maternity leave with pay and a guarantee of employment when the mother is ready to return to work.

Distribution of the Soviet Population

A national population density figure is meaningless in a country the size of the USSR. The Soviet population is very unevenly distributed (Fig. 15–4). In the western Ukraine, rural population densities are over 250 per square mile (100 per square kilometer), whereas large tracts of tundra and desert are essentially unpopulated. Almost three-quarters of the population live in European USSR, which contains about one-quarter of the national area. Conversely, about one-quarter of the population lives in three-quarters of the territory made up of Siberia and central Asia. In European USSR, areas with densities of at least twenty-five people per square mile (10 per square kilometer) are bounded by the international

boundary in the west and by the 60th parallel in the north. The northern limit of this zone continues just to the east of the Urals, where it extends southward, then westward to the bend of the Volga River near Volgograd. Most of the northern Caucasus and Trans-Caucasus regions are included within this population density zone. The areas of highest population densities are in the southwestern Ukraine, in the west and eastern lowlands of the Trans-Caucasus, in the vicinity of industrial regions and large cities such as the eastern Ukraine, the Moscow area, the middle Volga lands, and scattered areas in the chernozemic soil area.

Outside the European part of the country, the heaviest population concentrations are found in the foothills and valleys of the central Asian mountains and along principal rivers and irrigated areas. Scattered islands with population densities exceeding twenty-five per square mile (ten per square kilometer) are found in the southern portions of Siberia. Here

15–4 USSR: Population Distribution

One dot represents 100,000 people

Table 15–3   USSR: Regional Population Growth, 1940 to 1975

	Population in Millions			Percent Change	
	1940	1959	1975	1940–1959	1959–1975
USSR	194.1*	208.8*	253.3*	8	21
European USSR	159.8*	163.3*	189.6*	2	16
Northwest	11.2	10.9	12.7	− 3	17
Central Industrial	27.0	25.7	28.3	− 5	10
Volga-Vyatka	8.8	8.3	8.3	− 7	less 1%
Central Chernozem	9.1	7.8	7.8	−15	less 1%
Volga	15.6	16.0	19.0	2	19
North Caucasus	10.5	11.6	15.0	11	29
Trans-Caucasus	8.2	9.5	13.3	16	40
Urals	10.5	14.2	15.3	35	8
Baltic	5.9	6.6	8.0	13	21
Belorussia	9.0	8.1	9.3	−11	15
Donetsk-Dnieper	16.4	17.8	20.7	8	16
Southwest	20.0	19.0	21.3	− 5	12
South	5.0	5.1	6.8	2	33
Moldavia	2.5	2.9	3.8	17	30
Asiatic USSR	34.3*	45.5*	63.7*	33	40
West Siberia	9.2	11.3	12.4	23	10
East Siberia	4.9	6.5	7.8	32	20
Far East	3.2	4.8	6.4	53	33
Kazakhstan	6.1	9.3	14.2	51	53
Central Asia	10.1	13.7	22.9	25	67

*Totals do not always equal sums of parts due to rounding off.
SOURCE: USSR Central Statistical Committee, *Narodnoye Khozyaystvo SSSR v 1974* (Moscow: Statistika, 1975).

the Trans-Siberian Railroad has been an important contributing element in agricultural and industrial development.

Throughout most of the taiga, tundra, and desert zone, densities are two or less per square mile (one per square kilometer). Some sections of the tundra, ice-covered northern islands, and the central Asian desert are essentially unpopulated.

Changes in the Regional Distribution of the Population

Significant variations have occurred in the regional population growth of the country. An awareness of these changes introduces us to some population dynamics and aids in assessing Soviet attempts at planned economic development. Furthermore, an awareness of the recent regional population changes and the factors accounting for these changes can assist in speculation on future patterns.

From 1940 to 1959, the greatest single factor influencing the regional growth of the Soviet population was World War II. Several areas in the western part of the country had fewer people in 1959 than in 1940 (Table 15–3 and Fig. 15–5). In addition to very heavy civilian losses, large numbers of the people were relocated eastward beyond the war zone. Furthermore, relatively lower birth rates among the Slavic and Baltic people contributed to a slower recovery. The greatest declines occurred where a large rural population experienced high rural to urban migration rates. The urban areas to which they moved frequently lay outside the region.

15–5 USSR: Regional Population Growth, 1940 to 1959

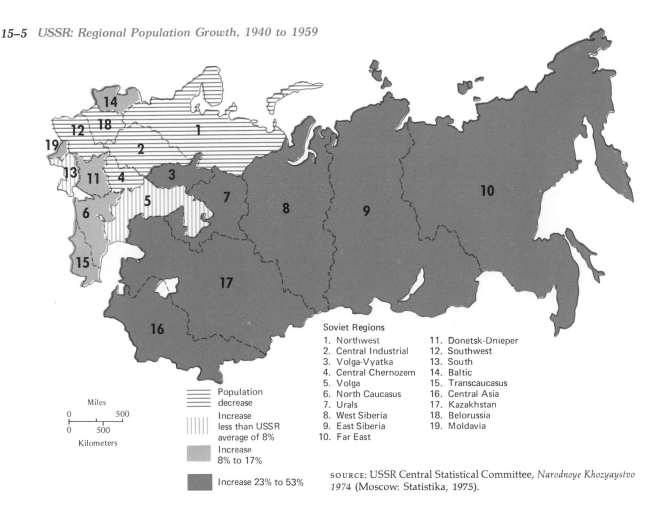

Soviet Regions

1. Northwest
2. Central Industrial
3. Volga-Vyatka
4. Central Chernozem
5. Volga
6. North Caucasus
7. Urals
8. West Siberia
9. East Siberia
10. Far East
11. Donetsk-Dnieper
12. Southwest
13. South
14. Baltic
15. Transcaucasus
16. Central Asia
17. Kazakhstan
18. Belorussia
19. Moldavia

Population decrease

Increase less than USSR average of 8%

Increase 8% to 17%

Increase 23% to 53%

Miles
0   500
0   500
Kilometers

SOURCE: USSR Central Statistical Committee, *Narodnoye Khozyaystvo 1974* (Moscow: Statistika, 1975).

The other regions in European Russia all increased their population in the 1940 to 1959 period. Two of these regions grew at a rate less than the national average, reflecting partially war casualties and the deportation of ethnic groups. The growth of the remaining regions of European Russia all equaled or exceeded the growth of the country as a whole. The reasons for the greater increases in these regions are varied: for example, both the Baltic and Donets-Dnieper regions were greatly affected by the war; however, the desire of the Soviets to settle the strategic Baltic lands with Russians spurred population growth. The need to reconstruct the industries in the Donets-Dnieper region, Russia's foremost zone of heavy industry, led to a rate of population increase equal to that of the USSR.

Increased industrial activity in North Caucasus, and particularly the Urals, stimulated population growth, whereas the more traditionally rural populations of Moldavia and the Trans-Caucasus expanded largely as a result of high rates of natural increase.

In Asiatic Soviet Union, all regions grew at least three times faster than the country as a whole. In-migration, quickened by the war and consequent industrial and agricultural development, accounts for much of the increase in Siberia and Kazakhstan. High birth rates, coupled with low death rates uninterrupted by the war, contributed to the rapid population increase for the Central Asian republics. The eastern regions were also the destination of millions of political prisoners and national groups deported from European USSR.

15-6   USSR: Regional Population Growth, 1959 to 1975

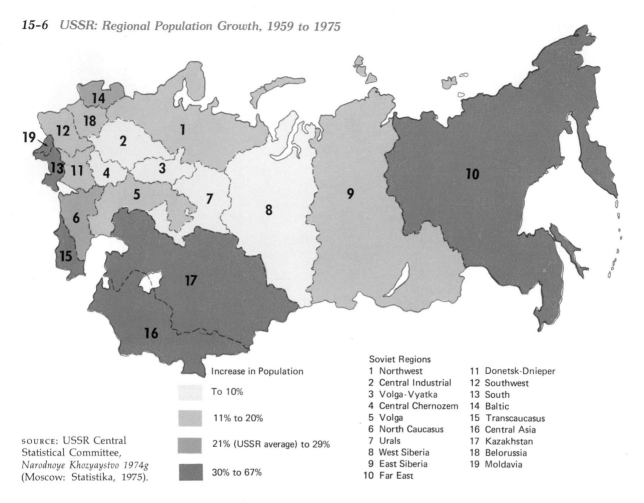

Increase in Population

To 10%

11% to 20%

21% (USSR average) to 29%

30% to 67%

SOURCE: USSR Central Statistical Committee, *Narodnoye Khozyaystvo 1974g* (Moscow: Statistika, 1975).

Soviet Regions
1 Northwest
2 Central Industrial
3 Volga-Vyatka
4 Central Chernozem
5 Volga
6 North Caucasus
7 Urals
8 West Siberia
9 East Siberia
10 Far East
11 Donetsk-Dnieper
12 Southwest
13 South
14 Baltic
15 Transcaucasus
16 Central Asia
17 Kazakhstan
18 Belorussia
19 Moldavia

The spatial variations in population growth from 1959 to 1975 differed in some important respects from the period 1940 to 1959 (Fig. 15–6). All regions experienced growth; however, the slowest growing section remained European Soviet Union, particularly the agricultural regions. The southern regions in both the European and Asiatic parts witnessed the fastest growth of population. A significant share of this population increase is attributed to high rates of natural growth. In-migration, because of economic expanse, however, was also an important contributing factor in some of these rapidly growing regions. Furthermore, the return of peoples deported during World War II also contributed to the growth of the North Caucasus.

The rate of growth of the Urals and Siberia con-trasts sharply with the increases they experienced during the 1940–1959 period. The lower growth rates for these regions indicate the difficulty in attracting and keeping people in these more remote areas. Increased mining activities in the Far East help account for its larger population gains; however, a general slower rate of economic expansion in the Urals and Siberia has taken place during the 1959 to 1975 period.

Urbanization

It has been pointed out several times that the economic development of a country, particularly the

Table 15–4 *USSR: Growth of the Urban Population, 1897 to 1975*

	Urban Population (in millions)	Rural Population (in millions)	Percent Urban
1897	18	106	15
1913	29	131	18
1940	63	131	33
1959	100	109	48
1961	108	108	50
1970	136	106	56
1975	153	100	60

SOURCE: USSR Central Statistical Committee, *Narodnoye Khozyaystvo SSSR v 1974g* (Moscow: Statistika, 1975).

growth of industrial activities, leads to urban growth and an increasing proportion of the population living in cities. This thesis holds true for the Soviet Union. Despite increased industrialization and commercialization in tsarist Russia, only 18 percent of the population lived in cities in 1913 (Table 15–4).

During the next decade the percent of urban population declined because of the chaos created by World War I, the revolution, civil war, economic dislocation, and famine. It was not until the Soviet Union embarked on an all-out program of industrialization in the late 1920s that urban growth began to experience its most rapid increase. By 1928, the proportion of the population living in urban areas

Leningrad (formerly St. Petersburg) was Peter the Great's gateway to the West and is one of the Soviet Union's major cities (4.3 million inhabitants). (Sovfoto)

Moscow, capital of the Soviet Union, with a population of 7.6 million is the political and economic heart of the USSR. During the Soviet period broad streets and high-rise buildings have been constructed yet housing is still in short supply. (TASS from Sovfoto)

had attained the pre-World War I level; by 1940, this share had more than doubled to 33 percent of the population.

World War II temporarily slowed the pace of urban growth. The return to peacetime conditions stimulated urban growth. By 1961, as many people lived in cities as in rural areas in the Soviet Union. Thirteen years later approximately 60 percent of the population was urban. During the seventy-seven years from 1897 to 1975, the urban population increased more than eightfold, from 18 million to 153 million. The rural population, on the other hand, was 6 million less in 1975 than in 1897. During the intercensus period 1959 to 1970, 36 million people were added to the urban population. Of this number, 16 million resulted from a net in-migration to cities from rural areas. As a result of converting rural settlements to urban settlements, 5 million people were added to the urban lists. Natural increases within urban centers accounted for another 15 million people.

Regional variations in city growth reflect overall regional population trends. Widespread rapid growth of cities from the Volga to the east occurred between 1940 and 1959. From 1950 to 1970, however, the most rapid increase in the urban population occurred in those parts of the Soviet Union that were in the early stages of the urban and demographic revolutions. These areas are found along the western border, southern Ukraine, Central Chernozem region, Armenia, and Azerbaydzhan. In the Asiatic part of the Soviet Union, Kazakhstan and central Asia experienced rapid urban growth. Cities with chemical industries and administrative functions grew at a faster pace than others.[2]

Since 1959, the cities in the Donbass, the Moscow area, the Volga region, the Urals, and Siberia grew more slowly. Professors Lydolph and Pease conclude that these latent trends indicate:

. . . that a recent spurt in growth has taken place primarily in less developed areas and less developed cities within the European part of the country, the Caucasus and Central Asia.

They continue by remarking that these vagaries in the urban growth rates:

. . . might also indicate a trend away from the concentration on heavy industry to increased emphasis on diversified lighter industries. It appears that the formerly less developed portions of the western part of the country are perhaps receiving the bulk of new plants associated with growth industries (for example, chemicals in Belorussia).[3]

The distribution of cities, like the distribution of the population as a whole, is uneven throughout the Soviet Union. The map (Fig. 15–1) of large cities (100,000 or more) shows that the greater number lie in European Soviet Union. Two hundred forty-five cities belong to this group. Of these, sixty-four are located in Siberia, Kazakhstan, and central Asia. The remainder (181), including eleven in the Caucasian republics, are in the western portion of the country. The distributional pattern of these cities forms a wedge with a broad base stretching from the Baltic to the Black Sea. Eastward, this wedge narrows, excepting the line of cities along the eastern shore of the Black Sea through the northern Caucasus. In Siberia, large cities are concentrated primarily along or near the Trans-Siberian Railroad. The Soviet Union has twenty-eight cities with populations between one-half million and 1 million. Of these, nineteen are in the European part and nine in Siberia and central Asia. The thirteen Soviet cities with populations of 1 million or more are Moscow (7.6

²Chauncy D. Harris, "Urbanization and Population Growth in the Soviet Union, 1959–1970," *Geographical Review,* 61 (1971), 103–107, 124.

³P. E. Lydolph and S. R. Pease, "Changing Distributions of Population and Economic Activities in the U.S.S.R.," *Tijdschrift voor Economische en Sociale Geographie,* 63 (1972), 155.

million), Leningrad (4.3 million), Kiev (1.9 million), Baku (1.4 million), Kharkov (1.4 million), Gorki (1.3 million), Sverdlovsk, Kuybyshev, and Minsk (1.1 million inhabitants each) and Tbilisi and Odessa (1.0 million) in European Russia, Novosibirsk (1.3 million) in western Siberia, and Tashkent (1.6 million) in central Asia.

Further Readings

Several excellent articles on the Soviet population are available as follows:

Harris, Chauncy D., "Urbanization and Population Growth in the Soviet Union, 1959–1970," *Geographical Review,* 61 (1971), 102–124. Harris provides much valuable information for the reasons behind urban growth and its regional variations.

Leedy, Frederick A., "Demographic Trends in the USSR," in Joint Economic Committee, Congress of the United States, *Soviet Economic Prospects for the Seventies* (Washington, D.C.: Government Printing Office, 1973), pp. 428–484. A detailed account is given of the growth of the Soviet population from 1950 to 1972, its age structure, sex composition, birth, death, and growth rates. Leedy also discusses Soviet population policy, regional and urban population distributions, and gives projections of future Soviet population growth.

Lewis, Robert A., and Rowland, Richard H., "Urbanization in Russia and the USSR: 1897–1966," *Annals, Association of American Geographers,* 59 (1969), 776–796. This article is most useful for examining the growth of the urban population particularly during the Soviet period.

Lydolph, P. E., *Geography of the USSR* (New York: Wiley, 1970), 683 pp. Lydolph has a very good account of Soviet population problems (pre-1970 census) and the nationality situation in Chapter 13, "Population, Nationalities, Manpower and Employment," pp. 335–397.

Lydolph, P. E., "Manpower Problems in the USSR," *Tijdschrift voor Economische en Sociale Geographie,* 63 (1972), 331–344. Lydolph presents an excellent survey of Soviet population characteristics as they relate to problems of the availability of workers, the reasons behind these variations, and possible solutions to the problems created by these variations.

Lydolph, P. E., and Pease, S. R., "Changing Distributions of Population and Economic Activities in the USSR," *Tijdschrift voor Economische en Sociale Geographie,* 63 (1972), 244–261. This paper examines regional growth patterns in the Soviet Union during the period 1959 to 1970. Both natural population increase and migration are discussed along with economic factors and governmental policies relating to regional population changes.

Book-length studies on the Soviet population include the following:

HARRIS, CHAUNCY D., *Cities of the Soviet Union: Studies in Their Functions, Size, Density and Growth* (Skokie, Ill.: Rand McNally, 1970), 484 pp. This volume contains a detailed analysis of Soviet cities. Maps, tables, and graphs are abundantly used. A most valuable bibliography of Soviet urbanization is also provided.

LORIMER, FRANK, *The Population of the Soviet Union: History and Prospects* (Geneva: League of Nations, 1946), 289 pp. A near classic on the population during the earlier part of the Soviet period.

Many books are available on Soviet ethnic groups. Noteworthy of these are as follows:

ALLWORTH, EDWARD (ed.), *The Nationality Question in Soviet Central Asia* (New York: Praeger, 1973), 217 pp. A collection of essays focusing largely on the contemporary position and problems of the people of Kazakhstan and Soviet central Asia.

CONQUEST, ROBERT, *The Nation Killers* (London: Macmillan, 1970), 222 pp. An interesting account of the ethnic groups deported by Stalin during World War II. Included is a discussion of the status of these peoples as of the late 1960s.

GOLDHAGEN, ERICH (ed.), *Ethnic Minorities in the Soviet Union* (New York: Praeger, 1968), 351 pp. This work is a collection of scholarly articles on selected nationality groups and general political, economic, and language problems of Soviet nationalities.

The USSR: Economic Activity

16

Planned Development

The Soviet Union's most significant achievement during its sixty years of existence is clearly its rapid industrial development. Industrialization has progressed to the point that the USSR has emerged as a major world power second only to the United States. The Soviet Union points with great pride to its success in working toward the attainment of its two principal goals: surpassing the United States in the production of goods and thereby becoming the world's greatest economic power; and raising the material level of life in the USSR to the highest levels. By past accomplishments and anticipated growth, the Soviet leaders believe they will prove their system superior to any of the capitalist varieties. At a time when so many poor nations are searching for ways to raise their own standard of living and attain a higher level of modernization, the Soviet challenge to the Western industrial powers cannot be ignored.

The Soviets contend that the key to their economic development lies in the application of the "truths" of social and economic development expounded in Marxist-Leninist philosophy. This body of ideas bestows on the handful of people (or on a single person like Stalin) the power to control and plan virtually every segment of the economy through a large bureaucracy.

The state planning agency (Gosplan), following Communist party directives, works out detailed plans. A plan is devised for a five-year period and for each separate year of the period. The complexities of these vast economic blueprints invariably lead to numerous problems. Frequently overoptimistic production goals have had to be reduced. The inability of one producer, for example, the coal industry, to meet its stated goals reverberates through other industries such as steel and thermal electricity. Shortages in these industries in turn affect industries they supply. Because the plan targets have usually been expressed in quantity of production, the Soviet economy has been faced with perennial problems involving the waste of resources, the inefficient use of labor, and poor-quality goods. Furthermore, Soviet politics are inseparably linked with the economic plans. For example, the invasion of Czechoslovakia in 1968 led to the diversion of capital, material, and labor and caused disruptions in the five-year plan.

Despite these and many other shortcomings, the Soviet Union has achieved notable success in its overall economic growth. This growth, however, has largely been the result of the development of heavy industry. The decision to build the industrial and military strength of the USSR was made in the late 1920s and has been the priority of every plan since 1928. Sectors of the economy such as agriculture and light industry (notably consumer goods) have suffered from low rates of investment, low priority status, and consequently low rates of growth compared to heavy industry. To illustrate this point, Soviet official statistics indicate that if one assumes the value of producer goods, consumer goods, and agricultural production in 1922 equal to "1," then currently the value of these indices would be 770 for producer goods, 95 for light industry, and only 5 for the value of agricultural goods.

The Soviet Union has made rapid industrial growth by emphasizing heavy industries such as iron and steel making and machinery. Here is a synthetic rubber plant in which hydrocarbons from petroleum are the principal raw material. Production of synthetic rubber lessens Soviet dependence on imported natural rubber. (Novosti from Sovfoto)

The Level of Economic Development in Old Russia

Soviet writers stress the point that the economy they inherited was poorly developed. They point out that Russia was a "backward" state, no match for its more modernized and industrial competitors such as the United States, the United Kingdom, Germany, and France. The economic achievements of the country, they contend, are first and foremost the product of the system imposed on the state by the Soviet leaders. It is true that over 80 percent of the labor force of tsarist Russia was employed in agriculture. Russia did, however, rank fifth in the world in terms of total industrial output on the eve of World War I. Russia's per capita levels of production, though, placed Russia behind such states as Sweden, Spain, and Italy. Russian prerevolutionary industry was one of great contrasts. Small, traditional craft industries were found along with large, new factories employing the latest technological innovations of the West.

Large-scale industrial production was concentrated in European Russia. The largest concentration was centered on Moscow. Here textile, engineering, and metal industries were prominent. Metalworking and engineering industries were also located near Russia's then largest city, St. Petersburg (Leningrad). The southern Ukraine was the most important region of heavy industry, iron and steel, coal production, machine building, and engineering industries. Metallurgy was also a principal activity of the Polish provinces, which were then part of the Russian Empire. In the Urals an older, less productive iron industry existed with the mining and processing of numerous other minerals. The area around Baku was the principal petroleum-producing area.

The most notable feature of Russia's prerevolutionary economy was the newness of its industrial economy. If one were to set a date for the inception of the Russian "industrial revolution," it would have to be in the early 1890s. During the last decade of the nineteenth century, Russia's industrial production doubled. Spurred by a great expansion of the rail network in the country, a modern large-scale iron and steel industry developed in the Ukraine. From 1890 to 1900, the production of pig iron increased by more than three times. Coal extraction increased by more than 2.5 times, as did petroleum production. Russia's industrial growth was greater than that of the Western industrial powers. Although the growth of Russian industry slowed somewhat in the first decade of the twentieth century, on the eve of World War I total industrial output was more than three times that of 1890.

The foundation of an industrial society had been laid under the old regime. Tsarist Russia was developing a modern industrial economy. The economist Marshall Goldman succinctly points out the significance of this prerevolutionary development to those nations who see the Soviet example as the key to their own successful modernization.

Those in the developing countries who look to the Soviet Union as a model for themselves often are unaware of just how much progress had been made. In comparison to Russia of 1913, the countries of Africa and Asia are at a much lower economic level. If nothing else the Russian population prior to World War I was about one-fourth the population in present-day India and one fifth that of China. Consequently, on a per capita basis, Chinese and Indian production levels in 1950 and even in 1957 and 1960 were considerably below Russia's in 1913.[1]

Recent Trends in Soviet Economic Planning

The Soviet economy, its planning and administration, are neither a "chaotic mess" nor a model of perfection. The Soviet leaders, like their counterparts in the satellite states of Eastern Europe, are concerned with increasing economic efficiency and reducing waste. Unlike the governments of Hungary and Czechoslovakia, however, Soviet economic reformers do not yet appear willing to relinquish the hold the authorities in Moscow have on production in favor of the enterprises' managers. Nevertheless, concerned by decreasing rates of growth during the late 1960s and early 1970s, and the unchanging gap between Soviet GNP and United States GNP, the Soviet leaders are searching for ways to increase the efficacy of their economic planning and administration. The present leadership has introduced a complex series of reforms into the system. Basically these reforms are aimed at developing better and more scientific means of directing the economy, for example, greater use of computers. It is the Soviet planners' goal to reduce the waste of materials and increase the productivity of the economy by expressing production goals not in quantitative terms but in relation to costs and profitability of production. The present leadership has also increased investment in the production of consumer items and agricultural production. These sectors of the economy, however, still lag far behind heavy industry.

A significant change in Soviet economic policy is an increased willingness to forsake a traditional policy of autarky (economic self-sufficiency) in favor of increased cooperation with Western industrial nations. This change does not only mean expanded trade such as grain purchases but also exchange of

[1]Marshall I. Goldman, *The Soviet Economy* (Englewood Cliffs, N.J.: Prentice-Hall, 1968), p. 12.

technical information and even Western investments in developing new industries where the Soviets are reluctant to commit their capital. Many obstacles exist before this economic interaction can become fully operational. Countries such as the United States and Japan have expressed interest in participating in the development of the oil and gas fields of western and eastern Siberia. The exchange of technical information, capital, and materials may be beneficial to both sides.

Industrial Resources

A critical factor contributing to Soviet economic development is its diverse and rich natural resource base. The Soviet Union ranks among the world's leaders in reserves of oil, natural gas, coal, iron ore, timber, copper, and chromium. The application of the Soviet model of development to any other state must inevitably be modified to adjust for the deficiency of its industrial resources compared to the Soviet Union's.

With the decision to industrialize, the Soviet Union embarked on a vast program of geological surveys and evaluation of the mineral resource base of its immense land area. Not only were efforts made to search out the potentials of the European part of the country, but attention was also concentrated on the great unknown of Siberia. Needless to say, the Russian geologist occupies a position of prestige and importance in Soviet society. Throughout most of the Soviet period, one could characterize the Soviet view of its resources as that of an unlimited supply for the service of the state. With increased awareness of excessive production costs and wastes, decreasing quantity and quality of important materials, and environmental pollution problems, Soviet planners and leaders are stressing more effective management of their raw materials.

Also, it should be repeated that the existence of a natural material does not guarantee its availability as a resource. Much of the Soviet Union's resource potential lies in remote sections of the country, accessible to the market areas only at tremendous cost. This high cost results from expensive long hauls by railway or even from the need to build transport facilities to the site of the deposits. Many deposits

are located in areas of harsh environmental conditions. It is difficult to attract workers to these sites. Furthermore, the costs of building the required housing and transportation facilities and the mining operations themselves may push the expenditures beyond practical economical levels.

Fuel Resources

Since the end of World War II, the most significant development in the Soviet fuel industry has been the growth of oil and gas production. From 1945 to 1972, the production of oil increased almost twenty-one times. Today, the Soviet Union is reputed to be the world's largest oil producer. During the same period, the extraction of natural gas, of which the Soviet Union claims to possess the largest reserves in the world, increased sixty-five times. Although the proportion of coal in total fuel production has declined, the actual output of coal has almost quadrupled. On the other hand, the amount of wood used as a fuel has declined by about 10 percent.

The growth of the oil and gas industry is attributed to several factors. Soviet officials estimate that the cost of oil production is only one-sixth that of producing an equivalent amount of coal. Production costs of gas are even cheaper; roughly one-thirteenth those of coal. Overall, it is more economical in the long run to pipe oil and gas than use coal, peat, and wood, even though they may be located closer to the points of consumption. Soviet oil pipeline construction has not, however, kept pace with oil consumption. About 40 percent of the petroleum is moved by trains at a relatively high cost. Increased demand for aviation and automotive fuels and the rapid growth of the petrochemical industry in the 1960s greatly stimulated petroleum production.

Approximately 80 percent of Soviet oil production comes from the rich Volga-Urals Field (Fig. 16–1). These deposits are well situated in relation to the principal consuming regions of the Soviet Union. Pipelines feed crude oil to the heavily populated area centered on Moscow, the Leningrad and Baltic areas, and across central European Russia to Belorussia and Eastern Europe. Lying just to the east of the Volga-Urals Field is the Ural industrial complex.

Although it is unlikely the Volga-Urals Field will soon lose its prominent position in oil output, decreasing production in several wells has prompted an awareness of the need to expand petroleum

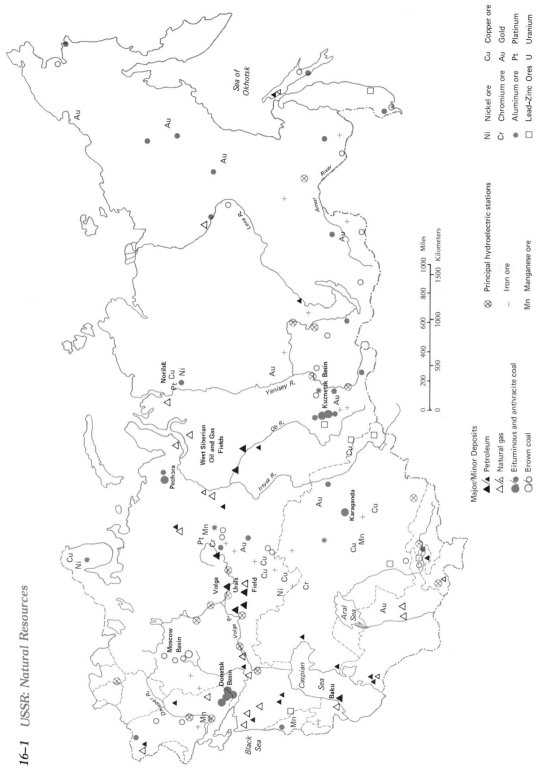

16–1 *USSR: Natural Resources*

Major/Minor Deposits

▲△ Petroleum
△▲ Natural gas
●● Eituminous and anthracite coal
○○ Brown coal

⊗ Principal hydroelectric stations
— Iron ore
Mn Manganese ore

Ni	Nickel ore	Cu	Copper ore
Cr	Chromium ore	Au	Gold
※	Aluminum ore	Pt	Platinum
□	Lead–Zinc Ores	U	Uranium

Sea of Okhotsk

Amur River

Lena R.

Norilsk
Pt Cu · Ni

West Siberian
Oil and Gas
Fields

Pechora

Yenisey R.

Ob R.

Irtysh R.

Au

Kuznetsk Basin

Au

Cu

Karaganda

Cu

Cu Mn

Moscow
Basin

Volga

Urals
Field

Cu Cu

Ni Cu

Cr

Dnieper R.

Donetsk
Basin

Volga R.

Mn

*Black
Sea*

Mn

*Caspian
Sea*

Baku

*Aral
Sea*

Au

Pt Mn
Cr

Au

Cu
Ni

1000 Miles
1500 Kilometers
800
1000
600
400
500
200
0
0

Petroleum produced in the Volga-Urals field and along the Caspian Sea is piped to industrial centers. Pipelines are also being extended to Eastern Europe. Once the pipes are welded together, the line is then insulated. (Novosti from Sovfoto)

Lena River in eastern Siberia will help to ensure future petroleum needs.

The increasing desirability of using natural gas as a fuel has led to opening new fields. The Soviets have concentrated on a program of gas pipeline construction. As with oil, the majority of Soviet gas production is west of the Urals. Unlike oil, however, the Volga-Urals Fields account for only about 16 percent, slightly less than the fields of the north Caucasus region. The Ukraine fields account for about 30 percent of total Soviet production. The greatest expansion of Soviet gas production has occurred in western Siberia. Production here increased almost fivefold during the five-year plan ending in 1975. These fields now supply about 15 percent of total Soviet production compared to only 4 percent in 1971. Several United States companies have expressed interest in the Siberian fields and have proposed financial and technical assistance in exchange for gas. Japan also has indicated an interest in assisting development of gas deposits in eastern Siberia in exchange for gas exports to Japan. Important gas fields are also located in central Asia. These latter fields are destined to become increasingly important to the Soviet Union.

Not only is the Soviet Union the leading coal producer in the world, but its coal reserves also rank among the world's largest. With estimated reserves of over 8.6 trillion tons, Soviet coal can last 13,000 years at present rates of extraction. More than 93 percent of these reserves are located in Asian Soviet Union with the majority in Siberia. Coal production, however, is concentrated in the western USSR. Soviet Europe (including the Urals) accounts for about 60 percent of the extracted coal. The Donetsk Basin, in the southern steppes, alone produces 35 percent of total Soviet output. Extraction costs of Donetsk coal are relatively high due to thin seams and depth of the deposits. Nevertheless, the Donetsk Basin remains the Soviet's largest producer because of proximity to heavily populated industrial regions in the Ukraine and central European Russia. The Kuznetsk Basin is the Soviet's second major source of coal, with output about one-half that of the Donetsk Basin. This west Siberia deposit meets local needs and is shipped as far west as the Urals. Two other notable but smaller fields are the Karaganda Basin in Kazakhstan and the Pechora Basin in northern European Russia. Karaganda coal supplements Kuznetsk coal in both

production elsewhere. Most of the growth in petroleum production during the 1970s has come from the extensive deposits in western Siberia, particularly in the area of the Ob River. Production here is hampered by the harshness of the environment, especially permafrost, which complicates drilling and the laying of pipelines. West Siberian oil, however, is now transported by pipeline east to Irkutsk and west to the Urals and Moscow.

Most notable among the remaining oil-producing areas of the USSR are several fields in the vicinity of the Caspian Sea and scattered deposits in the Ukraine and northern European Russia. Small fields on Sakhalin Island serve local needs, and recently discovered deposits along the upper reaches of the

The Soviet Union is the world's leading producer of coal. The Donetsk Basin accounts for 35 percent of the nation's total. Donetsk coal is of high quality and suitable for coking. The Abakumov mine shown here is one of the largest in the Donetsk Basin. (Novosti from Sovfoto)

Kazakhstan and the Urals and is used in central Asia. Coal from the Pechora deposits is consumed in northern European Russia.

The lower-grade fuels (peat and oil shale) are significant locally primarily as energy sources for small thermal electric stations. Oil shales are mined largely in Estonia and around Leningrad. Peat is extracted throughout much of European USSR. With the increasing availability of natural gas and petroleum, the less efficient fuels of peat and shales are losing their comparative cost advantages.

Hydroelectric Power

The hydroelectric potential of the Soviet Union is enormous, estimated at twice that of the United States and second only to China's. Nevertheless, only about 17 percent of all electricity generated in the USSR comes from hydroelectric stations. Most electricity is generated by thermal stations utilizing coal, peat, oil, and gas. Atomic power plants produce only a very small share of the Soviet Union's electricity.

Only about 11 percent of the Soviet Union's hydroelectric potential is actually utilized. A major handicap is that about 70 percent of the total Soviet potential lies within Siberia, far from the centers of demand. From the late 1950s to 1970s, several large stations were constructed on the Yenisey River and its tributary, the Angara. With a capacity of 4.5 to 6.5 million kilowatts, these hydroelectric stations are the largest in the world. The Soviets now concentrate on thermal stations within the consuming areas rather than continued development of the remote Siberian hydroelectric resources.

European Soviet Union's hydroelectric potential is more fully developed, particularly the Volga and its tributary, the Kama. Much of these rivers has been converted to a string of large lakes behind nine hydro stations that supply electricity to the Volga cities, the Urals, and the Moscow area. The Caucasus and central Asian mountains have also experienced an expansion of their hydroelectric capacity since 1960.

Metallic Ores

Complementing its energy resources, the Soviet Union possesses an ample and diverse base of metallic resources. Tungsten and tin are the only materials in short supply. Soviet iron ore reserves (according to Soviet estimates) are the largest in the world, representing 40 percent of all known reserves. The most important of the producing iron-ore deposits are in the west, whereas the majority of the known reserves are in Siberia and Kazakhstan. Approximately 50 percent of the iron ore extracted in the Soviet Union comes from the Krivoy Rog deposits in the Ukraine. These ores have been mined since the late nineteenth century. Although ore quality is declining and extraction costs are high, they remain the Soviet Union's major source of iron ore because of proximity to principal consuming industries. These ores supply not only the country's largest concentration of iron and steel works in the Donetsk Basin–Dnieper Bend area but also other areas in western USSR and are exported to Eastern Europe.

The second major area of iron ore mining is in the Urals. In the eighteenth century, the Urals was the Russian Empire's major producer of iron, smelting local ores with charcoal. With the advent of the coke smelting process in the late nineteenth century, the

Urals rapidly lost its prominence to the Ukraine. The Soviet program of rapid industrialization and the desire to disperse heavy industry for strategic purposes injected new life into the Urals' iron and steel industry. The otherwise mineral-rich Urals is deficient in coking coal. Ural industrial expansion, therefore, depended on coking coal imported largely from the Kuznetsk Basin 1,000 miles (1,600 kilometers) to the east. In exchange, Urals iron ore is carried to the Kuznetsk Basin for processing in a newly established metallurgical complex. Decreasing quality and availability of Urals iron ore and the high cost of transport, however, have led increasingly to utilizing closer iron ore deposits for Kuznetsk industries.

The Urals iron and steel industry has turned to local lower-grade iron ores and ores from northern Kazakhstan. Kazakhstan ores are also shipped westward to Karaganda where there is a small ferrous metallurgy industry. Increasing quantities of coal from Karaganda also are shipped to the Urals to reduce the expensive, long-haul shipments from the Kuznetsk Basin. The higher quality of Kuznetsk coal, however, and its low mining cost continue to make the basin the Urals' major source of coking coal.

Among the diverse metal resources, the ferrous alloy metals such as manganese, nickel, and chromium are found in sufficient supplies. Most noteworthy are those deposits accessible to the principal steel-producing centers, such as the numerous ferrous alloys extracted in the Urals and the mining of manganese at Nikopol in the Ukraine.

The Soviet Union possesses a wide array of other mineral products. Copper extraction, smelting, and refining are carried on in northern Kazakhstan, in the Urals, in central Asia, and in the Kola Peninsula. Copper is found in conjunction with nickel at Norilsk. The Soviets hold second place in copper mining and claim their reserves are the largest in the world. Lead-zinc ores are mined in Kazakhstan, central Asia, and the northern Caucasus. For the production of aluminum, the Soviet Union must rely chiefly on low-grade ores and aluminum imported from Hungary for processing at Volgograd. Siberia with its surplus hydroelectric energy accounts for about 65 percent of the aluminum metal production in the USSR. The Soviet Union is also a major producer of gold, platinum, and uranium.

Industrial Production

Basis of Soviet Regional Industrial Development

In a command-type economy such as the Soviet Union's, planning for regional distribution and development of industries is a major task. The difficulty is compounded by the nation's large area, imbalance in the distribution of natural resources and population, and diverse ethnic groups at different levels of economic development.

Industrialization Program Policies

From the start of the industrialization program, the Soviets have attempted to disperse their industrial production beyond the limits of the European part of the country. The justification for such a policy was found in ideological as well as strategic reasons. Marxist-Leninist doctrine contends that because capitalism involves the exploitation of mankind by mankind, one can also expect the exploitation of a region by another region. Theoretically, therefore, an equal and balanced level of economic development throughout a country is possible only in a socialistic society that has eliminated capitalism and therefore exploitation. The goal of rapid industrialization was to build "socialism in one country" to use Stalin's phrase. This goal would then enable the Soviet Union to withstand any attack from the Western capitalist powers, which Stalin believed to be imminent. The development of industrial production in the more defensible areas, the eastern portions of the country, seemed advisable. With much fanfare, the Soviets embarked on expansion of the Urals industrial base and the development of the Kuznetsk metallurgical base.

Although Soviet economic planning policies emphasize the balanced distribution of economic activities, including the development of the less economically advanced areas, they also stress production close to raw materials and markets to minimize transportation costs; the development of specialized forms of production in areas best suited for them; and creation within each region of adequate production to meet the basic needs for the population.

World War II was the greatest stimulus to increase industrial production in the east. The Germans occupied an area of European Soviet Union that

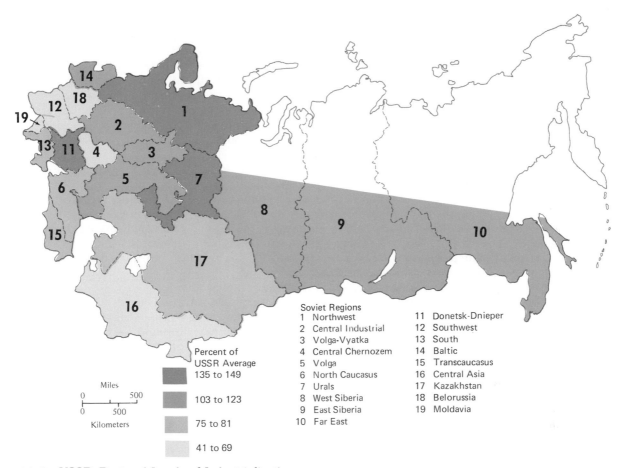

Soviet Regions
1 Northwest
2 Central Industrial
3 Volga-Vyatka
4 Central Chernozem
5 Volga
6 North Caucasus
7 Urals
8 West Siberia
9 East Siberia
10 Far East
11 Donetsk-Dnieper
12 Southwest
13 South
14 Baltic
15 Transcaucasus
16 Central Asia
17 Kazakhstan
18 Belorussia
19 Moldavia

Percent of
USSR Average
135 to 149
103 to 123
75 to 81
41 to 69

Miles
0   500
0   500
Kilometers

16–2 *USSR: Regional Levels of Industrialization*

SOURCE: Reproduced by permission from *Annals* of the Association of American Geographers, Volume 62, 1972.

included 40 percent of the population and the bulk of heavy industry (62 percent of coal production and 58 percent of steel production). New factories were established in the Volga, Urals, and Siberian regions. More than 1,300 factories were relocated in the Urals and points east where industrial production increased rapidly during the war years. From 1940 to 1943, for example, industrial production increased 3.4 times in western Siberia. During the post–World War II period, the west was reconstructed and resumed its dominant role in industrial production but at a level slightly below previous times.

Governmental regional development policy now seems to favor increased investment in the west, particularly medium-sized cities in areas with large labor supplies. Siberian development is slated to continue at a rate slightly higher than the national average. The high cost of attracting and sustaining labor in the east, however, will mean that development will be more technological than labor intensive. On the other hand, the increasing availability of oil and natural gas and even hydroelectricity means that previously energy-deficient regions will have greater access to fuels for industrial expansion.

Regional Pattern of Industry

Regional variation levels of industrialization are seen in Figure 16–2. The index of industrialization, based on computations by Dienes,[2] is expressed for each region in terms of its percent of the USSR average. This map shows industrial development in the Soviet Union is far from evenly balanced regionally. The lowest levels of industrialization are found in the Central Asian republics (No. 16) and the less developed agrarian sections of European Soviet Union: southwest Ukraine (No. 12), Central Chernozem (No. 4), Belorussia (No. 18), and Moldavia (No. 19). At the other extreme, the highest index values are found in the Donetsk-Dnieper area (No. 11), Urals (No. 7), and Northwest, dominated by the city of Leningrad (no. 1).

There are six areas of prime industrial significance: the Center, Ukraine, Leningrad, Mid-Volga, Urals, and Siberia (Fig. 16–3). The Center, with the Soviet Union's most populous and the largest single industrial city, Moscow, owes its industrial prominence to a number of factors including a large market, an ample supply of labor both trained and unskilled, and excellent transport accessibility to all parts of the country. The area was the core of the Muscovy state. The growth of Moscow's political influence and the increase in its population aided in the development of crafts and industrial activities throughout the centuries. Before World War I, approximately 27 percent of all industrial production in the Russian Empire was found in the districts surrounding Moscow.

The Center

The preeminence of this area was fortified by improving the transport routes from Moscow to other parts of the empire. Canals were constructed linking the navigable rivers together. By the early Soviet period, Moscow was accessible to the Caspian, Azov, Black, Baltic, and White seas. Railroad construction in European Russia likewise reflected the prominence of Moscow and the surrounding areas. Even at the present time, the radial nature of the railroad network emanating from Moscow is an evident characteristic of the railroad pattern. Electric lines from Volga power stations and gas and oil pipelines from the Ukraine, north Caucasus, the Volga-Urals Fields,

[2]Leslie Dienes, "Investment Priorities in Soviet Regions," *Annals, Association of American Geographers*, 62 (1972), 439. Index of Industrialization equals the index of industrial employment for each region added to the index of per capita industrial fixed assets, divided by two.

central Asia, and western Siberia supply the Center with important sources of energy. These same growth factors continue to assure this area's prominent position in industrial production, although its share of total industrial output may slowly decline.

The industrial resource base of the Center is weak. Energy resources of brown coal and timber are inferior. There is a small iron ore deposit, phosphorous for fertilizer, and some building materials. Textile manufacturing is prominent. Around Moscow, 30 percent of all industrial workers are employed in the manufacture of linen, cotton, wool, and silk fabrics. Other major industries include metals, machine construction, engineering industries, chemicals, food processing, and woodworking. Important production centers in addition to Moscow are Gorki, Yaroslavl, Tula, Ivanovo, and Ryazan.

Ukraine Industrial District

The Ukraine Industrial District is the principal heavy manufacturing area. The availability of Donetsk coal, Krivoy Rog and Kerch iron ore, and Nikopol manganese facilitates major iron and steel production including various forms of steel, heavy machine construction, and the coal-based chemical industry. The industrial resource base of the Ukraine extends beyond coal and iron ore. Energy resources are supplemented by gas fields to the north and by the gas and oil fields of the northern Caucasus. The base of raw materials also includes such minerals as salts, potash, mercury, and brown coal. Furthermore, the high productivity of agriculture within the district and throughout the southern steppes of western European Russia has stimulated the development of extensive food-processing industries and the production of agricultural equipment.

The largest concentration of cities within this industrial area is on the coal fields of the Donetsk Basin. Donetsk, Makeevka, Gorlovka, and Voroshilovgrad are the larger cities of this conurbation. The Donetsk Basin is not only a major producer of iron and steel, and principal coal mining district, but it also manufactures railroad equipment, machinery, plastics, food products, and clothing.

Outside the Donetsk Basin, iron and steel production is prominent in three other locations: at the intermediary site between the iron ore and coal deposits on the Dnieper in the cities of Dnepropetrovsk, Dnieproderzhinsk, and Zaporozhe; near the

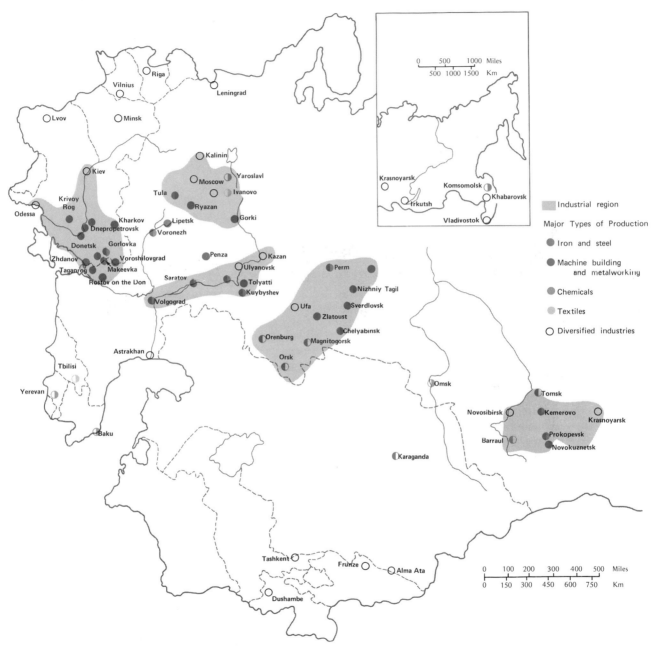

16–3 *USSR: Industrial Regions and Important Industrial Cities*

iron ore mines at Krivoy Rog; and on the northern shore of the Sea of Azov where Kerch iron ore is used in conjunction with Krivoy Rog ores and Donetsk coal.

The two largest cities of the Ukraine, Kiev and Kharkov, are located on the northern margins of the Ukrainian manufacturing area. In addition to its political function as capital of the Ukraine, Kiev is a major diversified industrial city. Machinery, textiles, and food processing are major industries in both these cities. To the southwest is the diversified industrial port and shipbuilding city of Odessa.

Leningrad

Leningrad, the second largest city in the country, occupies a position similar to that of Moscow. It lies in a region with a deficient resource base. Aside from hydroelectric stations, local energy resources are limited to peat deposits and oil shales. Leningrad industries include machine tools, equipment for hydroelectric plants, and shipbuilding.

Mid-Volga Area

The industrial strength of the Mid-Volga industrial area rests primarily on the extensive energy resources. Not only is the major petroleum-producing field of the Soviet Union (Volga-Urals Field) found here, but there are important gas fields and surplus hydro-

The Volga River system is the USSR's largest water transport route. Not only is the river an important economic asset to the nation, but it is also closely linked to the country's cultural heritage. (Novosti from Sovfoto)

The Volzhsky Motor works at Tolyatti built jointly with the Italian Fiat company. The plant was completed in 1972 and incorporates assembly-line techniques. (Novosti from Sovfoto)

electric power generated by some of the country's large dams on the Volga River. During World War II, this area experienced rapid industrial and population growth. It was located east of the front for security but was readily accessible to areas of need in the west. From the late 1950s, industrial production has increased in the Volga area at a rate greater than that of the Soviet Union as a whole. Expansion has been based on the development of its energy resources, the growth of the petrochemical industry, and the attraction of industries that are large energy consumers, for example, aluminum.

The Mid-Volga area has several advantages in addition to its fuels and hydroelectric power. It is connected by the Volga River and its tributaries to large areas of the western Soviet Union. The Volga River system is the Soviet Union's major water route, carrying over 60 percent of all the freight transported by river. Principal commodities shipped on the Volga are northern timber for the southern steppes and grain, Caspian fish, building materials, and petroleum products (from Baku and the Volga-Urals Fields).

Accessibility by water, rail, and pipeline not only has stimulated the development of the Mid-Volga's energy resources but also has allowed the expansion of industrial activities as a whole. One major example is the large automotive plant built as a joint venture with Fiat, the Italian firm, at Tolyatti. The plant is to produce 600,000 cars a year, 48 percent of the planned Soviet total automobile output. Principal industrial cities include Kuybyshev, Saratov, and Volgograd (formerly Stalingrad). Oil refining, metalworking, and chemicals are prominent activities in this area.

The Urals

In terms of overall industrial production, the Urals ranks third behind the Center and Ukrainian areas. Urals industry depends chiefly on the rich and varied local mineral deposits. In addition to its important iron and steel industry, the area is known for smelting and refining of copper, zinc refining, and production of both alumina and aluminum. Moreover, the Urals accounts for slightly more than one-quarter of the total mineral fertilizer production.

Sverdlovsk, a major railroad center, is the largest of the Urals cities. Here machine construction, especially for the mining industry, dominates; ferrous metallurgy is also a prominent industry. Iron and steel centers also include Chelyabinsk, Magnitogorsk, and Nizhniy Tagil. Perm in the western portion of the district is noted for its oil refining, chemical, and woodworking industries.

Siberia

Siberian manufacturing stretches from Novosibirsk to Krasnoyarsk and includes the metallurgical region of the Kuznetsk Basin. Along with coal mining and iron and steel production are two large integrated plants at Novokuznetsk; industries of the Kuznetsk Basin include coke, coke-chemical industries, metal-working, and machine construction. A large aluminum plant is located in Novokuznetsk and another in Krasnoyarsk where energy is supplied from a large hydroelectric station on the Angara River. The largest city is Novosibirsk lying to the west of the Kuznetsk Basin. Novosibirsk is both a major transport center located on the trans-Siberian railway and a river port on the Ob River. It is also a major industrial city with some ferrous and light metallurgy, machine building, textiles, and food and chemical industries.

Agriculture

Soviet agriculture has not experienced the same successes as industry. From the late 1920s to the early 1950s, agricultural output barely kept pace with

Table 16–1   *Comparisons of USSR and United States Agriculture*

	US	USSR
Size of farm labor force	4.4 million	36.9 million
Share of labor force employed in agriculture	4%	31%
Output per farm worker	$7,746	$834
Number of persons supported by one farm worker	46	7
Total sown area	301.4 million acres (122.0 million hectares)	521.6 million acres (211.1 million hectares)
Value of crop production	$19.4 billion	$16.8 billion
Value of livestock production	$21.2 billion	$16.0 billion
Number of planted acres (hectares) for each tractor	64 acres (26 hectares)	258 acres (104 hectares)
Trucks per 1,000 farm workers	665	34
Crop yields, bushels per acre (kilograms per hectare)		
Spring wheat	28 (1883)	14 (942)
Winter wheat	33 (2219)	26 (1749)
Corn	69 (4330)	35 (2197)
Sugar beets, tons per acre (tons per hectare)	18 (40)	10 (25)
Ginned cotton, pounds per acre (kilograms per hectare)	442 (495)	784 (879)

SOURCE: F. Douglas Whitehouse and Joseph F. Havelka, "Comparison of Farm Output in the US and USSR, 1950-1971," in Joint Economic Committee, Congress of the United States, *Soviet Economic Prospects for the Seventies* (Washington, D.C.: Government Printing Office, 1973).

population growth. Since the early 1950s, however, production of agricultural goods (both crops and livestock) has grown almost three times faster than population. Despite this recent growth in agricultural output, Soviet farm production still lags behind that of the United States. For every American farm worker, the Soviet Union has eight. About 31 percent of the Soviet labor force is engaged in agriculture compared to 4 percent in the United States (Table 16–1). The area sown to crops is 73 percent greater in the USSR than in the United States, yet Soviet crop production is equal to only 87 percent of the United States and 75 percent of the United States livestock output. Table 16–1 also compares the level of farm technology of the United States and the USSR. These data readily demonstrate the lower level of technology and the greater labor utilization in Soviet agriculture. Each farm worker in the United States produced $7,746 of farm output compared to only $834 per Soviet worker. The United States agricultural worker feeds forty-six people, whereas his Soviet counterpart feeds only seven. In the United States, there is one tractor for every 64 acres (26 hectares) of cultivated land compared to one tractor for every 258 acres (104 hectares) in the USSR. The number of trucks per 1,000 workers in the United States is almost twenty times greater than in the USSR. Furthermore, American yields are generally higher than those of the USSR, but a notable exception is cotton (Fig. 4–3).

The reason why Soviet agriculture has developed less successfully than industry probably lies within the organization of Soviet agriculture. Institutional restraints have restricted agricultural growth.

Position of Agriculture and Soviet Development

Early in the postrevolutionary years, Soviet policies were directed against the large, more efficient producers. With the victory of the Bolsheviks in the November Revolution and the subsequent nationalization of land, the countryside was in chaos. The peasant farmers, land hungry and landless, directed their wrath against the large estates and prosperous landowners. Lands were confiscated and turned into small peasant holdings controlled by the traditional peasant village system. Agricultural productivity consequently decreased. The large landholdings had advanced most in the use of machinery), whereas on the increasing number of small peasant holdings,

the old tools, the sickle, and even the wooden plough, became the standard of Russian farm technology. War was declared against the *Kulaks,* the better-off peasant farmers. Although they may have had only a couple of horses and cows, a *Kulak* occasionally hired a laborer to aid in production of a surplus for sale. Food supply decreased as a result of these chaotic conditions.

The disruption of the agricultural economy was temporarily alleviated with the introduction of a new policy that encouraged private trade and the more productive peasants to increase their efforts. At the same time, it was hoped that the small landhold peasants would be shown the advantage of cooperation in farm work and voluntarily join large communal organizations.

According to Stalin, the decision to embark on an all-out program of rapid industrialization in the late 1920s, however, necessitated the collectivization of agriculture. There were several reasons for this "second revolution."

First, the peasant class represented a capitalist or latently capitalist element that was ideologically unacceptable to the regime. Second, by forcing the peasants into large collectives, agricultural prices and wages could be controlled at low levels to allow accumulation of capital for industrial expansion. Third, it would be more efficient to control the peasantry grouped in large farms rather than scattered in smaller units. Fourth, this control would facilitate the flow of foodstuffs to the cities to feed the growing industrial labor force. Fifth, the envisioned large scale units would not only make it possible to mechanize agriculture, but also the planned increased productivity of agriculture would free labor for the growing industrial activities. Class war raged in the countryside, and the better-off peasants and opponents to the collectivization process were exiled, imprisoned, or murdered. Livestock herds were decimated, as peasants slaughtered their animals rather than surrender them to the new collectives. By 1940, virtually all peasant households were part of the collective agricultural economy.

Two forms of farm organization emerged: the collective farm, *Kolhoz,* and the state farm, *Sovkhoz.* The collective farm is a group of workers responsible for seeing that their obligations to the state are met. After the needs of the collective farm are satisfied (capital for farm repairs, taxes, and seed for the next

A modern collective farm (Kolkhoz) village in Lithuanian SSR. This collective specializes in dairying. (TASS from Sovfoto)

An older collective farm (Kolkhoz) near Moscow. In the back of each house is the farm family's private plot. These private plots provide cash income to the farmer and supply badly needed produce, meat, and milk to the nearby urban settlements. (USSR Magazine from Sovfoto)

season), the remaining crop is divided among the workers as their share of the profit. The peasants are, therefore, residual claimants to the farms' production. The state farm (*Sovkhoz*) workers, on the other hand, are paid a set wage, and the total costs of the operation are underwritten by the state. Needless to say, the cost of operation for the state was greater for the *Sovkhoz* than for the *Kolkhoz*. The *Kolkhoz* system was greatly favored by the Stalin regime. Another control system of the collective farm during the Stalin regime was the Machine Tractor Station (MTS). All machinery was held by the MTS, which in turn performed such tasks as plowing and combining for a share of the crop. Not only was the MTS an additional means of increasing the state's share of the *Kolkhoz* production, but it also enabled the party to keep an eye on the collective operation. During the Stalin regime, *Kolkhoz* efficiency suffered because of required low-priced deliveries of agricultural products. The capital thus provided for the operation of the *Kolkhoz* was insufficient and failed to provide work incentives. The state's investment in fertilizers, machinery, and other necessary technological improvements was woefully inadequate. All these factors contributed to stagnated agricultural production through the early 1950s.

The critical difference between survival and starvation was the private sector of Soviet agriculture: those products raised on the one-half to 2-acre (.2 to .8 hectares) plots allowed the collective and state farm workers and some industrial workers. For example, in 1953 private plots made up only about 4 percent of the cultivated land in the country but produced 72 percent of the potatoes, 48 percent of the vegetables, 52 percent of the meat, 67 percent of the milk, and 84 percent of the eggs. The majority of the collective farmers' incomes was generated by sale of products from these private plots in the free market.

The Post-Stalin Period

The dire situation of agriculture at the death of Stalin forced his successors, notably Khrushchev, to turn their attention to the needs of the agricultural economy. A multipronged attack on the problem resulted initially in a significant increase in agricultural production. The program called for the expansion of cultivated lands in the dry steppes of the Trans-Volga region. This virgin and idle lands program brought 122 million acres (47 million hectares) of new land under cultivation mostly in western Siberia and northern Kazakhstan. In these new lands precipitation ranges from 16 inches (41 centimeters) in the north to 9 inches (23 centimeters) in the south, a marginal farming area. Anticipated produc-

tion levels have not been attained, and production variation occurs from year to year. Nevertheless, the increased wheat area has freed land in the moister areas of European Russia for other crops. Furthermore, inadequate precipitation in one area is often offset by adequate rainfall in another.

With the opening of the virgin wheatlands, Khrushchev embarked on a vast program to expand corn planting. The objective of this effort was to increase the supply of livestock feed in order to promote meat production. Other changes implemented at this time included the abolishment of the MTS and the granting of loans to collectives for the purchase of their own machinery. The system for procuring collective farm products was simplified, and the prices paid for these goods were raised to provide greater economic incentive. The state also increased its investment in machinery and fertilizer production. These measures, along with good weather, promoted an increase in agricultural production. The gains were short-lived. Increased prices of goods obscured the increased wages of the collective farmers. Furthermore, the

Dryland wheat planting in the dry steppes of central Asia. About 122 million acres (47 million hectares) of this type of land have been put to cultivation in the virgin land program fostered by former Soviet Premier Khrushchev. (Novosti from Sovfoto)

state reduced some of the investment programs and began to pressure the state and collective workers to turn over their private livestock to the *Sovkhoz* and *Kolkhoz*. Bad weather also contributed to declining harvests in the late 1950s and early 1960s. The Soviet Union was forced to import grain from the West, and political pressure on Khrushchev culminated in his ouster in the fall of 1964.

The leadership of Leonid Brezhnev and Aleksei Kosygin inherited the agricultural problem which has come to be called by some Western critics the Soviet's "permanent crisis." The seven-year plan which ended in 1965 fell far short of its goals, and the new leaders took steps to improve conditions. They reinstated a more tolerant attitude toward the private plot, increased prices on purchases from the collectives, and introduced a guaranteed minimum wage for the collective farm worker. Prices on machinery and fertilizer were reduced, and the state indicated the right of collective farms to participate more in the planning procedure. The last half of the 1960s was marked by notable success in Soviet agricultural production. Imports ceased, while the output of agricultural products increased by 23 percent from 1965 to 1971.

The need of the Soviet government to import large quantities of grain in 1972, 1974, and 1975 underscored the fact that Soviet agriculture is still beset by numerous problems. Despite increased wages, the average collective farmer makes less per year than the average industrial worker or even the average state farm employee. Soviet agriculture still suffers from inadequate mechanical equipment and deficient storage and transportation facilities. The more productive young workers leave the farms because of the low salaries, restricted opportunities for advancement, and a scarcity of amenities that are available in an urban life. Furthermore, central decision making in the Soviet system contributes to interference by party bureaucrats and inefficient use of agricultural resources.

In 1950, Soviet per capita agricultural output was only 49 percent of that of the United States. The gap closed to 68 percent of United States per capita output by 1971. The total output of Soviet farms was about 80 percent that of the United States. The 1971–1975 plan called for equaling United States output, but this goal was not attained. The possibility of achieving the same levels of the United States agricultural production appears highly remote for any time in the foreseeable future.

The *Kolkhoz, Sovkhoz,* and Private Plot Today

The trends during the post-Stalin years were to increase the size of the *Kolkhoz*, conversion of *Kolkhozs* to *Sovkhozs,* and the establishment of state farms on newly opened lands. Some basic comparisons are made between the *Kolkhoz* and *Sovkhoz* in Table 16–2. Both are large, containing hundreds to several thousands of people in each unit. There are approximately twice the number of *Kolkhoz* than *Sovkhoz*. The amount of land sown to crops is approximately the same in each. The average total land of the *Sovkhoz* is considerably greater than that of *Kolkhoz*. There is also a greater number of workers on the average state farm than on the average collective. In general, the *Sovkhoz*

Table 16–2 USSR: *Comparison Between the Sovkhoz and Kolkhoz*

	Kolkhoz	Sovkhoz
Total number of units	31.4 thousand	17.3 thousand
Total number of workers	15.9 million	9.8 million
Average number of workers per unit	506	566
Amount of sown land	243.6 million acres (98.6 million hectares)	257.0 million acres (104.0 million hectares)
Sown land per worker	15.3 acres (6.2 hectares)	26.2 acres (10.6 hectares)
Average amount of sown land per unit	7,758 acres (3,140 hectares)	14,855 acres (6,012 hectares)

SOURCE: USSR Central Statistical Committee, *Narodnoye Khozyaystvo SSSR v 1973g* (Moscow: Statistika, 1973).

produces greater output per worker, which reflects the availability of more and better equipment on the *Sovkhoz.*

The private plots, though small (average size 1.3 acres or .5 hectare), are still an important source of income for the workers and a major source of food for the country. Although the dependence of the *Kolkhoz* workers on the income earned from the private plot has decreased in recent years, in some poorer collectives it may account for one-half of the worker's income. It is estimated that these plots, which comprise only 3.2 percent of the cropped area of the Soviet Union, account for about one-third of the total food production in the country. From these small parcels of land come about 65 percent of the potatoes, 40 percent of all vegetables, 35 percent of milk and meat, and 50 percent of eggs. These figures demonstrate that small socialist economic incentives, though they may be improving, make these very intensively worked private plots significant suppliers of foodstuffs and important sources of supplemental income.

Agricultural Regions

Soviet agriculture suffers from the institutional restraints of its organization and from physical environmental handicaps. It is, of course, possible that improved strains of plants, land amelioration programs, or even domes with controlled environments will someday reduce nature's unfavorable aspects. It is also true that even under similar environmental conditions in the United States and Canada, agricultural productivity surpasses that of the Soviet Union. Within the context of contemporary Soviet organization and technology, however, agriculture faces more risks from climatic elements than those faced by the United States. Although the land area of the Soviet Union is two and one-half times that of the United States, its area suitable for crop cultivation is only one-third greater. Only about 11 percent of the Soviet territory is arable.

Figure 16–4 shows generalized zones of agricultural use of the Soviet Union. The areas with no agriculture or widely scattered small farms clearly occupy the majority of the Soviet lands. These are areas of too short a growing season and poor soils covering northern European Soviet Union and virtually all of Siberia except the south. On the other hand, there is the dry desert occupying the bulk of central Asia.

To these generally nonagricultural zones must be added the scattered mountains of the west, central Asia, and Siberia. The remaining area comprises the agricultural zones of the USSR, but even among these lands are areas too cool, too moist or too dry, which would be judged marginal in the United States.

The agricultural zones indicated in Figure 16–4 are found mainly south of the 60°N. parallel in the European part of the country and south of the 57°N. parallel in Siberia. In the south the agricultural area is limited by high evapotranspiration rates.

Zone I, bordering the southern limit of the taiga forest, is an area of scattered agricultural development and stretches from the Baltic to the Urals. Within this zone, dairying is a prominent activity. Flax and hardier grains such as rye and oats do well in the short moist summers. Potatoes, long a staple of the European Soviet diet, are also a prominent crop, along with a variety of other vegetables. Some sugar beet cultivation is carried on in the west. Swine, present in all farming areas except the Moslem regions, is a prominent livestock product.

In Zone II, the frequency of cropped land increases, and grains (rye, oats, barley, and wheat) with potatoes, flax, and hemp are most characteristic. Dual-purpose (milk and meat) cattle dominate over cattle raised exclusively for meat or milk. Sugar beets are the most important industrial crop in Zone III. Higher temperatures than in the north balanced by adequate precipitation favor sugar beet cultivation, together with grains (wheat, corn, barley, rye, and oats) and potatoes. Cattle raising is primarily for meat products.

The subhumid steppe with widespread occurrence of fertile chernozemic soils, Zone IV, is the most important wheat-growing region in the Soviet Union. Winter wheat is grown largely in the European section, while spring wheat dominates the east. Corn has become an increasingly important crop in the west, and plantings of this crop have expanded into the drier steppe region of the east. Sunflowers for oil are a major crop, particularly in the western sections. A variety of grains and other crops (sugar beets, potatoes, flax, and barley), along with milk and meat livestock and sheep raising, round out the characteristics of the agriculture of this region.

In addition to this principal agricultural wedge, there are two regions noted for their specialized crop production. One is the Trans-Caucasus (Zone V),

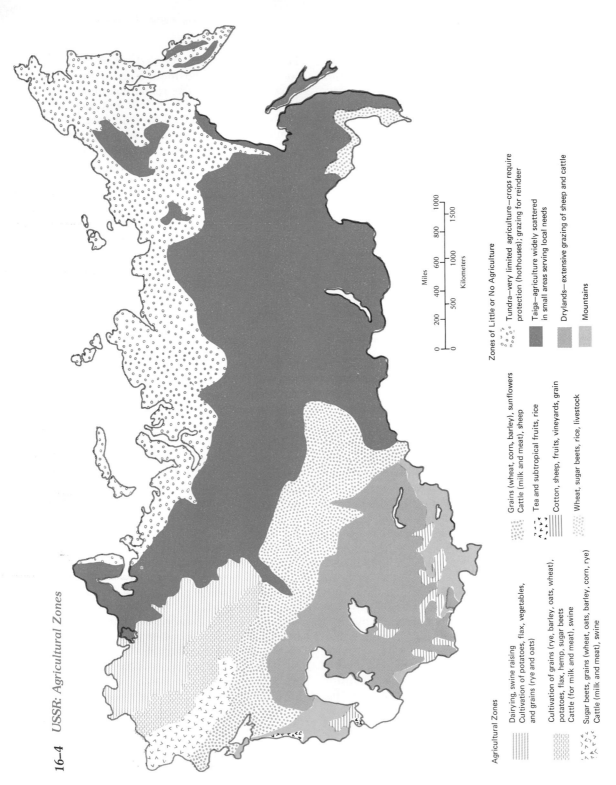

16-4 *USSR: Agricultural Zones*

Agricultural Zones

|||| Dairying, swine raising
Cultivation of potatoes, flax, vegetables,
and grains (rye and oats)

|||| Cultivation of grains (rye, barley, oats, wheat),
potatoes, flax, hemp, sugar beets
Cattle (for milk and meat), swine

Sugar beets, grains (wheat, oats, barley, corn, rye)
Cattle (milk and meat), swine

Grains (wheat, corn, barley), sunflowers
Cattle (milk and meat), sheep

Tea and subtropical fruits, rice

Cotton, sheep, fruits, vineyards, grain

Wheat, sugar beets, rice, livestock

Zones of Little or No Agriculture

Tundra—very limited agriculture—crops require
protection (hothouses); grazing for reindeer

Taiga—agriculture widely scattered
in small areas serving local needs

Drylands—extensive grazing of sheep and cattle

Mountains

Miles
0   200   400   600   800   1000
0   500   1000   1500
Kilometers

where citrus crops and tea thrive in the well-watered, protected western portion. In the drier eastern Trans-Caucasus area, tea is grown along with cotton and rice in irrigated regions. The other region is the principal cotton-growing area of the Soviet Union in the irrigated valleys of central Asia. Rice is the staple food of central Asia and is also widely grown in the irrigated areas along the rivers.

The USSR in the Coming Years

At present, the USSR's position as a major world power is unshakable. The development of Soviet economic power has indeed been impressive. Many claim this growth has occurred not because of but in spite of the political system of the country. The almost single-minded drive toward industrial and military development and the affluence of raw materials and borrowed Western technology have enabled the country to achieve high levels of economic growth. This progress has been costly. Natural resources have been wasted and labor inefficiently used. Most important, personal liberty and material well-being have been sacrificed for the state's developmental plans.

The Soviet Union is faced with a number of problems, the first of which is a slowing of economic growth. Efforts to remedy this difficulty have focused largely on reducing waste and developing a more efficiently run economy, not on fundamental organizational changes.

The shortages and subsequent high prices of raw materials, particularly energy resources, that have threatened the Western industrial states have been used to the USSR's advantage. The rich natural resource base of the Soviet Union gives it an enviable position. This position means that not only is the USSR's own economic growth provided for but also that the country can acquire a greater role as a supplier of raw materials to deficient industrial states. This fact cannot help but enhance the USSR's political and economic position in the world. To maximize this potential, however, the Soviet Union needs Western technological assistance to exploit its far-flung resources. Thus, there occurred in the early 1970s what

some observers viewed as a major turning point in Soviet economic strategy: its expanded commercial relations with the United States and other states of the industrialized West. The full impact of the potential of these new trade agreements may materialize in the near future. There still remain, however, strong suspicions that Soviet objectives are aimed at expanding its economic power for ultimate political and military purposes.

Agriculture remains the major bottleneck in the Soviet economy. Despite significant improvements in production during the last decade, Soviet agriculture fails to provide the quantity and quality of foodstuffs promised by the government. Frequent crop failures require imports of food from abroad. Consequently, the ability of the USSR to increase its own supply of foodstuffs is important not only for its own well-being but also for a food-deficient world as a whole.

Further Readings

LYDOLPH, PAUL E., *Geography of the USSR* (New York: Wiley, 1970), 683 pp., is an excellent reference for further discussion on the Soviet Economy. Chapter 14 (pp. 398–459) deals with agriculture; Chapter 15 (pp. 460–550) with industry; and Chapter 16 (pp. 551–589) with transportation and domestic trade.

Many articles have been written on selected aspects of the USSR's economic geography. The following are of particular significance: Rodgers, Alan, "The Locational Dynamics of Soviet Industry," *Annals, Association of American Geographers,* 64 (1974), 226–240; Dienes, Leslie, "Investment Priorities in Soviet Regions," *Annals, Association of American Geographers,* 62 (1972), 437–454; and JACKSON, W. A. DOUGLAS, "The Virgin and Idle Lands Program Reappraised," *Annals, Association of American Geographers,* 52 (1962), 69–79.

The diverse aspects of Soviet economic life have been the subject of many excellent books which includes CONNOLLY, VIOLET, *Beyond the Urals, Economic Development in Soviet Asia* (London: Oxford University Press, 1967), 420 pp. This valuable volume traces the economic development of central Asia and Siberia from the times of the tsar to 1966.

DEMKO, GEORGE J., and FUCHS, ROLAND J. (editors and translators), *Geographical Perspectives in the Soviet Union: A Selection of Readings* (Columbus: Ohio State University Press, 1974), 742 pp. The bulk of this collection of works by Soviet scholars is devoted to economic geographical themes—namely economic regionalization, resource management, agricultural, industrial, and transportation geography, as well as population and economic geography.

DIBB, PAUL, *Siberia and the Pacific: A Study of Economic Development and Trade Prospects* (New York: Praeger, 1972), 288 pp. A rather thorough study on the economic problems and future of eastern Siberia and the Far East.

GERASIMOV, I. P., ARMAND, D. L., and YEFRON, K. M., *Natural Resources of the Soviet Union* (English edition edited by W. A. DOUGLAS JACKSON, translated by JACEK I. ROMANOWSKI) (San Francisco: W. H. Freeman, 1971), 374 pp. This volume is a collection of fifteen essays on Soviet views of selected environmental problems in the USSR and the utilization of some of its natural resources (for example, water, soil, climatic, vegetative, and animal resources).

LYASHCHENKO, PETER I., *History of the National Economy of Russia to the 1917 Revolution* (New York: Macmillan, 1949), 880 pp. This well-known book is an important source of information on Russian economic development in spite of its Marxist-Leninist interpretation.

NOVE, ALEC, *The Soviet Economy: An Introduction* (New York: Praeger, 1969), 373 pp. Nove's book provides very good coverage on the often complex nature of the structure and problems of the Soviet economy.

NOVE, ALEC, and NEWTH, J. A., *The Soviet Middle East: A Model for Development* (London: Allen & Unwin, 1967), 160 pp. The economic and social development of the central Asian and Trans-Caucasian Republics is examined in a concise yet readable manner.

SCHWARTZ, HARRY, *The Soviet Economy Since Stalin* (Philadelphia: Lippincott, 1965), 265 pp. A valuable survey of economic developments in the USSR from the death of Stalin (1953) to the downfall of Khrushchev (1964).

SHABAD, THEODORE, *Basic Industrial Resources of the USSR* (New York: Columbia University Press, 1969), 393 pp. Shabad presents in this volume a very thorough discussion of the Soviet Union's very plentiful industrial resource base.

SYMON, LESLIE, *Russian Agriculture: A Geographic Survey* (London: G. Bell, 1972), 348 pp. This study is particularly useful for its treatment of the spatial aspects of Soviet agriculture; however, other pertinent features such as Soviet policy, politics, manpower problems, and the overall organization of the agricultural economy are very well treated.

VOLIN, LAZAR, *A Century of Russian Agriculture: From Alexander II to Khrushchev* (Cambridge, Mass.: Harvard University, 1970), 644 pp. The economic history of Russia and Soviet agriculture from the early 1860s to the early 1960s is the subject of this scholarly volume.

Japan and Australia–New Zealand

Jack F. Williams

PART FIVE

Japan: The Resource Foundation

17

A T THE PRESENT TIME, Japan remains the only non-Western country to enter the relatively exclusive club of rich nations. True, a number of other states have made remarkable progress in the last two decades, including a few in Asia such as Taiwan, South Korea, Singapore, and Hong Kong. None, however, can yet be classified truly as rich. Japan unhesitatingly is deemed a rich nation by whatever standards one wishes to use. Its GNP is over $400 billion, the third largest in the world, and the per capita income is more than $4,400. One has only to travel in Japan to see the physical signs of affluence: a modern, thriving society with well-dressed, well-fed people. Japanese products, businessmen, and tourists have become as ubiquitous around the world as Anglo-Americans. It is almost a running joke that if one looks at the label on nearly any product, it will read "Made in Japan."

The nation is a key member of important international financial and economic organizations. Many nations of Asia and other parts of the world depend on Japan for substantial amounts of their imported manufactured goods as well as investment capital. Japan has, in fact, succeeded so well that unflattering names, such as "Japan Incorporated"; and "economic imperialism," have become increasingly directed at the Japanese from many parts of the world.

Uniqueness of the Japanese Experience

The mere fact that Japan still is the only non-Western nation to reach such a high stage of affluence, the fifth or "high mass consumption" stage in Rostow's theory (see Chapter 4), makes the Japanese experience unique in the history of the modern world. It is all the more exceptional when seen in relation to the resource foundation of Japan. The question that has intrigued outsiders for decades is how Japan succeeded in developing, while so many other nations with far richer endowments of natural resources failed or floundered in their development attempts. What factors have accounted for the uniqueness of the Japanese experience?

Impact of the Environment on Japan's Development

Japan is often used as an example to disprove the old theory of environmental determinism because of the way the Japanese have succeeded seemingly in spite of the natural environment and poor resource endowment. Yet a closer analysis reveals a situation more complex. The various elements that make up the nation's physical environment—location, size, topography, climate, soils, arable land, and mineral resources—have had positive as well as negative effects on Japan's development.

Location and Insularity
Japan's unique role in East Asian civilization can be attributed in part to the nation's relative isolation off the east coast of Asia. The country consists of four main islands (Hokkaido, Honshu, Shikoku, and Kyushu), plus many lesser islands, stretching in an

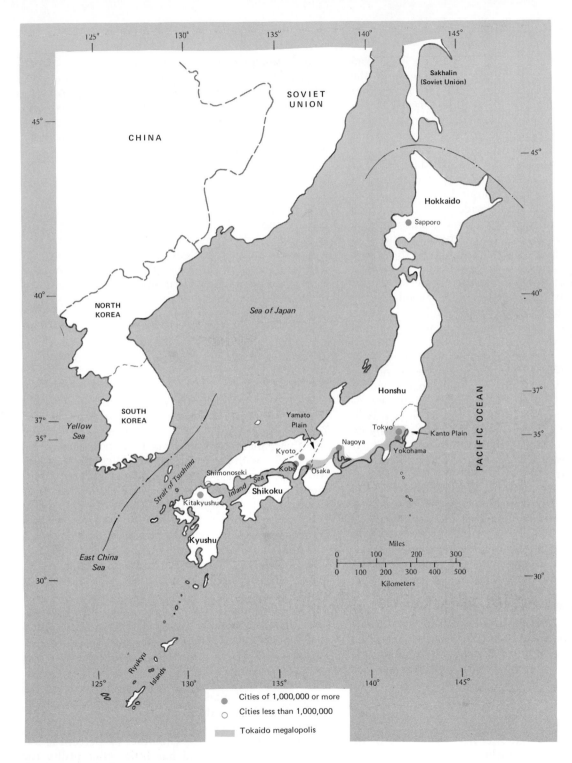

17–1 *Japan: General Locations*

Map labels:

125° 130° 135° 140° 145°

SOVIET UNION

Sakhalin (Soviet Union)

CHINA

45° 45°

Hokkaido

Sapporo

40° 40°

NORTH KOREA

Sea of Japan

Honshu

PACIFIC OCEAN

37° Yellow Sea

SOUTH KOREA

Yamato Plain

Tokyo Kanto Plain

35° 35°

Kyoto Nagoya

Yokohama

Shimonoseki Kobe Osaka

Inland Sea Shikoku

Strait of Tsushima

Kitakyushu

Kyushu

East China Sea

30° 30°

Miles
0 100 200 300
0 100 200 300 400 500
Kilometers

125° 130° 135° 140° 145°

Ryukyu Islands

● Cities of 1,000,000 or more
○ Cities less than 1,000,000
  Tokaido megalopolis

arc about 1,400 miles (2,253 kilometers) long from around 45° to 31°N. latitude (not counting the Ryukyu Island chain, which continues in another arc southward to near 24° N. latitude) (Fig. 17–1).

Comparisons are often made between Japan and the British Isles in their development experience and the roles they have played in history, but in fact the Japanese experience has been quite different. For one thing, Japan has been more isolated than Britain. The Japanese islands are separated from the Korean peninsula by the 115-mile-wide (185 kilometers) Strait of Tsushima, compared to the 21-mile-wide (34 kilometers) Strait of Dover between France and Britain. Moreover, in premodern times, there was still the mountainous Korean peninsula to traverse before arriving at the great culture center of China. The advantages of this isolation were several. On the one hand, Japan had natural protection from invaders and unlike Britain was never successfully invaded in its history. The Mongols tried it twice, in 1274 and 1281, and failed; the United States was poised to do it at the end of World War II. On the other hand, the relative proximity to China enabled Japan to adopt many aspects of Chinese culture on a largely voluntary and selective basis and at a slower rate than that experienced by Korea and Vietnam. The Japanese were able to adapt these innovations to their situation and create a truly unique and brilliant culture on their own.

Japan's proximity to China and the Soviet Union in modern times has contributed to the increasingly important role that Japan is playing in trade relations with those two states. For example, Japan may help China and the Soviet Union develop some of their natural resources: oil, gas, coal, and timber in Soviet Siberia, and oil in eastern China.

A Temperate Land

Japan's long latitudinal sweep within the temperate zone, combined with insularity, has benefited the country in terms of its climate, which is roughly comparable to that of the east coast of Anglo-America from New England to northern Florida (Fig. 17–2a). The maritime climatic patterns mean that Japan has no real dry season, unlike Korea and north China which frequently suffer from drought. There is sufficient rainfall throughout the year for crop growth, ranging from about 40 inches (102 centimeters) in Hokkaido and the Inland Sea area to over 100 inches

(254 centimeters) in the wettest sections of the south (Fig. 17–2b). There is a general progression from long cold winters and short mild summers in Hokkaido to long hot summers and short mild winters in subtropical Kyushu. Likewise, the growing season ranges from around 150 days in Hokkaido to over 260 days in Kyushu (Fig. 17–2c).

A significant dividing line is roughly along 37°N. latitude. South of the line double-cropping is possible, usually paddy rice in the summer and a dry crop in the winter. North of the line the winters are usually too long to permit double-cropping. This basic fact has been reflected in Japan's historical development in that settlement north of the line, in northern Honshu and Hokkaido, came much later than in central and southern Japan. Even today, most of Japan's population is found south of 37°N. The northern lands remain less densely populated, with larger farms, compared to the south. Hokkaido is still a frontier region in Japan with a much less developed appearance than the southern parts of the country.

A Crowded Land

Japan is a small country when compared with the other great powers of the world. Japan has a mere 143,000 square miles (370,370 square kilometers), slightly smaller than California. Perhaps a fairer comparison is with the nations of Western Europe. Although smaller than France, Japan is larger than the British Isles, Italy, or a combined East and West Germany. Japan's land problem stems from too many people on too little land. In terms of the population-land ratio, Japan is indeed a "small" country.

Compounding the problem is the rugged nature of Japan's terrain. The islands are actually the summits of immense submarine ridges thrust up from the floor of the Pacific Ocean. The island chain is but one part of the unstable orogenic (mountain-making) zone that encircles the Pacific Ocean. The Japanese islands rise abruptly from the deep waters of the Pacific on the east and the Sea of Japan on the west. Hence, Japan is an unstable part of the earth's crust, with hundreds of volcanoes scattered along the archipelago and earthquakes a common occurrence.

In this geologic setting, low, level lands are in short supply. Only 25 percent of the total land area has slopes under 15 degrees. The other 75 percent is too steep for cultivation and has little other utility for human occupancy. What level land does exist is

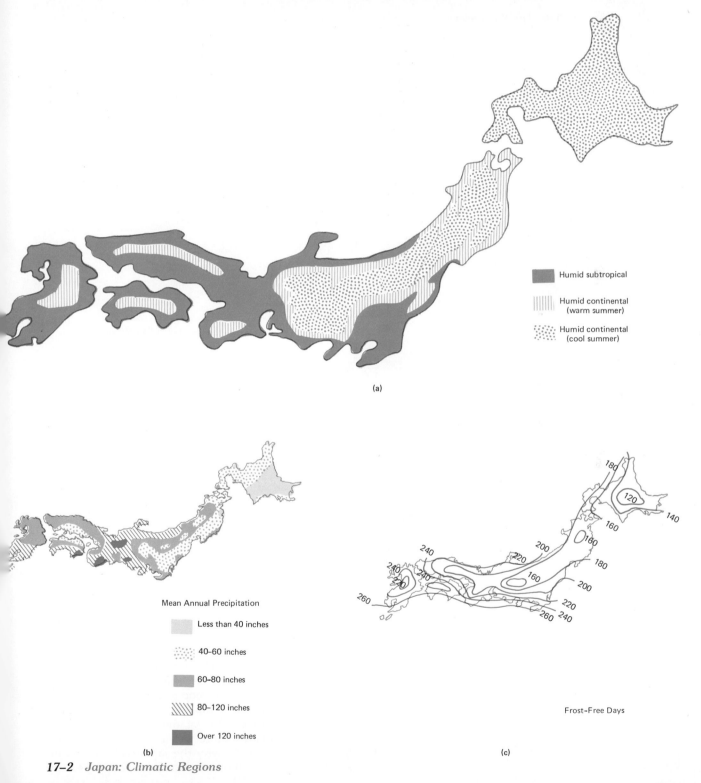

(a)

Humid subtropical

Humid continental
(warm summer)

Humid continental
(cool summer)

Mean Annual Precipitation

Less than 40 inches

40–60 inches

60–80 inches

80–120 inches

Over 120 inches

(b)

Frost-Free Days

(c)

17–2 *Japan: Climatic Regions*

337

17-3   *Japan: Population Distribution*

One dot represents 100,000 people

found in narrow river valleys and alluvial coastal plains separated from one another by stretches of rugged hills. Thus, Japan's more than 114 million people are actually concentrated in a land area slightly smaller than the state of Indiana. The population per square mile of arable land is well over 4,000 (1,544 per square kilometer), one of the highest in the world. Even the density of population per square mile of total land area, which comes to nearly 800 (306 per square kilometer), is ten times that of the United States. Japan is thus one of the most crowded human landscapes in the world (Fig. 17–3).

The Japanese have been able to support such a dense population because quite early in their history they adopted an intensive form of irrigated agriculture from China. This agriculture, almost like gardening, produces relatively high yields per land unit, and can be found in many parts of Asia. The chief distinction of Japan's variant of this agricultural system is that the Japanese, in modern times, developed productivity to a degree that has not yet been matched by any other Asian country.

A Maritime Nation

In premodern times, land communication was difficult in Japan. Thus, the surrounding seas provided links among the islands and along the coast, as well as contact with the outside world. Most of Japan's population is still located close to the sea. Not surprisingly, then, the sea has always played an important role in Japan's national life. The rugged coastline, with countless bays and inlets, the most important of which is the great Inland Sea separating Honshu from Shikoku, made fishing a major activity of the Japanese early in their history. Stimulated by the shortage of land for livestock raising, the fishing industry is one of the country's most important activities today. The Japanese continue to obtain the major share of protein for their diet from fish and other sea products. In modern times, the fishing industry has become a global enterprise, with vessels roaming the four corners of the world in search of seafood for Japan's large population. On another level, it is easy to see why the Japanese moved into shipbuilding early in their national life, but especially

Japan is a maritime nation. Fishing is a traditional occupation, and fish and other sea products are important elements in the Japanese diet. Today their fishing industry is among the largest, with Japanese vessels working the sea's resources throughout the world. (Consulate General of Japan, New York)

with the industrialization effort at the close of the nineteenth century. The shortage of domestic supplies of raw materials was the chief stimulus for shipbuilding. The result was that Japan became one of the major maritime powers of the world by the 1920s, a position it has held to this day. In fact, in the post–World War II period, Japan has surged ahead as the leading shipbuilder in the world, far outdistancing any of its rivals. The Japanese merchant fleet also continues to expand at a rapid rate.

Other Agricultural Resources

Forests still cover over 65 percent of Japan's land area, one of the highest proportions for any of the world's rich nations. Unlike the Chinese, who were not particularly conservation-minded through most of their long history, the Japanese have treated their forest lands with greater care in spite of the heavy population density and the demand for agricultural land. This concern was partly due to the fact that the forests are found primarily on the steeper slopes unsuited for agriculture, and partly because wood products are important in the Japanese culture. Even today, the majority of Japanese live in wooden houses, and wood and charcoal are still important domestic fuels. Also, the great relative importance of hydroelectric power and the need for abundant irrigation water for paddy rice growing have long encouraged the Japanese to respect their forest lands. It should be noted, however, that demand for forest products far exceeds domestic production, so that forest products are among the major import commodities of Japan.

The preservation of the forests has helped protect Japan's valuable water resources. The mountainous terrain, abundant rainfall, and swift streams provide power for one of the world's largest hydroelectric industries, in addition to irrigation water for agriculture. The rivers, however, are not without prob-

In the post-World War II period, Japan has become the world's principal shipbuilder. Nearly all the materials needed must be imported, yet well-organized operations with highly skilled labor have overcome the shortage of local raw materials in this labor-intensive activity. (Consulate General of Japan, New York)

Wood plays an important role in Japanese culture. The Japanese have practiced forest conservation for many decades. Most homes are constructed of wood. The demand for wood is so great that forest products are a major import item. (Consulate General of Japan, New York)

lems. Because of their shortness, steep gradients, and relatively small drainage basins, flash floods often follow heavy rains. These floods can result in large discharges of sand, gravel, and even boulders into the valleys and can be very destructive to agricultural land.

Within the limited areas suitable for agriculture, the chief resource of importance is, of course, soil. In general, the quality of Japan's soils is not conducive to high agricultural productivity. Partly, this circumstance is the result of many centuries of intensive use by Japanese farmers, and partly it is due to the country's geologic and topographic character. The most important soils are the alluvial deposits of floodplains, deltas, and alluvial fans. These are the most productive soils in Japan and are used primarily for rice cultivation. Nevertheless, Japan's high agricultural productivity is directly dependent on large inputs of chemical and organic fertilizers, plus very careful cultivation techniques.

Mineral Resources

Japan was cheated by nature in terms of the distribution of the world's mineral resources. The country is practically devoid of significant mineral deposits. As a result, the Japanese have had to rely

Japan is dependent upon imports of raw materials and food products and exports of manufactured goods. Major port facilities such as those of Yokohama have been built to further trade. The accumulation of materials at port sites has stimulated manufacturing in the local area. (Consulate General of Japan, New York)

primarily on imported raw materials to supply their industrial development. Japan's present dependency on imports of significant minerals ranges from 100 percent for aluminum and nickel to 45 percent for lead. Japan imports 98 percent of its iron ore, almost 100 percent of its petroleum, 75 percent of its coking coal, and 50 percent of its zinc and copper requirements. Among the important industrial raw materials needed, the country is self-sufficient only in limestone and sulfur.

The demand for mineral raw materials has mushroomed in the postwar period. Raw materials for industry now account for over 50 percent of Japan's total imports. Over 40 percent of Japanese investment overseas is for securing raw materials. Hence, foreign trade assumes a special importance for Japan, compared with other major industrial countries such as the United States and Soviet Union, which are far more self-sufficient. The desire to secure permanent sources of raw materials was one of the reasons for

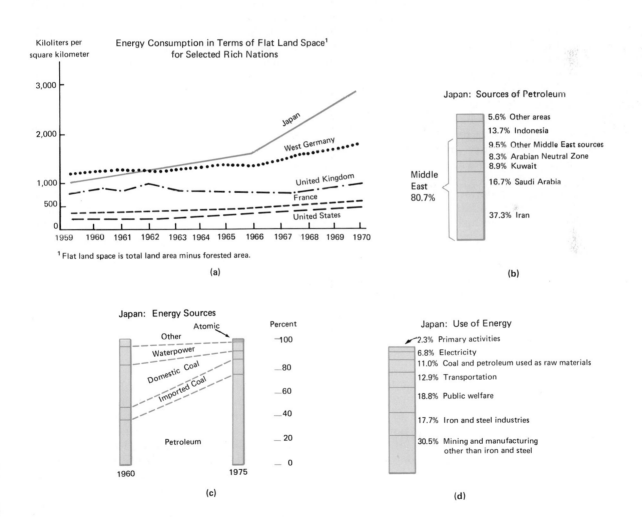

17-4 *Japan: Energy Characteristics*
SOURCE: *Japan Report* (December 1973; February 1974).

Japan's imperial expansion before and during World War II. An important lesson of that era is that Japan will do whatever is necessary to maintain secure overseas sources of raw materials.

The magnitude of this dependency on overseas sources is best illustrated by the case of petroleum. Oil is now Japan's largest single import commodity, accounting for 15 percent of total imports. In the early 1970s, Japan annually imported nearly 250 million tons and spent over $4 billion for them. Moreover, until the oil crisis of 1973–1974 erupted, Japan's rate of energy consumption was increasing each year, even faster than that of the United States (Fig. 17–4a). Unfortunately for Japan, oil accounts for three-fourths of its total energy output, and nearly 81 percent of the oil comes from the Middle East. Presently, 37 percent of Japan's oil comes from Iran, 43 percent from the Arab states, and 19 percent from elsewhere, mostly Southeast Asia (Fig. 17–4b).

Japan does have alternative sources of power. Coal still provides about 18 percent of total energy supply (Fig. 17–4c and d). Coal's relative importance, however, has declined markedly, and the domestic coal industry is one of the least efficient sectors of Japan's economy. Waterpower is important, presently providing about 7 percent of the country's power needs. Rising construction costs made further development of waterpower economically unattractive during the days of low world oil prices; however, this situation may now be changed. Nuclear power is also being developed rapidly, but at best it is estimated that atomic power can meet no more than 10 percent of the nation's needs by the end of the 1980s. Moreover, nuclear power is meeting increasing opposition from the Japanese people, as is the case in the United States. Japan's total energy situation points up the extremely vital role Japan plays and will continue to play in the global oil industry, with all of its political, economic, and even military implications.

In balance, the Japanese have responded to their physical environment by channeling their national development to take maximum advantage of their opportunities and by overcoming environmental deficiencies. For example, the Japanese have made up for the shortage of arable land by developing a highly intensive form of irrigated agriculture and by developing the fishing industry and water transportation to take advantage of the country's insularity. Yet much of this progress might not have been possible without the remarkable characteristics of the Japanese as a people and culture.

Role of Human Resources in Japan's Development

It would not be stretching things to say that Japan has succeeded so well in its development and modernization primarily because of the quality of its human resources. The Japanese people have a high degree of racial and cultural homogeneity. It is one of the country's great strengths, because it has helped foster a sense of national identity and cohesiveness that multicultural societies often find difficult to achieve. It is unity and purpose that have fueled Japan's modernization in the last century.

The Emergence of the "Japanese"

The Japanese already had achieved their distinctive physical and linguistic identity 2,000 years ago. The Japanese language is a polysyllabic, highly inflected language, similar to Korean and the Altaic languages of northern Asia. Although it continues to use a large number of Chinese characters in the written form, the Japanese language is actually very different from Chinese and was one of the factors that helped Japan to preserve its cultural distinctiveness.

Although the Japanese are a homogeneous group today, many racial strains have been blended into the people over a long period. Neolithic peoples inhabited the islands for several millennia. These were hunters, fishermen, and gatherers, and had only partial Mongoloid racial features. The Ainu were another important element in the creation of the modern Japanese. The Ainu were a proto-Caucasian people who inhabited much of northern Japan and are still found in small numbers on the island of Hokkaido. Mixture with the Ainu is believed to account for the fact that some Japanese today have more facial and body hair than the Koreans or Chinese. Other traits may have been acquired from peoples of Southeast Asia. This interchange was possibly the result of diffusion via coastal southern China, rather than actual migration of Southeast Asian people to Japan (Fig. 17–5a).

The real beginnings of the Japanese as a distinctive people and culture can be traced, however, to south-

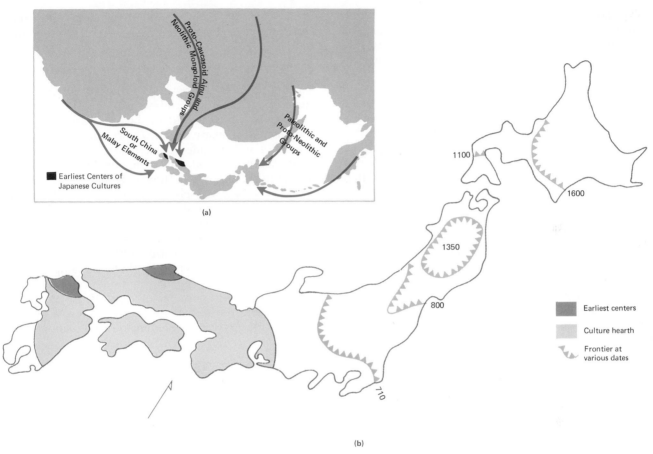

17–5 *Japan: Origin and Spread of Japanese Culture*

SOURCE: Adapted from Joseph E. Spencer and William L. Thomas, *Asia: East by South,* 2nd ed. (New York: John Wiley & Sons, 1971).

west Japan in the early centuries just before the Christian Era. At that time a culture known as Yayoi developed in north Kyushu, spreading eastward into the Inland Sea area by around 200 B.C. Yayoi culture evolved from Mongoloid peoples who migrated from the mainland under the push of the expanding Han Chinese (the Han Dynasty, 206 B.C. to A.D. 220). These peoples brought with them paddy rice agriculture and the use of bronze and iron.

Southwest Japan was a logical place for Japan's culture hearth. Aside from the fact that it was the closest part of Japan to Korea and China, the climate there was much milder than places farther north in the islands. Moreover, the sheltered coastal fringes of the Inland Sea area provided reasonably good habitats for these peoples to establish themselves (Fig. 17–5b).

By the third century A.D., Yayoi expansion resulted in the shifting of the center of their culture from north Kyushu to the Yamato Lowlands at the eastern end of the Inland Sea, near present-day Nara and Kyoto. This area remained the focus of Japanese culture for more than 1,500 years. As the evolving Japanese multiplied through migration and population growth, they began to fill up the lowlands of southern Japan. By the sixth century, they had begun to push northward. As they reached the northern half of Honshu, they encountered the culturally very different Ainu.

The clash of cultures and demand for land resulted in military struggles along the frontier with the Ainu that were similar to the clashes between European settlers and the American Indians. Most of the Ainu retreated to the island of Hokkaido. As the Japanese reached northern Honshu, they found themselves in a forested environment with a much cooler climate and few directly usable land resources. Clearing of the land for agriculture was difficult, and even then the shorter growing season prevented a large population density. Hence, the major thrust of development of Japan as a nation was focused on the area south of about 37° N. latitude.

During the Tokugawa period (1615–1867), the focus of power shifted farther eastward to Tokyo (then called Edo) in the Kanto Plain, where it has remained ever since. Today, Japan's core area remains essentially in the same area in which the nation's foundations and early growth were laid. This core is a belt between approximately 34° to 36° N. latitude, from Tokyo on the east to Shimonoseki at the western end of Honshu and encompassing the Inland Sea. The major share of Japan's population, cities, industry, and modern economy remain concentrated in this zone (Fig. 17–6).

National Character and Social Attitudes

Much has been written and said about the character and attitudes of the Japanese people. As is true of most peoples of the world, the Japanese are extremely complex, with many seemingly contradictory characteristics and attitudes. Nevertheless, a few of these traits can be singled out as having had particular impact in shaping the destiny of Japan. Each of these characteristics is complex in itself. Hence, they can be introduced here only in a very generalized sense. Most of these characteristics have evolved from centuries of national development. Some were not firmly stamped on the national character, however, until the Tokugawa period, perhaps the single most important period in Japan's premodern era. During the Tokugawa period, Japan's rulers isolated the country from the rest of the world. It was a period of consolidation and refinement of the Japanese cul-

17–6 *Japan: The Core Region*

SOURCE: Adapted from Joseph E. Spencer and William L. Thomas, *Asia: East by South,* 2nd ed. (New York: John Wiley & Sons, 1971).

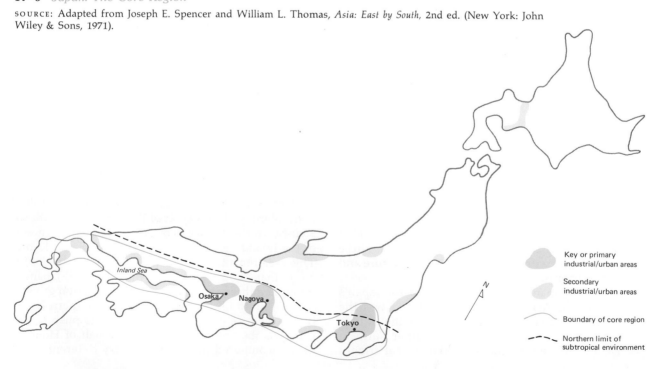

Key or primary industrial/urban areas

Secondary industrial/urban areas

Boundary of core region

Northern limit of subtropical environment

ture as a distinct entity. Moreover, within the feudal economy and society of Tokugawa Japan, several important characteristics of Japanese society became entrenched.

Of all the characteristics of the Japanese people, perhaps the single most important one to help explain the developments of the last century is the remarkable adaptability of the Japanese. Just as they adapted to their peculiar natural environment, they have been willing and able to borrow culture and technology from others when they clearly saw the superiority of something foreign. This ability was evidenced in their adapting aspects of Chinese culture in the premodern era and in their adapting Western technology in the modern era. Certainly contributive to this quality have been the innate intelligence of the Japanese and their deep respect for education. Even before the Meiji Restoration, Japan had the highest proportion of educated people of any country in Asia. In the course of the last century, the Japanese have evolved from the master imitators of Western technology to technological innovators of the first rank. Also important has been the great capacity of the Japanese for thrift and hard work, qualities undoubtedly instilled by the relatively difficult natural environment in which they had to create a nation.

Underpinning these qualities have been the basic homogeneity and unity of the Japanese already referred to, although political unity has been a more recent phenomenon. Out of that homogeneity grew a deep and unshakable consciousness of being Japanese, a clear recognition that they are a unique people. With the success of modernization in the last century, a great sense of national pride has been fostered.

The role of the individual in Japanese society has also been critical. In many respects, Japanese society is totally the opposite of Western—especially Anglo-American—society where the individual is paramount. In Japan, the group is all-dominant. Everything in life including work, study, and play is done largely within a group context. An individual never asserts himself to stand out from the crowd, to be different in any way from his peers. Moreover, society is extremely hierarchical, with clearly defined social classes and modes of behavior within classes and between members of different classes. Yet there is social mobility in that it is possible to move upward in society through education, hard work, and intelli-

gence. Mobility, however, is not as great as that in the United States and in some other Western countries.

The Japanese have long had a strong sense of national and personal self-discipline, a strong sense of duty to country and society, and a strong respect for authority and willingness to obey orders. The subordination of personal desires to national needs has been a key element in the success of Japan's government and business leaders in setting goals for the country's development and being able to count on the total commitment of the people to those goals. The defeat in World War II perhaps weakened some of these qualities, some more than others, but the Japanese remain today without question one of the most vigorous and purposeful societies in the world. They could never have achieved so much without their exceptional human resources.

Further Readings

ACKERMAN, E. A., *Japan's Natural Resources and Their Relation to Japan's Economic Future* (Chicago: University of Chicago Press, 1953), 655 pp. Somewhat old now but still a useful discussion of an important topic.

ALLEN, G. C., *A Short Economic History of Modern Japan*, 2nd ed. (New York: Praeger, 1972), 262 pp. This classic, first published in 1946, remains one of the best studies of Japan's first century of modern development.

BEARDSLEY, R. K., and others, *Village Japan* (Chicago: University of Chicago Press, 1959), 498 pp. Detailed study of a typical Japanese village in the Inland Sea area.

DORE, R. P. (ed.), *Aspects of Social Change in Modern Japan* (Princeton, N.J.: Princeton University Press, 1967), 474 pp. A collection of essays on the accelerating social changes since 1945.

DORE, R. P., *City Life in Japan: A Study of a Tokyo Ward* (Berkeley: University of California Press, 1958), 472 pp. An interesting account of a typical urban neighborhood in modern Tokyo.

EMBREE, J., *Suye Mara: A Japanese Village* (Chicago: University of Chicago Press, 1939), 354 pp. A classic anthropological study of village life in prewar Japan; still useful.

GUILLAIN, ROBERT, *The Japanese Challenge: The Race to the Year 2000* (Philadelphia: Lippincott, 1970), 352 pp. An experienced French journalist looks at the Japanese "miracle" in a fascinating way, although events of the early 1970s cast doubts about some of the author's bolder predictions.

JANSEN, M. B. (ed.), *Changing Japanese Attitudes Toward Modernization* (Princeton, N.J.: Princeton University Press, 1964), 546 pp. A collection of essays on Japanese reactions to the process and effects of modernization in twentieth-century Japan.

JOHNSTON, BRUCE F., *Agriculture and Economic Development: The Relevance of the Japanese Experience* (Stanford, Calif.: Food Research Institute Studies, 6, 1966), 312 pp. A brief but useful analysis of the role of agriculture in Japan's modern development.

JONES, F. C., *Hokkaido: Its Present State of Development and Future Prospects* (New York: Oxford University Press, 1958), 146 pp. A look at Hokkaido's historical development and potential as the last frontier in Japan.

KAHN, HERMAN, *The Emerging Japanese Superstate, Challenge and Response* (Englewood Cliffs, N.J.: Prentice-Hall, 1970), 274 pp. One of America's foremost futurologists explores the reasons for Japan's economic growth and makes bold predictions for the country's role in the world in the coming decades; as with Guillain's book, some of Kahn's predictions no longer seem realistic.

LOCKWOOD, W. W., *The Economic Development of Japan, 1868–1938* (Princeton, N.J.: Princeton University Press, 1954), 603 pp. A fine treatment of Japan's prewar development experience, especially interesting because of comparisons the author makes with the experience of Europe and India.

NORMAN, E. H., *Japan's Emergence as a Modern State: Political and Economic Problems of the Meiji Period* (New York: International Secretariat, Institute of Pacific Relations, 1960), 254 pp. Contains a wealth of information about the political and economic history of Japan's early modern period.

SMITH, T. C., *The Agrarian Origins of Modern Japan* (Stanford: Stanford University Press, 1959), 250 pp. A treatment of the transition from a feudal to industrial society, and the relationship with agriculture.

YAZAKI, T., *The Japanese City* (New York: Japan Publications, 1966), 105 pp. A study of the rapid growth of Japan's urban areas, with a special section on Tokyo.

Modern Japan: The Transformation

18

still has a dichotomous nature, a curious mixture of the traditional and the Western.

In a manner of speaking, Japan has gone through three transformations since 1868. The first consisted of the early period of modernization and industrialization in the late nineteenth century, followed by the move to heavy industry and militarization that reached its peak with World War II. The second transformation was the reconstruction and return to international power in the postwar period. The third transformation, now in its formative stages, involves the seeking of new directions in a world of increasing scarcity of raw materials and discontent with past growth strategies.

Japan's First Transformation: The Meiji Period and Its Aftermath

The Japanese experience illustrates the kind of cooperation and balance needed between the government and people in the economic development of a nation. Japan's leaders recognized the military superiority of the West. They also saw with alarm what was happening in China when that country failed to respond adequately to the challenge of the West. Japan's leaders were determined that a similar fate not befall their country.

The Japanese adopted a very pragmatic approach to transforming Japan step by step. As a consequence, they achieved spectacular and rapid results. Within a mere fifty years, the leaders created a sound and modernized economy. Japan achieved a position of national security and the international equality it had long sought. This position was attained in part by the government's providing a proper environment for development. Many measures were used. Feudal restrictions were removed on trade within the country and on individual activities. Internal stability was assured. Sound currency, adequate banking facilities, a reasonable tax system, and efficient government services were provided. The government took a direct role in industrial development by pioneering many industrial fields, and by encouraging businessmen

*I*n 1868, the feudal Tokugawa rule of Japan was overthrown by a group of revolutionary young Japanese. These revolutionaries responded to the challenge raised by the arrival of the Americans in the 1850s. Following that event, a series of shock waves reverberated throughout Japan and led to the beginning of the transformation of Japan into a modern nation.

In the more than one century that followed, Japan was physically transformed almost beyond recognition, yet some of the transformation was superficial. In a sense, it consisted of changes in the economy, in the use of the country's physical resources, in the makeup of the cities, in modes of transportation, and in habits of dress. Many of the basically feudal characteristics of Japanese society, however, were retained. Although the social transformation has been more rapid since the end of World War II, Japan

to move into new and risky ventures. The government also helped fund many ventures, providing private entrepreneurs with aid and privileges. Active cooperation between government and big business that would be viewed as collusion in the United States worked well in Japanese society. That kind of cooperation continues to be a basic characteristic of Japan's economy.

The role of the people was equally important in making a success of the transformation. Thousands of individual Japanese responded eagerly to the new economic opportunities. In the long run, it was this private initiative operating within the context of Japanese society that produced the bulk of Japan's economic modernization.

A critical aspect of this response, however, was the emphasis in the first two decades of the Meiji period on development of the traditional areas of the economy: agriculture, commerce, and cottage industry. There was no attempt made to build modern industry immediately and directly on a weak local economy, as many developing nations in the postwar period have mistakenly tried.

Growth of Industry and Empire

The modern industries deemed important by the government were those on which military power depended. Hence, the government led the way in developing shipbuilding, munitions, iron and steel, and modern communications; the first railway was built between Tokyo and Yokohama in 1872. At the same time, as the need for raw material imports grew, export industries were strongly encouraged, particularly silk and textiles.

In a quantitative sense, Japan's take-off period really did not begin until after the Russo-Japanese War of 1905 (see Chapter 4). At the end of the nineteenth century, the country still had a small industrial base. Growth after 1900, however, was great. Between 1900 and the late 1930s, the production of manufactured goods increased more than twelvefold. Export trade grew twentyfold in the same period; manufactured goods accounted for most of the increase. The Japanese excelled at producing inexpensive light industrial and consumer goods cheaper than many other countries. It was a production approach that has continued to serve the Japanese well. Foreign markets, however, played a less important role in this export strategy than is commonly believed. Japan's economic growth in the early twentieth century was largely self-generated. Foreign trade accounted for a smaller percentage of Japan's total economic activity than was true of most European countries during the same period.

Two decisive events that shaped the course of Japan's development in the twentieth century were the victories over China in 1895 and over Russia in 1905. These two wars had a number of consequences. For one thing, they started Japan on a course of imperial conquest that ended in the disaster of World War II. This course was partly an imitation of what was then an accepted practice of modern Western nations. It was also partly a quest for secure sources of raw materials and markets for industrial goods. The two wars also greatly expanded Japan's territorial control to include Taiwan, Korea, and parts of northern and coastal China, including Manchuria. By the end of World War I, Japan was a fully accepted imperial world power.

Growth of Agriculture and Population

A significant change that accompanied the modernization of Japan after 1868 was an upsurge in population growth. During the latter half of Tokugawa rule, Japan's population had stabilized at about 33 million. Stabilization was due, in part, to the prevalence of infanticide and abortion. Moreover, the population had increased to the maximum that could be supported by the technical levels of the feudal economy. Between 1868 and 1940, however, Japan provides the classic illustration of the interaction of economic and demographic factors (demographic transformation). As industrialization and urbanization proceeded, birth and death rates both declined. Population began to increase, but the rate averaged only about 1.5 percent a year up to 1940. Still, that growth rate was sufficient to double the population to just over 73 million by 1940. In spite of this considerable growth in population, the development of the economy was rapid enough to produce substantial increases in the level of living.

At the close of Tokugawa, three-fourths of Japan's population consisted of peasants, and about four-fifths of the labor force was engaged in agriculture, forestry, and fishing. From 1868 to the present, the rural population has declined steadily in proportion to the urban population. In absolute terms, though, agricultural population did not start declining until just before World War II. Today, agricultural workers constitute about 20 percent of the total labor force.

Japanese agriculture in the last century has been characterized by several important trends. Until World War I, there were great increases in production. These increases resulted mainly from improved farming methods, including more efficient irrigation, better crop strains, pest control and, above all, a lavish application of fertilizers. Agriculture succeeded in supplying all but a small part of the enlarged demand for rice that accompanied the growth in population and the rise in per capita consumption. As Japan's population continued to increase, however, the country's ability to feed itself steadily declined. As a result, reliance on the colonies of Taiwan and Korea for rice and other foodstuffs increased.

Another trend was that as industrialization progressed, the gap between urban and rural standards of living increased. Even at the turn of the century, the average Japanese farm was extremely small; one-third of the farms were under one-half *cho*[1] and two-thirds between one-half and one *cho*. Productivity was dependent on heavy inputs of labor on tiny fragmented fields often scattered some distance from one another. Moreover, the incidence of tenancy increased after 1868. By 1910, about one-third of the farmers were full tenants, two-fifths owned their land, and the remainder were part tenants, part owners. Tenancy was not substantially reduced until after World War II when a compulsory land reform program was initiated under the American occupation forces.

Another important development arising from the pre–World War I period was a growing tendency for rural people to seek part-time employment in secondary occupations. Reliance on rice alone could not bring the farmers enough to sustain an acceptable standard of living. Hence, many farmers engaged in silk-cocoon raising, and large numbers of rural people, especially young girls, sought employment in silk mills. This movement out of agriculture into industry was hastened by the world depression of the 1930s. Large numbers of rural people forced to seek work in industry competed with urban dwellers for a decreasing number of jobs. The transference of

workers from agriculture to industry and from rural to urban trades resulted in keeping down industrial wages. Low wages combined with technical improvements during this period enabled Japanese industry to limit production costs and remain highly competitive in world markets.

Japan's Second Transformation: Rise from the Ashes of War

When Japan surrendered in August 1945, the nation was prostrate. Destruction from the war had been catastrophic, especially in urban and industrial areas. The nation was stripped of its empire and consisted only of the main archipelago. The future of the 72 million Japanese seemed bleak indeed. Yet within a decade Japan was thriving; most of the physical destruction of the war was erased and the nation well on its way to regaining a position of economic might in Asia and in the world.

This revival is attributable in substantial part to the resilient fiber of the Japanese people. Physical destruction from war does not destroy the inherent qualities of a strong society. Japan, like Germany in 1945, still had a solid base of educated, technically proficient people. Its able administrators and entrepreneurs were eager to seize the reins of rebuilding the nation just as fast as the United States occupation authorities would allow them. The "Cold War" in the latter 1940s, and the fall of China to the Communists in 1949, led the United States to return the reins of government to the Japanese much faster than might otherwise have been the case. The United States needed a strong Japanese ally and played a critical role in the rebuilding of Japan.

The American Role

American assistance took several forms. Financial aid was especially critical in the immediate postwar years. It consisted of billions of dollars in foodstuffs, military procurement orders during the Korean War (1950–53), and other aid. Of even greater long-range benefit was the policy of relatively open doors for Japanese exports to the United States. By 1970, the

---

[1]One *cho* is equivalent to 2.45 acres or about 1 hectare.

United States was buying about one-third of Japan's total exports. This huge market was and still is of great importance to the Japanese. Another benefit was military protection by the United States which enabled Japan to spend less than 1 percent of its GNP annually for military expenses. Japan's defense budget constitutes a smaller proportion of its GNP than that of any other major nation. Also beneficial to Japan was American technology. With their own industry nearly leveled to the ground by the war, the Japanese bought American technology at most attractive prices and revitalized their industry. This approach gave Japanese industry a competitive advantage over many other countries, including the United States, whose plants and technology were much older and could not be replaced so easily.

Development Strategy of the Japanese

The policies of the Japanese government and businessmen were probably even more important than the American role. The close cooperation between government and business continued and even grew. Cooperation was particularly important in financing the modernization process. Banking credit was backed by the government and made heavy capital investment possible. The economy was geared to a high-growth-rate strategy that relied on large increases in productivity to provide surpluses to pay back capital debts. Commercial banks lent out a high percentage of their funds with the backing and guidance of the government-run Bank of Japan. Two-thirds of the capital requirements of the average Japanese company were met by loans from banks and only one-third by stock, just the reverse of the system in the United States. The strategy worked. The growth rate of the GNP averaged 8.6 percent annually from 1951 to 1955, 9.1 percent from 1955 to 1960, 9.7 percent during 1960 to 1965, and 13.1 percent during 1965 to 1970.

To a far greater degree than in the United States, the Japanese government, with great skill, has guided the economy much as in a socialist state. Growth industries are determined and supported with generous assistance of many kinds, including high depreciation allowances, cheap loans, subsidies, and low taxes. The results of research carried out in government laboratories are turned over to companies for commercial development. As Japan's economy has grown during the last century, industry has gradually progressed from labor-intensive light industrial production to capital-intensive heavy industry. In the 1970s, the official government stimulus shifted more to such industries as automobiles, precision tools, and computer electronics, an indication of the great sophistication of contemporary Japanese industry.

On the other hand, inefficient or nongrowth potential industries are dealt with rather ruthlessly. The attitude of many Japanese, including the government, is that uncompetitive industries should be forced to the wall by governmental financial practices. Some of the resources of those uncompetitive industries can thereby be freed for more efficient enterprises. Transference of resources from less to more efficient sectors is a key ingredient of economic progress. Japan's vigorous pursuit of this transference is in marked contrast to many other developed and underdeveloped countries, where inefficient industries are protected. The Japanese recognize the importance of the principle of comparative advantage in allocating the country's resources.[2]

At the same time, Japan has been protectionist in the postwar period. High tariff barriers have been raised against thousands of products from foreign countries. Foreign investment in Japan is very restricted. The rationale for a protectionist policy was that Japan's economy was too weak to withstand uncontrolled imports and foreign investment. This position was generally accepted by the United States and other nations in the 1950s and 1960s. Following the events of the early 1970s, however, Japan has been forced to start lowering its protective barriers.

The *Zaibatsu*

The corporate structure also has played a key role in Japan's development. When any traditional country industrializes, there is a shortage of capital, skilled labor, and technical resources. To obtain rapid growth, resources must be concentrated. Since there was no model of socialist development in the late nineteenth and early twentieth centuries for Japan to follow, it was logical for concentration of resources to fall into private hands. Hence, the *zaibatsu*, or "financial cliques," emerged out of the close relation-

[2]The principle of comparative advantage, simply stated, means that a country (or region) should produce those goods for which it is best equipped, in terms of resources, labor, capital, and technology.

ship between government and business. By the 1920s, the *zaibatsu* controlled a large part of the nation's economic power. The *zaibatsu* worked through vertical and horizontal integration of the economy. Thus, a single *zaibatsu* might have control of an entire operation from obtaining raw materials to retailing the final product. The nearest equivalent in the West are the giant conglomerates or the large international petroleum companies.

It is generally agreed that the *zaibatsu* were very efficient. Certainly they provided the entrepreneurial strength that led to the modernization process. Moreover, although there were monopolistic tendencies, the importance of foreign markets and raw materials helped keep *zaibatsu* prices competitive. Also, although much of the nation's wealth became concentrated among a few immensely rich and powerful families, enough profits from industry filtered down to establish a substantial and growing Japanese middle class.

Efforts by the United States occupation authorities to break up the *zaibatsu* were not very successful. Many have reemerged in the last two decades. These cliques, joined by many other giant corporations outside the *zaibatsu* system, have played a vital role in Japan's second transformation.

The Double Structure of the Japanese Economy

The resurrection of the *zaibatsu* has not surprised the Japanese. They tend to view it as a natural development. This attitude is related to the peculiar double structure that has long characterized the economy. Basically, the structure consists of a handful of giant combines, thousands of tiny workshops, and relatively few medium-sized firms. This structure had fully emerged by the 1930s and was in part the result of a "split technology." The leaders of modern industry followed Western technology, contrasted with the owners of small shops rooted in the traditional ways of old Japan.

The giant modernized companies, because of large outlays for advanced techniques, have succeeded in greatly increasing the productivity of their labor forces. The medium and small firms have relatively little capital outlay and rely on cheap labor to make their products competitive. The relationship between the two levels of the economic hierarchy is close. The larger companies job out substantial parts of their production to the smaller firms, because it is

cheaper to do it that way. In the postwar period, rapid industrial growth accompanied by a sharply decreased rate of population growth have caused a decline of workers in small industries and agriculture. Wages are rising, even in the small firms, and the double-structure aspect of the economy is diminishing. Nevertheless, about 70 percent of Japan's industrial labor force is still employed by small and medium-sized firms; only about 30 percent work in the large firms.

The double structure has contributed to social inequality in Japan. The employees in the larger firms have reaped the greatest benefits from Japan's growth. A paternalistic relationship between workers and management is very strong in these larger businesses. Workers tend to stay with a firm for life and to identify with a particular company rather than a skill. If a person's skill becomes obsolete, the company provides retraining with no loss in pay. Unions thus do not resist new technology. Employers in turn have great freedom to shift workers from one job to another and can invest huge sums for training without worrying that the employees will leave the company. Labor mobility remains extremely low. In return for their employees' absolute loyalty and hard work, the large companies reward them with lifetime security, relatively modest salaries, and generous fringe benefits. These benefits bring the level of living of these workers to levels comparable to those of workers in many Western countries. The majority of workers in the small firms has not shared as much in the modernization process. Their job security and fringe benefits are far poorer. This inequality remains one of the major social problems to be resolved in Japan.

Population Stabilization

The rising standard of living of the Japanese in the postwar period can also be attributed to the gradual stabilization of population growth (Fig. 18–1). In 1945, Japan faced the specter of eventually becoming so densely populated that the quality of life would seriously deteriorate. To control population growth, the government passed a Eugenics Protection Law in 1948 that legalized abortion for economic as well as medical reasons. Public and private efforts were also made to spread birth control practices. Government propaganda stressed the advantages of a small family. Many other factors, including more years of higher education (Fig. 18–2), later marriages, and the two-child family, also contributed to a steady decline in the rate of population growth. Although the total population is now over 114 million, the annual growth rate is 1.1 percent, one of the lowest in the world. Japan may be one of the first nations in the world to attain zero population growth.

Growth of Urbanization

One of the most dramatic developments of the postwar era, and one with profound consequences for Japan, has been the rapid increase in urbanization and its concentration in a small fraction of the country. In 1950, there were 6.2 million farm households in Japan, almost the same number as during the early Meiji era. By 1970, the number had declined to 5.3 million. Farm population as a percentage of total population declined from 85 percent in early Meiji to about 50 percent in 1945. It is now under 20 percent. The farmers and rural people have migrated to the cities.

Today, over 50 million people live in the three great metropolitan regions centered on Tokyo, Nagoya, and Osaka. Tokyo already has over 26 million people in its metropolitan region, Osaka over 16 million, and Nagoya over 9 million. Tokyo is believed to be the largest urban area in the world in population. In these regions, the density of popula-

18–1 *Japan: Population Characteristics*
SOURCE: *Japan Report* (June 1974).

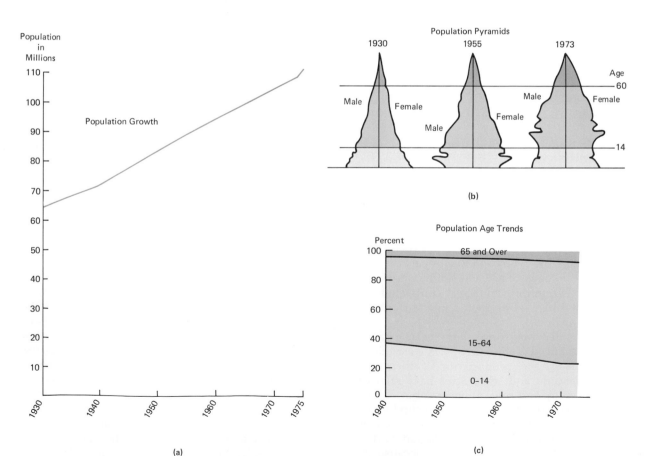

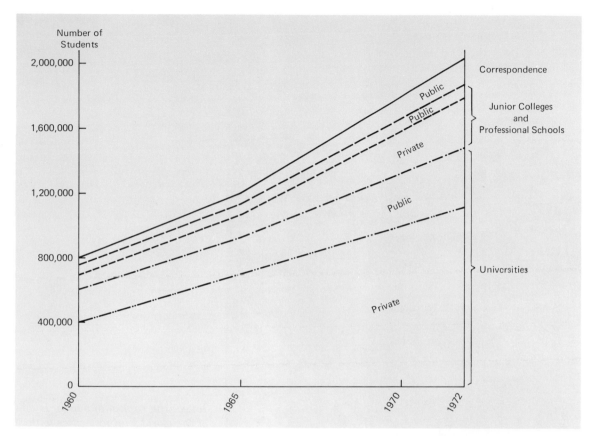

Number of Students

18–2 *Japan: Higher Education Enrollment*
SOURCE: *Japan Report* (February 1974).

tion in places exceeds 10,000 per square mile (3,861 per square kilometer). A major share of the remaining 65 million Japanese are found in other parts of the core region, between Shimonoseki and Tokyo. Movement into these zones was phenomenal in the last two decades. Tokyo alone has received 300,000 migrants each year. If migration and natural population growth rates were to continue, by the year 2000, 80 million Japanese would be located in the "Tokaido Megalopolis" (named after the ancient Tokaido highway) stretching between Tokyo and Osaka.

The reasons for urbanization in Japan are much the same as in other countries of the world. In the case of Japan, however, certainly an important additional factor was the desire of Japan's industrialists and government to concentrate industry in a few areas, most of them near the coast. Concentration was use-

ful to take advantage of economies of scale.[3] Concentration of industry near the seashore also made it cheaper to handle large quantities of imported raw materials, such as iron ore, coal, and oil. Where industry exists, so do the jobs that draw the migrants.

It appears that the growth rates of the giant cities of Japan have slowed in recent years. A new trend has appeared, termed "U-turn migration." The suburbs and satellite cities of the major metropolitan centers, such as Tokyo, are still growing rapidly. But many migrants are now moving to medium-sized cities, normally in the area from which the migrant came, and usually either a prefectural or regional

[3]"Economies of scale" means it is usually cheaper to produce goods on a large scale in a few factories rather than on a smaller scale in many small factories.

Tokyo, the capital of Japan, is one of the world's largest cities with over 26 million people in the metropolitan region. Tokyo is the social, economic, and political heart of the nation. (Consulate General of Japan, New York)

center. Hence, the fastest growing urban areas today are provincial cities in the 200,000 to 300,000 range.

The Consequences of "Japan, Incorporated"

As Japan proceeds through the 1970s and 1980s, a host of problems that need urgent attention are pressing on the leaders and people. These problems are the consequence of the development policy pursued since World War II that emphasized growth in the GNP at the expense of social welfare. This policy succeeded in catapulting Japan to the forefront of the industrial nations. The price, however, was high.

Regional Imbalances

There is no country in the world that is without regional imbalances in the distribution of population and levels of economic development. Nevertheless, Japan exhibits some of the sharpest contrasts of any Asian country. As urbanization and industrialization grew in the core area, another kind of double structure occurred. There is an increasing concentration of the population and modern economy on the outward or Pacific side of the nation at the expense of the inner side, which looks toward the Sea of Japan. There is an increasingly marked geographical division between a modern, industrialized, urbanized, densely populated Japan, and a "backwoods" Japan that is underdeveloped and not much changed from the rural Japan of many decades ago.

Urban Ills

An assortment of urban ills has been another by-product of economic growth. The term *kogai,* meaning "environmental disruption," is much on the minds of millions of people in the core region. *Kogai* has taken many forms. One form stems from the decision in the postwar period to opt for an automobile society. Starting in the early 1960s, the auto was vigorously promoted before roads, safety systems, and driver training programs were developed. Rapid increases in per capita income in the 1960s led to an explosion in demand for cars and other vehicles. By 1973, over 5 million new cars were sold in one year, making Japan the second largest auto consumer in the world. There are today more than 20 million motor vehicles on Japan's crowded roads (not counting over 10 million motor bikes), compared to only 1.4 million in 1960. Heavy traffic is an ill common to practically every major city in the world, but in Japan it is compounded by the haphazard way in which the cities have grown. Most of these cities have grown without serious attempts at zoning.

It must be admitted, however, that the Japanese have not neglected public transportation. The Japanese, in fact, have been among the pioneers in development of high-speed express trains and electrified railways, such as the famous Tokaido Bullet Express between Tokyo and Osaka on the Shinkansen Line. The route is now being extended westward through the core region (Fig. 18–3). The major cities also have excellent subways and bus systems; but because of the large populations in the urban areas, the demand for public transportation far exceeds capacity. During the rush hour in Tokyo, for example, the commuter trains are commonly crowded to 300 percent of capacity.

Many social problems have also arisen in the cities. Some of these problems are physical and tangible. These include the high price of housing, the extremely crowded living conditions (average per capita living space in Tokyo is only 74 square feet [7 square meters]), and the lack of modern sanitary facilities (which helps account for the continued popularity of public bathhouses). Moreover, there are problems of air and water pollution, noise, and the generally unaesthetic appearance of Japan's major cities, including the garish neon displays in the central business and entertainment districts. Some of the bleakest industrial slums in the world are found in Japan's cities, such as around Osaka. These slums remain one of the paradoxes of modern Japan, because appreciation of nature and beauty in architecture and life have been quintessential parts of traditional Japanese culture for centuries. Japanese artistic values have also had a profound impact on art and architecture in other countries, especially the West.

Other problems are less immediately tangible but no less important. Many stem from the vast social changes that have swept Japan since World War II. Included are the breakdown of the family, the increasing independence of children from parental authority, the rising desire of young married couples to live alone away from parents and relatives, and juvenile delinquency and crime in general (still much below the levels of the developed Western countries). Other changes are reflected in the increased freedom of women in a society where women's liberation has been the slowest to develop, the trend toward pursuit

18–3 *Japan: The Shinkansen Line—Past, Present, and Future*

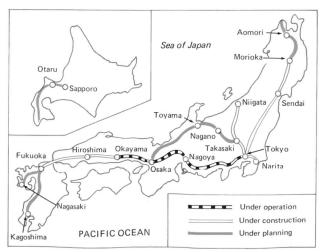

The Tokaido Bullet Express is the world's fastest train, covering the 320 miles (512 kilometers) between Tokyo and Osaka in three hours, In the background is Mt. Fuji, Japan's highest peak. (Consulate General of Japan, New York)

of happiness, the weakening of the Spartan work ethic, and the neglect of old people in a society where social security has always been provided by the family, not the government. Most of these problems are inseparable aspects of industrialization and urbanization in any country, but in Japan they have taken on their own special nature.

Pollution of the Environment

Of all Japan's problems, one of the best known is environmental pollution. Until very recently, Japan had done the least of any major industrial nation to protect its natural environment from the effects of uncontrolled industrial development. The seriousness of the problem was finally recognized by the government when it officially declared six areas in Japan as dangerous to human health. A total of forty-six major areas of pollution have been officially pinpointed. In actuality, no city or prefecture in Japan has been unaffected to at least some degree.

Most serious have been air and water pollution. A survey was made in one part of Tokyo where houses are sandwiched between many small factories. It was found that dustfall averaged 130 tons per square mile (50 tons per square kilometer), and that 35 percent of the residents examined suffered severe lung trouble. Traffic police must take periodic breaths of pure oxygen while on duty in Tokyo's crowded streets. Cases of *minamata* disease (named after the Minamata area where it was first noticed), or organic lead poisoning, are widespread. Also widespread are cases of *itai itai* (literally "ouch, ouch") disease, or cadmium poisoning. Reported cases of the effects of air and water pollution on human health are widely believed to be merely the tip of the iceberg. Much less easily measured or quantified are the effects of noise on human health in Japan's major cities.

Rural Problems

Rural areas have not escaped the effects of modernization either. For political and social reasons, farming has been one of the last sectors of the economy to modernize. The farmer's natural conservatism was reinforced in 1946 to 1949 by the land reform program that awarded tenant farmers the small plots they had tilled for generations. These smallholders, farming an average of only 2.5 acres (1 hectare), have been loyal supporters of the postwar conservative governments. In return, the government has offered the farmers the highest subsidies and support prices for their rice crop of any other Asian country. These policies have contributed to the high food prices urban consumers pay. A paradoxical situation now exists of Japan's having the highest rice yields per unit of land of any country in Asia, but produced at three times the cost of rice grown in the United States. Yet farming by itself is not profitable enough to sustain a farm family using large amounts of chemical fertilizers and mechanization. By 1970, the income of the average farm family had risen to $3,880, surpassing that of urban worker families, but about 60 percent of this amount came from nonfarm sources. About 80 percent of Japanese farmers are now part-time farmers. Agriculture has reached the point where further gains in productivity can be achieved only by removing the marginal farmer from the land and consolidating landholdings through such measures as cooperative farming or larger private farms. Progress along these lines has been slight because of the reluctance of farmers to part with their land, unless they are fortunate enough to be in the path of urban-industrial sprawl.

Japan's food situation has been further complicated by changes in diets. Food habits have been partly Westernized, especially in the large cities. The Japanese have developed a great liking for items such as beef and other meats, dairy products, sugar, tropical fruits, as well as products made from wheat and soybeans. The country cannot produce sufficient quantities of these commodities to meet demand. As a result, imports of foodstuffs have risen dramatically in the last two decades. Japan now imports about one-quarter of its total food supply, and the proportion is increasing. The government has ambitious plans to increase domestic food production and to reduce imports. It remains to be seen, however, how successful the program will be unless significant changes are made in other aspects of Japan's economy.

Foreign Trade and Aid: International Ill Will

Another problem emerging from Japan's success has been a rising tide of antagonism toward Japan's foreign trade, aid, and investment policies. Japan's exports still account for only slightly over 10 percent of the GNP, a smaller figure than for many countries of Western Europe. The total volume of exports, however, has risen enormously because of the high growth rates of the economy. Since domestic demand has not risen as rapidly, an increasing percentage of Japan's total production has had to be exported. Because Japanese products have been marketed at relatively low prices and are generally of high quality, they have sold extremely well.

Nowhere is this trade advantage more apparent than in the United States, which rapidly developed an enormous trade deficit with Japan in the late 1960s. The deficit reached a peak of over $4 billion by 1972. It is sometimes said that the United States has become an economic colony of Japan. The United States supplies Japan with foodstuffs and industrial raw materials and in return buys vast amounts of manufactured goods. Cries for protectionist policies in the United States have led to strained relations that heralded a new era in United States-Japanese relations.

Similar complaints are raised by other major trading partners of Japan, particularly in Southeast Asia. There the problem is complicated by the memories of wartime experiences that have been revived by the sometimes undiplomatic behavior of Japanese businessmen and tourists who have flooded the region. In many respects, Japan's economy has become a global one (Fig. 18–4). Not only does Japan buy raw materials and sell finished products, but also it has set up overseas operations, sometimes on a joint-venture basis with Anglo-Americans and other industrialists. Labor costs have continued to rise in Japan. It has been found more profitable to move labor-intensive industry, such as assembly plants for electronic goods, to foreign areas where labor

Japanese agriculture is largely small-plot horticulture with rice as a major crop. Mechanization is confined to small machines such as garden tillers and the rice-threshing machine shown here. (Consulate General of Japan, New York)

Japan today is a blend of traditional and Western life-styles. (Consulate General of Japan, New York)

18–4 *Japan: Foreign Trade, Direction and Value*
SOURCE: *Japan Report* (April 1974).

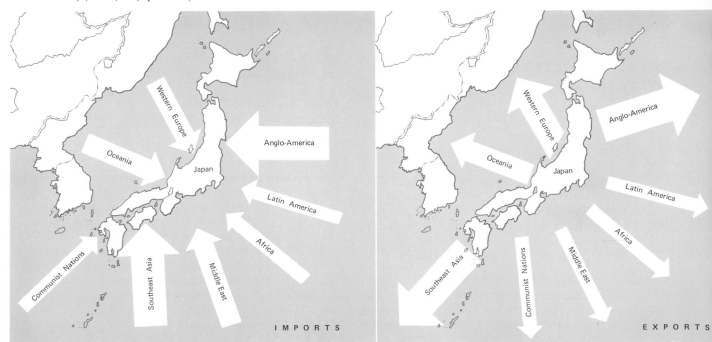

IMPORTS

EXPORTS

costs are much lower. Attractive countries for investment have included Taiwan, Hong Kong, South Korea, the Philippines, Thailand, and Indonesia. Although many benefits accrue to these countries, Japanese investments have met with a mixed response. Some local critics who espouse the notion of economic nationalism have been very vocal in expressing their displeasure. Critics contend that Japan has succeeded in creating economically what it failed to create militarily during World War II: the so-called "Greater East Asia Co-Prosperity Sphere." That is, Japan is the headquarters of an Asian economic system dependent on Japanese capital and leadership. The other Asian countries provide the cheap labor, raw materials, and markets for industrial manufactures in a neocolonial relationship. Although the Japanese vehemently deny this allegation, with some justification, there is no escaping the fact that the economies of the countries of East and Southeast Asia are increasingly tied to Japan.

Even in the realm of foreign aid, Japan has gained more brickbats than praise. Although Japan has been more generous in recent years than the United States in terms of aid as a percentage of GNP, critics complain about the relatively harsh terms of the aid agreements and that the terms tie the economies of the recipient countries even tighter to that of Japan.

Japan's Third Transformation: Correcting the Mistakes of the Past

As Japan moved into the decade of the 1970's, it was clear to many Japanese that the path followed since World War II would have to be changed. The international oil crisis of 1973–1974 provided the shock that jolted the country out of its complacency to a far greater degree than the shocks generated by United States' curbs on textile imports in the early 1970s, or the United States' détente with China.

Remodeling the Japanese Archipelago

It remains unclear, however, just how sincere Japan's government and leaders are in their professed desires to correct the problems created by an expansionist economy over the last two decades. For example, one of the most visionary proposals to be aired in the early 1970s was the government's design for social and economic change in Japan: "Remodeling the Japanese Archipelago." The plan was aimed first to relieve the congestion in the Tokaido megalopolis and to end the economic stagnation of some rural areas. Industries would be relocated in less developed areas of the country designated by the government's "Corporation for the Relocation of Industry." Various financial incentives would be used to lure industry out of the core region and to make the new industrial areas a more desirable place to live. Major improvements in highway and railway transportation would accompany the creation of the new industrial zones. The ultimate benefits, theoretically, would be a major redistribution of population and modern industry, a reduction of pollution in existing industrial areas, and considerable improvement in the quality of life for millions of Japanese.

Serious problems arose in association with the plan. A wave of land speculation swept Japan after the plan's announcement. There also were complaints that the plan would merely spread pollution around the country. Many rural people simply do not want industry in their areas. Moreover, the $1 trillion estimated cost of the 12-year program would be highly inflationary at a time when Japan is suffering from already high inflation. Nevertheless, even if the plan is never implemented in its original form, it has functioned as a catalyst to make the Japanese think about alternative directions of development.

Other Approaches

Even without the plan for remodeling the Japanese Archipelago, Japan appears to have begun significant shifts in priorities. The economy is becoming increasingly service oriented, like that of the United States. The government hopes that by de-emphasizing the industries most responsible for pollution and exports, it can alleviate Japanese discontent and lessen international concern about the country's aggressive trading practices. Japanese leaders can see that the old route to economic growth may no longer suffice. For one thing, the heavy industries have reached capacities exceeding domestic demand; surplus production can no longer be easily disposed of overseas. Moreover, if Japan does not lessen its emphasis on heavy industry, it could face severe shortages of raw materials and increased dependency on supplier nations. Industrial workers are already in short supply because of past rapid economic growth. They are likely to decrease further in numbers as the country's population pyramid continues to change, with a declining number in the productive age group of 15–64 years (Fig. 18–1). Workers are also demanding higher wages.

Japan, in short, is now going through a transition to a still more mature economy, perhaps a sixth stage beyond Rostow's theory. This stage is more strongly oriented toward services, with a reduced demand on the world's resources, and a less lopsided trade balance. Yet, in spite of serious attempts to balance trade, Japan still has an embarrassingly large trade surplus, especially with the United States. The Finance Ministry is revamping its policies to discourage the outflow of Japanese capital and encourage the inflow of foreign capital. A $10 billion pollution control program was put into effect in 1971 with the establishment of the Japan Environment Agency; attention is to be focused initially on the Tokyo area. Another law passed in 1970 is designed to assist over 1,000 areas in Japan where a sharp fall in population in the last decades has made the maintenance of a normal life difficult for the residents (Fig. 18–5). These are but a few examples of the seeking of new directions in a world vastly different from the one of the early 1950s.

The Japanese Experience as a Model

In reviewing the last century of Japan's development experience, one important question remains unanswered. Can Japan's experience be used as a model for other developing countries? Or is Japan's experience truly unique and nonduplicative? Certainly the Japanese experience has been unique in the sense that Japan will always have the distinction of being the first non-Western country to have achieved a high level of development. Japan, however, is almost certain to be joined by other countries within a few decades. Many of these countries have adopted development strategies with strong parallels to those used by the Japanese in the past. These strategies have worked, as evidenced by the remarkably strong

18–5 *Japan: Depopulated Areas*
SOURCE: *Japan Report* (January 1974).

Sea of Japan

Depopulated Areas

growth of nations such as Taiwan, South Korea, and Singapore. Obviously every country has its own peculiar mix of population and resources, and every country's experience will be unique. Yet the basic factors that have been the backbone of Japan's development could be viewed as essential to any country's development strategy. These factors include such things as national unity, a sense of purpose, strong and effective government, hard work, thrift, and a sense of pride. Without these basic requisites, it is questionable how successful any development policies can be.

From another point of view, it might be asked whether or not other countries ought to model their development after that of Japan. One is reluctant to recommend that a country follow directly in Japan's footsteps to the tune of promoting a high growth rate strategy with minimal concern about social welfare and the natural environment. Yet, to a disturbing degree, that is exactly what many developing countries in Asia and elsewhere are doing. This trend does not appear to be due to a lack of awareness of what Japan has inflicted on itself, or to a lack of concern about allowing such problems to grow. Rather, there seems to be a feeling that the environment and social welfare are secondary in importance to increased GNP and per capita income.

Further Readings

BEARDSLEY, R. K., and HALL, J. W. (eds.), *Twelve Doors to Japan* (New York: McGraw-Hill, 1956), 649 pp. An excellent collection of essays by specialists from various disciplines designed to introduce the student to the history and culture of Japan.

BENEDICT, RUTH, *The Chrysanthemum and the Sword* (Boston: Houghton Mifflin, 1946), 324 pp. Written during World War II by an American anthropologist, this analysis of traditional and modern Japanese culture of the prewar period has become a classic.

HALL, ROBERT B., *Japan: Industrial Power in Asia* (New York: Van Nostrand Reinhold, 1963), 127 pp. A brief introductory survey, somewhat out of date now but still useful.

LEONARD, JONATHAN N., *Early Japan* (New York: Time-Life Books, 1968), 191 pp. A beautifully illustrated, well-written introduction to Japan's historical foundations, in the Great Ages of Man series.

NOH, TOSHIO, and GORDON, DOUGLAS H. (eds.), *Modern Japan: Land and Man* (Tokyo: Teikoku-Shoin, 1974), 146 pp. A translation of the Japanese edition by Noh and others. A compact, well-illustrated, traditional type of geography, especially suited for the layman or beginning student of Japan.

REISCHAUER, EDWIN O., *Japan, The Story of a Nation* (New York: Knopf, 1970), 345 pp. One of America's foremost authorities on Japan, Reischauer in this book brought up to date and expanded his definitive work, *Japan: Past and Present.* A fine historical introduction to Japan.

REISCHAUER, EDWIN O., *The United States and Japan* (New York: Viking Press, 1965), 394 pp. Reischauer's other classic work on Japan, emphasizing the history of Japanese-American relations. An excellent book for the beginning student to read.

SEIDENSTICKER, EDWARD, *Japan* (New York: Time, 1965), 160 pp. A handsomely illustrated, popularly written introduction to Japan for the layman, in the Life World Library series.

TREWARTHA, GLENN T., *Japan: A Geography* (Madison: University of Wisconsin Press, 1965), 652 pp. The classic basic geography of Japan, still useful, especially for the treatment of the physical landscape of Japan. The statistics can be updated by the reader from other sources.

Australia and New Zealand

19

Australia and New Zealand are as unique members of the community of rich nations as is Japan. Their uniqueness stems in part from their historical development and in part from the stark contrast they exhibit to their densely populated, underdeveloped Asian neighbors. The "secret" of the development of Australia and New Zealand lies in the successful transplantation of Western society and economy to virgin territories that, due to the accident of geography and history, had been largely unknown and untouched by the peoples of Asia.

Australia and New Zealand share many common characteristics. Both were founded around the same time in the late eighteenth century as British colonies of white settlement. Both have large land areas in proportion to their population, high standards of living, and much closer ties with the United Kingdom and the United States than with most of the Asian countries. Both Australia and New Zealand developed in their early periods as sort of "supermarkets" for Britain; that is, they provided many of the foodstuffs Britain and the British Empire needed. That function is still important, but no longer dominant. New Zealand remains the more pastoral and agriculturally based of the two countries, with its high level of living dependent on abundant production of dairy products, meat, wool, and other animal products. Australia's wealth is more diversified, with rich deposits of minerals, coal, natural gas, and a bountiful agricultural basket of meat, dairy products, wool, wheat, and sugar, and increasingly industrial manufactures.

Because of their small populations, Australia and New Zealand depend on trade with the industrialized nations to maintain their high standards of living. For many decades, this trade was directed primarily toward Great Britain and the British Commonwealth. Because of tariff and other trade privileges, it was profitable for Australia and New Zealand to market their products thousands of miles away while largely ignoring the nearer but poorer Asian market. Since World War II, however, and particularly since Britain entered the European Common Market, the overseas relations of Australia and New Zealand have undergone a metamorphosis. Ties with the British have gradually weakened, whereas those with the United States and with Japan and other countries of Asia, especially Southeast Asia, have assumed new importance. This changing overseas outlook of Australia and New Zealand has evolved out of necessity, not choice, on the part of Australians and New Zealanders.

Australia

A Vast and Arid Continent

Much of Australia's development experience is related to the continent's physical environment, particularly its isolation, vastness, aridity, and topography. With nearly 3 million square miles (7.8 million square kilometers), including the offshore island of Tasmania, Australia extends 2,400 miles (3,862 kilometers) from Cape York in the north at 11°S. latitude to the southern tip of Tasmania at 44°S. latitude, and stretches for about 2,500 miles (4,023 kilometers) east-west (Fig. 19-1). Although this land area is approximately equal to that of the conterminous United States, Australia's population is far smaller, at just under 14 million. From this point of view, Australia is actually one of the smaller countries of the world.

The sparseness of the population and its concentration in a relatively small part of the continent are associated particularly with aridity and low elevation (Fig. 19–2). The continent extends east-west in subtropical latitudes. In addition, the whole landmass is relatively flat except along the east coast where a mountain range lies across the major rain-bearing winds. Hence, Australia has an immense area of insufficient and irregular rainfall in the center and in the western half of the continent (Fig. 19–3). Aridity is the greatest physical obstacle to development of the continent. Only 11 percent of the area gets more than 40 inches (102 centimeters) of rain, while two-thirds has less than 20 inches (51 centimeters). Australia is easily the most arid of the continents. In addition, the variability of rainfall is high for almost all of the continent, making agricultural development in marginal lands even more precarious.

Four major natural regions are distinguished on the basis of climate and relief (Fig. 19–4). The core region of Australia is the humid eastern highlands, which

19–1 *Australia–New Zealand: Location Map*

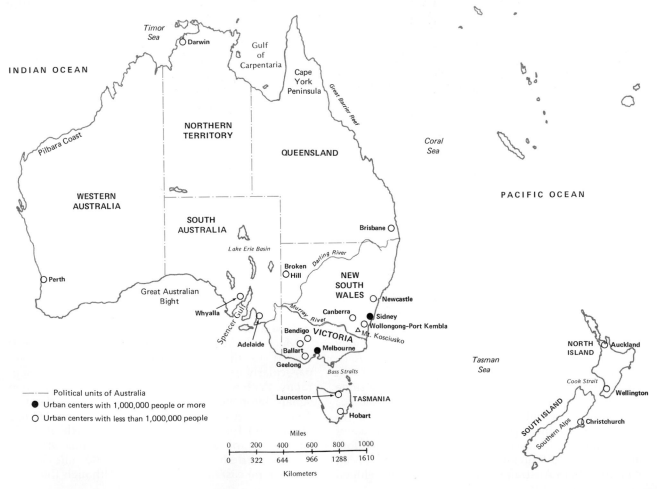

extend in a belt 100 to 250 miles (161 to 402 kilometers) wide along the east coast. The highest elevation, Mt. Kosciusko, only reaches to 7,316 feet (2,230 meters). The narrow and fragmented coastal plains along the base of the highlands provide the only part of Australia not subject to recurrent drought. Most of Australia's population, major cities, agriculture, and modern industrial economy are concentrated in this coastal fringe.

The three other natural regions have various disadvantages for human settlement, and land use is confined largely to mining and livestock raising. Population density is very light in these regions.

Along the northern fringe of Australia are the tropical savannas, where the monsoonal climate of six months of heavy rain followed by six months of almost total dryness make settlement and agriculture extremely difficult. Soils are poor, and the lack of highlands prevents the existence of large perennial streams for irrigation in the dry season. The southwestern corner of Australia and the lands around Spencer Gulf have a Mediterranean or dry summer, subtropical type of climate. This region is the second major zone of population concentration, particularly around the cities of Perth and Adelaide, but the total population density is still very sparse. Agricultural possibilities

19–2 Australia–New Zealand: Population Distribution

One dot represents 100,000 people

Miles

| 0 | 200 | 400 | 600 | 800 | 1000 |

| 0 | 322 | 644 | 966 | 1288 | 1610 |

Kilometers

19–3 *Australia–New Zealand: Annual Precipitation*

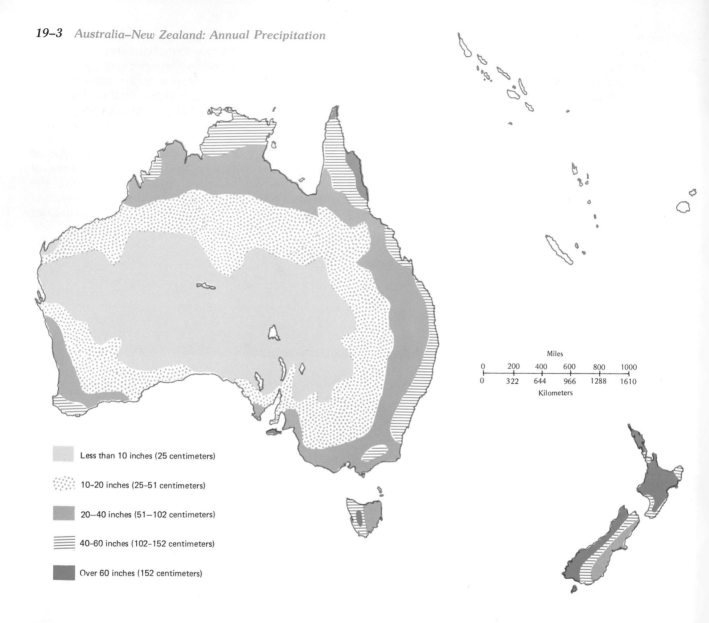

Miles

0	200	400	600	800	1000
0	322	644	966	1288	1610

Kilometers

Less than 10 inches (25 centimeters)

10–20 inches (25–51 centimeters)

20–40 inches (51–102 centimeters)

40–60 inches (102–152 centimeters)

Over 60 inches (152 centimeters)

are again limited by the lack of highlands to catch moisture and supply irrigation water to the lowlands. The huge interior of Australia is desert surrounded by a broad fringe of semiarid grassland (steppe) which is transitional to the more humid areas around the edges of the continent. These dry areas cover more than half the continent and extend to the coast in the northwest and along the Great Australian Bight in the south. The western half of Australia is in fact a vast plateau of ancient rocks with a general elevation of only 1,000 to 1,600 feet (305 to 488 meters). The few isolated mountain ranges are too low to influence the climate significantly or to supply perennial streams for irrigation. The Lake Eyre Basin, an area of interior drainage and salt flats in the state of South Australia, is the driest part of the continent

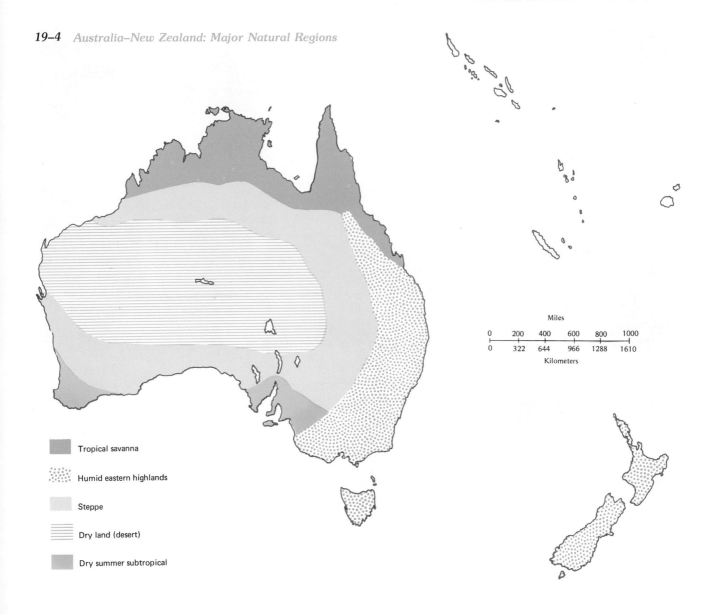

19–4 *Australia–New Zealand: Major Natural Regions*

Miles

0	200	400	600	800	1000
0	322	644	966	1288	1610

Kilometers

Tropical savanna

Humid eastern highlands

Steppe

Dry land (desert)

Dry summer subtropical

with an annual rainfall of less than 5 inches (13 centimeters).

Shortage of Arable Land

Australia actually has a remarkably small amount of arable land for such a large landmass. Fully one-third of the continent in the arid interior region has been found totally unusable for agricultural purposes, including livestock raising. Another 40 to 42 percent of the land area to the north, east, and west of the arid interior receives only enough rain to support cattle and sheep. The remaining land area, about 25 percent, receives sufficient rainfall to support agriculture, but mountainous terrain and poor soils further reduce the truly arable area to no more than about 8 percent. Much of the arable land is already

The Darling River near Broken Hill is only a chain of waterholes during most of the year. The line of gum trees marks the river's course and forms the only break in a semi-arid grazing zone. (Australian Information Service)

used for intensive sheep and cattle raising and dairy farming. In reality, less than 2 percent of Australia's total land area is devoted to crop cultivation at the present time, an indication of the very extensive use of the land resources of the continent. Australia could, if desired, support a much larger population. For example, the Murray-Darling Basin, the only river system of sufficient size to warrant large-scale development, is in the process of being developed as the major region of irrigated agriculture on the continent.

Settlement and Population Growth

The long delay in the discovery and settlement of Australia was due to many factors, including the vastness of the Pacific Ocean, prevailing winds and currents, and lack of any sign that the continent possessed resources worth having once it was sighted. Until 1788, Australia was inhabited only by Aborigines, numbering perhaps 300,000. These dark Negrito-type peoples of complex origin had been in Australia for thousands of years living a very primitive existence as hunters-gatherers. In 1770, James

Cook became the first known European to reach the east coast of Australia, the one part of the continent that appeared suitable for settlement.

One of the unique characteristics of Australia's early development was its use as an overseas prison for exiled convicts from Britain. The first prison group arrived in 1787, at what became Sydney, the major city of Australia. Exploration and settlement by adventurers, emancipists (convicts who had served out their sentences), and others continued into the nineteenth century. Immigration from Britain was encouraged by land grants in the opening continent, but the total population remained very small. One of the greatest stimuli to development and immigration was the gold rush of the 1850s, which brought large numbers of prospectors and settlers. In 1901, the six Australian colonies—New South Wales, Victoria, South Australia, Queensland, Western Australia, and Tasmania—were federated into the Commonwealth of Australia. The central government obtained certain limited, defined powers, and the states were left with the residual powers,

The Murray River area in New South Wales is in the process of being developed for irrigation agriculture. Much of the area, however, is still used for dryland cropping and livestock raising. (Australian Information Service)

somewhat analogous to the United States federal system.

One of the most important developments of the nineteenth century was the strengthening of the "White Australia" policy as the first trickle of nonwhite immigrants set foot on the continent. These were some Chinese and Indian laborers who entered the colonies in the 1830s and 1840s to meet a labor shortage. Their numbers, particularly of Chinese, increased rapidly thereafter and became the focus of often bitter disputes between various factions in the colonies over the question of nonwhite immi-

Australia is an urban nation. Sydney, with nearly 3 million people, is the largest city and an important port. Across the bay and to the left is the city's famous opera house. (Australian Information Service)

grants. The supporters of the White Australia policy generally won out. Successive Australian governments have recognized the dangers of a relatively small white population controlling such a large land area so close to the overpopulated regions of Asia. Immigration into Australia thus has been a major concern of governmental policy. In general, this policy has been characterized by three main features: alternating support for and opposition to large-scale immigration, closely linked to periods of domestic prosperity; strong preference for immigrants of British origin; and, since the late nineteenth century, exclusion of nonwhite immigrants with only a few exceptions.

The main phases of large-scale immigration were in 1852–1858, 1876–1891, 1909–1913, 1921–1925, and post-1948. Britons predominated in this immigration pattern until World War II and were aided by Australian government assistance. After World War II, the supply of Britons declined, and the Australian government changed its policy and accepted other European and Anglo-American immigrants as long as they were white. The response has been impressive and has given Australia an increasingly cosmopolitan character, at least in the big cities. These new immigrants have helped create an increasingly distinct "Australian" character for the country, as opposed to the previously dominant British mold of Australian society. Australia continues to cultivate immigration with the aim of maintaining net annual migration as nearly as possible at 1 percent of the total population.

The White Australia policy—officially termed "Restricted Immigration Policy"—though still enforced, has been softened somewhat as Australia's focus has been forced to shift more toward Asia. Efforts were made in the late nineteenth and early twentieth centuries not only to prevent nonwhites from entering Australia but also to deport as many as possible. Now, however, Australia freely admits nonwhite tourists, students, and businessmen on temporary permits. Thousands of Asian students now study at Australian schools and universities. Since 1956, nonwhites in certain categories, such as those with special skills, have been admitted for indefinite residence and may apply for naturalization after five years. In spite of the liberalization in immigration policy, however, Australia remains a white bastion in the Asian-Pacific world and is resented by many of the region's governments and peoples. Of Australia's current population of about 14 million, whites account for 99 percent, and about 94 percent are of British origin. The coming decades are likely to see Australia severely tested in regard to its immigration policies if the country continues to increase trade and other relations with the countries of Asia and the Pacific.

An Urbanized Society

An unusual characteristic of Australia, for a country with so much land and so few people, is the high degree of urbanization. About two-thirds of the population live in cities of over 100,000. Around 40 percent of the people live in the two great metropolitan areas of Sydney and Melbourne, each with populations of 2.5 to 3 million. There are several reasons for this pattern. Although the production of agricultural products is still very important to the economy, these activities are extensive.[1] They employ relatively few people (less than 10 percent of the labor force) and in most cases are highly mechanized. Moreover, especially since World War II, Australia has encouraged industrialization as a means of supporting a population increasing at nearly 2.5 percent a year, providing more home-produced armaments for defense, and securing greater economic stability from a more diversified and self-sufficient economy.

A further characteristic of Australia's urban population is that all of the five largest cities (in decreasing size, Sydney, Melbourne, Brisbane, Adelaide, and Perth) are seaports, and each is the capital of one of the five mainland states of the Commonwealth. This uniqueness is due to the fact that much of the country's production is exported by sea, and much internal trade is also conducted by coastal steamer. Moreover, before federation in 1901, each state built its own rail system focusing on its chief port. These rail systems were often of different gauges, and each state operated almost as an independent economic unit. The integration of the states into a national body has been the focus of attention of the national government for the last seventy years and has included installation of a standard-gauge railway network for the whole country.

[1]An "extensive" agricultural activity is one in which limited amounts of labor and capital per unit area are expended on a relatively large area, as, for example, in wheat farming.

Adelaide, one of Australia's five largest cities, is the capital of South Australia, a major port, and a road and railway center. (Australian Information Service)

The Success of Australia's Development

Australia's high standard of living can be attributed to a small population and a reasonably well-developed and diversified economy dependent on production of agricultural, mineral, and industrial goods (Fig. 19–5). This trilogy provides a solid base on which to build a prosperous economy. The future prospects for the country are extremely bright so long as world trade remains healthy.

Agriculture is dominated by sheep, cattle, and wheat—extensive forms of agriculture well suited to Australia's environment. Sheep ranching became the first mainstay of the economy in the nineteenth century and provided wool for Britain's textile industry. By 1850, Australia was already the world's largest supplier of wool, a position it has never lost. Wool still accounts for about one-quarter of Australia's total exports, and the country produces over one-quarter of total world production. Sheep thrive in the broad crescent belt of territory coinciding with the steppe surrounding the arid interior. New South Wales has over 40 percent of the nation's sheep and Victoria 15 to 20 percent. Sheep ranches, or "stations," are usually quite large, some encompassing

19–5 *Australia–New Zealand: Rural Land Use and Minerals*

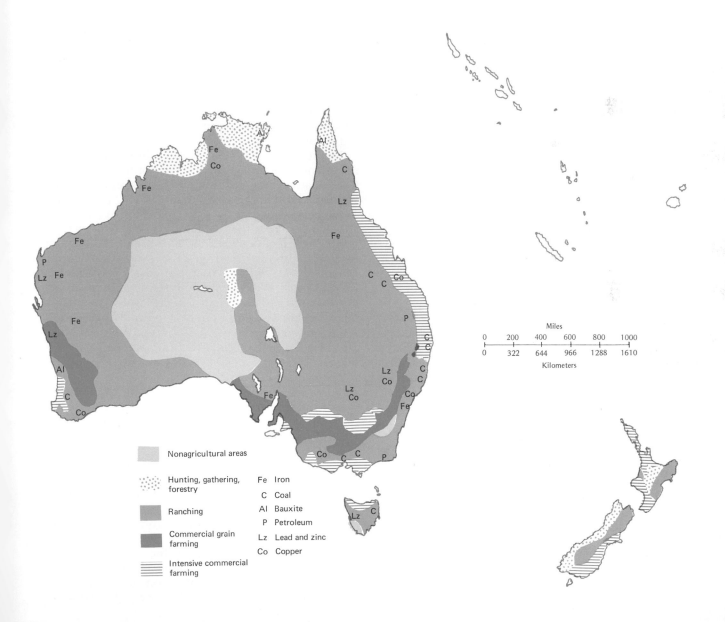

Nonagricultural areas		
Hunting, gathering, forestry	Fe	Iron
	C	Coal
Ranching	Al	Bauxite
	P	Petroleum
Commercial grain farming	Lz	Lead and zinc
	Co	Copper
Intensive commercial farming		

Miles
0   200   400   600   800   1000
0   322   644   966   1288   1610
Kilometers

thousands of acres, on which motor vehicles and airplanes are important equipment. Cattle raising has been particularly stimulated since World War II. The use of refrigerated ships enables Australia to supply northern hemisphere markets with meat and dairy products. Export of mutton has been helped by this development also.

Generally, cattle have been relegated to those areas not suitable for sheep, and the cattle industry remains secondary in importance, although it is growing rapidly. The dairy industry has also seen strong growth since World War II, including increased exports of dairy products. Dairy farming is confined largely to the eastern and southeastern coastal fringes. Much of the increased world demand for wool, beef, and dairy products has come from the countries of East and Southeast Asia, especially Japan, where rising standards of living are changing consumption habits. It is very much in Australia's interest to cultivate this growing Asian market.

Wheat production has also benefited by modern technology, with the introduction of mechanization in the twentieth century that permitted the extensive cultivation of the crop. About 19.8 million acres (8 million hectares) are now devoted to wheat alone, some 60 percent of the total cropland in Australia. Along with Canada and the United States, Australia has become one of the great breadbaskets of the world. The country exports nearly three-fourths of its annual wheat production. Given the existing world grain shortage and extremely strong demand, the only limitation to continued increased production of wheat in Australia is the physical environment. Most of the wheat is grown in areas having 12 to 24 inches (30 to 61 centimeters) annual rainfall, the same region of semiarid and subhumid lands where sheep production is concentrated.

Australia also produces many other crops and is virtually self-sufficient in foodstuffs. Among these, sugar is one of the most important. Sugar cane is grown along the northeastern coastal fringe. Because of high labor costs, the industry is highly mechanized and is protected by a high tariff. Australia annually produces nearly 3 million tons of sugar. Most is exported to Japan and to other Asian markets, making Australia a major participant in world sugar trade. Other important crops include a wide variety of temperate and tropical fruits for both domestic consumption and export markets.

Besides agriculture, mining is a mainstay of the economy. Australia is blessed with a bounty of mineral deposits. The development of the Australian mineral industry began with the discovery of copper in South Australia in 1842, followed by gold in the eastern states in 1851. In the 1870s and 1880s came the discovery of copper, tin, and lead-zinc in the eastern states and Tasmania. Since those early finds of the nineteenth century, successive discoveries have revealed a richness of minerals. These minerals are sufficient to make Australia a major provider of raw materials for the mineral-hungry industrialized countries of the northern hemisphere, especially Japan. Japan now buys the bulk of the island continent's 64 million-ton annual iron ore exports, 36 million tons of coking coal, and a host of other minerals, including lead, zinc, silver, copper, bauxite, and natural gas.

The locations of mineral deposits have played important roles in the development of the continent. Coal deposits, for example, are concentrated near the coast in New South Wales and Queensland. These deposits led to the development of Australia's steel industry near the coal mines at Newcastle to the north of Sydney and at Port Kembla to the south. Iron ore is brought in from distant deposits, which fortunately are located near the coast, particularly the mines at Whyalla in South Australia, itself a growing iron and steel center. In the last two decades, major new finds of iron ore have been discovered in the remote northern coast of Western Australia and are being developed primarily to supply the Japanese market.

These and other mineral finds, such as rich natural gas deposits on the Northwest Shelf off the Pilbara Coast, and copper, nickel, and aluminum, promise to make a significant transformation of Western Australia. The government has plans to create a modern industrial complex in Western Australia that would bring in people and industry to the mining sites and port towns. Projects such as these, and there are likely to be more, have particular appeal for two reasons. First, they fit in with the desire of the government to decentralize industry away from the crowded southeast. Second, they would give Australia a big opportunity not only to increase the value of exports but also give Australians a larger share in the development of their natural resources and industries. Heavy foreign investment by Americans, British, and

Sheep raising became Australia's first economic mainstay, and wool still accounts for about one-quarter of the nation's exports by value. (Australian Information Service)

Only a small part of Australia's land area is cropped. Wheat is the most important crop and is cultivated on large farms using mechanized equipment as illustrated by these combines. (Australian Information Service)

increasingly Japanese have aroused fears in some Australians that control of some sectors of the economy will fall into the hands of foreigners.

Petroleum is one of the few important mineral deficiencies of Australia and constitutes a serious obstacle to further industrialization. The continent's limited oil production provides only a small fraction of the country's needs. Intensive exploration continues, however, and future discoveries are possible. Self-sufficiency in petroleum would greatly enhance industrialization hopes.

The development of industry has always been hampered by the relatively small population of Australia. Industrial establishments find it difficult to produce and distribute on a large enough scale to minimize unit costs and use labor efficiently. Nevertheless, industry has grown appreciably, fostered by such successive events as federation in 1901, with centralized customs and tariff control; a single tariff policy toward other countries and internal free trade; the extension of tariff protection during the 1930s; and the stimulus of World Wars I and II. Spectacular growth took place during and after World War II in oil refining and the manufacture of motor vehicles, steel, chemicals, plastics, and many other products.

Manufacturing is concentrated in the state capitals, which have three-fourths of all factory workers as opposed to somewhat over one-half the total population. Each of the capitals has attracted a dominant share of the state's industrial activities because of the availability of markets, labor, fuel, business and government contacts, and access to overseas and internal transport systems. Much of the remaining industrial activity is located at a few large provincial centers on or near the coast, such as Newcastle, Wollongong-Port Kembla, Geelong, and Launceston, and at some inland urban areas, such as Ballarat and Bendigo. The leading industrial state is New South Wales followed by Victoria.

A dominant force in Australia's industry is the Broken Hill Proprietary Company (BHP), which began in the nineteenth century as a developer of the lead, zinc, and silver mines at Broken Hill in the desert of western New South Wales. Later, the company branched into steel production, which it now monopolizes, and into other fields. The BHP is the backbone of the entire Australian industrial system, controlling a major share of all the capital invested in private manufacturing companies in Australia.

The effects of industrialization have been felt in the country's trading composition, which has seen declining imports of finished consumer goods and increased imports of capital equipment. Manufactures (including processed foodstuffs) have increased to 15 to 20 percent of total exports. Although dependence on imported manufactured goods has been lessened, Australia's economy is still essentially based on the export of primary products in exchange for manufactured goods from the industrial countries of the world. A truly fundamental change in this pattern must await a number of developments: further increases in Australia's population; greater efficiency in Australian industry; further development of the continent's resources; and greater trade ties with the countries of Asia, which could serve as markets for Australian manufactured goods. The latter development, however, will require the ability of Australian manufactures to compete effectively with Japanese goods, which have already captured a lion's share of the markets in East and Southeast Asia.

New Zealand

New Zealand, lying over 1,000 miles (1,609 kilometers) southeast of Australia, consists of two main islands: North Island with the smaller area but two-thirds of the total population, and South Island, plus a number of lesser islands. The country is located entirely in the temperate zone from about 34° to 47°S. latitude. Like the Japanese islands, New Zealand is a section of the circum-Pacific mobile crustal belt and forms the crest of a giant earth fold rising sharply from the ocean floor. The country is mountainous, hilly, and corrugated. One-half the surface has mountain landforms, and another quarter consists of steep and broken hill country.

South Island consists of nearly three-fourths mountainous terrain, dominated by the Southern Alps rising to elevations above 10,000 feet (3,048 meters). North Island is less rugged, but many peaks still exceed 5,000 feet (1,524 meters). New Zealand has a wet temperate climate commonly known as marine west coast, with summer temperatures averaging

60° to 70°F. (15.5° to 21.1°C.) and winter temperatures 40° to 50°F. (4.4° to 10°C.) in the lowlands, with more severe weather in the highlands, including glaciers on both islands. The climate is produced by the location of the islands in the southern hemisphere belt of westerly winds and by the pervasive maritime influence.

A Pastoral Economy

Settlement of the islands is confined largely to the fringing lowlands around the periphery of North Island and along the drier east and south coasts of South Island. A large section of the country is completely unproductive, except for a growing tourist industry in the mountains.

The Dutch explorer Tasman was the first known European to sight the islands in 1642, but it was not until Cook arrived in 1769 that exploration and settlement really got under way. Hence the development of New Zealand closely parallels in time that of Australia. As with Australia, the climate is ideal for growing grass and raising livestock, and New Zealand

Livestock raising, especially sheep, is the backbone of New Zealand's economy. Only about 3 percent of the land area is cropped, mostly for animal feeds. (Consulate General of New Zealand)

has specialized in this activity from the very founding of the country in the eighteenth century. Some 200 years later, New Zealand's economy is still extremely dependent on specialized production of animal and dairy products. No other country has so swiftly, completely, and successfully been converted from a pre-European forested land into a land of productive pasture.

The country has one of the highest national proportions of livestock in relation to human population in the world, a ratio of 25:1. Pastoral industries completely dominate exports, and because of the country's small population and high standard of living, New Zealand stands near the top among the world's countries in per capita trade. New Zealand is among the top two or three exporters in the world of mutton, lamb, butter, cheese, preserved milk, wool, and beef. In exchange for huge volumes of these products, New Zealand receives most of its manufactured goods and considerable quantities of food, since the country grows relatively few basic foodstuffs. The 3 percent of the land area that is cropped is devoted in large part to the growing of animal feeds.

The Need for Industry and Diversification

With such a heavy dependency on trade and a very narrow economic base, New Zealand is far more vulnerable to the vagaries of world economic conditions than Australia. As long as New Zealand had Commonwealth ties, especially with Britain, over-specialization in pastoral production was not exceptionally dangerous. With Britain's entry into the Common Market and the weakening of Commonwealth ties and tariff privileges, however, New Zealand will likely lose its position as the meat and dairy market for Great Britain. Although the United Kingdom remains the principal trading partner of New Zealand, the United States, Japan, and Australia are increasingly important.

Attempts at diversification, primarily by industrialization, have not been very successful. For one thing, New Zealand does not have the rich mineral resources that Australia has. Coal, iron, and a few other minerals are present only in very modest quantities. Considerable potential for hydroelectric power development exists. Forestry also offers much potential. A manufacturing industry is not easy to develop for a number of reasons. First, the local market is too small and dispersed. Large-scale production and

efficient marketing are thus restrained. Second, the cost of skilled labor is high. Third, competition from overseas producers, such as Japan and the United States, can be severe, as Australia has found out. Most of the present manufacturing industries are high-cost producers surviving under tariff protection.

Given the much less developed manufacturing industry in New Zealand, the degree of urbanization is not strong. Under one-half the population live in urban centers of over 100,000. The country's largest city, Auckland, has only about 600,000 people. Sydney, Australia, has almost as many people as all of New Zealand.

Population and Society

A significant difference between New Zealand and Australia lies in the way in which the indigenous population has fared under the rule of the Europeans. When Cook arrived in the eighteenth century, he found the islands inhabited by a few hundred thousand Maori, a Polynesian people believed to have migrated to New Zealand from the Society Islands in the fourteenth century. Unlike the Australian Aborigines, the Maori had a relatively well-developed culture. During the first 100 years of contact with the white settlers, the Maori declined sharply in population because of disease and social disintegration. By the late nineteenth century, however, the Maori began to increase once again, recovering earlier population levels, and now account for about 8 percent of the total population. This turnabout was partly the result of accommodation to European culture and partly the result of conscious efforts of the white New Zealanders to integrate the Maori into the national fabric. Intermarriage has been a major contributor to this integration process. By contrast, the Aborigines of Australia have fared far worse and remain a badly neglected and tiny minority within Australia.

In spite of New Zealand's policies toward the Maori, the country follows immigration policies similar to those of Australia in severely restricting nonwhite immigration and in providing assistance to immigrants of British or European origin. Notwithstanding these programs, however, emigrants from New Zealand have exceeded immigrants in many years, probably because of the limited economic opportunities, particularly as compared with Australia.

Wellington is New Zealand's second largest city and a principal port. Most of New Zealand's population live in small towns and on farms. (Consulate General of New Zealand)

Further Readings

BLAINEY, G., *The Tyranny of Distance* (New York: St. Martin's Press, 1966), 365 pp. An interesting analysis of Australia's vastness and isolation and their effects on development.

CONDLIFFE, J. B., *The Economic Outlook for New Zealand* (New York: Praeger, 1969), 318 pp. A well-known economist looks at the general economic picture and presents some possible trends for the future.

CUMBERLAND, K. B., *Southwest Pacific: A Geography of Australia, New Zealand, and their Pacific Island Neighbors*, 4th ed. (New York: Praeger, 1969), 423 pp. A basic text.

DAVIDSON, B. R., *The Northern Myth: A Study of the Physical Limits to Agricultural and Pastoral Development in Tropical Australia* (London: Cambridge University Press, 1966), 283 pp. Attempts to answer the questions and speculation regarding the potential development of northern Australia.

MEINIG, D. W., *On the Margins of the Good Earth: The South Australian Wheat Frontier, 1869–1884* Skokie, Ill.: Rand McNally, 1962), 231 pp. An important study of the early period of wheat farming and grazing on the frontier of the steppe.

PIKE, D. H., *Australia: The Quiet Continent,* 2nd ed. (London: Cambridge University Press, 1970), 243 pp. A well-written, concise history of Australia.

ROBINSON, K. W., *Australia, New Zealand, and the Southwest Pacific* (London: University of London Press, 1964), 340 pp. A general geographic survey.

SINCLAIR, K., *A History of New Zealand* (Harmondsworth: Penguin Books, 1959), 320 pp. A short and interesting introduction to New Zealand's history.

SPATE, O. H. K., *Australia* (New York: Praeger, 1968), 328 pp. A broad review of all aspects of Australia by a leading geographer; likely to become a classic text.

TAYLOR, T. G., *Australia, A Study of Warm Environments and Their Effect on British Settlement* (New York: Dutton, 1951), 490 pp. An older work but still of considerable value.

WARD, R., *The Australian Legend* (Melbourne: University of Oxford Press, 1966), 262 pp. A study of the Australian character by a social historian.

WATTERS, R. F. (ed.), *Land and Society in New Zealand: Essays in Historical Geography* (Wellington: A. W. Reed, 1965), 216 pp. A collection of studies on the growth of New Zealand as a nation and society.

Latin America

Don R. Hoy

PART SIX

Latin America: The Physical and Colonial Legacy

20

W HEN COLUMBUS REPORTED HIS FAMOUS "New World Discovery" in 1492, a chain of events began which transformed the western hemisphere. Age-old American cultures, some of high-attainment levels, were shattered almost overnight. European nations established colonies and introduced their value systems and their political, economic, and social organizations. All mother countries employed the mercantalistic philosophy toward their colonies which, along with usurpation of native-held lands and enslavement of Indians, formed the basic foundation of the colonial economy. Latin America, that part of the western hemisphere south of the United States, felt the brunt of the European conquest a century or more before European presence was significant in Anglo-America. To a large degree, the present differences between Latin America and Anglo-America are a result of differences in colonial mother countries, British and French mainly in Anglo-America, and Spanish and Portuguese in Latin America.

The Fate of Pre-Columbian Civilizations

At the time of Spanish and Portuguese conquests, some 75 to 100 million Indians inhabited Latin America. Most lived within the densely settled realms of four high-culture groups: Aztec, Maya, Chibcha, and Inca (Fig. 20–1). These groups had a well-developed social, political, and economic organization, although the Mayan civilization was declining and had lost some of its internal cohesiveness. Agriculture formed the livelihood base, and advanced land-management techniques were common. Irrigation, land drainage, fertilization, terracing, and crop and land rotation were widely practiced. Crops used by these and other groups within Latin America apparently had been largely developed independently from the outside world. The more well-known cultigens included maize (corn), manioc, sweet and white potatoes, pineapple, cacao, tomato, avocado, cotton, tobacco, and peanuts. To this list is added a number of plants such as beans and squash varieties which were also cultivated in many other parts of the world. Few domestic animals were found in Latin America. The dog was ubiquitous, and the Incas had the guinea pig as a pet and food source, the llama as a beast of burden, and the alpaca as a source of fiber.

Agriculture was so productive that part of the labor force was freed for other pursuits. Soft metals (silver, gold, lead, and zinc) were fashioned into jewelry and other ornaments. Craftsmen such as potters and masons formed specialized labor groups. The Aztecs had a class of professional traders who traveled far outside the limits of their empire. All had a well-organized governmental structure and a priestly class. The priests were more than religious leaders, for they pursued the sciences of astronomy and mathematics and assisted in record keeping. Both the Mayans and Incas had developed precise calendars based not only on the sun and moon but also on nearby planets. Conversely, the high-culture groups did not have the concept of the arch, a functional use of the wheel, nor the alphabet.

High Cultures

Aztec and
related groups

Maya

Chibcha

Inca

Other groups shown by
principal linguistic
families

Miles

| 0 | 200 | 400 | 600 |

| 0 | 322 | 644 | 966 |

Kilometers

20–1 *Latin-America: Pre-Columbian Indian Cultures*

NOTE: Chichimecs is not a linguistic group but the Aztec name applied to the
nomadic peoples of northern Mexico.

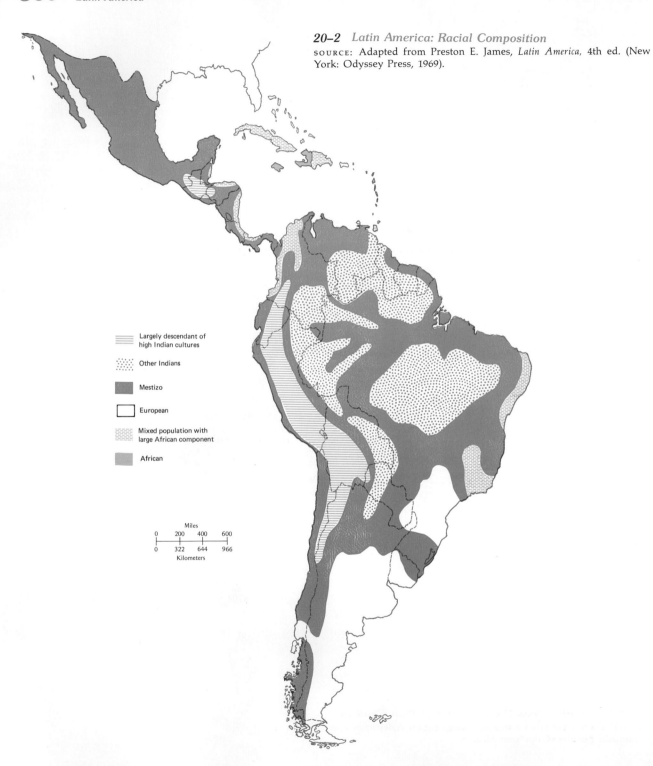

20–2 Latin America: Racial Composition
SOURCE: Adapted from Preston E. James, *Latin America,* 4th ed. (New York: Odyssey Press, 1969).

Largely descendant of
high Indian cultures

Other Indians

Mestizo

European

Mixed population with
large African component

African

Miles
0   200   400   600

0   322   644   966
Kilometers

Outside the regions of high culture lived numerous small, scattered, and loosely organized groups ranging in economic attainment from total dependency on hunting, fishing, and gathering to simple agriculture. These groups can be linked together by their common languages.

The Conquest and Its Aftermath

The Spanish and Portuguese conquest of Latin America was both rapid and devastating. The Spanish quickly carved out a vast colonial empire extending from California to Tierra del Fuego. In the space of sixty to seventy years, they explored, established sovereignty, and organized politically and economically an area of about 9.5 million square miles (24.6 million square kilometers). The Portuguese moved less quickly onto the east coast of South America, where they found few Indians.

The location of large, high-culture Indian groups was especially attractive to the European conquerors. These groups possessed the elements needed to satisfy the conquerors' wants: accumulated stores of gold and silver, an abundant and well-trained labor supply for work in the fields and mines, and a large population for the church to convert to Catholicism. Smaller Indian groups had neither the wealth nor the work skills required by the Europeans. Moreover, the diseases brought with the conquest severely depleted all-Indian groups, but in the high-culture areas a sufficient labor residue remained. Smallpox, measles, typhus, and probably yellow fever and malaria were among the diseases introduced from Europe and Africa. By 1600, epidemics had reduced the Indian population by as much as 90 percent. The Aztecs of central Mexico were reduced in number from 25 million to 3 million in less than 100 years. Most of this decrease has been attributed to disease, although malnutrition and warfare were contributory causes.

After the first flush of the conquest, and faced with the decimation of the Indian labor force, the Spanish imported African slaves. Later the Portuguese, and in the seventeenth century the English and French, followed Spain's example. In the Antilles where the Arawak and Carib Indians were destroyed early in the conquest, the descendants of these slaves make up a large part of the present population. The use of African slaves in the mainland of Latin America was neither very successful nor widespread, since the slaves could escape to the backlands, areas the Spanish and Portuguese did not effectively control.

The early Portuguese and Spanish rarely brought their families. Indeed most Europeans apparently intended to return to their European homes. Miscegenation between Spanish and Portuguese men and Indian and African women was widespread, contributed to retarding population decline, and gave rise to two new racial groups: the mestizo (part European, part Indian) and the mulatto (part European, part African). Of these two groups the mestizo became the more numerous. One lasting effect of racial mixing has been the lack of significant racial prejudice in Latin America. These racial groups varied in their distribution and have given rise to distinct differences among the various parts of the region (Fig. 20–2).

Contributions of Pre-Columbian Civilizations

Although the cultural bond of most Indian groups was destroyed by the Europeans and the Indians soon became subservient to their new masters, not all aspects of their culture were lost to the world. Maize has become one of the three most important cereals in the world and is cultivated from the tropics to the higher mid-latitudes. Plant geneticists have improved the yielding qualities of maize to the point where it is one of the highest-yielding crops in the world. Cacao, from which chocolate and cosmetics are made, is now an important commercial crop in west Africa. The white potato (Irish or Idaho) is a main feature in the diet of rich nations and is particularly important in northern Europe and the Soviet Union. Tobacco has spread worldwide, and peanuts, high in protein, are increasingly important in food-deficient areas and in other areas where they are also used for their oil and pulp. No part of the world today, except perhaps the polar zone, fails to use crops of Latin American origin.

Old World Contributions to Latin America

Early in the colonization of Latin America, a number of Old World crops and animals were introduced and eventually gained wide acceptance. These contributions expanded the resource base. For example, sheep, goats, and cattle could forage on alpine meadows and in grassland areas that formerly were marginal for Indian crops. Wheat, a plant that can grow in relatively dry areas, extended the cultivable area. Imported cotton varieties outproduced native cotton, and the latter type all but disappeared. Sugar cane and bananas, raised in heretofore sparsely populated areas, and coffee have become major export crops throughout tropical America. Rice and a number of vegetables were added to the subsistence farmers' repertoire lending diversity to the diet. Pigs and chickens were also readily adopted on small farms where they scavenged for food. Even the grape and olive were adapted in the more temperate areas, and the famous Captain Cook brought the breadfruit tree from Polynesia which became an important food source for slaves in the Antilles. Finally, the horse, mule, and donkey proved superior beasts of burden to man and the llama, enabling outlying areas to carry on trade with the population centers.

Equally important was the myriad of iron and steel tools far superior to the wooden and stone implements of the Indians. The digging stick gave way to the hoe and plow and the stone and obsidian ax to the machete. Digging tools provided a means to expand mining, and the cart made transport cheaper and more efficient.

The Europeans brought a completely new life style to Latin America. Landownership was essentially unknown to the communal Indians. Although used for decoration, gold and silver had no intrinsic value to Indians. Urban living and towns built along European lines far surpassed the limited pre-Columbian market and religious centers. Trade and manufacturing were greatly expanded over the barter, short-distance marketing, and handicrafts to which the Indians were accustomed. Catholicism was imposed, and Indian ways of worship were destroyed. European languages became the official tongues, and those unable to speak one of them were destined to servitude. The Europeans imposed their will on all others and attempted with considerable success to establish their value systems and cultural institutions.

By 1600, only a few areas remained outside European domination, and these were not considered valuable by the Spanish and Portuguese. The Lesser Antilles, largely bypassed by the Spanish, were occupied by the French and English who all but eliminated the Carib Indians by 1650. Later, these northern European nations along with the Dutch settled the coastal fringe of the Guianas. In tropical South America, much of the interior was outside effective Spanish and Portuguese control, and even today the Amazon Basin and the western part of the Brazilian Plateau are sparsely settled. Finally, in southern South America, the fierce Araucanian Indians held back the Spaniards from southern Chile, while the grasslands of the Argentine Pampas and the arid lands of Patagonia were dismissed as worthless by the early conquerors.

The Physical Environment

The Europeans in the conquest of Latin America considered some parts valuable but others essentially worthless. Part of this appraisal was based on the number and qualities of Indian inhabitants, but also important was the physical environment. The location of gold and silver deposits attracted Europeans who established towns near mining sites and brought Indian laborers from other areas to work the mines. Most of the crops and domesticated animals brought by the Europeans, and those indigenous to the region, had definite environmental limitations beyond which they could not feasibly be grown or raised.

Early Spanish chronicles vividly describe what happened when Old World plants and animals were introduced into a drastically different environment. Wheat, oats, barley, olives, grapes, and stone fruit (peach, plum, cherry, and apricot) were planted in the Greater Antilles but died soon after sprouting. Sugar cane and numerous vegetables, however, did well in the Antilles. The Spanish, like it or not, had to change their diet and eat a combination of New and Old World plants. Sheep could not withstand the rigors of the warm, humid tropics, but the pig soon almost overran the islands. Cattle and horses adapted less well, but in time they too proliferated greatly.

In the highlands, where the high-culture Indians existed, climatic and other environmental conditions were more suited to most Old World livestock and cultigens. There a dual system of agriculture soon appeared. The Indians kept their traditional crops and land-management systems but added some Old World vegetables, such as onions and lettuce, and

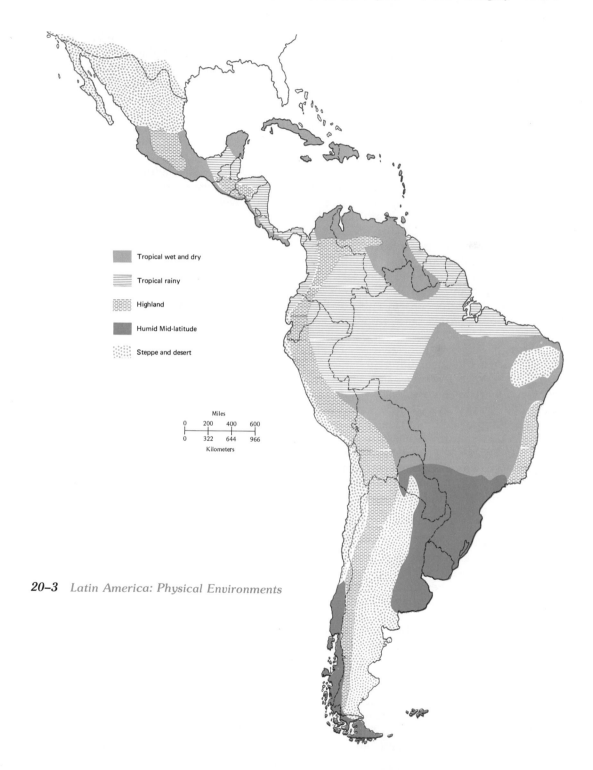

Tropical wet and dry

Tropical rainy

Highland

Humid Mid-latitude

Steppe and desert

Miles
0   200   400   600
0   322   644   966
Kilometers

20-3 *Latin America: Physical Environments*

poultry, pigs, and sheep. The Spanish established large landholdings devoted principally to cattle, horses, and wheat.

The highlands also were considered a desirable area by the Spanish because many of the important mineral deposits were located nearby. In the Inca, Chibcha, and Aztec areas a number of important silver and gold deposits were located within and adjacent to the major population clusters. In the highland and lowland Mayan areas, however, few significant deposits were found, and the Spanish consequently placed less emphasis there.

Southern South America, a mid-latitude environment, has become one of the most productive areas of Latin America but in the early days was considered of little value and consequently avoided for many years. The area's isolation, lack of minerals, and sparse Indian population did not hold much interest for the Europeans in spite of the fact that most European crops and livestock were easily adapted.

Latin America exhibits a diversity of physical environment. Each area of different physical elements possesses its own resources and obstacles that confront mankind in the struggle to obtain necessities and fulfill desires. As we have just seen, the early Europeans appraised the various parts of Latin America based partly on the environment. This evaluation continues. When the present situation of Latin America is viewed, knowledge of the environmental base is necessary for proper appreciation of opportunities and problems. Five major environmental types cover Latin America (Fig. 20–3).

Tropical Wet and Dry

The Spanish and Portuguese and later the British and French encountered the tropical wet and dry environment when they first settled in Latin America. This environment has a climate with a year-round growing season. Temperatures rarely fall below 60°F. (15.6°C.), which inhibits or precludes successful cultivation of most mid-latitude field crops. Precipitation is highly seasonal with abundant rainfall in the summer and an almost rainless winter. Consequently many tropical plants can be grown only with irrigation. In the Antilles, and to a lesser extent along the west coast of Central America, severe tropical storms are characteristic from June to December.

Soils vary considerably and depend partly upon the nature of the surface rock formations. In the Antilles and the Yucatan Peninsula, soils derived from limestone are normally quite fertile but tend to be shallow; most of Yucatan has shallow soils. Where deep, however, the soils are among the most thoroughly used. Soils of volcanic origin are generally deeper and typify the Brazilian Plateau and the higher elevation of the Antilles. In southern Mexico through Central America and the Llanos of Venezuela, alluvial soils, usually quite fertile, predominate. In Central America, they are mainly devoted to crops. In Venezuela, the Llanos is susceptible to flooding and used primarily for cattle raising. Most tropical wet and dry areas are plains at low elevations. Only the Brazilian Plateau is of intermediate elevation and of more rugged topography. Even on the plateau, however, extensive areas of level land are located on the plateau surface. Important mineral deposits are known only on the Brazilian Plateau. There, large quantities of iron ore and ferroalloys are located near the surface.

Highland Environment

The Spanish conquest came early to the highlands. The Spanish soon recognized several different altitudinal zones within the highlands based on temperature, and their terminology describing these zones remains with us (Fig. 20–4). Typically at the base of the highlands is the *tierra caliente* (hot land) extending to about 2,000 feet (610 meters). The *tierra caliente* has the basic characteristics of the tropical wet and dry or tropical rainy areas. Next is the *tierra templada* (temperate land) between 2,000 feet (610 meters) and 6,000 feet (1,829 meters). The *tierra templada* has temperatures 6°F. to 18°F. (3.3°C. to 10°C.), lower than those of the *tierra caliente,* giving it a springlike character. Most of the highland population of Mexico and Central America live within this zone. Above the *tierra templada* is the *tierra fria* (cold land), which extends to 12,000 feet (3,658 meters). This zone describes the upper limit of crop cultivation, and its alpine meadows often are used for grazing. Most of the *tierra fria* is in the Andean mountain system and supports a large Indian population. Only a few places in Mexico and Central America have such high elevations. Finally, above 12,000 feet (3,658 meters) is the *tierra helada* (ice land), an area where frost occurs almost nightly and at higher elevations is covered with permanent ice and snow. Soils within the high-

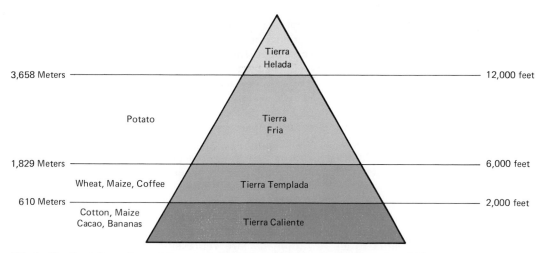

20-4 *Vertical Zonation of Tropical Highland Climates with Associated Crops*

lands are mostly of volcanic parent material. Their quality varies greatly, and no general statement can be made regarding fertility.

The highlands of Mexico and the Central Andes contain much of the gold and silver deposits from which vast quantities were removed to the coffers of Spain. Even today these areas remain one of the world's leading gold- and silver-mining centers. Other minerals have been added to the list, including copper, tin, lead, zinc, and various ferroalloys.

One of the principal problems of the highlands is erosion. The Indians had developed a well-organized system of terracing and constant soil improvement. During the conquest and later with the destruction of Indian government and decimation of the population, most of the terracing systems fell into disuse. Within the twentieth century, population growth has been rapid, and more and more sloping land has been used for crops. Without the organization and management of old, severe erosion has occurred.

Steppe and Desert

Four major areas of steppe (semiarid) and desert environment exist within Latin America. All are moisture deficient, and availability of water or its scarcity are the keys to their use. The Spanish pushed northward out of the highlands of Mexico with their herds of cattle and horses to establish large ranches in the steppe and desert area of northern Mexico and the United States Southwest. Settlement northward

was also stimulated by discovery of silver deposits in the Sierra Madre. More recently northern Mexico has been the scene of numerous irrigation projects. Similarly the steppe and desert area of northeastern Brazil was encountered by the Portuguese as they expanded westward. The dry area of northeast Brazil held no important minerals, nor have its water sources been successfully harnessed. It remains a poverty-stricken area. The dry area of Argentina (Patagonia and western Argentina) was ignored by the early European settlers, and even today much of the Patagonia is little used or developed. Western Argentina saw some development during the colonial period as a transport corridor between Chile and Bolivia. Later, irrigated agriculture along the piedmont brought considerable prosperity to the area.

The last dry area is along the west coast of South America, from Ecuador on the north to mid-Chile in the south. This narrow strip between the Pacific Ocean and the Andes is extremely arid. The Atacama Desert (northern Chile) is reputed to be one of the driest deserts in the world, and for a period was a major factor in world affairs. Within the Atacama was the world's only significant source of sodium nitrate, an important fertilizer and formerly used as an explosive. Foreign corporations vied for the privilege of exploiting the mineral, and for a while it accounted for much of the Chilean government's revenue. The value of the nitrates led to one of Latin America's few wars: the Pacific War of 1879 to 1883 in which Chile defeated Peru and Bolivia and took over all nitrate deposits; Bolivia lost its access to the sea as a result. The nitrate deposits are a good example of a resource that has almost lapsed to an inert phenomenon. Nitrate is now extracted from the air, and the demand for Chilean nitrate has dramatically declined. The Peruvian part of the western coast desert, aside from some petroleum in the north and iron ore in the south, is devoid of significant minerals, but numerous rivers flow out of the Andes onto the alluvial coast and supply water for intensive irrigated agriculture. Lima, Peru's largest city, is located along one of these rivers.

Tropical Rainy

Tropical rainy areas have been a dilemma to all European powers in Latin America. On the one hand, the luxuriant broadleaf-evergreen forest with its hundreds of different tree species and a warm, always moist climate is much like a greenhouse promising great productivity. In Latin America, the promise is yet to become a reality. Today, the tropical rainy areas are sparsely populated and of limited economic value, yet in the Asian tropics large populations are supported. Along the east coast of Central America, United States fruit companies established plantations devoted largely to bananas for export but tap only a small part of the area. The Amazon Basin has intrigued developers from colonial time, but successful use has been limited. Brazil is now engaged in a massive road-building and colonization program in the Amazon.

Soils in tropical rainy areas are often very infertile, but near surrounding mountain flanks and along rivers, rich alluvium may be found. It is the alluvial areas that are best for productive agriculture. All tropical rainy areas in Latin America possess potential mineral wealth. Petroleum deposits have been worked in eastern Ecuador and Peru. Exploratory wells have hit oil-bearing strata in Brazil's part of the Amazon, although no commercial production has resulted. The same circumstances pertain to the west coast of Colombia and the east coast of Central America. Also in the Brazilian Amazon, quantities of ferroalloys have been found.

There is increasing evidence that in pre-Columbian times the tropical rainy areas of Latin America were more densely settled than now. The lowland Mayan of Guatemala and Mexico lived largely within the tropical rainy environment. Elsewhere archeological evidence is accumulating that points to substantial early populations. Why the Mayans and other groups declined in population is not surely known, but introduced diseases such as malaria and yellow fever were certainly contributory causes. With development of pesticides like DDT and preventive and curative medicine, many of these diseases are not the scourge they once were. Latin American nations with part of their area in tropical rainy environments are looking to these environments as colonization and development zones.

Humid Mid-Latitude Environment

Aside from the highlands where mid-latitude environments are represented by *tierra templada* and *tierra fria,* only part of southeastern South America and middle and southern Chile falls within the humid mid-latitude environment. Like other parts of southern South America, these areas were not favored by early Spanish colonists. Middle Chile did become an outpost of Spanish settlement, and here the Spanish found environmental conditions like those of part of Spain. Crops such as olives, grapes, and wheat could be readily grown and were widely sought by Spaniards in other parts of South America. Southern Chile is colder and more humid and even today is sparsely settled, although large potentially commercial forests and abundant hydroelectric sites exist. Southeast South America's humid mid-latitudes are a combination of grasslands (pampas) and open forest on a plain surface underlain by alluvial soils. The soils in the east are particularly fertile, whereas those westward are of coarser material and more droughty. The growing season is long, although freezing temperatures do occur in the south and occasionally extend northward into southern Brazil. No significant minerals, except some petroleum, are known to exist in the area.

The Colonial Period

The conquest of Latin America was rapidly completed, and a new era began, lasting for the most part to the early nineteenth century. After decimation of the Indians, so few people were left within such an immense area that great difficulties were encountered

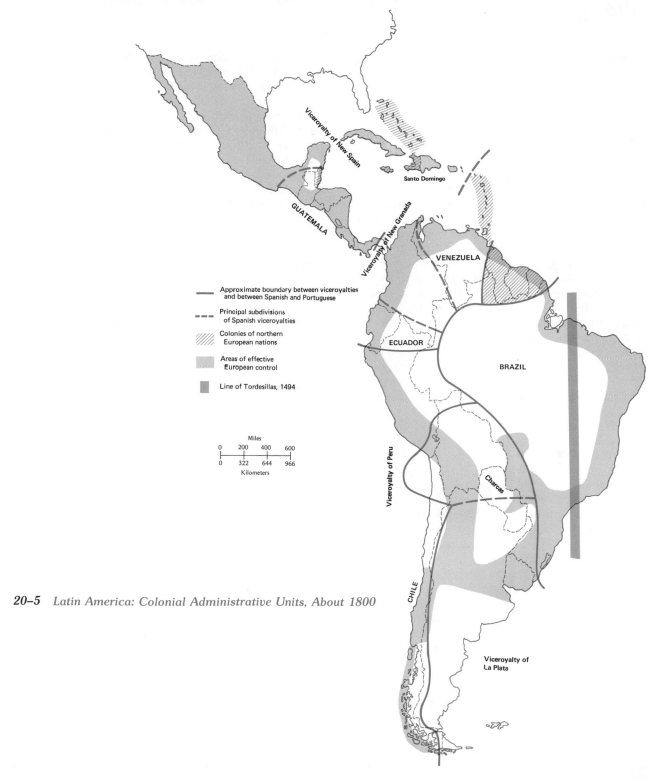

Legend:

——— Approximate boundary between viceroyalties and between Spanish and Portuguese

– – – Principal subdivisions of Spanish viceroyalties

/// Colonies of northern European nations

Areas of effective European control

Line of Tordesillas, 1494

Miles
0   200   400   600
0   322   644   966
Kilometers

Viceroyalty of New Spain

GUATEMALA

Santo Domingo

Viceroyalty of New Granada

VENEZUELA

ECUADOR

BRAZIL

Viceroyalty of Peru

Charcas

CHILE

Viceroyalty of La Plata

20–5 *Latin America: Colonial Administrative Units, About 1800*

in building a viable political, economic, and social system. Many of the characteristics developed during the colonial period have persisted and are integral aspects of Latin America today.

Political Organization

To a large degree the present-day political map is a legacy of the colonial period. Spain and Portugal found communications so difficult and tenuous that the respective crowns named local administrators to act in their stead. Spain divided its territory into viceroyalties, and each viceroy subdivided his area of responsibility and designated lieutenants to administer part of the territory under his jurisdiction. On the eve of independence (early nineteenth century), four viceroyalties and their subparts represented the crown in Latin America (Fig. 20–5). Portugal followed a system similar to Spain's. In Brazil, the Portuguese settled first in the northeast and gradually spread inland and southward along the coast. These settlements were distributed among several *Capitanias* that functioned in a manner like Spain's viceroyalties. The boundary between Spanish and Portuguese administrative units usually passed through sparsely settled areas, and most units possessed a single center of dense population.

In the Caribbean Sea area, Spain's dominion was disputed by the British, French, and Dutch. Continued inroads by these northern European nations eventually led to their control over the Guianas, Lesser Antilles and Trinidad, Jamaica and Haiti in the Greater Antilles, and Belize (British Honduras) in Central America.

Economic Organization

All mother countries attempted with varying degrees of success to operate their colonies with the mercantilistic philosophy. Trade was limited to that between colonies and the mother countries, and intercolonial trade was severely restricted. As a result, few linkages among colonies of the same mother country existed, and there was little feeling of commonality. Rather, competition was encouraged. This characteristic of mother-country policy is evident in modern Latin America where nations of common cultural heritage are often at odds with one another.

The mercantilistic policy also strongly affected economic activities. No colony was permitted to produce anything in competition with the mother country. Manufacturing was especially curtailed. For Spanish and Portuguese colonies, this policy meant that sheep raising, wheat, olive, and grape growing were suppressed. The British, French, and Dutch colonies were in such a different environmental realm that the agricultural economies of colonies and mother countries were inherently complementary rather than competitive. In the early nineteenth century, however, competition did develop between the sugar cane of the colonies and sugar beets of northern Europe. Since the colonial sugar cane industry was already firmly established and no alternative activities could be found, the industry was permitted to continue, but tariffs on imported cane sugar were increased to the advantage of local beet sugar producers.

Early in the colonial period, a dual economy began to evolve. Production for export to mother countries was largely the function of large landholdings controlled by Europeans. In the colonies of northern European nations, the colonial plantation typified this activity. African slaves were introduced into the Antilles as laborers. Management was carried out by Europeans, although the owners often lived abroad. Large blocks of the best land were set aside for these plantations, and the entire economy and life in general revolved around the plantation. Not all plantation land was devoted to export crops, basically sugar cane, for some land was set aside for provisioning the slaves and for pasturage of oxen. In less accessible areas and on poor-quality land, crops for the local market were raised on small properties. In Spanish and Portuguese colonies, the large landholdings were typified by three land units: the *hacienda,* the *estáncia,* and the *fazenda;* the first two terms are Spanish and the last-mentioned term Portuguese. The *hacienda* and *fazenda* usually refer to land units devoted to crops and animals, whereas the *estáncia* normally refers to a livestock operation, that is, cattle and horses. These large landholdings were not so export-oriented as the plantation, since neither the Spanish nor Portuguese market was large enough to accept the vast productive potential of their colonies. Also, in contrast to the plantation, labor was obtained from the local Indian supply, and the owners usually lived on the property or in a nearby city. Elsewhere in less accessible areas and on poor lands, small Indian subsistence economies persisted, using age-old practices.

Although neither the Spanish nor Portuguese had

scruples against using African slaves, blacks were important only in the Spanish Antilles (Cuba, Puerto Rico, and the Dominican Repulic) and in the Brazilian northeast. The relatively easy escape to the interior of the mainland made slavery less attractive as an economic institution. Nevertheless, Brazil did not abolish slavery until 1888.

The patron system and debt peonage were used to assure a continuing supply of Indian and mestizo laborers. The patron system is both an economic and social system. The patron, the large landowner, functioned both as an employer and a "godfather." Indeed, he was literally a godfather to many of the peons' (workers') children. He practiced a paternalistic policy toward his peons and in return gained their allegiance. In all things the patron's word was law, and few thought to dispute his prerogatives. The patron-peon relationship thus engendered the patron as almost a feudal lord and the peons as serfs.

Debt peonage further reinforced the patron system. Under the peonage system, the patron's power over his workers was given legal sanction. Wages paid to the peons were minimal, but housing and garden plots, and often some food, were provided by the patron. For other needs the patron maintained a store that granted credit to the workers. Over time, the peons usually got increasingly in debt, and these debts were inherited by the workers' heirs. Under debt peonage, no worker could leave his patron without approval. The patron, therefore, had a legal weapon to use over dissatisfied peons. Should a peon pay off his debts, there were still strong social ties causing him to remain with his patron. The *hacienda, fazenda,* or *estáncia* was a cultural institution as much as an economic organization. The peon's family and friends lived on the *hacienda,* and the peon usually felt a strong attachment to his patron. Neighboring patrons, in any event, rarely accepted a peon from another *hacienda,* and few alternative employment opportunities existed.

Social Organization

Society in colonial Latin America was organized almost on the caste system; that is, a person could rarely rise above the station in which he was born. In the northern European colonies, the African slaves formed the great lower class and had few privileges. Above them were the freed men who usually were mulattos. Mulattos were the craftsmen and gang bosses. Some had been educated and functioned as secretaries and accountants. Just above the mulatto were a small number of Europeans of limited wealth who lived as small farmers and competed with the mulattos for jobs on the plantations; this group has almost died out in the Antilles. At the top of the social order were the Europeans of wealth. They were the government officials, major merchants, plantation owners, and managers.

In the Spanish and Portuguese colonies, a similar stratification based on race prevailed. At the top was the small class of Europeans born in the mother country who held all the major governmental, military, and religious posts. This peninsular-born group was proud, exclusive, rigid in religion, and highly conservative. Just below them were the creoles (Europeans born in the New World), about whom life in the colonial cities revolved and who owned large rural landholdings. The creoles generally were the wealthy class but prohibited by royal decree from the most sensitive government positions. Three means of livelihood were socially acceptable to creoles: the military, in which they could serve as junior and medium-level officers; the priesthood, serving in offices generally below the rank of bishop; and as the landed aristocracy. It was as landed aristocracy that the creoles obtained their wealth and power. The creole measured wealth not only in terms of money but more important in landownership. In this attitude toward land we see one of the most fundamental characteristics of colonial Latin America and, although somewhat diminished, one that continues.

The Spanish word *caballero* has a double meaning, "gentleman" and "horseman." In essence, a gentleman is not a gentleman unless he is also a horseman. It is little wonder that creole military officers preferred the cavalry to all other army branches. Nor is it strange that the creole landowners preferred livestock raising (horses and cattle) to crop production. During colonial time the best lands were devoted to ranching and the poorest areas left to cropping; in some respects in the Anglo-American west a similar experience was the portrayal of cowboys as more glamorous and gallant than the homesteader. Even today, in spite of heavy population pressures, some of the best land is used for livestock rather than food production.

Below the creole was the mestizo, who on ranches became the *vaquero* (cowboy) and on farms the overseer. In the army he was the noncommissioned officer, and occasionally he was the village priest. In town he was the salesclerk and craftsman. Finally came the Indian, who performed the various jobs requiring manual labor (farming, mining) and who often farmed a small plot of land for his own needs. In the high-culture areas, many Indians were able to maintain their communal lands but were forced to work a certain number of days for local patrons. Others had their lands taken from them and became peons under a patron.

The Independence Movement

By the end of the colonial period in Spanish and Portuguese America, the basic social and economic systems had become firmly imbedded in the fabric of Latin American life. Independence did little to change the lives of most people.

Independence came quickly to most Spanish colonies in the period 1815 to 1825. Spurred by the philosophies of the United States and French revolutions, the capture of Spain by the French, and by creole demands for greater control of local affairs, numerous insurrections spread throughout the Spanish mainland colonies. Early recognition by the United States and the Monroe Doctrine of 1823, tacitly backed by the British fleet, encouraged and supported the independence movement. By 1825, independence was assured, and new nations were formed from the former Spanish viceroyalties and their parts. The boundary lines among those nations were essentially those of the colonial administration. One notable exception was the creation of Uruguay as a buffer between rival Argentina and Brazil.

In the Spanish Antilles, the independence movement followed a different course. The Spanish part of Hispaniola was annexed by Haiti in 1822, only to gain independent status in 1844. The last Spanish possessions were lost following the Spanish-American War when Cuba was made independent, and Puerto Rico became a dependency of the United States.

The Brazilian experience was somewhat different. Portugal, too, was overrun by the French armies, but the Portuguese king fled with his court to Brazil. For a time Brazil was the seat of the Portuguese empire. In 1821, the king returned to Portugal leaving his son, Dom Pedro, regent. Shortly afterwards, in 1822, Dom Pedro proclaimed Brazil independent with himself as Pedro I, Emperor of Brazil. From then until his death in 1888, Dom Pedro held Brazil together and laid the foundations of nationalism which have kept Brazil one nation.

The colonies of the northern European nations have followed a different course. Aside from Haiti, where a successful revolt against the French led to independence in 1804, the northern nations held onto their colonies during the nineteenth century. Only after World War II did some colonies move toward a sovereign status, and most did so reluctantly. Britain has followed a policy of gradual disengagement. In 1958, Britain sponsored the creation of an independent West Indies Federation that included the British Lesser Antilles, Trinidad, and Jamaica. The Federation, however, lasted only a short time. In 1962, Trinidad and Jamaica became separate and independent nations, and the British Lesser Antilles returned to colonial status, although with greater self-rule. In 1966, Barbados was granted independence, and in 1974, Grenada also became a sovereign state. On the mainland, Guyana gained independence in 1966. The French and Dutch have moved to integrate their colonies into their respective nations. Today, the former French colonies (Guadeloupe, Martinique, and French Guiana) are departments (states) of France and as much a part of France as Hawaii and Alaska are parts of the United States. The Netherlands Antilles are similarly integral parts of the Kingdom of the Netherlands, but Surinam is independent.

Summary

In this chapter we have traced some of the social, economic, and political processes at work in Latin America from pre-Columbian times to independence. Many of the events of the past (decimation of Indian populations; introduction of cultigens, animals, and tools; social, political, and economic controls and attitudes) have placed an indelible imprint on the present. To a large degree, colonial settlement patterns reflected the pre-Columbian population distribution and culture and the different physical environments encountered. In some environments, introduced crops and animals did not fare well; but where proper conditions were found, the resource base was greatly expanded. Colonial Latin America's economy and political structure were determined largely by the mother country. The economic policies were formed for the mother country's benefit and were one cause for the revolts leading to independence. Independence led to the formation of nations whose borders followed the colonial administrative divisions.

Further Readings

BRADING, D. A., and CROSS, HARRY E., "Colonial Silver Mining: Mexico and Peru, *Hispanic American Historical Review*, 52 (1972), 545–579. A chronology of silver mining in two of the most important producing nations emphasizing the role of technology in mining and refining, labor system, government, and production.

CROSBY, ALFRED W., Jr., *The Columbian Exchange: Biological and Cultural Consequences of 1492* (Westport, Conn. Greenwood Press, 1972), 268 pp. An interesting and informative account of the flow of crops, animals, and disease to and out of Latin America and the effects of these exchanges.

DORST, JEAN, *South America and Central America: A Natural History* (New York: Random House, 1967), 298 pp. A superbly illustrated book with many full-page color photographs emphasizing animal and plant life.

GOVEIA, ELSA V., *Slave Society in the British Leeward Islands at the End of the Eighteenth Century* (New Haven, Conn. Yale University Press, 1965), 370 pp. A study of the political, economic, and social organization of a "slave society." Although the study is principally of a few small islands, the ideas and thoughts are of much wider application.

INNES, HAMMOND, *The Conquistadores* (New York: Knopf, 1969), 336 pp. Innes, a novelist of considerable note, has prepared a beautifully illustrated and easily readable account of the Spanish conquest of Latin America. This book is an excellent companion piece to William H. Prescott's work and other early histories.

JAMES, PRESTON E., *Latin America,* 4th ed. (New York: Odyssey Press, 1969), 947 pp. A comprehensive geographic survey of Latin America and long a standard text.

PRESCOTT, WILLIAM H., *History of the Conquest of Mexico and History of the Conquest of Peru* (New York: Modern Library, no date), 1288 pp. These two books, now under one cover, were originally published over one hundred years ago and are considered classics. Prescott explores the state of the Aztec and Inca Indians just prior to the conquest and then follows both Cortez and Pizarro in their struggle to conquer these and other Indian groups.

STEWART, JULIAN H., and FARON, LOUIS C., *Native Peoples of South America,* (New York: McGraw-Hill, 1959), 869 pp. A companion piece to Wauchope's volumes, listed below.

WAUCHOPE, ROBERT (ed.), *Handbook of Middle American Indians* (Austin: University of Texas, 1964–1972), 12 volumes. A thorough investigation of the Mesoamerican area from the physical resource base to ethnohistory. A valuable data source.

WEST, ROBERT C., and AUGELLI, JOHN P., *Middle America: Its Lands and Peoples,* 2nd ed. (Englewood Cliffs, N.J.: Prentice-Hall, 1976), 449 pp. Like James, West and Augelli are well-known Latin Americanist geographers who present a geographic survey of the Caribbean, Mexico, and Central America.

Latin America: Processes of Change

21

For THE MAJORITY of the population, freedom from colonial rule did little to change their life styles. Throughout much of the nineteenth century the new nations experienced the growing pains of self-rule. Internal disunity led to stormy periods of near anarchy and strong, personalized, dictatorial rule, to battles over the location of the national capital, and to border conflicts. Even in Brazil under the rule of Dom Pedro I, great difficulties were encountered in keeping the fledgling nation together. The seeds of change were planted, however, and the processes that are presently modifying Latin America had their origin in the nineteenth and earlier centuries.

Population

Population Growth and the Population Pyramid

For most of Latin America's history, population growth has been slow. High birth and death rates were characteristic. In a few nations (Argentina, Brazil, and Chile), large-scale immigration from Italy, Germany, Spain, and other European countries in the second half of the nineteenth century added a new dimension to the population. Yet Latin America did not equal its pre-Columbian population until about 1920 (Fig. 21–1). During this century, population has increased by five times the 1900 figure, and by the year 2000 the population is expected to be ten times the 1900 level. All parts of Latin America are predicted to grow substantially during the rest of the century, but tropical South America and mainland Middle America will account for the greatest increases. Latin America's present rate of growth, about 2.7 percent annually, is one of the highest in the world. As a consequence, Latin America's percentage of total world population has increased. In 1920, Latin America had 5 percent of the world's population, but by 2000 it is projected that it will have 10 percent. Such a great increase is one contributing factor for Latin America's status as a poor or less developed region.

By far the most important factor in population growth is the natural (internal) growth rate; although pertinent in the past, immigration is not important today. Two characteristics of population are responsible for the rapid population increase: a high birth rate and a declining death rate, particularly in infant mortality. Traditionally high birth rates have prevailed and were necessary to maintain a viable population, since death rates were equally high. During this century and especially since about 1920, the death rate has dropped dramatically (Fig. 21–2). The six nations shown in Figure 21–2 are representative of the birth and death rate structure of Latin America. Guatemala, Mexico, and Ecuador have continued high birth rates, even though some decline is noted in more recent years. Most of mainland Middle America and much of tropical South America have the same pattern. Venezuela is at a variant position, since its birth rate has increased, but it is now declining. Barbados, like much of the Antilles, has had a more moderate birth rate, and one that has decreased in the last twenty years. Argentina, typical of much of southern South America, has a continued moderate birth rate.

Death rates of most nations shown in Figure 21–2 are now comparable with those of the United States (0.9 percent) and other rich or developed regions;

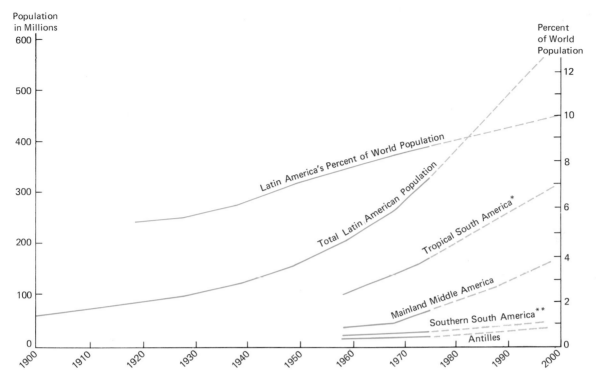

21–1 *Latin America: Population Growth, 1900 to 2000*

*Brazil, Bolivia, Colombia, Ecuador, Guianas, Peru, Venezuela. **Argentina, Chile, Paraguay, Uruguay.

SOURCE: Latin American Center, *Statistical Abstract of Latin America* (Los Angeles: University of California, 1976); and *World Population Situation in 1970* (New York: United Nations, 1970).

a few are lower. The decrease in infant mortality illustrates that death rates can be rapidly reduced. Moderate death rates presently prevail only in Guatemala, Honduras, Nicaragua, Bolivia, and Haiti.

As we saw in Chapter 4, poor nations have youthful populations. Latin America is no exception (Fig. 21–3). As noted in the population pyramid for Latin America, nearly one-half of the population is outside the most productive age group; 42 percent are under fifteen years old, a factor partly explaining low per capita productivity. Such a large youthful population indicates a great potential for further growth. In fact, as the population matures, a greater number will reach childbearing age. It is for this reason that with the trends in birth and death rates it is projected Latin America's population will double during the last thirty years of this century. Variation in the population age structure does exist. Mexico exemplifies a very youthful population, and Argentina shows a more mature distribution.

Although the population explosion is a factor in the low per capita incomes, it is also a reflection of greater productivity. Most people in Latin America do eat enough calories and protein, and most have a relatively long life expectancy (Figs. 4–9 and 4–10). These facts lead us to the proposition that the resource capacity of Latin America is sufficiently large to accommodate a rapid population increase but perhaps not at a level commensurate with the developed world. In further support of the proposition, some nations readily welcome immigrants; Brazil, for example, has a national policy favoring internal growth to occupy the western Brazilian Plateau and the Amazon Basin effectively. On the other hand, few acute observers would deny that heavy population pressures exist in some areas, for exam-

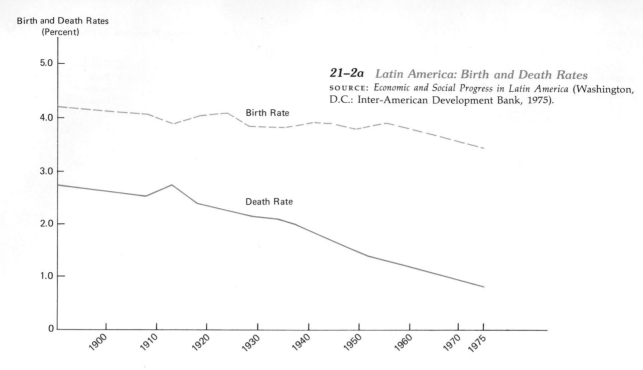

21–2a *Latin America: Birth and Death Rates*
SOURCE: *Economic and Social Progress in Latin America* (Washington, D.C.: Inter-American Development Bank, 1975).

21–2b *Birth and Death Rates for Selected Latin American Nations*
SOURCE: Latin American Center, *Statistical Abstract of Latin America* (Los Angeles: University of Los Angeles, 1976), and *Population Data Sheet* (Washington, D.C.: Population Reference Bureau, 1977).

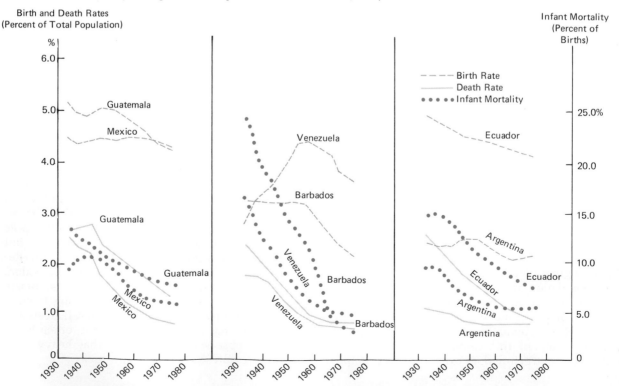

21-3 *Latin America: Age and Sex Pyramids*
SOURCE: *Economic and Social Progress in Latin America* (Washington, D.C.: Inter-American Development Bank, 1974).

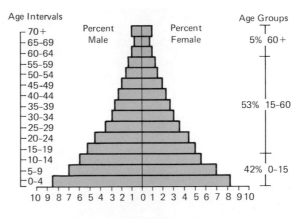

Latin America

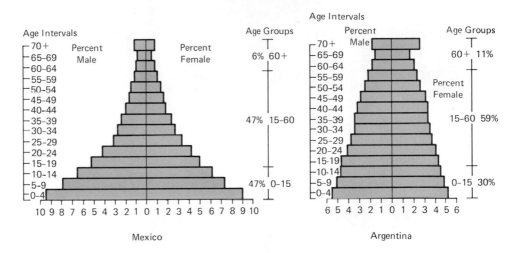

Mexico

Argentina

Urbanization

ple, the Antilles and areas of pre-Columbian high civilizations.

Latin America's population is dynamic not only from the point of view of growth but also in the urbanization movement. Migration from rural to urban areas and from small villages to large towns is progressing rapidly. Presently about 55 percent of Latin America's population live and work in towns and cities. By 2000, 75 percent are expected to be urban dwellers, a figure comparable to that projected for the rich or developed world. All parts of Latin America should participate in the urbanization movement, but southern and tropical South America will have the greatest part of their populations in urban areas (Fig. 21–4). Everywhere in Latin America the urban population will soon outnumber its rural counterpart.

Urbanization is perhaps more graphically demonstrated if we view projected absolute values. Figures 21–5 and 21–6 show the expected urban and rural populations for Latin America and its principal subdivisions. In the space of forty years, 1960 to 2000, Latin America's urban population is expected to increase from 100 million to 500 million, one-half of this number in tropical South America, principally Brazil. The rural population is also expected to increase but at a much slower rate; in southern South America it is projected to decrease.

Opposite: São Paulo, Brazil, is one of the world's largest and most dynamic cities. São Paulo has grown from a town of 600,000 in 1935 to over 8 million in 1970. By 1983, it is expected that the city will have more than 20 million inhabitants. (Varig Brazilian Airlines)

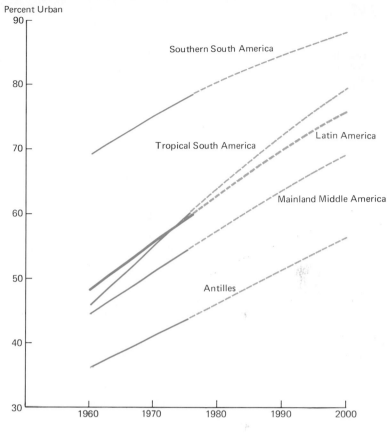

21–4 *Latin America: Projected Urban Percentage of Total Population*
SOURCE: *World Population Situation in 1970* (New York: United Nations, 1970).

Rapid urbanization in Latin American cities through rural to urban migration has led to large "shantytowns." It is difficult to provide sanitation and other services to these rapidly growing areas. (United Nations)

403

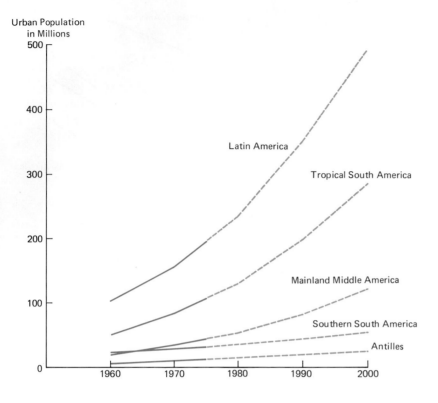

21–5 *Latin America: Urban Population Projected to Year 2000*

SOURCE: *World Population Situation in 1970* (New York: United Nations, 1970).

21–6 *Latin America: Rural Population Projected to Year 2000*

SOURCE: *World Population Situation in 1970* (New York: United Nations, 1970).

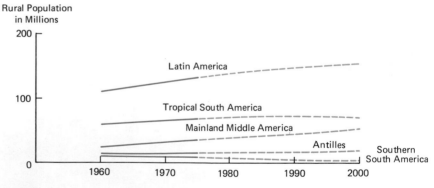

Urbanization is more than living in a town or city. It also means urbanism: a distinctive life style, different economic activities, labor specialization, and economic interdependence. These characteristics are reflected in changing social and political attitudes. A person born and raised in a rural area, especially of a subsistence background, finds the city an almost unfathomable complexity; yet nearly one-half of the urban dwellers are migrants. Few successfully make a direct farm-to-city transition. It seems more likely that movement from farm to regional town to large city is more common, often taking at least two generations. Moreover, a selective process occurs in the migration with the more educated, wealthy, and economically motivated participating in the movement. In many Latin American areas, the remaining rural population is almost devoid of a class of incipient entrepreneurs.

Economic Development

During the twentieth century and especially since 1950, a number of economic developments have begun to alter Latin America's economic structure. Overall, the economy has grown rapidly, although the high rate of population growth has diminished per capita values. The economy has become diversified with manufacturing and tertiary activities providing more and more jobs and contributing greatly to the growth of the gross national product. Foreign investment, largely from the United States, has increased greatly in spite of a growing local spirit of nationalism and actual nationalization of some foreign corporations.

Economic Growth

The gross national product for most Latin American nations has grown rapidly. Much of this growth has resulted from increased productivity, especially in manufacturing and service industries. In fact, some authorities place Latin America in an intermediate category between the rich and the poor nations groups. Since 1970, however, high rates of inflation have swept across the world and have altered the trends in GNP. Figure 21–7 shows the trend in per capita GNP for several Latin American nations. All nations show a very rapid increase in per capita GNP values, but that of Venezuela is especially high. Venezuela, one of the world's major petroleum exporting nations and a member of the Organization of Petroleum Exporting Countries (OPEC), has benefited greatly from increased oil prices. Per capita GNP for Latin American nations is not a good measure of income for most of the population, since wealth is concentrated in the upper-class stratum and increasingly in urban areas. For much of the lower-class and rural population, incomes have remained low. Moreover, inflation (loss of buying power) has reduced real gains in income. For nations such as Venezuela, total and per capita gains result largely from governmental petroleum revenues which go to the population only indirectly.

Economic Diversification

Agriculture remains the basic source of livelihood for much of Latin America's population, but other activities are of increasing importance. A comparison

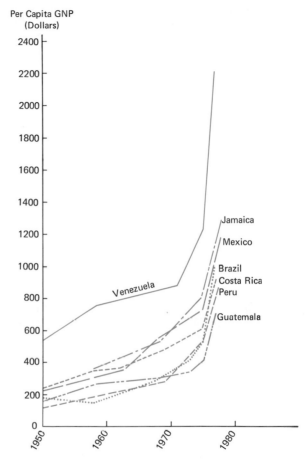

21–7 *Latin America: Per Capita GNP of Selected Nations*
SOURCES: *Statistical Yearbook* (New York: United Nations, 1974); and *Population Data Sheet* (Washington, D.C.: Population Reference Bureau, 1977).

of some sectoral characteristics illustrates the change in emphasis on economic activity (Fig. 21–8). The percent of the labor force engaged in primary activities (basically agriculture) decreased from 47 percent in 1960 to 39 percent in 1975, although the total number of primary workers actually increased. In contrast, secondary activities (manufacturing) increased from 18 to 26 percent. Tertiary activities (services and construction) remained unchanged. It is estimated that by 1980, the labor force will have 35, 29, and 36 percent in primary, secondary, and tertiary activities respectively. In 1940, 70 percent of the labor force was in primary activities.

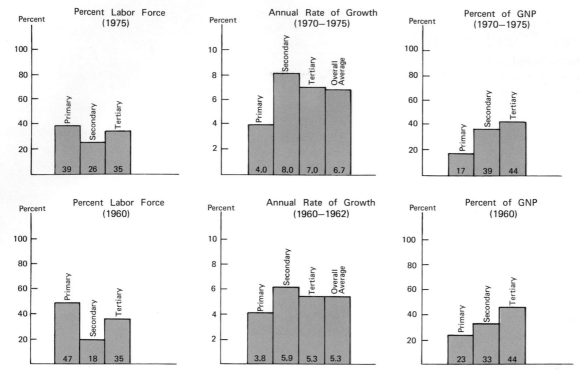

21–8 *Latin America: Sectorial Economic Characteristics*
SOURCE: *Economic and Social Progress in Latin America* (Washington, D.C.: Inter-American Development Bank, 1974).

21–9 *Latin America: Value Added Growth Rates*

SOURCE: *Economic and Social Progress in Latin America* (Washington, D.C.: Inter-American Development Bank, 1974).

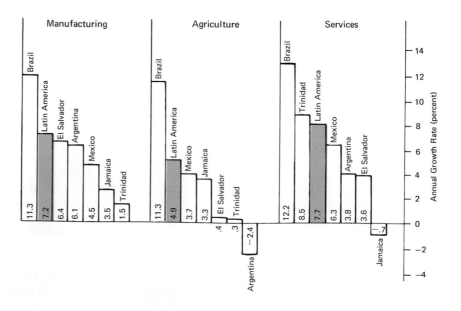

The annual growth rate of Latin America's economy has increased. In 1960–62, Latin America's GNP increased at a rate of 5.3 percent annually and by 1970–75, 6.7 percent. The growth rate varied among economic sectors. The growth rate in primary activities increased from 3.8 percent to 4.0 percent. Secondary activities' growth rate increased from 5.9 percent to 8.0 percent and tertiary from 5.3 to 7.0 percent. Even in 1960, primary activities accounted for but 23 percent of the gross national product and decreased to 17 percent in 1970–75. This decrease was compensated for by growth in the secondary sector.

Figure 21–8 also illustrates the difference in per capita productivity among the various sectors. Workers in primary activities in 1960 were only half as productive as workers in general; 47 percent of the labor force accounted for only 23 percent of the GNP. By 1970 to 1975, primary workers were relatively even less productive. Again, in contrast, secondary and tertiary workers had productivity ratings significantly higher than average. If we assume income is related to productivity, then the urban laborer has an income about twice that of the rural worker. It is little wonder that urbanization is increasing rapidly.

A second set of measures of sectoral differences is the value-added growth rate. Value added is calculated by subtracting all production costs (labor, energy, materials, depreciation of equipment) from the total value of the product or service. Value added is in essence the net profit of an activity. Figure 21–9 represents the value-added growth rates of manufacturing, agriculture, and services for several Latin American nations. Again we can see that Latin America's agricultural growth rate is much less than the growth rate of manufacturing and of services, and that there is great variation in rates among the nations shown. Brazil's economy is booming in all three sectors. The other countries presented are growing at lesser rates.

Foreign Investment

Independence of former Spanish and Portuguese colonies opened the door to economic interests in the United States and northwestern European nations. Britain early invested heavily in banking, railways and utilities, mining, meat-packing plants, and stock-raising operations. These activities provided the British Isles not only with income but also with products for the home market. In the twentieth century, British and other European investments have declined as United States corporations and more recently those of Japan have moved aggressively into Latin America.

Latin America has been particularly attractive to United States companies. From the time of the Monroe Doctrine, the United States has considered Latin America within its sphere of influence. United States government policy has protected private investment and has assured a degree of security from attempts to control foreign operations. Only since about 1960 have Latin American nations begun to exert a significant influence over foreign investments within their national territories.

History of United States Private Investment

Around the turn of the century, United States investment amounted to $320 million, about 47 percent of all its foreign investments (Fig. 21–10). Most of the investments were in mining, railways, and crops for export and were concentrated mainly in Mexico. The Spanish-American War peace treaty gave the United States Puerto Rico and a "protectorship" over Cuba. Investment in the Cuban sugar industry was rapid. In 1914, on the eve of World War I, United States private investment totaled $1.6 billion (46 percent of United States foreign investment). Mexico and Cuba were by far the most important investment areas. After World War I, investment in Latin America continued to grow, and by 1930, $5.2 billion were invested (33 percent of total United States foreign investment). The area of investment expanded; Cuba and Mexico remained the most important investment areas, but petroleum in Venezuela and Colombia and mining in other South American nations received substantial investor company interest. Although private investment in Latin America continued to increase, its share of total United States foreign investment declined as companies expanded into other parts of the world. The depression of the 1930s and World War II caused a drop in foreign investment. By 1943, the heart of the war period, investment in Latin America had declined to $3.4 billion (34 percent of total foreign investment).

Billions of Dollars

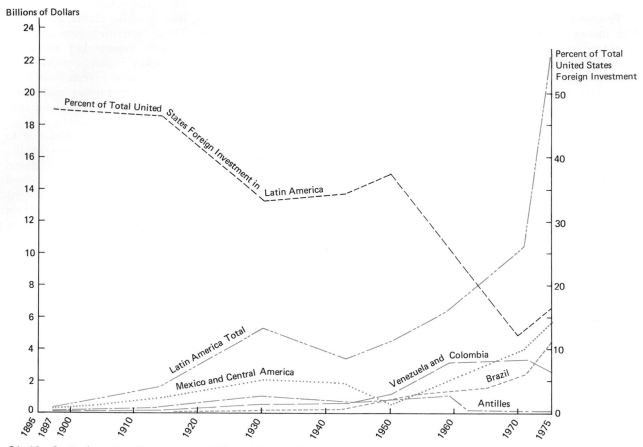

21-10   *Latin-America: Investment of United States Capital*

SOURCES: J. Fred Rippy, *Globe and Hemisphere* (Chicago: Henry Regnery Co., 1958); Department of Commerce, *U.S. Business Investment in Foreign Countries* (Washington, D.C.: Government Printing Office, 1960); and *Survey of Current Business* (1976).

After World War II, private investment entered Latin America at an unparalleled pace. By 1950, 4.4 billion (37 percent of total foreign investment) had been invested; by 1959, $6.5 billion (27 percent of the total), and by 1971, $10.5 billion (12 percent of the total). In the period since the war, mining, manufacturing, and service activities have received most of the investment interest. The sharp decline in percent of all United States foreign investment in Latin America reflects increased investment activity in Europe and elsewhere.

Foreign Investment: A Mixed Blessing

Foreign investment is a mixed blessing to Latin American nations. Corporations have provided badly needed capital and equipment, managerial and technical skills, and transport and marketing facilities. These assets have produced jobs, tapped heretofore unused resources, and created infrastructural improvements in the form of roads, banking, communications, and energy. In Central America, companies opened up little used, coastal tropical rainy areas for export banana production. Thousands of jobs were created not only directly in banana operations but also to service the industry and workers. Transport facilities (railways, roads, and ports) were built which all could use. Revenue was generated and taxes were paid to the government for use in other parts of each nation. In Venezuela, petroleum companies mapped, explored, and partly tamed parts of the Maracaibo Basin and have contributed greatly to Venezuela's large per capita gross national product. Cuba was probably most greatly affected by foreign investment. Within the space of a few years after the Spanish-American War, the island was the scene of feverish sugar cane development. New lands were opened, processing plants built, and transport systems established. Later, tourism provided jobs in Havana and other resorts. These are but a few examples of the benefits of foreign investments.

The other side of the story is not so rosy. Much of the wealth (profits) generated by foreign investment has been "exported" to the companies' home nations. Petroleum and other mining operations have traditionally sent 50 percent or more of their profits home. The same situation existed in the sugar industry of Cuba. In some countries, the private corporations were so strong they influenced national elections and local governmental policy. Banana companies in Central America are alleged to have "bought" politicians and may have sponsored revolutions to overthrow governments. In 1973, the International Telephone and Telegraph Company was accused of trying to prevent Salvador Allende's election as president of Chile and, when that failed, of attempting to overthrow his government. Where investment was particularly great, for example, in Cuba, the economy became heavily oriented to the export market and the country an "economic colony" of the investor country.

Changing Position of Foreign Investment

The position of foreign investment in Latin America is changing. Most Latin American nations have reassessed their policies toward foreign corporations and have moved or are moving to exert greater measures of control. Some forms of control are designed to integrate these traditionally export industries with local economic developments. Mexico, for example, encourages the processing of products before export by variable taxation. Raw materials (lead and zinc bars) are taxed at a higher rate than finished products (batteries). Mexico and other nations also have lower taxes for companies with majority ownership by nationals. Many countries require employment of nationals at all levels of management; technical and managerial skills are consequently nurtured. Still other nations limit the amount of profits that can be taken out of the country; substantial earnings of the petroleum industry in Venezuela are reinvested in other or related activities.

Finally, there is a growing trend for partial government ownership and nationalization of foreign corporations. Ecuador is in partnership with petroleum companies developing its eastern oil fields. Chile, before nationalization of its copper industry, held 51 percent of the stock. Nationalization is becoming more common. In 1954, Guatemala expropriated some of the foreign-owned banana lands. Between 1959 and 1961, Cuba nationalized its sugar industry and just about every other externally owned business. Chile and Peru in the early 1970s nationalized most of the foreign mining enterprises; additionally, Peru expropriated some of the larger corporate-owned agricultural lands. In contrast to other nations, Venezuela has proceeded slowly, gradually gaining control of petroleum operations and eventually nationalizing the industry completely.

The wave of increasing control over and nationalization of foreign investment marks a change in attitude on the part of Latin American governments and populations. Nationalism, the desire for greater control over resources, and improving ability for internal resource management are primary considerations for this change. Except for Cuba and Chile under Allende, nationalization has been directed primarily at agricultural and mining operations. Manufacturing and service activities are not materially affected. Indeed, many nations encourage investments in these areas.

Agrarian Reform

The preceding charts (Figs. 21–8 and 21–9) clearly demonstrate that agricultural development has not kept pace with other sectors of the economy. A rural-urban dichotomy is arising as many agriculturists continue traditional farming practices emphasizing hand tools, while in the city modern technology is rapidly replacing craftsmen and artisans. In recent years nearly every Latin American nation has announced an agrarian reform policy by which the fundamental structure of the agricultural economy may be changed. These reforms have two objectives. One objective is rural modernization, basically to improve overall productivity, especially improved yields per unit of land and per worker. The second objective is "social justice," which is interpreted differently from country to country but in general means the opportunity for each worker to own land, to enjoy greater income, and to participate in the benefits of modern society such as electricity, schools, and medical care.

Cause of Rural Poverty

Most of Latin America's rural problems are traced to the colonial system of land tenure and social organization. From almost the first day of European conquest, a process of land alienation began by which the conquerors and those who came later acquired large landholdings (*haciendas, fazendas,* and *estáncias*). In time, the countryside came to suffer from both *latifundia* and *minifundia.* Latifundia are large landholdings held by a small percentage of the population. Minifundia are small landholdings held by most of the people—the small farmers. The owners of large estates usually had the most productive land, used it inefficiently, and had a paternal relationship with their workers. In contrast, most minifundias were on poorer land and were not large enough to provide full-time employment. The small farmer was constantly faced with day-to-day problems of adequate food supply and had neither the capital, knowledge, nor opportunity to apply advances in technology should he have even learned that such technology existed.

Over time, as the rural population grew, the pressures on the land increased; but for the small farmer there were few alternative occupations. Through inheritance, the already small land parcels were further subdivided until they could be divided no more. Some of the younger generation had to migrate. Some went to sparsely settled areas, but they were often owned by a large land owner, were isolated, and if in the tropical lowlands were disease laden. Others, in search of work, went to the large

landowners, but oversupply of workers meant low wages even if jobs were available. More fortunate were those lucky enough to find employment in mines or in the latter nineteenth century on foreign-owned railway construction and on coastal plantations. In the twentieth century, urban development has provided an additional outlet for rural over-population.

Disparity in Income

For most of Latin America, the rural scene remains one of unemployment and underemployment. Great disparities in income exist. In 1969, the Economic Commission for Latin America concluded that 68 percent of the rural workers had an average yearly income of about $275, 30 percent earned $875, and 2 percent averaged $6,000.[1] There probably has been no significant change in the interim. The 2 percent owns more than 50 percent of the farmland, which averages about 1,000 acres (405 hectares) per person. At the other extreme, 25 percent of the rural workers hold but 2.4 percent of the farmland, about 5 acres (2 hectares) per family. It is estimated that each small-farm family has a labor force equal to two man-years. The Commission suggests that these farm families need at least 25 acres (10 hectares) to utilize the available labor supply fully and to provide an income of $600 to $800.

Serious income disparity in the rural sector where a majority of the population scarcely participates in the economy is detrimental to the economy as a whole. Other economic sectors have limited market outlets, since the purchasing power of a large part of the population is low. Low incomes also retard economic growth in agriculture itself. Lack of capital precludes the use of chemical fertilizers and high-yielding seed. Labor-saving devices such as mechanical equipment cannot be purchased, keeping yields per person low. Without alternative employment, labor-saving devices might very well accentuate unemployment problems.

Characteristics of Agrarian Reform

Agrarian reform has been heralded as the answer to the plight of the rural population. Most reform

[1]Economic Commission for Latin America, *Second United Nations Development Decade: Agricultural Development in Latin America* (New York: United Nations, 1969).

programs have social as well as economic aspects. Sometimes these two objectives are in conflict. This conflict is seen in the most widespread and popular part of reform, land redistribution. If economic aspects were the sole determining factor in improving agricultural productivity, land redistribution would probably play a minor role. Small land parcels would be consolidated into far larger units so that mechanization and capital could be accumulated. Also, it is easier to spread technologic knowledge among a small group of entrepreneurs than to a large mass of almost illiterate peons. Indeed, where land redistribution has been accomplished, productivity has often declined. Socially, however, land redistribution is most important. There is an egalitarian motive that the poor have a right to land control. Moreover, it is hoped that ownership of land may facilitate other changes that in the long run will be advantageous. In any event, political and social pressures demand land redistribution. Failure to do so would lead to political unrest and armed revolts. Indeed, armed revolution has already occurred, for example, in Mexico, Bolivia, and Peru.

Another aspect of agrarian reform is government sponsorship of rural cooperatives. These cooperatives form the contact point between the government and the small farmer. Often credit is provided individual farmers through the cooperative for seed, fertilizers, and insecticides. The cooperative may purchase a tractor and other equipment that any single farmer could not afford or use effectively. The cooperative is a distribution point where technologic information is passed on to the farmer. Such information may be in the realm of how to farm better, how to build a better house, how to dig a water well, or, in the area of health, how to deal with food preservation and nutrition. Finally, most cooperatives act as agents for their members. By buying in bulk, lower costs per unit often are obtained, and by selling in bulk, a better price is secured for farm commodities. In many respects the cooperative lessens the dependence on the large landowner and makes the farmer more independent. Lack of trained personnel and funds, however, has hampered cooperative effectiveness.

The reform movement is changing the relationship between the large landowner and his workers. The patron system, in which the employer provides housing, a plot of land, and other noncash allowances, is giving way to a wage system. Debt peonage and in-dentured work arrangements have been largely eliminated, and unionization of farm labor is increasingly important. To many workers the destruction of the patron-peon relationship has resulted in loss of security. The obligations of the owner to care for his sick and old workers no longer exist. Indeed, the necessity to pay wages has encouraged many landowners to mechanize their operations, reducing the need for manual labor.

Improvement of the infrastructure is another aspect of agrarian reform. Cooperatives are one type of infrastructure. Almost everywhere farm-to-market roads are being constructed, and new roads are pushing back the agricultural frontier. New and improved roads reduce the cost of moving goods; the farmer gains accessibility to a larger market area, and in return products become available to him at lower cost. In some countries, for example, Mexico and Peru, irrigation projects have increased the arable land. In others, land drainage has brought new land under cultivation. Nearly everywhere rural education programs have been instituted.

The Status of Reform

Most reform programs are designed for gradual implementation. Yet so slow has been the pace of agrarian reform that in many countries it exists more in name than in reality. Technical problems such as surveying property lines, building roads, the high cost of reform, and the resistance of the landed gentry have all acted to retard agrarian reform. Where reform policy has been implemented, it is often a matter of land redistribution.

Land for distribution has come from three sources. First, some land has been expropriated from foreign ownership; in many cases, however, the lands are located away from the areas of heavy population pressure. Second, some of the locally owned large landholdings have been nationalized and given to the workers. In Peru, Bolivia, Mexico, and more recently Chile, land for redistribution has come from the local gentry. Finally, most countries have an agricultural frontier beyond which is unused land (Fig. 21–11). Colonization of these sparsely settled areas is supported by the landed aristocracy, who see the frontier as a safety valve for surplus populations; they hope by encouraging colonization to hold onto their own lands. The peons see the frontier as a place of opportunity. The government views colonization

21–11   *Latin America: Agricultural Frontiers*

Miles

0	200	400	600
0	322	644	966

Kilometers

as desirable from two standpoints. First, it partially satisfies the demand for agrarian reform without economic or social unrest. Second, the sparsely settled areas if developed can contribute to the economic growth of the nation and may lead to the use of non-agricultural resources.

Colonization has some disadvantages. It is expensive. All kinds of infrastructure must be provided: roads, schools, towns, and banks. Often the area colonized has an environment different from what the settlers previously experienced. Different crops and land management techniques may be required that are unfamiliar to the settler. Most colonization zones are in the rainy tropics. Unless carefully controlled, endemic diseases can spread rapidly. The malaria-carrying mosquito once very susceptible to DDT is now increasingly resistant to the chemical. Some authorities believe destruction of the covering forest may lead to serious and irreparable damage to the soil.

During the last decade, over 1 million landless families obtained title or access to land through the agrarian reform movement. These families represented about 10 percent of the total number of potential beneficiaries in 1965. Unfortunately the rural population, in spite of migration, is growing at a rate faster than land is being distributed. Agrarian reform will be a continuing process in Latin America for many years and a point of internal conflict within each nation.

Nationalism and Multinational Integration

Nationalism, the allegiance to the nation-state, and the move toward multinational economic integration are, in some respects, complementary forces. In other aspects, however, these two forces are contradictory. Both are of increasing importance in Latin America and are shaping the region's development.

Nationalism

A spirit of national identity has been a missing prop in most Latin American countries. Cultural pluralism and a diffuse and illusive national allegiance remain regional characteristics. In some coun-

tries, however, nationalism has asserted itself. Mexico, since the 1910–1917 revolution, has successfully pursued a policy of national identification and unification. Argentina, under Perón's first tenure as chief of government, experienced a swell of nationalistic fervor for a time. Since about 1950, every country in Latin America has sought to subordinate local and provincial loyalties to national interests. In many cases they have been at least partly successful.

Nationalism takes several forms. In Mexico, a determined blending of Spanish and Indian cultural heritage has been the cornerstone of a nationalistic philosophy. A similar sentiment has characterized a segment of Peruvian and Bolivian political thought but with less success. In Brazil, national unity is fostered by a frontier spirit, the challenge to tame the west. Movement of Brazil's capital from Rio de Janeiro to interior Brasília is symptomatic of frontier nationalism. In Cuba, nationalism is attained by intensive indoctrination programs and economic and political reorientation.

Nationalistic attitudes are heightened in many countries by real or imagined threats from the outside and the need for internal unity. Often the United States is characterized as an adversary, and the international Communist movement is considered by some to pose a threat to national security. Recently the question of territorial control of offshore resources has become a rallying cry for national interests. Many Latin American nations have tried to extend their territorial sea limits to 200 miles (320 kilometers) from the shore. Finally, nationalism is expressed in the desire to have greater control over national resources. Foreign ownership of the natural resource endowment is being attacked, and a very conscientious effort is being made to assure that foreign corporations cannot exert much influence.

One goal of nationalism is self-sufficiency. No nation today can be completely self-sufficient and still provide the material benefits demanded by its citizens. A balance is usually struck between the desire for higher standards of living facilitated by trade and the desire for independence of external ties. Industrialization, particularly heavy manufacturing, is embodied in the spirit of nationalism. In Latin America, as in many other poor regions, the symbol of industrialization is an iron and steel mill. Mexico, Chile, and Brazil have both the raw materials and the market to support a steel plant. It is less clear if Argentina, Colombia, and Peru can economically support a steel industry. Few steel men believe a plant is economically feasible in Panama, Guyana, or Bolivia, yet serious discussions within the respective governments have revolved around ideas of such a plant. A self-sufficiency goal seriously inhibits multinational economic integration.

Multinational Integration

The European Economic Community (EEC) clearly has demonstrated that regional specialization can foster rapid economic development. Tariffs among member nations were abolished allowing the easy movement of finished products and raw materials to all member nations. Each nation can concentrate on producing those things for which it has an advantage and send its surplus to other member nations in exchange for goods it cannot readily produce. In Latin America, the success enjoyed by the European Common Market is viewed with great interest. The various governments reason that an organization similar to the EEC might spur development, yet the desire for each country to supply its own needs limits the application of regional specialization. Three multinational economic unions have been formed: the Latin American Free Trade Association (LAFTA), the Central American Common Market (CACM), and the Caribbean Common Market (CARICOM) (Fig. 21–12).

The Latin American Free Trade Association was formed in 1961 and presently has the following members: Argentina, Bolivia, Brazil, Chile, Colombia, Ecuador, Mexico, Paraguay, Peru, Uruguay, and Venezuela. LAFTA's basic goal has been to reduce tariffs among member nations. Progress has been slow, and the future of the Association is clouded. Trade among member nations has increased but not to the degree hoped. Two subgroups of LAFTA have been formed: the Andean Group and the La Plata Group.

The Andean Group has taken over many of the functions of LAFTA and is now the major catalyst in Latin American integration. Formed in 1969, members now include Bolivia, Colombia, Ecuador, Peru, and Venezuela. Nations within the group are committed to a multidimensional development program. Not only are there scheduled tariff reductions at rates faster than those of LAFTA, but also a common position is sought on several other aspects of develop-

ment. An industrialization program has begun in which certain industries are encouraged to develop or expand in specific countries with access to markets in all member nations. Agreement in principle is sought on treatment of foreign investments, an overall transport policy, and other means of strengthening the economic and social bonds among members.

Less progress has occurred in the La Plata Group. Formed also in 1969, its members comprise Argentina, Brazil, Uruguay, and Paraguay. This group is oriented toward cooperation in the use of natural resources, especially water, and improvements in the regional transport system. The goal of the La Plata group is a giant river basin development program for the watershed of the Rio de la Plata. Tariff reduction is not part of the group's goals.

The Central American Common Market (CACM) was established in 1960 and includes Guatemala, El Salvador, Honduras, Nicaragua, and Costa Rica. Patterned after the European Common Market, CACM has reduced tariffs among member nations and set up a common tariff structure on imports from other nations. The Central American Common Market nations, along with the United States, have funded the Central American Bank for Economic Integration to provide capital for a wide variety of development projects; transport and industrialization receive particular attention. Other treaties have been signed on common educational curriculums and texts, defense, and future formation of common economic and political policies. The 1969 "Soccer War" between El Salvador and Honduras disrupted the integration movement, and only in the mid-1970s have prospects for continuation of the CACM improved.

The Caribbean Common Market (CARICOM) was established in 1973 and is composed of former and presently associated British territories of the Caribbean. Members include the independent nations of Guyana, Jamaica, Trinidad, Barbados, and Grenada, and the colonies of Antigua, Dominica, Montserrat, St. Vincent, St. Kitts-Nevis-Anguilla, and St. Lucia. To date, a number of policy statements have been agreed upon and tariffs among members almost abolished. Aside from petroleum products from Trinidad, however, little trade occurs among members.

Lastly, the major exporting nations of bananas, coffee, and cacao have banded together in an attempt to control production and prices. By controlling the

LAFTA Members

Andean Group

La Plata Group

Other LAFTA Nations

CARICOM

CACM

Drainage Basin of the Rio de la Plata

Miles
0   200   400   600
0   322   644   966
Kilometers

21–12 *Latin America: Multinational Economic Organizations*

amount exported from each country, the producing nations can set their selling price. A recent example of this idea is the Organization of Petroleum Exporting Countries (OPEC) which has greatly increased oil prices throughout the world. Producers of tropical export crops point out that prices for their exports have not changed much in the last twenty years, but the finished products they import have doubled or tripled in price. These tropical export crop organizations have members not only in Latin America but also in Africa.

Summary

Latin America is in ferment. Many of the processes of change have become important only in the last twenty-five years, and the rate of change is accelerating. Change always brings uncertainty and confusion. As old systems are challenged, individuals and group relationships are disrupted. It is not clear what characteristics these new relations will have. The peon freed of patron control must make decisions he never before faced. Foreign corporations must reassess their investment opportunities. Nationalists must weigh the relative advantages and disadvantages of foreign investment in their country and how much interdependence they should permit in the multinational economic integration movement. Fundamental to these changes are population growth and urbanization-industrialization. How extensive these changes are and what direction they will take are questions that we can only ponder.

Further Readings

Department of Economic and Social Affairs, *Foreign Capital in Latin America* (New York: United Nations, 1955), 164 pp. A well-documented history of private investment in Latin America to about 1952. Department of Commerce, *Survey of Current Business* (Washington, D.C.: Government Printing Office). This monthly publication contains a wealth of statistical data on the United States economy. Each year, usually in the August or September issues, data are presented on United States corporate investment abroad. See also Swansborough, Robert H., "The American Investor's View of Latin American Economic Nationalism,"

Inter-American Economic Affairs, 26 (1972), 61–82; and Pinelo, Adalberto J., *The Multinational Corporation as a Force in Latin American Politics: A Case Study of the International Petroleum Company in Peru* (New York: Praeger, 1973), 180 pp.

Furtado, Celso, *Economic Development of Latin America* (translated by Suzette Macedo) (Cambridge: University Press, 1970), 271 pp.; and Prebisch, Raul, *Change and Development: Latin America's Great Task* (New York: Praeger, 1971), 293 pp. Furtado and Prebisch are two of Latin America's foremost development economists. Each traces the history of Latin America's economic trends and problems and analyzes the prospects and means of economic change.

Torres, James F., "Concentration of Political Power and Levels of Economic Development in Latin American Countries," *Journal of Developing Areas*, 7 (1973), 397–410; and Kleine, Herman, "The Evolving Pattern of Development," *Inter-American Economic Affairs*, 26 (1972), 92–96. These two short articles review some of the principal processes of economic development in Latin America. See also Scott, Robert E. (ed.), *Latin American Modernization Problems: Case Studies in the Crises of Change* (Urbana: University of Illinois Press, 1973), 356 pp.

Trewartha, Glen T., *The Less Developed Realm: A Geography of Its Population* (New York: Wiley, 1972), pp. 1–44. Trewartha begins his discussion of population with an overview of the less developed countries and then proceeds to a regional treatment. In the chapters on Latin America, he examines immigration, fertility rates, urbanization, and population characteristics of its various parts. Department of Economic and Social Affairs, *The World Population Situation in 1970* (New York: United Nations, 1971), 78 pp. This short but factually packed book is the first of a series on the world's population. Based on the most recent data available, projections to the year 2000 are made for population growth rates, total population, fertility levels, urban-rural distribution, education levels, and labor force.

Nationalism is expressed in several ways. See Stephansky, Ben S., *Latin America: Toward a New Nationalism* (New York: Foreign Policy Association, 1972), 63 pp.; Whitaker, Arthur P., "New Nationalism in Latin America," *Review of Politics*, 35 (1973), 77–90; and Edmunds, Dale C., "The 200-Mile Fishing Rights Controversy: Ecology or High Tariffs," *Inter-American Economic Affairs*, 26 (1973), 3–18.

The literature on agrarian reform is extensive. Some of the more recent studies include the following: Barraclough, Solon L., "Agricultural Policy and Land Reform," *Journal of Political Economy*, 78 (1970), 906–947; Branco, Raul, "Land Reform: The Answer to Latin American Agricultural Development," *Journal of Inter-American Studies*, 9 (1967), 225–235; Crossley, Colin J., "Continuing Obstacles to Agricultural Development in Latin America," *Journal of Latin American Studies*, 4 (1972), 293–305; Lockhart, James,

"Encomienda and Hacienda: The Evolution of the Great Estate in the Spanish Indies," *Hispanic American Historical Review*, 49 (1969), 411–429; Fedor, Ernest, *The Rape of the Peasantry: Latin America's Landholding System* (Garden City, N.Y.: Anchor Books, 1971), 304 pp.

The move toward international economic integration is the subject of the following: Balassa, Bela, "Regional Integration and Trade Liberalization in Latin America," *Journal of Common Market Studies*, 10 (1971), 58–77; Carnoy, Martin, *Industrialization in a Latin American Common Market* (Washington, D.C.: Brookings Institution, 1972), 267 pp.; Economic Commission for Latin America, *The Process of Integration Among the CARIFTA Countries* (New York: United Nations, 1971), 43 pp.; McClelland, Donald H., *The Central American Common Market: Economic Policies, Economic Growth, and Choices for the Future* (New York: Praeger, 1972), 243 pp.

Latin American Regions: The North

22

El Salvador, Honduras, Nicaragua, Costa Rica, and Panama (Fig. 22–1). All of these nations except Belize were under Spanish domain in colonial times. All except Belize and Panama have been independent for more than 150 years; all have rapidly growing populations.

Diversity characterizes many aspects of the region. Several racial groupings prevail ranging from the descendants of Aztec and Mayan cultures to whites in Costa Rica and blacks of European culture along the Caribbean coast. Many Indians continue the ways of their ancestors little changed four centuries after European contact. Cultural pluralism is one inhibiting factor in economic development. Economic diversity is exemplified by the mestizo farmer who cultivates a small hillside patch of maize, beans, and squash with hand tools, while a short distance away his children live in the city, wear European clothes, work in a factory, and buy their food in a store. Physically, Mainland Middle America varies from the steppe and desert environments of northern Mexico to the rainy tropics of southern Central America. Most of the region's population live in the highlands, although in this century other environments have witnessed an influx of people (Fig. 22–2).

So FAR WE HAVE VIEWED Latin America as if from a short distance from earth. From that perspective, we could see the broad patterns of the physical environment, trace mankind's occupation of the land, and perceive some of the more dominant contemporary trends. We now look more closely at Latin America's regions to appreciate differences from one place to another and to understand some of the development problems and potentials each part possesses. For our purposes Latin America is divided into six regions: Mainland Middle America, the Antilles, northern South America, Andean South America, southern South America, and Brazil.

Mainland Middle America

Mainland Middle America includes Mexico and the Central American nations of Guatemala, Belize,

Mexico

Mexico was one of the first Latin American nations to undergo many of the processes of change described in Chapter 21. Agrarian reform was instituted in the 1920s. Large amounts of foreign investment have been received, and Mexico early began a program of control. In some industries, such as the railways, the government bought out private companies and now operates them as national utilities. Mexicanization (majority ownership of corporations by Mexicans) rather than nationalization, however, has been the dominant trend. Industrialization and urbanization have progressed rapidly, and secondary and tertiary activities now produce the bulk of Mexico's wealth. Mexico has reached Rostow's "take-off" stage of development (Chapter 4).

In the area of social development, Mexico has made some dramatic advances. For more than fifty years, education has been a principal policy of the government. A large part of the government's revenue has

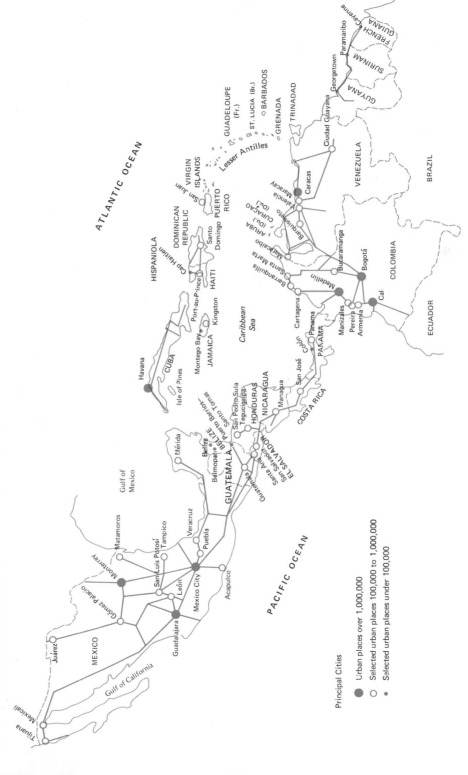

22-1 *Northern Latin America: Cities and Major Highways*

Principal Cities

● Urban places over 1,000,000

○ Selected urban places 100,000 to 1,000,000

· Selected urban places under 100,000

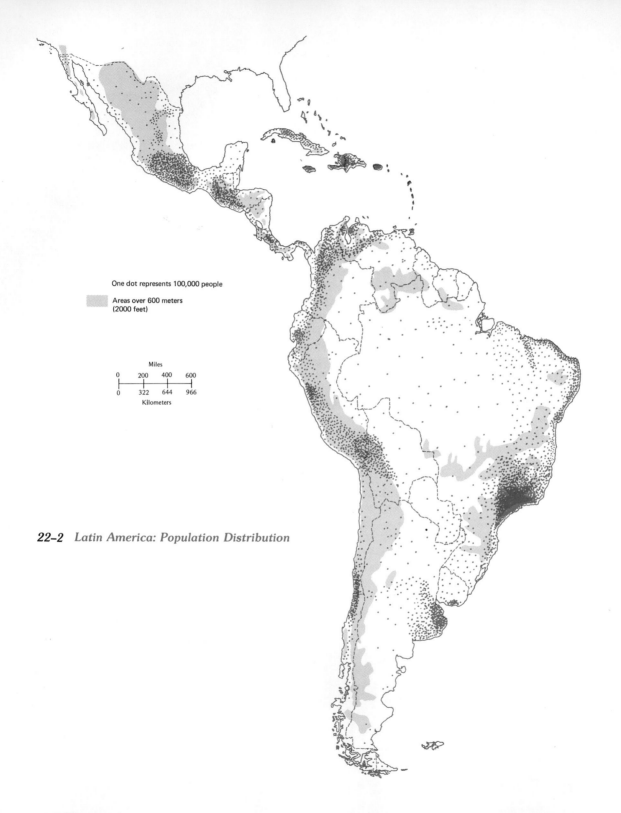

One dot represents 100,000 people

Areas over 600 meters
(2000 feet)

Miles
0   200   400   600
0   322   644   966
Kilometers

22-2   Latin America: Population Distribution

gone to education, 20 percent in recent years. Presently, 76 percent of Mexico's people are considered literate; in 1950, 56 percent were so classed. Another 5 percent of the government's expenditures is for public health. Improved sanitation including potable water, sewage facilities, and pre- and postnatal care are widespread, partly explaining the decline in death rates (Fig. 21–2).

In spite of a continued expansion of economic activity (7 to 9 percent yearly since 1960), increases in population have diluted economic gains on a per capita basis. Mexico's population has grown rapidly, especially since about 1930; it has doubled in the last twenty-five years (Fig. 22–3). By 1985, population is projected at 84.4 million. The rate of population growth has also increased greatly and is now one of the world's highest (see also Fig. 21–3).

Mexico's rapid rate of economic and social change has its roots in the late nineteenth and early twentieth centuries. Porfino Diaz came to power in 1876 and for thirty-four years ruled autocratically. Under the Diaz government, efficiency improved, and economic development was stimulated. Foreign investments, particularly from the United States, were encouraged. Mining and landownership laws were changed to benefit the foreign investor and the local aristocracy. Railway construction and public work projects provided the infrastructure to increase production and commerce. Diaz' policies favored the upper class, but the lot of the lower classes worsened. Indian communal lands and small properties were usurped and formed in large haciendas. Debt peonage was common, and education was reserved for the elite. By 1910, latifundia was highly developed, and the wealth of the nation was concentrated in the hands of only 3 to 5 percent of the population.

From 1910 to 1917, Mexico was racked by revolution and Diaz finally overthrown. More important, however, the revolution marked a fundamental change in national direction. The victorious forces committed themselves to a new national constitution that provided for universal suffrage and education, restriction of foreign and church ownership of property, a minimum-wage law, arbitration of labor

22–3 *Mexico: Growth of Population*

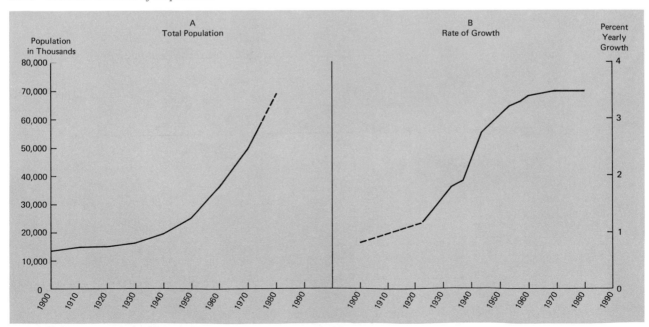

SOURCES: Latin American Center, *Statistical Abstract of Latin America* (Los Angeles: University of California, 1976); and *Economic and Social Progress in Latin America* (Washington, D.C.: Inter-American Development Bank, 1975).

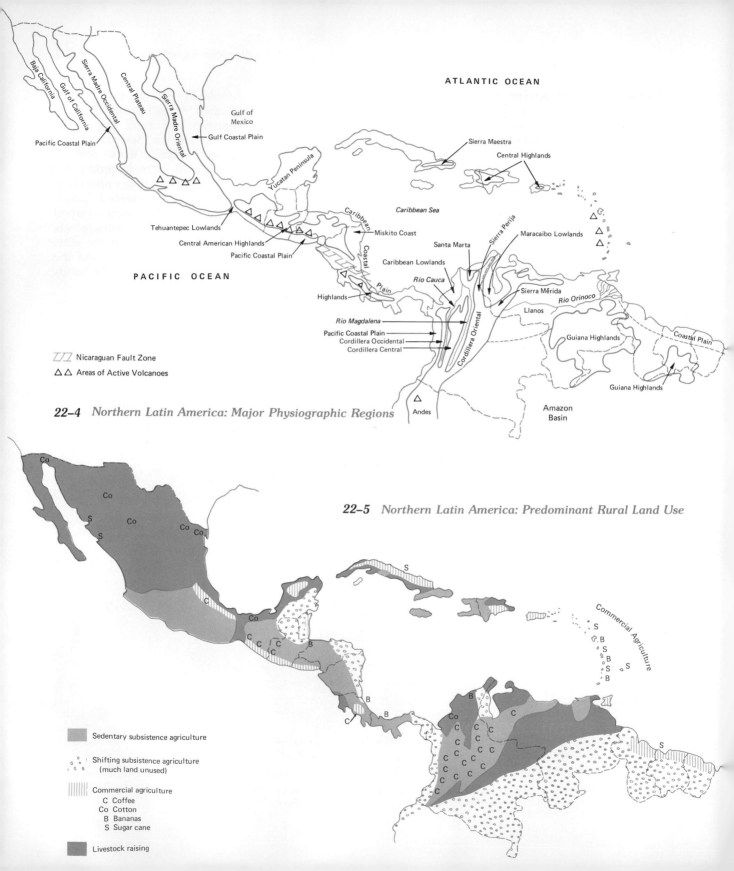

ATLANTIC OCEAN

Baja California
Gulf of California
Sierra Madre Occidental
Central Plateau
Sierra Madre Oriental
Pacific Coastal Plain
Gulf of Mexico
Gulf Coastal Plain
Yucatan Peninsula
Tehuantepec Lowlands
Central American Highlands
Pacific Coastal Plain

Sierra Maestra
Central Highlands

Caribbean Sea

Caribbean
Coastal
Plain
Miskito Coast

Santa Marta
Sierra Perija
Maracaibo Lowlands
Caribbean Lowlands
Rio Cauca
Sierra Mérida
Rio Orinoco
Llanos
Coastal Plain
Guiana Highlands

Highlands

Rio Magdalena
Pacific Coastal Plain
Cordillera Occidental
Cordillera Central
Cordillera Oriental

Guiana Highlands

Andes
Amazon
Basin

PACIFIC OCEAN

⧄⧄ Nicaraguan Fault Zone
△ △ Areas of Active Volcanoes

22-4 *Northern Latin America: Major Physiographic Regions*

22-5 *Northern Latin America: Predominant Rural Land Use*

Co
Co
S
Co
S
Co Co
C
Co
C C C
C
B
C C C
C
C
B
C
B
C
S
B
S B
S B
S B S
B
Co
C
C C C
C C C C
C C C C
C C C C
C C C
C
S

Commercial Agriculture

Sedentary subsistence agriculture

Shifting subsistence agriculture
(much land unused)

Commercial agriculture
C Coffee
Co Cotton
B Bananas
S Sugar cane

Livestock raising

disputes, and agrarian reform. Without exception, succeeding governments have adhered to these concepts.

National Unity

One of Mexico's most basic development problems has been the creation of national unity. The large unassimilated mass of Indians, one-half of the population in 1910, were looked upon as inferior by those of European culture. The postrevolutionary governments sought unity by heavily investing in basic education, an improved transport and communications network, and by extolling the virtues of Indian heritage. Mexican art and literature of the 1920 to 1940 period revered the dignity of manual labor and Indian ancestry. A large degree of national unity and allegiance is now an accomplished fact.

Agrarian Reform

Agrarian reform has become the keystone of Mexico's economic and social development programs. On the eve of the 1910 revolution, 95 percent of Mexico's rural families were landless. Land for the landless was one battle cry of revolution, and in the succeeding years land redistribution became a focal point of government policy. Fundamanetal to the agrarian reform movement was the *ejido* system. Under the *ejido* system, tenure rights for land expropriated from *haciendas* were given for communal land to villages. In turn, plots of village land were allocated to individuals. The alloted land was not privately owned, but the occupant had the right of use. The user could improve the land, build on it, work it, and will it to his heirs. This land, however, could not be mortgaged or sold. In a few areas, especially in irrigated areas, it was not feasible to subdivide an already operating unit. In these cases the *ejido* was operated as a unit with members sharing in the work and profits. Presently, about one-half of the population live and work on *ejidos*.

In recent years the agrarian reform movement has done more than redistribute land under the *ejido* system. Since 1950, various colonization projects have been started in the rainy tropical areas of the Gulf and Pacific coasts. In these projects private landholdings are encouraged. In northern and central Mexico, numerous irrigation projects have brought large areas of desert and steppe environment into agricultural production. Some of the larger projects

are near Gomez Palacio, Matamoros, Mexicali, and San Luis Potosí. Agricultural credit, farm-to-market roads, improved technology in the form of seed, fertilizers, and pesticides, and water supply, sewage disposal, and housing are all part of the movement. In the early 1970s, some *ejidos* have been oriented to provide facilities catering to tourism and small-scale manufacturing.

Not all hacienda land has been expropriated and turned into *ejidos*. In both northern and southern Mexico, some large land units remain. Even where *ejidos* have been organized, the hacienda owner was permitted to retain at least 100 hectares (247 acres) and often more, so private property is still a very important part of Mexican life.

Agriculture

Mexico's numerous physical environments facilitate the production of many different crops (Figs. 20–3 and 22–4). Governmental policy also supports crop diversity. The Aztec crop trilogy of maize, beans, and squash remains the principal land use for much of central Mexico and provides the bulk of the rural population's diet (Fig. 22–5). These cultigens are also important subsistence crops in other parts of Mexico. Commercial agriculture varies regionally. Wheat and barley are important land uses in central Mexico. In nothern Mexico, cotton is grown on irrigated land and is one of the nation's leading exports. Elsewhere in the north most of the land is used for cattle. Along the Gulf Coast, rice, citrus, bananas, and sugar cane along with cattle raising are major commercial activities. On the eastern flanks of the Sierra Madre Oriental and in the Chiapas Highlands, coffee is a major export crop.

The Yucatan Peninsula, around Merida, is one of the world's leading suppliers of henequen, a hard fiber used in bagging, carpeting, and twine. Much of the peninsula and adjacent southern Mexico is still a frontier area.

Mining

During the Diaz period, foreign investment in mining was stimulated. Old mines were revitalized and new mines developed. Since the revolution, mining has continued to play an important economic role, and ownership by national companies has been encouraged. Today, Mexico is of world importance as a producer of lead, zinc, silver, sulfur, copper, and

In northern Mexico irrigation projects have opened formerly desert areas to crop production. Ejidos in this part of Mexico are often operated as a single unit rather than in small individual plots characteristic elsewhere. (United Nations)

a number of minor minerals. Coal (near Sabinas) and iron ore deposits (near Chihuahua) support a growing iron and steel industry (Monterrey, Monclova, and Mexico City) that supplies most of the national demand (Fig. 22–6). Mexico's steel industry began in 1901 during the Diaz regime, financed by private capital. Today, the industry is largely owned by the government. During the Diaz period, Mexico became a major petroleum-producing nation, much of which was exported. After nationalization in 1938, production greatly decreased but in succeeding years has recouped its former level, making the nation self-sufficient. It is expected that large quantities of oil will be exported by 1980.

Industrialization and Urbanization

Much of Mexico's rapid economic growth is attributable to industrialization. Like most other nations,

Mexico's manufacturing was initially based on processing of agricultural commodities such as textiles, milling, and furniture making. Cottage or household manufacturing was particularly important. Although these activities are still important, processing has moved from the home to the factory where machines have replaced hand labor. Since about 1940, other, more complex forms of manufacturing have made their appearance. Today, in and around the large cities of Mexico City, Guadalajara, Monterrey, Juárez, Puebla, and Leon, many modern factories produce a multitude of consumer and capital goods. Some examples of consumer goods include auto assembly, pharmaceutical products, refrigerators, washing machines, and television sets. Capital goods include iron and steel bars and sheets, chemicals, cement, machinery, and fertilizers. The existence of many raw materials, adequate energy resources, and

Petroleum
Zone

Coal

Iron

Bauxite

Other Metals
(Silver, Lead, Zinc, Copper)

S   Sulfur

● Selected urban places over 1,000,000

○ Selected urban places 100,000 to 1,000,000

• Selected urban places under 100,000

22–6 *Northern Latin America: Principal Mineral Production Areas*

an abundant labor supply is a prime factor in the rise of manufacturing. Manufacturing now contributes more to the GNP than agriculture, although more people are employed in agriculture.

Urbanization has progressed rapidly. Greater Mexico City now has a population of over 8.5 million and is growing at a rate of 3.5 percent annually. Guadalajara and Monterrey are each over 1 million and growing at rates greater than that of Mexico City. Along the United States–Mexican border the cities of Juárez, Tijuana, and Mexicali have over one-quarter million populations and are growing at rates of about 5 percent annually. Many of these cities are now facing urban problems similar to those found in the industrialized world. Mexico City, for example, has problems of water supply, sewage disposal, and smog.

With urbanization have come economic interdependence and service activities. Service activities are the most rapidly growing sector of the economy.

Manufacturing in Mexico contributes more to the national economy than agriculture. Many large and modern plants, such as this paper factory, are found in major cities and occasionally in the nearby countryside. (United Nations)

Service activities also have grown in response to increasing tourism. Tourists, largely from the United States, each year spend over 1 billion dollars. Acapulco is heavily oriented to tourism as are many areas along the highways from Mexico City north to the border.

Present Status

Mexico is one of the most advanced of all Latin American nations. The development process began more than fifty years ago. The population has been largely unified and educated. A variety of natural resources and their integrated use have led to diversity of production and multiple exports; Tampico and Veracruz are the nation's principal ports. In contrast to most other Latin American nations, no one or two products dominate the export list. The economy is becoming increasingly commercialized. A well-developed infrastructure largely in the north and central sections has encouraged commercialization. An estimated 40 percent of the people, largely urban dwellers, live in reasonable comfort and form a growing middle class. These achievements have been attained under a stable government committed to economic and social change in an orderly manner.

Today, there are three Mexicos: (1) in rural areas there is still a large group of poor peons who live in the old ways; (2) in the urban areas are middle- and upper-class citizens who live as other members of a modern, industrialized society; and (3) there is the transitional group of villagers and rural migrants who live in urban shantytowns and are caught between the traditional and modern worlds.

Central America

Like Mexico, Central America is an area of rapid population increase (3.2 percent). Nearly 21 million people inhabit the isthmus, and it is projected that by 1985 the population will grow to 28 million. Unlike Mexico, mineral wealth is limited, and manufacturing is important only in the larger cities; agriculture is the principal source of employment. Small size both in area and in number of inhabitants is a limiting factor in development. Moreover, agrarian reform and modernization of the economy are only weakly supported. In Rostow's economic growth stages, Central America is somewhere between stage one (traditional society) and stage two (preconditions for take-off).

Central American Common Market

Ever since independence from Spain, there have been occasional attempts to reunite the colonial Captaincy-General of Guatemala. The most recent unification move is the Central American Common Market (CACM) of which Guatemala, El Salvador, Honduras, Nicaragua, and Costa Rica are members.

To many, CACM is a logical step to overcome the difficulties of each nation's small size. They reason that many economic activities are precluded by the limited market within individual countries. By eliminating tariffs among member nations, manufacturing plants can be constructed to serve all members. There is a minimum number of customers (a threshold) for every activity. Consequently the more customers there are, the greater number of activities is possible. Looking back to the discussion on resources in Chapter 1, we can see that by expanding the market, resources may be created. Similarly tariff elimination can lead to regional specialization. Each nation no longer needs to strive for self-sufficiency but can produce those things for which it has an advantage and trade for whatever it cannot readily produce itself. Theoretically the common market should stimulate economic activity and spur growth, sorely needed in the light of a rapid rate of population growth.

From a practical standpoint, however, the common-market idea in Central America has certain problems. One problem is the outlook of the population. Few have a supranational point of view seeing Central America as a unit and economic integration as a useful goal. Many are highly nationalistic and place their own country's welfare above the goals of the common market. Governmental policies are often protectionist and contradictory to common-market goals. Other Central Americans have a still more restricted view and pay primary allegiance to their local community. The large Indian population of Guatemala represents this latter view, and similar feelings are found in many outlying areas of all Central American nations. A second problem is that the economies of the member nations are more competitive than complementary. Only a few products make up the bulk of the exports for each country, and

Table 22–1 *Principal Exports of Mainland Middle America*

Country	Commodity	Percent of Total Exports by Value
Mexico	Cotton	6
	Sugar	6
	Coffee	6
	Shrimp	4
Guatemala	Coffee	33
	Cotton	11
	Bananas	6
Honduras	Bananas	34
	Coffee	20
	Wood	14
El Salvador	Coffee	45
	Cotton	10
Nicaragua	Cotton	23
	Meat	16
	Coffee	13
Costa Rica	Bananas	28
	Coffee	28
Panama	Bananas	98
	Refined petroleum	18
	Shrimp	12
Belize	Sugar	45
	Citrus	17

SOURCE: Latin American Center, *Statistical Abstract of Latin America* (Los Angeles: University of California at Los Angeles, 1976).

those exports are generally the same (Table 22–1). Central American nations have little surplus to trade among themselves. Third, rivalry among nations has greatly hampered common-market development. The "Soccer War" in 1969 between El Salvador and Honduras completely halted the movement of goods among member nations and greatly weakened the already troubled market. Also, Guatemala and Costa Rica, the two strongest members, have continued to duel over leadership.

Guatemala

Guatemala is the largest of the Central American nations (6.4 million) but is also one of the poorest. It is really four nations in one. The northern one-half is sparsely populated and forms the agricultural frontier. Into this area roads are being constructed and colonization projects established. The highlands which parallel the Pacific Ocean contain the bulk of the population. In the western highlands live most of the Indian population who follow traditional ways of life and still possess many customs of their Mayan ancestors. In and around Guatemala City, mestizos and whites follow European life styles. Guatemala City (825,000) is the only true modern city of the nation and is the social, economic, industrial, educational, and political center. In 1976, Guatemala City and much of the highland area experienced severe damage from a series of earthquakes. Some 22,000 people were killed and whole villages were leveled. Along the southern coast and slopes of the adjacent mountains lies the main commercial agricultural zone. Cotton, sugar, coffee, and cattle are widely grown and raised. Formerly this area also produced bananas for export, but this land has been redistributed and formed into small farms. Bananas are now raised for export only near Puerto Barrios–Santo Tomas on the Caribbean.

El Salvador

El Salvador is the most densely populated of all Central American nations. The results of heavy population pressures are seen in eroded hillsides, the flow of rural migrants to shantytowns in the principal cities of San Salvador (340,000) and Santa Ana (100,000), and the migration of Salvadorans to less densely settled areas in adjacent Guatemala and Honduras. The occupation by these migrants in disputed territory along the El Salvador–Honduras border was the cause of the 1969 war between these countries. As in other Mainland Middle American nations, population growth is rapid (3.2 percent). El Salvador's population of 4.3 million is expected to be 6.0 million by 1985. In contrast to other Central American nations, El Salvador does not have an agricultural frontier. Increased agricultural production must involve higher yields per unit of land and not expansion of cultivated areas. The traditional landholding patterns have not been broken, and little has been done to improve agricultural efficiency.

Rather, emphasis is on industrialization, and consequently the success of the Central American Common Market is vital to El Salvador. El Salvador, however, does not possess many attributes for manufacturing. No minerals of significance are known; power supply must be imported; the only local raw materials are those of agriculture; and the labor supply, although abundant and cheap, is not skilled.

Honduras

Honduras, with an area six times that of El Salvador, has a population of only 3.3 million. Some parts of Honduras are very sparsely settled. Outside the major cities of Tegucigalpa (225,000) and San Pedro Sula (120,000), where services and limited manufacturing provide employment opportunities, agriculture is almost the sole occupation. In the highlands, subsistence agriculture, livestock raising, and commercial coffee growing are the dominant activites. Along the coast around San Pedro Sula, foreign-owned banana plantations contribute much of the nation's exports. Aside from these plantations the rural scene is characterized by low productivity, poor infrastructure, uneven land distribution and use, limited storage, and few credit facilities. Except for San Pedro Sula, which is a thriving, growing city, other urban places, including Tegucigalpa, present an impression of stagnation.

Nicaragua

Nicaragua has had a long and stormy history. As recently as 1972, a disastrous earthquake destroyed over 80 percent of Managua (population 400,000), killing over 6,000 persons and leaving 200,000 homeless. A conservative estimate of the physical damage is 850 million dollars. In past times Britain tried to gain control of the Miskito coastal zone, and in the nineteenth century an American adventurer, William Walker, briefly captured the capital. In the early part of the twentieth century, United States marines occupied part of the nation for a time. Internal conflicts between liberals and conservatives contributed to national instability. Stability finally came to Nicaragua in 1937 when the Somoza family gained control of the government and has held power ever since.

Nicaragua is basically an agricultural nation with most of the production coming from the lowland area in the Managua area. Much of eastern Nicaragua is sparsely populated and underlain by sandy, unproductive soils. Banana plantations, so common along the isthmus' east coast, have never been important in Nicaragua.

Costa Rica

Costa Rica differs from other Central American nations in several aspects. More than any other Central American nation, Costa Rica has developed a spirit of national unity. Over 85 percent of the population is literate; the infrastructure is well developed; politically the government is democratic and forward-looking with a history of stability. Agriculture employs nearly one-half of the labor force, and large landholdings exist. Yet productivity is relatively high, and small- and medium-sized, owner-operated farms are widespread. Within Costa Rica's small area, there is considerable regional specialization. In the highlands around San José, high-quality coffee is grown; on the fringe of the highlands dairying is also important. Along the coast are banana plantations and cacao, and in the northwest beef cattle are raised. Centered in and around San José, manufacturing is limited but expanding. As with other Mainland Middle American nations, population growth is rapid. In Costa Rica many rural people are moving into the rainy tropics and to the San José area. Costa Rica presents a picture of progress, yet the high population-growth rate spells the need for continued economic progress.

Panama

Panama owes its existence to United States support in its revolution of 1903 and its economic viability to the Panama Canal, built and controlled by the United States. The ten-mile-wide (16 kilometers) Canal Zone divides Panama. The eastern part (Darien) is little developed. The western part contains numerous banana plantations along the coast and beef cattle, rice, and staple food crops in the interior. Tertiary activities in and along the Canal Zone contribute more than 55 percent of the total GNP and are the most rapidly growing sector of the economy. Panama City (420,000) on the Pacific Coast and Colón (70,000) on the Caribbean Sea are the major urban centers. Both cities cater to tourists and transit passengers by offering duty-free goods.

The area in and around the Canal Zone is truly a "crossroads of the world." English and Spanish are spoken by a majority of the population, and United

The Panama Canal along with the United States-controlled Canal Zone divides Panama into two parts. The canal provides the major focus of Panama's economic activities. (United Nations)

States currency is used everywhere. Since the canal acts as a funnel, finished products and raw materials can be brought together in Panama for transshipment elsewhere and for processing. Manufacturing, however, is only slightly developed. Many of Panama's development problems revolve around its dispute with the United States over the canal and the lack of an infrastructure, particularly roads, in other parts of the nation.

Belize

Belize, a British colony, is scheduled for independence soon, but its viability is in question. Its popula-tion numbers only about 130,000 living in an area of 8,866 square miles (22,963 square kilometers). Much of the land is little used. For many years the nation's principal exports were wood and other forest prod-ucts. Within the past twenty years, however, sugar and citrus fruit have come to dominate the export market. The English heritage of the predominantly black population is evidenced in language and Prot-estant religion. The city of Belize (about 35,000) is the principal urban center. In 1961, Hurricane Hattie struck the town, killed over 250 people, and destroyed much property. Shortly afterward the government decided to build a new capital in the interior; this new center is Belmopan, and its location is designed to stimulate occupation of the backlands.

The Antilles

Aside from being islands with a heritage of conflict among Spain, France, and Britain, the Antilles differ from mainland Middle America in three important ways. First, except for Haiti, export production dominates the economy of all islands. The Antilles, more than any other part of Latin America, are dependent upon trade with the industrialized world, not only as buyers of Antillean goods, but also as suppliers of products needed. Inflation, recession, and good economic times in the industrial nations are quickly reflected in the Antilles. Although many islands of the Antilles are politically independent, they remain in essence economic colonies. Second, the pre-Columbian Indian population was destroyed early in the European colonization period and has left little imprint on the present landscape. African slaves were introduced as a labor supply, and their descendants now form the principal population element in the Lesser Antilles, Jamaica, and Haiti, and a minority but significant group in Trinidad, Cuba, Dominican Republic, and Puerto Rico. East Indians also form an important segment of Trinidad's society. All of the population except for Haiti's has a European culture base; Haiti has a mixture of African and European cultures. Third, the Antilles are at the mercy of hurricanes. These strong tropical storms begin in the Atlantic somwhere east of the Lesser Antilles, move generally westward increasing in strength, and once in the Caribbean Sea generally curve northward through the Greater Antilles. From June to December, the danger of hurricanes is very real. Many of the rural dwellers build small, inexpensive homes accepting nearly a total loss if caught in the path of a hurricane. Crops grown in this area of mainly tropical wet and dry environment are usually those that can bend with the wind (sugar cane) or can be quickly brought back in production (bananas).

The Greater Antilles comprise the islands of Cuba, Hispaniola, Jamaica, and Puerto Rico. Each has developed economically in different directions, but only Puerto Rico has made substantial progress in increased per capita GNP. According to Rostow's stages,

Haiti is in stage one (traditional society); Cuba, the Dominican Republic, and Jamaica are in stage two (preconditions for take-off); and Puerto Rico is in stage three (the take-off).

Cuba

At the turn of the twentieth century, Cuba was a poor colonial remnant of the once great Spanish Empire. For more than thirty years sporadic revolts had disrupted the economy and had made much of the population destitute. Land had little value, and much of it was used for cattle ranching or only occasionally was cultivated. United States protectorship quickly changed Cuba. Foreign investors bought large sections of land and transformed Cuba into the world's leading exporter of cane sugar. Cuba's proximity to the United States also fostered tourism, and Havana became a major vacation spot. From 1900 to 1959, Cuba was an economic "colony" of the United States. On the island a well-developed infrastructure was built, sanitation facilities were greatly expanded, and average per capita incomes were substantially increased. Much of the wealth produced, however, was exported as corporate profits, and a substantial disparity in income between the rural poor and the urban upper class existed.

In 1959, the insurgent army of Fidel Castro, with much popular support, gained control of the country. From that time Cuba has undergone a social and economic reorientation. Following a break in diplomatic relations with the United States, Castro turned to the Soviet Union for assistance and trade. Foreign investments were nationalized; most locally owned property was confiscated. The economy was directed along Marxist lines with a central planning council to direct Cuba's economic and social objectives. Today, its goals are to increase export crop production, mainly sugar, and expand agricultural production for the domestic market to make the nation more self-sufficient. The council also supervises construction of rural and urban housing and expansion of electricity and potable water for Cuba's poor. Schooling is emphasized for both young and old.

There is little doubt that many in Cuba are now better off than before Castro took over. The former upper and middle class, the skilled and entrepreneur groups, however, have lost much, and many have escaped or emigrated elsewhere. Overall, economic growth has not kept pace with population growth.

The degree of success experienced by the Castro-Marxist program in Cuba is being watched closely by other Latin American nations.

Puerto Rico

Puerto Rico has followed a different socioeconomic development track. Until the 1940s, Puerto Rico had progressed little since becoming a dependency of the United States. Unlike Cuba, with its large area of level land and excellent soils for sugar cane, Puerto Rico received United States corporate attention only along the level fringe of the island. Interior Puerto Rico is hilly and mountainous.

In the post–World War II period, Puerto Rico began a period of continuing economic growth through a program called Operation Bootstraps. Operation Bootstraps is a three-pronged development plan. The first prong is industrialization. Like other Caribbean islands, Puerto Rico contained few raw materials other than those of agriculture, few power sources, a small market, and a cheap but unskilled labor supply. Industry was attracted by tax exemptions, government-training programs for labor, and by a strenuous governmental advertising effort to attract the industrialist to the island. So successful was the industrialization movement that by 1956, manufacturing produced more wealth than agriculture. Much of the manufacturing is centered around San Juan.

Agricultural improvement was the second prong of the program. Rural experiment stations were established which with the help of soil conservation agents introduced better land-management practices. Dairying and truck gardening, new kinds of land use, competed with the traditional crops of coffee in the highlands and sugar cane in the lowlands. Agricultural productivity was greatly increased. In more recent years, rural development has centered on improving housing and bringing water, electricity, and schooling to the farm family.

22–7 *Puerto Rico and Cuba: Per Capita GNP*

SOURCES: Latin American Center, *Statistical Abstract of Latin America* (Los Angeles: University of California, 1974); and *Population Data Sheet* (Washington, D.C.: Population Reference Bureau, 1977).

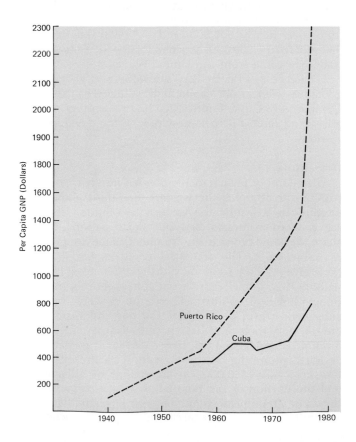

	Haiti	Dominican Republic
1. Annual per capita GNP ($)	180	720
2. Population density (per sq. km.)	165	105
3. Population growth rate (%)	2.0	3.5
4. Infant mortality (%)	15	9.8
5. Life expectancy at birth (years)	50	58
6. Percent of population literate	26	68
7. Annual per capita GNP growth rate (%)	0.4	2.5
8. Percent of population urban	22	41
9. Percent of labor force in agriculture	77	61
10. Daily per capita food supply (cal)	1,730	2,120

Table 22–2 Selected Differing Characteristics Between Haiti and Dominican Republic

The third prong is tourism. Again the government played a major role by sponsoring hotel construction and by an extensive promotion campaign. In 1953, 118,000 persons visited the island; by 1963, tourists numbered nearly 500,000; and by 1973, the figure had grown to 1,000,000. Much of Puerto Rico's tourist trade has come from the United States.

The success of Puerto Rico's development program is readily seen in per capita GNP values (Fig. 22–7). Compared to Cuba, Puerto Rico has made great progress, and Puerto Rico is now part of the developed world.

Hispaniola

The island of Hispaniola is divided into two countries. Haiti occupies the western one-third and the Dominican Republic the eastern two-thirds. There are marked contrasts between these nations that share the same physical environment; some are given in Table 22–2. The statistical comparisons of the table give only a partial view of the differences. Haiti with its Afro-French culture but almost entirely black population is a country of extreme poverty, small subsistence farm plots where the hoe and machete are the only tools, and has an aura of decay. Deterioration is evidenced in the roads and irrigation systems, plumbing and water supply in the city of Port-au-Prince (350,000), in the erosion of hillsides, and the decline of the population's social organization. The Dominican Republic has a Hispanic flavor; and although there are areas of hoe and machete subsistence agriculture, there are also areas of large farms with the plow and tractor. Road transport is especially good, and the infrastructure is improving.

Haiti under the French was a major exporter of cane sugar and coffee. Large landholdings under metropolitan ownership with mulatto overseers and black slaves were widespread. With independence, the French residents were expelled or killed, and their lands were progressively subdivided into smaller and smaller plots as the population increased. Over the years land originally devoted to export crops has gradually given way to subsistence agriculture. Alternative periods of anarchy and dictatorship prevented economic growth from keeping pace with population growth. Today the nation is faced with a dense population, moderate population growth, little capital, few resources, and a government unable to reverse a downward spiral of economic decay. Haiti is a good example of downward circular causation (see Chapter 4).

The Dominican Republic for most of the twentieth century has had a stable government. United States marines and government officials ran the nation from 1916 to 1924, and from 1930 to 1961, the dictator Trujillo ruled the nation as his own private fiefdom. Under Trujillo economic productivity was greatly expanded. Roads were built to facilitate production and open sparsely settled areas; foreign investment was encouraged, and colonization, particularly along the Haitian border, undertaken. At the same time all political opposition was ruthlessly crushed, and much of the nation's wealth went to Trujillo and his friends. Most of the Dominicans have remained poor and rural; Santo Domingo (675,000) is the nation's principal city. Since Trujillo's assassination, the Dominican Republic's political situation has been uncertain.

Jamaica

Jamaica became an independent nation in 1961, and is now a member of the fledgling Caribbean Common Market. For many years Jamaica supplied laborers to many parts of the Caribbean: the Panama Canal, railway construction in Central America, and on the isthmus' banana plantations. Others went temporarily to Cuba for the sugar cane harvest and as permanent emigrants to the United Kingdom. These outlets of economic opportunity and release of surplus population are no longer available. Jamaica's population must now look inward for support.

Along Jamaica's coast, plantation bananas and sugar cane have been the main exports of the island. In the interior, subsistence agriculture provides employment for a large segment of the population. Since 1950, two other economic activities have also played a major role in Jamaica's development. Jet-set tourism centered around Montego Bay has expanded to include less wealthy visitors. More important still has been the exploitation of bauxite, the ore from which aluminum is made. Today, Jamaica produces 25 percent of the world's bauxite, which is mined by United States and Canadian companies using local labor. Much of the government's revenue comes from taxes on the industry and is used to stimulate other activities such as manufacturing in the Kingston area (500,000). Like other parts of Latin America, Jamaica is demanding and getting a greater share and voice in the mining industry.

Trinidad and the Lesser Antilles

Trinidad's situation is very similar to that of Jamaica except that petroleum is the mineral exported, and Trinidad has a large East Indian population. Trinidad is also a member of the Caribbean Common Market (CARICOM). The Lesser Antilles are composed of numerous small islands. Barbados and Grenada are independent, and others are attached to the United Kingdom or France. These small islands were formerly important sources of sugar cane for their mother countries. Large plantations with African slaves dominated economic production. Abolition of slavery and the development of the European sugar beet industry led to disastrous economic results for the Lesser Antilles. Sugar on the low-lying islands and bananas on the mountainous islands provide the bulk of the exports and are the lifeblood of the population.

Today, high population densities and limited resources place these islands in a delicate position, and other economic opportunities such as tourism are being sought. The Lesser Antilles are in a potentially explosive situation, since the rising expectations of the population are not being fulfilled. One hope is the economic integration movement as represented by the Caribbean Common Market of which the British-related islands are members.

Northern South America

Along the Caribbean fringe of South America, the three Guianas (French Guiana, Surinam, and Guyana), Venezuela, and Colombia make up an amorphous but large Latin American region. All were at one time part of the Spanish Viceroyalty of New Grenada, but the Guiana area was never effectively occupied by the Spanish and subsequently was colonized by the British, Dutch, and French.

Guianas

French Guiana is sparsely populated and has experienced little economic development. Only along the coast, especially around Cayenne, is there some commercial agriculture specializing in sugar cane, pineapple, and rice; inland along some of the rivers, logging operations provide timber for export. Elsewhere, small populations live on a subsistence basis with little contact with the outside world.

In newly independent Surinam, the former Dutch government permitted foreign corporations to exploit large bauxite supplies located a short distance inland from the coast. Bauxite is the principal source of the government revenues and accounts for 80 percent of the exports. Along the coast commercial plantation agriculture using workers of Indonesian descent provides sugar cane, citrus fruit, and cacao. Also along the coast small farmers grow rice and operate dairy farms. In and around Paramaribo are blacks who form the professional class, run small stores, and control the political life of the country. The presence of these numerous ethnic groups, with differences in language and religion, has created much internal conflict. Much of the interior is not under effective governmental control.

Independent Guyana's economy is similar to that of Surinam. Bauxite is an important export from the interior, and along the coast sugar cane and rice are raised for local consumption and export. Like Surinam, ethnic conflict has hampered development. East Indians make up about 50 percent of the population and Africans and mulattos about 43 percent. Both these groups have formed political parties along racial lines and have waged bitter verbal and armed battles. As a result, there is a serious problem of maintaining a cohesive and viable economic and political structure. These problems have been increased by the government's policy of nationalizing the foreign-owned bauxite mines and the consequent decrease in inflow of development capital.

Venezuela

Venezuela has one of the highest per capita GNPs in Latin America, much of which comes from the petroleum industry. Oil and oil products account for over 90 percent of the country's exports, for 25 percent of the GNP, and for more than 60 percent of the government's revenue. Petroleum and development are synonymous terms in Venezuela.

The modern oil industry of Venezuela began in the early part of the century when the dictator Vincente Gomez encouraged foreign petroleum companies to develop the nation's reserves. The petroleum fields of the Llanos and the Maricaibo Basin became major sources of oil entering international trade. For many years Venezuela exported crude oil which was refined on the nearby Dutch islands of Aruba and Curacao and in the United States and in northwestern Europe. Since World War II, refineries have been built in Venezuela, the by-products providing raw materials for tires, synthetic fibers, medicines, and a host of other industries. Additionally, the government, in partnership with the petroleum companies, has obtained a greater share of the industry's profits. In 1976, Venezuela nationalized the industry. Nationalization is but the final step in Venezuela's move to regain control over its single most important economic activity.

Iron ore along the northern fringe of the Guiana Highlands has also contributed substantial exports for Venezuela. Originally developed by United States corporations, the operations were nationalized in 1975. The ores were located in a little-developed part of Venezuela and required large capital investments and an extensive infrastructure. Once developed, the area received additional investments from the Venezuelan government with the goal of creating a national center of heavy industry. Hydroelectric power and imported coal are now used to make iron and steel and aluminum. Ciudad Guayana (population over 100,000) is the center of this ambitious development program.

For many years Venezuela has had a policy of "sowing the oil." Governmental profits derived from petroleum and iron ore have been reinvested to stimulate other sectors of the economy and to provide a better level of living for the population. Funds have been used for highway construction and farm-to-market roads, low-cost housing and education, agricultural improvement and colonization, and for industrialization. By these means the wealth produced by mining is allocated to the population; the mining industry itself employs only about 3 percent of the labor force.

Petroleum provides much of Venezuela's export earnings. Developed by foreign corporations, largely those of the United States, the oil industry was nationalized in 1976. Shown here are storage tanks and wells of the Lake Maracaibo field. (United Nations/V. Bibic)

Venezuela is one of the most urbanized nations of Latin America; nearly 80 percent of the population is considered urban. Of Venezuela's 12 million people, 2 million live in the metropolitan area of Caracas; Maracaibo (650,000), Valencia (370,000), Barquisimeto (335,000), and Maracay (225,000) are the four next largest centers.

Most of Venezuela's population is located in the Cordillera de Merida highlands. Away from the urban centers, small-plot agriculture is characteristic. Population density is high. South of the highlands stretches the broad, flat, and sparsely settled Llanos, devoted mainly to cattle raising. Along the highland-Llanos fringe, however, some agricultural colonization is taking place. South of the Llanos, except for the Ciudad Guayana area, lies a large area outside effective government control. Venezuela is in Rostow's take-off stage.

Colombia

Colombia's 25 million citizens are distributed in several separated population clusters centered in the western one-third of the nation. The eastern two-thirds, sparsely inhabited and little developed, is a continuation of the Llanos and a small section of the Amazon Basin. In the west the great Andean mountain system is divided into three parallel ranges: the Cordilleras Oriental, Central, and Occidental. In the Cordillera Oriental are several intermontane basins in which the pre-Columbian high-culture Chibchas lived. Many of the early Spanish colonists settled in the same area with Bogotá (2.8 million) eventually becoming the largest city and focal point of the nation; Bucaramanga (320,000) is similarly situated. In the Cordillera Central, other settlers established the town of Medellin (1.5 million) and moved southward along the mountainside. Along the Caribbean Coast, population grew around the ports of Barranquilla (820,000), Cartagena (600,000), and Santa Marta (140,000), and expanded inland. Some of this expansion extended up the Cauca Valley and centered on the cities of Manizales (275,000), Pereira (245,000), Armenia (220,000), and Cali (1,000,000).

The various population clusters are poorly interconnected; east-west movement is particularly difficult. The Rio Cauca and Rio Magdelena provide a somewhat better line of communication in a north-south direction, but even along these routes transport is difficult. Limited interconnection among population clusters has led to strengthening of regional loyalties, lack of economic integration, and political factionalism.

In spite of numerous large cities, Colombia is basically rural. Coffee provides some 63 percent of the nation's exports, and 40 percent of the labor force is engaged in agricultural pursuits. Efforts to diversify the agrarian sector and lessen the poverty of most rural inhabitants have not been very successful. As a consequence, migration rates to the city are high, and unrest among the rural population remains high; in 1970, there were an estimated 270,000 squatters in the shantytowns of Bogotá. Agrarian reform in Colombia has been mainly concerned with land redistribution and formation of rural cooperatives. Since 1961, some 12 million acres (4.9 million hectares) have been redistributed to farm families.

Colombia's seven largest cities have about 31 percent of the nation's population. Within the cities is produced a full range of manufactured products from textiles and foods to metalworking and chemicals. Metalworking is facilitated by a steel plant located in the Cordillera Oriental at Belencito. Medellin has had a tradition of textile processing, and in recent years several other kinds of industries have located in the area. Cali, in 1912, had but 27,000 inhabitants; in 1950, 200,000; and today, over 1,000,000. Much of Cali's growth has come from industrialization by both national and foreign corporations.

Colombia has a diversity of natural environments in which the growth of a wide variety of crops is feasible. Moreover, coal, iron ore, and petroleum are, if not abundant, at least adequate for further development. The lack of a viable political organization and an adequate infrastructure has hampered economic modernization attempts.

Further Readings

The *ejido* system has received much attention. One of the most recent works is Barrett, Elinore M., "Colonization of the Santo Domingo Valley," *Annals, Association of American Geographers,* 64 (1974), 34–53. In the La Laguna district the *ejido* system developed differently from elsewhere in Mexico: Landsberger, Henry, and Hewitt de Alcantara, Cynthia, *Peasant Organizations in La Laguna, Mexico: History, Structure, Member Participation and Development* (Washington, D.C.: Inter-American Committee for

Agricultural Development, 1970), 144 pp. In recent years Mexico has attempted regional development using river watersheds. See, for example, Barkin, David, and King, Timothy, *Regional Economic Development: The River Basin Approach in Mexico* (London: Cambridge University Press, 1970), 262 pp. An overall view of Mexican development is Flores, Edmundo, "From Land Reform to Industrial Revolution: The Mexican Case," *Developing Economies*, 7 (1969), 82–95.

Some of the studies of the Central American Common Market include the following: Wynia, Gary W., *Politics and Planners: Economic Policy in Central America* (Madison: University of Wisconsin Press, 1972), 227 pp.; and Seligson, Mitchell A., "Transactions and Community Formation: Fifteen Years of Growth and Stagnation in Central America," *Journal of Common Market Studies*, 11 (1973), 173–190. The United States government has prepared country studies for most of the world's nations which are quite useful. See, for example, Blutstein, Howard, and others, *Area Handbook for Costa Rica* (Washington, D.C.: Government Printing Office, 1970), 323 pp. Belize's new capital is examined by Kearns, Kevin, "Belmopan: Perspective on a New Capital," *Geographical Review*, 63 (1973), 147–169.

For Cuba's radical change in development direction see the following: Mesa-Lago, Carmelo (ed.), *Revolutionary Change in Cuba* (Pittsburgh: University of Pittsburgh Press, 1971), 544 pp.; and Nelson, Lowry, *Cuba: the Measure of a Revolution* (Minneapolis: University of Minnesota Press, 1972), 242 pp. A pre- and post-Castro study of Cuban sugar is given in Dyer, Donald D., "Sugar Regions of Cuba," *Economic Geography*, 32 (1956), 177–184; and "Cuban Sugar Regions," *Revista Geografica*, 57 (1967), 21–30. For a different development story see Wells, Henry, *The Modernization of Puerto Rico: A Political Study of Changing Values and Institutions* (Cambridge, Mass.: Harvard University Press, 1969), 440 pp. For Hispaniola see Moore, O. Ernest, *Haiti: Its Stagnant Society and Shackled Economy* (New York: Exposition Press, 1972), 281 pp.; and Wiardi, Howard J., *The Dominican Republic: Nation in Transition* (New York: Praeger, 1969), 249 pp. Tourism in the Caribbean was examined in Mings, Robert C., "Tourism's Potential for Contributing to Economic Development in the Caribbean," *Journal of Geography*, 68 (1969), 173–177. The problems of the Lesser Antilles were investigated by O'Loughlin, Carleen, *Economic and Political Change in the Leeward and Windward Islands* (New Haven, Conn.: Yale University Press, 1968), 260 pp.

Ethnic and economic change are examined by Mandle, Jay R., *The Plantation Economy: Population and Economic Change in Guyana, 1838–1960* (Philadelphia: Temple University Press, 1973), 170 pp. The new city of Ciudad Guyana is examined by McGinn, Noel, and Davis, Russell G., *Build a Mill, Build a City, Build a School: Industrialization, Urbanization, and Education in Ciudad Guyana* (Cambridge, Mass.: MIT Press, 1969), 334 pp. For Colombian development see Bird, Richard M., *Taxation and Development: Lessons from the Colombian Experience* (Cambridge, Mass.: Harvard University Press, 1970), 277 pp.

The contrasts among areas of the Caribbean are expressed by Augelli, John P., "The Rimland-Mainland Concept of Cultural Areas in Middle America," *Annals, Association of American Geographers*, 52 (1962), 119–129.

Latin American Regions: The South

23

THREE FINAL REGIONS COMPLETE the broad realm of Latin America: Andean South America, southern South America, and Brazil.

Andean South America

Andean South America includes Ecuador, Peru, and Bolivia (Fig. 23–1). These nations have several common characteristics and face similar development problems. All are in Rostow's stage one of development (traditional society), although Peru is nearing stage two (preconditions for take-off). All are predominantly rural; more than one-half the labor force in Bolivia (56 percent) and Ecuador (51 percent) is engaged in agriculture. Peru has 45 percent of its labor force in agriculture. Manufacturing accounts for only 10 to 20 percent of the labor force in each country. Per capita GNP is low, varying from $320 in Bolivia to $550 in Ecuador and $810 in Peru. More significantly, all three nations have plural societies with the descendants of the Incas making up a large part of the population, although ineffectively integrated into the national economy. There is a great gulf between the rural Indian and the Western-oriented urban dweller. Cultural pluralism is also evidenced in a dual economy. There is a sharp line dividing economic activities for local consumption and those for the world market. These nations, along with Colombia, Venezuela, and formerly Chile, form the Andean Group of the Latin American Free Trade Association.

Andean Group

To date, LAFTA has not been very successful. The limited tariff reductions have benefited the large nations but have done little to spur economic advancement among the smaller nations. Over the last ten years, Bolivia, Ecuador, and Peru have had an annual growth rate in total GNP of 5.2 percent. When measured on a per capita basis, the growth rate shrinks to only slightly over 2 percent. The limited opportunity and growing population pressure have created much unrest, and the respective governments have searched for a means to stimulate the economy. Out of the search has come the Andean Group's economic integration plan.

Under the plan most goods are scheduled to move freely among the group by 1980. Manufacturing is especially encouraged. Integration of tourism promotion and education is also a group goal. To finance these and other integration programs, an Andean Group development bank has been formed (*Corporación Andina de Fomento*). One of the unique characteristics of the Andean Group is their common rule on foreign investment. Foreign investments are welcome if, in the group's opinion, they promote economic development. These investments are government controlled and subject to regulations on profits. All ventures are subject to gradual transformation to local ownership in which the foreign investor may retain a 49 percent interest. To date, the policies of the Andean Group have not been implemented to any significant degree. A structure has been established, however, by which a major economic reorientation is possible.

Cartagena
Barranquilla
Santa Marta
Maracaibo
Barquisimeto
Maracay
Caracas

Medellín
Bucaramanga
Valencia
Pereira
Manizales
Ciudad Guyana
Armenia
Bogotá
Cali

Quito
ECUADOR

VENEZUELA

Georgetown
Paramaribo
Cayenne

GUYANA
SURINAM
FRENCH GUIANA

COLOMBIA

Guayaquil

Iquitos

Manaus

Belém

Chimbote

PERU

BRAZIL

Fortaleza

Lima

Recife

Cuzco

BOLIVIA

Salvador

Arequipa
La Paz

Cochabamba

Santa Cruz

Brasília

PACIFIC OCEAN

Antofagasta

CHILE

PARAGUAY

Belo Horizonte

Asunción

São Paulo
Río de Janeiro

Tucumán

Curitiba
Santos

Córdoba

ATLANTIC OCEAN

Valparaíso
Mendoza

Rosario
Sante Fe

Porto Alegre

Santiago

Concepción

Buenos Aires
URUGUAY

La Plata
Montevideo

Temuco

ARGENTINA

Mar del Plata

● Urban places over 1,000,000
○ Urban places 100,000 to 1,000,000
• Urban places under 100,000

0 100   300   500 Miles
0   200 400 600 800 Kilometers

23–1 *South America: Principal Cities and Major Highways*

Ecuador

Ecuador has long been a nation beset with social and economic problems. During the colonial period, Ecuador was largely neglected. The few Spanish who did settle there gained their livelihood from the sweat of Indian farm workers on large estates in the highland basins of the Andes. The Spanish themselves lived in Quito, the capital and former Inca center, and in smaller regional centers. For many of the Indians, European conquest had little impact except for introducing new crops and animals. East of the Andes, the Oriente, fierce Amazon Basin Indians, and a paucity of resources of interest to the Spanish kept the area free from colonial administration control. To the west, the Pacific coast and Guayas Basin were largely neglected (Fig. 23–2). Part of the area was disease ridden (malaria and yellow fever), and part was quite dry.

The Pacific coast, especially the fertile Guayas Basin, is Ecuador's major center of growth. Guayaquil (860,000) is now the nation's largest city and functions as the principal port and manufacturing center; it is growing at a rate of 6 percent yearly. Control measures have freed much of the coast's rainy tropical areas from disease, and since 1940 a vigorous road-building program has opened large areas for settlement and commercial agriculture. Ecuador is the world's leading exporter of bananas, all of which come from the coast. Fishing in adjacent Pacific waters has contributed substantially to the nation's wealth. Ecuador claims a 200-mile (322 kilometers) territorial control of the bordering Pacific, a reflection of the importance of fish and of budding nationalism. Natural gas resources offshore of Guayaquil are now being developed. Throughout the coast there is the aura of rapid economic growth.

In contrast, the Andean segment is a stagnant area little changed since colonial times. Population is confined to several intermontane basins and surrounding hillsides. Quito (560,000) is the urban center of the region and serves as the national capital, home of the landed aristocracy, and as a regional service center. Agriculture, largely for local consumption, is the major economic activity of the area. The best lands are in large farm units; these estates specialize in dairying and maize. The poorer lands are divided into thousands of small Indian subsistence plots with wheat and barley cultivated on dry lands, potatoes at higher elevations (*tierra fria*), and pasture sheep on the alpine meadows. The Andes of Ecuador present a classic example of minifundia and latifundia.

The Oriente is a sparsely settled area inhabited largely by Amazon Indian groups such as the Jivaro who possess few modern techniques of production. In past times, the Oriente has received temporary influxes of people from the Andes in search of balsa wood, rubber, and chinchona (quinine used for malaria). Each time, however, the gathering of these wild products has lost out to commercial production elsewhere or to synthetic substitutes. In 1967, a major oil field was discovered in the Oriente, and Ecuador may soon become a leading oil exporter, second only to Venezuela in Latin America. By 1980, it is expected that 500,000 barrels per day will be shipped by pipeline to a Pacific terminal. Petroleum production may spur a more effective occupation of the Oriente.

Ecuador's 7.5 million inhabitants are about evenly divided between Andean Indian farmers and Spanish and mestizos who live in the cities and on the Pacific coast. These two groups differ greatly in their cultural outlooks and economic well-being. There is a minimum of communication between them. The Indian is oriented to the local group rather than to the nation, is illiterate, and has an income far less than the national average. Prospects for economic gain are dim, for population pressure has become increasingly intense. The Indian's worldly possessions are few: a one-room hut, a sleeping mat, and the clothes on his back. The city dweller, of Spanish descent or a mestizo, has a more worldy outlook, is nationalistic, and has an income much greater than the national average. Most of the latter class are literate, ascribe to Western dress and ways of life, and have alternative economic opportunities. These two groups present the nation with a problem of cultural pluralism.

Bananas, oil, and fish provide Ecuador with valuable exports, but national development is hampered not only by cultural pluralism but also by isolation. The Andes are a formidable barrier to internal transport. Few all-weather roads connect the intermontane basins of the highlands or the highlands to the coast. Roads are almost nonexistent in the Oriente. Ecuador as a nation is isolated by its west coast location. Only with the opening of the Panama Canal have the markets of eastern Anglo-America and Europe become available to Ecuador, but even so the nation is off the major shipping lanes. Since about 1950, with the rise of Japan as a major trading nation and popu-

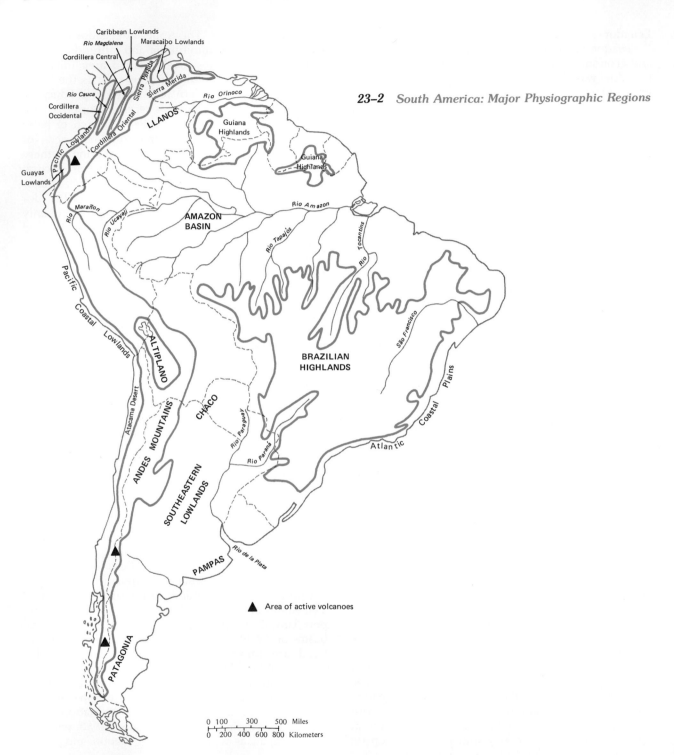

Caribbean Lowlands
Rio Magdalena   Maracaibo Lowlands
Cordillera Central
Rio Cauca
Cordillera
Occidental
Pacific Lowlands
Guayas
Lowlands
Sierra Perijá
Sierra Merida
Cordillera Oriental
Rio Orinoco
LLANOS
Guiana
Highlands
Guiana
Highlands
Rio Marañon
Rio Ucayali
AMAZON
BASIN
Rio Amazon
Rio Tapajos
Rio Tocantins
Rio
Pacific Coastal Lowlands
ALTIPLANO
Atacama Desert
ANDES MOUNTAINS
CHACO
Rio Paraguay
Rio Parana
BRAZILIAN
HIGHLANDS
São Francisco
Atlantic Coastal Plains
SOUTHEASTERN
LOWLANDS
PAMPAS
Rio de la Plata
PATAGONIA

▲ Area of active volcanoes

23–2   *South America: Major Physiographic Regions*

0   100   300   500 Miles
0   200 400 600 800 Kilometers

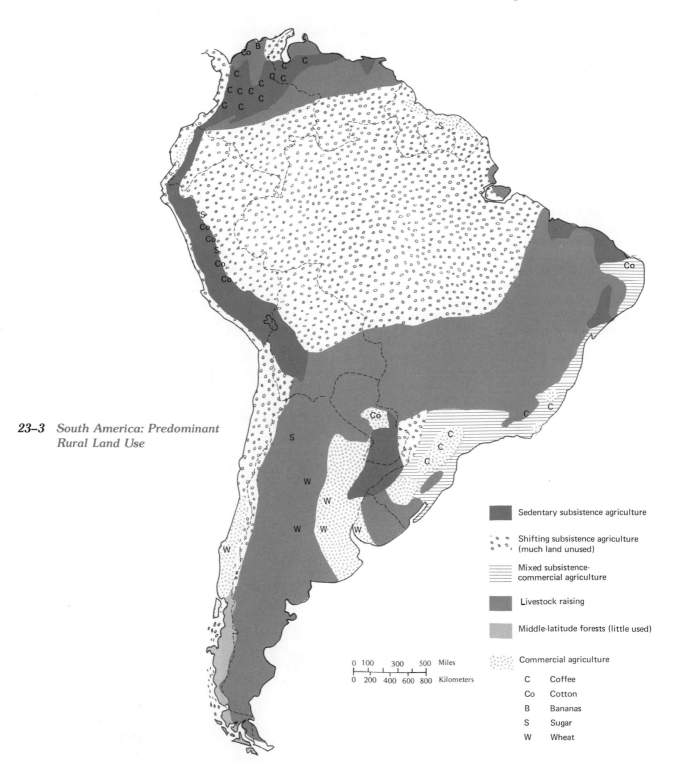

23–3 *South America: Predominant Rural Land Use*

Sedentary subsistence agriculture

Shifting subsistence agriculture (much land unused)

Mixed subsistence-commercial agriculture

Livestock raising

Middle-latitude forests (little used)

Commercial agriculture

C   Coffee
Co   Cotton
B   Bananas
S   Sugar
W   Wheat

0   100   300   500   Miles
0   200   400   600   800   Kilometers

lation growth along the United States Pacific coast, Ecuador's relative position to markets has improved. Japan particularly has become an important trading partner with Ecuador and other west-coast South American nations.

Peru

Like Ecuador, Peru has three major areal units: the coast, the Andes, and the east. Peru, however, has a more diversified economy. Much effort has gone to incorporate the Amazon lowlands into the national system, and in recent years Peru has pursued a policy of national unity. In many respects, Peru is in the forefront of Andean development.

The Peruvian coast is a narrow ribbon of desert climate. Settlement is confined to some forty rivers that rise in the Andes and cross the coastal zone to the sea. Most of the nation's commercial crops are grown in the river oases, cotton and sugar cane being the most important export crops (Fig. 23–3). Formerly these export crops were grown on large estates, many owned by foreign corporations, but in the late 1960s, most estates were nationalized and formed into worker cooperatives. The oases near the Lima-Callao area (population 3,350,000) are oriented to truck gardening for the urban market. The southern oases have a subsistence economy, but some grapes and olives are grown for the national market and food crops for the nearby regional center of Arequipa (200,000). Since 1950, the government has begun many projects to supply greater amounts of water to expand the irrigated area. One of the most ambitious projects, still in the planning stage, is to divert part of the Rio Marañon to the coast.

The coast is endowed with some significant mineral deposits. In the north, petroleum has been pumped since the late nineteenth century. Developed largely by United States and British capital, the oil fields and a refinery are now operated by the government. In the south are copper and iron ore. The iron ore is shipped to Chimbote where it is processed into iron and steel sheets, rods, and ingots.

Lima with its port Callao is the nation's social, political, and economic focal point. Founded by Pizarro, Spanish conqueror of the Incas, Lima soon became the center of Spain's South American empire. While preserving many of its colonial traditions, the city has become increasingly industrialized. Historic buildings on narrow streets vie with modern architecture and broad avenues. Around the city fringe and also on formerly vacant land near the city center are shantytowns populated by the urban poor and migrants from the countryside and smaller regional towns. The Lima-Callao area is experiencing a population growth rate of nearly 6 percent yearly, about double the national growth rate.

As in Ecuador, the Peruvian Andes are inhabited

Shanty towns are common sights in large Latin American cities. This one is in Lima, Peru. Rapid population growth and urbanization have led to an urban population explosion. Many of the urban poor must use whatever material they can find to build their homes. (United Nations)

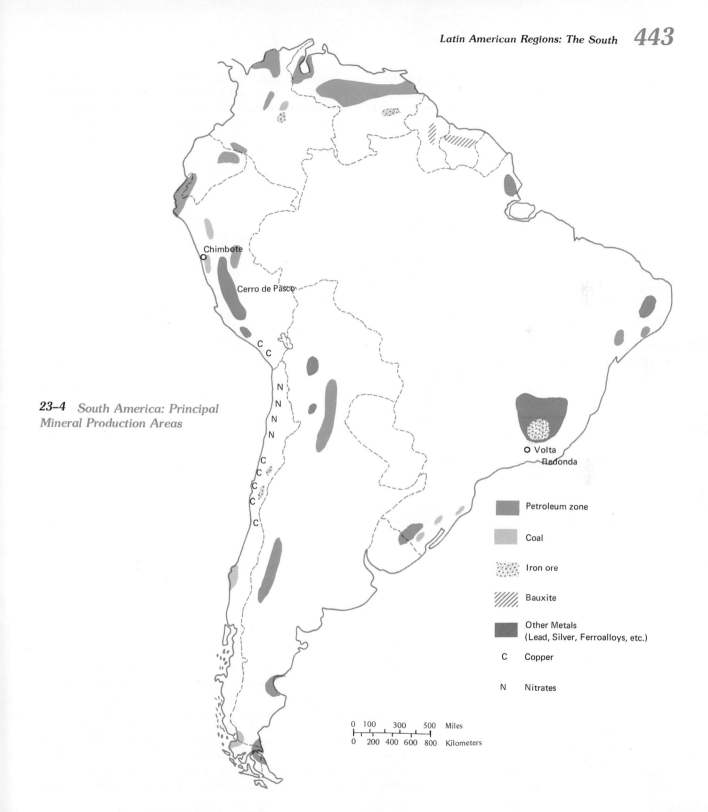

Chimbote

Cerro de Pasco

C C

N
N
N
N

C
C
C
C
C

O Volta
Redonda

*23–4 South America: Principal
Mineral Production Areas*

Petroleum zone

Coal

Iron ore

Bauxite

Other Metals
(Lead, Silver, Ferroalloys, etc.)

C Copper

N Nitrates

0 100   300   500 Miles

0   200 400 600 800 Kilometers

mainly by peasant Indians of Inca descent in contrast to the European-mestizo inhabitants of the coast. Population in the Andes is dense, and many peasants migrate both temporarily and permanently to the coast and to a lesser degree to the east in search of work. Land tenure in the highlands remains much the same as in colonial times. Large estates control the best lands. On the poorer lands and in isolated pockets, many small farms, often on communally owned land, dot the landscape with the poverty-stricken inhabitants living much as in pre-Columbian times. The old Inca center of Cuzco (110,000) is the only large city of the Andes. In the post–World War II era, the Peruvian government has attempted to break the landed aristocracy's hold on the land and peasants. Some of the large estates have been broken up and allotted to the peasants who worked them. These actions are part of a larger policy of national unification by integrating the Indian into the national Spanish-mestizo culture.

The Andes of Peru are also noted for extensive mineral deposits (Fig. 23–4). Coal is mined in several locations along the western flank of the Andes. The most famous mining district is Cerro de Pasco, an old colonial silver center redeveloped by foreign capital, which produces copper, silver, gold, lead, zinc, and bismuth. Present governmental policy is to gain greater control over all mining operations, and the government has sponsored higher wages for the miners, better housing, and other social services.

Peru more than any other South American nation has made repeated attempts to tame its large eastern lands of the Amazon Basin. For over 100 years the nation has sponsored colonization projects in the area, and these attempts have greatly intensified in the last forty years. Roads have been built across the Andes into the eastern plains, and along each road have come highland peasants in search of new economic opportunities. Illustrative of the interest in the east is Peru's proposed marginal highway projected to extend along the eastern flank of the Andes (Fig. 23–5). Peru has actively counseled its neighbors to assist in constructing a highway all along the eastern Andean margin from Venezuela to Argentina. Much of the east, however, remains sparsely populated and is inhabited by traditional Amazon Indians. River orientation is still characteristic of much of the area. Iquitos is the regional urban center and Peru's major port on the Amazon River. The existence of

petroleum deposits has been known for some time, but production is limited by inadequate transport. Active exploration for oil continues, and hopes remain high that vast reserves will be found.

In many respects Peru's development problems, like those of Ecuador and Bolivia, are similar to the problems formerly encountered by Mexico. In recent years the Peruvian government has played and continues to play an active role in directing change, both economic and social. Many of Peru's policies mirror those long instituted in Mexico: usurpation of political power from the landed gentry, unification of the society, an emphasis on education, agrarian reform, and increasing socialism.

Bolivia

Bolivia is the poor man of South America. The $320 per capita GNP figure fails to tell the true extent of the nation's poverty, since a large part of the population earns far less. In 1972, Inter-American Development Bank analysts reported that rural per capita income was about $75, while annual urban income was estimated at nearly $450, a sixfold difference. Bolivia's development problems are numerous. The nation's landlocked position inhibits international trade and communication with other nations. Bolivia has a history of political instability, and few elected leaders have been able to complete their terms. Since independence, no government has ever controlled all parts of the nation. Without a stable and effective government, all forms of development are most difficult.

Internal cohesion is lacking. Transport facilities and other forms of infrastructure are poorly developed. The population, grouped in scattered clusters, is poorly interlinked. Moreover, many of the Indians (53 percent of the population) do not speak Spanish but continue using Quechua (the Inca language) or Aymara (spoken by a large group living around Lake Titicaca).

Bolivia's heartland is the dry Altiplano, a series of high intermontane basins flanked by ranges of the Andes. It was in this area that most of Bolivia's pre-Columbian population lived, and it was here that the Spanish settled. The Spanish divided land among themselves in large estates and found large quantities of silver in the nearby hills and mountains. From colonial times to the present, minerals have made up the bulk of Bolivia's exports. Today, minerals

23–5 *Amazon Basin: Proposed Highway System and Connections to Other Regions*

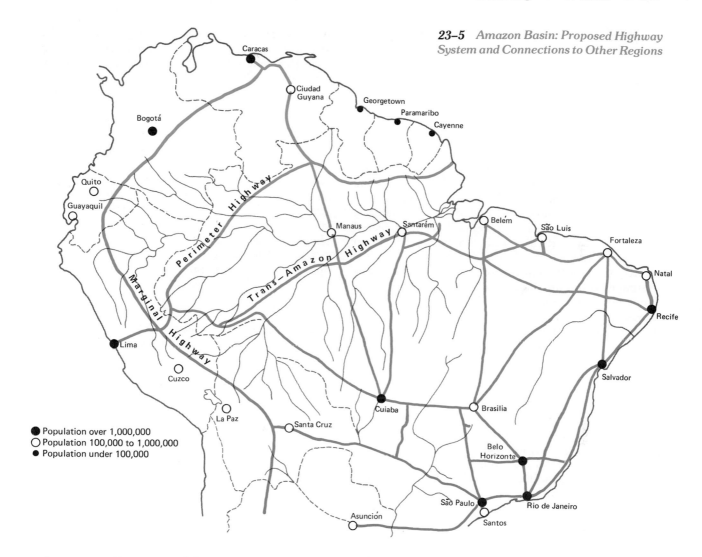

Population over 1,000,000
Population 100,000 to 1,000,000
Population under 100,000

make up over 80 percent of the nation's exports with tin accounting for nearly one-half of the total exports. In this century the mining industry became concentrated in the hands of a very wealthy few. The condition of the miners and their families was wretched, while the owners lived in luxury outside the country.

In 1952, the government of Bolivia was overthrown, and a fundamental change in political, social, and economic institutions began. Although there have been several subsequent heads of government, the basic tenets of the 1952 revolution have continued. The larger mines were nationalized and the large rural estates abolished. In succeeding years many other reform policies have been announced, but little real progress is evident.

One policy has been occupation of Bolivia's sparsely settled, eastern lowlands, both to relieve population pressures in the Altiplano and to mobilize the resources of the lowlands. Santa Cruz (100,000) is the urban center of the east and is growing at the rate of 5 percent yearly. La Paz (500,000) and Cochabamba (140,000) by comparison are growing at 2.5 and 3 percent, respectively. The presence of oil in the lowlands has long been known, and small quantities are produced for the national market. It is agriculture, however, that holds promise to employ large numbers, and it is people with agricultural backgrounds who make up the bulk of the Altiplano people who migrate to the lowlands.

Lowland colonization is inhibited for three reasons. First, the potential migrants, the rural poor, are an extremely conservative group with strong family and local ties who migrate only when forced by dire need. Second, many Altiplano dwellers live at elevations of 11,000 feet (3,353 meters) or more. Over the centuries these Indians have developed large lung capacities; many are unable to adjust physically to the oxygen-rich lowland air. They are also more susceptible to lung infections and diseases. Highland Indians throughout Latin America have long feared the lowlands, particularly the rainy tropics. From Mexico to Bolivia, the lowlands are often avoided. Third, many of the cultivation techniques and crops common to the highlands are inappropriate to lowland agriculture. New farming practices and new crops must be mastered. Unless adequate instruction is available, the migrants' crops produce poorly, and the land resources are destroyed.

Southern South America

Southern South America is composed of Chile, Argentina, Uruguay, and Paraguay. The first three nations share several common features that separate them from surrounding areas. Chile, Argentina, and Uruguay have relatively high per capita GNPs, $760, $1,590, and $1,330, respectively. Each is highly urbanized, has a well-defined middle class, and has a very high literacy rate. The state-idea is well established, and the populations are culturally unified. Measured by most standard indicators, these three nations fall in or near the rich side of the poor-rich continuum of the world's nations. This alignment with rich nations is accentuated further by demographic features. A review of birth, death, and growth rates shows that they have followed the course of the demographic transformation model (Chapter 1) and are now nearing the model's final stage. Birth and death rates are low, and the population growth rate ranged from 1.1 percent to 1.6 percent. In contrast to other Latin American countries, the population pyramid shows a more mature population (Fig. 21–3). According to Rostow's stages of economic develop-

ment, Argentina is probably in stage four (the drive to maturity); Rostow stated Argentina began stage three (the take-off) in 1935 (Table 4–1). Chile and Uruguay fall within stage three.

Paraguay is included in southern South America because its Guarani Indian population has intermarried with the Spanish immigrants to form a unified mestizo society, and Paraguay's connection with the outside world is largely through Argentina. Moreover, the physical environment within which Paraguayans live is an extension of the environment of adjacent Argentina. Finally, Paraguay, along with Argentina, Uruguay, and Brazil, is part of the Rio de la Plata watershed regional development group. Should this regional development proceed along the established tentative policy lines, Paraguay's development will be still more closely interlinked with southern South America.

Chile

Chile has achieved cultural unity in spite of its long (2,600 miles or 4,184 kilometers) and narrow (100–250 miles or 161–402 kilometers) shape. Central Chile, between 30° and 42° S. latitude, is the heart of the nation, and it is within this area that a cohesive society was formed. To the north is the bone-dry Atacama desert, sparsely populated but possessed of nitrates, copper, and iron that provide most of the nation's exports. To the south is a very humid, rough, and rugged area also sparsely inhabited but with abundant forest and water power potential. In the far south some coal and petroleum are found, and on the windy and wet tip of the continent is sheep raising with wool destined for export.

It is within central Chile that most of the population lives and where most of the industry and agricultural lands are located. Chile is an urban country (75 percent), and most of the cities are in the middle section. Santiago (2.7 million) is the largest center, with Valparaíso (480,000), Concepción (460,000), and Temuco (115,000), the second, third, and fifth principal metropolitan areas of the nation, respectively. Chile's fourth largest city is Antofagasta (125,000) in the north. All these cities except Valparaíso are growing at rates substantially above the national growth rate.

Chile's unified society is of recent vintage. In colonial times the economy revolved around rural estates with a patron-peon organization. The peon class was derived from mestizos (Spanish-Araucanian Indians); many Araucanians, however, retained their societal integrity until early in the twentieth century. The patron's hold was not strong, however, for population pressure was low, and to the north, south, and east across the Andes in Argentina were frontier areas. From 1879 to 1883, Chile successfully pursued a war against Peru and Bolivia over nitrate deposits in the northern Atacama area, which united the Chilean population in a common cause. Nitrates for fertilizer and explosives became the nation's major export, and mining it provided an alternative economic opportunity for the peons. Later, copper mining provided more jobs. Early in the twentieth century,

especially during World War I when goods from the industrialized world were difficult to obtain, industrialization for the local market began. In Chile, the landed aristocracy invested in manufacturing activities and services more readily than in Andean South America. The alternative economic opportunities in the cities provided another and increasingly important outlet for the rural farmer.

Cultural unity, national allegiance, valuable export products, lack of heavy rural population pressure, and diversification of the economy by manufacturing and service have played major roles in explaining Chile's development. To this list is added education; about 86 percent of the population is literate. Also of basic

Copper is an important export of Chile. Formerly developed by foreign (largely United States) corporations, the copper industry was nationalized in the 1970s. Here copper ingots are being processed for both export and local needs. (United Press International Photo)

importance are government stability and execution of development policies. Throughout most of Chile's history, political stability has been characteristic. In recent years political power has passed to urban-oriented parties who have campaigned for economic and social reform. These reforms are pointed along two lines. The first is greater control of foreign investment, especially the mining industries which provide most of the nation's exports. During the regime of Eduardo Frei (1964–1970), the government, for example, purchased majority ownership of part of the copper industry. The second major reform is in the rural sector and involves land redistribution. Large estates were purchased by the government and divided into small farms, although Frei's government was criticized for not moving reforms rapidly enough.

In 1970, Frei was succeeded by Salvador Allende, the first self-acknowledged Marxist president of Chile. Allende was elected by a minority of the electorate with barely over one-third of the vote in a three-man race. The Allende government attempted to reorganize Chile's economic and social structure very rapidly. In the process the mining industry was nationalized, land redistribution accelerated, and most banks and the communication media were taken over by the government. A true socialist economy was in the offing. This restructuring was not without problems and serious opposition. Allende almost bankrupted the nation and alienated a large part of the population. Congress repeatedly attempted to rescind Allende's actions; strikes of protest were common among workers and in the middle class. In 1973, the protest was brought to a head, and the military revolted. Allende was killed, and a military government was formed to rule the nation. Today, Chile is a politically divided nation with a disrupted economy.

Argentina

Argentina, in its early independence period, had to struggle for national identity. In colonial times Argentina rested on the fringe of the Spanish empire in Latin America. Along the Andean flanks in the west and northwest were the most developed areas, and commerce was directed through Lima. The small fort at Buenos Aires and the surrounding scattered cattle and horse *estáncias* were but outposts to defend against Portuguese penetration southward. With

independence came a period of internal conflict, and a reorientation from the west to the east began.

Argentina's modern development began about 1860 with the settlement of the Pampas, the broad, flat, and fertile grasslands surrounding Buenos Aires. Prior to 1860, the Pampas had been divided into large ranches devoted to cattle and horse raising. The cattle were valuable only for hides and tallow, since there was no nearby market for beef. In many respects the Pampas was like the American West with cowboys (gauchos) and nomadic Plains Indians. About 1860, immigrants from Italy, Spain, Germany, Switzerland, Britain, and other European countries began coming to the Pampas in large numbers; by 1930, some 6.3 million immigrants had come to Argentina largely through Buenos Aires. Some stayed in Buenos Aires, but many went into the country where they worked as tenants on the *estáncias.* The normal owner-tenant agreement was one of work for land. The tenant raised alfalfa, built barbed-wire fences, and sunk water wells powered by windmills, and in exchange could cultivate some of the *estáncia's* land on which he raised cereal grains, especially wheat, and garden crops for his family's needs. After cultivating the land for five to seven years, the tenant planted alfalfa or pasture grasses and moved on to another piece of land. Under this system many tenants were able to accumulate wealth which they used to buy land. In time the Pampas came to be one of the most agriculturally productive areas in the world, and by the early 1900s, the value of crops exceeded livestock. Livestock, however, remains very important, and Argentine beef is exported around the world.

Today, Argentina is one of the most developed nations in Latin America. Its per capita GNP rivals that of the rich world. Its population is both literate (94 percent) and urban (81 percent), and population growth is a moderate 1.3 percent. The nation's 26.1 million people are European and have a well-developed sense of national unity. Political strife is now rampant, however, over the direction of the nation's future. Buenos Aires and the Pampas form the country's heartland and are by far the nation's most important area. Greater Buenos Aires (Buenos Aires and its satellite cities) has a population of 8.5 million people, making it one of the largest urbanized areas in the world. The other principal cities of the

Cattle raising has been a traditional form of livelihood on the Argentine pampas. Meat and meat products are important exports. The Argentine cattle industry is noted for its efficiency and use of modern technology. (United Nations)

Pampas are Rosario (800,000), La Plata (470,000), Santa Fe (270,000), and Mar del Plata (300,000). Buenos Aires is the heart of the nation. A well-integrated road and railway system radiates outward from the city, and goods funnel through the city's port destined for the world market. From the Pampas come beef, maize, and wheat which are exported largely to the developed world. Imports are mainly manufactured goods from the United States and Western Europe. The development of an internal transport net, steam-powered ocean vessels, and refrigeration were vitally important to the economic advancement of Argentina and particularly the Pampas. Without these innovations, the export sector of Argentina's economy would be severely hampered.

Most of Argentina's manufacturing and service activities are in Buenos Aires; some 45 percent of the nation's industrial labor force is located there and about 50 percent of all those engaged in tertiary activities. From 1946 to 1955, when Juan Perón ruled the country, the government regulated and redirected the national economy toward manufacturing under a broad policy of creating national self-sufficiency;

Perón was also president from 1973 to his death in 1974. Metal fabricating, petrochemicals, and plastics came to vie with the more traditional agricultural processing industries of meatpacking, milling, and textiles. Government investments in manufacturing, an increasing bureaucracy, and numerous social projects such as housing and water supply were paid for by taxes on the agricultural sectors. Eventually the tax burden became too great, and the entire economy foundered, which forced Perón to resign.

Outside the Pampas are several regions that provide products complementary to the nation's economy. In the west near the Andes flank are the old cities of Córdoba (800,000), Mendoza (400,000), and Tucumán (340,000) which have become important growth centers. The west, a dry area, has an agricultural base with the dry lands devoted to cattle raising and irrigated zones to sugar cane and grapes for the national market. In the extreme northwest are a few small mines of asbestos and several metals. Also some petroleum is produced from a giant structural trough located just east of the Andes. To the northwest of the Pampas is a low, humid area be-

449

tween the Río Paraná and the Río Uruguay. The location of this area between rivers has created problems of access, particularly since the rivers are wide and subject to great variation in water flow. The Río de la Plata watershed development program in which Argentina, Paraguay, Uruguay, and Brazil are participants may make the Argentine "Mesopotamia" a more productive area, but even now the area is noted for its tea, maize, flax, cattle, and sheep. In the south is Patagonia, a sparsely populated, dry, wind-swept plateau, most of which is devoted to sheep raising, but in the lower, well-watered valleys are irrigated alfalfa fields, cattle raising, vineyards, and mid-latitude fruit. In this century some mineral resources have been exploited. Petroleum is found along the Andes, a continuation of the trough extending into Bolivia, and in the south near the coast. Also in the south near the coast are small quantities of coal and iron ore that are shipped to Buenos Aires for iron and steel processing.

Argentina's outlying regions are primarily product suppliers for the Pampas and its center, Buenos Aires. Much of the Pampas' population is supported by these regions, and their development is reflected in the improved well-being of the nation's heartland.

Paraguay

Paraguay's 2.8 million people are poor in material wealth, but because of low population pressure and productive land, they are able to live rather easily. Most of the inhabitants make their living from the soil. Asunción (450,000), the capital, is the only truly urban center, and located there are the nation's few manufacturing plants, basically meatpacking, and most of the tertiary activities. Paraguay's population is culturally uniform, and most are mestizos, a mixture of Spanish and Guarani Indian. Although Spanish is the official language, the Guarani language is spoken nearly everywhere.

Most of the population is located east of the Río Paraná which is the most humid part of the nation with fertile alluvial soils. West of the Paraná lies the Chaco, a broad, level plain with a drier climate. Soils in the Chaco are more sandy and gravelly and consequently droughty. Most of the Chaco is used for cattle raising and for forest products. In this century some foreign agricultural colonies from the United States, Canada, and Germany have been established in areas of better soil and water conditions.

Paraguay's limited development is partly explained by the nation's traditional policy of self-sufficiency, two disastrous wars, and its landlocked location and dependence on Argentina for an outlet to the sea. Some of the early leaders actively sought to keep the nation rural and limit contact with the outside world. This policy was not always followed. Paraguay attempted to extend its territory to gain access to the Atlantic. From 1865 to 1870, Paraguay fought the armies of Argentina, Brazil, and Uruguay. By war's end, fewer than 30,000 adult males remained in Paraguay. Some estimates give Paraguay a population of 1.3 million before the war and less than 250,000 immediately afterwards. From 1932 to 1935, Paraguay battled Bolvia over its western border. Casualties were fewer but still considerable. Paraguay still lacks an effective access to the sea. Although the advent of the Río de la Plata regional development program should give the nation better external transport facilities, Paraguay will still depend on good relations with Argentina.

Uruguay

Uruguay owes its existence to Brazil and Argentina. Its origin was one of a buffer state between these two South American giants. Uruguay's population is European and basically urban. Montevideo (1.4 million), the capital, is the only large city and has nearly one-half of the nation's total population. Uruguay is a flat, almost featureless plain on which livestock are raised on large estates. The bulk of the nation's wealth comes from meat, wool, and hides, the three principal exports.

Until the twentieth century, Uruguay was torn internally by policial strife. Unification was achieved under José Battle, who moved the country from agrarian feudalism to a democratic state. In succeeding years Uruguay progressively has socialized its economy and society. As part of the socialization process, liberal retirement benefits and pensions were set up for everyone including rural workers. Family allowances, unemployment insurance, sick-leave benefits, and guaranteed annual vacations became available to all. High taxes were levied to support these programs, and as long as the economy grew, the system worked. Unfortunately, the price of Uruguay's exports has risen but slowly, while those of its imports have climbed steadily. Since Uruguay possesses no minerals of significance, trade is vital and manufacturing confined basically to agricultural processing. As the economy has faltered, governmental revenues have not kept pace with the demands to pay benefits to which the population by law is entitled. Presently the economy remains weak, and the government is under continued pressure to meet its obligations.

Brazil

Brazil is by far the largest of all Latin American nations. Indeed, it is the fifth largest in the world in terms of area and seventh in population. Of the seven most populated countries, Brazil's rate of population, 2.8 percent, is the highest. If present rates of population growth continue to the year 2000, Brazil's population will increase to 215 million.

Brazil's large area is underlain by a diversity of geologic formations. Some of these structures contain rich and extensive ore deposits. The early Portuguese settlers discovered gold and diamonds and established mines to extract them. The more important minerals for manufacturing such as iron ore, bauxite, and ferroalloys are only now being extensively developed. It appears that Brazil holds vast quantities of many minerals.

There are several different physical environments within which many different crops are grown. In the north is the Amazon Basin with its rainy tropical climate and luxuriant rainforest. So far, much of the basin remains under traditional Indian land use. Just to the south is a large area of tropical wet and dry cimate. Some of the wet and dry area is located at elevations above 2,000 feet (610 meters) with moderate temperatures that permit a combination of mid-latitude and tropical crops to grow. Only the eastern part of the tropical wet and dry area is extensively used; the western part, like the Amazon Basin, is beyond the agricultural frontier. Southern Brazil has a mid-latitude environment and is almost fully settled with mid-latitude crops and animals dominating the agricultural scene.

Sleeping Giant

Brazil has been called the "sleeping giant," for in spite of its size and resource base, it has never attained the status of a world power. Apparently the giant is now stirring; not only is the population increasing rapidly, but so is the economy. The total GNP is growing at the rate of 9.3 percent yearly with a per capita rate of 6.8 percent. Industrialization has contributed much to the nation's economic growth; manufacturing accounts for 35 percent of the GNP and is growing at the rate of 13 percent yearly. Brazil's industry is diverse and has many branches that use high levels of technology and automation. Examples of modern factory products include electrical equipment such as motors and generators; transport equipment such as cars, trucks, and ships; and iron and steel production. In 1971, 500,000 motor vehicles were manufactured, more than double the 1967 level. Iron and steel, the basis of modern manufacturing, have grown as rapidly; steel production in 1970 was 5.4 million tons and by 1980 is expected to climb to 20 million tons.

Brazil is increasingly urbanized. Presently, 60 percent of the population are urban dwellers; in 1940, 31 percent were urban and in 1960, 45 percent. Six urban areas have over 1 million inhabitants: São Paulo (8.1 million), Rio de Janeiro (7.2 million), Belo Horizonte (1.2 million), Recife (1.1 million), Salvador (1.1 million), and Pôrto Alegre (1.1 million). There are many other cities with over 100,000 people. The growth rate for most urban centers is between 4.5 and 6.0 percent; Brasília, the new capital, is growing at the rate of 11.5 percent. With urbanization have come the problems of traffic congestion, deficiencies in public services, and air pollution.

The rural sector of the economy has grown at a slower pace, 4 to 5 percent yearly. About 46 percent of the population is engaged in agriculture and supplies low-cost foodstuffs and industrial raw materials. Moreover, it is agriculture that provides the bulk of the exports: coffee, 27 percent; cotton, 5 percent; and sugar, 4 percent. Manufactured goods and minerals only account for 17 percent of all exports. The rural sector contributes 20 to 25 percent of the GNP and is receiving much attention in an attempt to improve production rates.

One example of Brazil's drive toward development is its push to occupy its territory effectively and to tap an enlarged resource base. In 1960, the national capital was moved from Rio de Janeiro to Brasília on the agricultural frontier on the Brazilian Plateau. This move stimulated the western push of settlement. More recently has come the planned occupation of the Amazon Basin with an extensive road-building program. Road building in both the plateau and Amazon Basin is opening these areas to an influx of settlers.

One reason for Brazil's limited development has been the exploitative nature of the economy. From Portuguese colonization to the present, emphasis has been on derivation of maximum wealth in minimal time without regard to long-term, stable growth. Several cycles of boom-and-bust economic activity have been experienced. Sugar cane in the Northeast was the first commodity of importance. Large plantations with African slaves yielded great profits to the owners in the sixteenth and seventeenth centuries. For a long time, the Northeast was Brazil's most important region. The sugar industry eventually collapsed under stiff competition from other parts of Latin America, accelerated by the inability or lack of initiative to apply improved levels of technology used elsewhere. The second cycle was the push inland to the Brazilian Plateau north of Rio de Janeiro where gold and diamond deposits were discovered. The exploitation of these minerals led to partial settlement in the interior. Moreover, the discoveries encouraged other colonists to come to Brazil. Yet after the many surface ores were removed, there was an exodus from the area.

Coffee

The third cycle occurred after 1850 with the search for areas suitable for coffee cultivation. The perennial coffee bush has a very restricted environmental range. It cannot withstand frost, nor does it do well in warm-to-hot areas. These restrictions make the *Tierra Templada* zone in the tropics one of the few suitable areas; another area is along the boundary between tropical and mid-latitude climates. In Brazil, the southern area of the Brazilian Plateau is within the *Tierra Templada,* and the plains just to the south are on the tropical mid-latitude climatic boundary. It is around São Paulo that coffee production reached its zenith; it is there too that the famous *terra roxa* soil exists that is ideal for coffee. Until early in the twentieth century, coffee commanded a high price, and the area devoted to it spread into areas environmentally marginal for its growth. When prices dropped, these areas were abandoned.

Coffee culture put Brazil "on the map," and for many years Brazil exported more coffee than all other nations combined. By the 1910s, world production exceeded demand, and prices rapidly dropped. The Brazilian government tried to protect its most important export by buying the harvest and holding back coffee from the market until an acceptable price was attained. Other Latin American nations, notably Colombia, undercut Brazil and took a large part of the market. Since then, Brazil has repeated its attempt to control coffee prices but each time eventually has lost out to other countries. In recent years other coffee-producing nations have joined Brazil in trying to limit production and control the price. Such endeavors, called valorization, can only work when producing nations are in full cooperation.

The rise of coffee in Brazil illustrates a method of land development that requires little money and is similar to the tenant-owner agreement used in the Argentine Pampas. At the time coffee production began its great expansion, thousands of immigrants from Europe entered Brazil; most came to the southern half of the country as tenants on large *fazendas* (farms) whose owners were land rich but money poor. Between the owner and tenant, a working agreement was made that involved little or no cash. The tenant leased a portion of the *fazenda,* often uncleared land, on which he planted coffee bushes provided by the owner. For a period of years, the tenant tended the young plants and cultivated his own crops between rows of coffee. After five to seven years, the coffee bushes began to bear fruit, and the tenant moved on to another plot. In this manner an entire *fazenda* could be put to coffee with little capital; and once harvesting began, cash-wage employment was available to the tenant and his family.

Land of Contrasts

Today, Brazil is a nation of contrasts. The western half of the country is very sparsely populated, but the east and along the coast are densely settled. There is both a traditional and modern Brazil. Traditional Brazil is the rural and agricultural sector with the controlling landowner group of European extraction and the workers of European, African, and Indian stock. The workers are poor, often illiterate, and cultivate their small farms or work on large *fazendas* using hand tools. Modern Brazil is urban and industrialized. Living levels are high, literacy is nearly universal, and machines have replaced muscles. In the south, modern Brazil includes mechanized grain-

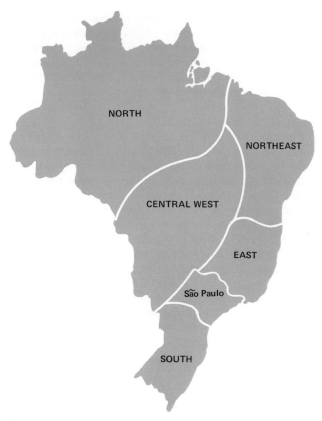

23–6 *Brazil: Regions*

São Paulo state is also the leading region in agricultural production. Coffee, beef, sugar, cotton, peanuts, truck crops, and rice are the principal farm products. Tractors and fertilizers are widely used.

Urbanization has progressed rapidly. The city of greater São Paulo has grown from a backland trade center to one of the world's largest urban areas. Growth has been rapid, particularly in the last half century; its population has increased from 600,000 in 1935 to 4.5 million in 1960, to 8.1 million in 1970, and is projected to be more than 20 million by 1983. São Paulo City alone accounts for 55 percent of the nation's manufacturing with almost 25,000 factories. It is the nation's leading financial center with the largest number of banks and the biggest stock exchange. Rapid growth has not come without problems. Many migrants from the countryside and other regions have been attracted to the city; most are illiterate, unskilled, and live in great poverty. Only a small part of this group has access to the city's water supply and sewage network. Transport, housing, public health, and schools are generally inadequate. The World Health Organization considers São Paulo's air pollution level near the dangerous point. Numerous planning agencies at the city, state, and national levels have been formed to cope with these problems. Other cities, including the port of Santos, have experienced similar growth and accompanying problems.

Why São Paulo state became the center of Brazil's economy and has developed so rapidly is not clear. In colonial times São Paulo seemed to have no inherent advantage in population, resources, or location. *Paulistas* (people of São Paulo) spread over large parts of Brazil in search of a livelihood. In the last half of the nineteenth-century coffee production began, and São Paulo eventually became the center of its cultivation. Perhaps the combination of propitious timing, wealth from coffee, and entrepreneurial skills of the *Paulistas* is the answer.

The South

Southern Brazil has its own distinctive flavor. In the nineteenth century large numbers of German and Italian farmers settled in the area, and many retained their national identity well into the twentieth century. German and Italian architecture is still noticeable. Agriculture is the basis of livelihood, but several cities of over 100,000, and growing at rates of about 4 percent yearly, are located on or near the coast; Pôrto

farming areas. Almost all modern Brazilians are of European origin. The difference between traditional and modern Brazil is striking. A third contrast is the regional differences. Areally there are great variations in population and economic activity. These differences are treated below (Fig. 23–6).

São Paulo

The state of São Paulo is Brazil's most modern and productive region. Per capita incomes are twice those of the national average, and nearly one-third of the national GNP is produced there. São Paulo accounts for almost two-thirds of the nation's industrial production. Practically all of Brazil's motor vehicles are produced there, and the state is the national leader in the manufacture of textiles, cement, shoes, paper products, soluble coffee, pharmaceuticals, and electrical goods. Petroleum refining, petrochemicals, and steel production are also important.

Alegre (1.1 million) and Curitiba (500,000) are the largest. Most of the region falls within the well-settled part of Brazil, but the western section is still part of a pioneer fringe. Transport facilities have been greatly expanded and have contributed much to the region's propserity. Exploitative agriculture is less prevalent in the South, perhaps because of the migrant farmers who have settled there. Cattle ranching and mechanized wheat farming characterize part of the agricultural zone. In the north, coffee raising has been pushed to its southern physical limit, since freezing temperatures penetrate from the south almost to the northern limit of the region. Rice, maize, and hogs are raised in the more traditional manner with the hogs associated especially with German settlements. Some Italian communities are noted for their grapes and wines.

Urban areas are basically regional service centers, but in the larger cities, such as Pôrto Alegre, industrialization based on processing farm and ranch products provides employment to a sizable part of the labor force. Milling, meat packing, tanning, textiles, and breweries are typical types of manufacturing.

Mineral resources of the south are limited. Small deposits of low-grade coal are mined to provide energy in the region and for shipment to the East and to São Paulo.

The East

The East is a region of contrast between modern and traditional Brazil. Large modern urban centers are in stark contrast to some of the nearby rural areas little changed in the last 100 years. The East is basically the hinterland (tributary area) of Rio de Janeiro, the former capital and second city of the nation. The region has experienced several waves of exploitation, first sugar cane, then gold and diamonds, coffee, rice, and citrus fruit. Between each boom period was a time of population decrease and reversion to a grazing economy.

Rio de Janeiro is the focal point of the East. Founded early in the colonization movement as a defensive position, its excellent natural harbor and location near the inland goldfields gave the city an early advantage. Later the capital was moved from the Northeast to Rio de Janeiro, and its functions consequently increased. As long as Brazil was a coastal nation, Rio continued to be a national focal point. Access to the interior, however, was difficult, since just behind the

narrow coastal plain is a steep escarpment. The advent of the motor vehicle and paved roads has partly offset this natural disadvantage, but other cities located inland have usurped some of Rio's former trade area. Rio itself faces a severe shortage of level land suitable for urban development.

The hinterland of Rio de Janeiro has both an agricultural and mining base. Most of the land is used for grazing; cropland is devoted to the cash crops of sugar cane, coffee, rice, and citrus, but much land is used for subsistence crops. Land rotation is a common farming practice; after cropping a plot for a few years, soil erosion and depletion make further cultivation uneconomical. The land is then allowed to "rest," and another plot is put to crops. Part of the East is one of the most mineralized areas in the world, the Mineral Triangle. Only within this century has large-scale mining been developed, and further development is in the offing. Gold, diamonds, and a number of precious and semiprecious gems have been mined for centuries. Industrial minerals such as manganese, chromium, molybdenum, nickel, and tungsten are of growing importance. Iron ore, however, is the most important mineral. The amount of iron ore reserves is unknown but is certainly one of the largest in the world. A conservative estimate is 15 billion tons of easily recoverable high-grade ore.

Volta Redonda is Brazil's iron- and steel-making center. Located close to the iron ore with nearby limestone and water and the São Paulo and Rio de Janeiro markets, the site is almost ideal. Coal is the one vital ingredient lacking and is brought in from the south and imported from the United States. The Volta Redonda plants began production in 1946, and since then have increased in capacity several times. Today, Brazil is the leading steel producer in South American and by 1980 is projected to be the world's tenth largest steel maker.

The Northeast

The Northeast was once the center of Brazilian culture, the location of the capital, and the most developed part of the nation. This favored position was based on sugar cane cultivation in the plantation system with African slaves and on the Northeast's location as the nearest part of Brazil to Portugal. Proximity was advantageous during the days of sailing ships but lost importance with the advent of steamships. Sugar cane is still grown in the region, but it is no longer a prosperous industry. Today, the Northeast is poverty-stricken, and the average per capita income is less than half the national mean. Thousands of people have migrated to other parts of Brazil, adding their number to unskilled laborers searching for work in the cities farther south and in the new capital of Brasília.

Population pressure is great and livelihood precarious even in the best of times. Over the years, settlement has moved inland away from the moist coast and into a drier environment. In the backlands, hillsides are cultivated with minimum regard for conservation, and pastures are overgrazed. Both practices have resulted in soil destruction and rapid water evacuation from the land. Exploitative agriculture is characteristic. The naturally semiarid environment has become still more moisture deficient by human actions. Droughts are frequent; and when they occur, there is a large movement of people from the interior to the already overcrowded coast and to other parts of the nation. Yet so strong is the tie to the Northeast that many persons forced to leave return when they can.

The Northeast has two cities of over 1 million people, Recife and Salvador, and one other over 500,000, Fortaleza. These port cities function as service centers. Manufacturing is limited.

Since about 1960, a number of government-financed projects have been directed to the Northeast and a regional development corporation has been established. Dams for hydroelectric power and irrigation are being built and roads improved and extended. Industries locating in the Northeast receive special tax rebates. All these efforts have yet to become effective.

Central West

The Central West region is part of Brazil's *Sertão*, the backlands where a frontier spirit and life prevail. Many parts of the region remain essentially unknown; most is very sparsely settled; and much remains outside effective control of the government. On the eastern edge of the Central West the new capital of Brasília was constructed to symbolize and encourage the occupation of the backlands. Roads have been built to connect Brasília with other parts of the nation and westward to new lands. Along the eastern fringe of the region, some commercial agriculture has evolved, but elsewhere subsistence agriculture, graz-

ing, small-scale lumbering, and mining are carried on. Throughout the region are small pockets of settlements, founded in earlier years by *Paulistas* and others who roamed the backlands searching for gold, and grass for their cattle.

In the pioneer zone, the frontier is gradually moving westward as new lands are opened for agriculture. Brazil believes its future lies in the west. Pioneering is at best a risky venture, but the farmers have discovered a relationship between vegetation and quality of land for cropping. The existence of this relationship is especially important, since there are few weather stations and soil surveys to provide more scientific data. Six vegetative types are recognized. In descending order of quality for agriculture, they are:

- *Mata da primeira classe* (first-class forest)
- *Mata da segunda classe* (second-class forest)
- *Campo cerrado* (mixed grassland and woodland)
- *Cerradão* (scrub woodland)
- *Campo sujo* (grassland with scattered trees)
- *Campo limpo* (grassland)

Mata vegetation is regarded as good-quality land suitable for cropping. *Campo Sujo, Campo Limpo,* and *Cerradão* are considered generally unusable for crops, although mechanized cropping might be feasible. Generally, however, these types of vegetation are used only for grazing. Use of areas covered by *Campo Cerrado* vegetation holds the key to occupation of the Central West, since this vegetative type covers about 75 percent of the region. *Campo Cerrado* soils are droughty and lacking in plant nutrients. Extensive mechanized agriculture emphasizing wheat has had some success in the eastern part, and alfalfa, a deep-rooted plant, offers another possibility. The main use of *Campo Cerrado* areas, however, is for grazing.

The Amazon Basin

In recent years much effort has been exerted in taming the Amazon Basin. Although the basin has long been considered an area of great potential and occupies over 40 percent of the nation's area, it has only 4 percent of the population. Only along the Amazon River and its principal tributaries have people of European culture been able to gain a foothold. Manaus (300,000) near the center of the basin and Belem (600,000) near the Amazon mouth are the major urban places. Away from the rivers live

The effective settlement and use of the Amazon basin have long been a dream of many Brazilians. This dream is now being realized through costly and ambitious governmental development programs. The key to these programs is the Trans-Amazonian Highway shown here. (Manchete from Pictorial)

an unknown number of Indians who, although touched by western society, still cling to their traditional ways; they gain their livelihood by shifting subsistence agriculture, emphasizing manioc, maize, and beans, and supplement their diet by hunting, fishing, and gathering.

In the nineteenth century, many parts of the Amazon were explored for rubber trees (*Heva brasiliensis*), the major source of natural rubber in the world. Tappers extracted the latex, smoked it into balls, and sold the balls to traders for shipment to the United States and Europe. Rubber first gained importance with the development of the vulcaniza-

tion process, but the demand greatly increased in the latter part of the century with the pneumatic bicycle tire and shortly afterward the automobile. Rubber prices skyrocketed, but still rubber was gathered only from the wild tree of Brazil. Finally, some seeds were smuggled to England, and these seeds became the foundation of the rubber plantations of Malaya, Sumatra, and elsewhere. Once plantation rubber came on the market, the prices dropped; and wild rubber gathering greatly decreased in importance.

Starting about 1950, two small attempts at taming part of the Amazon Basin have met with some success. The first has been the introduction of Japanese farmers who have been able to produce rice, jute (a hard fiber), and black pepper on a commercial basis. The second has been the mining of manganese ore in the extreme northeastern part of the basin. Many other minerals are known to exist in the basin but have yet to be mined. Of these, the search for petroleum has been the most extensive. Although oil has been discovered, deposits extensive enough for development have not been found.

More recently the Brazilian government has initiated a most ambitious development program for the Amazon Basin. In 1970, work on the Transamazonian Highway began and heralded the beginning of a massive road system throughout the basin and connection of the region to other parts of the nation (Fig. 23-4). With the road system is the expectation, already partly realized, that migrants from the Northeast and elsewhere will settle the area. Plans call for government-sponsored construction of urban areas, lumber and pulp plants, hydroelectric plants, and schools. To governmental officials and most Brazilians, the Amazon Basin development program is looked upon as vitally necessary to assure continued economic growth of the nation and a better life for its citizens.

This view, however, has been challenged by some ecologists who maintain that although short-term economic growth may result, disastrous results may occur over a long time. Many ecologists believe that destruction of the tropical forest may substantially change the area's climate, that drier and warmer conditions will prevail. Some estimates are that the forest could be cleared in less than 100 years. They also point out that many of the soils of the basin are low in plant nutrients and require careful handling to prevent their destruction. Whether or not the ecologists are right, Brazil is committed to Amazon Basin development.

Further Readings

For a general regional study, see Blakemore, Harold, and Smith, Clifford T. (eds.), *Latin America: Geographical Perspectives* (London: Methuen, 1974), 600 pp.; and Taylor, Alice (ed.), *Focus on South America* New York: Praeger, 1973), 274 pp.

The Inca civilization has placed a stamp on present-day Andean America. Some aspects of Inca culture are examined by Kelly, Kenneth, "Land Use in the Central and Northern Portions of the Inca Empire," *Annals, Association of American Geographers,* 55 (1965), 327–338. Agriculture remains a dominant way of life in the Andes. Some aspects are discussed by Blankstein, Charles S., and Zuvekas, Clarence, in "Agrarian Reform in Ecuador," *Economic Development and Cultural Change,* 22 (1973), 73–94; Wood, Harold A., "Spontaneous Agricultural Colonization in Ecuador," *Annals, Association of American Geographers,* 62 (1972), 599–617; Gitlitz, John S., "Impressions of the Peruvian Agrarian Reform," *Journal of Inter-American Studies and World Affairs,* 13 (1971), 456–474; Stearman, Allyn M., "Colonization in Eastern Boliva: Problems and Prospects," *Human Organization,* 32 (1973), 285–293. Other aspects of the economy are presented by Smetherman, Bobbie B., and Smetherman, Robert M., "Peruvian Fisheries: Conservation and Development." *Economic Development and Cultural Change,* 21 (1973), 338–351; Pearson, Donald W., "The Comunidad Industrial: Peru's Experiment in Worker Management," *Inter-American Economic Affairs,* 27 (1973), 15–29; and Andrews, Frank M., and Phillips, George W., "The Squatters of Lima," *Journal of Developing Areas,* 4 (1970), 211–224.

Southern South America has experienced different political philosophies. See, for example, Williams, John H., "Paraguayan Isolation Under Dr. Francia," *Hispanic American Historical Review,* 52 (1972), 102–122; Smith, Peter H., "The Social Base of Peronism," *Hispanic American Historical Review,* 52 (1972), 55–73; Porzecanski, Arturo C., *Uruguay's Tupamaros: The Urban Guerrilla* (New York: Praeger, 1973), 80 pp.; Petras, James F., "Chile: Nationalization, Socioeconomic Change and Popular Participation," *Studies in Comparative International Development,* 8 (1973), 24–51. Chile's recent experience in communism presented problems of international importance. See Gil, Federico G., "Socialist Chile and the United States," *Inter-American Economic Affairs,* 27 (1973), 29–47; and Vicuna, Francisco, "Some International Problems Posed by the Nationalization of the Copper Industry by Chile," *American Journal of International Law,* 67 (1973), 711–728. The Argentine Pampas received not only people but improved technology. See Jefferson, Mark S. W., *Peopling the Argentine Pampa* (Port Washington, N.Y.: Kennikat Press, 1971), 211 pp., a reprint of his 1926 book; and Winsberg, Morton D., "The Introduction and Diffusion of the Aberdeen Angus in Argentina," *Geography,* 60 (1970), 187–195.

Brazil's economic growth is traced by Alves, Marcio M., *A Grain of Mustard Seed: The Awakening of the Brazilian Revolution* (Garden City, N.Y.: Doubleday, Anchor, 1973), 194 pp.; and Worcester, Donald E., *Brazil: from Colony to World Power* (New York: Scribner, 1973), 277 pp. São Paulo's growth is traced by Dean, Warren, *The Industriali-*

zation of São Paulo, 1880–1945 (Austin: University of Texas Press, 1969), 263 pp.; and Morse, Richard M., "São Paulo in the 19th Century," *Inter-American Economic Affairs*, 5 (1951), 3–39. Sugar in Northeast Brazil is examined by Galloway, J. H., "The Sugar Industry of Pernambuco during the Nineteenth Century," *Annals, Association of American Geographers*, 58 (1968), 285–303. Problems of drought are noted by Brooks, Reuben H., "Human Response to Recurrent Drought in Northeast Brazil," *Professional Geographer*, 23 (1971), 40–44. Brazil's steel industry is presented by Rady, Donald E., *Volta Redonda: A Steel Mill Comes to a Brazilian Coffee Plantation* (Albuquerque: Rio Grande Publishing Co., 1973), 375 pp. Views on Amazon Basin development are given by Foland, Frances M., "A Profile of Amazonia: Its Possibilities for Development," *Journal of Inter-American Studies and World Affairs*, 13 (1971), 62–77; and Denevan, William N., "Development and Imminent Demise of the Amazon Rain Forest," *Professional Geographer*, 25 (1973), 130–135.

Africa and the Middle East

Douglas L. Johnson
Leonard Berry

PART SEVEN

Africa and the Middle East: Aspects of Development and Nation-State Formation

24

THE AREA ENCOMPASSING Africa and the Middle East is characterized more by diversity than uniformity. In location, it ranges from Turkey on the north to South Africa on the south and from Dakar on the west to Muscat on the east (Fig. 24–1). Environments range from thick tropical rainforests to almost rainless deserts. Diversity is also a factor in the human scene. The area has one of the world's richest nations, Kuwait, with a per capita GNP of $11,510, and some of the poorest, including Rwanda ($90), Mali ($90), Upper Volta ($90), and Ethiopia ($100). Culturally there is more homogeneity with the Moslem religion and culture in north Africa and southwest Asia and black African traditional religion and culture south of the Sahara Desert. Each dominates a large part of the region, but Moslem and African cultures intermix and overlap along the Sahara's southern margin. Moreover, Western thoughts and institutions have been introduced everywhere but with varied degrees of success.

Unifying Elements

In such a large and diverse part of the world, there are both unifying and differentiating elements. Important unifying elements include spatial linkages, common colonial or neocolonial history, broad cultural and historical traditions, and the common belief that all countries in the area, except Israel, South Africa, and perhaps Turkey, consider themselves part of the Third World.

Africa and the Middle East are the home of three overlapping cultural worlds: African, Moslem, and European. In each of the region's countries, some elements of each world are found. Arab Moslem and African interactions have taken place across the Red Sea and across the land routes of the Sahara for many centuries. European colonialism brought governmental institutions, and sometimes settlers, to African and Middle Eastern countries. Saudi Arabia and Turkey were affected less than most, and direct colonial rule is a thing of the past. Most independent nations, however, still show a strong heritage of the European period.

The concept of the Third World grew on the wave of national independence in former colonial territories. As new nations struggled with the critical problems of world trade, the Cold War, and international political and economic dependence, so too grew the need for solidarity. The Organization for African Unity (OAU) is a multinational entity set up to work for cooperation within the continent, but in the late 1960s and early 1970s, Third World unity was more common in international forums. On many issues deliberated in the United Nations, African and Middle East countries have often joined forces with other Asian nations.

Differentiating Elements

Within the framework of shared attitudes and positions in the international forum and a background of similar colonial experience, many differentiating elements separate the region's nations. Five elements are most important: (1) oil and mineral wealth; (2) religion and culture; (3) degree of economic diversity;

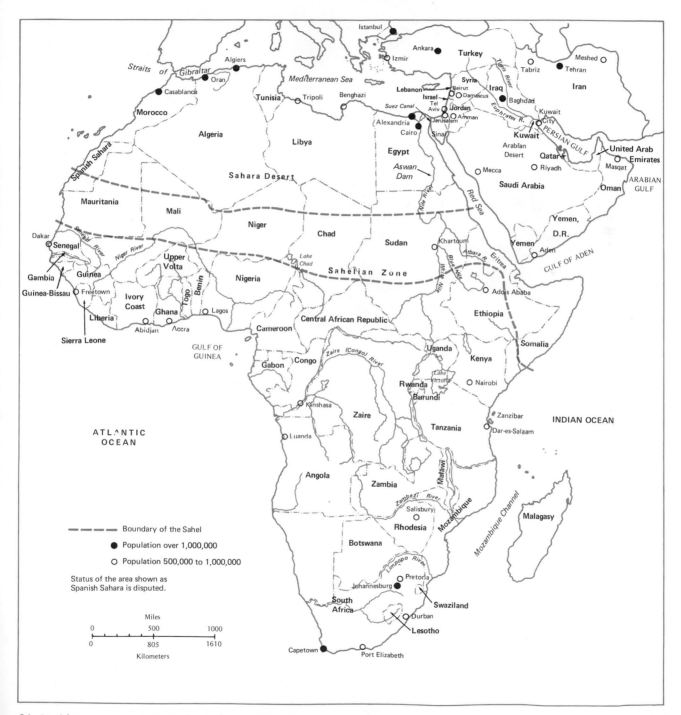

24–1 Africa and the Middle East: General Locations

The recent increases in oil prices have brought great wealth to petroleum exporting nations. Some of the North African and Middle East nations are major oil producers. Active exploration continues as shown here in Libya. (Photo Library, Mobil Oil Corporation)

(4) environment and location; and (5) external ties and dependencies.

Oil and Mineral Wealth

Oil and mineral wealth have played an important, sometimes overwhelming, role in the economic development of a number of the nations in Africa and the Middle East. Throughout the area, oil and minerals have been exploited mainly for export. The Western industrial world and Japan have had a rapidly growing need for fuel and minerals. Although prices for minerals have fluctuated, there generally has been a good economic return from petroleum production. In the case of Middle East countries such as Saudi Arabia, Iran, and Kuwait, oil wealth has transformed the economy. Since the Organization of Petroleum Exporting Countries (OPEC) raised prices for petroleum and petroleum products substantially in late 1973, some of the highest per capita GNPs in the

world are found in the small oil-rich Arabian states.[1] Similarly the GNP of Algeria, Libya, and Nigeria in Africa has greatly increased due to oil wealth. Most of these countries are part of a group of states with high actual or potential prospects for economic growth.

The effect of other mineral deposits has not been so great, although Zaire (copper, diamonds, and tin) and Angola (diamonds and iron ore) are examples of countries that have gained substantially from mineral exploitation. Countries without such resources often have had to rely heavily on agricultural products. In the past twenty years, agriculture has not

[1]OPEC, the Organization of Petroleum Exporting Countries, is a cartel composed of thirteen oil-producing countries. It is the most successful of the Third World organizations that attempts to control the price of raw materials essential to the economies of the rich industrialized consumer nations. By controlling the supply of oil, OPEC tries to raise, and to maintain at a high level, the price paid for petroleum.

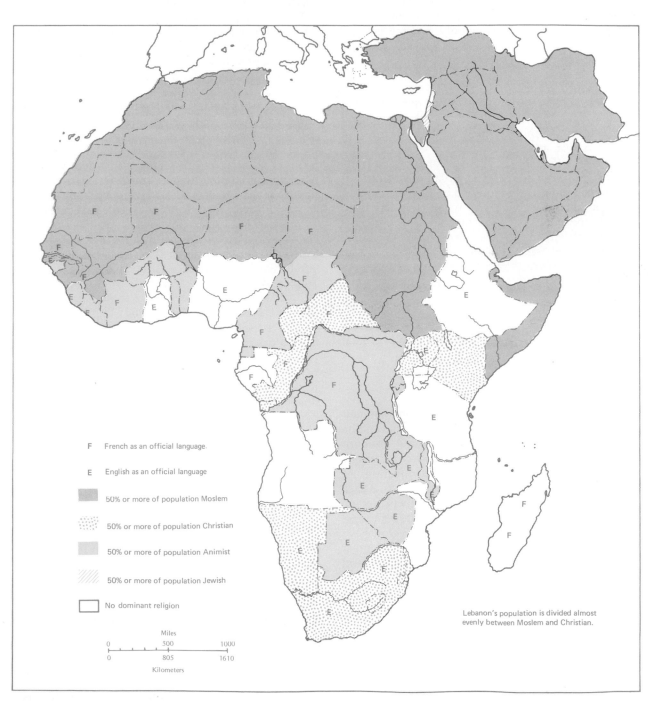

F   French as an official language.

E   English as an official language

▨   50% or more of population Moslem

⠌   50% or more of population Christian

▨   50% or more of population Animist

▧   50% or more of population Jewish

☐   No dominant religion

Miles
0   500   1000
0   805   1610
Kilometers

Lebanon's population is divided almost evenly between Moslem and Christian.

24–2 *Africa and the Middle East: Language and Religion*

brought returns comparable with those from mineral resources. As a result, those countries relying heavily on agriculture or possessing only modest mineral resource endowments have experienced difficulty generating rapid economic growth.

Religion and Culture

Religion and culture are strong differentiating elements. The Moslem religion and the associated Islamic culture dominate the Middle East, most of northern Africa, and significant areas of western and eastern Africa (Fig. 24–2). The rest of Africa is characterized by remnants of former animist cultures[2] which have been greatly modified by Christianity and European cultural beliefs. The whole area has been subject to the effects of European colonialism, including white settlement in some areas. Pockets of European settlement still remain, although their size outside South Africa is rapidly declining. More pervasive is the economic, cultural, and institutional impact of the former colonial powers. The unifying effect of a common colonial history is distorted by colonially initiated language and cultural differences among countries and by the relatively small colonial input in Middle East countries compared with much of Africa.

Economic Diversity

Economic diversity is also an important element differentiating a few countries from the rest. The region contains no highly industrialized modern state. Saudi Arabia and some other oil-rich nations are attempting to move rapidly in this direction. At present, however, Egypt, Turkey, and South Africa have the greatest range of economic activity. Israel, Algeria, and Iran have a significant but lesser range. In contrast, most of the countries of the region have little manufacturing industry. Large population, great internal purchasing power, availability of human and physical resources, capital, governmental policy and stability, and entrepreneurship are factors that contribute to the ability of some countries to develop a modern, diversified industrial sector.

[2]Animist cultures are characterized by the conviction that natural objects such as trees, stones, and rivers have conscious life and souls. This veneration of spirits present in the natural world often is associated with the belief that prayer and ritual can placate the spirit world and bring positive benefits such as increased rain or a better harvest.

Environment and Location

Environment and location also differentiate countries in the region. Middle Eastern countries have tended to be remote from Africa, although there have long been links across the Arabian and Red seas and through Egypt. Within Africa, problems of distance and access to world economic systems have tended to work against the landlocked states such as Chad and Mali. Possessing arid and semiarid climates, limited agricultural resources, and poor communications, the Sahelian countries, for example, have been at a disadvantage compared with the neighboring coastal African countries. It is no accident that the region's least developed countries are those afflicted with locational disadvantages and environmental constraints. Most of the countries characterized by rapid economic growth rates possess a diversity of environments or are located in more productive dry summer subtropical and humid-tropical settings.

External Ties and Dependencies

Lastly, the patterns of colonial dominance have been replaced partly by new links and alliances with other parts of the world. France has maintained close ties with its former colonies in West Africa. Former British colonies display a wider range of degrees of dependence. Countries like Zaire, Tanzania, and Zambia have each in different ways sought new alignments since independence. The bonds holding Moslem Middle East countries together have remained particularly strong.

Up to this point, we have noted briefly the diversity and unity within Africa and the Middle East mainly from the point of view of the current environmental, economic, and sociopolitical situations. We now turn to the economic and cultural history of the region to understand from the background of history how the present patterns of development evolved.

Historical and Economic Background

To characterize the common features of Africa and Middle East history and economy is no easy task. The region is so vast, its historical depth so great, and its cultures so rich and varied, that common denominators at first glance seem impossible; yet common themes are present. These themes, four in number, have shaped the main elements of the contemporary scene. The growth, often under difficult conditions, of indigenous cultures and economic systems, the impact of intrusive colonialist societies, the economic and cultural links and relationships that emerged during the colonial era, and the state of economic well-being and development that has resulted from these historical processes are the themes that give identity to the region.

Indigenous Systems: Contributions to the World

Perhaps as a result of the trauma of the colonial era, and the residual patronizing attitudes of Westerners toward "Darkest Africa" and "heathen" sections of the non-Christian Middle East, it is often suggested that little cultural and technological development took place in the region before the eighteenth and nineteenth centuries. Nothing could be further from the truth. The civilization of the European culture realm is based upon a variety of inventions developed in or near the African and Middle East shores of the Mediterranean since 8000 to 10,000 B.C. Barley and wheat, two cereals of great importance in the diets of today's rich nations, were first domesticated by Neolithic farmers in the fertile hills of Iran, Turkey, and Palestine. Sheep, goats, and certain varieties of cattle were also first domesticated in this region. The alphabet, arch, wheel, metallurgy, the first cities, astronomy, mathematics (including a system of counting by sixes that underlies our division of the hour into sixty minutes and the minute into sixty seconds), the potter's wheel, and numerous other inventions appear to have first emerged near the junction point of Africa and Asia and then spread elsewhere.[3]

If cultures south of the Sahara had less impact on Europe than did those of the Middle East, they were no less inventive. A different set of crops, often revolving around sorghum and millets in the drier areas and root and fruit crops in the tropics, formed the basis of local economies. These crops still supply basic subsistence foods for much of the region, and some, such as Akee (*Blighia sapida*), a fruit that is an essential ingredient in Jamaica's national dish, have become important in the New World as a result of the slave trade. Blocked from easy contact with distant cultures by long ocean journeys and the inhospitable Sahara Desert, much of sub-Saharan Africa's inventive energies were expended in local cultural and artistic developments, such as wood carvings.

Indigenous Systems: Two Traditions

Thus, two separate cultural traditions can be identified in Africa and the Middle East. On the one hand are the intertwined cultural streams that ultimately produced the Islamic realm. On the other hand are the animist traditions south of the Sahara. Both cultural traditions are held together by long-existing contacts via the Nile, the Saharan caravan trails, and the sailing routes of the East African coast. They overlap significantly in the grassland regions south of the Sahara where numerous animist groups have been converted to Islam. In the Islamic culture region of Africa and the Middle East, Arabic speech and culture are dominant. There are significant minorities (Berbers, Turks, Iranians, and Kurds), however, who retained footholds in remote or mountainous areas after the Arab conquest in the seventh century. South of the Sahara, blacks speaking a variety of Bantu and Sudanic languages dominate, although Cushite, Chadian, European, and Click-speaking language minorities are regionally significant. In both regions residual elements of Christian influence (Lebanon, Ethiopia, South Africa, and Rhodesia) are important, although their importance has declined greatly in the last few decades.

Two underlying cultural traditions have particular importance for economic development in Africa and the Middle East. The first is a set of driving forces that produce political integration and the formation of empires and embryonic nation-states. Movements of a national character predate European colonial occupation, and aspirations of and agitation for independent existence and identity remained strong throughout the colonial era. The second is the set of forces such as tribalism which tends to fragment national development and is divisive in nature.

Indigenous Systems: National Forces

In West Africa, extensive empires date back at least to the Ghana Empire of A.D. 1000 (Fig. 24–3). A series of empires, including the Mandingo Empire with a

---

[3]Charles Singer, E. J. Holmyard, and A. R. Hall (eds.), *A History of Technology*, Vol. 1, *From Early Times to Fall of Ancient Empires* (Oxford: Clarendon Press, 1954), pp. 327–803.

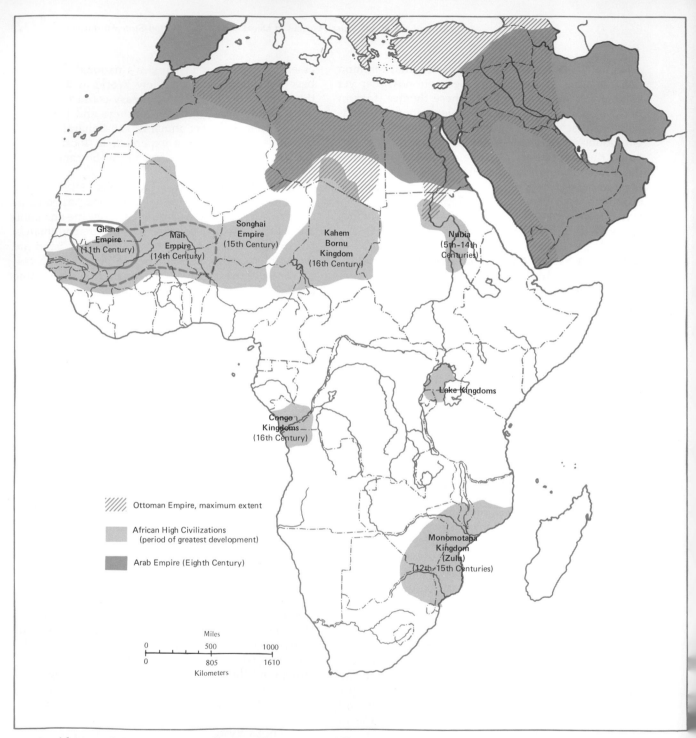

24–3 *Africa and the Middle East: Former Kingdoms and Empires*

SOURCE: Adapted from Regine N. Van Chi-Bonnardel, *The Atlas of Africa* (New York: Free Press, 1973); R. Roolvink, *Historical Atlas of the Muslim Peoples* (Amsterdam: Djambatan, 1957); and Harry W. Hazard, *Atlas of Islamic History,* 3rd ed. (Princeton: Princeton University Press, 1954).

capital at Mali and the Songhai Empire, encompassed a number of ethnic and tribal groups in the savanna-forest zone and experienced the same dynamic processes of growth and decay that characterized state and empire formation in Europe. Although written records from this period are generally lacking, impressive achievements in the arts, commerce, and administration were attained. With the collapse of the Songhai Empire, numerous smaller states emerged. Those along the coast, such as Dahomey, Oyo, and Benin, represented a local adjustment and response to the problems and opportunities presented by the slave trade. These states made a successful transition to an agricultural export economy when the slave trade was banned.

Farther inland in the savanna zone, where Islamic influence was greater, theocratic states organized by conquest and war (*jihad*) developed in the nineteenth century. The *jihad* of Usuman dan Fodio resulted in the enduring emirates of Sokoto, while more transitory states were developed by the *jihad* of al-Hajj Umar and Samori ibn Lafiya Toure. In the Middle East and along the Mediterranean shores of North Africa, the Ottoman Empire for centuries provided continuity and integration for the Islamic realm in the Eastern Mediterranean. At its height, the Ottoman Empire was the major military and political rival of the kingdoms of western and central Europe. Its collapse after World War I left a political void that the nations of the region are still struggling to replace with suitable and stable structures.

It would be a mistake to assume that either the past glories of ancient empires or the truncated aspirations of more recent national movements were eradicated by the colonial regime. On the contrary, they often provided the inspiration for the independence movements that flourished after World War II. The Somali, for example, although occupied were never really conquered, and Sheikh Muhammad Abdille Hassan's long rebellion (1900–1920) against outside political control constituted the country's primary political event in this century. In Libya, the Sayyid Muhammad Idris, the grandson of the Grand Sanusi and spiritual head of the Sanusiya, emerged as the symbol of national independence and became the first ruler of an independent state. Continuity with the past was claimed by the Gold Coast when it changed its name to Ghana upon attaining independence in 1957. The French Sudan underwent a similar transmutation

following independence, taking the name Mali in memory of the capital of the deceased Mandingo Empire. In these and numerous other instances, continuance of precolonial traditions as well as glorification of the resistance to colonial rule constituted a feature of the postindependence years.

Indigenous Systems: Divisive Forces

Tribalism and internal cultural divisions affect many of the region's new nations. Colonialism has often served to compound issues that already existed. Tribalism is a problem endemic to any area such as sub-Saharan Africa where more than 1,000 linguistic and ethnic entities are commonly identified. The situation, however, has been exacerbated because unlike and often mutually antagonistic tribal groups have been jumbled together without reason other than the convenience of colonial administration or accidents of history. Frequently, one tribe, such as the Kikuyu of Kenya, dominates the political process to the detriment of other smaller social groups. Often colonial administration preserved inequitable intertribal relationships such as that between the dominant Tutsi of Rwanda and Burundi and these countries' more numerous lower-caste tribes. The intertribal bloodbaths consequent upon independence in some countries developed directly from these unsatisfactory social and tribal traditions. Even where language and religion are shared, major conflict resulting from differing life styles and traditions, such as those separating Palestinian Jordanians from Bedouin Jordanians, can undermine the stability of the new postcolonial state.

Equally divisive are the traditions of feud and of clan or tribal protection of individuals, regardless of the actions such individuals may undertake. The assassination in 1975 of King Feisal of Saudi Arabia by a nephew over the king's refusal to support the nephew's family on a point of honor is a case in point. The paramount role played by familial or tribal loyalty in many countries reflects the tension existing between traditional patterns of behavior and the requirements of the modern state.

Intrusive Systems

European contact with Africa and the Middle East predates the massive technological and cultural impact of the nineteenth and twentieth centuries. It is, in fact, very old. Religious and economic contacts between the Middle East and Europe date back to the Greeks and the Romans. The Carthaginian Empire maintained trade relations with West Africa. One of its sailors, Hanno, not only established trading stations along the western African coast but also is reputed to have been the first person to sail completely around Africa. In all such interactions, it was the lure of exotic products possessed by the region, or controlled by the region as a transshipment point for the Orient, that attracted outside intervention.

Intrusive Systems: Slaving

One of the first resources sought was the slave. Black Africa had long been one source of slaves for the Mediterranean world, albeit a minor one, but the opening of trade routes to the Orient by sea made direct contact between Europe and sub-Saharan Africa possible. First, trading stations and staging points for the voyage to India were established. Attempts at colonization and settlement were minimal, and attention rapidly focused on the practice of collecting slaves from the interior tribes for shipment to labor-deficient regions. Spanish, Portuguese, and subsequently English and French colonial activities in the Americas provided the spur for slaving, and soon many nations had slave stations scattered around the coast of the Gulf of Guinea. Few countries, large or small, avoided involvement in this pernicious undertaking. The conviction that whites could not work in the tropics, and that blacks, physically vigorous and accustomed to tropical conditions, constituted a valuable resource for this purpose, led to the rapid growth of slave trading. An infamous triangular trade between Anglo-America and Europe, West Africa, and the West Indies soon developed. Manufactured products and trade goods were exchanged in Africa for slaves. Crowded into specially designed, speedy sailing vessels, the slaves were transported to the New World where an exchange for such agricultural products as rum and molasses, desired by Europeans and Americans, was consummated.

No one can be certain how many Black Africans, involuntarily involved in this movement, were killed as a by-product of expeditions and wars launched to accumulate slaves, or died during the voyage to the New World. Best estimates, however, suggest that approximately 9.5 million individuals were transported from coastal districts around the Gulf of Guinea.[4] An extensive trade in slaves also existed along the shores of the Indian Ocean with Arab slave traders penetrating far into the interior in search of suitable victims. The size of this East African slave trade is unknown, but it continued much later in the nineteenth century than did European activity in western and southern Africa and had an equally profound impact on those African societies within its reach.

Slaving was replaced in the last half of the nineteenth century by a scramble for territory on the part of European powers. Whereas the slave trade was confined to the coast, contact with the interior being largely in the hands of intermediaries, the pursuit of economic wealth in this second phase encouraged direct development of colonies and spheres of influence. The potential for conflict between rival powers was enormous, although the ground rules laid down at the Conference of Berlin (1885) controlled the situation to the satisfaction of most parties (Figs. 24–4a and 24–4b). Both the loosely organized kingdoms and tribal states of Black Africa and the declining Ottoman Empire succumbed to the colonial experience. Britain and France gained the largest blocks of territory, but other early comers such as Portugal or nonbelligerent minor states such as Belgium were also major gainers. Late comers to the game of empire, such as Germany and Italy, had to be satisfied with less attractive scraps of territory in less-productive districts. Initial penetration into the Ottoman Empire was often under the guise of protection of local Christian minorities, usually accompanied with trade concessions, and full control had to await the dissolution of the Ottoman state after World War I.

Intrusive Systems: Export Mining

Economic development during this colonial phase had both good and bad consequences. Since European interest focused on resources that would benefit the economy of the occupying power, emphasis inevitably was placed on developing primary resources unavailable in Europe. One direction taken was to concentrate on extractive mineral industries such as gold, diamonds, copper, and, more recently,

[4]Philip D. Curtin, *The Atlantic Slave Trade: A Census* (Madison: University of Wisconsin Press, 1969), pp. 86–93

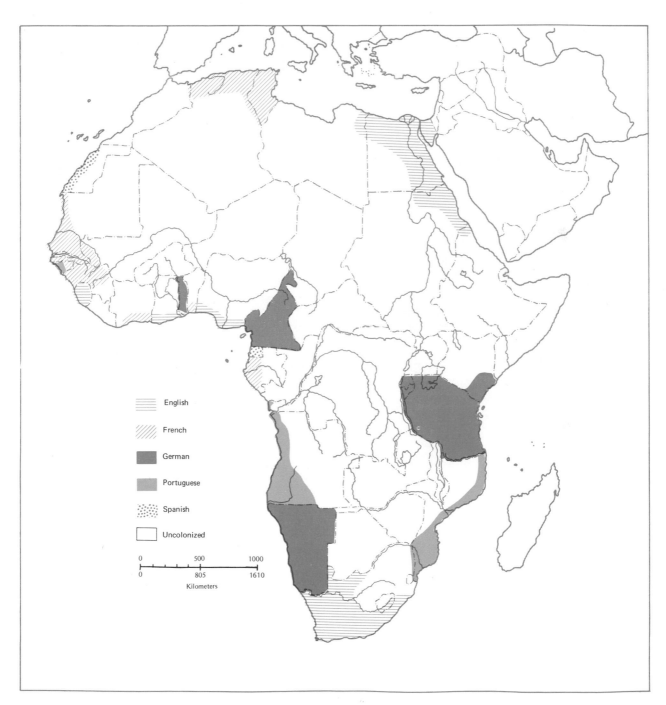

24-4a *Africa and the Middle East: European-Controlled Areas*
SOURCE: Adapted from Regine N. Van Chi-Bonnardel, *The Atlas of Africa* (New York: Free Press, 1973).

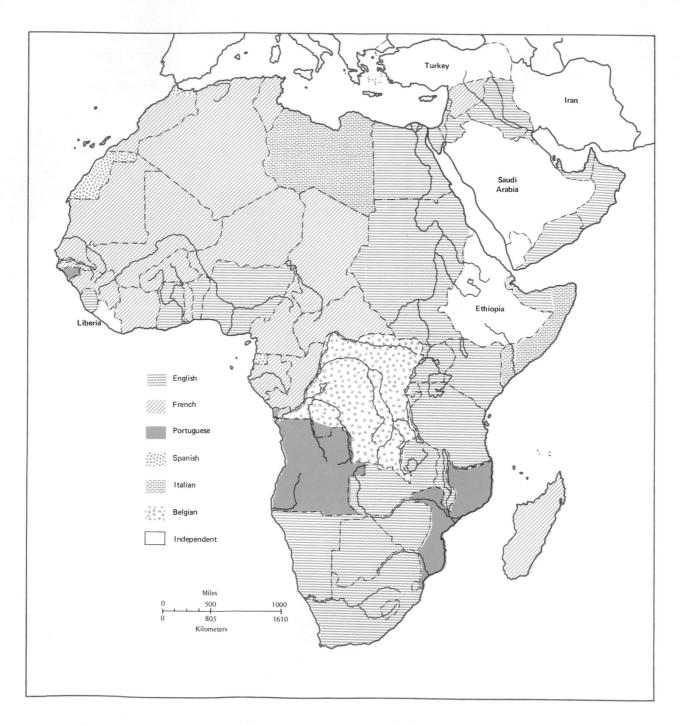

Turkey

Iran

Saudi
Arabia

Ethiopia

Liberia

English

French

Portuguese

Spanish

Italian

Belgian

Independent

Miles

| 0 | 500 | 1000 |

| 0 | 805 | 1610 |

Kilometers

24–4b *Africa and the Middle East: European Possessions and Mandates, 1925*

uranium. Minerals in countries such as Zaire and South Africa, possessing easily located reserves, were rapidly developed. The road, railroad, industry, and port facilities needed to process and move these products remain as an important legacy upon which contemporary economic development can be built. Unfortunately, since economic profitability criteria were the major factors determining the development of such infrastructure, many countries located far from the coast did not benefit. These bypassed countries emerged from the colonial era poorly developed. Niger and Chad are examples of interior states whose development was retarded by locational problems. Such countries were seldom left with anything other than unrealistic dreams with neither infrastructure nor trained personnel. Agriculturally marginal in an environmental sense, their only real hope for viability rests on the prospect that they, like Mauritania or Libya, will also discover mineral wealth.

Intrusive Systems: Export Agriculture

Agricultural development in the colonial period featured the introduction of new crops and the development of cash crop production for export. New crops such as maize (corn) and manioc (cassava) were introduced from the New World and became important subsistence crops, while other New World cultigens such as cacao were used by commercial plantations.[5] Export crop production focused on cacao, palm oil, peanuts, bananas, rubber, cotton, and sisal. Plantations were concentrated in the environmentally most favorable areas, especially those with relatively good transportation connections with the coast. After independence, commercial agriculture provided a basis upon which foreign exchange for development purposes could be generated.

Plantation agriculture, however, placed tremendous pressures on local and traditional resource use systems. The expansion of commercial agriculture often removed land from shifting cropping systems. The result was abuse and impoverishment of resources in the traditional sector. Removal of land for cash crops forced subsistence cultivation into more marginal areas. Often the shift to marginal lands was at the expense of pastoral activities in semiarid grasslands.

Ironically, while commercial agriculture usurped the best land, the peasant farmer was often made more vulnerable to drought and other natural disasters. The recent (1968–1973) Sahelian drought is illustrative of what can happen when heavy reliance is placed on a marginal semiarid environment. There an estimated 100,000 people died from drought-related causes, and 40 percent of the area's livestock perished. Marginal areas, such as the Sahel, provide few economic alternative pursuits and little margin for error in dealing with the environment.[6]

Intrusive Systems: Political

The impact of the colonial era was equally profound in the political sphere. In every case European colonial expansion and administration undermined and suppressed the indigenous forces of state formation and development. Arab nationalists, promised independence by the Allies in return for their help against the Turks in World War I, were frustrated when European-controlled mandate territories were carved out of portions of the prostrate body of the Ottoman Empire. Ashanti growth in central Ghana ran headlong into British imperialism whose interventionist activities undermined the authority and structure of the expanding state. In the Sahara, French activity ran directly counter to the expansion of the Sanusiya religious order. Organized by Muhammad ibn Ali as-Sanusi (the Grand Sanusi) and based in the interior oases of Cryenaican Libya, the Sanusiya movement spread rapidly along the caravan trails of the central Sahara before it was defeated by the French and forced into the role of a local nationalist opponent of Italian colonialism.

A heritage of antagonistic and ambivalent feeling toward colonial authority underlay the movement toward national independence characteristic of the post–World War II era. Only Liberia retained its independence throughout the colonial epoch, although Ethiopia's experience with Italian colonialism was very brief (1936–1941). As the pressure for full independence increased in the European colonies, the response of the colonial powers varied between adamant opposition and more reasoned willingness to grant independence expeditiously.

Where European investment was substantial and

[5]B. W. Hodder and C. W. Newbury, "Some Geographical Changes along the Slave Coast of West Africa," *Tijdschrift voor Economische en Sociale Geografie*, 52 (1961), 77–84.

[6]M. F. Thomas and G. W. Whittington (eds.), *Environment and Land Use in Africa* (London: Methuen, 1969), pp. 23–238.

where a large minority of white agricultural colonists had settled, opposition to independence was most extreme. It took the Mau Mau rebellion in Kenya to awaken British authorities to the need to set the colonies free. In Algeria, nearly a decade of guerrilla warfare by the National Liberation Front was required before France finally accepted Algeria's independence. France was so intimately attached to Algeria that 1 million Frenchmen had settled in a territory France considered an integral part of the metropolitan homeland. Similar guerrilla struggles during the last fifteen years were required to separate Angola, Guinea-Bissau, and Mozambique from Portuguese control, and analogous campaigns have arisen in the white-ruled countries of Rhodesia and South Africa. Other states such as Ghana, Nigeria, and Tanzania achieved independence in the 1950s in a more gradual and controlled transition. In some colonies, most notably Zaire (the former Belgian Congo), independence came so quickly and with such little preparation and training of indigenous personnel that chaos and civil war resulted. In most of the French West African colonies, independence was attained smoothly, and many links with France were maintained (Guinea was an exception).

Economic and Cultural Links

The newly independent, post colonial states of Africa and the Middle East are still linked in often intimate ways with their former colonial rulers. Although in a few cases such as the oil states, the wheel of economic and political power has turned to the point where former colonial mentors are now in a dependent relationship with their old colonies for some vital resources, many new nations are still closely linked to the old colonial power or are forced to struggle with conditions created in the colonial era. The survival of white-ruled states in Rhodesia and South Africa is a relic of the colonial era, a situation that has received much attention from the Organization for African Unity.

Boundaries imposed during the colonial era are an additional problem facing the new states. Many tribes were divided when boundaries were established on an arbitrary basis in unexplored and frequently inhospitable territory without reference to or knowledge of local ecological, social, and political conditions. For example, the frontier between Libya and Chad divides the Tibu people into two segments.

Problems of political control in Chad's northern Tibu districts are almost insoluble, given the antagonism between the Moslem north and the pagan-Christian south. The difficulty of precisely identifying the frontier, the ease with which Tibu guerrillas can withdraw to sanctuary in Libya, and the support given by Libya's government and Tibu population to rebellious elements in Chad further complicate the problem.

Somali-Ethiopian relations suffer from similar problems. Here political borders cut off the interior Ogaden desert from seasonal use by Somali nomads who traditionally spend the dry season in the coastal districts. The desert is ethnically and environmentally linked to the Somali Republic (as is the northern province of Kenya), but Ethiopian control is largely based on military power and on the expectation that valuable oil and mineral resources may be discovered in an otherwise inhospitable district. Although the principle that colonial boundaries should not be changed in the postcolonial epoch is agreed to by most African states, it has caused major problems where these boundaries run counter to the desires of the local population.

The impact of European culture has led to varying degrees of cultural pluralism. Traditional values have been challenged by a technologically sophisticated culture with often devastating effects on local institutions. Educational systems teaching subject matter more suited to Western Europe have replaced local knowledge and traditions. Behavioral values, styles of dress, political systems, technological innovations, and agricultural arrangements at variance with traditional local mores and expectations have arisen. Not all that is alien has been accepted, and the tensions generated by this culture conflict are often severe. Urban elites, often highly Europeanized, frequently are out of touch with the needs and aspirations of the average citizen. Differences in cultural values and inefficient political civilian leadership frequently have led to military coups with the armed forces seizing control, as the only power capable of holding the state together.

Most African states are still in the process of trying to establish a national identity and select those outside stimuli that are most suited to their own development. In this process they are often hindered by their colonial legacy. Economic ties to former mother countries remain strong because markets for agricultural and mineral products have been developed there, and preferential treatment is often extended by the former mother country. It is difficult to change these patterns, particularly if the alternative is to abandon export for outside markets and instead to concentrate on foodstuffs for home consumption.

Colonial economic ties are often reinforced by linguistic ties. Having developed an administrative elite whose second language is French, English, or Portuguese, it is difficult to move outside linguistic constraints to develop alternative sources of advice and assistance. Similarly, language barriers between former French and English colonies make regional cooperation difficult. Yet it is from such policies of interregional cooperation, whereby complementary environmental and cultural zones might be grouped together, that some of the best prospects for successful economic development may be derived.

State of Economy and Development

One consequence of the colonial era has been the shift in economic focus from interior portions of the continent to coastal districts. In the precolonial era most of the integrated states emerged in the savanna grassland zones south of the Sahara where they provided important entrepôt functions for caravan exchanges across the Sahara to the Mediterranean coast. The slave trade and the colonial era shifted the focus to coastal districts and left the interior zones as economic backwaters. It is no accident that the least developed countries in Africa are once prosperous interior states that suffer the disadvantages of a constrained environment and an unfavorable location.

African and Middle Eastern states today also suffer from many economic difficulties that are relics of colonialism. Often producers of agricultural products and raw materials, they have had their economic development skewed by the colonial process. They must struggle to overcome a century or more of colonial mercantilism that discouraged industrialization. For those countries that are still linked closely to the world market for primary products, price fluctuations and inflationary increases in the cost of materials can have catastrophic effects. The rising cost of oil is a doubly hard burden for the least developed countries, for they have the least budgetary flexibility in meeting cost increases. Those countries with monocrop economies, but more abundant resources, find themselves in an equally difficult position in attempting to diversify their economies. Even the spectacu-

larly wealthy, such as the oil states, encounter considerable problems in trying to overcome the disadvantages of an untrained population and in attempting to use sudden wealth most productively to plan for the day when the spigot will run dry. Only a few of the most diverse nations, such as Iran and Turkey, have managed to come through the colonial era with their social and political structure intact and with a headstart on the process of modernization and development.

Country Groups in Africa and the Middle East

To this point we have seen that Africa and the Middle East have both features of commonality and diversity. Our problem now is to classify the countries and colonies in a meaningful way for a more detailed study. Three characteristics are used in this classification: per capita GNP, environmental resources, and cultural-political institutions.

Per Capita GNP

Economic characteristics are commonly used measures of a country's "development." Economists, geographers, and others use a variety of measures, including per capita GNP, the percent of labor force in agriculture, and per capita energy consumption, to classify nations economically. Per capita GNP is the most frequently used single indicator of economic development. It is a measure of the production of a country in relation to its population. It is, however, criticized for several reasons. First, it is an average and like all averages does not express the wide variations in wealth that occur, particularly in developing countries. The validity of per capita GNP as a single indicator is probably greatest in highly developed countries and least in countries with a fast-growing urban or industrial sector together with a large subsistence sector. For example, the per capita GNP of Libya ($5080) is not an indicator of the level of economic well-being of many of the people. Similarly, Nigeria's recent greatly increased income from oil has not yet affected the well-being of most of the population. Despite these problems, per capita GNP is probably the best single indicator we can

Table 24–1 *Africa and the Middle East: Per Capita GNP (U.S. Dollars)*

Per Capita GNP $1,000 or More

Kuwait	11,510
United Arab Emirates	10,480
Qatar	8,320
Libya	5,080
Israel	3,580
Saudi Arabia	3,010
Bahrain	2,350
Gabon	2,250
Oman	2,070
Iran	1,440
South Africa	1,320
Iraq	1,280
Lebanon	1,070

Per Capita GNP $500 to $1,000

Turkey	860
Algeria	780
Tunisia	760
Angola	680
Syria	660
Zambia	540
Rhodesia	540
Congo	500
Ivory Coast	500

Per Capita $200 to $500

Morocco	470
Swaziland	470
Ghana	460
Jordan	460
Liberia	410
Guinea-Bissau	390
Senegal	370
Botswana	330
Egypt	310
Mauritania	310
Mozambique	310
Nigeria	310
Sudan	290
Cameroon	270
Togo	270
Uganda	250
Yemen, D. R.	240
Central African Republic	230
Kenya	220
Yemen Arab Republic	210

Per Capita GNP $200 or Less

Sierra Leone	200
Gambia	190
Lesotho	180
Tanzania	170
Malawi	150
Zaire	150
Benin	140
Malagasy Republic	140
Guinea	130
Niger	130
Chad	120
Burundi	100
Ethiopia	100
Somalia	100
Mali	90
Rwanda	90
Upper Volta	90

SOURCE: *Population Data Sheet* (Washington, D.C.: Population Reference Bureau, 1977.)

use.[7] In terms of per capita GNP, the countries of Africa and the Middle East fall into four groups: those with a per capita GNP of over $1,000, those in the range of $500–$1,000, those with $200–$500, and the remaining under $200 (Table 24–1).

Environmental Resources

Environmental resources such as land, climate, vegetation, and minerals are basic to economic acti-

[7]See also Chapter 4 for a further discussion of the use of per capita GNP.

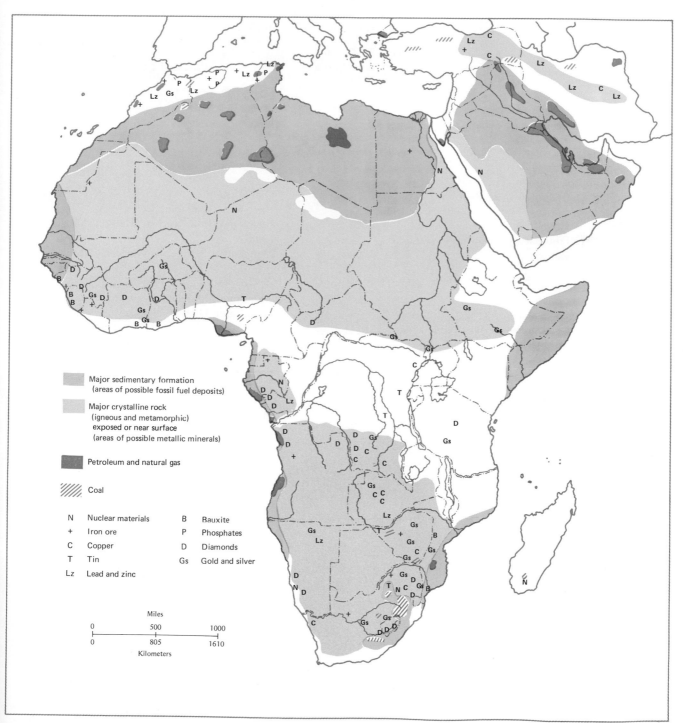

24–5 *Africa and the Middle East: Principal Mineral-Producing Areas*

Table 24-2   Africa and the Middle East: Major Mineral Resources

Country	Percent of Labor Force in Mining	Principal Minerals
Algeria	.9	Petroleum, Iron, Phosphate, Lead-Zinc, Coal
Angola	1.9	Petroleum, Iron ore, Manganese, Diamonds
Bahrain	.2	Petroleum
Benin	–	–
Botswana	.9	Diamonds
Burundi	–	–
Cameroon	–	–
Central African Republic	–	Diamonds
Congo	–	Petroleum, Lead-Zinc
Egypt	.2	Petroleum, Iron, Nuclear Materials
Ethiopia	–	Gold
Gabon	1.8	Manganese, Petroleum, Nuclear Materials, Iron
Gambia	–	–
Ghana	1.8	Aluminum, Manganese, Gold
Guinea	–	Aluminum, Diamonds
Guinea-Bissau	–	–
Iran	.3	Petroleum, Natural Gas, Chromite, Lead-Zinc, Coal
Iraq	–	Petroleum
Israel	.5	Phosphate, Copper
Ivory Coast	–	Diamonds, Aluminum, Gold
Jordan	2.4	Phosphate
Kenya	–	–
Kuwait	3.0	Petroleum, Natural Gas
Lebanon	.2	–
Lesotho	–	Diamonds
Liberia	3.5	Iron
Libya	3.0	Petroleum
Malagasy Republic	–	Coal, Nuclear Materials, Aluminum
Malawi	–	–
Mali	–	–
Mauritania	–	Iron
Morocco	.9	Phosphate, Copper, Petroleum, Lead-Zinc, Gold, Iron, Coal
Mozambique	–	Petroleum, Coal
Niger	–	–
Nigeria	–	Petroleum, Tin, Coal
Oman	–	Petroleum
Qatar	–	Petroleum
Rhodesia	–	Chromium
Rwanda	–	–
Saudi Arabia	–	Petroleum, Nuclear Materials
Senegal	–	–
Sierra Leone	5.1	Diamonds, Aluminum, Iron

Table 24-2 (Continued)

Country	Percent of Labor Force in Mining	Principal Minerals
Somalia	–	–
South Africa	10.5	Coal, Gold, Copper, Iron, Diamonds, Manganese, Pyrite
Sudan	–	Iron
Swaziland	–	–
Syria	–	–
Tanzania	–	Diamonds, Gold, Coal
Togo	–	–
Tunisia	2.2	Phosphate, Iron, Petroleum, Zinc
Turkey	–	Chromite, Coal, Iron, Copper, Petroleum, Lead-Zinc
Uganda	–	–
United Arab Emirates	–	Petroleum
Upper Volta	–	Gold
Yemen	–	–
Yemen, D.R.	–	–
Zaire	–	Copper, Manganese, Zinc, Diamonds, Tin
Zambia	4.2	Copper, Gold, Lead-Zinc

SOURCE: Bureau of Mines, *Minerals Yearbook* (Washington, D.C.: Government Printing Office, 1974); and *Yearbook of Labour Statistics* (Geneva: International Labour Office, 1974).

vities and life styles. There are a few countries (Israel) where the poor quality of these environmental resources has been compensated for by a high level of human resources in terms of education and skill or by advantages of position for trade. For most nations, however, the characteristics of the physical environment are important factors greatly influencing the pattern of life and state of economy. Arable land, pastoral land use, and geological resources are expressions of the nature of this resource base. Agricultural and pastoral resources are not just nature's endowment, as they are also influenced by technology and the laws of economics. Geological resources are known only to a degree, and they too vary depending upon technology and fluctuating costs and prices.

Africa and the Middle East are important world areas for two main types of geological resources: oil and natural gas and metallic minerals. Oil and natural gas deposits are found in sedimentary rocks in several highly localized areas. The older fields are located in Saudi Arabia and the Persian Gulf, whereas newer finds are located in North Africa (Algeria and Libya) and in the deltaic areas of Nigeria. Offshore areas are important in both the Persian Gulf area and in Nigeria.

Metallic minerals are related to quite different geological situations and are most commonly associated with the very old crystalline rocks of the African shield (Fig. 24–5). Metallic minerals are certainly more widespread than currently are known, but the deposits already being worked have provided an economic headstart for some of the African countries. Table 24–2 lists the known mineral resources and potential of the countries. As new discoveries are always being made, this table indicates only the current situation. The countries divide into oil producers, countries with diversified and important mineral resources, countries with a single significant metallic mineral resource, and those with few known mineral resources.

Arable land resources relate in part to a combination of climate and fertile soil, but the degree of exploitation of these resources is linked also to technology. For example, some of Africa's black clay soils are irrigated and intensively cultivated to produce important food and industrial crops. Other soils of this type are little used because technical and managerial skills and capital have so far not been applied to them. Moreover, there are many different agricultural systems within the region related as much to tradition and culture as to climate and soil. Table 24–3 lists the most important agricultural products of the various countries, dividing them into four major groups: countries with considerable large-

Table 24–3 *Africa and the Middle East: Agricultural Products and Labor Force*

	Countries with Significant Exports Mainly from Large Farm Units			Countries with Significant Exports from Both Large and Small Farm Units	
NATION	MAJOR AGRICULTURAL EXPORTS	PERCENT OF LABOR FORCE IN AGRICULTURE	NATION	MAJOR AGRICULTURAL EXPORTS	PERCENT OF LABOR FORCE IN AGRICULTURE
Liberia	Rubber Coffee Cacao Wood	76	Algeria	Wine Dates Citrus	61
			Angola	Coffee	64
South Africa	Wool Fruit	31	Egypt	Cotton Rice	54
			Ghana	Cacao Wood	58
			Israel	Citrus	10
			Ivory Coast	Coffee Wood Cacao Pineapple	85
			Kenya	Coffee Tea	82
			Lebanon	Citrus	20
			Malagasy Republic	Coffee Vanilla Rice Sugar	86
			Rhodesia	Tobacco	64
			Somali	Livestock	85
			Sudan	Cotton Peanuts	82
			Syria	Cotton Wheat	51
			Tunisia	Olives Citrus	50
			Turkey	Cotton Tobacco	68
			Zaire	Coffee Palm oil	80

SOURCE: Food and Agriculture Organization, *Production Yearbook.* (Rome: United Nations, 1975).

scale agriculture, countries with some large-scale agriculture and important smallholder production for export, countries with mostly smallholder products for export, and countries with largely subsistence farming and little or no agriculture for export.

There is a wide range in the level of economic activity associated with animal husbandry. In some places, ranching has been developed as a commercial enterprise; in other places, nomadic herding is the rule. Although a commercial orientation may exist

Countries with Exports Mainly from Small Farm Units			Countries with Mainly Subsistence Cropping or Herding with Little Agricultural Exports	
NATION	MAJOR AGRICULTURAL EXPORTS	PERCENT OF LABOR FORCE IN AGRICULTURE	NATION	PERCENT OF LABOR FORCE IN AGRICULTURE
Benin	Cotton Coffee Palm oil	50	Bahrain	7
			Botswana	87
			Burundi	87
Cameroon	Wood Cotton	85	Central African Republic	91
Congo	Wood Sugar	42	Chad	90
			Gambia	82
Ethiopia	Coffee	84	Kuwait	?
Gabon	Wood	82	Lesotho	90
Guinea	Pineapples Bananas	85	Libya	32
			Malawi	89
Guinea-Bissau	Peanuts	21	Mali	91
Iran	Varied	46	Mauritania	87
Iraq	Dates	47	Niger	93
Jordan	Tomatoes	34	Oman	—
Morocco	Fruit Vegetables	57	Qatar	—
			Rwanda	93
Mozambique	Cotton Sugar	74	Saudi Arabia	66
			Swaziland	81
Nigeria	Cacao Palm nuts Peanuts	62	United Arab Emirates	—
Senegal	Peanuts	80	Upper Volta	87
Sierra Leone	Palm nuts	72	Yemen, D.R.	65
Tanzania	Coffee Sisal Spices	86		
Togo	Cacao Coffee Palm nuts	73		
Uganda	Coffee, Tea	86		
Yemen	Coffee	79		
Zambia	Tobacco Peanuts	73		

Table 24–4 *Africa and the Middle East: Economic Role of Pastoral Activity*

Pastoral Activities Dominant or of High Growth Potential	Pastoral Activity Significant or Possessing Moderate Growth Potential	Pastoral Activity of Limited Importance and Growth Potential
Chad	Algeria	Angola
Iran	Botswana	Bahrain
Mali	Burundi	Benin
Mauritania	Cameroon	Central African Republic
Niger	Ethiopia	Congo
Somali Republic	Iraq	Egypt
Sudan	Jordan	Gabon
	Kenya	Gambia
	Kuwait	Ghana
	Lesotho	Guinea
	Libya	Guinea-Bissau
	Morocco	Israel
	Nigeria	Ivory Coast
	Oman	Lebanon
	Qatar	Liberia
	Rwanda	Malagasy Republic
	Saudi Arabia	Malawai
	Senegal	Mozambique
	South Africa	Rhodesia
	Swaziland	Sierra Leone
	Syria	Togo
	Tanzania	Zaire
	Tunisia	Zambia
	Turkey	
	Uganda	
	United Arab Emirates	
	Upper Volta	
	Yemen (Democratic Republic of)	
	Yemen (Republic of)	

Table 24–5 *Africa and the Middle East: Cultural-Political Attributes*

NOTES: [1]Data often incomplete.
[2]Protectorate or mandate.
[3]All German colonies made mandated colonies to allied nations after World War I.
[4]Full independence in 1956.

Country	Form of Government			
	MILITARY	REPUBLIC	ONE-PARTY STATE	MONARCHY
Algeria	X			
Angola				
Bahrain	X		X	X
Benin				
Botswana		X		
Burundi	X			
Cameroon			X	
Central African Republic	X			
Chad	X			
Congo			X	
Egypt			X	

within some nomadic societies generally, subsistence concerns are paramount. Table 24–4 lists the countries with large pastoral resources and notes the relative importance of the internal and export sectors of animal husbandry. Three major groups are found with pastoral activities either dominant, significant, or of little importance.

Cultural-Political Institutions

Cultural-political institutions are expressed in Table 24–5. Each country is classified by religious and colonial history, level of political independence, and political stability.

The Groupings

By combining the characteristics just discussed and applying them to Africa and the Middle East, three groups of nations are obtained, as follows:

Least Developed	Intermediate	More Developed
Benin	Angola	Algeria
Botswana	Central African	Egypt
Burundi	Republic	Iran
Cameroon	Congo	Iraq
Chad	Gabon	Israel
Ethiopia	Ghana	Kuwait

Gambia, Guinea, Lebanon
Guinea-Bissau, Ivory Coast, Libya
Lesotho, Jordan, Oman
Malagasy, Kenya, Palestine
Malawi, Liberia, Qatar
Mali, Muritania, Saudi Arabia
Niger, Morocco, South Africa
Rwanda, Mozambique, Turkey
Sierra Leone, Nigeria, United Arab
Somalia, Rhodesia, Emirates
Swaziland, Senegal
Togo, Sudan
Uganda, Syria
Upper Volta, Tanzania
Yemen, Tunisia
Yemen D. R., Zaire
Zambia

The first group, the least developed, are all small countries in terms of population, and most are characterized by difficult environments and no significant mineral resources. They are almost all African, though some are dominantly of the Moslem religion and culture groups. All have been colonial possessions (Fig. 24–6, page 484).

The second group, the intermediate, is more diverse and is largely composed of African countries, although Syria and Jordan are also included. These countries are characterized by a growth sector based on export crops or minerals and by a large subsistence

	Colonial History		Religion[1] (Percent of Total Population)			
COLONIZING POWER	DATE COLONIZED	DATE OF INDEPENDENCE	MOSLEM	CHRISTIAN	ANIMIST	OTHER
France	1830	1962	99	1		
Portugal	1483	1975				
Britain[2]	1820	1971	100			
France	1851	1960	13	15	65	
Britain	1885	1966			85	
Germany[3]	1898	1962		50	40	
Germany[3]	1884	1960	30	20	50	
France	1889	1960	5	68	27	
France	1900	1960	45	5	50	
France	1885	1960	1	50	49	
Britain[4]	1885	1922	92	8		

Table 24–5 (Cont.)

Country	Form of Government			
	Military	Republic	One-Party State	Monarchy
Ethiopia	X			
Gabon			X	
Gambia		X		
Ghana	X			
Guinea			X	
Guinea-Bissau		X		
Iran				X
Iraq			X	
Israel		X		
Ivory Coast			X	
Jordan				X
Kenya			X	
Kuwait			X	
Lebanon		X		
Lesotho				X
Liberia			X	
Libya	X			
Malagasy	X			
Malawi			X	
Mali	X			
Mauritania			X	
Morocco				X
Mozambique			X	
Niger	X			
Nigeria	X			
Oman				X
Qatar				X
Rhodesia		X		
Rwanda	X			
Saudi Arabia				X
Senegal			X	
Sierra Leone		X		
Somali	X			
South Africa		X		
Sudan	X			
Swaziland				X
Syria			X	
Tanzania			X	
Togo	X			
Tunisia			X	
Turkey		X		
Uganda	X			
United Arab Emirates		X		
Yemen	X			
Yemen, D.R.			X	
Zaire			X	
Zambia			X	

Colonial History			Religion[1] (Percent of Total Population			
Colonizing Power	Date Colonized	Date Independence	Moslem	Christian	Animist	Other
Italy	1936	1941	35	35	30	
France	1839	1960		46		
Britain	1816	1965				
Britain	1874	1957	12	43	45	
France	1849	1958	62	2		
Portugal	1446	1974				
—	—	—	98			
Britain[2]	1920	1932	97			
Britain[2]	1920	1948				88
France	1842	1960	25	12	63	
Britain[2]	1920	1946	94	6		
Britain	1895	1963	3	59	38	
Britain[2]	1899	1961	98	1		
France	1920	1944	50	50		
Britain	1868	1966		70	30	
—	—	1847			90	
Italy	1912	1951	100			
France	1896	1960		40		
Britain	1891	1964			67	
France	1904	1960	65	1	34	
France	1904	1960	100			
France, Spain	1912	1956	98	2		
Portugal	1500	1975				
France	1890	1960	85	1	14	
Britain	1883	1960	44	22	34	
—	—	—	100			
Britain[2]	1916	1971	100			
Britain	1890	1965				
Germany[3]	1898	1962				
—	—	—	100			
France	1861	1960	80	5	16	
Britain	1884	1960				
Italy	1889	1960	100			
Britain	1814–1877	1910	1	61		38
Britain	1899	1956	65	35		
Britain	1894	1968		57	43	
France	1920	1946	87	13		
Germany[3]	1890	1961	30	25	45	
Germany[3]	1885	1960	5	20	75	
France	1883	1956				
—	—	—	98	2		
Britain	1894	1962	6	50	44	
Britain	1880–1916	1971	100			
Turkey	1517	1918	100			
Britain	1839	1967	90			
Belgium	1908	1960		50	50	
Britain	1924	1964		30	70	

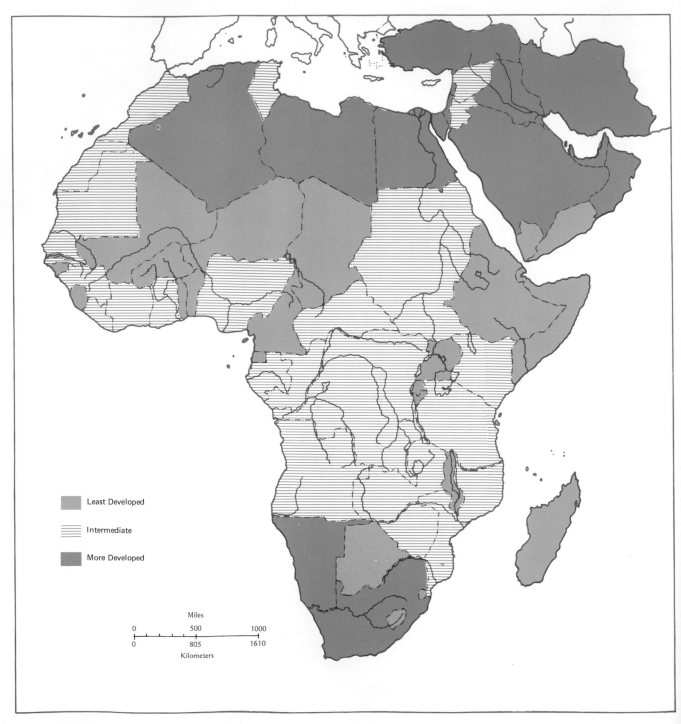

Least Developed

Intermediate

More Developed

Miles

0   500   1000

0   805   1610

Kilometers

24-6   *Africa and the Middle East: Grouping of Countries for Development Purposes*

sector still not greatly affected by economic change. Most are making moderate progress in fostering economic development or possess significant potential to do so.

The third group, the more developed, includes nations with a diverse economic base and those with such high earnings from oil that they are becoming wealthy and are now seizing the opportunity to diversify their economic structure. This group is mainly Moslem and Middle Eastern, although it includes South Africa and Israel. Both of these countries are economically within the group but have strong religious and cultural ties outside the region.

Our three-way division is made for ease of economic and spatial analysis and thus compromises on many of the basic divisions of the area. Arab-African, Moslem-Animist-Christian, French-English-speaking (with many other tongues), socialist-capitalist—these categories remind us of the range of cultural diversity.

Further Readings

A number of excellent general texts provide greater detail on the topics covered in this chapter. Coon, Carleton, *Caravan: The Story of the Middle East,* rev. ed. (New York: Holt, Rinehart and Winston, 1965), 376 pp., is a general introduction to the Islamic realm which evokes the flavor of the region with captivating flair; while the two-volume study, Sweet, Louise (ed.), *Peoples and Cultures of the Middle East* (Garden City, N.Y.: Natural History Press, 1970), is a fine collection of articles focusing upon specific cultural and ecological topics. Though dated, Cressey, George B., *Crossroads: Land and Life in Southwest Asia* (Philadelphia: Lippincott, 1960), 593 pp., remains a standard and lively geographical survey. A less exciting but eminently scholarly work is Fisher, W. B., *The Middle East,* 6th ed. (London: Methuen, 1971), 568 pp.

For Africa: Hance, William A., *The Geography of Modern Africa,* rev. ed. (New York: Columbia University Press, 1975), 653 pp., remains the best comprehensive introduction to the entire continent. Grove, A. T., *Africa South of the Sahara* (New York: Oxford University Press, 1967), 275 pp., adopts a more limited regional focus; and Hance, William A., *African Economic Development,* rev. ed. (New York: Praeger, 1967), 326 pp., concentrates upon a detailed examination of specific development schemes.

Atlases also provide a profitable source of information about the region's broad spatial patterns. The two best are Davies, H. R. J., *Tropical Africa: An Atlas for Rural Development* (Cardiff: University of Wales Press, 1973), 81 pp., and Ady, Peter H., *Oxford Regional Economic Atlas: Africa* (Oxford: Clarendon Press, 1965), 164 pp. Ginsburg, Norton, *The Atlas of Economic Development* (Chicago: University of Chicago Press, 1961), 119 pp., is the only atlas to attempt to quantify development variables and indicators systematically. The data are now out of date, but the regional patterns still retain their validity.

The history of trade contacts across the Sahara can best be grasped by the absorbing and magnificent work by Bovill, E. W., *The Golden Trade of the Moors,* 2nd ed. (New York: Oxford University Press, 1968), 293 pp. The scantier record of sailing connections between East Africa and the Arabian Peninsula has been examined by Villiers, Alan, *Sons of Sinbad* (New York: Scribner, 1940), 429 pp., and "Some Aspects of the Arab Dhow Trade," *Middle East Journal,* 11 (1948), 399–416. No examination of the slave trade can begin on firm ground without commencing with the work of Curtin, Philip D., *The Atlantic Slave Trade: A Census* (Madison: University of Wisconsin Press, 1969), 338 pp., a classic example of historical detective work. *Africa Remembered: Narratives by West Africans from the Era of the Slave Trade* (Madison: University of Wisconsin Press, 1967), 363 pp., is an equally rewarding examination of different facets of the slave trade issue. Also not to be missed is Davidson, Basil (ed.), *The African Past: Chronicles from Antiquity to Modern Times* (Boston: Little, Brown, 1964), 392 pp., a well-written general account that presents other aspects of African civilization. Studies of Islamic culture are legion, but Hitti, Philip, *History of the Arabs from the Earliest Times to the Present,* 10th ed. (London: Macmillan, 1970), 822 pp., continues to be the most detailed presentation of the historical processes molding the Islamic realm; von Grunebaum, Gustave, *Medieval Islam: A Study in Cultural Orientation* (Chicago: University of Chicago Press, 1961), 378 pp., is an examination of the origin of Islamic institutions and values. For the way of life associated with these institutions at the village level of social organization, see Fernea, Elizabeth W., *Guests of the Sheik: An Ethnology of an Iraqi Village* (Garden City, N.Y.: Doubleday, 1969), 333 pp. Numerous data sources for contemporary economic development problems are available. Published by the International Bank for Reconstruction and Development, *The World Bank Atlas* (Washington, D.C.), provides basic statistical data on an annual basis. Overviews of the salient problems in the region are given by Green, Reginald H., and Seidman, Ann, *Unity or Poverty: The Economics of Pan-Africanism* (London: Penguin Books, 1968), 364 pp., and Warriner, Doreen, *Land Reform and Development in the Middle East: A Study of Egypt, Syria and Iraq,* 2nd ed. (London: Oxford University Press, 1962), 238 pp.

Africa and the Middle East: The Least Developed Nations

25

T WENTY-TWO COUNTRIES comprise the group of least developed countries (Fig. 25–1). This group is characterized by limited wealth, significant environmental problems, a colonial heritage, and, in many cases, isolation. Although low GNP is a distinctive characteristic of all the least developed countries, the causes of low economic development levels are varied.

Two countries with similar low per capita GNP and environmental problems—Egypt, and Jordan—are not considered in this chapter because they are different from the others in several ways. Egypt is typical of a developed country in size, level of industrialization, and urban human resources, but the large poor rural population lowers the average per capita GNP. Jordan, different in part because of its human resources and special locational importance, is best considered with the other countries of the Eastern Mediterranean. Ethiopia, though included, needs

special mention because of its larger size and its particular culture and history.

Table 25–1 sets out basic data on the least developed countries. They are generally small, either in area or in population, or both (Fig. 25–2). They have a relatively low percent of their population in manufacturing and a high level of illiteracy. These are some of the dominant characteristics that UNCTAD III (United Nations Conference on Trade and Development) and other United Nations bodies have used to characterize the group of the least developed nations, sixteen of which are in Africa.[1] The designation of the twenty-five least developed nations has been adopted by many international bodies, and the World Bank has recognized their special need for assistance by setting aside special funds for their aid. Two African countries that fall within the United Nations least developed group are not included in our grouping. Both Sudan and Tanzania, in spite of a low per capita GNP, have a greater resource potential that places them in an intermediate cluster.

Environmental Constraints to Economic Development: The Arid and Semiarid Environments

The role of environment in the well-being of nations has been viewed very differently at different stages in history. In recent decades there has been a reaction from an earlier environmental determinism which, in its most dogmatic form, linked everything from human intelligence to economic growth with environmental factors. Such views gradually were rejected, and at one time it seemed as though mankind's technology and intelligence could overcome all limiting environmental factors. In the last few years, however, even in the technologically most

[1]UNCTAD III designated a group of nations as the twenty-five least developed countries in order to focus attention on the special needs of these states. Since the original designation, the effect of oil and commodity prices has changed the position of some other countries like Kenya, bringing them closer to the least developed states.

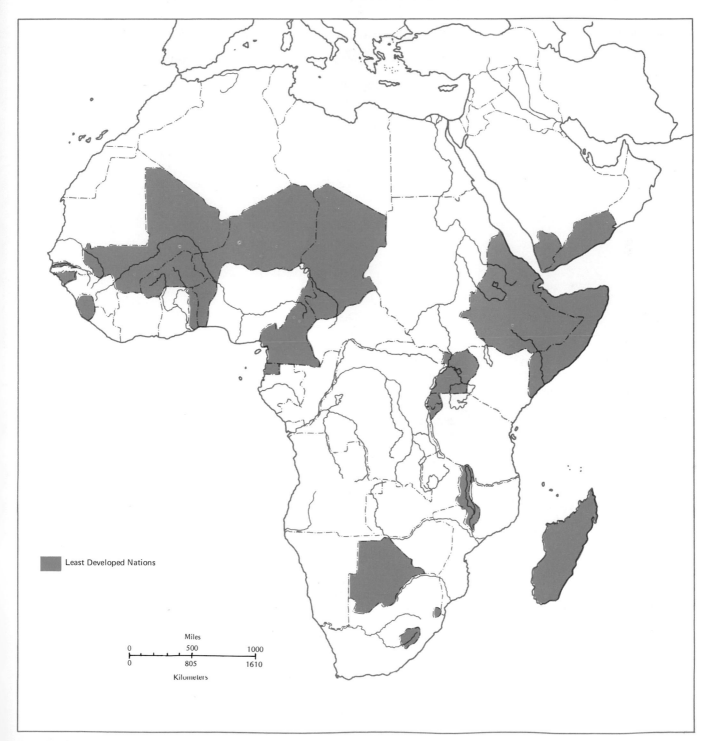

Least Developed Nations

25-1   *Africa and the Middle East: The Least Developed Nations*

Table 25–1 *Africa and the Middle East: Some Characteristics of the Least Developed Nations*

Nation	Area		Population (millions)	Population Density		Percent of Population Literate
	Square Miles	Square Kilometers		Per Square Mile	Per Square Kilometer	
Benin	43,484	112,624	3.3	76	29	20
Botswana	231,805	600,375	.7	3	1	20
Burundi	10,747	27,835	3.9	363	140	10
Cameroon	183,569	475,444	6.7	36	14	10–15
Chad	495,800	1,284,122	4.2	8	3	5–10
Ethiopia	471,778	1,221,905	29.4	62	24	5
Gambia	4,361	11,295	.5	115	38	10
Guinea-Bissau	13,948	36,125	.5	36	12	—
Lesotho	11,720	30,355	1.1	94	31	—
Malagasy	226,658	587,644	8.0	35	12	39
Malawi	45,483	117,801	5.3	117	45	22
Mali	478,655	1,239,716	5.9	12	5	5
Niger	489,200	1,267,028	4.6	10	4	5
Rwanda	10,169	26,338	4.5	443	171	10
Sierra Leone	27,699	71,740	3.2	116	45	5–10
Somali	246,210	637,661	3.4	14	5	5
Swaziland	6,705	17,366	.5	75	25	—
Togo	21,600	55,944	2.3	106	41	5–10
Upper Volta	105,869	274,201	6.4	60	23	5–10
Yemen	75,300	195,027	5.6	74	28	10
Yemen, D.R.	111,000	287,490	1.8	16	6	10

advanced countries, environmental problems have become increasingly apparent.

Importance of Environment

In the least developed countries, the "old-fashioned" scourges of environment have always been important. Most economies are vulnerable to fluctuating weather conditions. These fluctuations make for uncertainty both in terms of basic food supplies and export earnings. As agricultural products account for a large proportion of the total economy, weather fluctuations can have a significant impact on the overall economic situation. For example, during the Sahelian drought, the GNP of Mali dropped considerably.[2] Even in less severe years, countries may

show a 5 percent variation in GNP, depending on the weather. Environment-related problems such as the lack of good supplies of drinking water, insect infestation, and disease also take their toll on the health and vigor of the inhabitants. Although malaria is the most widely known of tropical diseases, yellow fever, bilharzia, and onchocerciasis (river blindness) are most important in Africa (Fig. 25–3). Animal diseases such as tsetse fly infections and east coast fever cause large areas of land to be removed from certain kinds of economic activity.

Ecosystem Uniformity

Among environmental problems of the least developed countries is the absence of ecosystem diversity (Fig. 25–4). Uniform environmental conditions often result in a narrow range of crops and other agricultural activity. Of the countries considered in this chapter, only Ethiopia and the Malagasy Republic show a wide range of diversity of ecosystem within

[2]Mali's GNP decline was due to a major drop in income from pastoral products and from grain production because of the Sahelian drought. Food relief imports were necessary to compensate for these losses.

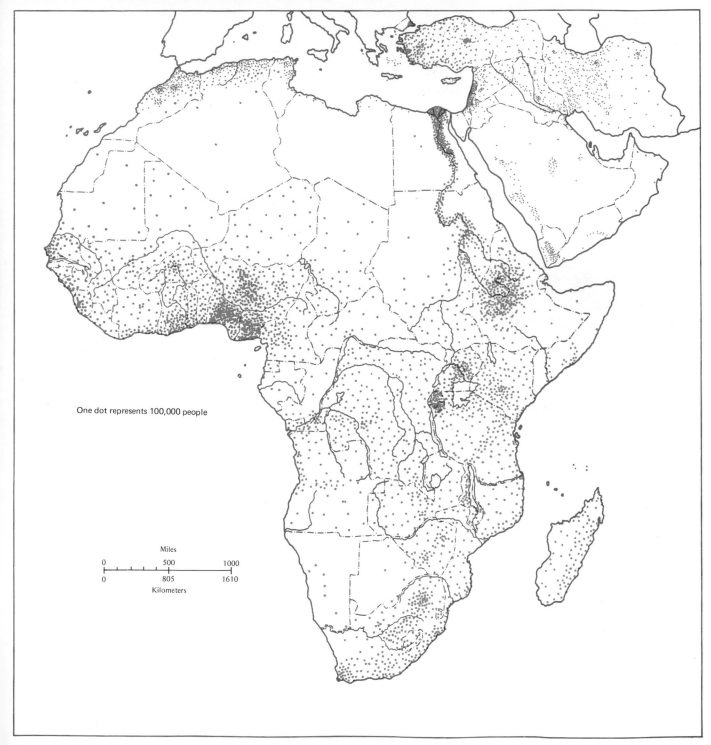

One dot represents 100,000 people

Miles			
0	500		1000
0	805		1610
Kilometers			

25–2 *Africa and the Middle East: Population Distribution*

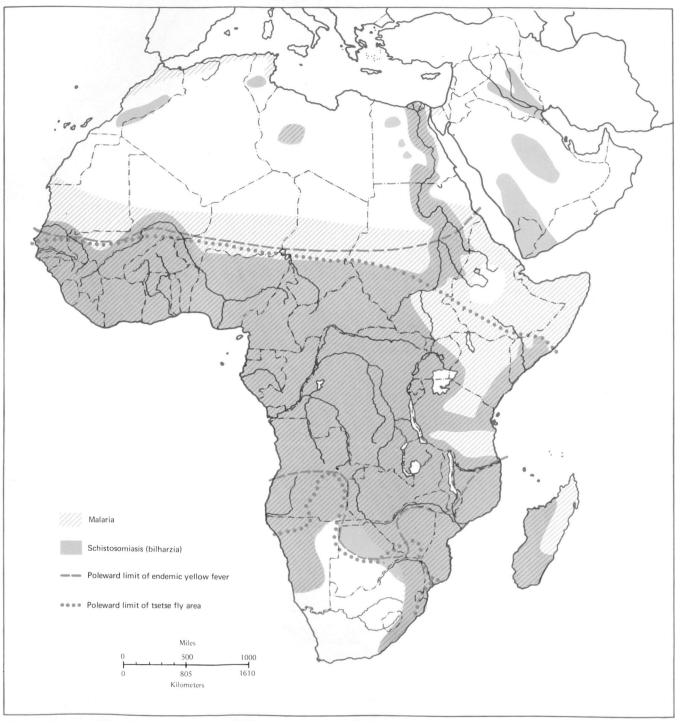

25–3 *Africa and the Middle East: Areas Infested by Selected Diseases*

SOURCE: Ernst Rodenwaldt, *World Atlas of Epidemic Diseases* (Hamburg: Fale, 1952–61), and *Atlas of the Distribution of Disease* (New York: American Geographical Society, 1950–55).

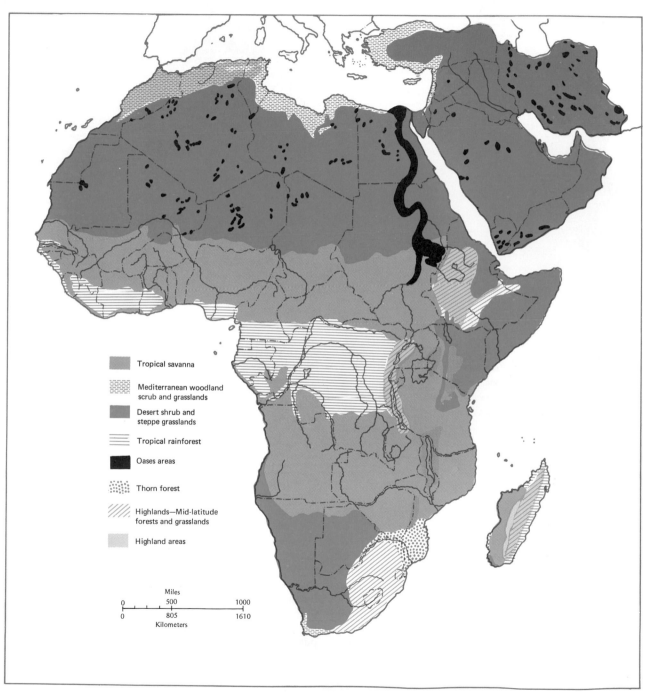

Tropical savanna

Mediterranean woodland scrub and grasslands

Desert shrub and steppe grasslands

Tropical rainforest

Oases areas

Thorn forest

Highlands—Mid-latitude forests and grasslands

Highland areas

Miles
0   500   1000
0   805   1610
Kilometers

25–4a *Africa and the Middle East: Natural Vegetation Regions*

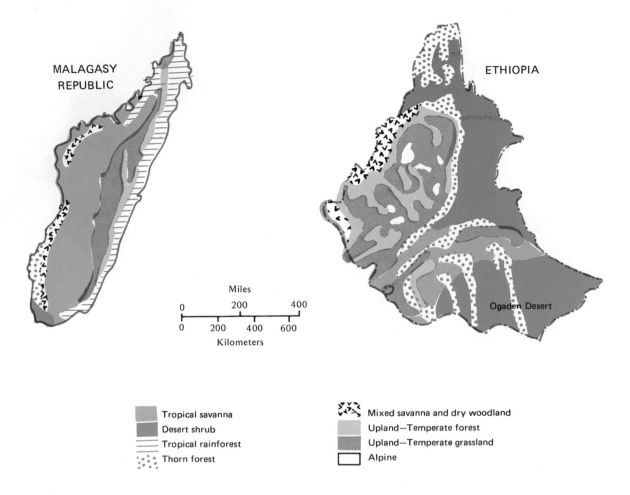

MALAGASY
REPUBLIC

ETHIOPIA

Ogaden Desert

Miles

0   200   400

0   200   400   600

Kilometers

	Tropical savanna
	Desert shrub
	Tropical rainforest
	Thorn forest

	Mixed savanna and dry woodland
	Upland—Temperate forest
	Upland—Temperate grassland
	Alpine

25–4b *Natural Vegetation Regions of Malagasy Republic and Ethiopia*

the country. In Ethiopia, the great altitudinal differences between one part of the country and another result in major differences in temperature, precipitation, soils, and vegetation. There are thus a variety of natural ecosystems and a considerable range in crops and land potential. In the coastal lowlands, an extremely arid environment and high temperatures are characteristic; on the Ethiopian mountains, over 80 inches (203 centimeters) of rain and year-round temperatures of 50° to 60°F. (10° to 15°C.) are common. In Malagasy, change in altitude, coupled with distance from the sea, provides a strong environmental

gradient, while seasonal differences in the effects of onshore winds result in cross-island contrasts.

The other countries are characterized by much more homogenous environmental systems, mainly of two types: arid or semiarid, and humid. Both of these ecosystem types have advantages and disadvantages. When a whole country is occupied mainly by one ecosystem, however, the lack of diversity places severe restraints on the range of agricultural and pastoral life-support systems that can be economically established.

Lack of diversity is evident in the crops and agri-

25–5 *Agricultural Systems of Niger, Guinea-Bissau, and Botswana*

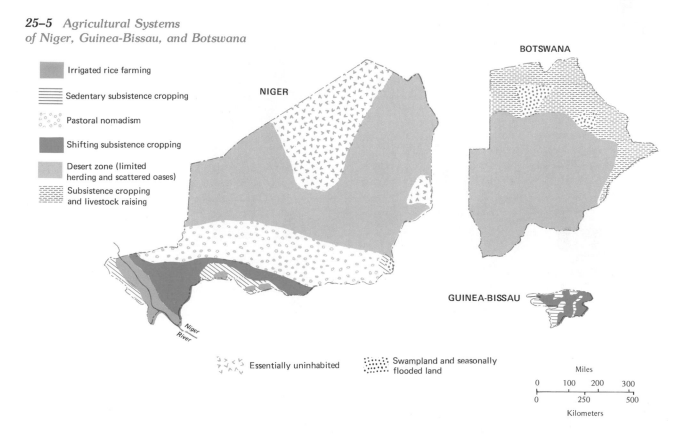

Irrigated rice farming

Sedentary subsistence cropping

Pastoral nomadism

Shifting subsistence cropping

Desert zone (limited herding and scattered oases)

Subsistence cropping and livestock raising

NIGER

BOTSWANA

GUINEA-BISSAU

Niger River

Essentially uninhabited

Swampland and seasonally flooded land

Miles
0   100   200   300

0   250   500
Kilometers

cultural systems of Niger, Botswana, and Guinea-Bissau (Fig. 25–5). Much of Niger is utilized by pastoral nomads, although in the southern parts of the country a range of grain and livestock-based agricultural systems are found. Some small diversity is provided by irrigation and garden watering from the Niger River. Botswana is dominated by pastoral activities with some high value mineral extraction constituting a recent growth point. At the same time, the relic hunting and gathering Bushmen survive by using extremely marginal resources unwanted by other groups. Guinea-Bissau, in quite a different ecological setting, is dominated by root crop agricultural systems.

The problems of single ecosystem countries have been dramatically highlighted when those ecosystems are placed under stress as, for example, in the recent Sahelian drought. The whole of the countries of Niger, Mali, and Chad were affected by drought, and

international help was essential to the survival of many of their people and for support of the economy.[3] Neighboring Nigeria and, to a lesser extent, Upper Volta were only affected in part by the drought, and resources were reallocated from the unaffected areas of these countries. Nigeria was particularly fortunate because its humid coastal zones escaped the drought. Niger was less environmentally diverse and so suffered disproportionately.

Arid and Semiarid Ecosystems

Many of the least developed countries are composed of arid or semiarid ecosystems. Niger, Mali, Chad, and Upper Volta lie in a belt on the southern margin of the Sahara known as the Sahelian zone.

[3]See, for example, Hal Sheets, *Disaster in the Desert: Failures of International Relief in the West African Drought* (Washington, D.C.: Carnegie Endowment for International Peace, 1974), 167pp.

The Sahelian drought of the late 1960s and early 1970s was disastrous to the people along the southern fringe of the Saharan Desert. Thousands of people died from drought-related problems and livestock perished by the tens of thousands. (United Nations)

Ethiopia, Somalia, Yemen, and Yemen DR are on the eastern extension of this zone, while Botswana is part of the arid interior of southern Africa. In all of these countries the rainy season is a short period (one to four months) of thunderstorms and showers associated with the movement north and south of the moist unstable equatorial air mass. The intertropical frontal zone between dry Sahara and humid equatorial air moves north and south in a regular way, but there is considerable variation in the amplitude of its movement and in the thickness and instability of the air mass behind it. There are consequently marked differences in the amount and distribution of rainfall from one year to the next. Because the precipitation occurs as localized storms, there is also a marked variation in the amounts of rainfall from place to place. Figures 25–6a and 25–6b show examples of this variability. The extreme uncertainty of adequate rainfall occurring in any one place makes all semiarid environments uncertain and potentially hazardous places in which to live.

Development Strategies

A variety of strategies to overcome these environmental problems are adopted by local people especially in the kind of agricultural-pastoral system they employ. In the more arid areas, various forms of nomadism are traditional ways of coping with problems of seasonal availability of grazing and water.[4]

[4]Y. Poncet, *Drought in the African Sahel: A Micro-Regional Study in Niger* (Paris: OECD, 1974), 51pp.

494

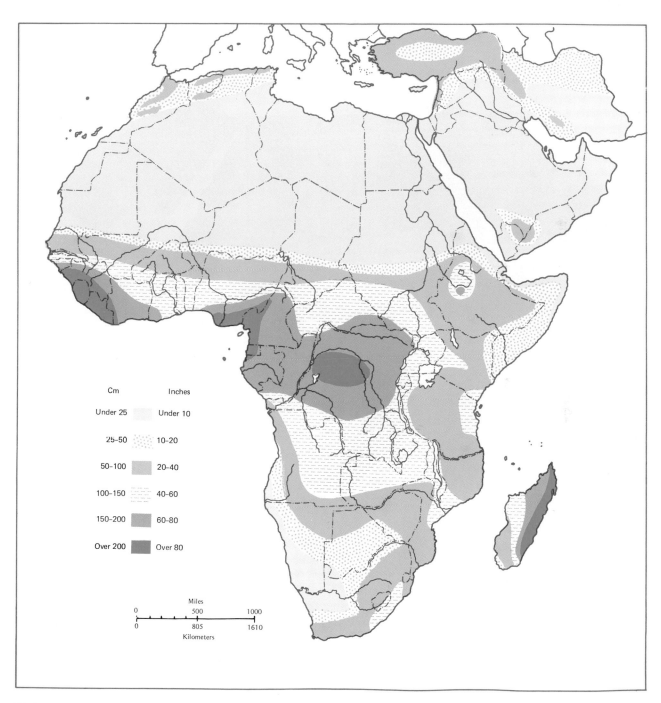

Cm   **Inches**

Under 25   Under 10

25–50   10–20

50–100   20–40

100–150   40–60

150–200   60–80

Over 200   Over 80

Miles
0   500   1000
0   805   1610
Kilometers

25–6a   *Africa and the Middle East: Average Annual Rainfall*

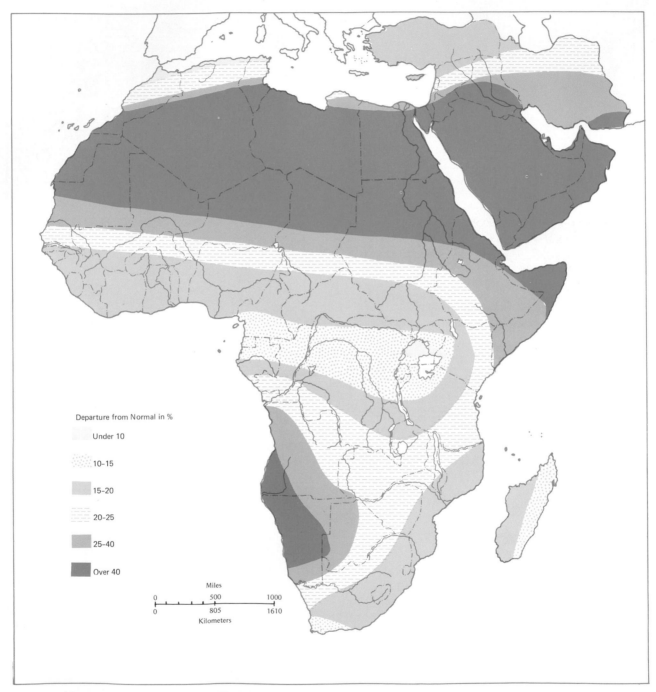

Departure from Normal in %

Under 10

10–15

15–20

20–25

25–40

Over 40

Miles
0   500   1000
0   805   1610
Kilometers

25–6b   *Africa and the Middle East: Variability of Precipitation*

The dry margins of the desert occupy vast areas and constitute a significant world resource in terms of their potential for livestock husbandry. This potential can only be realized under flexible systems of land use that allow every advantage to be taken of pasture and water at different seasons. Traditional nomadism is one response to this need, the huge ranches of western Texas another.

Nomadism, however, has become associated with a closely knit social and economic system. This system, effective in its mode of use of environment, is often difficult to fit into the fabric of new nation-states. Governments tend to favor settlement of nomadic communities. Taxes, security, ease of administration, and a modern national image are all contributing factors in this desire to settle the nomad. The flexible nomadic system has almost entirely broken down in the face of the combined effects of the severe drought that began in 1969, and the impact of political and economic pressures. A gradual increase in numbers of people and animals, especially in the comparatively wet years of the 1960s, led to overstocking of the range. When successive years of drought occurred, the result was a major reduction in stock levels, the need for massive food assistance, and concern about future patterns of life and land use in this area.

Problems arose not only because of the drought but also because of other changes affecting nomadic people. Growth in numbers of livestock, changes in the amount of land available because of increased settled agriculture, new national boundaries, and the effect of modern technology (new wells, medicines, and roads) have played a part in intensifying the effect of the Sahelian drought. A basic issue in the use of the resources of such dry areas is the way in which new patterns of life emerging in the post-drought years will be adapted both to the ecological needs of the environment and the socioeconomic needs of the people.[5]

Agriculture in Dry Lands

Even in very dry parts of the world, wherever water is available, some cultivation is practiced. Agricul-

tural life styles predominate in areas where underground or surface water occurs. In good years, when rainfall is above average, agriculture may move into marginal, drought-prone areas. In the very dry areas, cultivation is usually confined to the beds of seasonally flowing streams and may only take place during years in which rainfall is substantial. In average years, enough water can usually be obtained from shallow wells to support a tiny vegetable crop.

In good years, maize and millet are planted in the stream beds and wherever water has soaked into the ground. In areas of higher rainfall, cultivation is a regular part of the agricultural system with maize and millet the most favored grains. The pattern of cultivation follows closely that of the seasonal rainfall, and during the long dry season, animal herding supplements farm income.

Because people in this somewhat wetter zone live in fixed villages, and their grazing is limited to a small zone around their fields, they have also suffered a great deal from the recent drought. In the semiarid lands of Africa, the subsistence grains are cultivated alongside cash crops. Cotton is widespread on small and large farms in East Africa, and peanuts are especially popular in West Africa. Vegetables are almost always grown in small plots for family consumption, except near the large towns that provide a ready market for tomatoes, onions, and other more perishable items.

Prospects for major improvement in farm income from cultivation in this zone depend on much better water control as well as improved marketing. Although water availability is a major constraint to the production of a crop, quality and quantity of output are also influenced by soil type and farming methods. The soils of the arid and semiarid lands are not generally fertile, though there are pockets of high-quality soil. The major release from the constraints of aridity occurs where pockets of good soil can be supplemented by secure supplies of water as, for example, around Lake Chad where indigenous methods of *shaduf* irrigation can be supplemented by small pump irrigation. A *shaduf* consists of a bucket that is pivoted on a long pole. Also, on the terraces bordering the Niger River, pump irrigation is used for vegetable and rice cultivation, although currently on a small scale. In some areas, underground water may provide similar opportunities.

[5]UNESCO, *The Sahel: Ecological Approaches to Land Use* (Paris: UNESCO, 1975), 99pp.

Another traditional means of raising water for irrigation is the "Persian Wheel." Buckets are strung in series around a wheel. As the wheel turns, the buckets dip into the well and then raise the water for distribution in irrigation ditches. (United Nations)

Intrusion of Technology

The record of modern technology as it has been applied in semiarid environments is mixed. There has been a tendency to attempt large schemes of water control as a basis for major irrigation projects. The Gezera scheme in the Sudan, located south of Khartoum and using gravity-fed water to produce high-quality cotton for export, is an example of one successful scheme of this type. Others have met with many problems. These problems have, in part, been environmental, brought about by major changes in local ecosystems. Some of these changes include the spread of bilharzia (a debilitating intestinal infection) in irrigation water, salinity problems on irrigated land, the spread of water-borne plants such as water hyacinth (*Eichnornia crassipies*), and insect and pest problems on the growing crops. These insect and pest problems are particularly important in semiarid lands.[6] Recent ecosystem studies show that a high proportion of the total plant and animal life of semiarid ecosystems is concentrated in the insect stage of the food chain.[7] Newly irrigated areas within semiarid environments are thus risky, since they create ideal conditions for the proliferation of insects capable of doing great crop damage. The required human adaptation is an additional problem associated with large-scale water-control schemes. Farmers used to cultivating small plots under their own management

[6]For some examples of ecological problems resulting from irrigation, see Raymond F. Dasmann and others, *Ecological Principles for Economic Development* (London: Wiley, 1973), pp. 183–255.

[7]E. B. Edney, "Animals of the Desert," in E. S. Hills (ed.), *Arid Lands: A Geographical Appraisal* (London: Methuen, 1966), pp. 204–209.

have viewed the rigid irrigation scheme as a dubious alternative, particularly when production and marketing-pricing and management problems have added another element of economic uncertainty.

Water control on small farms has improved crop production in a number of localities in the zone, but once again there are many cases where water provision is seen as almost the only development need. Provision of new watering points without concern for the effects on grazing and other resource systems in the area has frequently resulted in an imbalance. Animal numbers are allowed to increase in relation to the water supply, and palatable grasses are soon consumed. A common sight is the water hole surrounded by a barren circle a mile or more in radius.

In the arid and semiarid countries development goals therefore need to be set with a clear idea of the environmental constraints and opportunities. In this area more than in most others, an ecological view of development is vital.

Throughout most of northern Africa and the Middle East, water is a major key to development. Many projects are underway to expand cropland areas by means of irrigation. (United Nations)

Environmental Constraints to Economic Development: The Humid Environments

A different set of problems is posed for small countries with humid tropical ecosystems. The natural vegetation, particularly in lowlands, is a thick forest with several layers of vegetation. This biomass (a term used to include all plant and animal life in an ecosystem) when cleared provides a good environment for the production of root crops like yams and cassava and has potential for cash crops such as palm oil, cacao, and rubber. Clearing is hard work, and soils are fertile only locally. The most common soil type in this zone is the deep-red latosolic which deteriorates quite rapidly under continual cultivation. Most of the nutrients of the humid tropical ecosystems are locked in the vegetation mass itself. Unless the natural vegetation can be replaced with a similar cultivated system, nutrients are lost through soil erosion.

Traditional shifting cultivation provides a mechanism whereby land is allowed to regenerate after a few years' use while a new plot of land is cleared. When plenty of land is available, shifting cultivation seems a reasonable economic use of labor and land from the point of view of the traditional cultivator. Clearing is effected economically by cutting and burning the smaller trees and litter. The ash provides a short-lived fertilizer, and the small plots prevent major soil erosion problems. After a few years, the vegetation is allowed to reestablish itself, and a new, nutrient-rich plot is cleared. This system is, however, impossible to sustain when land is in short supply. As population grows, more land formerly under shifting cultivation comes into permanent use with consequent problems of soil fertility and soil erosion. The small countries in the humid tropical ecosystems of Africa have yet to evolve productive small-farm systems that are alternatives to either shifting cultivation or plantation agriculture. In countries such as Ivory Coast and Zaire, plantation-type farms have been very successful in preserving soil and maintaining production levels, but only because of capital inputs in the form of fertilizers and machinery that are far beyond the capability of the small peasant farmer.

The large European-funded and managed rubber plantations and coffee farms are typical examples.

A further problem of the humid tropics is that animal husbandry, apart from pigs and chickens, is difficult. Cattle suffer from a variety of diseases in the wetter areas and are plagued by the tsetse fly in drier savanna conditions. The result is a general lower level of nutrition in some of the humid areas than in the arid zones where animal products are used much more.

The Mineral Resource Base

The role that mineral resources play in the economy of a country depends on several different factors. The state of development of the country, the state of knowledge of mineral resources, and the economics of development of a particular resource for internal use or for export, are all important. Other factors include distance from external markets, alternative supply sources available to the international market, availability of capital, and the perceived investment climate.

The least developed countries are the poorest nations, partly because they do not have exploitable mineral resources. The copper resources of Zambia, the iron ore resources of Liberia and Mauritania, and the oil of Libya have greatly altered these countries' overall national economic position, although great problems of wealth distribution within the countries remain. Table 24–2 illustrates the known mineral resources of the least developed countries. Malagasy has good bauxite deposits; Sierra Leone has an important diamond industry; Guinea is reported to have one-half of the world's bauxite supply; Lesotho exports some diamonds, and Rwanda some tin; 35 percent of Togo's exports are composed of phosphates; and Somalia (uranium), Cameroon (titanium), and Niger (uranium) have deposits of extremely rare and valuable minerals. Botswana has moved in the last few years from a national economy depending almost entirely on agriculture and livestock to one with great possibilities of mineral resource development; copper and diamonds are already being exploited.

None of the other nations are known to have

mineral deposits that are large enough to influence the path of economic development significantly over the next decades. Where deposits do exist, such as the placer gold or diamond gravels of Mali, they are too small and scattered to be worked by modern methods. Thus, while considerable numbers of people may subsist by working such deposits, they make only a limited contribution to the country's GNP. It is clear, however, that mineral exploration in many countries has been rather perfunctory. Recently, traces of oil have been noted near Gao in Mali, and there are reports that substantial oil reserves underlie the arid Ogaden district in Ethiopia. Confirmation of these reports is lacking. Even when located, they often have a rather unfavorable location with respect to possible export. Malawi, for example, is known to have deposits of bauxite, but they are too remote from markets and transport systems to be exploited economically at present. Similarly a deposit in Chad or Niger, far from the sea, would have to be very valuable to overcome the high overland transport costs to international markets.

Locational Constraints

Most of the least developed nations are isolated. Nine are landlocked; several are isolated because they have no significant port and little air traffic, while others are off the major trade and communication routes of the world. Many better-off developing countries gained the impetus for modern economic development through exports of commercial crops or minerals for European or Anglo-American markets. Because of their isolation, most of the least developed countries have been kept from fully participating in this global economy. Isolation, however, is not complete, and communication is well enough developed to make many inhabitants of the least developed countries well aware of the benefits of economic and social development to be found elsewhere and to make them displeased with their position.

The need to move goods through a neighboring country to reach the coast often means that not only long land distances have to be covered but also problems of external control over national trade flows

have to be solved. Niger has attempted to spread the risk by developing new routes through Nigeria as a complement to older routes through Upper Volta, Benin, and Mauritania. Uganda, however, still relies completely on Kenya for access to the sea, as routes through Zaire, Sudan, Tanzania, or Ethiopia are so difficult they do not present viable alternatives. Ethiopia, although it has a long coastline, has poor access to the sea. The coast, when reached, has few anchorages, and the central government is plagued by a large-scale revolt, supported by the Moslem population of the coastal province of Eritrea. In addition, since the closing of the Suez Canal in 1967, the area has been well removed from major trade routes, although the 1975 reopening of the canal may well change this situation. The Sahelian countries are in the unusual condition of once being important trading areas on the margins of the "Saharan Sea." But as times and modes of transport changed, the countries increasingly found themselves in an isolated backwater dependent on the former colonial powers or on the coastal states for their access to the world.

All of the countries considered in this chapter, except Ethiopia, were once colonies. The boundaries of all the states are thus determined more by the particular set of influences of the European colonial powers around the end of the nineteenth century than by the cultural, economic, and power relations that have generally governed the evolution of states. Many countries are peculiarly hampered by factors of shape and size inherited from the colonial period. Gambia and Togo are tiny states that were formed as the respective colonial powers gained control over narrow river basins. The major states of West Africa all cut across the boundaries of the old African kingdoms that were so important in the history of the area. Although the Organization of African Unity has agreed to maintain the status quo in national frontiers in Africa, many tiny states struggle to achieve minimum economic and political stability in the face of all but impossible physical and resource heritages. As Seidman and Green point out, if political unions are difficult, economic unions are vital for almost all the states of Africa.[8]

[8]Reginald H. Green and Ann W. Seidman, *Unity or Poverty? The Economics of Pan-Africanism* (Harmondsworth: Penguin, 1968), pp. 263–282.

Economic Structure

The least developed nations group together primarily because of their low levels of economic development. Although there is some diversity in the internal structures of these countries, there are common characteristics of relative poverty and relatively simple economic structures. The economic systems of these countries have a high proportion of their population involved in subsistence activity. They also frequently contain a small enclave of more advanced agricultural activities in the form of foreign-owned farms, state farms, or plantations and limited industrial sectors. In addition, relatively low levels of international trade reduce the potential for capital formation. In discussing economic structure, we deal first of all with some general points about the economy and illustrate these with reference to Mali and Ethiopia.

Subsistence Activities

There are few societies anywhere that are completely removed from the cash exchange economy, yet there are many countries in which a large portion of the inhabitants work first to feed themselves and their families and only then to earn cash for the extras needed in their lives. This subsistence attitude is a sound one in the light of the poorly developed marketing and communication systems in many areas. In these circumstances, a distinction between food crops and cash crops is a false one. The significance lies more in the emphasis the farmer places on his or her activity. Crops that are planted for sale are given second priority.

Goods that are sold—cattle, maize, cotton, peanuts, and the like—find their way into local and perhaps national and international markets. In some cases where the crop is specifically for export, government or cooperative marketing arrangements are made for the orderly movement of goods from local to national markets. For the small peasant farmers, however, located away from good communications or population centers, selling their crops is almost as big a problem as growing them.

Modern Agricultural Enclaves

The development of cash crops was encouraged by the colonial powers mainly to serve overseas markets.

Two main methods were utilized. The first was through native smallholder production of such crops as peanuts, cotton, and cacao. The second was the establishment of plantation enterprises both for these crops and others such as rubber, pineapple, and tea which were thought to need large-scale investment and experienced management. In some countries, expatriate farmers were encouraged to set up private plantations. In others, governments initiated state farms or joint ventures with overseas companies. In semiarid areas attempts have been made to introduce large-scale ranching as pilot projects to indicate ways of future modification of animal husbandry in these areas. Such ranches can undoubtedly produce meat and dairy products. Also involved, however, is the alienation of much land from the traditional pastoral economy, the investment of considerable capital, including imported animal stock, and considerable management skill.

In more humid areas, plantation enterprises dealing with palm oil, rubber, tea, and coffee are common. In most cases these plantations were started by foreign investors and later incorporated into the state structure. In both cases the small enclaves of large-scale farming have disproportionate importance in the export earnings compared with the peasant sector, but these activities are seldom effectively integrated with those of the small-scale farmer.

Industrial Sector

In the least developed countries, the industrial sector is small. Industries are typically concerned with first-stage industrial processing of mineral and agricultural products, such as peanut shelling or cotton ginning. Small-scale production of cement, cheap household goods, beer, soft drinks, shoes, plastic goods, and textiles, and heavier industry involving construction materials and products comprise the bulk of the consumer-oriented industrial establishment. As the internal market has low purchasing power, and many countries in this group are very small, market-oriented industry is indeed limited.

Links with the International Trading Network

Table 25–2 illustrates the pattern of trade for selected countries of the least developed group. The total volume of trade in relation to population size is low. These countries still have largely self-sufficient societies. The nature of the trade, however, places the

Table 25–2 *Trade Patterns of Selected Least Developed Nations in Africa and the Middle East*

Country	Major Sources of Imports and Percent of Total Imports by Value		Major Destinations of Exports and Percent of Total Exports by Value	
Cameroon	France	33	France	57
	Netherlands	24	United States	10
	West Germany	18	West Germany	8
	United States	11	Italy	7
Chad	France	45	France	72
	Japan	22	West Germany	7
	West Germany	10	Netherlands	6
Ethiopia	United States	55	Italy	20
	West Germany	9	United States	18
	Italy	9	Japan	18
Rwanda	Belgium	37	Belgium	28
	United Kingdom	29	Japan	28
	West Germany	15	West Germany	16
Uganda	United States	26	United Kingdom	31
	United Kingdom	23	Japan	16
	Japan	19	West Germany	12
Upper Volta	Japan	34	France	80
	France	27	West Germany	8
	Italy	19		

countries under much greater dependence than the figures suggest. Typically, exports consist of primary agricultural or mineral products, often in commodities for which there are many alternative sources of supply. Generally, these commodities have shown only slowly increasing price levels over the last two decades, and some, despite inflation, have shown a drop in value. On the other hand, the small volume of imports typically includes fuel oil vital for industry and communication, machinery, specialized goods of all kinds, and fertilizer. Most imports have shown steeply increasing price levels over the past two decades, and especially since 1970.

The least developed countries have been hard hit by fluctuating prices of commodities and the high cost of freight, and by far the best way the developed world could help would be through an improvement of their trading situation. One of the goals of the UNCTAD conference of the United Nations is to improve the trading situation.

Mali

Mali is a large country with a small population. The agricultural sector is the means of livelihood for 91 percent of the population, and three-quarters of all exports come from this sector. Food crops are millet and sorghum, the staple grains, an increasing quantity of rice (particularly popular in urban markets), and a wide range of crops such as maize, sugar cane, tobacco, tea, vegetables, and fruit. Cotton, peanuts, and livestock are the main exports, the first two mostly for European and Anglo-American markets, and livestock mainly for West African needs. Manufacturing is still in its infancy, although this sector is growing quickly, mostly under state ownership. Enterprises include eight cotton gins, a peanut oil mill, processing plants for rice, meal, sugar, soft drinks, and canned fruits and vegetables, as well as two textile plants, a shoe factory, and a bicycle assembly plant. There is no current national income generated from mineral deposits.

Per capita income in Mali is only about $90 and the total national budget only $70 million, much less than in many small towns in the United States. The Sahelian drought caused a major setback to the already slow rate of growth of national income.

Yet Mali, at present under a military government, is blessed with an energetic population and a government that is struggling to initiate suitable development programs for the people. It is more thoughtful than some other Sahelian countries about the future problems of the nation's large nomadic population and plans to develop this livestock sector steadily. The inland delta of the Niger is a great resource in this context, as the area provides precious dry-season grazing in an otherwise arid zone. The same delta is the site of an ambitious irrigation project, started by the French colonial government in the 1920s to provide vegetable oil and cotton for France. This overcapitalized effort indicates the difficulties of massive transformation solutions to the problems of the area.

In summary, Mali is a poor country, isolated, agricultural, with poor internal and external communications. Yet it has agricultural potential and prospects for slow economic growth in the future.

Ethiopia

Ethiopia is larger than most other countries treated in this chapter. It has a steadily growing population of about 28 million with a GNP of $26 billion and a per capita GNP of about $100. Agricultural products account for most of the exports: coffee is dominant (40 percent of total by value), followed by hides and skins, oils, seeds, cereals, and meat. There are several small-scale mining enterprises, of which gold is the most important. Ethiopia, however, is typical of the least developed group in that only 0.3 percent of the GNP is in mining compared with 84 percent in agriculture, 2.3 percent in forestry, and 12.5 percent in industry.

The industrial sector is bigger and more diversified than most other countries with low per capita GNPs, partly because of the larger market provided by the relatively large population. Many import substitution industries were established in the 1950s and 1960s. The success of industrial enterprises has been mixed. Textiles is a fast-growing industry with a small export trade, but a pulp and paper enterprise has not proven feasible. A modern tire factory opened in 1972 is capturing a large proportion of the local market.

In agriculture, there have been few changes, and the pattern and products of agriculture have been held constant by the many problems of land reform and rural development. Under the new socialist government, attempts are being made to break up the feudal system that prevailed into the 1970s in rural Ethiopia and to produce a more modern form of landholding, often cooperatives. This task has only just begun and is likely to prove long and difficult, given the isolation of one part of Ethiopia from another and the stronghold of the feudal systems.

In the long run, Ethiopia appears to be a land of substantial opportunity; there are rich soils in places, though terrain is rugged; the climate is varied; and a wide range of agricultural products are possible. Continued industrial growth is likely, although it will be mainly to serve local markets. Tourism is a possible growth sector of the future; yet before much of this potential is realized, many social and administrative problems must be overcome. Ethiopia has a long way to go before leaving the ranks of the least developed.

Summary

The least developed countries have special problems. In a few cases, these may be helped by new mineral resource discoveries or other sources of new income, but for the most part there is a long hard road to economic growth and improvement in living standards. There are, however, encouraging signs that both the world community and the leadership in the countries themselves are beginning to see ways of working together to this end.

Further Readings

Keeping current with the latest developments in the least developed and developing countries of Africa and the Middle East is no easy task. The job is made immeasurably easier by the existence of the monthly journal *African Development.* The journal provides both news items of immediate interest and an overview analysis of a specific country in each issue. The result is a scholarly, timely, and indispensable source of information.

Two serials that strive for a comparable level of contemporaneity are the *Arab Economist* and the *Africa/Middle East Business Digest.* Both feature short articles focused on specific aspects of a country's economy, and the *Arab Economist* publishes occasional supplements surveying a specific national economy.

Development issues in the least developed countries inevitably begin with the traditional life styles and economies of this group of states, since they form the baseline from which prospects for growth must be assessed. Traditional nomadic ways of life have received considerable attention, but the works of Cunnison, Ian, *Baggara Arabs: Power and Lineage in a Sudanese Nomad Tribe* (Oxford: Clarendon Press, 1966), 233pp.; Asad, Talal, *The Kababish Arabs: Power, Authority and Consent in a Nomadic Tribe* (London: C. Hurst, 1970), 263pp.; and Stenning, Derrick J., *Savannah Nomads: A Study of the Wodaabe Pastoral Fulani of Western Bornu Province, Northern Region, Nigeria* (London: Oxford University Press, 1959), 266pp., provide a representative introduction to the topic. Allan, William, *The African Husbandman* (Edinburgh: Oliver & Boyd, 1955), 505pp., is a valuable introduction to African subsistence production systems. Johnson, Bruce F., *The Staple Food Economies of Western Tropical Africa* (Stanford: Stanford University Press, 1958), 305pp., examines the implications of food production in a development context, and

deSchlippe, Pierre, *Shifting Agriculture in Africa: The Zande System of Agriculture* (London: Routledge & Paul, 1956), 304pp., analyzes a classic low technology economic system. A case study of patterns of change in a traditional culture due to increased contact with the outside world is presented by Knight, C. Gregory, *Ecology and Change: Rural Modernization in an African Community* (New York: Academic Press, 1974), 300pp.

An attempt to introduce high technology into the least developed economies in order to promote rapid economic growth is one potential solution for a resource-constrained setting. Some of the problems resulting from utilization of foreign, large-scale technology for development purposes are explored in Farvar, M. Taghi, and Milton, John P. (eds.), *The Careless Technology: Ecology and International Development* (Garden City, N.Y.: Natural History Press, 1972), 1030pp. with the articles on African diseases spread by technological development, pp. 69–101; salinization in Algerian oases, pp. 276–287; changes in East African soils, pp. 567–576; consequences of nomadic settlement, pp. 671–682; Bedouin settlement in Saudi Arabia, pp. 683–693; and range management in Masailand, pp. 694–711, being the most immediately relevant. The reason why technology can have adverse impacts is discussed in a very important paper by Holling, C. S., "Resilience and Stability of Ecological System," *Annual Review of Ecology and Systematics,* 4 (1973), 1–23.

Africa and the Middle East: The Intermediate Nations

26

T HE LEAST DEVELOPED countries, dealt with in Chapter 25, are clearly very poor nations with little hope of foreseeable rapid advancement; the oil-rich nations of the Middle East are, conversely, going through far-reaching changes in economic well-being. Between these two extremes there is a large group of nations, with quite varied characteristics, designated "intermediate" for several reasons (Fig. 26–1).

Some intermediate countries have growing but relatively undiversified economies; most gain the bulk of their foreign exchange by exporting one or two agricultural commodities. Dependence on a small range of agricultural exports is exacerbated by the general slow rise in price of agricultural commodities on the world market and the much higher rise in prices of other commodities and manufactured goods. Recently, countries heavily dependent on agricultural production have been major victims of the worldwide rise in the price of petroleum products. Fertilizers for farms and fuel for production and marketing

machinery have risen steeply in price. Another group of intermediate countries has begun to develop a more diversified resource base, mainly through the exploitation of mineral resources. Nigeria with oil and Angola and Zaire with a variety of minerals have experienced growing national incomes. None of these countries have yet been able to transform this new income into an improved standard of living for most of the people or into a wider range of production enterprises.[1] A few intermediate nations fall into our purview for different reasons. They, despite relatively high levels of current economic development, appear to have but moderate potential for the future. All nations dealt with in this chapter share the experience of a colonial inheritance. Nearly all are still poor countries in economic terms, but this low level of economic activity is accompanied by a moderate-to-high potential for future growth.

Table 26–1 presents the categories that form the basis for the discussion in this chapter. The first set of countries is dominated by a single economic sector. Most are grounded in agriculture, although Mauritania is heavily dependent on a single mineral resource, iron ore. The second set of countries has more diversity in their countries' national economies, some on the basis of a single mineral resource with an agricultural base, others on several mineral resources and an agricultural base.

Countries with Dominantly Single Sector Resources

In some respects, nation-states are not unlike ecosystems. Complex ecosystems are more stable and have less risk of destruction than simple ecosystems. Nation-states with a diverse economy are likewise less vulnerable than those whose economy depends heavily on only one or two products. Vulnerability increases when the country must compete with many others in the world market. States possessing diverse environments and a number of developed economic components stand a good chance to provide prosperous and stable conditions for their citizens. Some

[1]In the 1960s the overriding economic development concern was for growth in national income; it is now more widely appreciated that equity problems are equally important.

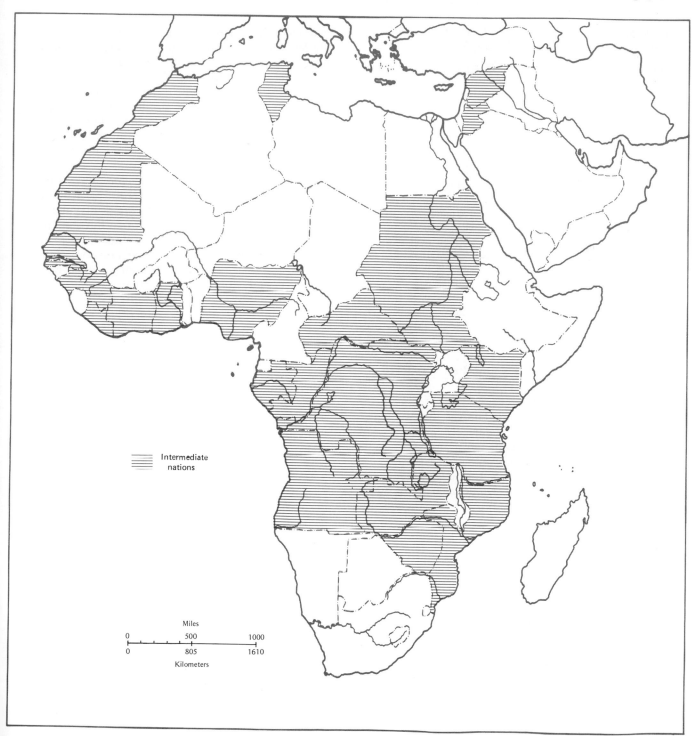

26–1 Africa and the Middle East: Intermediate Nations

Table 26–1 *Africa and the Middle East: Intermediate Nations, Level of Economic Diversification*

Single Sector Economies	Multisector Economies
Central African Republic	Angola
Guinea	Congo
Ivory Coast	Gabon
Kenya	Ghana
Mauritania	Liberia
Senegal	Jordan
Sudan	Morocco
Syria	Mozambique
Tanzania	Nigeria
Tunisia	Rhodesia
Zambia	Zaire

countries, however, such as Libya and Kuwait, have an abundant single resource of such value that it overcomes all other difficulties. Most nations are not so fortunate.

Problems Associated with a One-Item Resource Base

Many countries have development prospects in one sector of the economy. Tunisia, Ivory Coast, Tanzania, Syria, and the Sudan are examples of countries nearly totally reliant upon agriculture not only as a prime source of daily subsistence but also as a generator of foreign exchange for development and other purposes. In contrast, Mauritania is an agricultural and pastoral community that struggles to maintain a precarious existence, while mineral exploitation offers the best hope for future economic growth.

Dependence on only one sector to stimulate growth may cause serious problems. Some of these problems are internal; others are produced by the operation of international market mechanisms. Internally, concentration upon agricultural development often takes the form of cash crop production for overseas sales as a means of gaining foreign exchange. Agricultural developments of this type often place great strain on local socioeconomic systems. Frequently, capital generated from commercial farming benefits only a small section of the agricultural community. Inequitable distribution of wealth in turn may create social tensions where none existed before or may

encourage a drift of population away from economically less advanced rural regions toward the seemingly greater economic opportunities of urban areas. Commercial agriculture may also disrupt traditional agricultural patterns by encouraging a removal of land from subsistence food production or by encroaching upon fallow or pasture land. The growth of agricultural settlement on government irrigation schemes has in some cases encroached upon land formerly an essential part of a pastoral system.

Externally, concentration of development effort in the commercial agricultural sector, particularly when only one or two crops are exported, may make a country extremely vulnerable to price fluctuations on the world market. Cotton in the Sudan, cacao in Ghana and the Ivory Coast, coffee and cloves in Tanzania, and olives in Tunisia are all major foreign exchange earners for their respective countries. As long as commodity prices are high, the economy of these countries prospers. When prices fall, however, it can mean a major economic setback for the producers. Year-to-year weather conditions can also result in considerable differences in production and unforeseeable fluctuations in the national income.

Other problems exist in a one-resource economy, even if the nation concerned turns its back on the world economy and stresses self-sufficient patterns of development. With medical technology removing traditional constraints on population growth, rural population grows at rates greater than 2 percent yearly. The capacity of an agricultural economy to absorb the additional labor remains limited, and employment opportunities in the agricultural sector are often inadequate to absorb the new labor force. Both seasonal and permanent unemployment and underemployment result. In fact, the problem of employment is most important for developing countries, especially those where the economy has begun to grow. Available jobs are growing more slowly than the work force, and a latent potential for social and political unrest is frequently a consequence.

The economies of almost all of the countries considered have provided a very narrow base upon which to attempt to develop a manufacturing industry.[2] Agricultural processing industries frequently remain

[2]Reginald H. Green and Ann W. Seidman, *Unity or Poverty? The Economics of Pan-Africanism* (Harmondsworth: Penguin, 1968), pp. 52–78.

small in size, and most products are either consumed fresh or shipped in an unprocessed state to overseas markets. Even where a local mineral resource such as Mauritania's iron ore is developed, it is exported overseas after only minimal processing, and stimulates little indigenous employment.

As a result, there are great disadvantages facing any country that attempts to raise personal income significantly from an agricultural base. Yet in an age of nearly instantaneous electronic communications, citizens of the underdeveloped world are aware of better living conditions elsewhere. Not unnaturally, they would like a share, however small, of the good life as they conceive it. The governments of one-resource countries have tried various solutions in an attempt to foster economic growth. These alternative strategies may be ideological in nature, such as Tanzania's emphasis on instilling a spirit of *ujamaa* (socialist cooperation) and pride in the people. On the other hand, they may artificially attempt to increase the diversity of the natural ecosystems available by massive application of technology. Giant irrigation schemes such as Sudan's Gezira or Syria's Euphrates Dam project are examples of an attempt to create totally new agricultural environments as a spur to increased productivity. Still other strategies call for the creation of new sources of revenue, such as from increased cash cropping or from a tourist industry based on mild climate and attractive waterfrontage, or the development of known mineral resources. Let us consider some of these alternatives more closely.

Alternative Development Strategies

Tourism as a Development Alternative

All countries in the intermediate group gain some income from tourism, but Ivory Coast and Tunisia have placed special emphasis on the tourist industry as a means of accelerating economic growth. Tanzania has important tourist potential but has hesitated to commit resources to this sector. The Tanzanians have felt that tourism on a large scale may create more problems than it solves, though there are signs now

that more emphasis will be given to promoting tourist visits.[3] For Ivory Coast and Tunisia, there have been few such inhibitions.

The Ivory Coast is something of an economic miracle. For the decade 1965–1975, an economic growth rate of over 7 percent per year has given the Ivory Coast an economy more expansive than many in Europe. So properous is the Ivorian economy that 1.5 million workers from the surrounding countries of Ghana, Mali, Guinea, and Upper Volta have flocked to the region to share in the boom. Perhaps the most remarkable feature of Ivorian development is its nearly total reliance on growth in the economy's agricultural sector (Fig. 26–2). Exports of coffee, timber, and cacao account for 80 percent of all exports with such specialty crops as pineapples, peanuts, sugar cane, bananas, and rubber comprising most of the remaining earnings. Unlike the small industrial sector where management and ownership are concentrated in French hands, most of the Ivory Coast's cash crops are produced by peasant farmers. The direct infusion of funds into the hands of the bulk of the population (per capita GNP has doubled in the last ten years) plus great political stability have created a climate conducive to large amounts of foreign investment to complement government expenditures on ports, roads, and other infrastructural improvements.

Although the immediate future of the Ivory Coast's economy appears bright, there are problems. Basically, these are associated with the country's vulnerability to price fluctuations on the world market. In an attempt to reduce the risk of sudden market variations, the Ivory Coast has searched for alternative devices for extending its economic boom. Investment in the tourist industry has been adopted as a means of rapid future economic growth.

Large sums of money are being expended to increase the number of hotel rooms, increase accessibility to interior locations of cultural and environmental interest, and improve tourist facilities. By 1970, some 5,800 new rooms, one-half located in and around the capital, Abidjan, were completed. Visitor-nights now total 300,000 annually and are projected

[3]Tourism was thought to benefit mainly the foreign companies engaged in the business and to encourage "servant" attitudes among local peoples.

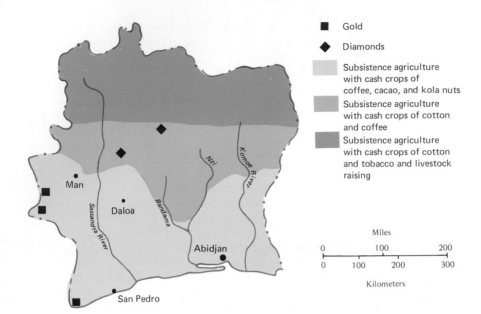

Gold

Diamonds

Subsistence agriculture with cash crops of coffee, cacao, and kola nuts

Subsistence agriculture with cash crops of cotton and coffee

Subsistence agriculture with cash crops of cotton and tobacco and livestock raising

26–2   Ivory Coast: Agricultural Systems and Mineral Deposits

to swell to over 1 million by 1980.[4] Lured by long sandy, palm-lined beaches, luxurious hotels, exotic game parks, and excursion air fares, tourists are expected to spend lavishly in Abidjan and several other target zones near San Pedro along the west coast and Man and Daloa in the interior. The resulting infusion of new jobs is meant to spur economic growth.

That this alternative is not necessarily a sure-fire solution is indicated by Tunisia. Although the economy is primarily agricultural, despite having valuable though small-scale iron ore and phosphate deposits, Tunisia also has a flourishing tourist industry (Fig. 26-3). Resort hotels line Tunisian beaches, and nearly 800,000 foreign tourists visit the country yearly. Government-sponsored hotels often are developed in abandoned Islamic schools or constructed in indigenous house-type styles, providing an authentic flavor that retains much of the uniqueness and charm of local culture. Many traditional craft industries survive by producing high-quality items at low cost for shopping tourists. The *suq* (market) of Tunis' old city is a magnificent setting in which to hunt for bar-

gains. Moreover, a system of government *artisanates* preserves traditional crafts, sets quality-control standards, helps the artisan develop traditional

26–3   Tunisia: Economic Activities

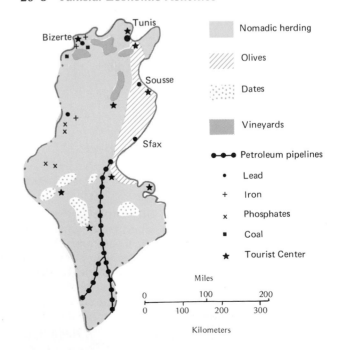

Nomadic herding

Olives

Dates

Vineyards

•—•—• Petroleum pipelines

•   Lead

+   Iron

×   Phosphates

■   Coal

★   Tourist Center

---

[4]Visitor-nights is a tourist-industry measure of the number of beds occupied by tourists in a given statistical period (generally a year). Since tourists usually remain in a country for more than one night, the figure drastically overstates the total number of tourists actually visiting a country.

In recent years Tunisia has built a number of hotels and initiated other projects to attract increased tourist traffic. Shown here is a modern hotel where the government sponsors training of hotel personnel. At the right is a restored fortress. (United Nations/B. Graham)

designs that appeal to European aesthetic tastes, trains new apprentices, and establishes fair-pricing practices. Proximity to Europe, together with vigorous government support, makes the tourist industry a major contributor to the Tunisian economy.

The tourist alternative, however, is not sufficient to provide adequate employment. While as much as 5 percent of Tunisia's population is directly employed in the tourist industry, nearly one-half million Tunisians are unable to find work locally, and many seek work abroad. Often it is low-cost Tunisian labor that performs the least desirable jobs in France, Tunisia's former colonial master. This pattern of labor migration mirrors a traditional specialized rural-urban labor flow in which specific villages provide workers for certain urban jobs. The newspaper vendors of Tunis, for example, come largely from the southern Berber village of Chenini. Extension of the system overseas is a measure of the inadequacy of tourism to diversify the Tunisian economy.

In addition, concentration of employment opportunities in the tourist sector inevitably gives the foreign visitor a false impression of local talents and personality. Confined to a subservient role, the servile image projected can become both demeaning and debilitating. The culture conflict inherent in a tourist industry is also threatening to local values and often accounts for the hostility that frequently accompanies a massive tourist influx. Thus, although tourism may help diversify a local economy, it is a mixed blessing; and because the tourist's ability to spend is linked directly to the health of the global economy, it may expose the tourist-oriented economy to many of the same vicissitudes that affect a cash crop concentration.

Application of Technology

Use of technology to overpower the environment and achieve a quantum leap in production is a Western strategy for accomplishing development

511

objectives. Diffusion of Western technology through-out the poor world has made it possible for under-developed countries to emulate the activities of the industrialized nations. For example, in the arid zones, large-scale agricultural development projects based on dam-fed irrigation systems have been a common pattern of development. Two countries, Syria and the Sudan, have adopted this strategy to promote growth in their agricultural sectors as well as provide hydro-electric power for projected industrial opportunities. Since this strategy is under consideration by a number of other nations with as yet underdeveloped river basin systems, examination of the Syrian and Sudan-ese experience is instructive.

Irrigation is an ancient technology in Syria. The *shaduf* (lever-operated bucket), *kanat* (underground water-collection tunnel), and *noria* or *dulab* (water-wheel) have long been used to raise water from streams or convey water from distant sources to agricultural fields. After World War II, major efforts were made to develop the irrigation possibilities of the Orontes ('Asi) River using modern technology. The gains from these development projects, although important, were minor when compared to Syria's total arable area. Attention was then turned to the Euphrates Valley which represented the last major source of water available for irrigation (Fig. 26–4). Work began in 1968 on a giant dam project at ath-Thawra (Tabqa) twenty miles south of Meskene. By creating a 50-mile-long (80 kilometers) artificial lake, the new dam will provide water with which to irrigate 1.5 million acres (607,000 hectares). This project will more than double Syria's irrigated area and dramatically increase crop yields, as well as produce more cotton for export.

The resulting improvement of Syrian standards of living could prove considerable. The Euphrates Dam's hydroelectric power capacity of 12 million kilowatts will exceed the power potential of the Aswan High Dam in Egypt and should help compensate for Syria's lack of coal or oil energy resources. Construction of the dam will, however, necessitate the resettlement of more than 60,000 farmers presently farming rich floodplain areas that will be inundated by the dam's large lake. Although plans call for resettling these displaced villagers in new modern communities, not all farmers are happy with the prospect of relocation. The social costs attendant upon such compulsive dislocation and social upheaval, together with pos-

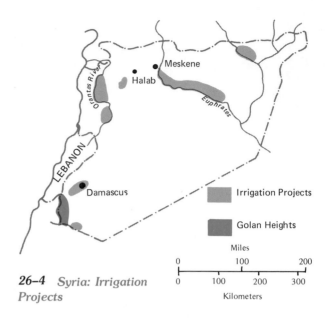

26–4 *Syria: Irrigation Projects*

sible unforeseen ecological side effects of the project, constitute the major hidden costs of this vast, techno-logically sophisticated scheme. The expectation is that local problems will be compensated by substantial gains at the national level, and that the potential threat of conflict with Iraq over distribution of Euphrates River water can be avoided. There is little doubt that most Syrians, seeing a chance to seize control of their own destiny and in dramatic fashion take a giant step toward their society's long-range goal of economic betterment, would count such problems as minor distractions on the road to prog-ress and development.

Sudan has used a similar technology, irrigation, but the schemes have evolved over longer periods. Also, the programs place less reliance on single large structural improvements to trigger growth (Fig. 26–5). Begun in 1925 with the construction of the Sennar Dam on the Blue Nile, the first stage of development brought irrigation to over 1 million acres (405,000 hectares). Much of this newly irrigated land lies in the Gezira ("island" in Arabic), the triangular area between the White and Blue Niles. Formerly a semi-arid plain used primarily by pastoral nomads raising animals and farmers cultivating a precarious rainfall crop of *dura* (a millet), much of the Gezira has been converted into more intensive and productive agri-

26–5 *Sudan: Irrigation Projects*

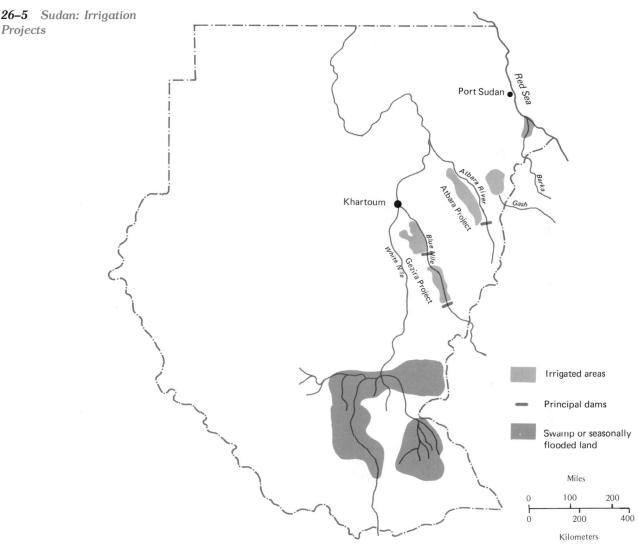

cultural land. The primary product in the Gezira region is long-staple cotton; this variety is used to make fine cloth, and its export is the major foreign exchange earner for the country. Unfortunately, growth of the Lancashire, England, fine cotton industry, for which much of Sudan's cotton crop was originally intended, has stagnated. Much of Sudan's cotton crop of 1.2 million bales (850,000 bales produced in the Gezira) now is marketed in Eastern Europe and the Soviet Union in return for military and technical assistance, although this political and economic alliance is weakening.

The Gezira is not only Sudan's major cash crop region, but it is also an important source of food for internal consumption. One-half of the country's wheat is produced in the Gezira, and substantial amounts of other crops are raised in the region's coordinated and integrated irrigation system. Its importance is such that major expansions have taken place. The Manaquil Scheme has extended irrigation to an additional 800,000 acres (324,000 hectares), and the Roseires Dam (finished in 1966) upstream from Sennar waters has added another 300,000 acres (121,000 hectares) in the El Rahad project. Much of

One of the irrigation outlets of the Gezira irrigation scheme in the Sudan. Waters of the White and Blue Nile are impounded and used to grow cotton for export and a variety of foodstuffs for local consumption. (*United Nations*)

this new cropland is planted in short-staple cotton in an attempt to avoid excessive dependence on the specialty cotton market and to provide an easier-to-process fiber that can meet the competition of artificial fibers. The Khashim El Girba Dam on the Atbara River also irrigates 740,000 acres (300,000 hectares) of cotton and food crops in another development project. These projects employ relatively simple technology and construction and are able to rely on an efficient and cheap gravity flow water system. Moreover, in each scheme the tenant farmers participate in the local development board's decisions. Sudanese irrigation agriculture has made a major contribution to the country's development.

A Mineral Alternative to Agro-Pastoralism

Mauritania is an interesting comparison to other nations in the intermediate group, for its one major activity suitable for rapid economic growth is the mining sector of its economy (Fig. 26–6). Although most of Mauritania's small population (1.4 million) is actively employed in agriculture, only 1 percent of the country's land is arable. Most of this land is

concentrated in the extreme south in the Senegal River valley, and the remainder of the country is composed of the low productivity ecosystems of the semiarid Sahelian grassland and the Sahara Desert. Except for the desert oases of Tagant, Adrar, and Assaba in the center and north that produce dates, and the millet zones along the Senegal, most of Mauritania is suitable only for livestock raising. It is no accident that nomadic pastoralism continues to be the major concern of 70 percent of Mauritania's population. Although efforts to improve all sectors of the traditional agro-pastoral economy are being made, the low productivity of the land presents many problems. Drought such as that which has blighted crops and decimated herds in West Africa's Sahelian zone is the most important of the environmental problems. The risk of repeated severe drought is the prime reason why the opportunities for major economic growth in the agricultural sector are limited.

From an economic development perspective, Mauritania's situation would be truly desperate were it not for the existence of several rich mineral deposits. These deposits permit hope that the constraints of a one-resource agricultural environment can be avoided to at least some degree. The existence of rich iron ore deposits in northern Mauritania around Fderik has been known since 1935. Only since 1960 has exploitation of these iron reserves, in many cases with a metal content of 60 percent, been possible. A great deal of infrastructure, including construction of a railroad to move the ore to the coast and port facilities at Nouadhibou to speed handling, had to be built before major exports became possible. In 1963, exports from deposits at Taxadit and Fderik were begun and rose rapidly to a total of 8 million tons in 1969 as additional deposits at Fderik and Rouessa were brought into production. Continued growth in the iron ore industry has taken place. Copper deposits at Akjoujt also promise to generate substantial revenue, and a search for petroleum is under way. Although no oil has as yet been discovered, development of local energy supplies would be of tremendous significance in an age of increasing energy costs.

As a result of royalties and taxes generated by the mining industry, Mauritania's foreign trade balance has improved from one of small deficits to a modest net favorable balance. This favorable balance is not without its problems, for it has proved difficult

26–6 *Mauritania: Economic Activities*

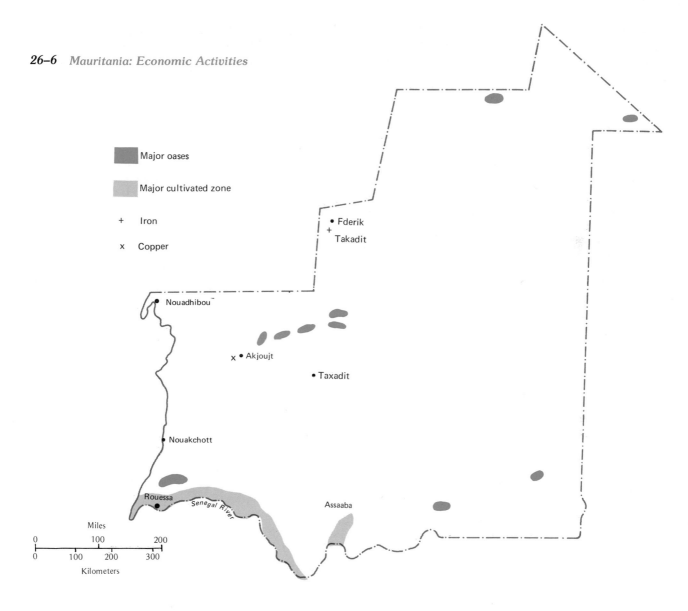

- �as Major oases
- ▱ Major cultivated zone
- + Iron
- × Copper

Fderik
+ Takadit

Nouadhibou

× Akjoujt

Taxadit

Nouakchott

Rouessa
Senegal River
Assaaba

Miles
0   100   200
0   100   200   300
Kilometers

to distribute the newly found wealth, modest though it is, equitably among Mauritania's citizens. Today some 15 percent of the population, largely those with managerial and professional skills or positions in the government bureaucracy, control 65 percent of the country's wage income. Such inequities are bound to have future political and social repercussions. Control of the major extractive industry is also a potential problem. The Mauritanian government owns only 5 percent of the stock in MIFERMA, the international company developing the iron ore reserves. A majority of the stock is held by public and private French companies and the remainder is shared by steel firms in Britain, Germany, and Italy. Although efforts have been made to develop new market outlets, most notably in Japan, the ownership of Mauritania's major revenue generator by industrial and financial interests that constitute the basic market for the product tends to hold down the price paid for Mauritanian ore. Most of Mauritania's income comes from indirect taxes on the industry rather than from a major share of the profits made by the sale of the iron ore. There is, thus, some doubt about whether Mauritania is realizing as much income as it might from the iron mining industry.

Nonetheless, revenue generated, despite income inequality, is finding its way into constructive development projects. Most of these projects involve trying to use the ore export revenue to develop the agricultural economy that has languished because of insufficient capital, technology, or expertise. The most noteworthy project, which has met with limited success, has been to increase Mauritania's share of the yield from the offshore fishery. By forcing foreign fishermen, largely Spaniards and Greeks, to land at least part of their catch in Mauritania for processing and sale, the government hopes to improve the availibility of protein in the local diet as well as produce processed fish for reexport. A number of fishing vessels have also been purchased for use by Mauritanians, although teaching new skills to people unfamiliar with deep-sea fishing is a slow process. Other projects involve improved slaughtering and cold storage facilities to open new outlets for the meat produced by the nomadic sector of the economy. Agricultural projects include efforts, aided by China, to develop rice production and start sugar cane cultivation in order to reduce import requirements. Although few of these projects have been unqualified successes, they do mark an important beginning in a strategy designed to raise living standards by applying revenue generated in one sector of the economy to growth in another.

Colonialist and Quasi-Colonialist Solutions

Desirable though a complete break with Africa's colonial past might be, in many countries severance of linkages to the former European colonial power is impossible without great distortions of many features of national life. Nearly all aspects of life in the newly independent states of Africa have been altered to a significant degree by the colonial experience. The end of direct foreign control does not guarantee an automatic end to educational, economic, technical, and cultural relationships forged under a foreign administration. Indeed, one option available to African states is to continue to tie future economic growth to the market opportunities and needs presented by the European "motherland."

Cutting all ties with Europe is an option that no African state has followed with short-run economic success. In 1960, when Guinea accepted France's offer of independence and withdrew from the French monetary system, the economic consequences were considerable, not least because of the immediate French reaction to this decision. Most of the foreign administrators and technicians, together with much of their equipment, were withdrawn almost at once. Guinea had great difficulties in raising money in international financial circles, and the economy rapidly became unstable. The government then turned to the Communist bloc, particularly the USSR and China, for assistance but found that not only were their ability and willingness to respond limited, but also that uncertainty about the political direction of Guinea frightened off investment from capitalist countries. Only after a lengthy period of dislocation and discontinuity was Guinea able to strike a balance between international political forces, provide guarantees to the international financial world, patch up relations with France, and resume a balanced program of economic development. The Guinean experience has served as a warning to other independent African states and is one reason why the Portuguese colonies, now emerging from the trauma of colonial guerrilla warfare into independence, are likely to retain reasonably close ties to Portugal.

There are many other reasons why it is difficult to disown the colonial relationship completely. Often the language of the colonial power serves as the lingua franca integrating a large number of tribal dialects. Moreover, education is frequently closely linked to the European system with standardized exams for admission to a university following the European model in content and format. African elites also are fluent in the language of the former colonial powers. In many instances they have adopted some of the manners, thought patterns, and beliefs of Europeans. These European traits often result in internal difficulties because they increase the difference between the country's leaders and the mass of the people. The ability to speak another language, however, does make it possible to keep abreast of technological developments overseas and eases the transition to a different culture for students sent overseas for specialized training.

The economies of the colonial territories were closely tied to those of the occupying power. Usually they produced crops or minerals in short supply in Europe. The legacy of this relationship is a role as a primary producer that is difficult to avoid. Markets for the cash crop, be it cacao or cotton, already exist; and since foreign exchange is important to most African countries, it is difficult to eliminate plantation agriculture.[5] Moreover, cash crop agriculture provides employment to peasant farmers; an end to cash cropping would drastically affect the living conditions of the bulk of rural populations. Where minerals are the primary export, the position of the producer is more secure. Because such minerals as titanium, copper, and uranium are in short supply in the world market, and most producers are now independent, consumers do not have the same control over the producer they once had. Whenever producers, such as the oil-exporting countries, have been able to develop a firm and concerted policy, they have adjusted price and royalty arrangements in ways that are favorable to themselves. This situation is not without its dangers, however, for where several potential producers exist, the consumer can select among the competitors. Such was the case in Niger where France, dissatisfied with local conditions, stopped producing uranium at the mine in Arlit. Never a major producer and dependent on French economic help, Niger was forced to accept an unpleasant situation involving substantial loss of revenue. Thus, there are serious difficulties in pursuing a development policy that is too dependent on one market outlet and one source of economic assistance. Associate status with a trade grouping such as the EEC (European Economic Community), however, can provide preferential treatment for various products and result in substantial benefits for the country involved.

The difficulties of diversifying market outlets remain substantial. The newly independent African states continue to be dependent on foreign experts for technical assistance. Often poorly provided with competent indigenous technicians before independence, most African states must continue to rely on expert help provided from overseas. In most cases this situation means large salaries for foreign technicians and a continuing image of dependence for the new nation. Few countries have been willing or able to follow the example of Tanzania, where foreign technicians are paid salaries compatible with local wage scales and are not allowed to remit substantial portions of their salaries to their homeland.

Not only are technicians expensive, but in addition, the technology they bring is often not really suitable to the local situation.[6] Capital intensive in nature, this technology fails to provide numerous jobs for, and skills to, the local population. Moreover, technological dependence continues to lock the recipient into a system of language, education, and skill dependence that reinforces the original pattern of colonial domination. The fact that most of the newly independent nations continue in a type of commonwealth relationship with the original colonial country is symbolic of their continuing dependent status. Continued involvement with the former mother country is inevitable, since in many cases it is the only way in which progress can take place. Despite ambitious goals, independence politically is only slowly followed by economic self-assertiveness.

Ujamaa in Tanzania

Like all other countries in the intermediate group, Tanzania derives most of its resources from the agricultural sector (Fig. 26–7). Farming and animal husbandry are the activities of the bulk of the population. Exports of agricultural commodities such as coffee, cotton, cloves, and sisal are the country's major foreign exchange earners. All of these crops are affected by fluctuations in world-market prices, cloves having been especially hard hit by a steady decline in price over the last decade. This situation has sparked efforts to diversify the economy of Zanzibar and nearby islands, where most of the cloves are grown, by introducing cattle ranching and new crops such as rice, cacao, and coffee.

Tanzania does have a small diamond mine and reserves of coal and iron which until now have been too remote to be exploited. Although the diamonds do not exist in large quantities and will only be exploitable for a few more years, they have helped balance Tanzania's foreign payment situation. De-

---

[5]For some of the problems inherent in this dependent state, see Reginald H. Green and Ann W. Seidman, *Unity or Poverty? The Economics of Pan-Africanism,* (Harmondsworth: Penguin, 1968), pp. 38–51.

[6]Gunnar Myrdal, "The Transfer of Technology to Underdeveloped Countries," *Scientific American,* 231 (1974), 172–191.

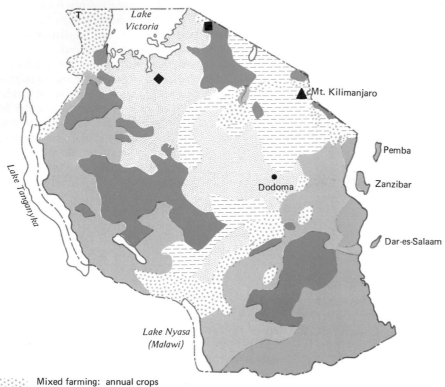

Mixed farming: annual crops
(maize, millets, cotton, rice,
oilseeds, tobacco)

Arable farming: perennial
and annual crops: (coconuts
cashew, maize, millets, rice,
oilseeds)

Mixed farming: perennial
and annual crops (coffee,
bananas, wheat, maize,
pyrethrum, tea)

Arable farming: annual crops
(maize, millets, cotton, rice,
oilseeds, tobacco)

Pastoral and semipastoral

Forest and game reserves,
national parks

◆ Diamonds

T Tin

■ Gold

26–7   *Tanzania: Agricultural Regions*

spite an excess in dollar value of imported machinery and manufactured goods over agricultural exports, Tanzania has been able until recently to maintain its foreign reserve holdings. The precariousness of this aspect of Tanzania's economic position is indicated by the rapid decline in foreign reserves in 1973–1977 as the price of oil skyrocketed and the need to replace drought-depleted grain stocks became acute. At present, Tanzania is in severe financial difficulties because of the drought and high fertilizer and oil prices. Unfortunately, neither drought nor petroleum prices are under Tanzanian control, and the best that can be done is to cope with them in the short term while designing long-range strategies that reduce exposure to the vagaries of nature and the international marketplace.

Tanzania is unusual with respect to the strategy selected to deal with the development problems. Tanzania has not adopted many aspects of Western technology. Instead, emphasis has been placed on developing the motivation and aspiration of the people in ways that seem authentically Tanzanian and African. *Ujamaa* is the word used to express this philosophy. The connotation of *ujamaa* is one of pride in self, of commitment to self-improvement, and development of the individual, region, and nation along communal and cooperative lines. This type of African socialism is seen by Tanzanians as preserving the best features of the African tribal, group-conscious social system. Also, by involving the people at grass-roots level in the development process, generally good results have been obtained.

Originally the government experimented with building a series of new villages with modern amenities in an effort to promote rural development. The experiment was not successful as costs were high, and many Tanzanians were reluctant to move to the new settlements. Emphasis shifted to introducing development projects into existing villages, the so-called *ujamaa* villages. Local people were encouraged and often exhorted to undertake communal projects in return for government assistance. Cooperative agricultural activities have been promoted with marketing and credit arrangements handled by the cooperative. The Chagga coffee cooperative that has been operating for many years on the slopes of Mt. Kilimanjaro is a good example of a successful cooperative. Government-controlled price structures are also an important device in encouraging farmers to produce crops in

the amounts required. For example, a small upward change in the price-support structure for rice encouraged a spurt in rice cultivation as farmers rapidly adjusted to the improved economic returns available in the alternative crop.

Great efforts are also being made to utilize technology that is appropriate to the local context. Tanzania has stressed improvements in its existing peasant technology, for example, producing stronger and more efficient ox plows rather than the introduction of tractors and other machinery. With foreign exchange reserves declining, this policy also conserves badly needed cash. The break with the colonial era is also symbolized by Tanzania's decision to move its capital from the coastal city of Dar es-Salaam, which was the center of colonial administration, to a site at Dodoma in the central part of the country. The cost of such a move is tremendous, both in terms of abandonment of existing facilities and infrastructure along the coast and in creating a new city in the water-deficient central interior. As a symbol of *ujamaa* and as an expression of a determination to make the government more accessible to the people, the move is more than an empty gesture. It is a vital expression of a strategy for galvanizing popular involvement in the development process and is a central ingredient in Tanzanian hopes for a better life.

Countries with Diversified Resources

Table 26–2 portrays seven countries studied in this section in terms of their agricultural and mineral resource use. Most of the countries have important agricultural resources, either potential or current. In each case, except perhaps Gabon, there is within the country some diversity of national ecosystems. This ecosystem diversity is matched with potential diversity in agricultural production. Nigeria has a wide range of environmental resources ranging climatically from the Sahelian zone in the north to a humid coastal environment in the south. Most other countries have only part of this range. Culturally the group is diverse. Politically it includes three former British colonies, Nigeria, Ghana, and Zambia; two former French colonies, Guinea, and Gabon; a former Belgian colony, Zaire; and a former Portu-

Table 26–2   *Africa and the Middle East: Intermediate Nations, Mineral and Agricultural Resource Development*

		Mineral Resources	
		SINGLE	MULTIPLE
Agricultural Resource Potential	HIGH	Nigeria	Zaire
		Guinea	Gabon
			Angola
	LOW	Zambia	
		Ghana	

guese colony, Angola. Current political status is equally diverse ranging from dictatorial military rule to socialism.

A focus for economic growth in all of these countries is mineral resource development, though for each this development has a different pattern and potential. The economic position of Nigeria has been transformed since 1967 with the increasing importance of oil. The fourfold increase in oil prices in 1973–1975 reemphasized this change. A major problem for Nigeria and many others in this group, however, is how to invest the increased wealth in productive activities benefiting a wide section of the nation. Guinea has the world's second largest bauxite deposits, Zambia important copper deposits, and Ghana major bauxite resources. In each case the exploitation of these resources by international companies has brought increased national income but limited growth in other sectors of the economy.

Zaire, Gabon, and Angola possess diversified mineral resources: copper, diamonds, and cobalt in Zaire; petroleum, manganese, natural gas, uranium, and iron ore in Gabon; diamonds, petroleum, and iron ore in Angola. In every case the minerals are the leading edge of economic development. Agriculture in Angola is also important, though most of the export crops are in foreign hands. Since independence, Angola's export crop sector has been unstable, as many Europeans have fled the country. In Zaire and Gabon, agriculture remains of great potential despite only moderate actual achievement. Gabon, with a

population of just over one-half million, has increasingly concentrated effort on the lumber industry and mining, and only a small part of the country is in agriculture. Zaire and Angola have a more mixed economy, though in Zaire at least, development of agriculture still takes second place.

Each of the nations in the intermediate group is grappling with the problems and opportunities of development using externally derived capital. In each country the multinational company is a fact of life. The general sentiment before 1973 was that the multinational corporation had an overwhelming strong hand in dealing with small African countries. In most cases the companies had alternative countries in which they might concentrate their investment, and the host country was left with the need to provide incentives for investment. Tax holidays, remittance guarantees, and special privileges regarding work permits and trading arrangements were common features of concessions made to multinational companies by African countries.

This situation has not changed completely, and for some commodities and countries it is still basically the status quo. But the solidity of OPEC (Organization of Petroleum Exporting Countries) at least through 1973–1977 has allowed oil-producing countries in Africa to take a much stronger stand. Nigeria and Gabon have begun to renegotiate arrangements with oil companies working in their countries. Renegotiation will undoubtedly lead to an increase in taxes and participation for host countries, in addition to price increases. In this situation the balance of power is changing.

Zambia: The Copper State

The problems of Zambia are a good example of one kind of situation related to mineral development. Zambia with a population of 5.2 million and a per capita GNP of $540 is a rich country compared with others in Africa. It has a favorable balance of trade. Still, ten years after independence, copper accounts for 98 percent of foreign exchange earnings. It is directly responsible for 29 percent of government revenue and contributes additional indirect sums. Cobalt, zinc, and tobacco are other minor export earners. Zambian towns appear on an average to have as high a standard of living as anywhere in Africa and higher than some European countries. The countryside, however, is a different matter. Zambia has considerable agricultural potential, yet imports 40 percent of its food.

There are many reasons why Zambia is food deficient. For some years after independence, little investment occurred in agriculture, and the proportion of people on the land dropped from 70 to 75 percent in 1963 to 60 to 65 percent in 1975. The apparent high standards of living in the towns compared with the countryside were a major factor in a rural-to-urban migration. Despite the wealth generated by copper, however, employment opportunities are not being created comparable to the urban growth rate. Large ranges in wealth even among working

Gabon's economy differs from its neighbors. Subsistence agriculture is of minor importance, and lumbering and mining are major forms of livelihood. (United Nations)

adults occur in the towns, and still more discrepancies in standards of living occur between town and country. At the same time, ironically, much wealth is being lost to the country because there is not yet a sufficient reservoir of trained technical and managerial workers to run the copper mine and other industries. Zambia thus has an employment problem and a need to employ many foreigners at high salaries.

Major efforts are beginning on both problems, though there is a great range of relative concern with these two problems. Some, at ministerial level, welcome skilled foreigners who are "valuable to the economy"; others are worried about the present labor structure in economic, social, and political terms. Through the University of Zambia and a range of formal technical education programs, the Zambian government is attempting to create a reservoir of skilled manpower quickly. Training is relatively easily acquired; job experience unfortunately takes much longer.

In agriculture the position is beginning to change. One success story is in sugar where over the past fifteen years there has been a steady improvement from 1960 when all needs were met by imports to 1974 when Zambia became self-sufficient. Prospects in the future are good for cattle. If World Bank plans materialize, Zambia could become a major exporter of meat and maize—for which Zaire is a large potential market—and of fruit, particularly pineapple and mango.

Zambia has currently begun to define a policy of diversification of resources and greater spread of wealth. It will take some years for this policy to become fully effective.

Gabon: A Timber and Mining Economy

Gabon is a large country in area but small in population. It is a country where subsistence agriculture has been replaced as the livelihood for rural people, first by the timber industry and more recently by a rapidly growing mining industry (Fig. 26–8). Although in terms of climatic and general environmental resources, there is a moderate long-term potential for production from the land at present population levels, this potential is only being realized in terms of lumber. In 1972, less than 0.5 percent of the land area was under cultivation. Of this, two-thirds was for subsistence and the rest was used for cacao, coffee, and palm oil for export. One result

26–8   Gabon: Economic Activities

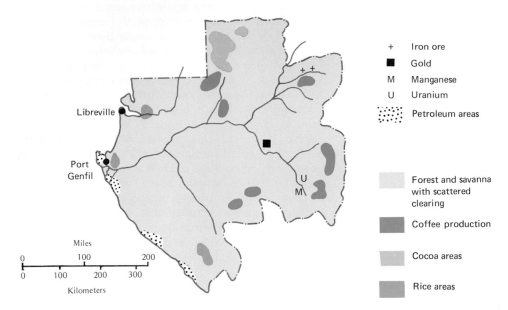

of this situation is a high import bill for foodstuffs (20 percent of total imports). Forests cover 56 million acres (23 million hectares), or 25 percent of the country, and in 1960 timber accounted for 73 percent of total exports. Timber is exploited by large European countries and increasingly by smaller Gabon enterprises. A great advantage of timber production is that processing industries can be established without massive capital investment. Timber processing accounts for over 50 percent of Gabon's industry and for much local employment.

In the 1960s, mining became increasingly important under license to international companies. Petroleum and natural gas have become economically dominant because of world-price increases, but manganese, uranium, iron ore, and gold provide a diversified and expanding mining economy.

Gabon has great potential, but major future constraints exist in the human resources in the country. Wealth is being generated rapidly from mining, yet many of the rural population still live at low economic levels.

Zaire: Minerals and Hydroelectric Power as a Development Focus

In Table 26–2, Zaire is placed as a multiple, mineral-resource nation with considerable agricultural potential. Like Zambia, Zaire has relied heavily on copper as the main basis of export earnings. Copper accounted for 60 percent of export earnings in most years. Cobalt and diamonds provide an important subsidiary mining income and rank with coffee and palm oil in export values. Zaire is the world's largest producer of industrial diamonds. Additionally there is a range of other as yet little-developed mineral resources including silver, iron ore, manganese, and possibly oil.

Zaire has considerable agricultural potential, though agriculture is still a struggling sector. Coffee, cacao, tea, cotton, and tobacco are produced for export, but with the increasing rate of urban population growth in Kinshasa (currently about 1.75 million) and the relatively high degree of employment in mining within the country, food import needs continue to grow.

Zaire has the great long-term advantage of one of the world's great hydroelectric power sites at Inga near the mouth of the Zaire River (formerly Congo). Initially 300 megawatts were developed, and in 1976 another 1,000 were added. This energy resource is one key to Zaire's development prospects, increased by the growing price of other energy sources.

A number of industrial projects are being considered to use hydroelectric power, including a steel complex, a caustic soda factory, a polyvinyl (PVC) factory, a nitrogen fertilizer factory, and an aluminum smelter. The steel complex is already under way, and the caustic soda and PVC plants are the next priorities. Power will also be used for the copper-producing area in Shaba (Katanga).

Zaire, with a large urban population and major mineral and power resources, may well become a nation with a major manufacturing section. Manufacturing is one conscious focus of government development plans. Much of the industry will be capital intensive, and there is a great need for massive training of skilled workers and an upgrading of employment possibilities in the rural and service sectors.

Nigeria: Black Africa's Most Powerful Economy

Nigeria, with a population of 67 million and a GNP of $19.5 billion, has the largest economy in Africa south of the Sahara outside South Africa. Much economic growth has occurred since 1968 because of the discovery and exploitation of oil in the Niger delta area. In 1968, exports were valued at $321 million with oil accounting for only $58 million. In 1972, before the abrupt rise in oil prices, exports were $2.2 billion with oil representing $1.8 billion of the total. In 1974, oil revenues soared to $7 billion. This growth has revolutionized the whole economy of a country which, when oil began to be important, was recovering from the effects of a long and devastating civil war.

Before the oil boom the main staples of the export trade were cacao, peanuts, and tin. Nigeria also has a great variety of indigenous production systems. Cattle and peanuts are dominant in the north and palm oil, rubber, timber, and cacao in the south. Agriculture in Nigeria, however, is reported to be in an indifferent state; output in 1971 was only about 36 percent of the 1960 level.[7] Perhaps the major factor accounting for slow growth in agriculture has been the social and economic differential between towns and surrounding countryside. Wages, standards of health, and diversity of opportunity have all become greater in the city, and young people are reluctant to stay on farms in isolated rural districts.

[7]Alan Rake, "The Decline and Fall of Agriculture," *African Development,* 8 (1974), 55.

With all data indicating that the country will maintain a rapid growth of population, increased food production is important, unless the growing oil wealth can be distributed more effectively to the population.

Oil mining is already making a major difference in some aspects of internal investment in the country. Nigeria has always placed a high priority on investment in higher education; there are five universities and many technical colleges. The country has a variety of research institutions of national and international character. There are signs that the oil money is being used to reinforce and broaden some of these institutions.

Some of the oil money is, as in other newly rich countries, being used to improve Nigeria's power in the West African region (Fig. 26–9). Nigeria has engaged in such humanitarian and prestige-enhancing projects as contributing to drought relief in Mali and sponsorship of the 1975 All Africa Games. Efforts to improve regional commercial ties have also emerged. Nigeria has invested heavily in Guinea's iron ore mines in expectation that much of the ore can be used in its own planned iron and steel industry. Transit taxes have been removed on goods crossing the border of some of its neighbors in an effort to promote regional trade. Additional efforts to promote regional development through the Lake Chad Basin Commission and through cooperation with CEAO (Francophone West African Community) have also been undertaken. These attempts by Nigeria to promote a West African economic community are viewed with distrust by some countries, for any such community would inevitably be dominated by Nigeria. The addition of oil wealth reinforces this traditional pattern of dominance, but the road to a successful economic union is laced with pitfalls. An integrated regional system lies many years in the future.

Nigeria has long had a varied industrial sector. Little modern industry exists, however, and most of the industrial sector is made up of thousands of small units working at the craft or local level. This sector continues to flourish and dominate many parts of the country. In the 1960s and increasingly in the 1970s, investment in Western-type industry and services has been heavy. As export earnings grew, imports of food (more than doubled in 1969–1972), machinery, and equipment grew phenomenally even as the import of manufactured goods also exploded.

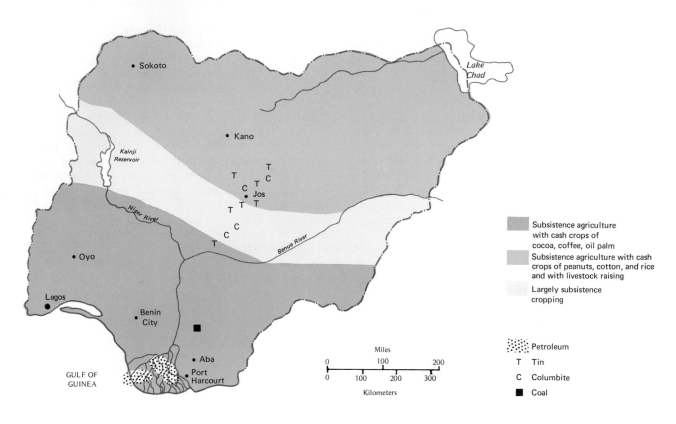

26–9 *Nigeria: Agricultural and Mineral Economy*

The rapid increase in world oil prices helped pay for this import expansion. Attention, however, has turned recently toward investing oil revenue in infrastructural improvements, such as paved roads and new railway equipment, and in large-scale industrial enterprises. Attention now focuses on increasing oil refinery capacity and beginning an iron and steel industry. If the new oil wealth can be invested in successful projects of this type, dependence on a high volume of imports can be somewhat reduced.

Ghana: A Mixed Economy

Ghana, one of the wealthiest states in tropical Africa on a per capita basis, has long had a mixed economy. Before independence in 1957, Ghana exported cacao, gold, and timber, though cacao accounted for nearly 60 percent of export earnings. The efficient farmer-level production of cacao brought a widespread growth of income to Ghana accounting for a large part of the high per capita income the country enjoys. Nkrumah, the first leader, saw great dangers in a reliance on one crop that was steadily falling in value and was linked to an uncertain demand in Europe and Anglo-America. While maintaining cocoa production, he advanced ambitious schemes to move Ghana ahead as an economic and political force in West Africa. Bauxite, diamonds, and manganese production were expanded, but the main thrust was to turn Ghana into an industrial nation. Major investment in the Volta River Dam was planned to produce large quantities of electricity, to be used in part for a large aluminum smelting plant to process Ghana's bauxite. The electricity is also used to help the development of other industries and for railway and port infrastructure.

Unfortunately, massive debts were incurred in

this major program and in the other projects undertaken by Nkrumah; Ghana is still saddled with the need to repay these debts. Now a new pattern of economic growth is emerging, involving new attempts to diversify the agricultural base and to develop small-scale industries for local and regional markets.

Conclusions

The set of intermediate countries ranges from relatively poor countries with moderate development prospects to countries with clearly defined growth sectors. Countries with strong growth points often exhibit a marked disparity in income and prosperity between one part of the country and another. Two overall problems occur. In the first group of countries there is an urgent need to diversify the economic base because of a dominant agricultural economy, or, in the case of Mauritania, because of a dominant mineral resource economy. This diversification, however, is necessary at a "grass-roots" level. The other groups of countries are dominated by the need for a spread of wealth to other parts of the economy from the leading sector. In both cases, there is an urgent need for the development of exchanges within the economy in many different ways. In most countries there is the parallel need to consider economic unions, as by far the largest proportion of these countries appears too small to develop a completely industrialized economy, at least for a very long time. Lastly, there is a ubiquitous need for human resource development coupled with major efforts to meet the employment problem.

Further Readings

The diversity of countries found in this intermediate category precludes a comprehensive listing of works dealing with their problems. The complex and continuing problems generated by the colonial era are provocatively discussed by Amin, Samir, *Neo-Colonialism in West Africa,* translated by McDonagh, Francis (Harmondsworth: Penguin, 1973), 298pp.

The role of irrigation in fostering development is placed in global perspective by Cantor, Leonard M., *A World Geography of Irrigation* (London and Edinburgh: Oliver & Boyd, 1967) 252pp., who devotes a chapter to Southwest Asia (pp. 131–149) and Africa (pp. 178–206).

An integrated examination of the Tunisian economy by Duwaji, Ghazi, *Economic Development in Tunisia: The Impact and Course of Government Planning,* (New York: Prager, 1967), 222pp., is a fine example of the difficulties encountered by a state with a limited natural resource base and an economy that was slowed in its development by the impact of colonialism.

Grove, A. T., *Africa South of the Sahara,* 2nd ed. (London: Oxford University Press, 1970), 280pp., gives a comprehensive survey of much of the region and emphasizes ecological and historical factors as well as future directions.

The role of specific development projects and mineral resources as economic growth factors are examined by Hance, William A., *African Economic Development,* rev. ed. (New York: Praeger, 1967), 326pp.

Seidman, Ann, *Planning for Development in Sub-Saharan Africa* (New York: Praeger, 1974), 382pp., provides an overview of obstacles to and prospects for economic growth.

Stamp, L. Dudley, and Morgan, W. T. W., *Africa: A Study in Tropical Development,* 3rd ed. (New York: Wiley, 1972), 520pp., presents a broad overview of many obstacles and prospects of development.

Africa and the Middle East: The More Developed Nations

27

THREE TYPES OF NATIONS are considered in this chapter: (1) countries that are now benefiting from a major influx of capital from oil revenues; (2) countries with a few natural resources but unique kinds of human and locational attributes; (3) states that vary in their levels of prosperity but are characterized by integrated economies. All of these countries are economically more developed than those considered in the previous two chapters, but for most, basic problems of income distribution and levels of literacy remain (Fig. 27–1).

The Oil States

No region of the world is more intimately associated in the popular mind with oil than the Middle East. This image is not unreasonable, because more than one-quarter of the world's known reserves are found in the region. Yet this wealth of "black gold" is unevenly distributed within the Middle East, and some nations, by accidents of geology and of discovery, have much greater oil reserves than others. So abundant are the proven reserves and present output of such oil producers as Libya, Kuwait, and Saudi Arabia that they deserve consideration in a separate category. The oil industry dominates the economy of these states and pours more money into national treasuries than the countries can spend, in the short run, on productive internal enterprises. With prosperity so closely tied to one resource, the oil states must grapple with a complex array of developmental prospects and problems.

Characteristics of the Oil States

With some exceptions, the oil states share certain common features. First, unlike their larger oil-rich neighbors of Iran and Algeria, they possess small populations. Iraq is the only exception to this generalization, although it shares most other characteristics. These populations are small not only with respect to absolute size but also are generally found concentrated in a few centers. Thus, most oil states have large relatively uninhabited areas and overall low population densities. This relationship is shown in Table 27–1. Most of Libya's 2.7 million inhabitants are concentrated on less than 3 percent of the total land in a narrow strip of land along the Mediterranean coast. Similarly, localized population clusters in uplands, oases, or coastal districts, usually wherever water can be found, characterize the other oil countries.

Second, until oil development created new employment opportunities and sparked massive population movement to a few urban centers, much of this small population was engaged in traditional livelihood systems. Oasis agriculture and pastoral nomadism constitute the two main traditional systems, although fishing, particularly for oyster pearls, is important in the Persian Gulf. Piracy also flourished in the gulf and along the north African coast. Piracy's traditional importance is an economic indication of the paucity of opportunity. Urban traditions generally are weak in oil states, and the educational and technological sophistication of the bulk of the population is limited. Skills lie almost entirely in the traditional sector of the economy and are unsuited to employment in many aspects of the oil industry. This shortage of

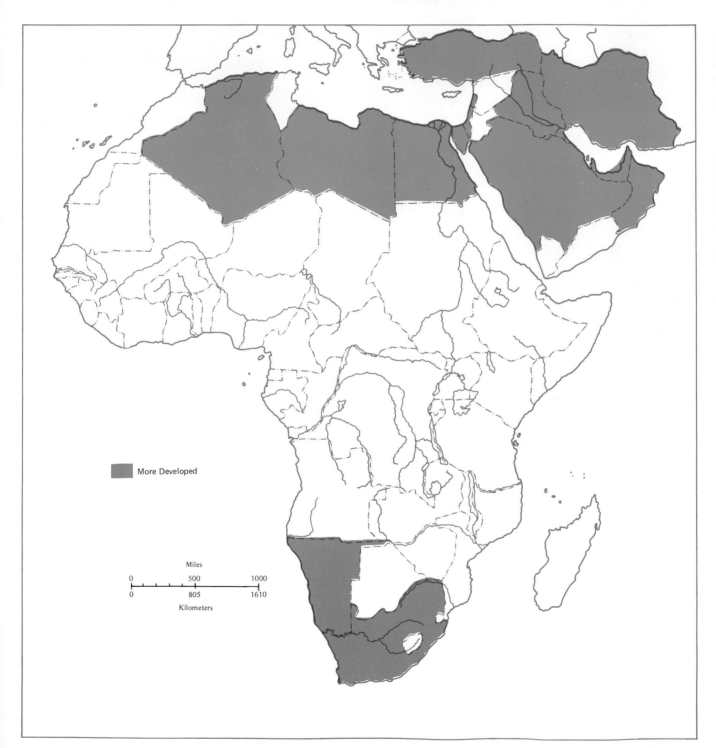

27–1 *Africa and the Middle East: The More Developed Nations*

Table 27–1   *Africa and Middle East Oil States: Population Density and Petroleum Production*

Country	Population Density		Petroleum Production 1973	
	PER SQUARE MILE	PER SQUARE KILOMETER	THOUSANDS OF BARRELS*	BARRELS PER PERSON
Algeria	18	7	400,000	24
Bahrein	1299	502	24,946	83
Egypt	97	37	62,282	2
Iran	85	33	2,139,269	65
Iraq	66	25	706,275	64
Kuwait	171	66	1,044,781	950
Libya	3	1	791,530	344
Oman	10	4	106,925	134
Qatar	12	5	208,023	2,080
Saudi Arabia	11	4	2,677,146	297
Syria	103	40	51,000	7
Tunisia	90	35	29,962	5
Turkey	132	51	24,846	1
United Arab Emirates	6	2	Not Available	—
Abu Dhabi	2	1	474,968	10,325
Dubai	39	15	80,207	1,359

*One barrel contains 42 gallons (159 liters).
SOURCES: *Population Data Sheet* (Washington, D.C.: Population Reference Bureau, 1975); and *Bulletin, American Association of Petroleum Geologists,* 58 (1974).

human skills suitable to the modern sector of the economy has serious political consequences. It encourages the importation of skilled personnel both from the industrialized countries and India, Pakistan, Iran, and Palestine to fill specialized job niches unoccupied by other groups. Dissatisfaction with these conditions has led most oil states to engage in vigorous educational and job-training activities in an effort to replace foreign experts with local personnel as soon as is practicable.

A third characteristic is the overwhelming importance of wealth derived from oil in the local economy. The size of oil deposits is staggering. More than two-thirds of the known oil reserves outside North America and USSR–Eastern Europe are found in the Persian Gulf. Saudi Arabia alone may possess one-quarter of the world's total oil reserves, and it is already the world's third largest producer after the United States and the USSR.[1] Moreover, the

[1]Elizabeth Monroe, *The Changing Balance of Power in the Persian Gulf* (New York: American Universities Field Staff, 1972), pp. 78–79.

dominance of the Middle East is increasing despite efforts to discover and develop new fields elsewhere. Beginning with the discovery of oil in the Iranian hills at the head of the Persian Gulf (1908), the pattern of location and development has spread southward through the coastal regions of the Arabian Peninsula. Not only have new petroleum deposits been located but also knowledge of the areal extent and extractable quantity of oil in existing deposits has grown rapidly. Coupled with improvements in the technology of recovery and the traditional tendency of oil companies to understate reserves or otherwise obscure the reserve capacity of their concessions, the continued dominance of the oil states in the world petroleum economy seems assured for the foreseeable future.

Most oil states share a growing desire to participate directly in oil operations. This trend reverses the conditions of the early years of development and gives the countries more overt control over their own resources. In the past, foreign companies, often operating in a consortium such as Aramco (partly owned by several American oil firms), received concession rights to explore defined areas. Most revenues were generated from concession sales and royalties and taxes paid on oil extracted and exported. Local governments had few skilled oil administrators; oil companies possessed a monopoly on extractive technology and on market distribution in the industrialized states. Moreover, the impoverished and politically weak states of the region lacked the capital to develop the oil resources.

Gradually, this situation has changed. As administrative and technical skills accumulated and as better understanding of the intricacies of oil economies developed, Middle Eastern countries demanded more favorable agreements with the oil companies. Tough new leaders appeared on the scene. Increased percentages of profits from the oil fields were returned to national treasuries, and greater pressure was placed on oil companies either to develop their concessions or relinquish them to someone who would. In Libya, the government has assumed control of 51 percent of the stock of local enterprises exploiting the oil deposits. "Libyanization" ensures governmental control over a vital resource. It also faces the oil companies with the choice of abandoning their remaining investment to Libya without compensation or of continuing to provide leadership and technological skills in order to realize a return on the remaining capital equipment and stock. Saudi Arabia plans to purchase full control of Aramco by 1980, and the process has been followed elsewhere. This trend has been reinforced by an altered world market setting in which demand for oil has outstripped supply. This situation has had the double consequence of quadrupling oil prices and of making it impossible for oil companies to cut off access to market outlets in retaliation for "creeping nationalization."

Impact of the Petroleum Economy

The discovery and development of vast petroleum deposits have had a drastic impact upon the economies and sociopolitical systems of the oil states. A large petroleum output plus soaring oil prices have moved such former backwater states as Kuwait, Abu Dhabi, and Oman to international prominence. Formerly impoverished, the oil states now face the task of coping rationally with an embarrassment of riches. How to dispose of the petro-dollars is a crucial issue. Although abundance is the watchword of the present, the long-term future is by no means assured. Small states such as Oman have sufficient revenue for the present but must consider how best to use the money wisely in order to prepare for the day when oil revenue is no longer available. Equally important is the need to maximize the current high price for oil before competition from new supply points elsewhere in the world can drive down prices or capture present markets, as is possible in the case of Chinese and Siberian oil and gas deposits. Twenty-five percent of Japanese oil comes from the Middle East. Loss of such a large share of the Japanese market to Asian supply sources with lower transport costs would seriously affect Middle Eastern prosperity. The need to diversify an oil-based economy, to develop industrial processes that are not dependent on oil, and to invigorate an often sluggish agricultural sector before there is any decline in oil revenues are obviously high priority items for the oil states.

Internal Investment of Petroleum Income

One outlet for the petro-dollar is internal investment. Large sums are currently being expended to create infrastructure improvements. Ports such as Doha in Qatar have been built where none existed before. Investments in airports, highways, sewer systems, water systems, and pipelines have drastically changed the face of most oil states. Housing projects are also common. Oman is in the process of rebuilding its capital of Masqat. In constructing modern housing units for a growing urban population, Oman is destroying most of the traditional architectural style, but to most citizens this change is seen as an inevitable and progressive aspect of modernization. Libya has actively introduced housing units into rural areas in an attempt to encourage settlement of the nomadic Bedouin and as a way of slowing the rush of people to the cities.

Substantial sums are also being invested in agricultural improvements. In Kuwait, desalinization plants are now in operation whose output is used not only for drinking water but also for irrigation of vegetables to help feed the burgeoning population of Kuwait City. This type of agricultural operation is not economic in the strict sense. Rather, it is justified by a desire for a reduced volume of food imports in a setting of nearly unlimited financial resources. In Saudi Arabia, where date palms provide the traditional staple crop, vigorous attempts have been made to diversify agricultural production. Encouragement of cereal and vegetable production promises to diversify the country's agricultural base. Production increases should be forthcoming through the use of such technological developments as the drilling of artesian wells to expand the agricultural area in the oases of Al-Hasa and the construction of the Wadi Jizan Dam in Asir.

In Libya, where less than 3 percent of the country can support nonirrigated agriculture, attention focuses on the Kufra Basin. Here, near the oasis of Al-Jawf, plans call for tapping vast underground water reservoirs in order to bring a large area of barren desert sands under cultivation. This project is largely designed for pasturage for sheep. The animals then will be transported by truck to distant coastal market towns such as Tripoli and Benghazi. Since Libya must now import most of its meat supplies, this type of development could substantially ease its food production situation. Tremendous emphasis is also being placed on revitalizing the former Italian agricultural plans in the coastal districts. Planners hope that full development of these farms will result in agricultural self-sufficiency by 1980.[2]

In Oman, investment of oil revenues in agricultural development is especially crucial, since depletion of existing reserves late in this century is anticipated, and the prospects for additional discoveries are discouraging. Because agricultural land is limited by a low and unreliable rainfall and great expansion of farmholdings is not possible, Oman aims to achieve food self-sufficiency by improving the efficiency of existing agricultural systems, and by developing underground water resources (including the *qanat* system introduced centuries ago from Persia). Control of plant diseases, improvement of marketing and transport facilities, discovery of new and more productive plant strains, and the use of fertilizers and mechanized equipment where they meet the needs of local farming traditions are some of the projects presently under way. Concentration on the agricultural sector is a feature of development in the oil states

and is needed if a firm base is to be laid for economic viability in the future.

Industrial investment also is taking place. First priority is frequently given to the construction of cement factories to support the construction of industry. Plants based on refining oil products, such as fertilizer, chemicals, plastics, and gas liquification, are common. So too are consumer-oriented enterprises including soft-drink bottling plants, flour mills, fish-processing establishments, and textile-weaving firms. Often craft traditions form a focus for small-scale industrial development that caters to the tourist and export trade. The uniquely designed silver jewelry of Oman is one example of this type. In many oil states, rug-making, slippers, pottery, brass, copperware, and leather products such as hassocks represent a nexus for productive development as well as cultural continuity.

Few resources to support heavy industry are known as yet. Mineral exploration, however, is in its infancy, and prospects for major discoveries exist. Reports circulate of major finds of gold, silver, nickel, and copper in the Arabian Shield areas of mountainous western Saudi Arabia. If confirmed, Saudi Arabia may be able to translate its oil wealth into a diversified economic base that includes indigenous mineral-based industry.

Attempts to overcome skilled manpower shortages are also being made. Nearly every country has invested in modern universities and a rapidly developing primary and secondary school system. To staff these systems, numerous foreign teachers, often recruited from more educationally advanced countries such as Egypt, have been imported. Large numbers of students, however, also study abroad under governmental contract. Upon their return they are funneled into decision-making positions at all levels of the government. It is their newly acquired expertise that underlies the rapid replacement of foreign personnel in the larger oil states of Iraq and Saudi Arabia and the tougher and more aggressive posture of governments in dealing with the international oil companies.

Petro-Dollar Investment Overseas

Faced with an inability to absorb all of their oil revenue in internal investment projects, oil states have begun to seek investment opportunities overseas. Kuwait recently purchases a 14 percent interest in

---

[2]In a recent survey of the agricultural potential of Egypt, Israel, Jordan, Lebanon, Syria, and Iraq, it is estimated that production can be increased by a factor of from two to ten in all countries except Israel. Marion Clawson, Hans H. Landsberg, and Lyle T. Alexander, *The Agricultural Potential of the Middle East* (New York: American Elsevier, 1971), p. 4.

Germany's Deimler-Benz company and has invested $17.4 million in a projected luxury resort project on Kiawah Island near Charleston, South Carolina. A further $100 million loan offer to Lockheed Aircraft Corporation was recently rejected. Indeed, opposition of environmentalists to the Kiawah project and the Lockheed refusal are symptomatic of the ambivalent attitude of Americans and their government toward a massive infusion of petro-dollars into the United States economy. Long accustomed to treating with disdain the apprehensions of foreigners about overseas investment by American multinational corporations, considerable uneasiness arises in financial and political circles when outside capital reverses the process. Although American industry recognizes that a reverse flow of investment capital is essential to balance capital exports in payment of petroleum purchases, the threat of foreign domination of American industry sets up nationalistic currents in both public and private sectors.

In the face of such uncertainty, Arab oil investors have preferred to place their oil wealth into short-term, high-interest loans that can be moved rapidly to take advantage of shifts in the international monetary market. This action has tended to unsettle the world monetary system and increase fears that a major overhaul of the monetary structure is overdue. For the oil states, long-term investment in Western industry and technology represents a major device for increasing their financial security. Monies so invested could support their economies when oil revenue is no longer a viable generator of economic growth.

Changing Political Power Relationships

Oil wealth has altered the balance of power in several ways. Most of the oil states are now able to use their rapidly accumulating wealth to purchase arms with which to modernize their military forces. Possession of sophisticated weaponry has made Iran the dominant political force in the Persion Gulf. This situation, in turn, has sparked demands for enhanced military preparedness among other states in the region. Arms dealers are as common a sight in ar-Riyadh as are agricultural development specialists. Moreover, the region's dominance in the OPEC (Organization of Petroleum Exporting Countries) group gives it substantial force on the world political scene. Proposals to raise prices or threats to embargo oil shipments to supporters of Israel send tremors racing through the industrialized world. The oil states are not insensitive to their new position of influence, and there seems every reason to believe that those states with oil are not afraid to use their power. The psychological satisfaction derived from this change from an inferior to a preponderant political position explains much of the contemporary behavior of the oil states. They can no longer be ignored or dominated but rather must be dealt with as equals of great power and influence.

Changing Internal Conditions

Developmental growth fueled by oil wealth has worked massive changes in the societies of the oil states. One important change has been rapid urbanization.[3] Attracted by new job opportunties, social welfare programs, and a milieu of excitement and diversity, rural folk have flocked to the rapidly growing urban centers sparked by the oil boom. Benghazi, Libya, has grown from fewer than 30,000 at the end of World War II to over 150,000 at present. Doha, the capital of Qatar, leaped from insignificant village status to over 80,000 inhabitants in the space of twenty years. Similar growth has taken place in the capitals of most oil states and sheikdoms; Baghdad now exceeds 2 million, and Tripoli, Libya, and ar-Riyadh, Saudia Arabia, both top the 200,000 mark. Much of this growth has of necessity been accompanied by shantytowns and other temporary housing on the outskirts of the urban area. The result is a set of sanitary and health problems because of excessive crowding that is gradually being removed as oil money is invested in improved housing and infrastructure.

Centers of economic growth and change, the expanding cities are also the scenes of cultural conflict. For rural migrants, adjustment to urban life is often difficult. Moreover, it is in the urban milieu that traditional Islamic and customary values are receiving their strongest challenge. Bombarded by a variety of exotic stimuli, ranging from the material possessions of the industrialized West to the clothes,

[3]Saad E. M. Ibrahim, "Over-urbanization and Under-urbanization: The Case of the Arab World," *International Journal of Middle East Studies*, 6 (1975), 29–45.

Table 27–2 *Characteristics of Three Eastern Mediterranean States*

Lebanon	Israel	Palestine
Small Population (2.8 million)	Small Population (3.6 million)	Small Population (ca. 3.5 million)
Little Agricultural Resources	Some Agricultural Resources	Little Agricultural Resources
Minor Mineral Resources	Small Mineral Resources	Minor Mineral Resources
Major Service Function	Some Industry	
Location and Human Resources Important	Location and Human Resources Important	Human Resources Important
Financial Remittances from Overseas Important	Financial Remittances from Overseas Very Important	Financial Remittances from Overseas Very Important
European Outlet for Arab World	Asian Inlet of Western World?	Displaced Portion of the Arab World

values, movies, and behavior patterns of the foreign employees of oil and construction firms, the citizens of the oil states have been forced to adjust to a new social context. In many cases the shift has involved moving physically and conceptually from the Middle Ages to the twenty-first century. Equally profound are differences in work habits and social priorities. Western notions of the sanctity of time are seldom accepted by the new urban arrivals, and financial gains are often applied to social needs such as bride price or kin obligations rather than savings or job advancement. Tension between workers and foreign employees because of inadequate understanding of respective cultural values is often a result.[4]

Great income and status disparities have also appeared as a result of the oil boom. Although some of the newly found wealth has filtered down to lower social echelons, much of the income thus gained is eroded by an inflationary spiral triggered by oil development and accelerated imports. Individuals close to the import trade, those serving as local representatives of foreign firms, and professionals of all sorts have benefited most from petroleum growth. The unskilled have been left behind. Some states such as Kuwait have instituted a progressive policy of free social services to improve living conditions, but others have been less foresighted. The gap between the ruling elite and the bulk of the population is often wide.

[4]Frederic C. Thomas, Jr., "The Libyan Oil Worker," *Middle East Journal*, 15 (1961), 264–276.

Internal Political Situation

Social unrest stemming from culture change and income inequities shadows the political future of many of the oil states. Opposition to the central government of many of the oil states comes from a variety of sources. In Iraq the Kurdish population of the north was for years engaged in armed rebellion against the Arab-dominated central administration. Although a desire for greater regional autonomy was the ostensible motive, increased access by Kurds to revenue from the northern oil fields in or near Kurdish territory was an underlying motivation. Although recently suppressed when Iranian aid to the rebels was withdrawn, the Kurds remain a disenchanted minority in Turkey and Iran as well as in Iraq.

In Oman, where the obstructionist and repressive traditionalist policies of the late Sultan Said bin Taimur resulted in considerable unrest, armed rebellion has existed in the southern province of Dhufar since 1963. This guerrilla campaign, managed by the People's Front for the Liberation of Oman and the Arabian Gulf and supported by the neighboring People's Democratic Republic of Yemen, ultimately culminated in the deposition of the Sultan and his replacement by his son, Qaboos.

Security conditions in Oman deteriorated to such a state that Iran, at the request of Qaboos, dispatched troops and helicopters in a successful attempt to sustain the new Sultan's position. Iran's intervention underscores its determination to promote stability in the region. Concerned for the security of the vital oil export route to Europe and Japan, the Shah has served notice that Iran's new and massive military might well be employed unhesitatingly whenever vital Iranian interests appear threatened. The implications of Iran's emergence as a regional power are numerous, and the prospects for future confrontations over the political future of the Gulf region are very real.

Equally likely are governmental changes by military coups. The young officer corps emerging in many of the oil states possess nationalistic and Pan-Arabic ideals. Impatient with the slow rate of economic progress, the pronounced social and economic inequalities in many oil states, and the corruption that frequently characterizes a boom economy, they are apt to use their position to seize power. Concern for excessive social permissiveness and employment of foreign nationals are also likely to spark unrest in the lower levels of the military hierarchy.

The energy of the petroleum economy is changing many features of the political, economic, and social fabric of the oil states. Prospects for considerable future instability, exacerbated by the uncertainties of the Arab-Israeli conflict, are very real, as the traditional life styles of the oil states struggle to adjust to the forces of change and to harness their petroleum wealth to promote the development of their economy.

Small States of the Eastern Mediterranean: Unique Economies

Two countries on the eastern margin of the Mediterranean, together with one increasingly recognized *de facto* state organization (Palestine Liberation Organization), present a quite different set of issues for discussion than is usual for essays on development (Table 27–2). In this case, the agricultural and mineral resource factors, which are basic factors for many countries, are of secondary importance. They are superseded by questions of human resources, of service functions, and of various kinds of inflows of support for the nations concerned.

Lebanon, Israel, and Palestine are the three states involved. In this group Palestine, or more specifically the Arabic population of Palestine represented by the PLO, is unusual in being a state without a defined national territory. This condition ironically mirrors the situation of Israel some years ago. Despite its uncertain status, it seems important to include Palestine here because of the separate and fundamental issues involved. If we accept the three as political states and attempt to erect a national profile in each case, it is clear that there are some strong similarities, although individual differences are also important.

The eastern end of the Mediterranean has gained a reputation as a cockpit of the world and as a strategic area controlling important nodes of connection and trade. Routes linking Africa and Asia and the Arabian Gulf and Europe pass through the area, and control of these routes has long been important. For a time Beirut was a vital air link between Europe and the East, although now it is partly superseded by the new generation of long-distance jets.

The religious significance of the area is remarkable. Here is the seedbed of the ancient Jewish religion, and out of that tradition grew the various Christian denominations. The area is also significant to Moslems. While Mecca is the central holy place of Islam, Jerusalem, the site of the Prophet's ascension into heaven, is a pilgrimage center of major importance. Medieval maps showed Jerusalem as the center of the world with Asia, Africa, and Europe all focused toward it. Present power politics have greatly modified but not totally destroyed this view. From a religious and a geopolitical viewpoint, the area is still central in the world's attention.

Israel: Asian Inlet of the Western World

Israel is, of course, the major focal point. It is unique among countries in that it became established territorially on the basis of 2,000-year-old claims of the Jewish people to the area and as a result of the persecution of Jews as minorities in many countries of the world. The establishment of a nation in disputed land long occupied by indigenous peoples unsympathetic to the Israeli national cause was bound to provoke conflict. The small state of Israel, set up in 1948, was difficult to defend against surrounding Arab pressures. As a result of the 1967 war, Israel was able to "rationalize" its frontiers but at the territorial expense of Jordan and Egypt.[5] The Palestinians,

[5]The consequences of these changes for Jordan are analyzed by B. P. Birch, "Jordan's Geography after the 1967 War," *Tijdschrift voor Economische en Sociale Geografie*, 62, (1971), 45–52.

who had been dispossessed of part of their lands in the 1948 hostilities, found all their traditional territory in Israeli hands after the 1967 war.

Apart from the continued struggle to maintain and expand its territory, what kind of state is Israel? Statistics give some clues. In an area where per capita GNPs are generally low, Israel's is $3,580. In one sense Israel is a comparatively rich country in a poor part of the world. Some of the wealth comes from the industry of the people. The farming cooperatives of Israel are well known. The *Kibbutz* and the *Moshave* are two main types of agricultural production and marketing cooperatives that help Israel to maximize production from land that is fertile only in the narrow coastal plain. Efficient use of water, moved southward from Lake Kinneret along the national water carrier, assists in the intense cultivation of citrus fruits, vegetables, and some grain crops (Fig. 27–2). High-quality farm exports make a significant contribution to the economy, although only a small percent of the population work on farms.

Israel is basically an urban country; Tel Aviv alone houses 20 percent of the country's population. Industry is an important part of the economy, and diamond cutting, textiles, manufacture of woolen goods, and many other small industries provide a livelihood for the bulk of the population.

Another striking feature of the country is the high level of education. There are four main universities in Israel, but in addition many immigrants are highly qualified academically and technically. Israel is one of the few countries that has difficulties because of the general high level of manpower resources. Moreover, in some fields the quality of Israel's technical and academic achievement is unsurpassed. Israel's work on hydrology, on horticulture, and in some electronic fields is particularly important. As a result, significant Israeli aid has been given to developing countries. With its high living standard, technological sophistication, large, well-educated elite, and substantial European-derived population, Israel is more European in its values and orientation that it is Middle Eastern. Thus, Israel is doubly at odds with its neighbors, for political conflicts are reinforced by a clash of cultural values that increases the difficulties of attaining mutually suitable compromise.

Even with such rich and varied resources, it appears that Israel would not be a self-sufficient state were it not for the large flow of capital sent to Israel from

27–2 *Africa and the Middle East: The Small States of the Eastern Mediterranean*

High levels of technology have been applied to agriculture in Israel. Here, near the Gaza Strip, wheat is being harvested by combines on a kibbutz. (United Nations)

Jews around the world. The high burden of defense costs is one factor in promoting this capital inflow, but the maintenance of current levels of living might be difficult on internal resources even without the defense costs. Although Israel is a complex nation whose nationals possess high technical skills, its future will depend as much on the political skills both of Israel and others as on anything else.

Lebanon: Arab Window on the West

Lebanon, sharing the East Mediterranean coast with Israel, is likewise small both in area and population. It is a "country where economic theory is scarcely applicable and events regularly give lie to forecasts."[6]

Lebanese have for many years moved out from the poor agricultural lands of their country to become merchants and entrepreneurs in many parts of the world. The traders in many African capitals were until recently dominantly Lebanese; the Lebanese are the bankers of the Arab world and parts of Africa too. Large numbers of Lebanese moved to South America, and important Lebanese communities can be found in many American cities. Remittances from this overseas population have supported many of the rural villages at living standards well above the subsistence level their resource base would seem to justify.

This overseas activity has always been balanced by the steady growth of Beirut as the financial center of

the Arab world; and as more and more traders returned to their homeland, this role has intensified. Lebanon has no firm agricultural base, although bananas and citrus fruits grown in the coastal plain and deciduous fruits from the upland form the basis of an export trade with surrounding Arab states. Industry is only now beginning to expand. Metallic goods, processed foods, textiles, and pharmaceuticals are leading sectors, though tourism is also a vital source of income. The mixture of Christian and Moslem that characterizes Lebanon, together with a considerable veneer of French culture from the colonial era, make Beirut the most cosmopolitan city of the Middle East. Yet deep social, cultural, political, and religious antagonisms remain that can flare into serious conflict.

Lebanon is a major center for Arab tourism which continues to boom in the face of a general world recession. Seventy percent of the Lebanese national income is derived from the service sector, and 55 percent of the active population works in service-related jobs. Moreover, until recently oil sheiks were reluctant to invest in non-Arab countries. As a result, much of the capital that could not be employed in development projects at home was invested in Lebanese real estate. Thus, Lebanon is a classic crossroads economy whose location results in a good standard of living (per capita GNP $700) and a steadily expanding economy as the Arab hinterland grows in wealth.

Lebanon's long-term future remains cloudy. Torn by conflicting internal pressures, it suffers from a

[6]Special Report on Business and Finance: "Beirut," *The Times* (London), No. 59, 254 (Tuesday, November 26, 1974), 1.

major identity crisis. Thoroughly Arab, it is neither fully Christian nor Moslem, neither laissez faire nor socialist, neither pro- nor anti-Palestinian. Until these problems are solved, Lebanon will be unable to realize its full potential.

Palestine: A Nation in Search of a Homeland

The most volatile element in the politics of the Middle East is the Palestinian Arab population. Creation of the state of Israel in 1948 resulted in the suppression of Palestinian nationalism and a dispersal of Palestinian Arabs among neighboring states. Many fled to the Gaza strip, which fell under Egyptian control, to the West Bank (annexed to Jordan), or to Syria and Lebanon. Grouped into refugee centers, they subsisted on United Nations relief handouts. Unabsorbed into the structure and economy of their brethren Arabs, unreconciled to the loss of their homes and property, the Palestinian Arabs developed a national consciousness distinct from other Arab peoples. Neighboring governments have been unwilling to accept them as full members of their states, or in many cases Palestinians have refused to be absorbed because they wish only to regain their ancient land base. Their numbers were swelled by further refugees following the 1967 war (Table 27–3). Although it must be remembered that these figures only indicate the number of individuals entitled to receive relief rations, the actual Palestinian population is nearly three times as large as the 1.3 million official refugees.[7]

Large numbers of educated and skilled Palestinians have dispersed throughout the Middle East. They often staff the bureaucracies of the oil states and, like the citizens and associates of Israel and Lebanon, send remittances to their friends and relatives in the refugee camps. Consciously identifying with the goals of the PLO (Palestine Liberation Organization), they dream of one day returning to their homes. The resort to terrorism and guerrilla warfare is only one example of the frustrations engendered by an inability to establish the territorial homeland so ardently desired. The tragedy of the present Arab-Israeli conflict is that two nationalisms are competing for control of a geographic space that can only accommodate one. Even if the PLO attains its political

[7]Edward Hagopian and A. B. Zahlan, "Palestine's Arab Population: The Demography of the Population," *Journal of Palestine Studies,* 3 (1974), 61.

Table 27–3 *Palestinian Arab Refugee Population*

Area	31 May 1967	1968	1972
Jordan	722,687	747,434	829,867
Gaza	316,776	307,864	324,567
Lebanon	160,723	168,927	184,043
Syria	144,390	151,730	168,163
TOTAL	1,344,576	1,375,915	1,506,640

SOURCES: Gérard Chaliand, *The Palestinian Resistance* (Harmondsworth: Penguin, 1972); and Edward Hagopian and A. B. Zahlan, "Palestine's Arab Population: The Demography of the Palestinians," *Journal of a Palestine Studies,* 3 (1974).

objectives and establishes a national territory in Palestine, the problem is still far from solved. Creation of a new political reality does not provide economic viability, and the mobilization of skills and resources from outside the state would still be essential for its success.

The Integrated Economies

Five countries are discussed in this section: Egypt, Algeria, South Africa, Iran, and Turkey. At first, they seem dissimilar, but although each has unique features, they have many basic similarities.

Features of Similarity

First, in comparison with other countries in Africa and the Middle East, all of these states are relatively self-supporting. This high degree of self-sufficiency comes from a significant industrial sector based on primary processing of raw materials, diversified agricultural production, and a developed service sector. Second, countries with such diversified economies also show a marked concentration of population in urban settlements. Other countries in Africa and the Middle East have large towns, but they are usually single primate cities. In contrast, Egypt has 30 percent of its population in urban places with Cairo (5.1 million) and Alexandria (2 million) as major cities; Algeria has Algiers (1.8 million) and Oran (327,000); and South Africa includes Johannesburg (640,000), Durban (495,000), Pretoria (544,000), and Cape Town (691,000). Iran contains Tabriz

(465,000), Meshed (510,000), and Isfahan (520,000) in addition to Tehran (3.4 million); and Turkey possesses Istanbul (2.2 million), Izmir (521,000), and Ankara (1.2 million). Such cities serve complex functions; they are administrative, industrial, cultural, trading, and often religious centers. Third, the existence of these cities is made possible not only by the economic status of the five countries but also by their comparatively large populations. Turkey and Egypt are the largest with 42 and 39 million citizens, respectively, followed by Iran and South Africa, with 35 and 26 million, respectively, and Algeria with only 18 million inhabitants. Although size is not an absolute determinant, large populations can support a larger urban sector than can small populations despite a comparatively low per capita GNP as is found in Egypt.

Fourth, many factors contribute to the distinctiveness of these countries, but it is significant that each has appreciable mineral wealth. South African minerals are important world resources, as is Algerian and Iranian oil. But coal, copper, and chrome in Iran, coal, iron ore, and chrome in Turkey, and oil and phosphates in Egypt are important bases for economic growth.

Fifth, environmentally each country is characterized by a large area of semiarid or arid territory of little value to the agricultural economy. Agriculture thus tends to be concentrated in limited areas and to be more intense than in most other countries in Africa and the Middle East. Egypt with the narrow fertile strip of the Nile valley and delta is the extreme case.

Features of Diversity

Culturally, this group of countries falls into two categories. On the one hand, Turkey, Egypt, and Iran are states with a long, uninterrupted cultural heritage and a proud tradition of political and economic independence. In contrast stand Algeria and South Africa, where culture conflict and discontinuity are salient features.

Turkey, Egypt, and Iran share a common interest in their cultural richness and continuity which give an added dimension to their pursuit of economic development. It is this concern with the inheritance of the past that motivated the Shah of Iran's Persepolis fete in honor of the 2,500th anniversary of the founding of the Persian throne. The richness of Egyptian civilization was a wonder in Greco-Roman times, and the splendor of its scholarly institutions increased following the Arab conquest in A.D. 640. Rapidly developing a reputation as a paramount center of Koranic learning, at al-Azar University and other religious centers in Cairo, Arab scholars were largely responsible for preserving the writings of the classical world and transmitting them to the Christian West. At a period when the Dark Ages characterized Western Europe, the acme of scholarship and invention resided in the universities of the Middle East.

The Turks as the heirs of the Byzantine Empire assumed the mantle of empire in the Eastern Mediterranean. So successful were their military campaigns that Vienna nearly fell to their armies in 1529, and only when outstripped by the industrial development of the West did the Ottoman Empire lose its internal cohesion. Even then, Turkey retained sufficient vitality and creative drive to renew itself and escape colonial rule and a subservient status. Iran's path parallels Turkey's. The proud possessor of a rich and distinct literary, artistic, linguistic, and religious heritage, Iran traces its identity to the empire of the Medes and the Persians. It is this sense of distinctiveness and of continuity in the face of change that underlies the Shah's extravagant Persepolis birthday party and reinforces Iran's self-identity.

If Iran, Turkey, and Egypt can point to a historically rich cultural tradition of sufficient strength to survive external pressures, Algeria and South Africa present a contrasting case. Whereas the culturally rich countries were able to stave off the worst effects of colonialism or to avoid colonial control altogether, both Algeria and South Africa have been the scenes of major cultural conflict. In Algeria, this conflict began with the establishment of a French protectorate in 1830. Large-scale European emigration followed, and tremendous pressures were placed on local cultural institutions. Today, despite an end to French control, Gallic civilization maintains a very visible presence in Algeria. This French-Arab conflict mirrors the long opposition of Berber culture, protected by its rugged mountain retreats, to a succession of Phoenecian, Roman, and Arab invaders. It is perhaps ironic that guerrilla opposition to the French has done much to reduce the differences between Arab and Berber in Algeria and speed the process of Berber assimilation into the mainstream of Arab culture.

In South Africa, the culture conflicts occur in a

number of ways. There is the obvious black versus white dichotomy that pits the European-descended population against Zulu, Cape Colored, and Indian. South Africa's present apartheid policies are a consequence of this conflict and result in a very serious waste of valuable human resources that could make a significant contribution to the well-being of the country. Also present is a conflict between the Boers, the descendants of the original Dutch settlers in South Africa, and the English settlers who began to arrive after the Napoleonic Wars. No love is lost between these two groups, although in the last two decades the English and Dutch have been united by their mutual fear of black domination. Despite tremendous economic potential, the future of South Africa re-

mains uncertain until some acceptable solution to its culture conflict is achieved.

Algeria

Algeria has some similarities with other oil states. It has a rapidly increasing income from oil and is using this new wealth as the basis for economic growth. Like Egypt, however, Algeria has gained a position in the world far above that which might be suggested by its size and economic status. One reason for this position is related to the strength and maturity brought about through long cultural traditions; another is related to the good mixture of resources available to the nation (Fig. 27–3). The north part of the country includes large areas of fertile agricultural

27–3 *Algeria: Major Mineral and Specialized Agricultural Areas*

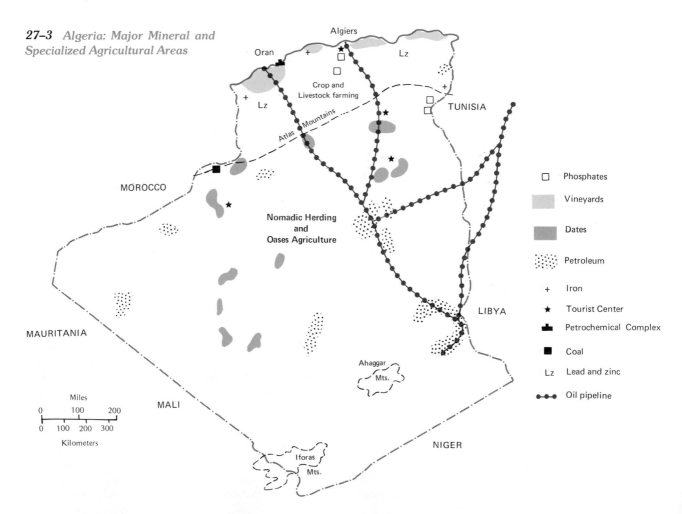

land, the granaries of the Roman Empire once more being used for wheat. The climate is ideal for citrus and other fruits and vegetables, and European markets are close at hand. Outside this commercial agricultural area, many peasants till poor soils for a low subsistence level of life. Oil and gas resources are backed by phosphates and major iron ore deposits and form a solid base from which to promote industrialization. Skills and infrastructure are well enough developed to allow the major development initiative to remain firmly in Algerian hands.

The major distinctive feature of Algeria is the single-minded political strength and acumen of its leaders. This feature has enabled the government to demand and get major internal sacrifices in return for the promise of long-term benefits. Luxury imports are strictly curtailed, and some 40 percent of the GNP in currently being reinvested in the economy. Such allocation of investment has enabled great headway to be made on a major iron and steel plant, a large plastics complex, and many smaller consumer industries. Algeria's current yearly growth rate of 7 to 10 percent is likely to continue and even grow over the next few years.

Algeria has been able to take a leadership role in the Arab group of countries, the OPEC group, and the Third World in general. There is good reason to predict that economic growth and political struggle will continue to be characteristic of this rapidly growing nation.

27–4 South Africa: Agricultural Systems and Mineral Deposits

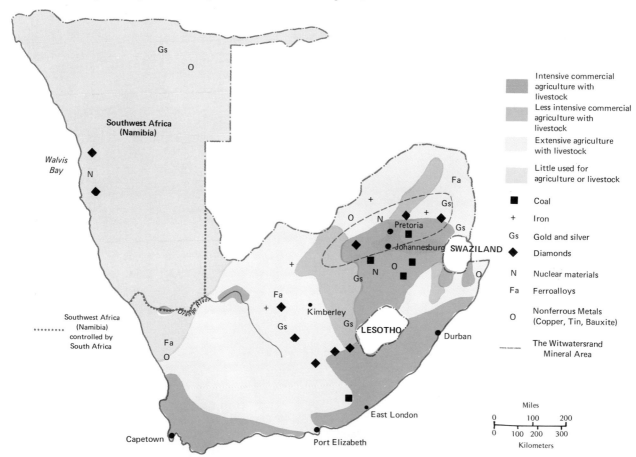

South Africa

South Africa is often in the news because many countries do not approve of the official apartheid policies of the government. It is also in the news because it is a leader in the production of a number of important minerals (Fig. 27–4). Its great economic strength may well be a major reason why other countries are unable to influence significantly internal South African policies.

South Africa is the world's leading producer of gold, antimony, platinum, and diamonds and the second leading producer of uranium, chromium, manganese, vermiculite, and vanadium. It is also important for its reserves of copper and nickel. Moreover, South Africa has over 90 percent of the continent's known coal deposits, and these deposits enable the nation to meet most internal energy needs and still export coal northward to Rhodesia and

The central business district of Johannesburg, the largest city of South Africa. Johannesburg is a major industrial and commercial center with gold mining nearby. (United Nations/Jerry Frank)

Zambia. The mineral sector is thus the leading edge in South Africa's economic growth, but the southerly location also provides a strong comparative advantage in the production of fruit and vegetables for northern hemisphere markets. Potentially at least, South Africa is a very prosperous country. Paradoxically, this very prosperity is beginning to cut away the country's archaic social structures. At present, the nation is not effectively using a large part of the available human resources, but there is some hope that more liberal economic and social views will begin to prevail.

Egypt

Egypt is a country of contrasts. Cairo and Alexandria are among the two largest cities in Africa, but peasants in upper Egypt live about the same as peasants in the area 1,000 years ago. The Egyptian economy is diversified, with a wide range of basic and processing industries and many large and small consumer-oriented enterprises. The country is a center of Moslem culture and tradition and is in the forefront of the modern Arabic world. For example, it manufactures most of the world's Arab language films. It is the leading political power among Moslem nations, skillfully organizing the diverse groups involved.

Egypt achieves all this power on a fairly limited resource base. Agricultural resources are limited by water available for irrigation (there is very little rain except along the Mediterranean coast) and by suitable soils. Clover for animal fodder, cotton, rice, maize, and wheat are the main crops, and sea island cotton is a major export crop. Land use is intense, and given the current socioeconomic situation, it is difficult to get major breakthroughs to new levels of produc-

tion. An attempt is being made to achieve a breakthrough by construction of a high dam at Aswan. By controlling the seasonal irregularity in Nile River flow, the Egyptians hope to convert much land to year-round cropland and to bring new, presently desert land into production. Both good and bad ecological and economic results have occurred since the completion of the High Dam, but as a symbol of Egypt's determination to achieve development, the dam has great psychological significance.

Industry is diversified and growing steadily, but even with the vast new resources of power from the Aswan Dam, there are major constraints on rapid expansion (Table 27–4). Cotton, oil, and phosphates remain the major industries. Cotton provides raw materials for some industries, but most other industries are based on imported materials. Oil and phosphates meet local needs but generate little employment and make a minimal direct contribution to local incomes. Perhaps, in the future, Egypt will develop more service-related industries for the Arab world that will actively utilize the nation's educated population and provide a wider range of job opportunities.

Despite these major constraints and considerable internal problems of improving well-being, especially among the peasants, Egypt is an important world power. Its strategic position and political sophistication are overriding assets.

Iran: The White Revolution

In the Third World, Iran is perhaps unique in that the force for radical economic and social change comes from elements of its traditional leadership. Convinced that the future of his throne depends on

Table 27–4 *Egypt: Principal Industrial Activities*

Activity	Number of Establishments	Labor Force
Food Processing	983	86,500
Cotton Ginning-Textiles	760	193,200
Other Raw Material Processing	693	91,900
Engineering-Machinery	508	63,000
Construction Materials	336	23,100

SOURCE: K. M. Barbour, "The Distribution of Industry in Egypt: A New Source Considered," *Transactions, Institute of British Geographers,* 50 (1970).

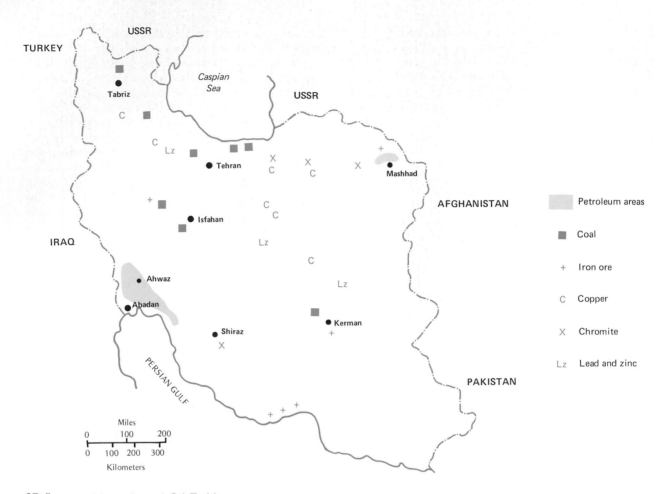

27-5   *Iran: Minerals and Oil Fields*

meeting the aspirations of his people for a better life, the Shah of Iran has pushed a variety of development programs including land reform and industrialization. Possessing a relatively limited agricultural base, for much of the center of the country is desert and many of the populated rim lands are mountainous, agricultural development also has been retarded by an archaic land-tenure system in which few farmers held their own land (Fig. 27-5).[8] The government, over the strong opposition of the landed rich and the traditional religious leadership, is attempting to redis-

tribute land to over 2 million peasant families. Secure tenure, it is hoped, will result in increased agricultural output.

The role of a dominant political leadership in accomplishing this change is crucial. This dominant central authority is also essential to plans for industrialization. Iran's rich oil reserves are producing huge revenues that are being invested in local enterprises such as petrochemicals, iron and steel, irrigation, copper and zinc development, and food processing. Large sums also are being used to purchase an interest in the industrial base of the developed countries including a substantial share in West Germany's Krupp Steel Company. Tremendous expenditures on

[8]Ann K. S. Lambton, *The Persian Land Reform: 1962–1966* (New York: Oxford University Press, 1969), pp. 112–130; 347–356.

Although the Iranian government has several development projects to improve the rural economy, most farmers are poor. Much of the arid environment is used for sheep pasturage. (United Nations)

a vast array of modern weaponry are making the Iranian military the most heavily armed and powerful in the Middle East. At the same time that Iran moves to assert its authority regionally and internationally, internal opposition is rigorously suppressed. In all these actions the personality of the Shah is paramount. Central to the enterprise, convinced of his historic mission, dedicated to the grandeur and progress of his country, the Shah is the personification of Iran's determination to achieve a level of economic growth comparable to that of the Western world. Time alone will tell whether revolutionary change imposed from the top of a political system can promote the changes required for development and whether that change can be confined to the economic aspects of Iran's life.

Turkey: Unrealized Potential

The end of World War I witnessed the demise of Ottoman Turkey and the emergence during the next decades of a new, smaller, but more homogeneous, Turkish state. Led by Kemal Ataturk, the forces of modernization and change dedicated to cutting ties to the discredited past assumed control. In this process the army played a crucial role and served as an integrating institution of national regeneration to replace the discredited Ottoman bureaucracy. Many old customs were abandoned; a secular state was created; and great impetus was given to industrial and agricultural development (Fig. 27–6). Tremendous progress has been made in agriculture with citrus, cotton, olive oil, and other Mediterranean crops being produced in the coastal district, while cereal grains dominate the drier regions of the interior. Mineral deposits are sufficient for internal needs, and coal, iron ore, and petroleum are the most significant. Important chromite, meerschaum, and manganese deposits enter world trade, but much of Turkey's mineral potential remains to be tapped.

A tough and disciplined people, a strong military, and a growing agricultural and industrial base made Turkey a powerful state. Yet this power is seldom

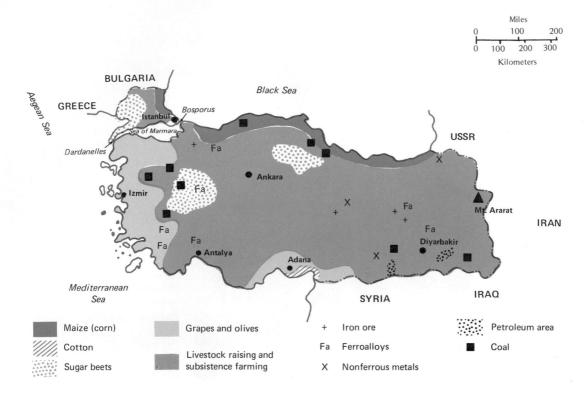

27–6 *Turkey: Mineral Resources and Specialized Agricultural Regions*

applied. The Turkish military is more apt to exert its influence in internal affairs than in the power politics of the Eastern Mediterranean. The Turkish invasion of Cyprus in 1974 is an exception to this rule but is understandable in terms of concern for the position of the Turkish Cypriot minority and for the future of a territory that Ottoman Turkey once ruled. Today's reluctance to engage in power politics stems in part from issues never fully resolved by Ataturk's revolution. While the state created is secular, 98 percent of the population is Moslem. With 40 percent of the population still living in rural agricultural settings, the influence of the religious leadership remains strong, and its proper role in the national structure remains unresolved. Turkey's ambivalent location between east and west also causes uncertainty as to the best foreign policy course to follow. While Turkish leaders remember traditional quarrels with Russia, a vocal minority continues to agitate for closer ties

with the USSR. Despite creation of a democratic framework after the Ataturk revolution, the army continues to exert a strong influence upon political concerns. Direct intervention in the political process is common and adds to uncertainty about future directions. When Turkey achieves agreement on a course of action, as in the Cyprus invasion, its regional effect can be enormous. Prospects for a consistently powerful regional role, however, are limited.

Further Readings

The impact of burgeoning oil revenues, prospects for economic improvement in both integrated and unique economies, and patterns of social and cultural change are topics introduced in this chapter but which deserve considerable elaboration. The reader interested in the

Istanbul, Turkey's largest city, stands on the gateway between Europe and Asia and between the Black Sea and the Mediterranean Sea. Istanbul possesses one of the world's finest natural harbors. (United Nations/M. Biber)

historical growth of the region's oil economy would do well to consult the following: Shwadran, Benjamin, *The Middle East, Oil and the Great Powers* (New York: Wiley, 1973), 630pp. Global implications of oil supply and demand are exhaustively examined by Adelman, M. A., *The World Petroleum Market* (Baltimore: Johns Hopkins University Press for Resources for the Future, 1972), 438pp. Melamid, Alexander, "Satellization in Iranian Crude-Oil Production," *Geographical Review,* 63 (1973), 27–43, analyzes some of the recent spatial changes in Iran's petroleum industry.

That large amounts of oil revenue place special strains on and present special opportunities to a developing country is a theme discussed by El-Mallakh, Ragaei, "The Economics of Rapid Growth: Libya," *Middle East Journal,* 22 (1969), 308–320. Clawson, Marion, Landsberg, Hans H., and Alexander, Lyle T., *The Agricultural Potential of the Middle East* (New York: American Elsevier, 1971), 312pp., give the best existing summary of the considerable pros-

pect for agricultural growth existing in the Middle East despite the region's severe aridity constraints. For those interested in attempts to institute agrarian reform, see Lambton, Ann S. K., "White Revolution," *The Persian Land Reform: 1962–1966* (New York: Oxford University Press, 1969), 386pp., which constitutes an essential point of departure. The best overview of the region's land tenure patterns continues to be Warriner, Doreen, *Land Reform and Development in the Middle East: A Study of Egypt, Syria and Iraq,* 2nd ed. (London: Oxford University Press, 1962), 238pp.; Lewis, Geoffrey, *Modern Turkey* (New York: Praeger, 1974), 255pp., affords a balanced perspective on economic development and patterns of modernization in a setting that predates the oil economy and still lacks the sudden infusion of oil-related capital.

The literature on social and cultural change is so extensive that any selection runs the risk of erring by omission. A study of modernization in the Levant, a good point at which to begin, is Lerner, Daniel, *The Passing of Traditional*

Society: Modernizing the Middle East (New York: Free Press, 1958), 466pp. Clarke, J. I., and Fisher, W. B., *Populations of the Middle East and North Africa: A Geographical Approach* (New York: Africana, 1972), 432pp., present the geographic context for contemporary change in much of the region covered by this chapter. An evocative account of the impact of modernization and the Aswan High Dam on the inhabitants of southern Egypt is Fernea, Robert A., *Nubians in Egypt: Peaceful People* (Austin: University of Texas Press, 1973), 146pp. The massive, widespread and rapid growth of cities as a result of rural-to-urban migration can only be hinted at by Harrison, Robert S., "Migrants in the City of Tripoli, Libya," *Geographical Review,* 57 (1967), 397–423. Finally, no account of change in the Middle East can be complete without at least passing reference to the Arab-Israeli conflict, which unsettles the region and drains capital and energy that might otherwise be applied to economic growth. Contrasting aspects of this fundamental political problem are admirably sketched by Elon, Amos, *The Israelis: Founders and Sons* (New York: Holt, Rinehart and Winston, 1971), 359pp.; and Chaliand, Gérard, *The Palestinian Resistance* (Harmondsworth: Penguin, 1972), 190pp.

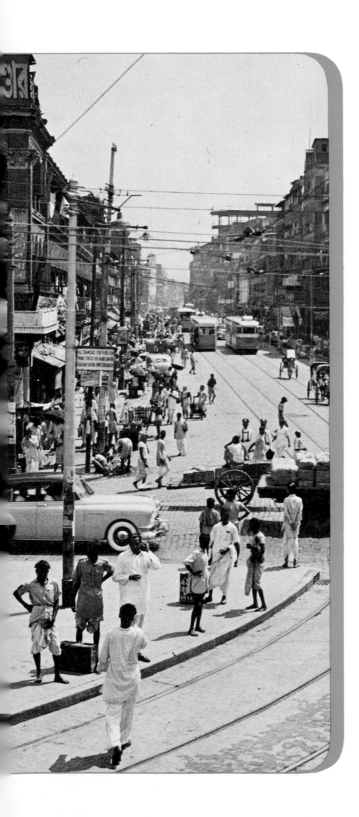

Monsoon Asia

Clifton W. Pannell

PART EIGHT

Monsoon Asia: Regional Unity and Disunity

28

THREE RELATED BUT distinctive regions form Monsoon Asia: South Asia (Indian subcontinent), Southeast Asia, and East Asia (Fig. 28–1).[1] More than one-half of the world's population lives in Monsoon Asia, many in great poverty. In the valleys of the Indus and Hwang Ho lie the remains of two of the world's important early centers of civilization and culture. Through time, many of the cultural influences and traits of these civilizations have spread to other parts of Monsoon Asia where they were modified in a variety of ways. More recently the introduction of European habits and practices has added new di-

mensions to the culture mosaic of this large part of the world.

The term Monsoon Asia implies a physical basis for regionalization in the seasonal pattern of summer rainfall common in the southeastern quadrant of the Eurasian landmass. Although the term originally was used to describe a part of the earth that shared generally similar rainfall regimes, it has come increasingly to be used as a more broadly descriptive areal term.

Recognition of South Asia, Southeast Asia, and East Asia as regions is based on a series of cultural and geographical features, many of which are examined in this and succeeding chapters. These features are both pertinent and significant, for it is the interaction of these different cultural features and environmental settings that underlies the development processes in Monsoon Asia.

India and China are of special importance. Both countries have gigantic populations, and both are important industrial states. Both, however, are poor, and their per capita output of goods and services is very low. Both are seeking rapid economic growth, yet each follows its own particular development path. Two alternative development systems are present, and in each are evident successes and failures.[2] Both are watched closely by other poor nations and sometimes are emulated by other smaller, developing states.

These two Asian nations in some ways parallel the American system and the Soviet system. Such comparisons imply competition, and during the decades of the 1950s and 1960s, this competiton was often clear and was violently expressed. In the 1970s, however, competitiveness has become more muted.

Common and Uncommon Themes

Several themes are common throughout Monsoon Asia. These themes are both physical and human, and

[1]Japan with its modern economy is not included in this treatment of Monsoon Asia. Yet its loss detracts little from the significance of the large region. In a region characterized by great contrasts, size, and large numbers, Japan is simply one of several medium-sized states both in population and area. Its remarkable economic growth of the last century, however, has distinguished it from other states in the region. Today, as a consequence, Japan serves as an excellent model of sustained progress and rapid development, highly visible to other Monsoon Asian states.

[2]S. Swamy, "Economic Growth in China and India, 1952–1970, A Comparative Appraisal," *Economic Development and Cultural Change,* 21 (1973), 1–4. Swamy has developed an analytical comparision of the two states and compared a number of quantitative indexes on growth.

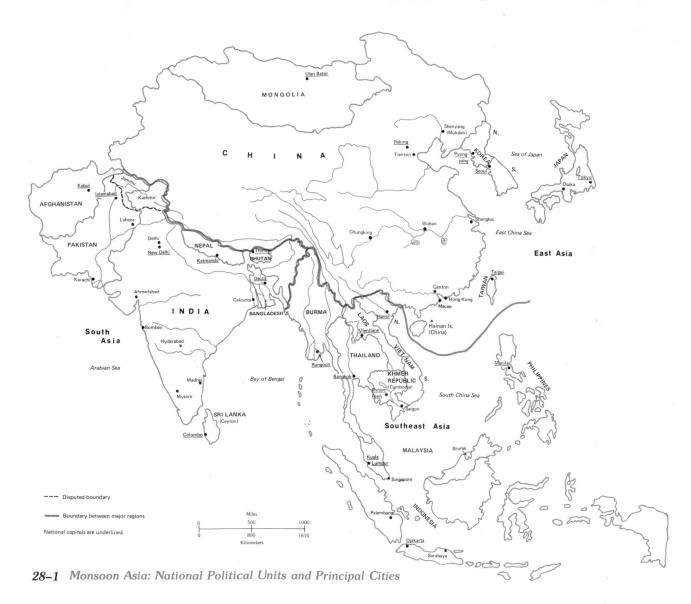

28-1 *Monsoon Asia: National Political Units and Principal Cities*

they provide a good springboard to introduce this diverse part of the world.

Environment: The Basis for Human Occupancy

The distribution of population in Monsoon Asia reflects an ordered relationship of people with particular physical environments. For example, the most dense population concentrations are in the great alluvial valleys and floodplains of the major rivers. The main clusters are located in the Indo-Gangetic plains of India, the Yangtze and Pearl River basins, and the northern plains of China, the Tonkin delta of Vietnam, and along the Irrawaddy, Menam Chao Phraya, and Mekong rivers in mainland Southeast Asia (Fig. 28–2). Even where there are no great rivers,

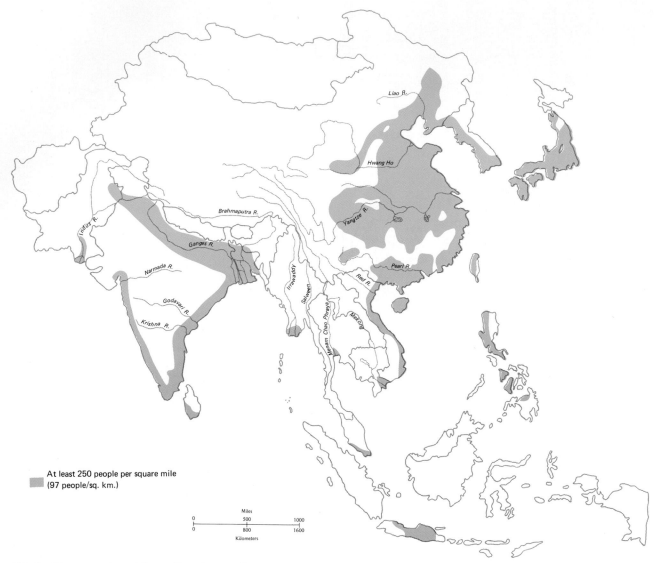

At least 250 people per square mile
(97 people/sq. km.)

Miles
0   500   1000
0   800   1600
Kilometers

28–2   *Monsoon Asia: Major Population Concentrations*

Asian peasants have clustered in the smaller alluvial basins and plains. There are some exceptions. Peninsular India, the island of Java, and the loess plateau of north China stand out as population centers, but special circumstances and historical events help explain these particular patterns.

The point is that in Monsoon Asia, people have gathered and proliferated most where the agricultural potential has been greatest. This concentration of population has meant that water has played a key role; and where water is available, agriculture generally has been oriented to wet rice (paddy). Although there are many different methods of growing wet rice, and yields vary considerably, rice is the principal food crop in Monsoon Asia. Where environmental conditions permit, two crops a year are customary in

China. In India, on the other hand, double-cropping is not widespread, even where the environment is suitable; thus India may have a considerable potential for increased crop production.

Accessibility and Population

Another facet of population clustering is accessibility. Clustering reflects the degree of accessibility as much as the nature of the environment and the agricultural potential. Perhaps the best example of this principle is seen in the Yangtze Basin of central China. Historically, areas within the Yangtze Basin where navigation was possible prospered and grew in population. Away from navigable streams, the economy stagnated, and population grew slowly. More than 1,000 years ago, food grains were transported by water from the Yangtze region to the ancient capitals in the plains of northern China, a pattern that suggests a remarkable degree of economic and spatial integration for parts of China. Accessibility and the ability to move foodstuffs cheaply have been very important in determining where people could and did concentrate.

The Monsoon and the Rhythm of Seasons

Monsoon Asia is used to describe that part of Asia that composes the southeastern quadrant of the Eurasian landmass. The term *monsoon* is believed to be of Arabian origin and has been used for centuries by mariners coasting the Arabian Sea to describe a seasonal reversal in wind direction. In simplest form, this reversal results from the different heating and cooling rates of land and water in the summer and winter months, coupled with latitudinal shifts in major air masses. Thus, during the summer in the northern hemisphere, southwesterly (onshore) winds dominate, but during the winter northeasterly (offshore) winds are customary.

The Indian monsoon has three distinct seasons, and the most important is the June-to-September rainy season. The warm, moisture-charged winds blow from the southwest and strike the southern coast of Sri Lanka (Ceylon) usually in late May (Fig. 28–3). By the first of June, the southern tip of the Indian peninsula has been contacted, and from there the monsoon progresses northward. The first occurrence of the monsoon is usually a brief transition period in which the extreme heat and aridity give way to light rains and more humid conditions. In some parts of India, afternoon convection thunderstorms precede the wet monsoon, and it is not always easy to distinguish showers from monsoonal rain.

The southwest monsoon in fact may be grouped into two divisions: an Arabian Sea branch and a Bay of Bengal branch. Winds of the former strike the sharp ridge barrier of the Western Ghats, are forced upward, and dump very heavy rains on the windward slopes of the Ghats. In the east, winds from the Bay of Bengal move rapidly northward and arrive at Calcutta about June 7. These winds blow north as far as Khasi Hills, where a piling-up effect forces them northwest up the Ganges Plain. By the end of June, the winds have reached Delhi; and in early July, Lahore and Rawalpindi.

In East Asia, the winter monsoon is characterized by cold and dry winds, sweeping out of Siberia. In northern China average winter temperatures are often 5° to 10°F. (3° to 6°C.) cooler than at comparable latitudes and locations in Anglo-America. The winds first move eastward, then southward toward the equator. In the movement south the air is warmed and gradually loses much of its Siberian character. Some winds flow across the warm Sea of Japan and the East and South China seas, and great quantities of moisture are added to the air mass. Where these warm, moist winds strike land areas, heavy precipitation can result. In some places—for example, the northern coast of Taiwan, the east coast of Vietnam and Malaya, and the northern and eastern sides of islands of Southeast Asia—winter brings the time of maximum rainfall. On Taiwan, the winter months are cool, wet, and dreary. More commonly, however, winters are dry and cool.

During the summer monsoon in East and Southeast Asia, the winds move in great arcs toward a center of low pressure situated in central Asia. Moisture is brought overland, and the pattern of a summer maximum of rainfall described for South Asia is characteristic with the monsoon, first influencing the Indonesian archipelago, then mainland Southeast Asia, and finally China.

Aridity and Sparse Population

Within Monsoon Asia is a large area of limited water availability. Much of western Pakistan, the

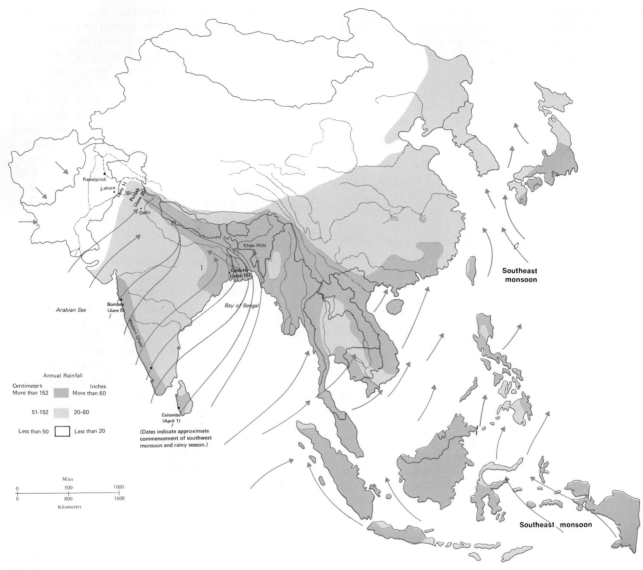

28-3 *Monsoon Asia: Summer Seasonal Winds and Annual Rainfall*

western two-thirds of China, and Mongolia are such examples. All of these places are dry, and the meager populations reflect an environment of limited food resources. Sedentary cultivation is difficult, and many of the inhabitants are pastoralists. Such an occupancy system does not support a dense population.

Monsoon Asia and Its Three Regional Parts

South Asia is a physical unit, the Indian subcontinent, but is also is a social unit based on the similar cultural origins and characteristics. East Asia has less physical unity, and some have argued that it has less cultural unity as well. Western China, for exam-

Although much of monsoon Asia is humid, part of the interior is dry. Here in Afghanistan the Hindu Kush mountains are barren and rocky. Shown here is the road connecting the city of Kabul to the Pakistan border. (United Nations)

ple, may be viewed as an exception to the case for regional unity. The efforts of the Chinese government to colonize this area, however, make the issue moot. It is today a political appendage and increasingly integrated into the state. Tibet and Inner Mongolia also have special administrative identities within China, but have close historical and ethnic ties with the Chinese.

Southeast Asia, composed of many smaller and medium-sized states, has less cultural homogeneity than its two neighbors. In fact, in many ways it is a buffer between Chinese and Indian civilizations and has received significant contributions from both. The unity of Southeast Asia, based on its tropical and maritime elements, gives it a distinctive character. Its regional coherence, however, has been recognized only recently. In many ways it is a region more for academic purposes than for its regional attributes.

The Cultural Basis for Regions and Regionalization

Peoples and Their Cultures

The cultural variety of Monsoon Asia is well expressed in the mosaic of ethnic and racial types and in the astonishing number of spoken and written languages. All three major gene pools, black, white, and yellow, have contributed to modern mankind in Monsoon Asia, although it is the Caucasoid (white) and Mongoloid (yellow) types that dominate contemporary racial patterns. More than 1,000 languages are spoken in Monsoon Asia, and languages such as Mandarin Chinese, Bahasa Indonesian, Hindi, Bengali, and Chinese Wu are spoken by millions of people. Religious diversity is equally broad; the dominant religions are mystical in character and recognize many gods. There are exceptions, however, such as the Philippines and Vietnam where Christianity is dominant or common and Moslem states such as Indonesia and Pakistan. China and other Marxist states officially deny or discount religion. In some areas there is little doubt that religious influences are declining in importance. What significance the neglect of religion will have on future events is hard to predict, but in some places it may have far-reaching effects as old traditions are cast aside.

Other distinctive cultural features are present in greater or lesser degree throughout Monsoon Asia. Architectural styles and practices are one example; agricultural systems and fertilization practices have been used as a means to define specific regions. Political and administrative institutions and land-tenure relationships also provide good measures of similarity

Monsoon Asia exhibits a wide range of economic development. Shown here are laborers in Taiwan leaving a modern manufacturing plant after work. The bicycle is a common means of transport. (United Nations)

and differences among the major regions or cultural realms of Monsoon Asia.

The nature of the development process in different areas is also a distinguishing characteristic. For example, a Chinese (Sinic) realm and an Indian realm are relatively simple to delineate and define by their distinctive development strategies. Common patterns in Southeast Asia, lying geographically and culturally somewhere between the two, are much harder to identify. Southeast Asia is a shatter zone that is fragmented into many political, cultural, and geographical subregions discerned on a map as national units. Whether such a shattering process retards or accelerates the speed of economic growth and national development is hard to estimate. One method of examination is to compare the development perfor-

mances of some of the various component states within the regions. Such a comparison may permit an evaluation of alternative means to national development.[3] At least it can provoke discussion and debate about the different approaches.

Throughout Monsoon Asia, the common objectives of governments and the populace are to improve levels of living and promote economic growth. Few disagree that these are worthwhile objectives, and most politicians, planners, and citizens share the

[3]Donald Fryer, *Emerging Southeast Asia, A Study in Growth and Stagnation* (New York: McGraw-Hill, 1970), pp. VII–XI. Fryer's geography, as the title implies, is focused on the theme of differential growth rates and performances. Monsoon Asia offers a number of excellent comparative situations.

vision of a better standard of living for all citizens. Disagreements arise over how this objective should be achieved.

Alternative approaches to development span the gulf between Asian Communist economies as exemplified by China and North Korea to free-enterprise economies such as exist in Malaysia and the Philippines. In between are a host of partly socialistic, partly free-enterprise economies. Political approaches offer an equally broad set of alternatives from highly authoritarian systems of control to Western-style democracies. Comparisons may be made between the many different approaches and the success or failures they may have achieved. Some of the more obvious examples useful for evaluative comparisons of economic growth performance are India-China, South Korea-North Korea, and Burma-Thailand. As the succeeding chapters are read, it might be well to keep in mind that comparisons of methods and approaches to development can be made and performance evaluated. In arriving at final conclusions, it is wise to note, however, that some economic growth goals in China or North Korea may have been achieved at some sacrifice of personal behavior and political freedom. Is it worth such a sacrifice? Is it better to have more food and better health care and less individual freedom? It may be that there is no definitive answer to such questions.

Regional Discordance and Difficulty

Political events since World War II have been unhappy for most of Monsoon Asia. Strife, war, famine, and hard times are common themes that run through the recent history of the region. These same themes occur in other developing areas. Yet such cataclysmic events as China's civil war, the end of colonialism in South and Southeast Asia, the Vietnam war, and the forging of Bangladesh appear to be more common and frequent in Monsoon Asia. Moreover, the magnitude of such events in Monsoon Asia involves so many lives that the events themselves take on special significance.

Regional Cooperation

All is not hopeless or disappointing in Monsoon Asia. A number of examples of cooperation among various states exist, and there are now institutions that seek to promote rapid economic growth. Perhaps the best example is the Asian Development Bank, a financial institution headquartered in Manila and directed by a Japanese president. The bank is largely Asian financed and staffed, and it seeks to promote economic growth and development, primarily in Monsoon Asia.

The point is that Monsoon Asia has hope and promise as well as tragedy and misery. The whole range of success and failure in the experience of developing states exists, and most of the approaches can be studied. The panorama of the less developed world is here, and we can review and learn from the range of human conditions under situations of poverty and growth, tradition and modernity, technological backwardness and innovative change.

Further Readings

American Geographical Society, *Focus* (New York: American Geographical Society, Monthly). *Focus* provides up-to-date geographical essays on topics and regions throughout the world. Many recent issues have examined development problems in Monsoon Asia.

Association for Asian Studies, *Annual Bibliography of Asian Studies* (Ann Arbor, Mich.: Association for Asian Studies, Annual). This is probably the most comprehensive up-to-date bibliography of English-language materials on the countries of Monsoon Asia.

Far East Economic Review, *Asia 1974 Yearbook* (Hong Kong: Far East Economic Review, 1974). This yearbook, which appears annually, provides up-to-the-minute coverage of political, economic, and social patterns in all countries of the region. Statistical summaries are also provided.

FRYER, DONALD, *Emerging Southeast Asia, A Study in Growth and Stagnation* (New York: McGraw-Hill, 1970), 486pp. Fryer's text focuses on Southeast Asia, but the issues he raises go far beyond and pertain to all of Monsoon Asia.

GINSBURG, NORTON (ed.), *The Pattern of Asia* (Englewood Cliffs, N.J.: Prentice-Hall, 1958), 929pp. Ginsburg's geography, although out of date, remains one of the best regional texts on the geography of Asia.

LEWIS, JOHN, *Quiet Crisis in India: Economic Development and American Policy* (Washington, D.C.: Brookings Institution, 1962), 383pp. Lewis poses the significance of India's economic development as a model for other states.

MCGEE, T. G., *The Urbanization Process in the Third World* (London: G. Bell, 1971), 179pp. Theoretical approaches to Asian urbanization form the main theme of this set of provocative essays.

SPENCER, JOSEPH E., *Oriental Asia: Themes Towards a Geography* (Englewood Cliffs, N.J.: Prentice-Hall, 1973), 146pp. This brief volume outlines the main geographical and cultural patterns of Monsoon Asia.

SPENCER, JOSEPH E., and THOMAS, WILLIAM, *Asia, East by South: A Cultural Geography,* 2nd ed. (New York: Wiley, 1971), 669pp. Without question, this text is the primary cultural text on the geography of Monsoon Asia.

SWAMY, S., "Economic Growth in China and India, 1952–1970, A Comparative Appraisal," *Economic Development and Cultural Change,* 21 (July, 1973), part 2, 1–84. Swamy presents quantitative comparisons of the recent economic growth rates of India and China.

United Nations, *Statistical Yearbook for Asia and the Far East* (New York: United Nations, Annual). The United Nations provides statistics of significant social, demographic, and economic topics.

United States, Department of the Army, *Communist China: A Bibliographic Survey,* 1971 ed. (Washington, D.C.: Government Printing Office, 1971), 253pp; *Peninsular Southeast Asia: A Bibliographic Survey of Literature, 1972* (Washington, D.C.: Government Printing Office, 1972), 424pp; *South Asia and the Strategic Indian Ocean: A Bibliographic Survey of Literature, 1973* (Washington, D.C.: Government Printing Office, 1973), 373pp. These bibliographies also contain maps and descriptive essays on the politics and economic matters of the various countries surveyed.

South Asia: Cultural Origins and Physical Environments

29

O N THE SOUTHERN FLANK of the Eurasian landmass, south of the Pamir-Himalayan mountain wall, lies a large region that juts almost to the equator. Known to the world as the Indian subcontinent, or South Asia, this region is today composed of three large countries, India, Pakistan, and Bangladesh, and three smaller and less significant states, the island state of Sri Lanka (formerly Ceylon), and the Himalayan states of Nepal and Bhutan (Fig. 29–1). To the northwest is Afghanistan, a transitional state with historic ties to both South Asia and the Middle East. The adjacent seas, the Arabian Sea, the Bay of Bengal, and the Indian Ocean, contain numerous islands and island groups that belong to India and also are part of South Asia. Despite the great human variety and the contemporary political fragmentation of the subcontinent, there are sufficient historical precedent and cultural homogeneity to justify regional characterization. Through an examination of the history and cultural origins of South Asia and a study of the contemporary situation, a better appreciation of this region may be derived. This examination can help us understand the nature of the development processes under way in South Asia and the likelihood of success or failure of these processes.

Origins and History

The Indian subcontinent is an ancient land with an old civilization rich in traditions of literature, art, technology, and religion. Although the early origins of the population are still in dispute, it is well established that sedentary cultivators were farming the Indus floodplain more than 5,000 years ago. Indeed the term "India" is derived from the Sanskrit word for river used to describe the Indus. The origins of South Asian civilization thus are rooted in an ancient society that took shape almost as long ago as the early development of mankind in the Middle East.

The ancient civilization that emerged in the Indus floodplain around 3500 B.C. is believed to have been composed of ancestors of the modern dark-skinned inhabitants of southern India, Dravidians. These people were the ancient city builders of the Indus Basin responsible for the construction of Mohenjo-daro and Harrappa (Fig. 29–2). A second distinctive human group entered the subcontinent from the northwest mountain wall of Baluchistan in successive migratory waves. These invaders were Caucasoid people, commonly described as Indo-Aryans, and they infiltrated the subcontinent over a long period of time. More recent historical examples of outside penetrations include the army of Alexander the Great, Islamic Moors, and lately Persians and Afghans.

All of these invasions have had some impact; but, as Brush has pointed out, none had the significance of the earlier blending of the ancient Dravidians and early Indo-Aryans.[1] It was the synthesis of these two groups that produced the Hindu-based civilization we today refer to as Indian.

[1]John Brush, "South Asia, Peoples and Culture," in Norton Ginsburg (ed.), *The Pattern of Asia* (Englewood Cliffs, N.J.: Prentice-Hall, 1958), pp. 458–482.

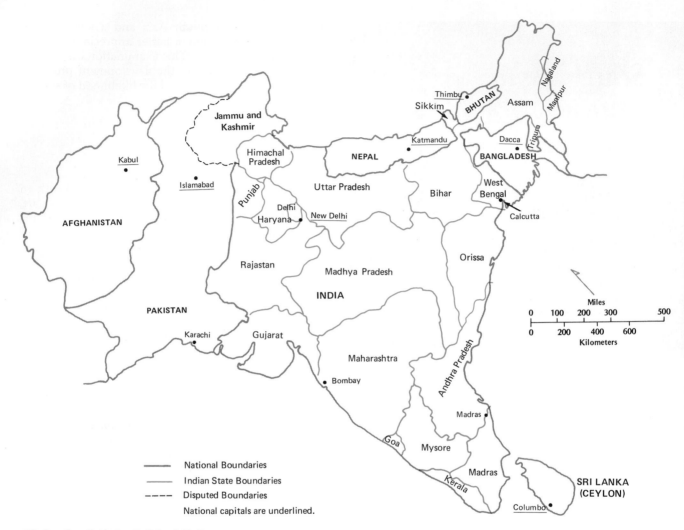

29–1 *South Asia: Political Units*

Religion and Its Impact

The formation of India and Indian civilization is tied closely to the Hindu religion. There emerged very early a formalized set of religious beliefs with societal and political ramifications. Local priests established themselves at the apex of the social strata and became a potent political force. A clearly identifiable social group (caste) emerged that was composed of this priestly society, the Brahmins. So influential was the Brahmin group that even today it is looked on as the capstone of modern Indian-Hindu society. Its members, despite the fact that many of them are not wealthy, continue to wield considerable power at the local level.

Other religious groups also emerged. Buddhism, a spin-off of Hinduism, appeared and under the Indian monarch Asoka enjoyed a flourishing but brief period of dominance. Buddhism spread throughout much of the rest of Monsoon Asia, but its impact on the subcontinent was reduced. Today, only in Sri Lanka and some of the northern mountain areas is it significant.

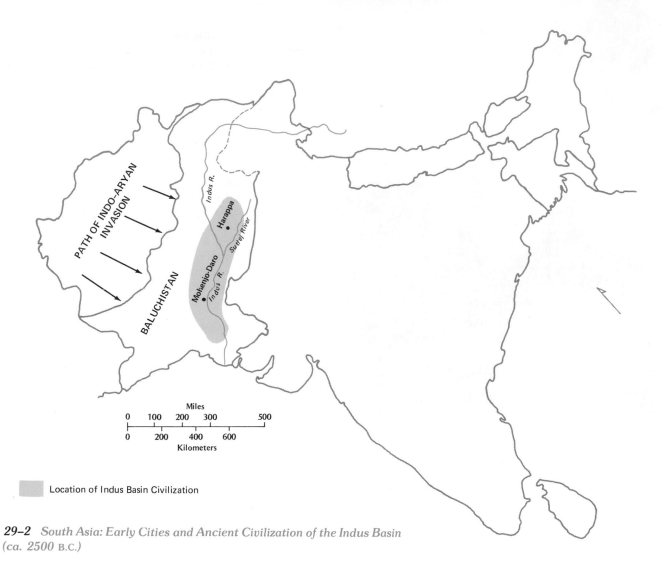

29–2 *South Asia: Early Cities and Ancient Civilization of the Indus Basin (ca. 2500 B.C.)*

Another major religious challenge to the dominance of Hinduism occurred about the tenth century A.D. with the introduction of Islam. Here was a Middle East religion that presented a supreme test to Hindu order and society. It offered identity and status to those excluded and downtrodden in the well-established Hindu system. Islam came from the northwest, the traditional direction of the Aryan invaders, and made itself felt most vigorously first in the Indus Basin and later in the Ganges drainage area. To some extent the invasion of Islam followed the main drainage systems. Thus, Islam did not spread rapidly in peninsular India, a plateau with greater impediments to movement.

Today, the impact of religion continues to shape the political geography of the subcontinent. Religious differences were responsible for the 1947 partitioning of the subcontinent into a Hindu component (India) and a Moslem component (Pakistan) from which East Pakistan broke off to form the state of Bangladesh in 1971. Although all three states have significant religious minorities, the Hindu-Moslem split has shaped and influenced political patterns and human activities more than any other. Other religious groups

Although Hinduism prohibits the eating of animals, animals are used for other purposes. Cattle are used as draft animals and as a source of milk. Additionally the bones of dead animals are collected and crushed for fertilizer. (United Nations/FAO/H. Null)

tions against taking the lives of animals. The cow is especially revered. In the absence of systematic approaches to the breeding and slaughtering of cattle, a great proliferation of animals has resulted, and most of these animals are undernourished. Many observers have criticized India's failure to use its bovine resources and have viewed the cattle population of India as essentially parasitic. Hindus, however, do use their cattle in several ways: dried cow dung for cooking fuel, milk as a source of protein, and for draft purposes.

Hindus not only revere cattle, but they also look upon them as a self-renewing, inexpensive resource. Although there is controversy over whether or not this approach is a wise and efficient exploitation of a resource, it is clear that the Indian method of using cattle is an excellent example of the impact of culture on the development process.

The Hindu Caste System

Hindu society with its privileged upper tier of Brahmins is tightly structured. Individuals are grouped according to occupation (castes). Members of each caste are restricted to a narrow range of groups with whom they may eat, marry, and otherwise associate. A highly formalized society has emerged which determines virtually all facets of life for an individual. Birth determines a person's destiny, for despite recent legal proscriptions to the contrary, a man's occupation, diet, wife, and daily habits have been largely foreordained. In village India, the home of at least three-quarters of the population, those who perform the mean tasks are still "unclean" and thus "untouchable." In the cities, however, the caste system may be weakening. It is within the urban areas that the opportunity for social mobility exists.

It has been postulated that stratification in Hindu society is to a large extent ethnically motivated. Centuries ago, the lighter-skinned Aryans set themselves up as Brahmins, whereas darker-skinned Dravidians were forced into the lower castes. Such an ethnic-racial background to the evolution of caste, however, has never been clearly established. Moreover, there is some degree of caste mobility, and examples exist of castes with vigorous, wealthy members which have advanced in the social hierarchy.

exist in relatively modest numbers, Jainists, Sikhs, Parsis, and Christians, but none has been effective in establishing a separate state based on their religion.[2]

Hinduism and the Case of the Sacred Cow

Religion and religious custom have played important roles other than political. For example, Hindu ideas about reincarnation have resulted in proscrip-

[2]Sikhs have attempted to form a separate political territory, and there is some recognition of the existence of the Punjab state as a Sikh administrative subunit of India. It likely will remain part of the Indian Union.

Other Cultural Aspects of Diversity

The Human Mosiac

India is a human mosaic of immense variety and complexity. There is no single India any more than there is a representative Indian; rather, there are many different Indias. Differences in religion, ethnic groups, language, and caste illustrate the enormous diversity and present difficult problems for development planners. Yet out of this diversity has emerged a common bond of allegiance to the Indian nation. It is this view of nationhood which today captures the essential spirit of modern India.

29–3 *South Asia: Major Languages*

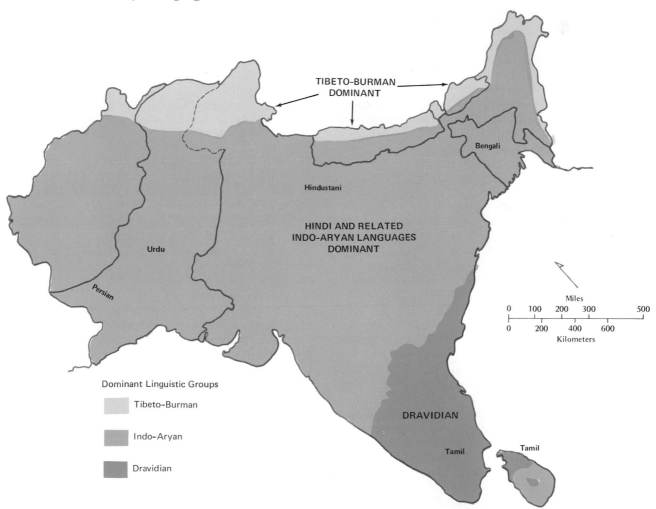

Dominant Linguistic Groups

- Tibeto–Burman
- Indo–Aryan
- Dravidian

Language

Social diversity in the subcontinent is matched by an equal plethora of languages. Although four or five languages dominate in South Asia, there are fifteen major regional languages, more than 150 official languages, and at least 400 dialects (Fig. 29–3). The major languages are Hindi (India's national language) and its two derivatives, Hindustani and Urdu (the main language of Pakistan), as well as Bengali and English. Hindi and Bengali are Indo-European and derived from the Indo-Aryan Sanskrit. In the south, several Dravidian languages such as Tamil are spoken, and they differ genetically from Indo-Aryan languages. The language difference, coupled with other distinctive traditions, is serious enough to give rise to separatist movements among the more than 100 million Dravidian speakers. The use of Urdu and Bengali in Pakistan and Bangladesh respectively provided one motive for political separation despite having a common religion, Moslem. English, despite its association with former colonialism, continues in use as a lingua franca and a semi-official language of administration. Inasmuch as it belongs to no group, its usage is sometimes more acceptable than Hindi. English to that extent serves as an important integrating element.

The Impact of Diversity on Territory

The political organization of the subcontinent expresses the impact of several cultural variables that have been able to assert themselves since British influence has waned. Evolution of the several political units of South Asia is perhaps the most obvious territorial expression of cultural differences.

British colonialism played a powerful role in providing numerous spatial linkages such as railways to tie together the territory of the subcontinent into meaningful, operating administrative and economic regions. Many of these links survive, three decades after the twilight of British political power. The withdrawal of the British in 1947 introduced a new formative phase in the political history of the subcontinent, a phase that is characterized by a high degree of stress and flux and one likely to persist for many decades in the future. That this condition has far-reaching implications for the pace and pattern of national development is obvious. The magnitude of the economic difficulties that face the recently created state of Bangladesh is one of the most recent examples of this problem.

South Asian Environments as a Resource Base

South Asia, like China and the United States, spans a large area and contains great physical variety. Most of the subcontinent is tropical; only at higher elevations in the north and northwest is frost common. Consequently the main climatic variable is the amount of precipitation and its seasonality.

The landforms of the subcontinent are divided into three major regions: (1) the southern peninsular massif; (2) the interconnected drainage basins of the Indus, Ganges, and Brahmaputra rivers; and (3) the northern mountains composed of the Himalayas, Karakoram, Pamirs, Suleiman, and Hindu Kush (Fig. 29–4). Considerable variety exists within each of these major units, but they stand apart as significant surface regions.

The Peninsular Massif

Peninsular India is an ancient massif composed of weathered crystalline rock. The peninsula is a plateau tilted up on the west to elevations exceeding 6,000 feet (1,829 meters) and sloping gently eastward. A narrow lowland fringes the coasts, but abrupt ridges (the Ghats) rise a short distance inland along both the eastern and western flanks. Drainage in the south is generally to the east, but north of Bombay the major rivers flow westward.

This peninsular massif with its ancient rock contains most of the mineral wealth of India, copper, iron, gold, lead, manganese, and coal. Soils vary in fertility, but paucity of water is a more serious problem to agricultural production than is quality of soil.

The Northern Mountains

Several mountain ranges compose the northern margins of the subcontinent, and this mountain wall forms one of the most imposing physical features on earth. The height and magnitude of the mountain system exceed all others, and the geology of the mountain region indicates compressional forces and stress at work that are almost unparalleled. Fossil

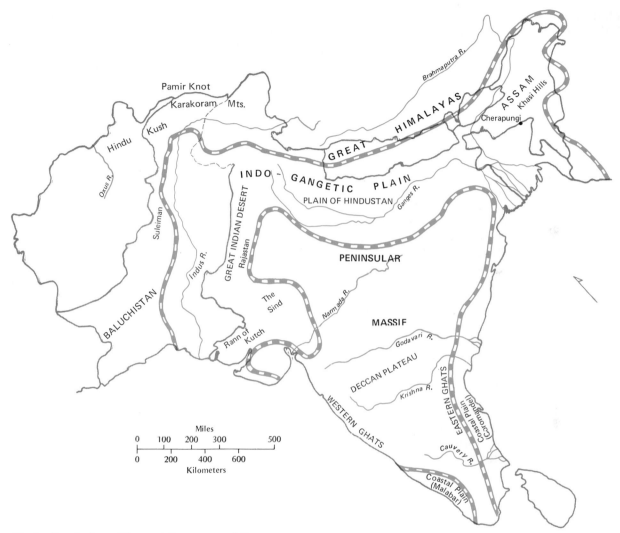

29-4 *South Asia: Physical Regions and Drainage Patterns*

evidence suggests a fairly recent uplifting of gigantic rock waves that formed the Himalayas. The high ridges are aligned parallel to the margins of the plains farther south.

To the west of the Pamir Knot radiate two major ranges, the Suleiman of eastern Afghanistan and western Pakistan and the Hindu Kush of Afghanistan. To the east are the Himalayas and the Karakoram which form the wall that has so long divided the Indian cultural realm from the Chinese. Other mountains have proved less of a barrier to movement and communications between the outside world and South Asia. India, for example, has long demonstrated a propensity for ties with the Middle East and Southeast Asia. One consequence of this directional tendency has been that the exchange of ideas and goods was made with the Middle East and Southeast Asia more than with East Asian civilization.

The Indo-Gangetic Plain

Of the three main physical divisions of South Asia, the contiguous plains of the Indus, Ganges, and Brahmaputra rivers are probably the most significant. Here is the heartland of ancient Hindu civilization; and Indians today look toward this region as the core of their nation. The drainage systems considered together compose the largest continuous alluvial plain on earth, altogether more than 300,000 square miles (482,790 square kilometers). The plain is a structured depression formed at the same time as the Himalaya Mountains. The main drainage of the subcontinent has been oriented to this depression, and the resultant floodplain has been covered with vast quantities of alluvial material. Farmers long ago found the plain a particularly attractive agricultural environment with its combination of productive soil and plentiful water. Today, the Indo-Gangetic Plain supports roughly one-half the population of the subcontinent and is the political focal point for the three major states of South Asia. The recent selection of Delhi, a city located in a commanding position in the upper Ganges Basin, as the capital of India reflects well the Indian view of the special significance of this region.

Mineral Resources

The resource base of South Asia, and India in particular, is impressive. Coal, iron ore, manganese, and chromite exist in large quantities, and smaller amounts of other valuable minerals are also present. Perhaps the most serious mineral deficiency that faces the subcontinent is the lack of petroleum to sustain a drive to industrialization. The continued increase in imported petroleum prices is a most serious matter for India, for future increases in food production are likely to be closely tied to increases in the production of petroleum-derived chemical fertilizers. Lack of petroleum could impede seriously the effort to improve the standard of living for millions of Indian peasants.

Land Resources

The agricultural potential of the subcontinent is not as poor as the commonplace notion often suggests. The Indo-Gangetic Plain is one of the largest, most productive alluvial lowlands found anywhere on the globe. The availability of water and the partial renewal of nutrient materials through irrigation and floodwaters have allowed this great floodplain to sustain large numbers of people for several millennia.

Peninsular India contains a high percentage of land suitable for farming, but the major problem is sufficient water to nourish crops. India is fortunate to have roughly one-half its area arable. Bangladesh has a still greater proportion of its land classed as arable. China, by comparison, has only about one-fifth of its area suitable for agricultural purposes. Only in the dry west—the Rajastan Desert, the Sind region, the Rann of Kutch, and the basins and plateaus of Baluchistan in western Pakistan—and in the high mountains of the north is the agricultural environment too harsh to sustain large numbers of people.

Human Adjustments to the Monsoon

The monsoon determines the agricultural calendar of most South Asian farmers. Notwithstanding the impact of irrigation, it is the arrival of the southwest monsoon that brings with it the surge in agricultural production and to which the peasants adjust many of their activities. During the extremely hot and dry season preceding the arrival of the summer rains, rural life has slowed. The earth is parched and cracked, and little can be done to prepare for the sowing. With rain, human activity resumes at an intense pace. Soil preparation, planting, weeding, and the many tasks associated with the growing season must be compressed into a relatively brief time span.

Almost three-quarters of India's rainfall is concentrated during the southwest monsoon between June and September. Those locations that receive the most rainfall are the windward slopes of the Western Ghats and the Khasi Hills of Assam (Fig. 28–3). Recordings along the windward slopes of the Ghats of 100 inches (254 centimeters) and more are common. The statistics from Khasi Hill stations, however, are the most impressive. Cherrapungi and some of the adjoining villages have average rainfalls of more than 400 inches (1,016 centimeters), and here is felt the full effect of the moisture-laden air moving off the Bay of Bengal over terrain barriers. Later the winds are funneled to the west, and their effect diminishes progressively. Thus, there is a steady decrease in the length of the rainy season and the amount of precipitation from southeast (Calcutta) to northwest (Delhi and Lahore).

By the end of September, the southwest monsoon is in full retreat. Fortunately, it is followed by a dry but comparatively cool period in which the activities of harvest and business may be carried out more comfortably. The late autumn and winter months are dry, cool, and pleasant for most of the subcontinent. The eastern two-thirds of southern peninsular India receives much of its rain at this time. The area is located in the rain shadow of the Western Ghats but is exposed to the northeast monsoon (Fig. 29–5).

Rain and season are so closely intertwined with agricultural activities that the agricultural calendar cannot be separated from climate. An intimate association has grown up between climate and the activities of the rural peasantry. It is difficult to understand rural life without a knowledge of the rhythm of the monsoon. That such behavioral patterns may have

29–5 *South Asia: Northern Hemisphere Winter Monsoon*

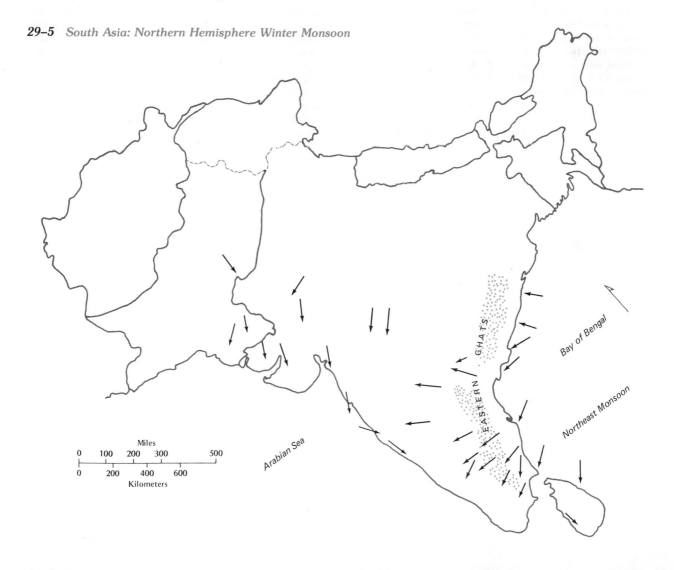

had implications on the speed and pace of economic growth and change in the past is conceivable, although it is difficult to pinpoint and evaluate in what manner such effects have occurred.

Further Readings

ANANT, SANTOKH, S., "Caste Hindu Attitudes: The Horizons' Attitude," *Asian Survey*, 11 (1971), 271–278. Caste, based on 2,000 years of Hindu social development, continues to play a profound role in contemporary Indian society. Caste is especially important in the more than half-million villages. This article examines critically the role of caste and focuses attention on recent relationships among caste groups and the condition of the Harijans (former untouchables).

BRUSH, JOHN, South Asian Chapters in Norton Ginsburg (ed.), *The Pattern of Asia* (Englewood Cliffs, N.J.: Prentice-Hall, 1958), pp. 458–678. Although this work is a little out of date, it is an excellent treatment of the geographic pattern of South Asia.

DUTT, ASHOK (ed.), *India: Resources, Potentialities and Planning* (Dubuque, Iowa: Kendall/Hunt, 1972), 144pp. This recent compendium is focused on the problems of India's modernization and economic growth.

NEALE, WALTER, *India: The Search for Unity, Democracy, and Progress,* (New York: Van Nostrand Reinhold, 1965), 128pp. Neale, an economist, has focused on all aspects of India's modernization.

SEN GUPTA, BHABANI, *The Fulcrum of Asia: Relations Among China, India, Pakistan, and the U.S.S.R.* (New York: Pegasus, 1970), 383pp. Sen Gupta identifies the main currents of political competition among the main powers that affect the Indian subcontinent. The relevance of the topic to development processes lies in the role that aid and political competition play in nation building.

SPATE, O. H. K., and LEARMONTH, A. T. A., *India and Pakistan: A General and Regional Geography,* 3rd ed. (London: Methuen, 1967), 877pp. If there is a standard geography of the subcontinent, it is this volume.

TINKER, HUGH, *India and Pakistan: A Political Analysis* (New York: Praeger, 1962), 228pp. This book provides a penetrating background to the political forces that have fragmented the Indian subcontinent. In exploring this topic, the book probes deeply into the cultural splinterings based on religion, language, and caste that are so fundamental to understanding this complex topic.

TIRTHA, RANJIT, "Population Problems in India: A Battle with Hydra," *Focus*, 24 (1974), 8. Tirtha has provided a good, current summary of the demography of India.

WHEELER, MORTIMER, *Early India and Pakistan to Ashoka,* rev. ed. (New York: Praeger, 1968), 241pp. A distinguished historian surveys the early history of the subcontinent.

WINT, GUY (ed.), *Asia, A Handbook,* rev. ed. (Middlesex, Eng.: Penguin, 1969), 856pp. This volume contains several good chapters on South Asian economic and political topics.

South Asia: Indian Economic and Social Systems

30

INDIA AND ITS South Asian neighbors are embarked on a course to promote national development and economic growth as rapidly as possible. In India's case, development and growth are being attempted within the framework of a democratic political system although with some authoritarian measures. This fact in itself is significant, for many poor nations look to the experiences in India and China for development strategies. Salient among the features of India's contemporary economy is that most of the employed population work on farms. From their toil and production must come much of the capital necessary to finance new investment in other sectors of the economy. Success in development of the agricultural sector is, therefore, critical both for food

supply and for growth of the industrial and service sectors.

Agriculture

Earlier it was pointed out that 50 percent of India's land is arable, a figure much higher than in most other countries of Monsoon Asia (Table 30–1). About 338 million acres (137 million hectares) have been under cultivation in recent years.

Bangladesh possesses a ratio of arable land to total area even higher than India. In this favorable agricultural environment has evolved an incredibly dense rural population, and the delta regions of the lower Ganges and Brahmaputra are as crowded as any place on earth.

Pakistan, by contrast, has a small percentage of land suitable for agriculture. The Indus floodplain, however, is a region of intensive cultivation. Where sufficient irrigation water is available, Pakistan also possesses an intensive agricultural system that supports a large population. Aridity, however, has placed limits on the agricultural potential of much Pakistani territory, but improvements in the irrigation systems may permit some expansion of the agricultural land.

The smaller states of South Asia are less favorably endowed with arable land. In most cases much of the surface terrain is too rugged, and significant agricultural development has taken place only in the level river valleys and basins. Nepal is fortunate in that several large valleys offer significant agricultural potential. Productive lowland plains comprise about one-third of Sri Lanka's area. Nowhere among these smaller states, however, are the arable land ratios as great as in India or in Bangladesh.

The distribution of agricultural systems and crops in South Asia corresponds closely to variations in the environmental setting and particularly to water availability (Fig. 30–1). In the dry west and northern mountains, herding is characteristic over much of the area. Most of the people, however, are supported by subsistence agriculture located in the irrigated valleys of the west and the mountain basins of the north. Estate agriculture (large-scale commercial production mainly for export) occupies some of the southern mountain slopes and nearby hills of northeast India. Tea in Sri Lanka and coconuts along India's Malabar

Region and Country	Number of People Per Land Unit		Number of People Per Cultivated Land Unit[1]	
	PER SQUARE MILE	PER SQUARE KILOMETER	PER SQUARE MILE	PER SQUARE KILOMETER
South Asia				
Afghanistan	80	31	649	251
Bangladesh	1,439	583	2,368	914
Bhutan	66	26	235	91
India	508	196	976	377
Nepal	244	94	1,727	667
Pakistan	215	83	995	384
Sri Lanka	564	214	1,844	712
East Asia				
China	230	89	1,733	669
Hong Kong	10,526	4,064	84,000	32,432
Korea (N)	355	138	2,284	882
Korea (S)	945	366	3,895	1,504
Mongolia	2.5	1	492	190
Taiwan	1,194	461	4,665	1,801
Southeast Asia				
Burma	121	47	435	168
Indonesia	182	70	1,958	756
Khmer Republic	116	45	1.142	441
Laos	39	15	954	368
Malaysia	98	38	927	358
Philippines	382	148	1,030	398
Singapore	10,450	4,035	44,000	16,988
Thailand	223	86	826	319
Vietnam	370	142	2,305	890
United States	59	23	294	113

Table 30–1 Monsoon Asia: Population-Land Ratios

[1]Includes cropped area and temporary pasturage. Does not include land in permanent pasture.
SOURCES: Food and Agriculture Organization, *Production Yearbook* (Rome: United Nations, 1975); and *Population Data Sheet* (Washington, D.C.: Population Reference Bureau, 1977).

coast are also estate crops. These estate crops are a legacy of past British rule and are grown in humid areas.

Over most of India, native subsistence and commercial agricultural crops are distributed largely in relation to water supply. In the humid areas rice and jute are dominant. In the drier areas wheat, millets, corn, cotton, and gram (a type of chick-pea) are characteristic. Throughout India, agricultural life revolves around the local village.

Traditional Indian Agriculture and the Village

More than 500,000 villages pepper India's landscape, and they contain more than 70 percent of the population. Village life, its economy, society, and politics, represents to a large extent the style and tempo of life for most Indians. The village, as the primary production and social unit, holds the key to many insights about the character of modern India and the forces at work leading to change and modernization.

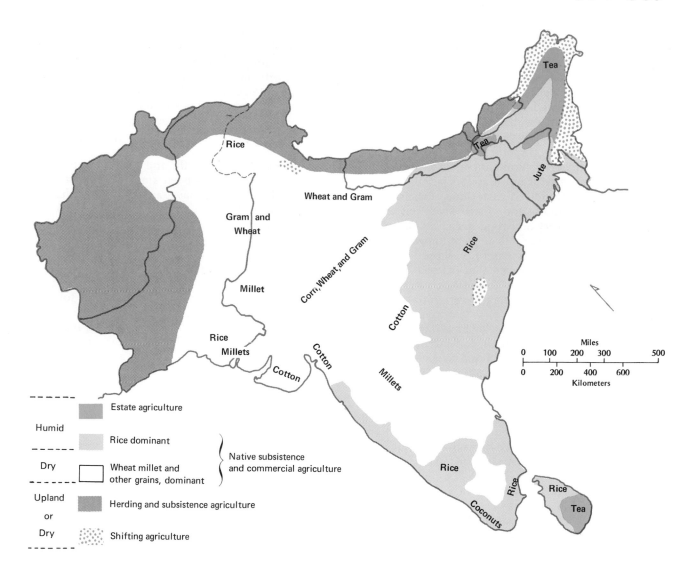

30-1 *South Asia: Major Agricultural Systems and Crop Regions*

Village life continues to be largely traditional, and the Indian peasant remains one of the most conservative members of human society. In recent years, however, demonstrations and popular movements in the lower Ganges delta region illustrate that these peasants can be politically volatile. Where conditions of rural poverty have become intolerable, rebellion and violence have flared, and local leaders who espouse extreme leftist Marxist programs have emerged. Prominent among these was a group from the small village of Naxalbari. So radical were the solutions they advocated and so violent their actions that they and their followers have been labeled "Naxalites," a political term that has come to stand for the most severe form of Marxist-inspired rural revolutionaries. For most Indian peasants, however, life continues along patterns that are familiar and predictable.

Traditional Village India

Production of essential foodstuffs, clothing, tools, and other commonly used goods has taken place at the village level for many centuries. A high degree of autonomy and self-sufficiency existed in which each village functioned as an independent operating and production unit. Paralleling this economic independence was a social system incorporating a strict ordering of social groups and families in which occupation was a major determinant of caste and thus social ranking. Many, if not most, social interactions and relationships followed caste lines. Inasmuch as most castes operated at the village level, this interaction yielded a strict and rigid social stratification. Individuals throughout a village were known to each other, and behavior was carefully scrutinized and often criticized. To the extent that external contacts existed, they generally were focused on neighboring villages, and intervillage linkages were formed along kinship lines of equivalent caste groupings. Exceptions existed, and in fact there have always been examples of social and economic advancement based on exceptional talent and industriousness. These exceptions, however, were rare, and the commonplace picture of rural India as conservative, tradition-bound, and slow to change is a reasonably accurate one. This characterization is slowly beginning to break down because of improved communications networks and the technical changes that are gradually being introduced to the farming systems.

The Indian Village in Transition

Some change in villages occurred with the imposition of British colonialism over the country and the improved transport system associated with this colonial enterprise. Prior to British rule, India had been compartmentalized into a number of social, economic, and political units. Regional and local isolation began to break down with the improvement of the transport and communication network. As the villages were affected and integrated more into the national framework, powerful social cleavages emerged between villages and the new elites in the urban centers. Such cleavages are characteristic of societies in transition from traditional to modern. New social and political institutions may be necessary to cope with the complexities of modernization and the consequences of national development.

Three prominent factors have significance for the long-term alteration of the traditional Indian patterns of economy and society. These changes—rapid population growth, land reform, and the Green Revolution—are interrelated and in some ways mutually reinforcing. Examining these factors and their implications will tell us much about how and why the traditional, rural patterns are changing.

Indian Population Growth and Food Production

Population growth has been among the most fundamental changes that have affected India during the last century. Both the rate of population growth and the huge absolute human increases are without precedent in Indian history. The implication and consequences of such rapid population growth are far-reaching and profound.

The increased number of people must be fed, clothed, and housed. The first need is food for the rapidly growing numbers, and enormous stress has been placed on the agricultural sector to increase food production continually. Although yields per unit area throughout most of India are still low by Western or Japanese standards, by and large production has increased and has managed to keep pace with the huge population growth. At least one economist, Ester Boserup, has suggested that rapid population growth is a positive factor because the intensified demand on the farming sector is the mechanism that has forced the agricultural sector to reform and modernize.[1]

Boserupian Thesis

According to Boserup, without the stress of increased demand, the traditional agricultural system simply continues without basic alteration. Population growth, however, has forced the agricultural system to begin a process of transformation and modernization in step with the increased demand for agricultural products. Boserup's thesis is interesting if for no other reason than it is contrary to the neo-Malthusian viewpoint of population growth (see Chapter 1). It should be pointed out, however, that Boserup's advocacy of population growth is qualified. She claims that the forced pace of rural modernization, brought on

---

[1]Ester Boserup, *The Conditions of Agricultural Growth* (Chicago: Aldine, 1968). See especially pp. 11–27.

30–2   Hypothetical Difference in Yields and Points of Diminishing Returns

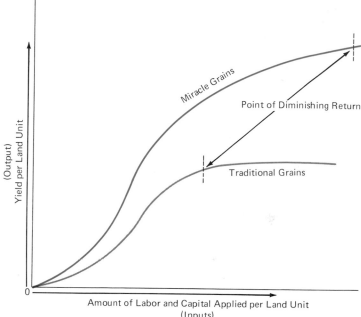

by rapid population growth, is a short-term proposition. Once the transformation process is well under way, population growth, she predicts, will begin to taper off.

Land Reform

One recent government effort is a program to restrict the size of rural landholdings and to encourage owner-operated farms. Holdings deemed too large are partially nationalized, and the excess land is redistributed to the laborers who have cultivated for the former owners. Despite the program, however, large landholdings persist, and the majority of India's rural people own little or no land. Presently about 40 percent of the rural farmers possess less than 2.5 acres (1 hectare), and another 22 percent are landless. On the other hand, about 4 percent of the rural landowners control over 30 percent of the area under crops.[2]

Such a pattern of landownership often leads to a two-tiered rural system. Cultivators of small plots remain subsistence-oriented, use traditional methods

[2] Ashok Dutt (ed.), *India: Resources, Potentialities and Planning* (Dubuque, Iowa: Kendall/Hunt, 1972), p. 58

and tools, and have little opportunity for social and economic betterment. Owners of large cropped areas generally posses wealth and knowledge to effectuate advanced farming methods and produce for the larger national and international market. In India, however, the large landholdings usually do not use modern methods. Nevertheless, the difference in wealth and social status is great, and increasing population pressure on the limited land resources of the small-plot farmer has led to increasing agitation and upheaval in the countryside.

Division of large land units into small plots may lessen, for a time, the frustrations of the Indian peasants. Small plot agriculture, however, presents problems of its own. Small plots often are not large enough to permit the farmer to save any of his income, and consequently he is unable to survive economically a poor agricultural year. Mechanization, even with garden-type equipment, is inhibited; purchase and use of chemical fertilizers are limited; and dissemination of improved techniques is difficult. In irrigated areas, water and drainage channels are tied closely with small field size. These channels and the paths necessary to provide access to each field greatly reduce the amount of land that can be cultivated. Thus, while fragmentation of landholdings may be desirable from a political and social point of view, it may be economically unsound.

India's Green Revolution

Far-reaching changes are under way in rural India, and much of this change is associated with the Green Revolution. Although not a panacea for all the problems, the Green Revolution can play a significant role in the process of development under way in rural India. Perhaps in this revolution lies the greatest promise for India's rural modernization program.

Slogans and catchwords often come into common usage, and their very simplicity and the promise of hope lead to abuse and misuse. The Green Revolution certainly fits such a mold. In brief, the concept of the Green Revolution is a thoroughgoing process of agricultural transformation with the goal of maintaining agricultural production at a level well ahead of population increases. There are two main elements of the Green Revolution. First are the introduction and use of recently developed high-yield strains of food grains, principally rice. Second is the establish-

ment of institutional supports to promote increased yields. These supports include irrigation projects; accessibility to fertilizer, insecticide, rodent controls, and herbicides; and improved cultivation techniques. Also included are education programs designed to stimulate the diffusion of new ideas and techniques in rural areas.

The introduction of new high-yielding varieties of grain, often called miracle grains, is the best-known aspect of the Green Revolution. A concerted effort was made in the 1950s and 1960s to improve the yielding qualities of rice, maize, and wheat. Plant geneticists eventually developed varieties that produced more seed and would respond better to the application of fertilizers. A particular characteristic bred into these miracle grains is that they are susceptible to more intensive cultivation; in essence, the point of diminishing return is delayed with increasing inputs. This feature of the new varieties is illustrated in Figure 30–2.

To date, these miracle grains, especially rice, are not widely cultivated in India. Nevertheless, total grain production has increased during the last two

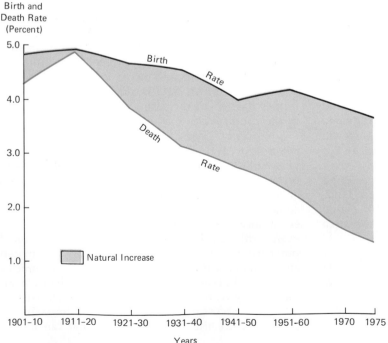

30–3 *India: Birth and Death Rate Trends, 1901 to 1975*

SOURCE: Adapted from Ashok Dutt (ed.), *India: Resources, Potentialities and Planning* (Dubuque, Iowa: Kendall-Hunt, 1972).

decades. The average yearly increase has been nearly 3 percent, a figure somewhat above the annual rate of population growth. Although the area under cultivation has expanded, much of this increase is attributable to improved yields per land unit. Yields, however, are still but one-half those of China and far below those of many other nations (Fig. 4–3).

There are four reasons for the difference in yields between China and India. First, China has a much higher proportion of land under irrigation (34 percent versus 18 percent). Second, China uses nearly two and one-half times as much fertilizer on about the same amount of cropland as India. Third, China's best land is generally used for grains, but some of India's best and irrigated land is used for other crops such as cotton and jute. Fourth, China double-crops its land much more than India. These differences indicate that India has a considerable potential for increasing grain production. If the use of miracle grains is expanded also, India's agriculture can be made much more productive.

Can India Feed Itself?

A program of rural transformation has been under way for some time in India. It could be argued that this transformation has both failed and succeeded. On the one hand, India's food production has, to a large extent, kept pace with the rapid population increase. Despite the frequent importation of foodstuffs, India has stood on the threshold of food self-sufficiency. Great gains are being made; unfortunately, these gains in production are offset by the increased numbers of people. Yields per land unit remain low, which suggests that the agricultural transformation is still in its early stages. The sheer size of India's peasantry reminds us of the monumental task of transforming traditional agriculture.

It is difficult to predict India's future rate of population growth. India's attitude at the 1974 United Nations Conference on World Population Growth disclosed a skepticism of Western views that the nation's population is too large and growing too rapidly. Yet it seems certain that if population growth continues unchecked, the outlook is not happy. Although India's population may be able to feed itself, the average peasant may remain impoverished and undernourished with no hope of ever achieving levels of personal consumption that equal those of the United States seventy-five years ago.

India's Population

Discussing population and population growth in India and South Asia is never easy, and for many Westerners the subject is most depressing. Second among the world's countries in population, India is estimated to contain nearly 623 million people. To this number, roughly 15 million new citizens are added each year. The latent potential for great human growth is being realized. Only China matches such huge annual increases in the population. The question is how long can such growth be sustained, and can India continue to feed such rapidly increasing numbers.

India is not alone in rapid growth. South Asia, as a whole, has almost 850 million people with an annual growth rate of 2.4 percent. To a large extent the subcontinent rivals the great concentration of people in East Asia. Indeed, if the high rates of growth continue in India, Bangladesh, and Pakistan, South Asia soon may contain more people than East Asia.

Earlier it was noted that approximately 70 percent of India's population is directly dependent on agriculture for its livelihood. Under present economic conditions, it is necessary for the agricultural sector to provide employment for huge numbers of new workers every year. At the same time, agriculture is attempting to modernize and become more efficient by substituting capital for labor to improve yields. A contradiction is obvious between the realities of India's current demographic situation and the nation's desire and capacity to initiate change in its traditional agricultural system. Of perhaps greater significance and concern is the assertion that India's per capita production of foodstuffs has declined by one-third in the first six decades of this century.[3] If this claim is true, India is in serious difficulty, and the urgency of curbing population growth becomes even more obvious.

Birth, Death, and Growth Rates

According to United Nations estimates, life expectancy for the average Indian is fifty years. The nation remains in the second stage in the demographic transition (Figs. 30–3 and 1–4). Despite a considerable

[3]Ibid., p. 96.

effort to control births, the birth rate is 3.5 percent. The death rate of 1.3 percent reinforces the suggestion that India has much to accomplish if the rate of growth is to be reduced.

Much public education and many inducements have been provided India's peasantry to practice various forms of birth control. Free transistor radios are given men who volunteer to have vasectomies, and animated cartoons depict the disadvantages of too many people in a family (Fig. 30–4). Government workers are provided special incentives for small families or penalized if they have too many children (more than two). In 1975 and 1976 serious consideration was given to the use of legal methods for controlling family size, but the use of law for such purposes is extremely difficult and always creates controversy. The problem appears to lie not so much in India's disapproval of birth control or neglect of family planning programs, but rather in the sheer magnitude of the task of spreading effective family planning concepts and techniques to the more than 500,000 villages. Family planning ideas have been provided in most cities and many adjacent villages, but the main task lies in spreading the work throughout rural India.

Continuation of the high population growth rate will give India a population of more than 800 million by 1990. Figure 30–5 offers several alternative population growth curves for India based on changes in fertility levels in the coming years. Based on what

30–4 *India: Family Planning Cartoon*

Family Planning Can Help You

The Indian government provides family planning instruction and contraception devices in most cities and many villages. Much public education and many inducements are provided to encourage birth control. (United Nations/ILO)

we know of India's cultural values and the esteem that Indian peasants place on large families, it is unlikely that comparable birth and death rates can be achieved in this century. Consequently, it seems likely that India's population will reach close to 1 billion by the turn of the century.

Population age and sex structure as evidenced in the population pyramid support the idea of continued growth (Fig. 30–6). The familiar conifer-shaped graph with the wide skirt at the bottom of the pyramid parallels the pattern found in most poor countries (Fig. 4–12). There is a marked concentration of people in the younger age groups. These young people provide an ever enlarging potential for future growth as more people enter the childbearing years. It is this

pattern, perhaps as much as anything about India's demography, which suggests little or no letup in the rapid population growth in the near future.

Population Growth and India's Future

The picture of India's population growth is not attractive, but there are some hopeful signs as well. For one thing, the speed and pace of change in today's developing countries are much greater than for comparable demographic experiences in Western history. New technologies and techniques of communication, education, and public health appear continually, and the rapidity of change in values and traditions has been much increased. Where the exigencies of the situation demand change, the pace of change may be rapid. If the information available from China is an accurate indicator of recent events there, it suggests that rapid and far-reaching demographic changes are possible even in large developing states. The prospect that similar events will happen in India is another matter. Of all the impediments to rapid economic growth and development, population growth is one of the most serious.

Where the People Are

The population of the subcontinent is unevenly distributed (Fig. 30–7). Several conspicuous regions of extremely dense clustering stand out. The middle and lower Ganges basin, the lower Brahmaputra basin, the coastal littoral along the eastern and western Ghats, and the Indus drainage area are the most densely inhabited regions. This pattern appears to reflect the attraction of people to the most favorable agricultural environments, for here are concentrated dependable sources of water and productive alluvial soils. Such a setting could be shaped and manipulated into a set of productive farming regions that support

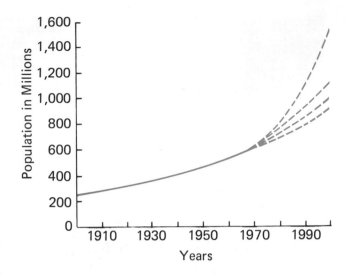

30–5 *India: Recent and Projected Population Growth*
SOURCE: Adapted from Ashok Dutt (ed.), *India: Resources, Potentialities and Planning* (Dubuque, Iowa: Kendall-Hunt, 1972).

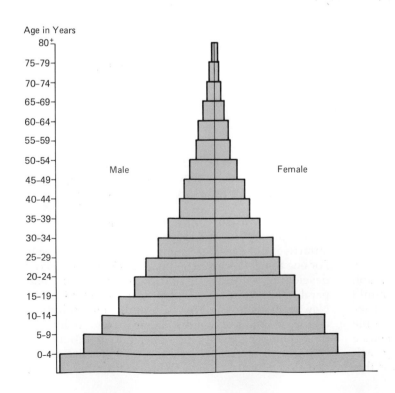

30–6 *India: Population Pyramid*
SOURCE: Adapted from Ashok Dutt (ed.), *India: Resources, Potentialities and Planning* (Dubuque, Iowa: Kendall-Hunt, 1972).

a large number of people and sustain frequent croppings of intensive food grains such as wet rice.

Peninsular India also supports a large population, but it is less dense than the rice-growing alluvial lowlands of the great river floodplains and the coastal littoral. For one thing, much of the peninsula is given over to crops such as wheat and millets. Yields per land unit are lower than in the rice-growing areas; thus, peninsular India cannot support as high a population density as is found in rice areas. In the drier desert and steppe regions of western India and Pakistan (Rajastan, the Rann of Kutch, and Baluchistan), populations are sparse, a reflection of the marginal nature of the environments.

Centralized Planning and Economic Growth

India, until Mrs. Ghandi's establishment of authoritarian control in 1975, provided an instructive model of a democratically based economic development program and of a growth strategy that was rooted in open legislative processes, a pattern that appealed to many Anglo-Americans and West Europeans. The diverse interests and social groups that contributed to and benefited from this program in many ways paralleled those that operate in the United States.

30–7 *South Asia: Population Distribution*

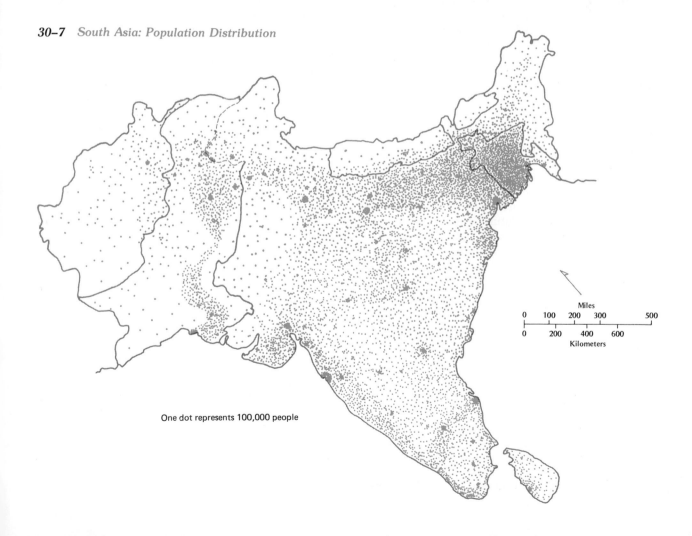

One dot represents 100,000 people

Miles
0   100   200   300   500

0   200   400   600
Kilometers

On the other hand, India's method of operation differed greatly from that of China's, which is regimented and authoritarian.

Since independence in 1947, India has developed its own form of central planning and government control to promote development. Such control has been expressed mainly in a series of five-year plans that have focused on the nationalization of key industries and services: iron and steel making, heavy machinery manufacturing, fertilizer and chemical production, irrigation systems, and transportation and communication networks. In general, this planning was done to channel large investments to these enterprises. The important thing to remember, however, is that the plans are made by a national planning commission, and this commission has acted within a democratic context in which local, regional, and national interests are considered. Such procedures are not entirely different from those that operate in the American federal political system, although the United States has no single national commission charged with overall economic planning.

The Five-Year Plans

Since 1951, controversy has revolved over the approach and the emphasis given the various sectors in the five-year plans. Controversy has centered on one principal issue: the relative attention and investment provided manufacturing as contrasted with agriculture. An important subissue is to what extent the manufacturing sector is to be nationalized. After more than two decades of economic planning, it is clear that the Indians have emphasized a heavy industry approach and invested heavily in this sector. Such a policy is similar to Soviet planning in the 1920s and 1930s and is aimed at making India self-sufficient in basic goods. Such an approach is termed import-substitution, and if successful should reduce reliance on external suppliers.

Agricultural Investment versus Industrial Development

Most Western planners have argued that India should focus more attention and investment on modernization of agriculture. Many Indians view an emphasis on agriculture, however, as extending the nation's dependence on foreign sources of basic equipment and industrial goods. It is, therefore, unappealing. Moreover, the Indians consider an agricultural emphasis will prolong India's political and military impotence in world affairs. Such a disagreement may have been a factor in the increased misunderstanding and mutual criticism that have characterized United States–Indian relations during the past two decades.

In recent years, rapid population growth and periodic droughts have shown the necessity of increasing agricultural production rapidly. Freedom from reliance on foreign manufacturers has not been achieved, and India has sporadically looked toward external sources of foodstuffs to meet internal deficiencies. One consequence has been increased investments in those industries and services that support agriculture. An associated policy, aimed at providing more rural jobs in community development, self-help, and education-related projects, has received increasing attention. Over the years it has become obvious that creating industrial jobs in large cities will not suffice to provide employment for the additional millions who enter the labor force each year. The real crisis for India's development program continues to lie in the countryside. Here are more than 70 percent of the population, and some means must be found to satisfy not so much their long-term aspirations for education and a better life but rather their immediate needs for food, clothing, and housing. Indian planners and politicians are coming to realize that urgent solutions are required in rural India if the more sophisticated goods of industrial growth and national development are to be delivered.

The Industrial Economy

Industrial development and expansion in India evolved out of a tradition of cottage manufacturing associated with metal smelting and consumer goods. British plans for India in the nineteenth century chose to neglect these native industries and concentrated on new industries that complemented manufacturing in Great Britain. British interests lay in introducing labor-intensive industries such as textiles or industries that relied on local commodities, for example, jute processing, sugar refining, and leather tanning. The British hoped to sell India the more sophisticated manufactured goods. British colonial rule ensured that such a policy did operate.

Britian's colonial policy in India restricted the range of production facilities, and most were associated with processing agricultural products. World War I with its attendant shortages brought some expansion in industries such as iron and steel, cement, and paper. World War II intensified these developments, for India was a rear supply area for the China theater. Despite these improvements, India on the eve of independence had few machine-building and chemical industries. The shortages of specific industrial enterprises meant continued dependence on foreign suppliers and the expenditure of scarce capital, a situation that many Indian planners believe seriously retarded economic growth.

Another consequence of British colonial policy was the concentration of industrial production in the three major port cities of Calcutta, Bombay, and Madras. The Indian government since independence has viewed this concentration as undesirable. A key feature of industrial planning since independence has been to spread production centers throughout the nation while continuing to strive for the least-cost location of such centers and industries.

Planned Industrialization, 1947 to the Present

A new policy of industrialization followed the establishment of an independent India in 1947. In line with the desire to achieve greater self-sufficiency, India has followed an import-substitution strategy. Planners have sought to develop industries to supply the entire range of producer and consumer goods except for highly specialized luxury goods. This range extends from the well-established textile, jute, and sugar-processing mills to locomotive, shipbuilding, and aircraft factories and even includes computer manufacture and the processing of materials for atomic energy. Almost the entire range of goods produced in the major rich nations are produced in India, although quantities are often modest.

Production of a wide range of goods is possible by virtue of the domestic market's large size and the ability of producers to achieve economies of a scale sufficient to keep prices low. India, like most other large countries, is reasonably well provided with resources and materials for industrial production. Thus, the nation does not have to rely on imports to provide most of the ingredients for its industrial enterprises.

Industrial Regions

Since independence, a major effort has focused on distributing new industry throughout the nation. The basic objective is to distribute industries closer to raw materials and markets and thus spread centers of production and growth more evenly.

Most prominent among the recently created industrial regions is the cluster of heavy manufacturing in the Damodar Valley region (Fig. 30–8). Symbolized by the planned industrial city of Jamshedpur, this center of iron and steel production is largely raw material oriented. Jamshedpur is close to the major iron ore and coal fields of the subcontinent. Most of the recently established heavy metals and associated chemicals and machining industries are also located here. With each new project, the raw material and assembly factors that favored this concentration have been reinforced by economies of scale and agglomeration. The growth and concentration of Indian industry in the Damodar Valley have been impressive enough to make this region one of the major metal and machinery production centers not only in Asia but also in the world.

Industrial Location and Clusterings

Some industries do not benefit through clustering. Many are oriented toward local or regional markets or are best located in association with agricultural or other raw materials that are dispersed more broadly. Textile mills are common in the cotton-producing region of Rajastan, and paper mills and sugar refineries are located in peninsular India near the areas of heaviest sugar cane production. At the same time, older centers of industry continue to attract new establishments, for they possess the advantages of large markets, good accessibility and transportation linkages, and large and relatively skilled labor pools. The dominance of older centers is increasingly challenged. The policies aimed at creating clusters of industry focused on smaller cities suggest some reorientation of industry in the future.

The Growth Pole Theory

If industrial dispersal succeeds over the long term, it should provide India with a more equitable distribution of economic activities and wealth. Redistribution may help reduce the criticism leveled against the concentration of wealth in a few great cities, a sharp contrast with the huge, impoverished, traditionally oriented rural hinterlands. India's desire to decentralize industry and encourage growth in many smaller cities corresponds with the growth pole development strategy. This strategy aims to promote growth in many centers in an effort to spread economic activities and benefits throughout the state.

By clustering new industries in selected centers over a wide area, jobs are created in depressed areas and provide alternative opportunities for labor. Once established, these new industries spur other economic activities (multiplier effect). These centers and their economic activities grow as the forces of economies of scale and agglomeration become increasingly important. Moreover, by careful selection of specific industries in a center, the planner can exert some control and direction over resource development and

30–8 *South Asia: Surface Transportation and Industrial Concentrations*

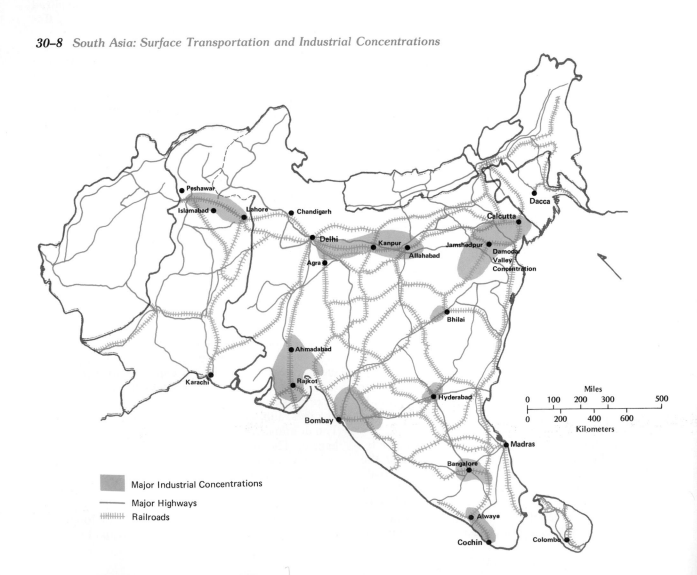

Jamshedpur, India's iron and steel center, is a planned industrial city and stands as a symbol of India's industrial growth. Iron and coal deposits are located nearby.

economic growth. Increased and diverse economic activity in turn becomes an agent of cultural change. Traditional behavior patterns are modified. In India's case, the caste system may be broken and opportunities for individual social mobility enhanced. Individual freedom of action is consequently improved. The effects of both economic and social change are then expected to be diffused throughout the surrounding countryside. By judicious use of the growth pole theory, uneven concentrations of wealth and production are avoided or at least reduced. In addition, decentralization is thought by some to speed the pace of economic growth and development.

The growth pole theory is controversial, and regional development experts have different opinions on how the theory should be used. For example, should poles be already existing centers, or should they be new towns? How large should the growth pole centers be? There are many such questions, and definitive answers are not yet available. Whatever the merits or liabilities of such a strategy, it is difficult to argue against the strategy's political benefits. In a fragmented political situation such as India's, with its diverse local and regional interests, there are benefits for most if not all regions. The economic benefits balance regional interests and help to oil the national political machinery and provide it with a greater chance for success.

Transportation Networks

India has possessed a large and dense railway system for more than a century. This system is the longest, most dense rail network in Asia, and it is one of the longest systems in the world. The usefulness of the rail system in promoting urban and economic growth is hindered, however, since it is not interconnected and was designed to further colonial policies of stimulating exports and imports rather than internal development.

Historically the rail system has focused on the four major urban centers: Delhi, Bombay, Calcutta, and Madras. These cities were linked together by a skeletal framework of broad-gauge lines. Each city, in turn, was tied to its large hinterland by a network of different gauge lines. Thus, a fully integrated rail system did not exist, nor was there a means to stimulate growth at smaller centers in the interior. Such a varied rail system typified British colonial practice and variations in British administrative and economic control.

Recently, attempts have been to standardize gauges where feasible and not too costly, but a long-term project of gauge standardization is yet to be devised. After independence, the system was divided into a series of regional railway networks under the direction of the government-owned Indian Railways. Despite problems of administration and operation, railways continue to be the most efficient and cheapest method for hauling goods and people throughout the nation.

The Highway System

In recent years attention has been given to highways as an alternative to rail transport. A long-term plan has been made for construction of an integrated system of all-weather roads. These roads may provide an effective means of linking together the many areas of India and provide an effective agent for stimulating growth and economic activity in smaller cities.

Long-term highway development plans are designed to integrate most of the settled area of India through a network of main and secondary highways. The network is to reflect population densities. In the thickly settled agricultural regions, a dense network of interlocking roads is to be built to place all villages within 2 miles (3.2 kilometers) of a secondary road and 5 miles (8 kilometers) of a main highway. In sparsely settled nonfarming areas, the network is to be thinner but still link the areas in a systematic manner.

A twenty-year highway plan is designed for completion in 1981. By that time, India hopes to have a highway network about one-half the density of that of the United States in 1970, and to have more than 50 miles (80 kilometers) of highway for every 100 square miles (259 square kilometers) of area. The system may appear insufficient by West European standards, but compared to most other developing nations, and especially large states, India will have an extensive network of highways.

Transportation and Development

The evolution of transportation networks commonly is seen as one of the primary infrastructural supports that underlie economic and political systems. Without access to its territory and effective linkages that integrate regions, a state cannot function effectively or promote the kinds of programs necessary for proper economic growth. A good transportation system is imperative in a large, developing state. For example, in China and the Soviet Union the expansion of transport networks has been critical to economic growth policies and to the political-military goals of national strength and power.

India offers a somewhat contrasting pattern. A fairly dense transportation network was one of the consequences of British colonialism. It provided a broad base on which to develop a locational strategy of economic growth. Such a base permitted India to concentrate investment on other problems and simply improve the transport system that already existed. In the search to identify bottlenecks and critical problems in India's development path, it is unlikely that transportation will loom largely as one of the key difficulties in the long struggle for economic growth and improved levels of living.

Urban India: Contrasting Points of View

Although perhaps only one-quarter of India's population resides in cities and towns, it should be remembered that this one-quarter includes close to 160 million people. India, by this standard, must rank as one of the world's leading urban states. These millions are contained in a variety of different-sized cities and towns. The prospects of future urban growth have given rise to several competing schools of thought.

Some planners and scholars have suggested that India is neglecting to influence urban growth and thus can look forward to huge concentrations in a few great metropolitan areas: Calcutta, Delhi, Bombay, Madras, and Hyderabad. These scholars also suggest this concentration is bad, as income levels will be inequitably concentrated, and increasing urbanization will destroy the traditional social patterns focused on small, more personalized villages. They point to future dense crowding in huge supercities with attendant social breakdowns and note that political stress already is experienced in some of the large cities. Calcutta is often described as a poignant example.

Calcutta and Bombay

Greater Calcutta with its 7.1 million people is probably the most frequently cited example of everything wrong with big-city life and activities in a developing state. The poverty and disease of Calcutta, and the congestion in its streets over which cars speed and sacred cows meander, have been chronicled so frequently that they need little repetition. Political rallies and urban violence also are no strangers to Calcutta's urban scene, and many sympathetic Westerners despair over how hopeless is this largest of India's urban centers.

Bombay, by contrast, is sometimes viewed as an example of the positive aspects of urban India. The city is smaller (6 million people) and surrounded by a less densely populated and more prosperous agricultural hinterland. Bombay has always seemed one of

India's most advanced and progressive places. Large and modern factories staffed by an industrious and relatively well-fed labor force create a large industrial output. These industries in turn support a thriving commerce and a growing middle and upper class of professional and entrepreneurial people. Cultural and educational institutions abound, and the city, with its outward gaze and booming economy, is among India's most cosmopolitan centers.

Such a view, with its typecasting of individual cities, is based as much on image as on fact and to some extent oversimplifies the realities. Both cities have good and bad points, and both have examples

Downtown Calcutta, a city of 7.1 million people. (United Nations)

Calcutta has been characterized as a city of poverty and disease. Beggars and street vendors line many of the side streets. (*United Nations*)

of growth and progress and of poverty and despair. Perhaps most pertinent is that two views exist on the role of cities in India's future growth strategy. These two cities, as is true of all of India's cities, will continue to grow.

Metropolis versus Town

The alternative to metropolitan growth is to concentrate new economic activities in smaller cities and towns dispersed throughout the country (growth poles). Some planners claim that the greatest production efficiencies are achieved in the larger cities with good transport systems and skilled labor forces. Others assert that towns and cities in the 250,000 and smaller size range are big enough to provide the economies of scale necessary for efficient production. The latter argue that India should plan for more of its future economic growth in the smaller centers and spread the wealth more equitably. Such greater dispersal also appeals to local and regional political interests and satisfies the sectional interests of the Indian political system.

It is too early to predict exactly what will happen in the future, but already it seems that India's gigantic population will require the use and development of both styles of urban growth. The great metropolitan centers can be used intensively for the production of specific goods, especially those that are export oriented. The smaller cities and towns can continue to satisfy local needs and demands, both commercial and industrial, as they always have. To the extent that the economy modernizes, these local and regional production centers will have to improve their levels of sophistication as they too are drawn increasingly into the national and even the international economy.

Further Readings

CHEN, KUAN-I, and UPPAL, J. S. (eds.), *Comparative Development of India and China* (New York: Free Press, 1971), 404pp. This volume, through a series of essays, compares several aspects of economic growth and development between India and China.

DEAN, VERA M., *New Patterns of Democracy in India* (Cambridge, Mass.: Harvard University Press, 1969), 232pp. This general volume on contemporary India covers the economy, society, and politics of India. It also contains a selected bibliography.

FRANKEL, FRANCINE R., *India's Green Revolution: Economic Gains and Political Costs* (Princeton, N.J.: Princeton University Press, 1971), 232pp. Frankel examines critically some of the economic and social implications of India's Green Revolution.

MEIER, RICHARD L., "Preparing Indian Coastal Metropolises for Accelerated Development," *Asian Survey,* 11 (1971), 506–521. Meier proposes a strategy to exploit the coming urban age in India.

MELLOR, JOHN W., and others, *Developing Rural India: Plan and Practice* (Ithaca, N.Y.: Cornell University Press, 1968), 411pp. This volume focuses on the modernization of Indian agriculture.

MOORE, RAYMOND A., "Military Nation Building in Pakistan and India," *World Affairs,* 132 (1969), 219–234. This article examines the role of the armies of Pakistan and India in promoting development in their respective countries.

MYRDAL, GUNNAR, *Asian Drama: An Inquiry into the Poverty of Nations* (New York: Twentieth Century Fund, 1968), 2284pp. A famous Swedish economist examines planning problems in economic development in East and Southeast Asia.

NEHRU, JAWAHARLAL, *The Discovery of India* (Garden City, N.Y.: Doubleday, 1959), 426pp. One of India's most famous twentieth-century statesman outlines his view and vision of the many faces of India.

SRINIVAS, M. N. (ed.), *India's Villages* (Bombay: Asia Publishing House, 1969), 221pp. This volume of essays contains descriptions of most aspects of traditional village life in India.

VENKATASUBBIAH, H., *The Anatomy of Indian Planning* (Bombay: Vora, 1969), 218pp. An Indian journalist appraises critically the nature of the Indian planning process.

South Asia: The Peripheral States

31

cultural heritage is associated more with Llamistic Buddhism than with the Hindu or Moslem religions more commonly associated with the subcontinent. Sikkim, formerly a semi-independent nation and similar to Bhutan, was incorporated into an Indian state in 1975. Afghanistan is on the periphery of South Asia and has long been a cultural transition area bridging the gap among South Asia, central Asia, and the Middle East.

Included within these six countries is a considerable range in levels of development. All are poor by Western standards, but at least one of them (Sri Lanka) has demonstrated some capacity to initiate programs designed to produce rapid progress. In other cases problems related to recent political disputes and military conflict have inhibited growth. In the case of Bangladesh most of the physical infrastructure was destroyed in recent fighting. Such destruction is serious for future growth; however, with the resolution of political problems and issues, both Bangladesh and Pakistan may now focus on the pressing problems of improving the levels of livelihood for their respective populations.

Afghanistan

INDIA DOMINATES THE South Asian subcontinent, but several of the bordering countries are themselves large. Two rank among the ten most populous countries in the world (Bangladesh and Pakistan), and three others (Sri Lanka, Afghanistan, and Nepal) are among the fifty largest countries in terms of population.

The creation of Bangladesh and Pakistan resulted from special political and religious problems. The politically divided region of Kashmir and Jammu, whose status is still ambiguous, is another illustration of the political issues that have fragmented the subcontinent. Nepal and Sri Lanka are closely akin to India in cultural makeup, but each has managed to maintain autonomy and independence. The small Himalayan kingdom of Bhutan is a border state that shares some of the physical characteristics of Nepal. The origins and ethnic composition of Bhutan, however, are more closely connected with Tibet, and the

Afghanistan is not climatically within Monsoon Asia. Its location is transitional between South Asia and central Asia and has long figured prominently in the political and military history of the Indian subcontinent. Since the time of Alexander the Great, the strategic location of Afghanistan with its command of the mountain passes that link eastern, southern, southwestern, and central Asia has attracted conquerors and rulers. The recent history of the country reflects the continuation of this competition between British and Russian interests in the nineteenth century and early twentieth century and more recently among the Soviets, Indians, Pakistanis, and Chinese.

The Physical Basis for Livelihood

Most of Afghanistan is high and enclosed, similar to the intermontane basin region of the western United States (Fig. 31–1). With the exception of the Kabul River and its tributaries which flow east into the Indus system, all other drainage in Afghanistan is internal (no outlet to the sea).

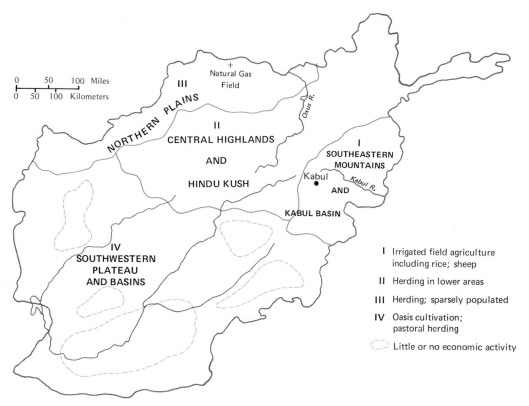

31–1 *Afghanistan: Physical and Agricultural Regions*

The dominant landforms of Afghanistan are the Hindu Kush which separates the plains of northern Afghanistan along the Oxus River from the rest of the nation. Despite this division, the country has managed to maintain a strong political coherence. South of the Hindu Kush are a great desert plateau and basin region in the southwest and south and a large alluvial basin centered on Kabul in the east. The Kabul Basin is the most densely settled part of the country and forms the heartland of Afghanistan.

Most of Afghanistan is dry, and precipitation averages 20 inches (51 centimeters) or less. This precipitation is concentrated in the winter and early spring, and much of it falls as snow. The seasonal distribution is the reverse of the monsoonal pattern found farther south and east. Daily temperature variations and seasonal ranges are great. These temperature ranges further illustrate the nonmonsoonal nature of the climate.

Isolation and a rugged land surface have impeded the introduction of technical innovations and the speed of economic change and growth. The difficulties of surface transportation and the sparseness of the road network illustrate this deficiency. There are no railways in Afghanistan, and the major roads serve only the main urban centers. Large sections of the country have no modern means of transportation, although in recent years a great deal of investment has been made in transport improvement. The impoverishment and extremely low level of economic development are not surprising in view of the impediments to internal movement of goods, ideas, and people.

Afghan Society

Afghanistan comprises a series of different tribal groupings and is dominated by the most numerous of these, the Pushtu-speaking Pathans, who are concentrated in the Kabul Basin. Of the nearly 20 million

Afghans, probably 90 percent are of Caucasian stock, and the remaining 10 percent are Mongoloid. Although strong dialectal differences exist among the major tribal groups, a common religion (Islam) provides a cultural cement that is very powerful in holding this traditional society together.

One of the most striking features of Afghanistan is its demography. The country possesses one of the highest annual birth and death rates in the world, 4.3 percent and 2.1 percent, respectively. Such a pattern suggests that Afghanistan remains in the initial stage of the demographic transformation. When coupled with a per capita GNP of $80, Afghanistan emerges as one of the poorest and least developed states on earth.

The Changing Economy

Subsistence farming of grains, fruit, and a few specialty crops, along with animal herding, are the traditional mainstays of the Afghan economy. More than 80 percent of the population is employed in these activities. In the Kabul Basin, the most productive region in Afghanistan, irrigation is commonly practiced with an intensive agricultural system that includes paddy rice. Recent agricultural innovations introduced from elsewhere such as new seed strains and better techniques of cultivation are expected to increase agricultural productivity in Afghanistan and meet the future food demands of a growing population. Although Afghanistan possesses reserves of several important minerals, only natural gas production is significantly developed. The Soviet Union has assisted with the production of natural gas and purchases most of the gas.

Economic Development and the Five-Year Plans

Since 1957, Afghanistan has followed a series of government-formulated five-year plans. These plans were designed originally to bring the country into an age of industrial modernization and to accelerate the pace of economic development. Initially the major focus was on capital improvements: roads, bridges, irrigation systems, and power production and transmission. Beginning with the third Five-Year Plan in 1968, however, more attention has been focused on social and educational problems. Funds recently have been funneled into improving literacy levels and into the commodity-producing sectors of the economy. Annual economic growth rates of 4 percent or better are sought.

These five-year plans have relied heavily on foreign financing, most of which has come from the Soviet Union, the United States, and West Germany. These countries and the World Bank also have assisted in arranging some of the foreign trade and special external purchases. Afghanistan also has helped pay for some of its modernization program by exports of cotton, wool, rugs, fruit, and natural gas. It seems likely, however, that the country must continue to rely on outside aid to finance economic development for some years to come.

Afghanistan is an extremely poor, landlocked, buffer state that has only recently begun the push toward modernization. The process of modernization

Kabul, Afghanistan, is that nation's largest city. The city is located in the country's largest agricultural area. In the background lies the Hindu Kush. (United Nations)

is long and difficult. Yet Afghanistan, with its 250,000 square miles (648,000 square kilometers), is not small, and it possesses a considerable resource base. Many problems remain such as how to bring into the twentieth century a society that until recently was largely feudal. How to prepare this society for the changes and challenges of an age of industrialization and technology and how to finance this shift are the basic questions that confront Afghanistan's contemporary leaders.

Pakistan

What began in 1947 as a political problem of partitioning the British colony of India's Moslems and Hindus into two nations has had serious economic consequences. The political divisions and economic troubles associated with the partitioning have continued, and violence has erupted periodically. Most such outbreaks have involved Indian and Pakistani forces fighting over control of Kashmir. The latest outbreak, however, involved the eastern territory of Pakistan, a Bengali-speaking region. Unfortunately for Pakistan, the linguistic and sectional differences among Urdu- and Bengali-speaking Moslems overcame whatever affinities based on religion existed previously. The result was that East Pakistan, with help from the Indian Army, established itself as an independent state. Pakistan was left as a single territorial unit focused on the Indus River floodplain.

Although the reasons for the political fragmentation of Pakistan into two independent states are many, the basic difficulty centered around the position of one group, the West Pakistanis. The Urdu-speaking West Pakistanis dominated the politics and military affairs of the nation, although their popular support was a minority position. The numerically dominant Bengali-speaking East Pakistanis were to some extent exploited and, as they saw it, oppressed. When new elections in 1970 were won by the East Pakistani Awami League Party, disagreements broke out, and fighting followed. The state ultimately fragmented into two political entities: Pakistan (the western sector) and Bangladesh (the eastern). Such fragmentation is another example of the diversity and variety of the Indian subcontinent with far-reaching political, social, and economic consequences for a large number of people.

The New Pakistan

As the President of Pakistan is fond of pointing out, the Pakistan that emerged from the political division is considerably different from the joint state that existed prior to 1972. The differences are political as well as economic. First, the new Pakistani state is patterned politically along British lines, to which is added an interesting blend of local religious feeling and socialist economic doctrine. The economy has become increasingly socialistic, with less freedom for the free enterprise sectors of industry, commerce, and finance. Such a socialist pattern, to some extent, may reflect Pakistan's attraction to China and its policies of state control throughout the economic system.

Contemporary Society and Fragmentation

Although separation from East Pakistan has led to the presumption of greater political and social coherence in Pakistan, other problems associated with a large area and a heterogenous population have become more apparent. The large, arid Baluchistan region west of the Indus floodplain, with its various tribal interests and political groupings, surfaces as a major domestic difficulty. In the northwest, tribal elements that look to Afghanistan or Iran for political support and cultural appreciation create other problems. Pakistan somehow must knit together these diverse ethnic threads into a strong social fabric if the progress it so much desires is to be achieved.

Kashmir and Jammu

No problem in contemporary Pakistan is of greater magnitude than the territorial dispute with India over Kashmir and Jammu. The cease-fire line drawn in the 1950s has become an effective boundary, and India occupies the most productive and densely populated parts of the region. The resolution of the territorial claims has not yet been found. In today's political climate the stalemate over Kashmir and Jammu is not likely to change. Since division into two nations, Pakistan's position has been weakened in economic and military strength vis-à-vis India. India emerged from the Pakistani-Bangladesh dispute as the dominant political and military force on the subcontinent, a position that is unlikely to be challenged in the near future. Pakistan is no longer a viable opponent; consequently, the cardinal tenet of Pakistan's foreign relations has been to improve its relationship with India. Such a policy would seem as much a matter of survival as of a desire simply to get along with one's neighbors.

Growth and Progress

Pakistan, like its neighbors, is poor, and its past political problems have aggravated its condition of impoverishment. Most of the population is concentrated in the watershed of the Indus, and two main productive regions stand out. One of these regions is focused on Karachi, the early capital and a large port and industrial center located at the mouth of the Indus. The upper Indus region, focused on Lahore and the new capital Islamabad, is the modern political center and region of greatest contemporary planning concern.

The economy of Pakistan continues to be largely agrarian. Even where irrigation is commonly practiced, as in the Indus Basin, weather patterns often play a key role in determining agricultural output and thus the size of the gross national product. Severe flooding in 1972, for example, seriously reduced crop production and the rate of economic growth.

Agriculture is based on three main crops: wheat, rice, and cotton. Cane sugar is also important as are pulses from which cooking oils are extracted. Government objectives, as implemented by various rural development programs, are to increase agricultural production through increased applications of chemical fertilizers, better seed strains, and improved farming techniques. Agricultural imports and exports have generally balanced, and recently there has been a modest export of agricultural commodities.

Industry accounts for a much smaller percentage of the GNP than does agriculture, although it is growing more rapidly. A broad range of industrial establishments exists, but the socialization of industry may alter the past patterns of growth by reducing levels of entrepreneurial investment. Traditionally, cotton textiles have been the most important foreign exchange earner, and a high rate of investment in these activities continues.

The years since separation from Bangladesh have witnessed more rapid economic growth, yet Pakistan continues as an impoverished country with a traditional, agricultural economic system. The large area and population provide many advantages, and industrial production can be geared to exploit the advantages of scale economies. With help from the Chinese and other external sources of investment capital, along with the end of the East Pakistan problem, Pakistan is able to focus more energy on its economy than was possible in past years.

Bangladesh

The creation of a new country is never simple or easy. In the climate of South Asian politics and society, it is especially troublesome to create permanent loyalties and symbols so necessary to the evolution of successful statehood. Events of recent years have been especially difficult and frustrating for the citizenry of the former territory of East Pakistan. It was against a background of political, social, and linguistic discrimination that Bangladesh emerged as a new nation-state. India's backing has been important in forging the new state, but it was with a sense of relief and satisfaction that the former eastern sector of Pakistan wrenched itself free from the confinement and pressure its coreligionists in West Pakistan had imposed on it.

The Delta Environment

Bangladesh is an unusual place. Its tropical, delta environment, limited territory, and large number of people present a host of problems. These features, and the failure of the former Pakistan government to recognize and cope with them, largely explain the desire of Bangladesh to seek its own destiny as an independent state. Bangladesh, with nearly 85 million people occupying 55,260 square miles (143,123 square kilometers), is one of the most crowded places on earth. Envision crowding 85 million people in the state of Illinois or Georgia for a reasonably close approximation of the population densities.

Most of the land area of Bangladesh is low lying and encompasses the intricate network of streams and distributaries that form the mouths of the Ganges and its confluent, the Bhramaputra. Much of the area is composed of delta lands and gives a common water character to the country. The extent of this delta environment is large and almost without parallel elsewhere on the globe.

Many of Bangladesh's people lost all their material possessions in the battles of the early 1970s that led to Bangledesh independence. Millions of people fled the fighting for refugee camps in India. (United Nations)

It is this delta environment that has permitted so many people to live on so little land. Almost two-thirds of the land area is farmed, and a very productive agricultural system has evolved based largely on the cultivation of paddy rice. Rice is not the only crop produced, but it is by far the most important. A variety of other food crops and some specialized cash crops such as jute are also produced in large quantity, but rice is the main provision of the great mass of people.

The rhythm of alternating wet and dry seasons already described for India also epitomizes agricultural life in Bangladesh; 80 or 90 inches (203–229 centimeters) of rain concentrated in the summer months provide much of the water necessary for rice cultivation. A finely developed system of irrigation supplements this rainwater and provides the controls for sustained high yields necessary to feed so many people. Unfortunately, tropical storms that originate over the adjacent warm Bay of Bengal disrupt life by causing periodic floods in many coastal areas. Destruction and loss of life are great as a consequence of the extreme crowding on the delta. One of the sad ironies is the vulnerability of so much productive agricultural land to the vagaries of natural catastrophe.

Prospects for Economic Growth

Bangladesh entered nationhood with a chronicle of suffering and disaster that appeared almost insurmountable in 1972. The war of independence caused great damage to the infrastructure. Bridges and transportation systems were wrecked; ports were clogged and not functioning; and power facilities were damaged. A great storm had caused enormous loss of human and animal life, and agricultural production had declined sharply. The United Nations sponsored relief efforts, and these were supplemented through additional donations from other foreign national and private agencies. Bangladesh was not without friends and sympathizers, and these external suppliers saw the nation through the immediate crisis.

The Future: Prospects

Shortly after independence, Bangladesh sought entry into the United Nations. Although denied entrance initially by the veto of China,[1] Bangladesh was permitted to join the various United Nations member agencies. These agencies initiated a series of projects designed to promote progress in the economy. It is still too early to predict the speed with

[1]China, because of its friendship and special relationship with Pakistan, chose to veto the admission of Bangladesh, a move very difficult to justify in view of China's claim as champion of the nonaligned countries.

which Bangladesh manages to accomplish the development goals that have been established. It is now clear, however, that the rate of recovery from the difficulties faced at independence has been slow. If these problems persist, it is only a question of time before serious political and social unrest erupt. The nation may be viewed not only as a country with many people but also as an example of a state that has failed to develop sufficient economic growth to counter the population growth.

Strategy for Future and Economic Growth

The immediate goal of restoring production on the eve of independence was a one-year plan. This plan was designed only to restore the country to its previous levels of production before the war. In 1973, a Five-Year Development Plan was adopted to promote future growth. An annual goal of 5.5 percent was set for production increases. This goal was unrealistic in view of the many natural and human problems. The goals aim mainly at the agricultural sector, an orientation that appears appropriate. The principal difficulty is the high degree of reliance on foreign capital sources. During its first Five-Year Development Plan, Bangladesh procured about 60 percent of its invest-

Creation of a new country is never easy. Bangladesh at independence was in chaos. The war for independence from Pakistan destroyed much of the infra-structure and capacity to produce. Thousands were homeless. Moreover, in 1970 a damaging tropical storm swept the region, destroying homes and crops. Many of the people of Bangladesh were literally forced to fight with the hogs for garbage. (United Nations/Wolff)

ment capital externally. It is a poor country and is not a particularly good credit risk. Nevertheless, it has found some capital assistance from sympathetic foreign sources and may rely on these sources at least for a short time. The difficulty has been that the capital has frequently been used to purchase food-stuffs, the most critical need, and there is little left for development projects.

The Himalayan States: Nepal and Bhutan

Lying athwart several of the main mountain passes that connect India with China's province, Tibet, are two Himalayan countries, Nepal and Bhutan. Nepal, the largest, is an old kingdom that has long survived both as an extension of Indian influence into the Himalayas and as a buffer separating Tibetan and Chinese influence to the north from the Indian subcontinent.

Although the Nepalese trace their history back many centuries, the forging of Nepal as a political entity dates from the late eighteenth century and the Gurkha conquest of the Katmandu Valley. In the nineteenth century, the Gurkhas extended the boundaries of Nepal far beyond today's 54,000 square miles (141,000 square kilometers). It was this expansion and territorial ambition that brought the Gurkhas into conflict with British interests. Out of this conflict the boundaries of the Nepali state were formed much as they appear today. Nepal was closed to outsiders for many decades but in recent years has opened its borders to foreigners and tourists. With the increasing number of visitors each year, the self-imposed isolation and established traditions are beginning to diminish.

Elevation and Environment

Everything about Nepal's physical environment suggests extremes, especially the dominating mountain wall that composes the northern two-thirds of the country, the Himalayas. The mountains are a series of parallel steps that ascend progressively from south to north. Along the southern edge of the state are low-lying alluvial tropical plains, the Terai, that are part of the Ganges floodplain (Fig. 31–2). The Terai is hot and humid, although in western Nepal it is seasonally dry.

North of the Terai rise, first, the Siwalik Hills, rounded foothills of the Himalayas, to an elevation of about 2,000 feet (610 meters). The hills give way abruptly to a series of sharp northeast-southwest ridge lines that cross the main structural alignments of the Himalayas. These ridges give a pattern of northeast to southwest flowing streams and associated valleys. It is in these valleys and the southeastern bordering Terai that most of Nepal's population are concentrated. Some of the larger alluvial valley bottoms are densely populated, for example, the core region of Katmundu. There the mild climate is pleasant, and little seasonality is expressed other than that of rainfall.

The Himalayas

Northward, the Lesser Himalayas rise sharply and average 5,000 to 15,000 feet (1,524 to 4,572 meters) in elevation. The highest mountains in the world lie just north of this middle range of the Lesser Himalaya. High peaks and serrated ridges are everywhere, and the common features of glaciation characterize the surface of this section of Nepal. Beyond, to the north, is the high Tibetan Plateau, a bleak and forbidding region with a special high altitude environment. In northern Nepal, people live only in the deeply incised river valleys that flow across the grain of the main mountain range. These village dwellers are composed of small clusters of tribal peoples ethnically related to Tibetans, apparently the original inhabitants of the Himalayan region.

Elevation and Rhythm of Life

Environment in Nepal correlates remarkably with different elevations, and several distinctive zones associated with elevation are noted. Closely attuned to these distinctive physiographic zones are styles of livelihood and occupancy. Such distinctions focus primarily on the character of the agricultural system and the crop associations.

The low-lying Terai in the south is a land of tropical forests; rice, sugar cane, and jute are the dominant crops. Double-cropping of food crops is common. In the large river valleys at 4,000 to 5,000 feet (1,219 to

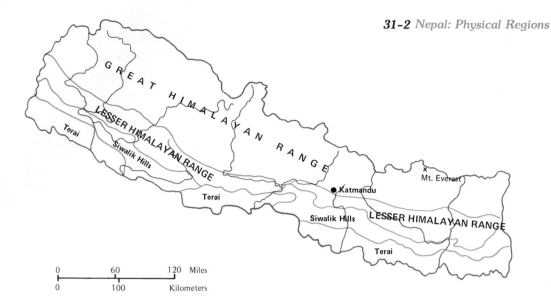

31-2 Nepal: Physical Regions

1,524 meters), rice continues to be the main food staple, although terracing of the fields is necessary. At higher elevations, where the soils are poor and stony and the winters long and cold, potatoes provide the staple food with corn and other grains as supplements. In all cases agriculture is largely subsistence, although as one descends south toward the Ganges, it is increasingly prosperous and commercialized.

Nepal's People and Society

Nepal's population is increasing rapidly and is still in the early stages of demographic transformation. With nearly 14 million people and an annual growth rate of 2.2 percent, there is no evidence that the high birth rate supporting this growth will diminish. At the same time it seems likely that the death rate of 2.0 percent will begin to drop and thus accelerate the rate of population growth in the coming years.

Most of these people are heavily concentrated in two broad bands. The Terai is the most densely populated section. Most of the rest of the people are concentrated along the alluvial valley floors associated with the major stream systems. In some of these basins, the conditions of crowding are extreme and rival those found in the floodplains of India. Nepal with only about 10 percent of its land suitable for intensive cultivation is crowded in its population-

arable land ratio. Such a statistic implies little optimism about the hopes for future growth.

Resources and the Potential for Future Growth

Only in recent years has Nepal opened its doors to foreign influences, and the reluctance to such intrusions lingers. The country, however, is being brought out of its largely feudalistic past, and the traditional isolation is being reduced. This change means a great deal to Nepal, for it not only brings ideas and new points of view, but it also provides investments to promote economic growth. Several foreign countries are eager to provide Nepal with money and technical assistance. They include India, China, the United States, and the Soviet Union. On the other hand, tourists bring in money, creating a new source of foreign exchange internally. In the past, Nepal has had to rely largely on wages and remittances made to Gurkha mercenaries employed in the British Army. In recent years, however, the British have reduced their support of the Gurkha battalions, and their number has been progressively decreased.

Nepal continues to be an impoverished country. Yet change is under way, and in terms of the pre-industrial economy and traditional society, the changes are almost revolutionary. It would be difficult to shut off the flow of new ideas and people now in motion. Such a flow suggests new patterns of society and more economic growth as Nepal struggles to modernize itself in step with its Asian neighbors.

Bhutan

Bhutan, the small Himalayan principality to the east of Nepal, lies between India and China and is most prominent for its buffer role. The native (Bhatia) population, of Tibetan origin, comprises most of the state's 1 million people. Although Bhutan has a long history of independence and autonomy, a special relationship with India exists. India manages the nation's foreign relations and sponsored Bhutan's entry into the United Nations. Bhutan exists as a client state of India.

Bhutan's physical environment in similar to that of Nepal. Most of the territory is at higher elevations, however, and the livelihood patterns are oriented toward subsistence farming and animal husbandry on the southern flank of the Himalayas. In recent years the United Nations, through its development programs, has initiated far-reaching economic changes in Bhutan, and a series of five-year plans has been fashioned. India has augmented the United Nations program; in fact, the United Nations commitments have followed the Indian investments.

Sri Lanka

Growth and Change

Of all the South Asian states, Sri Lanka (Ceylon until 1972) is perhaps the most modern and certain of the direction in which its society and economy are moving. Although poor, it has initiated a massive effort to cut birth and population growth rates and to promote improvements in the agricultural economy. Ceylonese agriculture, with its large plantation sector, has long been commercialized. The plantation sector, however, has been plagued with inefficiencies, and certain aspects of the agricultural economy have stagnated in recent years. Such agricultural stagnation has perpetuated low yields and intensive manual inputs. This economic bottleneck must be overcome before real growth can be achieved.

Land and Environment

Sri Lanka is a pear-shaped island state of 25,332 square miles (65,610 square kilometers) that lies 12 miles (19 kilometers) southeast of India. The island is a tropical land, although the hot and humid climate is moderated at higher elevations in the interior mountains. The climate is shaped primarily by a seasonal pattern of rainfall in which the southeast monsoon dominates between May and August and brings heavy rains to the southwest quarter of the island. From November to early spring, the wind direction is reversed, and the eastern flank receives most of the rain. Frost in unknown, and average monthly temperatures vary only slightly from the coldest to the hottest month.

About 44 percent of the island is forested, and much of today's plantation agricultural complex was formerly forest land (Fig. 31–3). Mineral deposits of graphite, quartz, feldspars, and gemstones are commercially significant, and exploitation for additional minerals continues.

Technologic Backlash

Sri Lanka, like many other developing nations, has applied some technological innovations. Occasionally these innovations can lead to unforeseen problems. The case of malaria and the insecticide DDT is one such case. During the 1950s and the 1960s, the use of DDT was widespread on the island, and the incidence of malaria was drastically reduced. In 1963, there were only seventeen reported cases of malaria in Sri Lanka. Lowland humid areas heretofore little used because of malaria mosquitoes were settled, cleared, and put to agricultural use. Within a short time, a sizable population lived and worked in these lowland areas. More recently the damages of using DDT became widely recognized and its use was severely curtailed. Without an effective means of mosquito control, the number of reported malaria cases rapidly increased. In 1974, over 315,000 cases of malaria were detected. Sri Lanka is now faced with the dilemma of either abandoning these new croplands, or using more DDT to control the increasingly resistant mosquitoes and accept other possible damage to human health (contaminated water supply) and to the environment, or funding other more costly control measures.

Sri Lanka is not alone in facing problems of technologic backlash. Similar DDT-malaria problems exist throughout the world's humid tropics and especially in Monsoon Asia. In India, for example, the number of malaria cases has increased from 1.3 million in 1972 to 2.5 million in 1974. Egypt, with its Aswan High Dam, has another type of technologic backlash. The Aswan Dam was constructed to control the flow and floods of the Nile River and to use the impounded water to improve and expand irrigated agriculture. The dam, however, reduces the amount of fertile silt deposited downstream, necessitating increased use of chemical fertilizers. The sardine fishing industry in the Nile Delta has been adversely affected because the organic matter in the river has decreased greatly. Also the incidence of disease, particularly the debilitating schizomycosis, has increased, reportedly because of the environmental changes wrought by the change in the river flow regime.

Contemporary Patterns of Development

Sri Lanka is confronted with a serious economic dilemma. Its socialist government made a number of campaign promises to provide weekly rice subsidies to consumers in 1973. There has been much difficulty in meeting those promises and in the continued ability to meet its development commitments. Sri Lanka has been caught up in a recent, serious inflationary spiral that has affected all of the world's

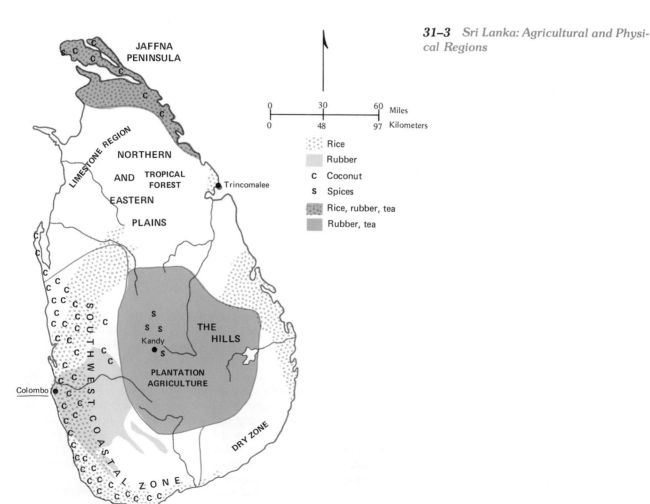

31–3 *Sri Lanka: Agricultural and Physical Regions*

economies. Export prices of tea, rubber, and coconuts, the principal cash crops, have boomed; but these have been offset by equally large increments in the cost of imports: rice, sugar, flour, and crude oil. The net effect has been to impede the rate of capital formation and thus restrict the rate of investment to promote economic growth.

Social problems also continue to plague the island. Of the 14 million inhabitants, 70 percent are of Indo-Aryan origin, Sinhalese, the ancient settlers. Twenty percent are more recently arrived South Indian Dravidians (Tamil speakers). The Sinhalese are Buddhists, and the Tamils are Hindus; thus, not only is there an ethnic, linguistic division but also a religious division. Violence between the Sinhalese and Tamils has been characteristic for several centuries, and as late as 1958 there were clashes between the two groups over the use of different languages for official purposes. More recently, newly arrived Tamil plantation workers have created additional stress, and the government has worked to repatriate some of these people to India.

Sri Lanka's government has followed relatively enlightened economic and social policies and energetically has promoted population control and economic growth. Political and social problems and unrest have been confronted and, in most cases, resolved. The economic problems of the small island state with a limited resource base, however, are serious and place Sri Lanka in a vulnerable position. Another small island Asian state (Taiwan) has been in a similar situation, and Sri Lanka may draw some inspiration from the remarkable growth and economic progress of Taiwan. The lesson learned is that such an effort requires disciplined effort from the people as well as investment capital. If these two criteria are met, Sri Lanka may have the same kind of success as Taiwan. In that case, the characteristics of small-island accessibility, a high level of administrative control, and an integrated economic system become advantages. These advantages may be developed to the point that they outweigh the liabilities of a limited resource base.

Further Readings

BARNDS, WILLIAM, J., *India, Pakistan and the Great Powers* (New York: Praeger, 1972), 388pp. This volume, published for the Council on Foreign Relations, presents an excellent treatment of the political issues that divide India, Pakistan, and Bangladesh.

Department of Census and Statistics, Government of Ceylon (Sri Lanka), *Ceylon Year Book* (Colombo: Department of Census and Statistics, since 1948). These year-books contain data on many aspects of the society and economy of the island republic.

FARVAR, M. TAGHI, and MILTON, JOHN P. (eds.), *The Careless Technology: Ecology and International Development* (Garden City, N.Y.: Natural History Press, 1972), 1030pp., presents numerous examples of technological backlash throughout the world.

KARAN, P. P., *Nepal: A Physical and Cultural Geography* (Lexington: University of Kentucky Press, 1960), 100pp.; and Karan, P. P., *Bhutan: A Physical and Cultural Geography* (Lexington: University of Kentucky Press, 1967), 103pp. Karan presents a general background to these two small little known mountain nations. His books are a good starting place for those interested in the region.

NYROP, RICHARD F., and others, *Area Handbook for Ceylon* (Washington, D.C.: Government Printing Office, 1971), 523pp.; Harris, George L. and others, *Area Handbook for Nepal, Bhutan, and Sikkim*, 2nd ed. (Washington, D.C.: Government Printing Office, 1973), 431pp.; and Smith, Harvey H., and others, *Area Handbook for Afghanistan* (Washington, D.C.: Government Printing Office, 1969), 435pp. These handbooks provide general background information on the areas covered and emphasize data on politics, society, and economy. They are excellent sources for factual information.

PAVITHRAN, A. K., *Bangladesh: Principles and Perspectives* (Homers Beach, Fla.: W. E. Garent, 1971), 108pp. One of the first studies that examine the forging of the new state.

ROSE, LEO E., *Nepal: Strategy for Survival* (Berkeley: University of California Press, 1971), 310pp. Rose examines the delicate political position of Nepal as a buffer state.

ROBINSON, HARRY E., *Monsoon Asia: A Geographical Survey* (New York: Praeger, 1967), 561pp. Robinson has compiled a traditional regional geography of the region. His book is a useful introduction to Monsoon Asia.

China: Origin and Development of Civilization

32

A̲ᴛ ᴛʜᴇ ʙᴇɢɪɴɴɪɴɢ of the section on Monsoon Asia, it was pointed out that two of mankind's oldest and richest civilizations had emerged in the region. The civilization that grew up in the Indus Basin and slowly evolved into Indian civilization has already been discussed. Farther north and east, in the drainage area of another great river, the Hwang Ho or Yellow, there gradually developed the second of Monsoon Asia's great cultures, the Chinese. Although little is known of the earliest origins of the Chinese, and the written record dates back only 3,700 years, enough archeological evidence has been found to suggest that the Chinese civilization emerged largely independently of external links and innovations. Thus, as in the case of the Indus Basin civilization, the Middle East hearth, and Mesoamerica, the north China loess plains and plateau country is believed to be one of five or six centers of early civilization. In north China, a society evolved out of late Stone Age

culture and developed techniques of sedentary agriculture, writing, formal religious, social, and political systems, and distinctive technological and architectural practices, which gave rise to the sizable concentrations of people that we call today urban centers or cities.

Man and Nature in China

The North China Culture Hearth

The origins of the Chinese go back a long time in human prehistory. China, like parts of North and South America, Europe, and Asia, experienced the influence of a series of ice ages. It is in the relics of these ancient times that fossils of the early Chinese have been found. Although China was not covered with an ice sheet, it was near enough to the great ice cap that covered much of western Siberia so that its climate and environment were much affected. Near the modern city of Peking, at a place called Choukoutien, extensive remains of early Stone Age hominids (humanlike creatures) have been discovered in a cave. These ancient people, called by some "Peking man" (Sinanthropus pekinensis), inhabited an area of the North China Plain, collected and hunted food, constructed crude tools and implements, and were able to build and use fire (Fig. 32–1).

Although these activities appear simple to us, they mark an important step forward in human progress. Archeological evidence from China also supports the idea that human beings began to evolve complex civilizations in a number of different locations scattered throughout the world. Recognition of multiple centers of origin may help us understand how the process of modern cultural differentiation and specialization came to take place.

Peking man did not survive; he was followed in time by true Homo sapiens, Neanderthal. Remains of this successor, who roamed over much of northwestern China 30,000 years ago, have been found in many places. Neanderthal was the early ancestor of the modern Chinese. The environment in north China at that time was cold and dry, and it is believed that this early human lived a nomadic existence gathering plants and hunting the large mammals associated with the last of the great ice ages. From residential and burial sites, enough tools and relics have been

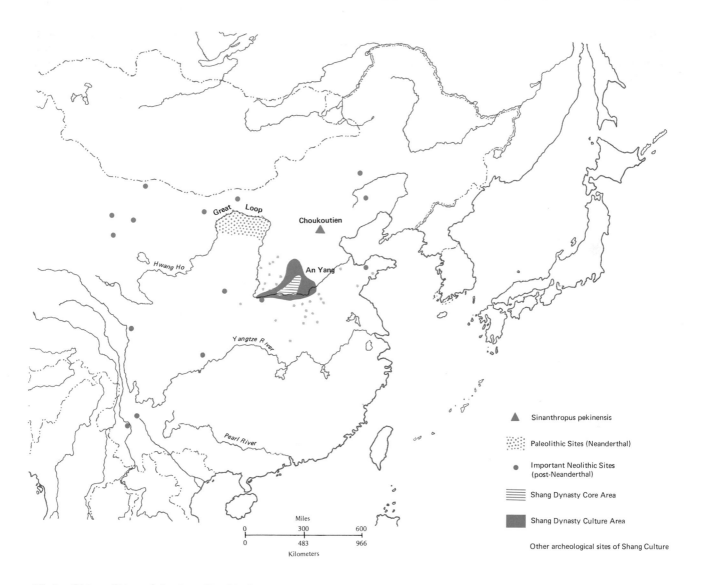

32–1 *China: Sites of Ancient Mankind*
SOURCE: Adapted from L. C. Goodrich, *A Short History of the Chinese People* (Harper & Row, 1959); and Paul Wheatley, *The Pivot of the Four Quarters: A Preliminary Enquiry into the Origin and Character of the Ancient Chinese City* (Chicago: Aldine Publishing Company, 1971).

found to indicate that life was rough and crude. The fossils suggest that improvements had been made over earlier techniques, and the stage was being set to develop better techniques for survival, techniques that would permit cessation of wandering and foster settlement in one place.

Loess Land and the Physical Environment

Much of north China is covered by a mantle of fine, wind-blown, dustlike soil material called loess or yellow earth by the Chinese (Fig. 32–2). Loess is common not only in China but also in places in Anglo-America and Europe and is usually associated with past glacial activity. So extensive has been the deposition of loess in north China that the material is as much as 500 feet (152 meters) deep in places. Although loess is of eolian (wind-borne) origin, erosion has taken place, and much of the material has been carried off the Loess Plateau by the Hwang Ho and redeposited as alluvium in the North China Plain.

Climate in north China is characterized by dry, cold winters and hot summers. Sandstorms are common in spring, and dry winds of Siberian origin sweep

Figure 32–2 *China: The Hwang Ho and Adjacent Areas*

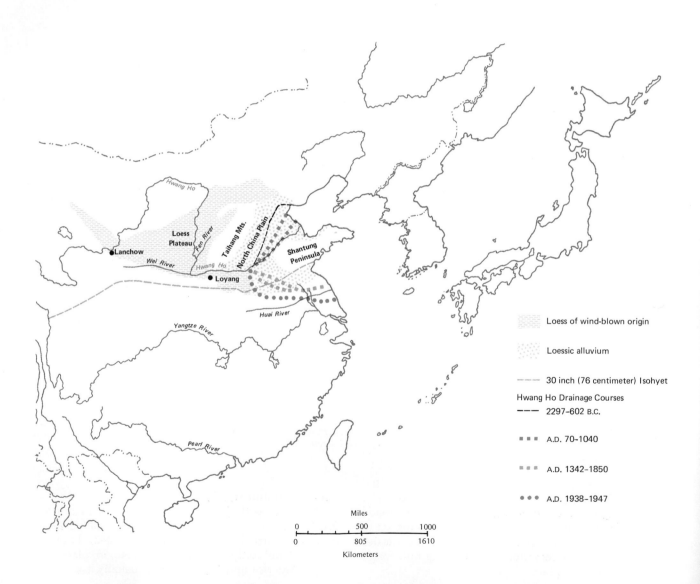

down from the north and west during the cooler months. On the Loess Plateau, annual precipitation is commonly between 10 and 20 inches (25 and 51 centimeters). On the North China Plain, precipitation is greater, between 16 and 30 inches (41 and 76 centimeters). South of the 30-inch isohyet (equal rainfall line), loessial materials dissipate and are not present in the Yangtse River Basin. Throughout north China, most of the rain is concentrated in summer; and although this concentration is good for cropping, the evapotranspiration is high enough to render the Loess Plateau and much of the North China Plain marginal for many crops.

It is interesting that such a marginal area is the site of the birth of Chinese agriculture. Yet, as Ho has noted, it was on the Loess Plateau and not on the North China Plain that Chinese agriculture first developed.[1] The explanation adduced is that the loessial soils were fertile, porous, and easy to work. The early Chinese with their primitive, wooden digging tools could operate much more effectively on such soils than on the dense, more compact alluvial soils of the low-lying floodplain farther east.

Little has been said about the environment other than to note it was cold, dry, and possessed an unusual soil condition. Yet this environment began to change; and as it did, succeeding conditions offered new opportunities. By 5000 B.C., much of the plain and plateau area of north China was no longer a cold and desolate steppe. Broadleaf forest vegetation had taken over the mountain and ridge areas, and some of the plains and plateaus are believed to have supported a mixed grass and forest cover. With increasing rainfall and more moderate temperatures, a greater ecological diversity was available, diversity that proved attractive and offered the early Chinese a broader range of options for occupancy.

In the loess lands at the southern end of the Great Loop of the Hwang Ho, where several major tributaries join the Hwang and the river turns sharply east in its plunge off the Loess Plateau, a remarkable combination of environments exists. Such a combination is called an interface or juncture, a place where different conditions come together. This interface,

with its adjacency of forested hills, watercourses, and grassy high plains and its diversity of conditions and opportunities, was the location of one of the most important events in Chinese history, the development of agriculture.

The Origins of Chinese Agriculture

The agricultural origins of China are ancient, complex, and controversial. For a long time it was believed that China, like so much of the rest of the world, had derived its agriculture through contacts with the Middle East and had build on agricultural dispersals and innovations that diffused out of ancient Mesopotamia. Recently such thinking has been revised, as it was recognized that Mesoamerican agriculture had contributed uniquely to world agricultural dispersals. Indeed, China too is believed to have developed and contributed its own techniques and crops. Support for this idea has come from several sources. Archeological remains and pollen samples suggest that crops not found contemporaneously in the Middle East hearth were commonly cultivated in the north China loess land. These crops include several types of millets and very early evidence of rice. Supported by subsequent historical writings, the net effect is to suggest that a number of crops such as foxtail millets, sorghum, soybeans, hemp, mulberry, and possibly some species of rice were indigenous to China.

This ancient stage of Chinese culture with its distinctive agriculture has been called Yang-shao after a village of the same name. Its origin dates back to 3000–5000 B.C. (Table 32–1). Its focus on dry field agriculture distinguished it from the water-oriented agricultural systems of the Middle East. Nonirrigated agriculture is the basis for establishing Chinese Yang-shao civilization as an event and a set of cultural processes independent of the Mesopotamian cultural hearth.

Yang-shao practices gradually spread eastward onto the North China Plain. Eventually on the North China Plain the Chinese did begin to irrigate, and wheat and barley entered into the agricultural complex of this ancient period. These events, however, were subsequent and added diversity to an already well-established and easily identifiable cultural base.

China, whether or not one admits that rice was indigenous, developed its own peculiar and special agricultural style and system. To a large extent this

[1]P'ing-ti Ho, "The Loess and the Origin of Chinese Agriculture," *American Historical Review*, 75 (1969), 31.

Period or Dynasty	Date
Yang-shao Culture	ca. 5000 B.C.–3000 B.C. (?)
Lung-shan Culture	ca. 3000 B.C.–(?)
Hsia (mythological?)	ca. 1994–1523 B.C.
Shang	1766–1122 B.C.
Chou	1122–221 B.C.
Ch'in	221–207 B.C.
Han	206 B.C.–220A.D.
The Three Kingdoms	220 A.D.–280
The Southern Dynasties	265–589
The Northern Dynasties	386–581
Sui	581–618
Tang	618–907
The Five Dynasties	907–960
Liao (Khitan Tartars)	916–1125
Sung	960–1279
Chin (Nangen Tartars)	1115–1234
Yuan (Mongols)	1277–1368
Ming	1368–1644
Ch'ing (Manchus)	1644–1911
Republic (Sun Yat-sen)	1912–1949
Peoples Republic (Mao Tse-tung)	1949 +

Table 32–1 *Chronologic Dynastic Table of Chinese History*

SOURCE: National Palace Museum (Taipei, 1971).

style was associated with an unusual environmental circumstance. The agricultural system was an obvious expression of the culture. The distinctive nature of the origin of Chinese agriculture thus has had profound implications on the nature and character of Chinese civilization as well.

Ancient Shang China

Chinese scholars traditionally have traced the origins of the Chinese state to the Shang Dynasty (ca. 1765–1123 B.C.), although Chinese tradition traces its origins to an apocryphal period of a preceding dynasty, the Hsia (Fig. 32–1). The Hsia Dynasty, it is claimed, dates from before 2000 B.C. Despite these earlier claims, Shang is a good place to begin a review of Chinese history, for Shang culture possessed a linguistic system of written characters, the forerunner of modern Chinese script. This writing, largely discovered on ancient bones (the Oracle bones), has told us much about the formation of Chinese institutions and civilization. From these bones a reasonable picture of ancient China has been constructed.

Shang civilization contained most of the cultural attributes assigned to China, attributes that distinguish China from the other cultural systems of equivalent age. Its priests, scholars, administrators, and artisans created and developed a society and culture that rivaled other contemporaneous civilizations. Settled agriculture, formalized religious, administrative, and legal systems, writing and literature, the tradition of a dynastic ruling house, sophisticated handicrafts and metallurgy, and the creation of a great city (An Yang) in the loess plains of north China—all existed in Shang times. To the extent that East Asia has evolved subsequently into a related, broader region with common historical antecedents, the regional roots may lie in Shang times. This period and its developments are a key to understanding all other East Asian cultures and societies, both historical and contemporary.

The Hwang Ho and Loess

The arid and semiarid environments of north and northwest China for a long time have been closely intertwined with the evolution of human activities and cultural practices. Within these environments several features have been important for the Chinese and their technological and economic systems. Chinese civilization developed in and around the drainage basin of the Hwang Ho. Although irrigation was not a vital part of the earliest agricultural settings in north China, the Hwang Ho has affected the Chinese in many ways. The difficulty of controlling and using this river has been a long chapter in Chinese history. The chronicle has not always been a happy history, and the phrase "China's sorrow" long ago gained currency as a means of describing the devastation brought on by this powerful and unusual river.

The Hwang Ho rises among the swamps and lakes found in the eastern part of the Tibetan Plateau. The river initially flows east and north and after 1,000 miles (1,609 kilometers) reaches the vicinity of the modern city of Lanchow. The river, with its steep gradients, has cut its way down to the western Loess Plateau, a drop of some 10,000 feet (3,048 meters). Once on the plateau, its gradient and velocity are suddenly reduced, and the river broadens and meanders. There the Hwang is broad and shallow. In the arid climate considerable amounts of its volume are lost to irrigation and evaporation, and the river is a sluggish and undramatic stream. At the top of the Great Loop, the river abruptly shifts direction and begins to flow south. As it changes directions, its character also begins to change.

Several factors account for this change. First, the river begins to flow through poorly consolidated and easily eroded loesslands. Second, numerous tributary streams join the main channel. Some of these tributaries are large; they too flow through the easily eroded loess and contribute enormous quantities of material to the main stream. Third, as the stream flows south, it descends more rapidly, and its capacity to carry a large silt load is increased. Fourth, there is more rainfall, and more water enters the Hwang and brings quantities of silt through the tributary network.

Together, these factors mean the stream is much larger and carries a very heavy silt burden. Conditions are further complicated by the seasonality of rainfall, concentrated in the summer months of July and August. Such a concentration of rainfall results in great changes in the volume of flow of the river. In fact, the Chinese claim the Hwang in a flood in the nineteenth century flowed at a discharge rate 250 times greater than its lowest flow, a record for the world's major rivers. Finally, the absence of heavy vegetative cover and the nearness of a number of important tributaries combine to increase the speed with which rainwater is funneled into the main stream, a speed that is analogous to water flows in irregular channels in desert areas.

All of these features combine to produce floods of devastating proportions. At the south end of the Great Loop, the river flows almost due east and within a short distance drops sharply off the Loess Plateau and enters the North China Plain near the city of Loyang. The same mechanics of stream flow operate again: high silt load; sudden reduction of gradient and velocity and thus loss of capacity to carry the high load; and seasonal concentration of precipitation. The results long have been of profound consequence to the great number of Chinese living in the North China Plain, for the annual floods have become legendary; and Chinese history is filled with accounts of the mighty river.

Most rivers flood, but the nature of the floods on the Hwang is spectacular. The character of the stream, perhaps coupled with mankind's ecologic abuse of the loess lands and the progressive degradation of local vegetation, has meant that flooding apparently has become worse over time. Moreover, the river is continually building up its bed with the dumping of its heavy sediment load, and the stream bed has become elevated throughout much of its path across the North China Plain. The Chinese have responded by building levees, but the levees must continue to be heightened as the river bed is elevated. Yet it is impossible to build a dike high enough to protect against all floods. Such a vicious spiral of progressive higher levees has resulted in elevating the bed high above the surrounding floodplain. Consequently, when major floods have occurred, the river has broken out of its bed and inundated enormous areas of the North China Plain and has been terribly difficult to re-channel.

Since 602 B.C., the Hwang Ho has changed its course fifteen times and flowed to the sea through different channels. Some of these channels have entered the sea at distances as much as 500 miles (805 kilometers) south of the present mouth (Fig. 32–2). At one time the flow was actually channeled south into the Huai River and ultimately into the Yangtze. Such a condition might be analogous to seeing the Mississippi River flood and alter its channel to the point that it actually flowed to the Gulf of Mexico through the Rio Grande.

Flooding of the Hwang Ho is of monumental proportions. When the river has broken out of its channel, the results have been catastrophic. For example, in 1938 the Chinese Nationalist government destroyed the dikes on the south side in an effort to slow the advance of the invading Japanese Army. More than 6 million acres (2.4 million hectares) were flooded and caused the relocation of an estimated 6 million refugees. As many as 500,000 people may have died as a result of this induced flood. Nine years passed before the river was rechanneled to resume its flow north of the Shantung Peninsula.

Mankind, Nature, and Environmental Change In China

The Chinese concern with nature and natural processes is an ancient one. Based on the marginal character of much of the Chinese land, inhabitants of China early began to respond in several different ways to these physical difficulties. First, the Chinese over time developed a powerful consciousness of nature and natural forces. In coming to grips with these forces, a set of practices and customs were developed that set the Chinese in apposition with rather than opposition to nature. The Chinese did not want to antagonize and battle the cosmic and worldly forces; rather, the best path was to attempt to establish a kind of harmony and compatibility with nature. Such harmony did not mean total surrender to forces and elements of nature but rather saw people placed in the position, sometimes uneasy, sometimes comfortable, of accommodation to the natural surroundings.

Construction Projects and the Great Wall

Such an approach to nature is witnessed in the sensitive landscape paintings of Chinese artists. Moreover, much of the literary tradition of the ages further entrenched this feeling, and the Confucian norm of good was a person not in conflict with his surroundings. Yet the Chinese have not always been passive when dealing with their environment. Some of their most notable achievements have been modifications of their environments or at least have been conditioned by the circumstances of the environment. Before the birth of Christ, the Chinese had begun work on a major construction project, one of the largest in human history. This project, the Great Wall, is commonly believed to have been a protective device aimed at keeping nomadic herders from the dry, cold north and west out of those parts of China that were more mild and humid and were occupied by sedentary farmers. More realistically, however, it appears that the Great Wall was primarily an attempt to demarcate the boundary between Chinese sedentary farming and Mongolian pastoral nomadism (Fig. 32–3).

Ecology versus Culture

Both Chinese sedentary farming and Mongolian pastoral nomadism have always been associated with certain environmental conditions. The nomadic herders of the north and west evolved their forms of livelihood based on animal husbandry in an area that afforded few other opportunities for human sustenance. The Chinese, in building the Great Wall, said in effect the sedentary Chinese should be distinguished and cut off from the footloose and presumably uncultured nomadic herders. Chinese culture and civilization were characterized by cities, specialized social and political systems, and settled farming. Pastoral herding did not have these attributes. In the Chinese view it was wise and proper to separate the two to maintain the distinction and the higher character of Chinese civilization.

Such an effort to divide cultural types by a wall failed to recognize that sharp environmental boundaries rarely exist in nature. Rather, climatic zones and their associated vegetative complexes are differentiated more often by broad transition zones as patterns blend into each other. Moreover, there is likely to be considerable variation in climate from year to year. In agriculturally marginal areas such as China's north and west, the attempt to define and separate the grazing range of the herder from the land of the millet-growers was difficult to achieve. In the dry years the herders encroached south and east in search of adequate pasturage for their animals. In years of plentiful rainfall, the settled farmers extended their cropland far into the steppe. Similar

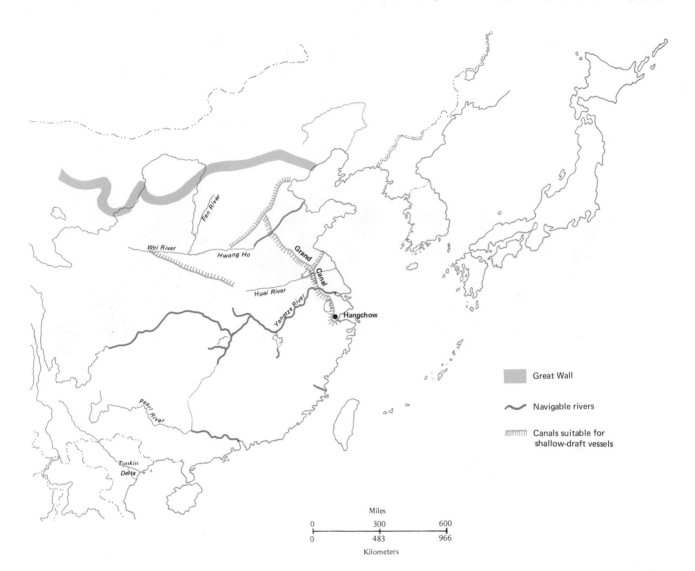

32–3 *China: Selected Major Construction Projects*
SOURCE: Adapted from Mark Elvin, *The Pattern of the Chinese Past* (Stanford: Stanford University Press, 1973).

conditions and practices occurred in the United States as "sod-busters" increasingly farmed the Great Plains of Texas, Oklahoma, Kansas, Nebraska, and the Dakotas in the late nineteenth and early twentieth centuries. The results were sometimes disastrous, for when dry years occurred, crops failed, and hard times followed.

In the case of northern China, the attempt to demarcate a cultural boundary along an ecological interface failed. It failed because the mechanisms that underlay the cultural systems, attuned carefully to specific environmental conditions, were little understood. The Great Wall, to the extent that it was designed to preserve a kind of purity of essence to

The Great Wall of China was built more than 2,000 years ago to separate the Mongolian nomads of the north from the Chinese farmers to the south. (United Nations/T. Chen)

things Chinese, may have succeeded. In the end, it was the aggressive, energetic push of Chinese culture with its trappings of sedentary farming that came ultimately to encroach on and dominate the nomadic herders on the northern periphery. It may be that the force of numbers overwhelmed the herders. Whatever the case, the remnant of a once magnificent empire of nomadic Mongols and associated tribes is today a political and military midget caught between the two giants of the Communist world, China and the Soviet Union.

Canal Building as Environmental Modifier and Spatial Linkage

Another and certainly more practical application of Chinese energy in grappling with the natural environment was in the construction of canals. Contemporaneously with the construction of the Great Wall, the Chinese built canals; the dimensions of some of these

canals were very impressive. Although their major use was to facilitate the movement of goods by barge in the early days, it appears that many such canals began based on the need to convey water from one location to another.

By the Sui Dynasty (A.D. 590–608), the Grand Canal had been completed that linked the North China Plain with the Yangtze Delta (Fig. 32–3). This canal was designed to provide a link whereby tribute grain could move from the south rice-growing areas north to the capital, basically a political-administrative objective. Yet the net effect of designing and operating such a canal was to demonstrate the possibilities of tying together a large empire through waterways. Barge canals and river transport spread rapidly and were most effective in the Yangtze drainage system and the river basins farther south. Such a network has long served China as a basic means of hauling goods and people even in the twentieth century. Although crude and slow, it was a cheap and effective means of transporting and assembling goods at central locations. Such a network of waterways enabled the Chinese to support the large and magnificent capital of Hangchow that Marco Polo described in 1273, and permitted the rapid growth of modern Shanghai at the turn of this century.

Diffusion Out of the North China Hearth

The Hwang Ho and Loess Plateau have meant many things to China and the people. Among them, one theme is common. The Chinese have long struggled with an environment over which they exercised little control and direction. The ecological conditions that confronted the Chinese as they developed sedentary agricultural practices in north and northwestern China were unusual and probably not dependable. It may be that the Chinese, through their various activities of cultivation and forest removal, rendered this environment even more intractable and uncertain. Whatever the causes, the Chinese began to migrate south away from the ecological rigors of the Loess Plateau and the North China Plain. In so doing, they were confronted with a substantially different environmental setting that offered new opportunities for agriculture and more dependable physical conditions for human sustenance. Migration south also extended the influence of cultural practices, as tribal people who inhabited south China were acculturated and absorbed into the vigorous and dominant Chinese way of doing things.

The March Toward the Tropics

As early as Lung-shan times (3000 B.C.), Chinese agriculturists were present in the lower Yangtze Delta. It is clear that by Shang times the Yangtze was within the orbit of Chinese civilization. It was not until 220 B.C., however, that the Chinese were able to extend their influence, culture, and control much farther south. The legendary emperor Shih Huang-ti, who first brought unified control and management to the diverse factions and regions of China between 221 and 214 B.C., led a military campaign to the south. This campaign resulted in adding the Pearl River drainage basin of southern China and the Tonkin Delta of Vietnam under the territorial control of China.

Central and southern China possess very different environments from the north. Once south of the Chin-ling Mountains, the climate patterns are more mild and humid. The effects of continentality, so strong that they promote steppelike conditions on the Loess Plateau, are softened in the south. Probably the most conspicuous differences are in the average daily temperature and amount of annual precipitation. Subtropical conditions are common, and in the far south these grade into tropical conditions with a growing season that lasts all year. With plenty of rainfall and long, hot summer days an environment is found well suited for the cultivation of wet rice.

It is not known for certain who first cultivated rice. Some argue that this crop originated in India, and recently fossil evidence found in northeast Thailand has indicated it was cultivated there as early as 3500 B.C. Rice thrives where it is warm and wet. If the plant is to survive in nature, a subtropical or tropical location appears logical. Consequently, if some species of rice were indigenous to China, they were probably cultivated in the south. As the Chinese moved south and interacted with native tribal people, an exchange of techniques, innovations, and crops took place. In the south, agricultural practices increasingly focused on wet rice cultivation, the crop practice that yielded most and was best suited to the environmental circumstance.

The Landscapes and Environments of South China

Southeastern China possesses a more intricate and complex surface geography than the broad plains and plateaus of the north (Fig. 32–4). The main rivers and their tributary networks dissect the uplands into a system of river valleys separated by rugged and sometimes high ridges. With the exception of the Yangtze and Pearl river drainage basins, there are no extensive low-lying floodplains. The consequence of the diverse and rugged landscape has been to restrict movement and communication, which in turn has led to isolation of local areas and the formation of sectional interests. Probably the most obvious and conspicuous example of isolation is found in the great number of spoken dialects in southeast China. Al-

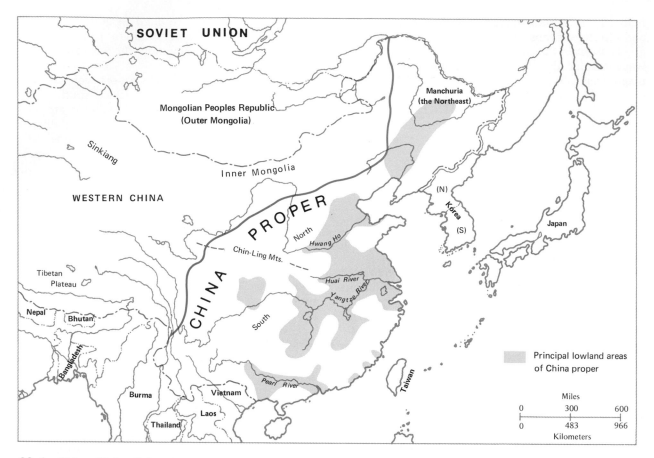

32–4 China: Major Subregions

though grouped into major families such as Cantonese, Fukienese, Hakka, and Wu (the lower Yangtze dialects), the variations are so great they are mutually unintelligible. Within short distances, from one river valley to another, local people often are unable to converse with one another. Such a condition has long been a fragmenting force in the political evolution of the Chinese state. Here the geographical foundations for this problem are easy to discern.

The Chinese took to the lowlands and began specialized rice growing. Native peoples who were not absorbed were forced into the rugged uplands. So rough was some of the terrain that these native peoples isolated themselves and preserved some degree of ethnic and cultural independence and integrity. In some cases ethnic separateness has con-

tinued to the present. The southern province of Kwangsi has been identified by the leaders of the new China as an autonomous ethnic region associated with the Chuang tribes. A number of other tribal groups, such as the Miaos, Meos, Thais, and Laos, are also present in south China. They exist in the purest form high in the uplands. Where they are found in lowlands, they invariably came into contact with Chinese agriculturists. Upon contact, there began that inexorable process of acculturation, whose end result was the conversion of aboriginal peoples to Chinese in thought, speech, and practice, a process known commonly as Sinicization.

Thus, it happened that the environments of south China too played a role in shaping the historical and social patterns that evolved. On the one hand, the

Chinese were drawn to this environment, for it proved more tractable and dependable and could support more people through the mechanism of water control and rice cultivation. Sustained food production permitted the sustenance of large numbers of people and provided a safety valve area when conditions became bad in the traditional north China hearth. On the other hand, the ruggedness and diversity of landscapes in the south promoted political fragmentation and regional interests. China, like the United States, has old and powerfully established sectional interests. Even under a Communist government, with its emphasis on centralization, sectionalism has not disappeared. The south is probably the best place to witness these sectional interests, and it may be that part of the explanation for this regionalism is to be found in physical geography.

The Outer "Non-Chinese" Regions

Little has been said about the western part of China or about Manchuria, the region the Chinese refer to as the Northeast. Several distinct areas exist outside what is commonly referred to as China proper. These are Tibet, Sinkiang (Chinese Turkestan), Mongolia, and Manchuria (Fig. 32–4). These regions, with the exception of the Manchurian Plain, all share some common physical characteristics. They are for the most part high and dry, with great extremes of temperature between winter and summer. Much of their mountain and basin area is characterized by internal drainage, salt marshes, and desert, and the environment has never supported a dense, productive agricultural system. Traditionally peopled by non-Chinese, these areas have long existed politically somewhere between central Asia and the Chinese. There people have been conditioned to their particular environments, and most livelihood is derived from herding, oasis cultivation, and subsistence farming. Today, these non-Chinese regions, except for Outer Mongolia, have once again been attached to China, but their features of physical geography and human population serve to distinguish them from China proper.

The Three Chinas

Based on broad environmental and ecologic factors, at least three distinct Chinas are discerned. To some extent these may be identified with human populations, but occupancy styles and activities are a better way to describe them. In brief, the three Chinas exist as large, ecologic regions. The first is western China—the dry, high China with its highly specialized, adaptive livelihood styles closely attuned to the physical conditions of environment. Its people are of Turkic origin, Tibetans, or Mongols. The other two parts of China lie in the east, and these two may be divided between north and south. The south begins south of the Chin-ling Mountain-Huai River line and includes the main agricultural lands of the Yangtze and the Pearl River drainage systems. It is mild and wet in climate, much like the climate of the southeastern United States. Its agriculture is mainly wet rice. Although included in China proper and typically inhabited by Han Chinese, it entered the Chinese orbit during the late formative period. Its dialects and human population are diverse enough to betray its more recent entry into the great Chinese cultural system.

Finally, there is north China, the source area out of which China was born and grew. North China includes the middle and lower Hwang Ho drainage basin, to which should be added Manchuria. This is a cold, dry region in winter, but the summers are hot and humid and support intensive Chinese-style agriculture. The people are largely Mandarin speakers; and although there are ethnic infusions of Manchu, Mongol, and other tribal elements, it was here that the proto-Chinese came to be. Based on the features of environment and occupancy, these three broad regions persist to distinguish the main areal parts of the Chinese state even in the face of strong central government efforts to counter this diversity. Environment and geographical features as much as politics continue to shape the face of modern China.

Further Readings

CREEL, H. G., *The Birth of China* (New York: Frederick Ungar, 1937), 402pp. Creel's study is a classic account of China's early history suitable for college students.

CRESSEY, GEORGE, *Land of the 500 Million* (New York: McGraw-Hill, 1954), 387pp. Cressey's geography is the standard classic of China on the threshold of revolutionary change. The chapters on north China and the Hwang Ho drainage basin are among his best.

DE CRESPINGNY, RAFE, *China: The Land and Its People* (New York: St. Martin's Press, 1971), 235pp. This study, by a historian, has an excellent chapter on the Hwang Ho and its hydrology. It also examines recent efforts to modernize the agricultural economy.

ELVIN, MARK, *The Pattern of the Chinese Past* (Stanford: Stanford University Press, 1973), 346pp. This new work focuses on the economic history of China's last 2,000 years. It is a lively and highly readable account.

GOODRICH, L. CARRINGTON, *A Short History of the Chinese People*, 3rd ed. (New York: Harper & Row, 1959), 295pp. A brief but lucid outline of China's long history.

HO, P'ING-TI, "The Loess and the Origin of Chinese Agriculture," *American Historical Review*, 75 (1969), 1–36. An interpretive statement of Chinese agricultural origins from the Chinese point of view.

LATTIMORE, OWEN, "Origins of the Great Wall of China: A Frontier Concept in Theory and Practice," *Geographical Review*, 27 (1937), 529–549. For those interested in an interpretive essay on the development and operation of the Great Wall, this work is excellent.

TRIESTMAN, JUDITH, *The Prehistory of China* (Garden City, N.Y.: Doubleday, 1972), 156pp. Triestman's text is a brief but worthwhile account of the archeological record of China's prehistory.

TUAN, YI-FU, *China* (Chicago: Aldine, 1969), 225pp. Tuan, a geographer, follows an ecological, environmental approach in describing the evolution of Chinese landscapes and human occupancy patterns.

WHEATLEY, PAUL, *The Pivot of the Four Quarters: A Preliminary Enquiry into the Origin and Character of the Ancient Chinese City* (Chicago: Aldine, 1971), 602pp. Wheatley's work is long and difficult but is the most definitive statement on the origins of Chinese cities and their place in the world's urban history.

WIENS, HEROLD G., *China's March to the Tropics* (Hamden, Conn.: Shoestring Press, 1954), 441pp. This study traces the migration of Chinese southward and the incorporation of southern indigenous people into the Chinese culture system.

WITTFOGEL, KARL, *Oriental Despotism: A Study of Total Power* (New Haven, Conn.: Yale University Press, 1954), 556pp. This work, like Wheatley's, is long and difficult. It contains Wittfogel's controversial "hydraulic theory" for the rise of Oriental civilizations.

China: Forces of Social and Political Unity and Change

33

ONE OF THE MOST interesting and unusual facts about China is its permanence and long political tenure. Despite changes in dynasties, occasional disruption by outside groups, and short terms of competing factions, China has managed to persist and to maintain control over essentially the same territory for some 2,000 years. For Europe, it would be as though the Roman Empire had remained virtually intact to the present, and its institutions of governance had, with certain modifications, survived.

Integrating Elements

The Administrative Structure

One reason for China's long-term cohesiveness has been a powerful and effective central political apparatus that prevented fragmentation. The traditional approach to governance was based on a large and centralized bureaucracy that operated with a surprising efficiency on a modest budget. The constant circulation of bureaucrats from one post to another reduced the growth of powerful local interests. Through these and related policies, the central administration was able to maintain a remarkable degree of control even in the absence of modern means of communication and transportation. When times were bad or if a dynasty were overthrown, a new dynasty emerged and the political system persisted and remained effective. In the absence of an integrating economic system, the centralized bureaucracy provided some of the glue that held the diverse parts of the empire together.

The Chinese Language

As noted in Chapter 32, the people of China spoke a variety of different dialects, many mutually unintelligible. In contrast, written Chinese based on ideographs (picture symbols that symbolize ideas, objects, and actions) was used in all parts of China. It was the language of government and provided the central means of communication among the various dialect groups. Here, then, was another aspect of Chinese culture that promoted integration of the various territories.

That the written language was known only to a small fraction of the total population was not a great disadvantage. The literate group—scholars, government functionaries, and the wealthy landed gentry—formed the leadership group at the local level. They were the decisionmakers, and it was this group with whom the central government communicated and to whom it addressed calls for taxes, laborers, and military conscripts. Both the central government apparatus and the use of a common written language became national symbols for the Chinese and important components of the emerging cultural system of the Chinese state.

Confucian Social Order

Another key integrating element was the Confucian tradition. Confucius and his many succeeding disciples and interpreters established for China a social order whose total impact has been pervasive for more than 2,000 years. Confucius outlined a set of social and behavioral patterns for ancient Chinese society.

These views were transmitted through an important segment of China's rich literature, and over time became embedded in the social system that ordered most, if not all, relationships among individuals, families, and the political system. In general, the Confucian precepts emphasized stability. In that fundamental change in the existing order of social, economic, and political relations was excluded from these precepts, Confucianism often has been viewed as rigid and conservative.

It is not surprising that Confucius has been blamed frequently in recent years for disallowing innovation and technical change and preventing China's economic modernization. Yet such criticism oversimplifies the nature of traditional China. Confucianism was but one of many underpinnings of traditional Chinese society. Nevertheless, in the last few years Communist leaders in China have waged a vigorous campaign to exorcise the ghost of Confucius from the new China in their drive to shift attitudes on such matters as how many children a family should have or what the nature of social stratification should be. Communist planners search and strive for a new order as they seek economic modernization and growth, and Confucian ideas have no place in their revolutionary society. Consequently, it seems likely the idea of Confucius as a great man of Chinese history and letters will continue to be challenged.

The Impact of the West

In the early part of the nineteenth century, new influences entered China along its coasts and rivers that challenged the traditional government system. The British Navy, for example, forced China to accept the superiority of Western power in the Opium War of 1838. The West had progressed greatly in the seventeenth and eighteenth centuries, largely because of new techniques and innovations associated with the Industrial Revolution. China had progressed little toward these new technologies and developed few modern scientific forms of inquiry. European powers forced China to open its doors to the outside world and to begin the long drive toward modernization.

In the middle of the nineteenth century, there occurred one of the great Chinese political upheavals, the Taiping Rebellion, a partial result of the new challenge from the West. The Taiping Rebellion was followed at the turn of the century by the Boxer uprising. Finally, the old dynastic order could stand no longer, and fledgling revolutionary forces succeeded in establishing the Chinese Republic. Through these years, China was increasingly set upon by external forces unlike any the nation had ever been exposed to previously. Not only were these forces alien and not susceptible to acculturation, but they also introduced modern weapons and industrial technologies superior to indigenous techniques. For the first time, China was confronted with an external force that asserted and promoted its own racial and cultural superiority and could support its claim with power, money, and technology. Traditional Chinese ethnocentrism and pride were challenged, and the Chinese were humiliated and beaten by foreigners whom they had considered barbarians. Reinforced by the aggressive efforts of Christian missionaries to make over the Chinese in the Christian image, the basis for the thoroughgoing reform and revolution in China in the late nineteenth and early twentieth centuries became increasingly apparent.

China's Revolution and the Rise of Chinese Communism

Among the ideas and innovations introduced into China in the nineteenth and twentieth centuries were the concepts and views of Western democracy and Marxism. Following the overthrow of the Ching Dynasty in 1911 and a brief period of near anarchy, the Chinese Republic was established in 1918 under the guidance of the revolutionary patriot, Sun Yat-sen. His party, the Chinese Nationalists, announced the "Three Principles of the People" by which there was to be reform in politics, society, and the economy. This reform looked toward the enfranchisement of individuals and their improved status and economic well-being within the state.

Upon Sun's death in 1926, Chiang Kai-shek took over the Nationalist Party and government, and except for brief periods, he ruled the Nationalists and their exiles in Taiwan until his death in 1975. Chiang had a more conservative political outlook than Sun, and he soon aligned himself with the urban, commercial interests and large landholders in the rural areas.

A number of young Chinese had been sent to study in Europe and the United States. Among the ideas they were exposed to was Marxism, a political philosophy of great interest as a result of the Communist revolutionary movements in Europe and Russia. With the success of the Russian Revolution and the establishment of the Soviet Union, a base of support existed, and Marxist ideas gained currency in China as a competing means to achieve the goal of reforming and modernizing China.

The various Marxist groups in China argued and fought bitterly among themselves. A major conflict existed as to whether the Communist party should focus on converting the urban workers (the proletarians in classic Marxist parlance) to communism as the main strategy for revolution or whether a rural strategy that focused on the peasantry would be more appropriate. Among those who supported the rural-peasant line was a young, articulate librarian in Peking. His name was Mao Tse-tung.

After a short period of unsuccessful competition with the Nationalists, the Communist urban strategy failed, and the Communists increasingly turned to the peasantry for support. In the 1930s, however, Chiang Kai-shek, aided by foreign advisors, attacked the rural Communists and drove them out of south and central China to a remote location in the Loess Plateau. The retreat of the Communists from the south to the isolated Great Loop of the Hwang Ho is now an epic in the lexicon of Chinese revolutionary literature, The Long March.

Driven into hiding but not destroyed, the Communists regrouped themselves and their movement. With the Japanese invasion in 1938, the Nationalists were preoccupied with fighting the Japanese and no longer were able to prosecute vigorously their campaign against the Communists. In a short time, the Nationalists managed to lose the political and military initiative. By war's end in 1945, the Communists had formidable military and political arms, and the Chinese Communist Party offered itself as a new and energetic answer to the plights of the poor, landless peasant. The issue was joined. The Communists increasingly established their power in north China, and the Nationalists no longer seemed able to compete for popular support. The results are well known; Chiang Kai-shek and the remnants of his army re-treated to Taiwan, and Mao Tse-tung and his followers established a new, revolutionary government. These actions were the culmination of events that began early in the nineteenth century when the West seriously began to challange the traditional Chinese system. The ascendance of a Communist government in 1949 was but another change in a century of change and turmoil for China. The Communists, now firmly in power, are carrying on the processes of change as rapidly as possible. One aspect of the Communist strategy is to destroy many of the nation's traditions and to build a new China, politically, economically, and socially.

China's Minorities

Minority groups have always been part of the Chinese experience. China's dealing with its minority peoples often have not been friendly, and two minority groups, the Mongols and the Manchus, conquered and ruled China. In all cases, however, there has been an assumption of the superiority of Chinese culture and civilization. If a traditional Chinese policy toward the minorities existed, it was always based on that assumption, and often took the form of Sinification of these barbarian peoples.

Within China today, about 50 million people are classified as minorities, about 6 percent of the population. In the early 1950s, fifty-four groups were designated as minorities, and thirty-six of them have been awarded autonomous status at various levels in the territorial hierarchy of administration. Among these, five areas exist at the first-order level, the autonomous region. The five autonomous regions are: Kwangsi, Tibet, Sinkiang, Inner Mongolia, and Ninghsia (Moslem) (Fig. 33–1). These autonomous regions, although sparsely populated, occupy a large part of China's total area. Such an administrative recognition of minorities suggests that the Chinese have followed the Soviet example in administering state territory. In practice, policy toward these minorites has been considerably different.

Initially the Communist government proceeded cautiously and allowed a liberal policy of local minority self-government and the study, preservation, and use of local languages, customs, and traditions. Since the late 1950s, however, the Communist government

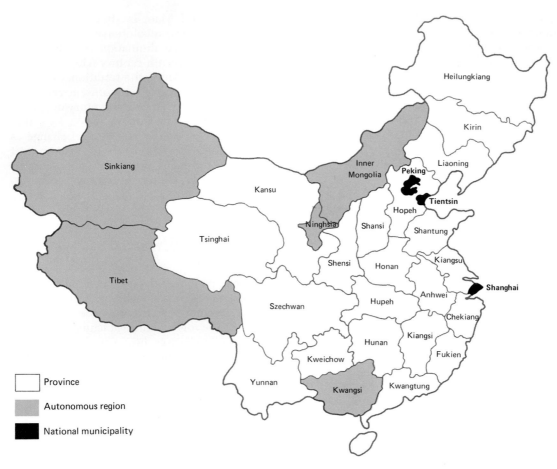

33–1 *China: Administrative Divisions*

has strongly curtailed the freedom of minorities. The most recent example occurred on the heels of the cultural revolution in 1970–1971 when a large amount of territory that belonged to Inner Mongolia was reassigned to three other provinces. This stricter policy is believed to result from two circumstances. First, China became increasingly uneasy about its frontiers, and those people who inhabited frontier areas were viewed with suspicion. For example, unrest and dissent among several groups in Sinkiang were thought to be inflamed by Russian provocation. At the same time, the Tibetans were distrusted, as they were considered to be slow in adjusting to Chinese Marxism in socializing their region. Second,

Chinese demand for increased agricultural output forced new policies of collective landownership and the conversion of pasturage to cropland in minority areas. The Mongolians, especially, resisted these policies, and the Chinese were prompted to force the adjustments.

Together, these circumstances resulted in a more conservative and strict approach to dealing with minorities. In support of this approach and perhaps looking toward ultimate assimilation, 3 to 5 million Han Chinese have been sent to minority areas since 1950. It appears likely that the central government has adopted a program of integration of its minorities and is willing to enforce this policy if necessary.

Territory and the Nature of Chinese Administration

A central government and bureaucracy existed throughout much of China's history, and its operation was vital to the formation of the state. In traditional China, with the exceptions of rivers and canal routes, transportation and communication were difficult and expensive. Economic interchange and integration among regions were poorly developed. The government was responsible for some commodity transfers, but these transfers were a form of taxation. For example, tribute grain was moved from the Yangtze Basin north along the Grand Canal to the food-deficient region in the North China Plain.

Outside the Yangtze Basin, economic intercourse and trade did not furnish the main ties that linked the traditional core area of eighteen provinces settled by the Han people.[1] Where economic linkages failed to hold China's territory together, the central administration developed alternatives through the establishment of an administrative system. To make the system work and to maintain communications at the local level, all the territory of the state was divided systematically and hierarchically, with well-defined responsibility and authority residing at all levels of administration. The system also recognized the different functions of administration between urban and rural areas and between Han Chinese and non-Han (other ethnic groups) areas.

Out of this administrative system there grew the political institutions and mechanisms that maintained control and ties between a central authority and its large supporting territory. This system prevented a fragmentation of territory according to local and sectional interest groups. So successful were the administrative system and the hierarchy of administrative units that they survived to the twentieth century. The Communist government has adapted the system to fit the changing needs of the modern Communist state.

[1]"Han people" is a name the Chinese use to refer to themselves (i.e., sons of Han or the Han Dynasty) to distinguish cultured Chinese from less cultured or "barbarian" people on the frontiers of the core area.

Territorial Structure of Contemporary Administration

According to the 1954 constitution, China's territory is divided into a three-tiered administrative hierarchy. At the top are (1) the twenty-one provinces (the eighteen provinces of the traditional core area and the three provinces of Manchuria) which were more recently settled by Han people; (2) the five autonomous regions; and (3) the three national cities of Peking, Shanghai, and Tientsin (Fig. 33–1). The middle tier in the administrative hierarchy is composed of districts (subunits of provinces) and non-national cities. At the bottom of the hierarchy are the rural villages and their surrounding lands. In this bottom group is the commune.

Rural Commune

The rural commune is the basic unit that the Communist government has used to initiate change. To some extent, it is a Chinese contribution to Marxist ideology in the countryside. Part of China's effort to reorganize and socialize agriculture in 1958 was the "Great Leap Forward." This effort was a radical movement to transform the traditional agricultural landscape and increase production rapidly. One of its basic features was consolidation of agricultural land and formation of cooperatives into large producing units, communes, as a means for transforming the traditional livelihood patterns. Some features of the "Great Leap Forward" have been abandoned because they did not work; but the structural unit of the commune has remained and today forms the main administrative unit of rural production and society.

Indeed, the territorial basis for the formation of the communes is rooted in the earlier administrative organization. No matter whether one considers the labor brigade, the producer cooperative, or the rural commune—all used the already established administrative territorial units of neighborhoods, hamlets and villages on which to build their own organization. It is a testimonial to the practicality of the old established structure that it should survive in a somewhat altered form. Often we are told how radical and changed the new China is from the old, traditional China. If one examines underneath the surface veneer of change and rhetoric and assesses the way things are organized, it is remarkable how much has survived, although with new names. Perhaps the key point is that although many facets of the structure and organization remain the same or are little altered, the operation of the administration system has changed greatly. Here, unquestionably, a far-reaching revolution has taken place.

China's View of a World Order

The Chinese internal political system has changed abruptly in response to new and powerful forces from the outside. In parallel fashion, China's dealing with other states has also changed over time to reflect new conditions and realities.

Several scholars of Chinese politics and diplomacy have sought to explain the political behavior of the modern Chinese state by referring to the map of Asia and examining recent behavioral patterns in terms of past patterns. Such an approach, based on map inspection of China's territory and neighbors, is termed a "cartographic approach." This approach is probably the most logical, for China has had a long time to develop a perception about its own and adjacent political areas.

The Chinese Point of View

China's view of a world order was well-established before extra-Asian contacts became important. The key point is that a Chinese-centered (Sinocentric) point of view has developed. The Chinese call their country *Chung-kuo* (The Middle Kingdom or Central State). Implicit in their name for their own country is the idea of the central point in a larger community.

33–2 *China: Its View of National Space*

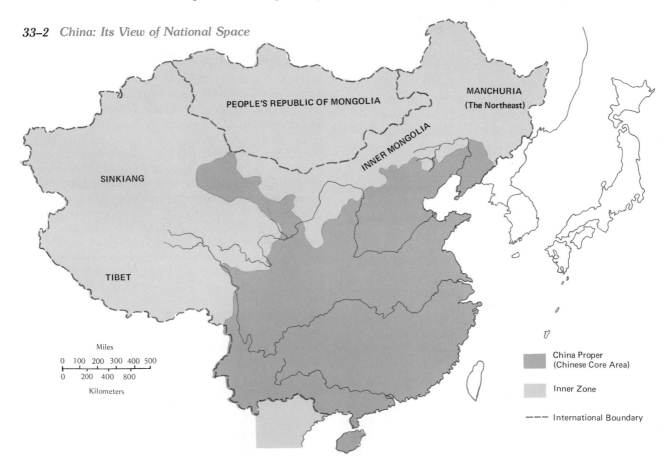

Moreover, there has always been a powerful element of ethnocentrism and felling of cultural superiority among the Chinese.

The Traditional Core and Inner Zone

In Han times, about the time of the birth of Christ, the Chinese empire stretched over much of central, southeast, and northeast Asia. China proper, or the eighteen provinces traditionally inhabited by Han Chinese, formed a cohesive spatial unit regarded by the Chinese as their heartland (Fig. 33–2). Adjacent to the heartland existed a large area over which the Chinese enjoyed a certain level of control. This controlled area included northern Vietnam, much of Tibet, Sinkiang, Mongolia, and Manchuria. The core of China is still essentially based on the eighteen original provinces. To these, however, should be added the three provinces of Manchuria and parts of autonomous regions of Ninghsia and Inner Mongolia to which Han Chinese have migrated in large numbers in the last century.

The Outer Zone

Associated with the adjacent inner zone was a series of flanking areas or an outer zone that traditionally had been under loose Chinese dominion; it was comprised of southern Vietnam, Korea, central Asia, and the trans-Amur region north of Manchuria. Included in this outer zone also were those tribute-bearing states that had long recognized Chinese hegemony (political dominance and leadership) in this part of the world. Examples include Afghanistan in the west and the Ryukyu Islands in the East China Sea (Fig. 33–3).

33–3 *China: Outer Zone and Beyond*

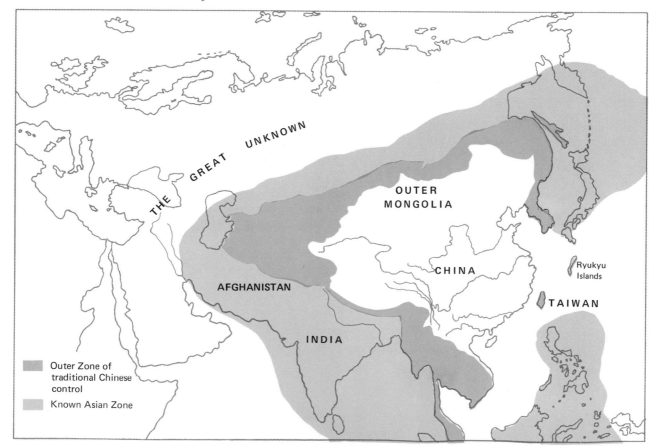

USSR

Trans-Amur Region

Sakhalin

People's Republic of Mongolia

Manchuria

Afghanistan

Pakistan

Korea

Japan

CHINA

Nepal

Bhutan

Ryukyu Islands

India

Bangladesh

Burma

Taiwan

Laos

Thailand

Khmer Republic

Vietnam

Philippines

Territory enclosed in broken colored line was under Chinese domination prior to the Opium War of 1838.

Malaysia

Indonesia

33–4 *China: Territorial Extent, 1838 and at Present*

Such was the outer zone. Its significance is that, although it was once regarded as Chinese-dominated territory, its value was less; it was a region peopled by barbarians, and its status was undetermined in areal or boundary terms. Indeed, one valuable insight into the Chinese view of a world order suggests that the Chinese had never defined political boundaries sharply but had viewed adjacent areas as frontiers of political influence and dominance. By the rise of the Ching Dynasty (1644–1911), China had established a perception of eastern Asia as a part of the world that it controlled and dominated. Although some of this territory was viewed as less important because it lay outside the core, nevertheless it was still Chinese or paid homage to the Chinese court. The frontiers of this Chinese part of the world included all of mainland southeastern Asia, several island groups in the South China Sea, Taiwan, the Ryukyu Islands, Korea, the Trans-Amur region of the Soviet Union including Sakhalin Island, Mongolia, much of Soviet central Asia, the Himalayan Kingdoms, and the Indian state of Assam.

Beyond the outer zone lay two further zones. One may be termed the known Asian zone or realm. This zone included kingdoms such as India and Persia (Iran) about which the Chinese knew a considerable amount and with whom they long had extensive contacts. Beyond was the "Great Unknown," a part of the world about which the Chinese traditionally had known very little and were presumed to care about even less. Europe and much of the Middle East were not well known to China or considered of much importance.

33–5 *China: Contemporary Core Area*

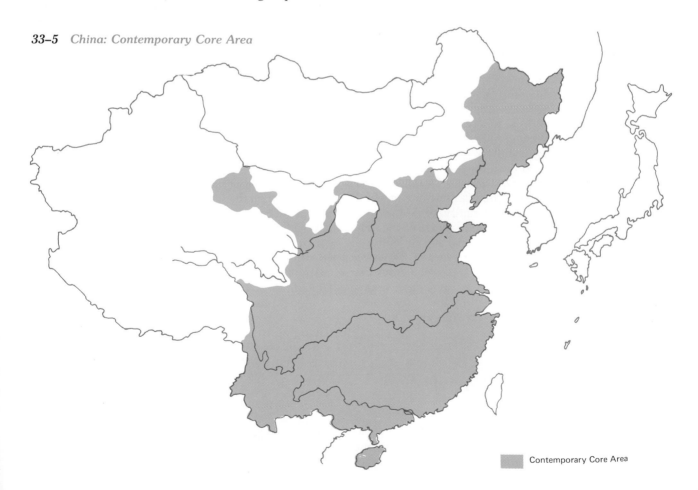

Contemporary Core Area

Boundary Problems and Territorial Conflicts

Although China has been gaining and losing control of territory for a long time, in the nineteenth and early twentieth centuries, the pressure placed on China by foreign powers was most serious (Fig. 33–4). In 1858 and 1860, China ceded to Russia the Trans-Amur region as well as territory in central Asia. In 1888, the Japanese successfully claimed the Ryukyu Islands, and in 1895 China ceded Taiwan to Japan after the Sino-Japanese War. In 1921, Mongolia was established as a separate state under Soviet tutelage. Territorial enclaves of colonial control were established by European powers, and the treaty ports became small islands of foreign power and influence in the giant sea of imperial China. By 1930, it appeared that Manchuria would be taken over by foreign control, either Soviet Russian or Japanese. In 1934, the issue was finally settled; Japan marched in and established a puppet state, Manchukuo.

Chinese colonization of Manchuria began in earnest in the late nineteenth century and provided a safety valve location for the rural poor of north China and resulted in the dense settlement of the Manchurian Plain. Early in this century, it is estimated that as many as 1 million farmers migrated yearly to Manchuria. This part of China has witnessed one of the great human migrations of recent history.

Rapid Chinese colonization of Manchuria, coupled with Japanese efforts to develop Manchuria and the rapid growth of industry, transport facilities, and cities, has made it one of China's most modern and progressive regions. The prosperity of Manchuria makes it important to the central planners of the new China as a major center for industrial and agricultural expansion. At the same time, in view of Taiwan's increasing political and physical separation from the People's Republic since 1949, it no longer seems appropriate to include Taiwan in the core. These recent changes are reflected on the latest map of China's core area (Fig. 33–5).

The Post–World War II Political Scene

Part of the settlement of World War II and the Potsdam agreement restored some lost territory to China. In all cases it was territory that Japan had seized. Thus, Manchuria and Taiwan were taken from Japan and restored to China. The most serious Chinese irredentist claims thus were satisfied.

Contemporary Border Problems

Since the ascendancy of a Communist government in China in 1949, China has embarked on an effort to modernize and develop the economy. Yet political uncertainties and antagonisms have interrupted and slowed the development process. Based on twenty-five years of hindsight, it is now clear that China's conflicts with its neighbors and other powerful countries have focused often on issues of territory and boundaries.

Since 1950, China has fought at least three mini-wars, and all appear to have centered on boundaries and territorial claims. These wars—United Nations–Korea (1951–1953), India (1963), USSR (1968–1969), and other incidents such as those with Burma and in Vietnam and the running dispute with the Soviet Union—illustrate how sensitive China is over matters affecting national territory. Given the long history and traditional manner in which China has viewed its own territory and that of its neighbors, some of the recent foreign affairs of China may make more sense. Thus, one can interpret China's willingness to fight when the United States and United Nations forces entered North Korea and posed a threat, according to the Chinese view, to Manchuria. In contrast, when the United States made clear its intention in 1965 not to enter North Vietnam, the Chinese never joined the Vietnam conflict as an active protagonist. It seems a historical lesson was heeded, and attention to the events of the past helped prevent an enlargement of the Vietnam War to the point that the United States and China did not confront each other in head-to-head fighting.

Changing International Relations

Recent events that affect the way in which China sees other powers in a world order and world community of states can tell us much about how China behaves, especially toward the United States. It may be argued that once the United States removed many of its troops from the rim of eastern and southeastern Asia, China perceived less of a threat from the United States. A major obstacle to United States–Chinese relations was thus removed. In 1971, Henry Kissinger journeyed to Peking, and President Nixon followed the next year. Diplomatic relations short of full recognition have been achieved and a small but active trade established. Thus, despite ideological conflicts between Marxism-Maoism and American democracy and free enterprise, the United States and China may be able to form a basis for cooperative coexistence. Probably the only thing restraining full diplomatic relations is U.S. continued recognition of an independent government on Taiwan. China believes Taiwan should be brought back into the Chinese fold. The United States now follows an ambiguous policy on the issue of one or two Chinas.

One far-reaching implication of China's new diplomacy, especially toward the United States, is the opportunity to increase internal economic growth and the development process in the People's Republic. The United States has an enormous range of scientific and technical knowledge that could help China. For example, China is attempting to exploit rapidly its large oil deposits but lacks the necessary sophisticated drilling technology. The United States could provide this technology if appropriate political and commercial linkages existed. In return, an alternative source of oil might become available to the United States. By establishing better political and trade relations, some of this knowledge could be shared. Moreover, new sources of capital are also involved. The United States could make valuable contributions to Chinese economic growth by providing low-cost financing for the purchase of this new technology. How far such contacts and exchanges go remains a major political question on both sides, but the potential contribution of the United States for promoting modernization and development in China is great.

Political realists may inquire why the United States should want to deal with China and what benefits might accrue to the United States. One reason is the opportunity to lessen tensions between the two major powers. Political relations may improve in step with increased economic interchange. A second reason is that trade can lead to a large, new market for American goods and technology. Recent sales of foodstuffs to China have resulted in a large trade surplus in favor of the United States. Such dealings could greatly assist the United States in overcoming a balance-of-payment problem and stimulate United States economic activity. Some authorities suggest that increased economic interdependence among nations can lead to more friendly political relations.

It seems that tensions between the United States and China have lessened. By contrast, if one were to predict future conflicts in Eastern Asia, it is likely the Soviet Union would be seen as China's chief antagonist. The territorial and border problems between China and the USSR appear to outweigh any affinity or trust based on similar political and economic systems. At the same time the potential for future Sino-Japanese differences exists as both countries strive for hegemony among other Asian states. China's relations with India are also soured by territorial claims and counterclaims, and no mutual agreement on these issues appears likely without major concessions on one side or the other.

Further Readings

Central Intelligence Agency, *Communist China, Administrative Atlas* (Washington, D.C.: Central Intelligence Agency, 1969), 42pp. A detailed atlas of China's administrative areas.

CHAO, Y. R., "Languages and Dialects of China," *Geographical Journal,* 102 (1943), 63–67. Chao is one of the world's foremost linguists. In this brief article he presents an accurate linguistic map of China.

CLUBB, O. EDMUND, *Twentieth Century China* (New York: Columbia University Press, 1967), 470pp. An interesting and readable account of China's recent political history.

GINSBURG, NORTON, "On the Chinese Perception of a World Order," in Tsou, Tang (ed.), *China in Crisis* (Chicago: University of Chicago Press, 1968), pp. 73–93. The author presents several cartographic models of China's view of Chinese and neighboring territory and urges that maps be consulted to determine China's view about the world in evaluating Chinese behavior.

HARRER, HEINRICH, *Seven Years in Tibet* (New York: Dutton, 1953), 314pp. Harrer's account is a narrative of his travels in Tibet and is one of the last detailed geographic descriptions of late traditional Tibet.

HERMAN, THEODORE, "Group Values toward the National Space: The Case of China," *Geographical Review,* 49 (1959), 164–182. Herman outlines the main Chinese attitudes toward their own territory and interprets policies in light of these values.

HSIAO, KUNG-CHUAN, *Rural China: Imperial Control in the Nineteenth Century* (Seattle: University of Washington Press, 1960), 783pp. Hsiao presents a detailed account of the nature and functioning of the local administrative system in nineteenth-century China.

LATTIMORE, OWEN, *Inner Asian Frontier of History* (New York: American Geographical Society, 1940), 585pp. This work contains a collection of essays on the ancient and modern history of the peripheral areas of China, Mongolia, Manchuria, and Sinkiang.

LATTIMORE, OWEN, *Studies in Frontier History* (London: Oxford University Press, 1962), 565pp. Using themes set forth in his *Inner Asian Frontier,* Lattimore focuses this work on minorities and the regions they traditionally inhabit.

McCOLL, ROBERT W., "The Development of Supra-Provincial Administrative Regions in Communist China, 1949–1960," *Pacific Viewpoint,* 4 (1963), 53–64. In this essay, political events are interpreted for the bearing they have on the designation of administrative territory.

SALISBURY, HARRISON, *War Between Russia and China* (New York: Norton, 1969), 224pp. A provocative work that suggests the inevitability of war between China and Russia over the question of territory.

SCHWARTZ, HENRY G., "The Treatment of Minorities," in Oxsenberg, Michael (ed.), *China's Developmental Experience* (New York: Praeger, 1973), 227pp. Schwartz presents a lucid statement of the past and present Chinese policies toward minority peoples.

China:
Changing Society
and Economy

34

A FTER 1949, with the ascendancy of the Communist government in China, the main objective of the new political order was to make over China completely. The new government sought to modernize China's economy and society as its political system had been made over during the years of revolution. Since 1949, radical new economic and social programs and policies have been promulgated. These revolutionary policies and the events associated with them form a modern phase of China's development. In this chapter modern China is contrasted with traditional China.

China's Population

China as long ago as the year A.D. 2 is believed to have had a population of 60–70 million (Table 34–1). Although the population may have fluctuated by more than 20 million during specific periods, ten centuries later it was still around 60 million. Such a demographic history, on the one hand, suggests that in early China a crude balance was achieved between the number of people and the ability of the agricultural system to support them. Without some innovation or technical change, the population did not grow.

On the other hand, it may suggest that the system of counting people was inaccurate, and that the census enumeration system of the time resulted in a serious undercount. Whatever the case may be, by 1400 the population had begun to grow, and by 1751 it had reached 207 million. A century later, on the eve of the Taiping Rebellion, China's population had grown to about 420 million. Perhaps some better method of enumeration was followed which gave a more accurate figure. More likely, better techniques of farming, technical innovations, and greater internal movement of large amounts of food grains led to substantial growth in the population.

The Last Century

The outbreak of the Taiping Rebellion in 1850 began a century of turmoil, war, and demographic uncertainty. The consequence of this uncertainty was a high death rate and a decrease in the rate of population growth. After the Communists assumed power, they reported a 1949 population of 542 million people, an increase of about 25 percent over the preceding century. Since 1949, population growth has been great.

Although subsequent estimates of China's population are rather crude, one thing is certain: even the most conservative estimates of China's population growth suggest that more than 800 million Chinese live on the mainland today. More generous estimates assert that there were 962 million Chinese on the mainland in 1975 (Table 34–2). The truth probably lies somewhere between these two figures and may be in the neighborhood of 900 million. In addition to these 900 or so million, there are another 40 to 45 million Chinese who reside in Taiwan, Hong Kong, Southeast Asia, and other overseas locations.

Population Structure, Composition, and Growth

A most significant act of the Communist government when it took over in 1949 was to institute a comprehensive program of state-assisted public

Table 34–1 *Estimates of China's Population, A.D. 2–1851*

Dynasty and Year	Population (millions)	Dynasty and Year	Population (millions)
Western Han		Sung (*cont.*)	
A.D. 2	71		
		1065	77
Eastern Han		1075	94
88	43	1086	108
105	53	1094	115
125	56	1103	123
140	56		
156	62	Chin	
		1193	123
Sui			
606	54	Ming	
		1381	60
Tang		1393	61
705	37		
726	41	Ch'ing	
732	45	1751	207
742	51	1781	270
755	52	1791	294
		1811	347
Sung		1821	344
1014	60	1831	383
1029	61	1841	400
1048	64	1851	417

Note: These figures should be considered as only rough guides to the demographic situation at particular points in China's history.

SOURCE: J. D. Durand, "The Population Statistics of China, A.D. 2–1953," *Population Studies,* 12 (1960).

health. This program led to a rapid decline in the death rate and a great surge in the annual population growth that began to subside only after 1960. At the same time, family planning has led toward greater balance between the sexes, for traditionally there were more males than females, and a low value was placed on female infants and children. Today's balance between males and females is in line with China's new social objectives that stress equality between sexes and a more important role for women.

Presently it is estimated that the annual birth rate has dropped to 2.7 percent. This rate is still high, but it is a significant decline from the 4.5 to 5.0 percent of traditional times. Offsetting the decline in birth rates is a continued decline in death rates to an estimated 1.0 percent (Fig. 34–1). Thus the annual growth of 1.7 percent per annum, if accurate, indicates that China is midway in its path along the demographic transformation (Fig. 1–4).

Additional statistics suggest that although China continues to be a relatively poor state in terms of per capita income, the capacity exists to reduce birth rates rapidly in step with the demographic transition from traditional patterns to modern. The point is well illustrated in the reports of recent visitors who claim they have seen statistics that indicate drastic drops in

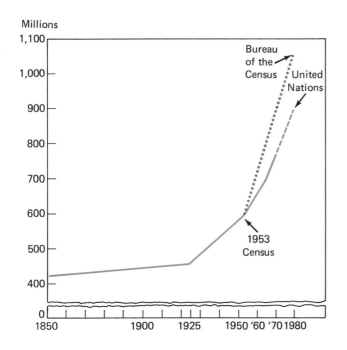

34–1a *China: Estimated Population, 1851 to 1980— Two Projections.*

SOURCE: "China: Population in the People's Republic," *Population Bulletin,* 27 (1971).

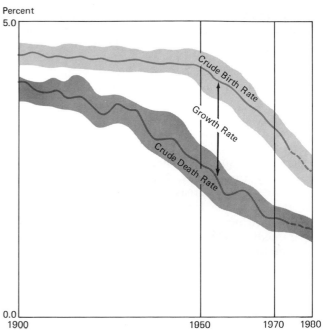

34–1b *China: Trends in Birth and Death Rates, 1900 to 1980*

SOURCE: Population Reference Bureau.

Table 34–2 *China: Alternative Population Estimates (in thousands)*

	U.S. Bureau of the Census		United Nations (Medium Variant)		Orleans	
YEAR	NUMBER	AVERAGE ANNUAL GROWTH RATE	NUMBER	AVERAGE ANNUAL GROWTH RATE	NUMBER	AVERAGE ANNUAL GROWTH RATE
1960	718,004		—		655,000	
		1.8		—		1.4
1965	782,555		695,000		701,000	
		2.3		1.8		1.5
1970	871,035		759,619		753,000	
		2.1		1.7		1.7
1975	962,480		825,821		813,000	
		2.0		1.6		1.7
1980	1,060,695		893,900		887,000	

SOURCE: China: Population in the People's Republic (*Washington, D.C.: Population Reference Bureau, 1971*).

the birth rate in urban precincts in Shanghai.[1] If these statistics, which are for a relatively small sample in the Shanghai area, reflect national patterns, China has made great strides toward getting its population growth into balance with its resources and production, an achievement unprecedented for such a populous country.

Most claims about China's birth and death rates are difficult to evaluate inasmuch as China is so large and family planning programs implemented in a large, modern city such as Shanghai may not have diffused to many rural areas. Nevertheless, enough information on rural areas has been made available to suggest that the family planning policies and programs are now being extended to rural areas. The paramedical teams, for example, function in part to disseminate knowledge and instruments of birth control. For this reason among others, some authorities have suggested that the Chinese birth rate has dropped more sharply and rapidly than suggested above. One spokesman has even gone so far as to suggest the Chinese birth rate, based on recent trends, will drop below the rate of birth in the United States within the next decade.[2] If such a trend is real and continues, the Chinese have indeed demonstrated the powerful role of central government direction in producing rapid social and demographic results.

Family Planning

Achieving a more moderate growth rate has not come easily. Traditionalists believed that large families were good, the natural order of things in Confucian terms. As in agrarian societies throughout the world, children, and especially sons, were seen as economic assets to look after their elders in their declining years. Although the Communist government was skeptical about birth control initially, it appears that after 1960, family planning was viewed with increasing favor and necessary to rapid economic progress. Except for the period of the Cultural Revolution (1965–1968), it seems that China has promoted vigorously family planning. The "pill" is the most commonly used method of birth control, although a variety of other practices are also employed, including sterilization, especially of females, and abortion.

The extensive public health and welfare system is an effective agent in promoting birth control, and the "barefoot doctor" (paramedic) is heavily involved in spreading the word about family planning into remote rural areas. With the improvements in medicine and greatly reduced death rates, parents may anticipate their children will reach adulthood, and there is no need to have more to ensure that some children will live. The idea the children will provide a support for their aged parents is also outmoded, for the state now cares for people too old to work who have no descendants to look after them.

Other measures may be even more influential in affecting birth rates in the countryside. Especially critical is the "work-point" system whereby total income of rural work teams and brigades is apportioned according to the total work points compiled by the working adults. Thus, the more children in a family, the smaller each individual's share of the family total will be.[3] Another device to reduce rural births is the allocation of private gardening plots to each family. Plot size increases with the number of members in the family to a maximum of four. Families with two children thus receive the same amount of land as those with six children. This system is used as a kind of negative tax. More Draconian measures are also employed; for example, young people are discouraged from marrying before a certain age (twenty-four for women, twenty-nine for men). In urban areas young married couples may be denied housing before they reach the required age.

The success of these programs and policies indicates that China has made remarkable progress in altering the value system of its citizens. A new ideal of a small family with two children has been proclaimed. If we can believe the available information, this ideal is being accepted rapidly. Such a change in values appears to place China in opposition to the established Marxist position on population growth. The Marxist position accepts any population growth as good in that the labor potential is increased. Such a

[1]Victor and Ruth Sidel, "The Delivery of Medical Care in China," *Scientific American,* 230 (1974), 19–27.

[2]A reply attributed to the Minister of Health of the People's Republic of China when queried as to China's 1974 rates of birth and death were 2.5 and 0.7, respectively. The reply was conveyed through a spokesman for the United States Department of State (June 1975).

[3]An excellent discussion of the operation of these policies is contained in Sterling Wortman, "Agriculture in China," *Scientific American,* 232 (1975), 21–22.

position, however, seems to conflict with policies of social enlightenment such as the emancipation of women from their traditional roles. Chinese Communists may perhaps justify family planning on the basis of its role in gaining greater status for women. Also, it may well be that the Chinese are not too concerned with the link between Marxist theory and Chinese practice as applied to their goals of modernization. Pragmatism and success may be more important.

Where the People Are

In China, people are concentrated where they have been for many centuries. Today, within the eighteen provinces of China's traditional core area more than 800 million people reside. One large area has been populated within the last century, Manchuria. Han Chinese recently have also migrated in large numbers to the region of the Great Loop of the Hwang Ho. If a line were drawn connecting Kunming in southern Yunan Province with Tsitsihar in western Manchuria,

34–2 *China and the Rimland: Population Density*

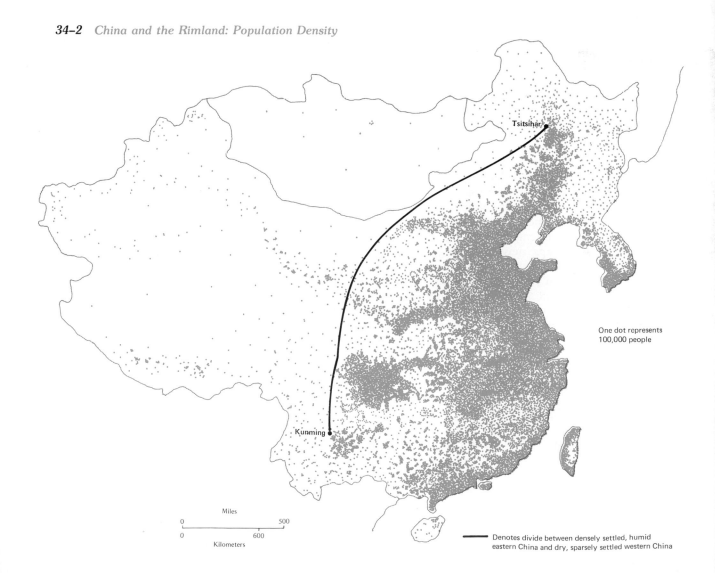

One dot represents 100,000 people

Miles
0   500
0   600
Kilometers

—— Denotes divide between densely settled, humid
eastern China and dry, sparsely settled western China

more than 90 percent of China's people would live east of that line. The eastern part of China contains less than one-third of the total area (Fig. 34–2). It is also true that eastern China contains most of the well-watered, productive agricultural land. Thus, there exists in China a coincidence between arable land and population. With the dense settlement of the central part of the Manchurian Plain in this century, the last major region of the Chinese state able to support a dense population was occupied. Future development of agricultural lands will depend on expensive and difficult irrigation and reclamation projects if population densities similar to the core region are to be supported.

The Dry West

West of the Kunming-Tsitsihar line, rainfall decreases to the point that settled agriculture is marginal. Perhaps for this reason most of the region has traditionally been the domain of pastoralists and specialized oasis farmers. Although two-thirds of China's territory is included, only about 60 to 70 million people live here, and many are members of China's minorities. These relatively empty areas of China would appear to invite migrants from crowded eastern China. The nature of the environment with its paucity of rainfall, however, is alien to the traditional methods of intensive farming common in eastern China. Perhaps new technologies, such as mechanized agriculture and improved irrigation techniques, will open much more of this region to denser settlement, but it seems unlikely that Han people will live in western China in the concentrations so common in the more humid east.

China's Changing Social Patterns

Changes in China's society have been dramatic in recent years. In part, these are the changes expected in any society transforming itself from traditional to modern, from rural to urban, and from agricultural to industrial. To these changes has been added Marxism, which seeks to transform the basic value system rapidly and thereby speed up modernization. The special Chinese brand of communism, Maoism, has contributed a distinctive flavor and demands radical approaches and great speed in the development drive.

The Traditional Pattern

Society in traditional China was focused for the most part on family relationships at the village level. Other social units and relationships existed, but for more than 2,000 years, a family-oriented social system existed based on ethical norms and values outlined by Confucius as early as 436 B.C. In this system, order was based on the primacy of the head of the household, and his desires ruled over the younger generation who lived in the same household. Male heirs, who could carry on the family name and line, dominated the system and ensured its perpetuation. Women were important for the work they performed but had little role in formal decision making and no proprietary rights. They were to some extent thus excluded from a meaningful role in affairs of the family and the village.

Beyond the family, there were established common surname associations or clans. These clans frequently acted to satisfy larger mutual interests and would temporarily side with other clans when some joint purpose was served. In addition to clans, other voluntary associations and groups emerged. Often these were focused around a religious objective, some common commercial, or special activity interest.

Such was traditional society. It centered on the village, and most of the relationships that grew up were village oriented and based on kinship ties. To a considerable extent, villages were self-sufficient. There was a tie between the economic system with its self-sufficiency and the inward-looking social system that was associated with it. It was the political system as much as anything that linked the village with the rest of China. Social links with the outside did exist, but for most peasants, these links were not well established and were rarely used.

Society in the New China

Social revolution is as much a part of the Maoist remolding of China as change in the political and economic systems. In fact, all three are related components of a broader process of development. The basis for the success of Chairman Mao's revolution was founded at the village level, and social reform in the village followed wherever the Communists were able to establish themselves.

The Gate of Heavenly Peace in Peking, the capital of China. The Tien An Men Square in the foreground has been the scene of many parades and demonstrations. (United Nations/T. Chen)

Among the most important consequences of the new social revolution was the shift in family dominance. Authority has shifted from the traditional head, the eldest male member of the family, to a younger man, usually the oldest son, who is also the most productive member of the family. Although this action may seem plausible, it is a reversal of the established Confucian norm and an indicator that social changes in China are fundamental and far-reaching.

Another abrupt change from established practice was the creation of new roles and rights for women. Women have been freed from their traditional subservience and lowered status and have been encouraged to spend time away from home, often as salaried members of the labor force. Men at the same time have been encouraged to view their wives as equals and to share in the chores of household work. Men also have had to recognize the rights of their wives to join in political activities and associations that promote the group interests of women. In this new set of

relations, a basic alteration in the established Confucian order has taken place.

Rural and Urban Society: The Role of "Hsia Fang"

In social relations, it is almost surprising that the part of China usually regarded as most traditional and conservative, the village, has been deeply involved in the Chinese revolution. Urban society too has been much changed, but that is expected, for cities everywhere have been agents of change and modernization. Changes in urban China have been even more far-reaching than in rural areas. For example, the life that revolves around the employment center and its associated dormitories suggests an aura of the brave new world. It may be that the urban quarters or districts function successfully as cohesive social units and provide services and social needs that are often unsatisfied in cities in other societies. It is clear that certain neighborhoods have developed collective manufacturing projects or have otherwise operated in a joint manner. Such collective behavior suggests an urban neighborhood extension of the old rural pattern of local production for local consumption.

One aim of the "Cultural Revolution" was to reform society and destroy the emerging "elitism" of the party members and educated groups. A special means devised for reeducating the elite was to transfer these people from their privileged jobs and positions in the cities and send them to rural communes to perform periods of farm labor. This transfer process is known as *hsia fang,* a practice that Chairman Mao believed was necessary to maintain the revolutionary ardor and purity of the urban masses. Although *hsia fang* was widely practiced in the late 1960s, it apparently was unpopular, and the urbanites for whom it was designed have not always accepted it fully as a good means for strengthening their revolutionary zeal.

The Changing Economy

China's contribution to the development process may be its emphasis on rural areas as economic growth centers. The government believes that the village has played a key role in revolutionizing China from its traditional patterns in less than a generation. It is this accomplishment, based on the special approach to and reform of traditional village China, that is believed to be a unique contribution of the Chinese to the theory and process of development and modernization.

The economy of China has always been rooted in and dependent on agriculture. Even after twenty-five years of Communist rule, China remains an agricultural country with more than 80 percent of its people engaged in farming and related activities. Despite the fact that farming continues to dominate employment activities, the nature of the rural economy has changed greatly in this century. Collectivization of land and organization of rural activities by communes have altered many traditional rural patterns and especially the uneven distribution of landholdings. At the same time, some of the old cropping practices and farming techniques linger.

Part of the difficulty in economic modernization that faces contemporary China is how to accommodate the 15 million or so new citizens that enter the labor force each year. Cities have grown and in total may hold as many as 200 million people. Yet industry and services cannot absorb the huge number of new workers. For example, in 1975 it was reported that between 680 and 756 million people in China depended on the agricultural sector for their livelihood.[4] The agricultural sector must provide most of the new jobs. In view of the labor supply, a serious question should be raised about the necessity and desire of Chinese planners to mechanize agriculture. The social consequences of reducing labor requirements on farms and releasing great numbers of rural dwellers to move to cities would place enormous added stress on cities that may already be too crowded.

Traditional Village Agriculture

For many years, villages in China were described as micro-units of production in which most things consumed in the village were produced locally. Food, seed, manure, tools, and clothing were the most obvious items. It was on these goods and their production that life in rural China depended.

In recent years a contrasting view of rural China has been depicted which asserts that in many parts of China, especially where water transportation was available, a significant degree of spatial integration of the economy occurred. Commercial linkages did exist between village and city and between local producers and the national economy. Although essential foodstuffs, cloth, and basic tools were produced locally, many other items were desired. Through a network of periodic and permanent market centers, goods from other places were sold locally, and local commodities were taken in exchange. A number of items of trade were involved: salt, luxury goods, specialty tools, equipment and metals, special edibles, and for a time opium. The degree of exchange appears to have been related to the accessibility of a place. The point is that most places did have some exchange and trade with the outside and to that extent were tied into a national economy. It has been further demon-

[4]Ibid., p. 13.

strated that social relationships followed the direction of the commercial and marketing patterns.

The two views of traditional village China may not be incompatible, for it may be that self-sufficiency in basic goods was a hallmark that coexisted with a modest, specialized trade in certain necessities and luxuries. Surplus agricultural commodities financed trade. Much of traditional China was not isolated, and spatial linkages did exist to ensure the flow and exchange of goods, money, people, and information.

Beginnings of Modernization

Much of the modernization of China began with the arrival of Europeans in the early nineteenth century. The establishment in the nineteenth century of European-dominated treaty ports along the coast and major rivers began a new era in the commercial evolution of China (Fig. 34–3). For the first time, China was forced by foreign powers to open up and establish commercial linkages. It is true that the influence of the new foreign ties was largely confined

34–3 *China: Foreign Penetration, ca. 1910*
SOURCE: Adapted from W. Gordon East and O. H. K. Spate, *The Changing Map of Asia* (London: Methuen, 1966).

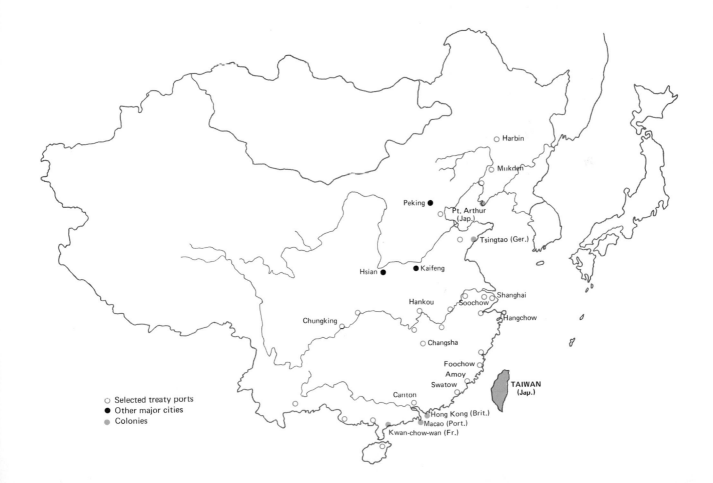

to the areas immediately adjacent to the treaty ports and did not penetrate much of the great interior of China. Nevertheless, a new age had begun. Technical innovations were introduced, and capital improvements in harbors, cities, transport systems, and factories were constructed. Equally important, the training of the human resources was accomplished rapidly, and an effort was launched to repair the centuries of neglect in the sciences and engineering.

One aspect of the treaty ports was their concentration in the most accessible locations associated with water transportation, the only cheap, effective means of transport in traditional China. Such a spatial concentration has left its mark on contemporary China. The main industrial and service centers and the largest cities continue to be concentrated along the coasts and rivers and in Manchuria, the latter a testimonial to the more recent efforts of Japanese colonialism. Such a pattern has led to calls to move the new centers of commerce and industry into the interior in order to break the "improper colonial" pattern of past development and to permit the defense of China's cities from invasion by sea.

Chinese Agriculture

Traditionally, China increased agricultural production through increasing inputs of organic nutrients and labor, multiple cropping, and through expanding cropland. This process was especially common in the south with its rice-growing environment and the steady southward expansion of the frontiers of Chinese settlement over the centuries. With the settlement of Manchuria in this century, no more attractive agricultural frontiers remain. The primary path to greater agricultural output in the future must come from increased per-unit yields.

The High-Level Equilibrium Trap

Traditional techniques of increasing crop output in China were simple: add more labor and night soil (decomposed human excreta) and intensify multiple cropping. Unfortunately, increased crop output led to more people. A greater population in turn reinforced the cycle by providing more laborers and more human waste. Gains in production were countered by

an increased population, and such a pattern meant that most people existed at a low level of subsistence. This system developed a kind of self-reinforcing chain reaction whose ultimate result could only mean a descending spiral of poverty for most people. There were no stimulants for technical breakthrough or investments in ideas or products that might have led to technological innovation. Innovation in turn might have lessened the traditional demand for labor and reduced the incentive for large families. The population continued to expand in step with expanded commodity output. Per capita wealth and demand remained stagnant and may have even decreased after the sixteenth century. It was this situation, described as a "high-level equilibrium trap,"[5] that resulted in increasing poverty and created so much stress on the traditional Chinese economy and society. It has been argued that this stress was the main precondition to the demands for the reform and modernization of China.

Peasant life in traditional China was frequently difficult, and even the village elites and main landlords were hardly wealthy by Western standards. Several levels of peasants existed: landlords, middle peasants who farmed their own land, and poor peasants. The poor peasants frequently hired themselves out or otherwise became indentured or tenanted and, thus, associated with one of the wealthier peasants. The landholding relationship and pressure for land became more urgent with increasing populaiton growth, for the effect of more sons was to fragment more and more the existing cropland and to reduce continuously the amount of farmland available to most peasant families.

Such a situation when coupled with the vagaries of natural catastrophes of drought, flood, earthquake, or locust plagues yields an unpleasant picture of traditional peasant life. This picture was especially true in north China where the natural conditions were more erratic and less subject to control. Floods and drought would lead to food shortages, and many families would flee to the south in search of a better situation. Landlords would squeeze poor families harder, and often the difficulties would lead to local violence. In some cases such violence may have been

[5]Mark Elvin, *The Pattern of the Chinese Past* (Stanford: Stanford University Press, 1973), pp. 203–319.

Table 34–3   *China: Food Grain Production,*
1949–1971 (in millions of tons)

Year	Estimated Output	Stated Output (Chinese Sources)
1950		125
1955		175
1957	200	185
1960	160	150
1965	190–195	200
1971	215–220	246

SOURCE: U.S. Congress, Joint Economic Committee, *People's Republic of China: An Economic Reassessment* (Washington, D.C.: Government Printing Office, 1972).

widespread enough to crystallize into larger movements with political objectives.[6]

Modern Agriculture

In addition to socializing agriculture through state ownership of land and the formation of rural communes, Chinese planners in recent years have attempted to raise crop yields and production. Many approaches have been followed: land reclamation, better land management, improved irrigation, and increased dry cropping. Perhaps the most promising of these new approaches are innovations in the form of chemical fertilizers, hybrid grains, insecticides, and improved irrigation systems. All of these techniques are forms of substituting capital for labor in the agricultural sector, a practice common in the West but only recently applied intensively in China.

Production and Output

In the early 1950s, the new government was fortunate to have several years of good weather, and agricultural output increased. The "Great Leap Forward" program of 1958, however, coincided with

[6]A chronicle of such events on the eve of Communist takeover may be reviewed in William Hinton, *Fanshen: A Documentary of Revolution in a Chinese Village* (New York: Vintage Books, 1966), 637 pp.

several years of poor weather, and crop production fell. Since then, except for some uncertainty during the years of the "Cultural Revolution," output of food grains is believed to have increased gradually in pace with the growing population (Table 34–3).

Despite increased crop production, China has frequently purchased grain (wheat usually) from several Western states. The major suppliers have been Australia, Canada, and the United States. Such sales suggest that China may not produce enough food to support its huge population. At the same time, however, China exports a considerable amount of high-cost foodstuffs (vegetables, pork, condiments, and specialty foods) to Hong Kong, Southeast Asia, and Chinese communities all over the world. China may well be a net exporter of foodstuffs, despite the large grain purchases. China has claimed self-sufficiency in food production since 1971.

Food exports notwithstanding, China's foreign grain purchases suggest a weakness in agriculture that makes modernization and increased productivity of the system ever more pressing. One major problem is that little good land is available for agricultural expansion, at least under present technologies. For example, the United States with only one-quarter the Chinese population, has 153 million acres (60 million hectares) more arable land. With only dry or marshy lands available for future expansion, China faces an increasingly difficult situation with its growing population. Statistics on rural land use show that cropland has not expanded significantly in fourteen years. Yet China depends largely on the agricultural economy to form capital in order to finance development projects. It is imperative that per-unit yields be increased rapidly if agriculture is to continue financing development and living levels be raised simultaneously.

An associated and equally urgent problem is that as the economy modernizes, it will increasingly urbanize. Like Japan, the cities are commonly found on the best agricultural land. As cities grow, good agricultural land is lost to urban land uses. Here is a built-in conflict, wherein modernization and economic development programs are likely to result in the loss of a significant amount of the best agricultural land. India, with its higher percentage of arable land and lower multiple-cropping index, does not yet face this problem to the same degree.

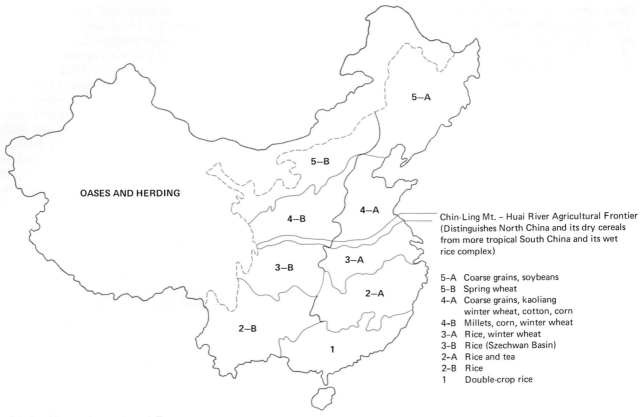

Chin-Ling Mt. – Huai River Agricultural Frontier
(Distinguishes North China and its dry cereals
from more tropical South China and its wet
rice complex)

5-A Coarse grains, soybeans
5-B Spring wheat
4-A Coarse grains, kaoliang
     winter wheat, cotton, corn
4-B Millets, corn, winter wheat
3-A Rice, winter wheat
3-B Rice (Szechwan Basin)
2-A Rice and tea
2-B Rice
1   Double-crop rice

34–4 *China: Agricultural Regions*

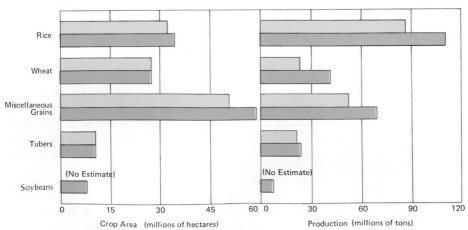

34–5 *China: Area and Production of Important Crops, 1957 and 1971*

SOURCE: Adapted from Sterling Wortman, "Agriculture in China," *Scientific American,* 232 (1975).

Crop Regions and Crop Types

Food grain (including rice) are the leading crops produced in China. As a general rule, rice is grown wherever environmental conditions permit. In the southeast, where frost is uncommon, double-cropping of rice is the usual pattern. Rice tends to produce the most edible grain on a given unit of land, provided that there is plenty of labor, nutrient material, and water. Thus, the alluvial valleys and basins of the southeast, where rice can be grown, are among the most densely inhabited parts of China (Figs. 34–2 and 34–4).

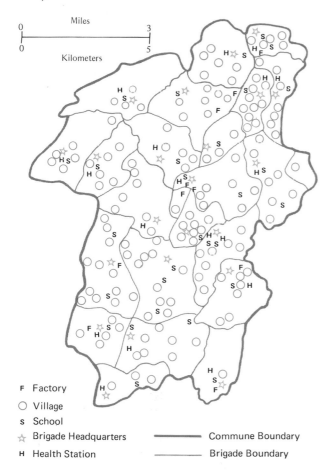

F   Factory
○   Village
s   School
☆   Brigade Headquarters
H   Health Station
──────   Commune Boundary
──────   Brigade Boundary

34–6   *China: Hwa Shan Commune, Division into Production Brigades*

SOURCE: Adapted from Keith Buchanan, *The Transformation of the Chinese Earth* (London: G. Bell, 1970).

Dry cereals, wheat, millets, and kaoliang (a type of grain sorghum) tend to dominate foodstuffs in the drier, cooler north, although cash crops such as cotton are also common. In the south the more common cash crops are tea, sugar cane, and tung. The most striking pattern of Chinese agriculture is the transition from wet rice in the south to dry cereals north of the Chinling Mountains–Huai River frontier. Production and area statistics for the most important crops are depicted in Figure 34–5.

Throughout China, a rich and interesting array of fresh vegetables is part of the agricultural scene. Today, much of this fresh vegetable production, like that in the Soviet Union, is produced on private plots and is either consumed at home or marketed privately. A large quantity of poultry and hogs is also produced on the small amount of land assigned as private plots to most peasant households for their own production and consumption.

The Rise of the Commune

Probably the most important economic unit in modern China is the commune. Although at one time the term commune was used to refer to urban neighborhoods and districts, it is in the rural areas that communes have functioned best as operating units. Today's commune is large in area (an average of 60 square miles or 155 square kilometers) and usually composed of the territory and fields of a number of small hamlets and villages. Farmers are organized in labor brigades and teams. The commune is organized hierarchically and composed of several brigades and many teams (Fig. 34–6). In addition, a central office conducts administrative affairs and maintains ties with other political, social, and economic units.

The Commune Structure and Operation

Communes first appeared in China in 1958, about a decade after the ascendancy of the Communists. Their formation was the achievement of a stage of communism outlined by Chairman Mao to promote modernization through one of the great movements that have swept revolutionary China. In this case, "The Great Leap Forward" was designed to push economic growth to a level of output equal to that of the United Kingdom within a decade. To accomplish the goals, it was necessary to rely on small-scale rural industries as well as the production of large factories. The communes were organized to support the new efforts to modernize, but they built on an extablished pattern of village production and social, political, and economic self-sufficiency. Not only were these communes to grow their own food, but also they were encouraged to build local industries to produce clothing, tools, and other necessities: local production for local consumption. At the same time any surplus could be sold or shipped to other regions. Return goods would complete the flow and suggest the modern adaptation of a well-established Chinese pattern of spatial organization and interchange.

The self-contained aspects of the communes are seen in the services available. Communes today generally contain schools, medical and dental facilities, market and retail establishments for necessities and provisions, a post office, a library, and recreational facilities and clubs. They are functioning communities and to some extent represent a modern version of the traditional Chinese village and rural township pattern.

Individual families are assigned to specific labor teams and as such compose part of a particular brigade within the commune. In general, all adult members and some older children of a farm family perform labor and are awarded work points for their labor. The work points form the basis for the share the family will receive from the total earned by the commune after expenses and taxes have been paid. In addition, the family works a small private plot of land assigned it by the commune and usually also owns a couple of pigs and a few ducks or chickens.

The private plots, as in the Soviet Union, are very important. Between 5 and 7 percent of China's arable land is composed of such plots, and their output of fresh vegetables is probably equal to 50 percent of total national production. Of equal importance, however, is the support they provide for swine production, a primary source of animal protein in the Chinese diet and of organic fertilizer for crops. As the Chinese discovered during the Great Leap Forward, a decline in swine production brought about through curtailment of the family-operated private plots has far-reaching consequences for agricultural output. The private plots were restored to individual families after a brief but disastrous interruption.

How much better off a family in one of China's rural communes is today than a poor peasant family of fifty years ago is not easy to evaluate. Most members of every family must still work long and hard.

Today, however, there is enough food for everyone, even if the government must import it. A basic educational opportunity is provided everyone in free schools, and a health service of national dimensions provides reasonable care to all at very low cost. Along with these benefits goes a certain amount of political education, and the state expects everyone to comply with its directives without question.

Recent Changes and New Policies

In 1961, the "Great Leap Forward" was abandoned as a revolutionary strategy. But some aspects of the "Great Leap" have survived. One of the most important aspects relates to the nature of the commune and the ideal of rural self-sufficiency. In recent years a new policy or movement called "Walking on Two Legs" has been cited frequently as the correct path to developing China. "Walking on Two Legs" is designed to use both large and small production units, both traditional and modern means of production, and both urban and rural areas to promote development. This policy demonstrates that the leaders and planners of new China have a pragmatic and flexible approach that includes building on established, workable patterns from the old China in their search to find solutions to the problems of achieving rapid economic growth.

Industrialization and Economic Growth

Location and size have been fortunate for China. One of the important consequences of China's great size and particular location is a rich and diverse resource base and accessibility to these resources. Both factors help promote industrialization and economic growth. The resource base includes not only a wide variety of environments for the production of food and cash crops, some of which (cotton, tobacco, soybeans, and tung nuts) are closely associated with industrial processing, but also substantial reserves of fossil fuels (coal and petroleum) and minerals (iron ore, tin, manganese, copper, and molybdenum). These resources serve the state by assuring a degree of economic self-sufficiency that requires less dependence on external sources. China, therefore, does not need to expend scarce capital on imported raw materials. Indeed, foreign exchange can be gained by the sale of some of these resources. Today, for example, China is able to finance the purchase of sophisticated machinery and technology from Japan through the sale of crude oil.

Accessibility, by contrast, is more difficult to assess, for in much of China the landscape is rough, and transportation has traditionally been slow and expensive. Water transportation is an exception, and based on the former treaty ports, links China increasingly with the outside world. In this century new emphasis has been placed on other forms of surface and air transportation, as China has sought to tie together the far-flung parts of its vast territory.

Pre-Communist Modernization

Although China has a long tradition of small-scale, cottage industry, manufacturing, based on innovations associated with the Industrial Revolution, arrived in China in the nineteenth century. This new form of manufacturing came to China when Europeans established treaty ports along the coasts and great rivers of China (Fig. 34–3). Later, Japan seized Manchuria and promoted its very rapid economic growth as a puppet state. Initially, Western powers wished to take advantage of cheap Chinese labor; therefore, they established industries based on textiles and other labor-intensive activities. Shipbuilding and munitions manufacturing also were developed.

Along with the development of industry, a revolution in commerce was set in motion in the treaty ports, and a group of Chinese emerged who served as middlemen linking the colonial interests with the native Chinese labor force and market. Banks, trading and shipping companies, insurance firms, and other commercial and financial institutions soon came into being. Many were Chinese owned. All of these new energies and activities were concentrated in the few peripheral treaty ports. The largest of these ports, Shanghai, serving all the hinterland of the Yangtze River Basin, soon came to stand for all the new changes introduced from the West. Through such cities, China was linked increasingly with the outside world and the vast wealth and trade of the international marketplace.

A controversy exists over how deeply these new impulses of commerce and industry affected the huge body of the Chinese economy, and no doubt many parts of the interior were little touched by the new growth. A start toward modernization was made, however, and China was forced to acquire new techniques and ideas. It is this modernization process, initiated through the existence of the treaty ports and their Western sponsors, that continues today. The consequences of these events are likely to extend far beyond anything anticipated by those early colonials and traders who set the process in motion.

Contemporary Patterns of Industrialization

After seizing power at mid-century, the Communist leaders and planners established a basic goal of industrializing the state as rapidly as possible. It appears that this policy and its implementation were based on the Soviet model of investing most in heavy industry (producer goods) to the relative neglect of consumer goods and agriculture. The Soviet Union initially provided a great deal of technical assistance in constructing new plants and provided, for a price, much of the equipment and machinery necessary.

At the beginning, things went well. Good weather provided large agricultural yields, and the Soviets cooperated and helped complete a number of important industrial and construction projects. Strains, however, soon began to appear between the two states, and in the late 1950s, the Soviets withdrew their assistance. Soviet engineers abandoned some projects in the middle of construction. It was at that point the Chinese changed direction in central planning and established their own path for industrialization and economic growth.

The "Great Leap Forward," "Walking on Two Legs," and Other Chinese Innovations

At the time of the "Great Leap Forward" in 1958, it was difficult to gauge the full nature of the new changes under way in industry. Subsequent events have permitted a better understanding of the new policies for economic growth. An established part of Communist policy was to develop the interior and eradicate the "evil" influence of the treaty ports and colonial cities as centers of industry and innovation. This new policy, according to the planners, would serve to develop backward areas, take industry to the raw materials, reduce transport costs, and provide better protection for the centers of production. A conflict existed, for it was recognized that the treaty ports as great industrial centers had good transportation access, were well supplied with skilled labor, already had established industrial and power facilities, and were low-cost centers for industrial production. All of the amenities and advantages of large cities already existed. As the Chinese planners knew, these assets are very difficult and expensive to create and build anew.

During the exuberance of the "Great Leap Forward," a new experiment was undertaken to short-cut the process of industrialization. This process focused on local small units of industrial production; the "backyard" iron furnace was one example. The Communists planners sought to reduce dependence on the great industrial centers by promoting local industries designed to produce goods required at the local level. In part, this new policy failed. Products from local industry were often inferior or useless and in some cases were not cheap. Apparently, the principle of local production for local consumption survived, however, for in the late 1960s, it reappeared in a new guise, "Walking on Two Legs."

As noted earlier, "Walking on Two Legs" refers to the use of both small and large units of productions: the traditional and modern sectors and the rural and urban locations. It might best be described as a policy of dual industrialization. Moreover, this policy has been strengthened by attaching its objectives of local industrial production and self-sufficiency to the functioning of the communes. The lesson learned is: small-scale industrial units have limited capacity and can produce only certain goods in limited quantities. Such a policy if followed carefully may have many far-reaching advantages and be well suited for China's current situation.

Industry, Cities, and Transportation

Despite the aim of China's new planners to shift the centers of production to the interior, the well-established industrial cities such as Shanghai, Tientsin, Shen Yang, Canton, Wuhan, and Peking continue to dominate as the great industrial cities. A few new centers have been developed. The Inner Mongolian city of Pao-Tou, with its new steel mills, is a frequently cited example. Many other older cities in the interior, such as Tihua, Chengtu, and Changsha, also have been given new industries. Nevertheless, the major centers of industry continue to be the cities of the coast, the Yangtze ports, and the great Man-

churian cities, a testimony to the efficacy of their choice initially.

This situation is neither startling nor illogical if one examines a transportation map of China (Fig. 34–7). These great industrial cities for the most part are those best served by rail and sea and are the most accessible. By virtue of their large populations and established industries, they are the centers of the greatest consumption of goods, fuel, and raw materials. Such processes and activities have a tendency to feed each other, to reinforce the growth already well established. This principle of the location of economic activity seems to operate universally enough to continue in China despite the apparent attempts of planners and policies to reverse it.

One plan to counter the influence of the treaty ports was to build new rail links throughout the interior. The Chinese have built and continue to build an impressive if not dense rail network throughout the interior. In time an expanded transport network may help reorient new economic growth to the interior centers. The new growth centers, however, will be hard pressed to rival the advantages of international trade linkages and sea transport shared by the former treaty ports.

China's industrial and transportation geography may be usefully compared with India's. India, it will be recalled, possesses a dense network of railroads focused on the major industrial and commercial centers, all of which were inherited from British colonial days. China, by contrast, contains two patterns: the indigenous interior urban and water transport network which has been supplemented with a foreign-associated rail system, and a coastal and northeastern pattern of cities (treaty ports) and transport artieries. The main industrial areas are associated with the latter pattern, although in recent years the two have been increasingly integrated.

Both China and India have used the preexisting network of transportation and cities in their recent development strategies. Altthough the Chinese have talked more about developing new areas and cities in the interior, it is difficult to evaluate which country has done more. China, in recently developing its rail system, has done a great deal, but it had a smaller rail base in 1949, whereas India inherited a dense and good rail network. Both countries have developed new industrial centers, and it is hard to know whether Pao-Tou is more important to China's economy than

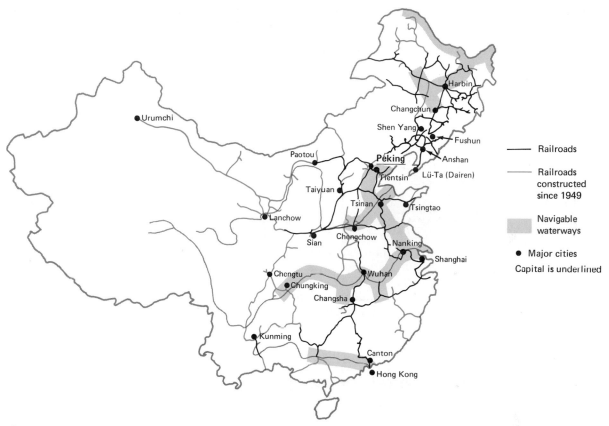

34–7 *China: Railways and Waterways*

is Jamshedpur to India's. Perhaps the most significant point is that both countries seek to develop new areas in the interior and to redistribute wealth away from the coastal areas associated with past colonial policies. Neither, however, is likely to follow too closely such a locational policy if it impedes the rate of economic growth and progress.

Oil: A Promising Development Opportunity

Oil is perhaps the best example of an industry that has been developed very rapidly under Communist guidance. Prior to 1949, although the presence of oil resources was known, little systematic exploration or exploitation of that oil had occurred. Prior to 1949, less than 3 million tons of oil had been produced over a half-century period of consumption. Following the establishment of the People's Republic in 1949, the Soviets provided both equipment and personnel for oil production. By 1955, annual crude oil production totaled nearly 1 million tons (Table 34–4). By 1960, production had jumped to 4.5 million tons, and estimates of production in 1970 were 20 million tons.[7] Much of the great new production was associated with the development of the Taching oil field in Manchuria. It was on the basis of production from this field that China is believed to have achieved self-sufficiency in petroleum production and to have initiated oil exports to Japan.

The exast size of China's reserves is still uncertain. Rough estimates based on proven and probable

[7]Tatsu Kambara, "The Petroleum Industry In China," *The China Quarterly*, 60 (1974), 699–719.

Table 34–4 China: Crude Oil Production (Including Shale Oil) and Oil Production Sufficiency

Year	Production (thousands of tons)	Self-Sufficiency (percent)
1950	200	27
1955	966	38
1960	4,500	58
1964	6,900	90
1965	8,670	100
1970	20,000	100
1973	40,000	100

SOURCE: Tatsu Kambara, "The Petroleum Industry in China," *The China Quarterly*, 60 (1974).

reserves suggest between 1 and 2 billion tons, enough to support China's estimated 1973 production of 40 million tons for twenty-five to fifty years. The point is that China has a lot of oil and is in no danger of depleting its reserves soon. If China is able to lay claim to and undertake oil development offshore in the Yellow, East China, and South China seas, its resources may be much larger than is currently assumed.

Offshore development may have to wait some years, for one of the serious problems confronting China is the level of oil drilling technology. Up to now wells have been shallow and fairly easy to drill using older methods. Future developments, especially those offshore, will require very large investments in equipment and very sophisticated equipment, neither of which China possesses at present. The future of Chinese oil production, then, may be tied closely to political events, for the Chinese may need assistance from other countries such as Japan, the United States, or other Western states. The potential for greater resource exploitation, coupled with the opportunity to capitalize on that resource for capital formation, would suggest the Chinese will consider carefully such factors in determining their future relations with foreign powers.

Future Trends

Today, China produces, in modest quantities, a full array of industrial and scientific goods: cars, planes, locomotives, computers, and atomic bombs. Iron and steel, cement, and electric production are about the same level as that of the United Kingdom. Fossil fuel production is large and climbing rapidly, based on new petroleum discoveries. Total industrial production, however, by United States or Soviet standards is small. When divided among China's great population, the per capita output is very small.

China's economy by almost any standard remains agrarian in nature and is likely to continue so for some time. In terms of the Rostow model outlined in Chapter 4, China would appear to be at the stage of "precondition" to economic take-off. It is conceivable that the recent shift in foreign policy and the efforts to buy new technologies will accelerate economic development. Both Japan and the Soviet Union industrialized quickly, and it may happen in China also. Special conditions and traditions, however, when coupled with the huge population, suggest that China's economic development will proceed in a somewhat different manner and at a more modest pace. How fast or slow is impossible to judge in a world where events occur with increasing rapidity and where time itself seems compressed in the rush to get things changed.

Further Readings

BUCHANAN, KEITH, *The Chinese People and the Chinese Earth* (London: G. Bell, 1966), 94 pp. Buchanan's book is a brief survey of China's changing geography after a decade and a half of revolution. The volume is a readable, introductory account.

Central Intelligence Agency, *Peoples Republic of China, Atlas* (Washington, D.C.: Government Printing Office, 1971), 82 pp. This atlas provides an excellent and up-to-date regional textual and cartographic coverage of China. Major topics treated are population distribution, physical geography, transportation, agriculture, linguistic and ethnic patterns, minerals and industries, and administrative structure.

HAN, SUYIN, *China in the Year 2001* (London: Penguin, 1967), 254 pp. This volume is a sympathetic account of recent developments in revolutionary China coupled with speculations about future trends.

HINTON, WILLIAM, *Fanshen: A Documentary of Revolution in a Chinese Village* (New York: Vintage Books, 1966), 637 pp. Hinton's study of revolution in a small village in the loess region of north China is something of a classic.

The book chronicles the social and economic revolution and change at the time of Communist takeover.

LINDBECK, JOHN M. H., *China: Management of a Revolutionary Society* (Seattle: University of Washington Press, 1971), 391 pp. Although Lindbeck's volume is a set of politically oriented essays, the aim is to demonstrate the manner in which the political system has organized society to promote national development.

OKSENBERG, MICHAEL (ed.), *China's Developmental Experience* (New York: Praeger, 1973), 227 pp. Oksenberg has edited a series of brief essays, many of which focus on aspects of the economy and society of contemporary China. The volume is readable and the essays are broad and introductory.

ORLEANS, LEO, *Every Fifth Child, The Population of China* (Stanford: Stanford University Press, 1972), 191 pp. Orleans' short book provides a brief but reasonable account of the growth of the Chinese population. It is most complete on demographic trends since 1950 and has sections on urban population and migrations.

TREGEAR, T. R., *An Economic Geography of China* (London: Butterworth, 1970), 276 pp. Tregear's economic geography of China is one of the few recent geography texts concerned solely with China. The text is straightforward and provides a clear explanation of the major patterns of economic development in recent years.

WILLMOTT, W. E. (ed.), *Economic Organization in Chinese Society* (Stanford: Stanford University Press, 1972), 461 pp. Willmott's volume is a collection of essays whose main theme is the linkage between Chinese society and economy. The essays, some of which use Taiwan as a Chinese setting, are serious and profound. This book is more suitable for serious students of Chinese affairs.

YANG, C. K., *Chinese Communist Society: The Family and the Village* (Cambridge, Mass.: MIT Press, 1965), 276 pp. The book contains two sections, one on the family and the other on the village during the first decade of Communist control. Although a bit dated, both are classics on China's changing social patterns at the most basic level.

China: The Rimland

35

S URROUNDING CHINA ON ITS eastern and northern flanks are several nations and colonies. For many years these lands have been linked either directly or through ancient cultural and historical ties to China (Fig. 35–1). In some cases, as in Korea, physical isolation and a degree of political autonomy have promoted a separate and distinctive culture. In other cases, Chinese territory and Chinese-speaking populations are involved (Hong Kong, Macao, and Taiwan). The Chinese government has made clear it considers these three areas integral parts of the Chinese state. The Chinese view is that at some appropriate time these three territories will be returned. Both Chinas, Communist and Nationalist, appear to agree on this eventuality.

Mongolia too is considered Chinese territory but of a somewhat different nature. Its status, in the Chinese view, is analogous to Tibet, Sinkiang, and Inner Mongolia. Although its inhabitants are recognized as non-Chinese, their history and destiny are so interlinked with China as to make them proper subjects for political integration with the People's Republic. Were it not for Soviet opposition and power, the Chinese probably would have forced the incorporation of the People's Republic of Mongolia into their state.

Hong Kong and Macao

At the mouth of the Pearl River estuary, just within the tropics along China's southeast coast, are two colonies that were established during an earlier age. On the west side of the mouth of the Pearl River lies Macao, a 400-year-old, 6-square-mile (15.5 square kilometers) Portuguese colony. Thirty miles (48 kilometers) east is Hong Kong, comprising 404 square miles (1,046 square kilometers) of the New Territories, Hong Kong Island and Kowloon.

Hong Kong was first established in 1842 as a result of British pressure on China. Initially established as a small-island colony, additional territory was later ceded or leased to the British. The present territory is composed of 32 square miles (83 square kilometers) of Hong Kong Island and additional land on Kowloon, which is theoretically British territory and to which have been added the leased New Territories and adjacent islands (Fig. 35–2). The ninety-nine-year lease on the New Territories is due to expire in 1997, and it is difficult to predict the Chinese response to the end of that lease.

Colonial Development

Traditional functions of Hong Kong were trade and commerce. From its beginning, the British developed Hong Kong as a major entrepôt and trading center in south China. The colony served this function well for a century, and many of the export goods that originated in south China transited the colony. Unlike Macao and Canton, Hong Kong's harbor is deep and spacious. The development of steam-driven ocean vessels with increasingly deeper drafts was an advantage to Hong Kong over its neighbor. Although British-financed and -controlled, Hong Kong functioned as an appendage of China's foreign commercial and trade structure and traditionally accounted for a substantial part of China's foreign trade.

35–1 *East Asia: Main Political Units*

After the ascendancy of the Communist government in mainland China, Hong Kong's future appeared bleak. Military defense of the colony against China was not feasible. Somehow the colony has survived, and indeed prospered, although there have been periods of great uncertainty and instability. For example, in 1967 during the "Cultural Revolution," businessmen and investors became very nervous. Land prices and the local market in stocks and securities plummeted; many observers foresaw the quick demise of Hong Kong. The colony did not disappear, but the events of 1967 illustrated how tenuous is its position and how directly events and

policies in mainland China affect the colony. Inasmuch as China controls the water supply for Hong Kong, the colony is hardly in a position to resist or even to debate policies and desires that emanate from the mainland.

Recent Economic Changes

Another recent change in Hong Kong brought about through its relations with China is the nature of the colony's economic structure. The traditional functions of trade and transshipment were interrupted with the establishment of a new Chinese government in 1949. New enterprises were necessary

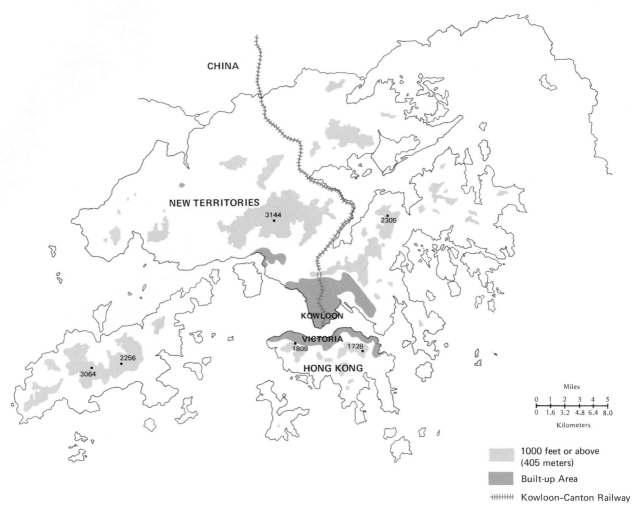

35-2 *Hong Kong: Urban Areas*

Table 35–1 *Hong Kong: Labor Force*

Activity	Total Work Force	Percent
Manufacturing	613,620	39
Services	375,440	24
Commerce	259,690	17
Construction	96,000	6
Agriculture, Forestry, Fishing	81,300	5
Communications & Utilities	121,810	8
Other	10,640	1
TOTAL	1,558,500	100

SOURCE: *Report of the Year* (Hong Kong: Government Printing Office, 1970).

Table 35–2 Hong Kong: Foreign Trade

	Source of Imports		Destination of Exports	
Country	Percent of Total Imports		Country	Percent of Total Exports
Japan	24		United States	42
China	16		United Kingdom	12
United States	13		West Germany	8
United Kingdom	9		Japan	4
Taiwan	5		Canada	3
West Germany	4		Australia	3
Remainder	29		Singapore	2
TOTAL	100		Sweden	2
			Netherlands	2
			Remainder	22
			TOTAL	100

SOURCE: *Report of the Year* (Hong Kong: Government Printing Office, 1970).

if the colony were to survive, and a series of activities were promoted. The most prominent was manufacturing. By 1970, more than 40 percent of the labor force worked in manufacturing (Table 35–1). Textiles and garments account for nearly 45 percent of the total work force in manufacturing and for 60 percent of the colony's exports. Other important industries are plastics (toys, recreational goods, and sundries), electronic components, wigs, and machine tools.

Banking, insurance, and other forms of commerce continue as important activities and make up the other major employment field. Although trade with China was interrupted in 1949, some has been reestablished. Hong Kong continues its role as an entrepôt, and approximately one-half billion dollars worth of goods are transshipped through the colony. This transshipment trade, primarily of diamonds, textiles, pharmaceuticals, watches, and foodstuffs, is increasing. About one-fifth of it goes to Japan. Singapore, the United States, Indonsia, and Taiwan are among the other important destinations of reexports. Tourism is also extremely important to the economic health of the colony. More than 100,000 tourists arrive in Hong Kong each month. The earnings derived from the tourist trade and from other invisible sources of foreign exchange, such as banking and insurance, help offset an unfavorable balance of trade that amounts to $100 million per month.

Hong Kong must trade to live. Although its population is less than 1 percent that of China, its foreign trade is larger than China's, a remarkable achievement for such a small place. Japan and China are the principal sources of imports (Table 35–2). By contrast, exports go mainly to the United States and West Europe.

Social Patterns and People

One of the revealing facts about Hong Kong's character is that about one-half of its citizens were born elsewhere, mostly in China. Some 98.5 percent of them are described as Chinese, based on their language and place of origin. At the end of World War II, Hong Kong's population was about 600,000. By 1960 this number had swollen to 3.13 million. Presently, the colony's population is over 4 million. Immigration has accounted for this remarkable growth in such a short period of time, and one must look to events in China for an explanation. China's recent political activities and history, such as the "Great Leap Forward" and the "Cultural Revolution," have apparently spurred a migration. The mainland government at times has allowed some of its citizens to leave. Moreover, some people leave China illegally, and Hong Kong is only a short swim or boat ride away.

With such a large population and modest land area, Hong Kong is really a city-state and is one of the most crowded places on earth. The colony's population density is nearly 10,000 per square mile (approximately 4,000 per square kilometer). This figure is even more astonishing when it is pointed out that 90 percent of the land area is either in rugged uplands and scrub vegetation or under cultivation. Only 10 percent of the area is urbanized, that part of the colony in which the 4 million inhabitants live and work. For example, in one urban district there are about 430,000 people per square mile (166,000 per square kilometer), unquestionably one of the most densely populated urban concentrations on earth.

Hong Kong's Future

It is almost impossible to predict what will become of Hong Kong in the future. For the short term, say, a decade, things may not change radically. For the year 2000, it is extremely difficult to forecast what changes and political events will have occurred in China. Yet, clearly, it will be impossible to separate political events and changes that occur in China from the destiny and political control of Hong Kong. It appears that the British government understands this situation and attempts to accommodate Chinese desires and demands to fit the interest of all parties concerned in the governance and operation of the colony.

It is in China's economic interest to maintain the colony, for it is estimated that Hong Kong returns more than $1 billion each year to China. Moreover, Hong Kong plays a valuable role as a point of contact and interchange with the outside world. As long as China remains relatively closed to outside contacts, a linkage and communications broker is useful. There is, however, a negative side for China, a leading anti-colonial force among developing nations, to permit the maintenance of a colony of Chinese on Chinese territory. The Chinese are embarrassed when other Communist countries criticize them for allowing the colony to thrive and grow.

Macao

Macao, the oldest surviving European colony in Asia, was first established by the Portuguese in 1542. The small colony is composed of a narrow peninsula and two islands. As Hong Kong grew and prospered in the nineteenth and twentieth centuries, Macao stagnated. Its harbor was shallow, and its shipping and commercial functions were taken over by its energetic British colonial neighbor.

In 1967, associated with China's "Cultural Revolution," Macao came very close to being restored to China. Curiously, after a series of demonstrations

and strikes when it appeared the Portuguese would be ejected, an agreement with China was worked out. Since then, Macao has prospered. Increasingly it has been drawn into Hong Kong's dynamic economic orbit. Today, linked to the adjacent British colony by large ferries and high-speed hydrofoils, it has become almost an appendage of Hong Kong. New investments in garment manufacturing and construction have given Macao and its 300,000 residents a new face. Although tourism and gambling are its leading earners, it exports yearly about $10 million worth of textiles and fabrics. Its 1,000 Portuguese colonials are increasingly less important in commercial and internal affairs, and the colony best describes the new Chinese pragmatism that recognizes and uses economic assets even in the face of ideological contradictions. How long such a contradiction can continue is a mystery to all, except perhaps the leaders of the new China.

Taiwan

All places on China's rim have somewhat uncertain futures, and Taiwan is certainly no exception. Although its economic growth and progress in recent years have been very impressive, its political future is cloudy. The island-state contains a variety of political factions. All are tied in one way or another to the island's historical evolution as a frontier territory of China and to China's pressing desire to restore and reintegrate Taiwan to China politically.

History and Sinicization

China has had ties with Taiwan for more than 1,000 years and, since the sixteenth century A.D. has sent a steady stream of immigrant farmers and adventurers, first to the Pescadore Islands and then on to Taiwan (Fig. 35–3). At various times this stream has been interrupted, and the island has been occupied by the Dutch, Spanish, French, and Japanese. Despite these non-Chinese occupations, ever since Chinese farmers began to arrive and cultivate the productive, alluvial western coastal plains and basins, the island has been Chinese in ethnic and linguistic composition. Although proto-Malay peoples originally inhabited Taiwan, these tribal groups were increasingly pushed into the high, rugged inter-

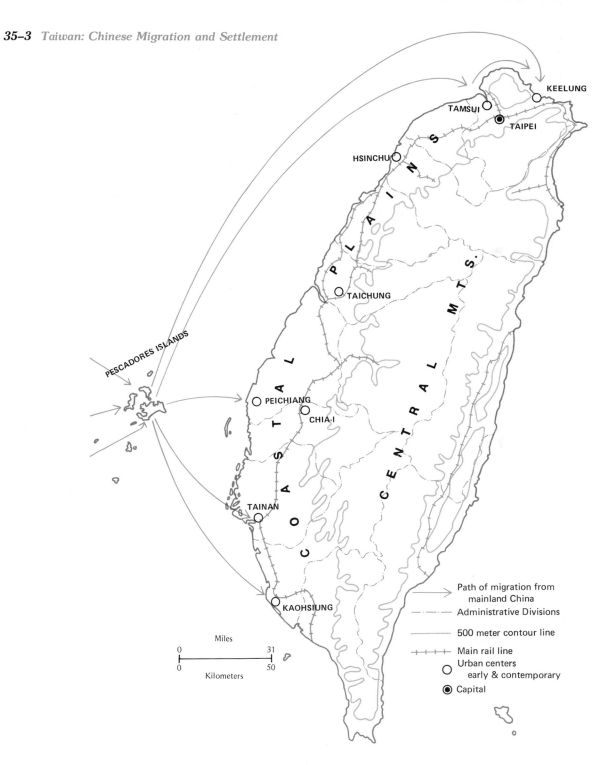

35–3 *Taiwan: Chinese Migration and Settlement*

KEELUNG

TAMSUI

TAIPEI

HSINCHU

P L A I N S

PESCADORES ISLANDS

TAICHUNG

C
E
N
T
R
A
L

M
T
S
.

PEICHIANG

CHIA-I

C
O
A
S
T
A
L

TAINAN

KAOHSIUNG

Path of migration from
mainland China

Administrative Divisions

500 meter contour line

Main rail line

Urban centers
early & contemporary

Capital

Miles

0   31
0   50

Kilometers

ior of the island. They still exist there in scattered remnants, but the inexorable force of Sinicization has exerted great pressure on them and has continued to absorb them into the Chinese cultural mainstream.

The Physical Environment

Physical features have aided these aboriginals in maintaining their separateness and identity, for the interior of Taiwan is composed of a rugged northeast-southwest trending cordillera. Peaks rise to 13,000 feet (3,962 meters). The eastern two-thirds of the island are high and rough. Except for a narrow valley and a small alluvial basin in the northeast, little suitable land exists for intensive farming. The western one-third of the island is composed of a series of alluvial plains and basins separated by broad, cobble-strewn river beds. A dense network of roads and rail lines link these plains and basins, and the densely inhabited parts of the island are well integrated spatially.

Taiwan lies astride the Tropic of Cancer 100 miles (161 kilometers) east of the China mainland. Its climatic patterns result largely from a combination of location, seasonal reversal of prevailing winds, surface configuration, and seasonal storms, especially typhoons. Climate is monsoonal, with most precipitation occurring in the summer. The northeast quadrant, however, displays a reverse pattern with heavy winter rains from the prevailing winter northeasterlies. The southwest is in a rain shadow in the winter and has mild, dry winters. Precipitation ranges from 55 inches to 220 inches (140 to 559 centimeters) and except at high elevations occurs as rain. Summer and autumn typhoons are an annual threat and often bring great devastation and flooding. Frost is unknown at lower elevations.

The combination of climate and alluvial lowlands has resulted in a productive paddy rice agricultural environment characteristic of southeastern China. Other crops include sugar cane, tropical fruit, tea, and a rich variety of vegetables. Intensive cropping of the alluvial area has led Taiwan to develop one of the most productive agricultural systems found anywhere. Aside from a modest quantity of coal and natural gas and small quantities of gold, copper, and less valuable minerals, Taiwan's only other natural resources are the forests and the hydropower potential of the mountain streams. Taiwan's development has been constructed on productive agriculture, loca-

tion, and the industriousness of the people. These assets have been supported by a recent history of foreign investment and management which has produced a remarkable record of growth and progress.

Early Modernization and Japanese Colonialism

One of the most significant events that has affected contemporary Taiwan took place in the late nineteenth century. In 1888, the Chinese Ching imperial government sent a new governor to Taiwan and made the island a province. This governor, Liu Ming-chuan, set about to modernize and develop the island as a showpiece province. Seven years later, the Japanese took over the island as a colony and accelerated the pace of investment and growth. The combined efforts of late Ching China and Japan to modernize Taiwan through the construction of roads, dams, irrigation systems, bridges, a railway system, new cities, and ports laid the stage for a transformation of the traditional economic system. Although the succeeding fifty years of Japanese colonial rule were designed to benefit Japan, long-term benefits have accrued to the contemporary inhabitants and government. The basis for the current economic prosperity and growth was laid initially almost a century ago.

Population and Society

Another consequence of Japanese colonialism was the promotion of improved standards of public health, a decline in the death rate, and rapid population growth. At the turn of this century, there were about 2 million people on the island. By 1920, population had climbed to 3.65 million. In 1949, when the Nationalist Government was exiled on Taiwan, there were more than 6 million inhabitants. To these were added more than 1 million soldiers, functionaries, and refugees from the mainland. Today about 14 percent of Taiwan's population is composed of this recently arrived group; rapid population growth has continued; and now there are 17 million inhabitants, an incredible number of people for an island of 13,892 square miles (35,980 square kilometers), roughly the combined size of Maryland and Delaware.

The restoration of Taiwan to China and the subsequent establishment of the exiled Nationalist Government introduced a new stage in Taiwan's checkered history. One aspect of this new stage was the further mixing of Taiwan's peoples. Traditionally the immigrants to Taiwan came from southeastern China, and these people spoke either the Fukienese or Hakka dialects. Most of the 1949 immigrants were Mandarin speakers from north and central China. Inevitably a division set in between Taiwanese residents and the Mandarin speakers, commonly referred to as "Mainlanders." Some of this division was transferred to politics, for the government, army, and police apparatus became dominated by mainlanders, and the Taiwanese majority was to a large extent excluded from decision-making processes. In recent years, with increased use of the Mandarin dialect and intermarriage between Mainlander and Taiwanese a common occurrence, this division has waned.

Recent Economic Growth and Progress

Associated with the exile of the Nationalist government was a new stage in the economic development of Taiwan. After a brief period of retrenchment, recovery, and little growth, the Nationalist government in 1955 seriously began to promote the rapid growth of Taiwan's economy. Part of the hesitancy to spur growth earlier was political, for the Nationalists have long claimed that Taiwan is only a temporary headquarters, and they will return to mainland China eventually. Such a pretense initially inhibited development on Taiwan, for it was not considered desirable to invest heavily in a temporary situation. Increasingly it became obvious that there would be no return to China, and the promotion of Taiwan became a more serious and desirable goal.

Another significant consequence of Nationalist rule on Taiwan was the fulfillment of a meaningful program of land reform, an achievement the Nationalist government was never willing or able to complete on the mainland. Indeed this failure was unquestionably one critical element in their loss of mainland China. On Taiwan, in contrast, the landlords were associated with the Japanese colonial government. The Nationalists had no ties or loyalties with them and were able to resist the temptation to ally with the traditional rural elite. Consequently, a comprehensive and profound land reform program was instituted in the 1950s in which land was given to those who tilled it. A program of government-supported, low-interest loans provided financial support. Land redistribution was accomplished, and output has climbed steadily since that time (Table 35–3).

Despite Taiwan's very high ratio of people to arable land, the island, with its double-cropping of rice,

Table 35–3 *Taiwan: Agricultural Production*

Index of Agricultural Products (1963 = 100)					
1959	1961	1963	1965	1967	1969
89.5	100.2	100	121.1	134.9	141.7

Rice Production (thousands of tons)					
1959	1961	1963	1965	1967	1969
1,856	2,016	2,109	2,348	2,414	2,322

SOURCE: *Statistical Yearbook for Asia and the Far East* (New York: United Nations, 1970).

abundant vegetable, grain, and fruit production, remains basically self-sufficient in food needs. During the last decade, it has exported large amounts of sugar, rice, mushrooms, pineapples, and pork. The relatively high cost of these items in relation to imports of other less expensive food grains has permitted Taiwan to remain a net agricultural exporter.

Industrialization and Rising Levels of Income

Since 1960, Taiwan has prospered increasingly as its economy has industrialized. In part, this prosperity is attributed to United States foreign-aid contributions, which were both large and well conceived during the 1950s and 1960s. So successful was the United States aid program that no new projects were funded after 1965. Growth was both rapid and self-sustained, and the annual increases in per capita GNP during that last decade gave dramatic proof of this growth, from $157 in 1959 to $260 in 1969 and $500 in 1975.

Associated with the remarkable economic progress has been a shift in the structure of the economy from a rural-based farm economy to an urban-oriented commercial and industrial economic system. Some 38 percent of the gross national product is accounted for by manufacturing. In contrast, agricultural production accounts for only 15 percent of the GNP, a decrease from previous times. Employment figures also confirm this trend; there were more people employed in industrial production than in agriculture, a strong indicator of the shift in economic structure from traditional and agrarian to modern and industrial.

Economic Growth and the Duty-Free Export Processing Zone

In the mid-1960s, economic planners in Taiwan pioneered the development of a duty-free export processing zone in the southern industrial, port city of Kaohsiung to encourage foreign investment. The zone, which was designed to use Taiwan's low-cost and skilled labor force for processed goods such as textiles, electronics, and toys, has been so successful that two other zones are now in operation with more than 270,000 workers. The concept of export-processing zones has spread to other Asian countries such as South Korea and India.

Per capita gross national product is about $500; the annual economic growth rates of recent years have been around 10 percent, a spectacular growth even by Japanese standards. Moreover, since 1966, Taiwan has had a trade surplus and has been able to collect a large reserve of foreign exchange. Much of this trade surplus results from Taiwan's large volume of exports to the United States. Such trade relations help illustrate a key dilemma facing modern Taiwan: the linkage between commercial and political alliances and patterns.

Politics, Economics, and Taiwan's Future

In 1972, Taiwan was voted out of the United Nations as the Chinese representative, and the mainland government took over the Chinese seat. Since then Taiwan has drifted in political uncertainty without aim or seemingly clear objectives. Taiwan's trade relations, wherein 42 percent of the exports go to the United States, suggest a very heavy dependence on United States policies and goodwill. Although there have been great economic advantages in this policy, changing United States policies have been very difficult for Taiwan; many Nationalists fear an ultimate "sellout" of Taiwan to China in some secret United States–China deal.

Taiwan is Chinese ethnically and linguistically, yet its politicians are of all persuasions, and its citizenry seems to reject the idea that it should be restored to a China governed by Maoists. Moreover, with increasing growth and prosperity and rising levels of living, the impediments and antagonisms to rejoining China are increasing with time. Almost everyone agrees that Taiwan is Chinese. Yet must it be politically attached to the mainland? Perhaps the most promising future for Taiwan lies as a politically independent Chinese state, free to develop the particular type of political, economic, and social system it believes will benefit its citizens best. Such a policy likely would be acceptable to all parties except those on mainland China.

Taiwan—with its great prosperity even in the face of limited resources, its thrifty and hardworking people, its balance of a sound agricultural economy paralleled by a booming commercial and industrial economy, its quiet villages offset by bustling cities and crowded highways—is a model of East Asian growth and change. The island is a good example of the success of a workable economic development policy. Here is a country with few resources other than its people which in a short time has been transformed into a thriving, dynamic state with one of Asia's highest per capita incomes. Politics, Maoism,

A shirt factory in Taiwan's duty-free zone. Materials are imported free of tariffs, processed using low-cost labor, and then exported to world markets. (United Nations)

and international relations cloud Taiwan's otherwise promising future.

Korea

In many ways Korea might have become a part of China. Through much of its history, it has existed as a loose political appendage of the Middle Kingdom and owes China a profound cultural debt. Much of what traditionally was Korean began as Chinese: religion, art forms, part of the written language, political and social institutions, agricultural practices, and land tenure relationships. Despite the many cultural threads and practices from China that have entered the Korean fabric, somehow an independent and separate entity has emerged. Modern Koreans, with their own distinctive culture, language, and history, see themselves as an individualized form of East Asian civilization. They have worked and suffered to establish this cultural integrity, a feature that may be vitally important in the evolution and acceptance of the idea that being a Korean means something profound and significant to those people who inhabit the peninsula.

Background

Koreans are predominantly of Mongoloid stock with perhaps some ancient proto-Caucasoid admixtures. Taller than the Japanese, they resemble most closely northern Chinese and the non-Han people of China's Manchuria to whom they are closely related. Although carbon-dating of artifacts suggests a Neolithic culture about 3000 B.C. in the Korean Peninsula, the tribal elements that inhabited Korea did not coalesce in a meaningful political form until the early Christian Era. Several tribal groups competed with one another until one, Silla, was able to dominate the others and to extend its control over much of the peninsula during the seventh century A.D.

As the first Korean dynasty, Silla rule resulted in great advances in promoting formalized religion (Buddhism) and ethical practices (Confucianism). Arts and crafts prospered, and new techniques for producing paper, silk, pottery, and crafts flourished. Trade linkages with Japan and China were established, and after A.D. 700, two-thirds of the peninsula emerged as the Silla state. By the ninth century, Silla power waned and was replaced by a period of even more intensive flowering of Korean cultural institutions, the Koryo Dynasty (A.D. 918–1392). Although never a time of conquest or military power, Koryo witnessed the maturing of many of those institutions and art forms initiated earlier and formed a logical bridge between a formative and more mature period of Korean history.

Grandeur and Decline: The Yi Dynasty

The great Korean dynasty that unified the peninsula was the Yi (A.D. 1392–1910). Early, the Yi flourished as it borrowed heavily from Chinese politics, law, society, and culture. In time the society became increasingly stratified, and its economic and political institutions ossified. The country increasingly turned inward and closed itself off from all except China. To a large extent, Korea became a political and cultural hermit. In the nineteenth century, Korea, like China and Japan, was pressed by the West to modernize and open its doors to the outside world. Modest beginnings were made, but a new shock soon set in, the shock of Japanese aggression.

Japanese Colonialism

Korea's location as a peninsula jutting off the Asian mainland toward Japan has made it a sensitive and strategic bridge. It has been so used by various warring groups for centuries. Yet traditionally it has commonly been under China's protection and influence. In 1905, following the Russo-Japanese War, Korea became a Japanese protectorate. Five years later it was annexed and became a part of the Japanese Empire. Although this colonial rule lasted only until 1945, its impact was far-reaching. The thrust of Japanese investment and development projects began after 1910 and initiated the transformation of a hitherto traditional economy and society. The process of transformation and modernization, begun more than a half-century ago, is only now achieving rapid motion. The point, however, is that without Japanese stage-setting contributions in capital investments such as roads, railways, ports, mines, factories, and irrigation systems, no matter how painful at the time, today's growth and progress would likely not be possible. Japanese colonial rule, however, left a legacy of distrust and dislike that continues to sour and rankle relations between the nations, but there were undeniable economic benefits.

The Resource Base and Physical Environment

Korea's 84,562 square miles (219,016 square kilometers), about the size of Minnesota, lie in the middle latitudes, and climatic patterns largely reflect the peninsula's location and position. Summers are warm and moist; winters are dry and cold, a reflection of the dominant air mass, the Siberian high, that controls much of the winter climate of eastern Asia. In North Korea, especially at higher elevations, winters are very cold and harsh. Farther south, the influence of the adjacent water bodies moderates the temperature and provides a better opportunity for moisture accumulation in the air. The southern part of the peninsula thus has a much greater potential for agricultural production (Fig. 35–4). With its less severe winter, longer growing season, and higher average daily temperatures, a cool season second crop is possible, and a dense agricultural population is supported.

Landforms reinforce the agricultural pattern, for the largest alluvial plains are in the south. Most of the peninsula is mountainous, focused on a north-south trending spine, with peaks rising to 9,000 feet (2,743 meters). Only about one-fifth of North Korea is suitable for cultivation. Although this rugged relief is a serious obstacle to intensive agriculture, the mountains offer other types of resources and economic opportunity. Hydropower production, timber, and mining are major activities in the mountainous regions, and Korea is well endowed with coal, iron ore, tungsten, graphite, magnesite, and a variety of other minerals. Most of the mineral wealth is concentrated in the north and to some extent has offset the greater agricultural potential of the south.

Complementarity and Independence
Between North and South

During colonial rule, the Japanese, in making their investments and constructing cities, factories, and transport lines, focused on the north. Over time much of the industrial output of the state became concentrated in the northern half of the peninsula. The south, with its more intensive wet-rice agriculture, supported twice as many people and was looked on as a breadbasket. Thus developed the geographic complementarity of the two halves of the country, an industrial north and an agricultural south.

Such a regional characterization has never been entirely fair or correct, and large cities in the south,

Cultivated Area

35–4 *Korea: Cultivated Area*

such as Seoul, Pusan, and Taegu, long have had important industrial functions. The north, although a focal point for much Japanese energy in building an industrial base, has been traditionally an agricultural land also. Recently the North Korean government has claimed further progress in agricultural investment, great improvements in farm production, and the successful socializing of agriculture.

Following the division of the country into two political halves—a Soviet-oriented north (population 17 million) and a Western-oriented south (population 36 million)—and the Korean War which hardened this division, new economic patterns have emerged. Each half has attempted to round out its economic structure in order to achieve greater self-sufficiency and less reliance on external sources of supply. To some extent, both have succeeded in promoting economic growth and in providing necessary production for their own economies. It has been suggested that over time, such policies will lead to an acceptance of the political division as permanent, inasmuch as the two halves no longer need each other.

The Contemporary Economy

Both North and South Korea have prospered in recent years, and both countries are increasingly looking toward the conversion of their economies to industrial, service-oriented systems from the traditional, agrarian patterns. Although per capita gross national product remains low in both cases (about $300), growth rates have been impressive in recent years. South Korea especially has had good success in exporting textiles, apparel, footwear, and electronics to consumer-oriented economies such as Japan and the United States.

In keeping with the plans for economic growth, both countries have invested heavily in new roads, railways, mines, irrigation systems, new power sources, and ports. For example, South Korea has constructed a four-lane expressway linking Seoul with Pusan. The highway even has modern Greyhound buses. Cities are growing at prodigious rates, and the capitals, Pyongyang in the north and Seoul in the south, are among the most impressive. Indeed, Seoul with close to 6 million inhabitants is one of the world's largest cities. The rate of economic expansion in both the industrial and agricultural sectors, coupled with the resource base and the industrious, educated,

Seoul, South Korea, has nearly 6 million inhabitants and is growing at a rapid rate. (United Nations)

and disciplined labor force, suggests a happier economic future for both Koreas.

Prospects for a Unified Korea

In 1971, after two decades of bitter antagonism, the South Koreans proposed direct contacts with North Korea. The political differences and the ideological divisions, however, between a Marxist and a free-enterprise economic system are too great to permit much more than talks, at least for the foreseeable future. Without the use of force, it is unlikely that these two fundamentally different economic and political systems can overcome their differences to permit reunification, despite a common culture and historical tradition.

Mongolia

The traditional Mongol heartland was focused on the Gobi, a great elevated, arid plateau (Fig. 35–5). Much of the Gobi is in that part of the Mongol domain called traditionally Inner Mongolia, today an autonomous region of the People's Republic of China. Although the Mongols under Genghis Khan and his successors at various times in history have controlled enormous areas of Asia and even parts of Europe, today the Mongolian state is a buffer and owes its existence to the geopolitical competition between the two giants of the Communist world, China and the Soviet Union.

The Mongolian People's Republic

A revolutionary government came to power with Soviet assistance in 1921, and the People's Republic of Mongolia was proclaimed in 1924. Although the state consists of only part of the domain of the Mongols, that part known as Outer Mongolia, it is a large country of more than 600,000 square miles (1,564,000 square kilometers). The country stretches 1,400 miles (2,253 kilometers) from east to west and has a latitudinal spread generally between 300 and 500 miles (483–805 kilometers). With a population estimated at 1.4 million, it is one of the most sparsely populated countries on earth.

Physical Environment and Aridity

The People's Republic of Mongolia occupies a great plateau at elevations ranging from 3,000 to 6,000 feet (914–1,829 meters). Although east-west trending mountains are common, they are generally low and rolling. The flank of the higher mountains slopes gently toward Siberia. Inasmuch as most of the precipitation originates from the north, the more productive part of the country lies closer to Siberia.

Throughout Mongolia, the key climatic control is moisture. Annual precipitation is greatest in the north at higher elevations. Precipitation decreases southward, and many parts of the Gobi receive no more

35–5 *Mongolia: Physical Features*

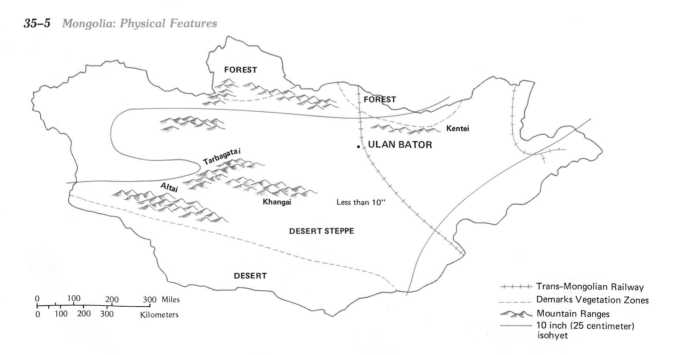

FOREST

FOREST

Kentei

• ULAN BATOR

Tarbagatai

Altai

Khangai

Less than 10"

DESERT STEPPE

DESERT

| 0 | 100 | 200 | 300 Miles |
| 0 | 100 | 200 | 300 | Kilometers |

+ + + + Trans-Mongolian Railway
- - - - Demarks Vegetation Zones
～ Mountain Ranges
——— 10 inch (25 centimeter) isohyet

than 2 inches (5.1 centimeters) annually. Most of the precipitation is concentrated in the summer, and dry farming of cereals in the north is possible. This continental climate of severe dry and cold winters and hot summers is similar to that of the state of Montana and the Canadian province of Alberta.

Pastoral Nomadism as a Way of Life

Mongols have traditionally engaged in herding as a means of using the resources of their marginal, arid environment. On the broad steppe grasslands and plains, they have developed extensive herds of sheep, goats, camels, and cattle. Today, there are roughly thirty domestic animals for each person, probably the highest such ratio in the world. The characteristic seasonal migration of these herders has long formed the main activity patterns of these people. Their well-established disdain for farming and other forms of sedentary activity has stood in the way of modernization. Traditionally only priestly service to Buddhistic Llamaism, borrowed from Tibet, formed an acceptable alternative to the wandering existence of the pastoralist.

Modernization and Change

Formation of a People's Republic under Soviet control has wrought great changes for Mongolia. The Soviet approach to modernization has been gradual and perhaps for that reason successful. Although 60 percent of the labor force is engaged in animal husbandry, it is state controlled and based on collectively organized farms and ranges. Associated with collectivization have been a scientific approach to animal raising and centralized rural organization complete with five-year plans. Other improvements in medicine, transportation, education, power production, and mining have followed. The influence of the traditionally powerful religious establishments, the lamaseries, has been eliminated.

Today Mongolia exists as a satellite of the Soviet Union. Its role is more appropriately that of buffer state separating Russian influence from Chinese. Its location and position are sensitive and strategic, and its role in not an entirely happy one. The viability of this buffer role is based on the strength and competition of its two giant neighbors. It is difficult to foresee the future of Mongolia in the event this uneasy balance is disturbed.

The Rim of China Reconsidered

The states and colonies that lie along the northern and eastern rim of China share at least one feature. All of them, to a greater or lesser extent, are dominated by the enormous and powerful presence of China. Whatever the established political systems that control Hong Kong, Taiwan, the Koreas, and Mongolia, the main fact of life for them is their relationship to China. The behavioral patterns of these states—economic, social, and political—are closely attuned to, and conditioned by, political events in China, although the effects often may be indirect and subtle. The point is that all of these places have long been associated with China and to some extent even today see themselves in a satellite role to China. Geopolitical realities have caused them to follow other political masters and make other friends. China's increasing military strength, however, makes it a most formidable political force. An increasingly powerful China is already seeking a more prominent role in political affairs of Monsoon Asia.

Further Readings

BARTZ, PATRICIA, *South Korea* (Oxford: Clarendon Press, 1972), 203pp. This text is a good, up-to-date account of South Korea's geography. It is well researched and has excellent maps and other illustrations.

CHEN, CHENG-HSIANG, *Taiwan, An Economic and Social Geography* (Taipei: Fu-min Geographical Institute of Economic Development, Research Report No. 96, 1963), 653pp. Without question, Chen's study of Taiwan is the best English-language geography of the island.

CLAVELL, JAMES, *Tai-Pan* (New York: Atheneum, 1966), 590pp. *Tai-Pan,* a novel, chronicles the rise of Hong Kong through the efforts of the great British trading firms that built the nineteenth-century China trade.

Far Eastern Economic Review, *Asia Yearbook* (Hong Kong: Far Eastern Economic Review). This annual yearbook presents statistics and summaries on all countries in Monsoon Asia.

HEDIN, SVEN, *Across the Gobi Desert* (translated from the German by H. J. Cant) (New York: Greenwood, 1968), 402pp. This book is a recent republication of one of the accounts of the well-known Swedish geographer-explorer who visited Mongolia in the early part of this century.

Hong Kong Government, *Hong Kong* (Hong Kong: Hong Kong Government Publications). The Hong Kong colonial government produces this annual yearbook of facts and figures on social and economic conditions.

Hsieh, Chiao-min, *Ilha Formosa: A Geography in Perspective* (London: Butterworth, 1964), 372pp. Hsieh's geography provides a useful introduction to the physical and human geography of Taiwan.

Hughes, Richard, *Hong Kong: Borrowed Place, Borrowed Time* (New York: Praeger, 1968), 171pp. A journalist looks at Hong Kong and its uncertain future.

McCune, Shannon, *Korea's Heritage, A Regional and Social Geography* (Tokyo: Charles Tuttle, 1956), 250pp. McCune's geography remains the standard English language work on Korea.

Nuttonson, M. Y., *The Physical Environment and Agriculture of Central and South China, Hong Kong, and Taiwan* (New York: American Institute of Crop Ecology, 1963), 402pp. This study is an excellent description of the physical and agricultural environment of the southeastern rim of China.

Rupen, Robert, *The Mongolian Peoples Republic* (Stanford, Calif.: Hoover Institution, 1966), 74pp. A political scientist's view of contemporary conditions in Soviet-controlled Mongolia.

Sih, Paul (ed.), *Taiwan in Modern Times* (New York: St. John's University press, 1972), 521pp. Sih has edited a large anthology of essays on all facets of Taiwan's modern history.

Southeast Asia: A Region Divided and United

36

P RIOR TO THE NINETEENTH-CENTURY European occupation of most of Southeast Asia, little was known in the West about this part of the world. Traditionally the region has been a zone of contact and interaction between India on the west, from whom it drew much of its early culture, and China to the north. For many centuries the Chinese had been interested in Southeast Asia and had at various times extended their culture and influence to what is now known as Vietnam. In the last 150 years, many Chinese have migrated southward to what the Chinese call Nanyang (the Southern Ocean or Sea), and several Asian countries today have large and significant Chinese minorities.

During the nineteenth century and the first half of the twentieth, the age of colonialism for Southeast Asia, many divisive and heterogeneous forces fragmented the region. World War II and invasion by Japan brought a new political awakening and change in traditional values to this corner of Asia and a new view of its regional character. This new view is based on the belief that despite a considerable degree of local variation, there exists sufficient uniformity of climate and physical patterns, of race and ethnic groups, of history and related origins to justify the identification of a Southeast Asian region.

Perhaps most important, World War II ushered in a new age that signaled the decline of European colonialism in Asia. The war exposed the weaknesses of European colonial control over most of Southeast Asia. Following the war, the British, French, and Dutch colonies emerged as independent nations. This transfer of power was not always easy. Often local groups fought among themselves for political power, and the polarization between Marxist-oriented movements and other approaches to government has been especially bitter. Political democracy has been challenged everywhere and has been under serious stress even in places as modern and "Western" as Singapore and the Philippines. In southern Vietnam, democracy was never taken seriously, and nowhere in the region does it thrive.

Today, with the emergence of Asian nationalism, the traditional view of the region as a cultural interface between China and India is declining. The future is likely to focus more and more on indigenous views and patterns that demonstrate national pride rather than on those features that have been borrowed from India or China. On the other hand, the military and political power of both China and India are geopolitical realities that all Southeast Asian states recognize and must consider in designing their own futures.

Geographical Outlines and Patterns

Despite the idea that Southeast Asia is a distinctive region, within the region there exist both cultural and physical contrasts and diversity. Perhaps most obvious is the division between the mainland: Burma, Thailand, Indo-China (Laos, Cambodia, and Vietnam); and insular and archipelagic Southeast Asia: the Philippines, Indonesia, Brunei, Singapore, and Malaysia. These two parts are separated by shallow waters that lie over the Sunda Shelf (Fig. 36–1).

Southeast Asia extends more than 3,000 miles

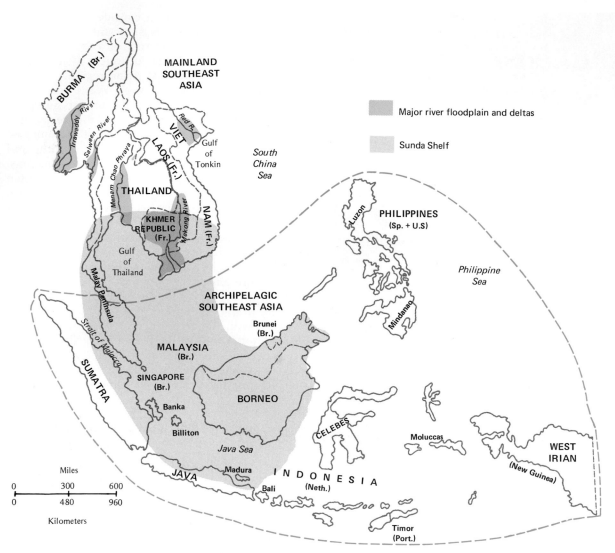

MAINLAND
SOUTHEAST
ASIA

BURMA (Br.)

Irrawaddy River
Salween River
Red R.

VIET
LAOS (Fr.)
NAM (Fr.)

Gulf
of
Tonkin

Menam Chao Phraya

THAILAND

South
China
Sea

Mekong River

KHMER
REPUBLIC
(Fr.)

Gulf
of
Thailand

Malay Peninsula

Strait of Malacca

Major river floodplain and deltas

Sunda Shelf

Luzon

PHILIPPINES
(Sp. + U.S)

Philippine
Sea

Mindanao

ARCHIPELAGIC
SOUTHEAST ASIA

Brunei
(Br.)

MALAYSIA
(Br.)

SINGAPORE
(Br.)

SUMATRA

Banka

Billiton

BORNEO

Java Sea

CELEBES

Moluccas

WEST
IRIAN

(New Guinea)

JAVA

Madura

Bali

I N D O N E S I A
(Neth.)

Timor
(Port.)

Miles

0   300   600

0   480   960

Kilometers

36–1 Southeast Asia: A Mainland and Archipelagic Realm (Former Colonial Associations Indicated)

(4,828 kilometers) from east to west (Burma to New Guinea). If the western half of New Guinea (West Irian[1]) is included, Southeast Asia contains about

[1]West Irian, the western half of New Guinea, formerly was occupied and controlled by the Netherlands. In 1962, the Indonesian Army seized West Irian. It was later incorporated into Indonesia with little objection from other states.

1.72 million square miles (4.45 million square kilometers), 807,000 square miles (2.09 million square kilometers) on the mainland, and 915,000 square miles (2.37 million square kilometers) of insular territory. Although this region lies near the equator, it stretches to almost 30°N. latitude in northern Burma, and a considerable part extends as far north as 20°N. latitude.

Areal Organization

One of the most conspicuous features of human activity and political territory in mainland Southeast Asia is the formation of national corelands around the major river basins: in Burma, the Irrawaddy; in Thailand, the Menam Chao Phraya; in Laos, Cambodia, in southern Vietnam, the Mekong; and in northern Vietnam, the Red River. It is these great rivers that provide the soil-enriching floodwaters, the water supplies for irrigation systems, and the primary corridors of transportation and access to the broad, adjacent alluvial floodplains. The large river basins contain the most productive agricultural environments and the most dense concentrations of people.

By contrast, archipelagic Southeast Asia has a different pattern of areal organization and formation of national territories. No single, clear pattern is obvious. In the archipelagos of Southeast Asia, the role of European colonial policies in shaping the patterns of development and political evolution was especially prominent. To a great extent, it was the activities of several European powers—the Spanish in the Philippines, the Dutch in Indonesia, and the

Human settlement is largely water oriented in mainland Southeast Asia. Along the rivers are fertile alluvial soils irrigated with river water. The rivers also provide a transport route. *(United Nations)*

British in Malaya and northern Borneo—that gave rise to the political boundaries that form the outlines of the modern states.

Marine Location and Accessibility

Among the most significant features of the geographical setting of Southeast Asia are its maritime orientation and location. Composed of a series of islands and peninsulas, no country, with the exception of Laos, is without a shoreline and adequate anchorage for seagoing ships. Moreover, the Strait of Malacca, separating Sumatra from the Malay Peninsula, is one of the great shipping corridors through which passes most maritime traffic sailing between Europe and eastern Asia. Accessibility ranks high among the assets of Southeast Asia, and the seas that separate the various countries and islands are much more a tie than a barrier.

This feature of maritime accessibility can best be

Singapore, a major entrepôt, controls the Malacca Strait. Coastal freighters bring cargo from throughout Southeast Asia for shipment by oceangoing vessels to the world's markets. (United Nations)

illustrated through the example of Singapore, the region's most important entrepôt. British colonials created this great city from a tropical island and mangrove swamp in the mid-nineteenth century. Its location, carefully selected by Sir Stanford Raffles at the southeastern end of the Malacca Strait, controls access to and through this part of the world and gave Singapore its growth and prosperous future. Like Europe, Southeast Asia has long benefited as much from its location and accessibility as from the resources of its physical environment. The difference is that the use of this access has developed more slowly than in Europe.

Mainland Southeast Asia

For most of mainland Southeast Asia, the dominant physical features are the rugged cordilleras that splay out from the Himalayan Mountains to the north and arc to the south. These mountains are underlain by an ancient crystalline mass of stable granite material. It is this mass and its subterranean extension, the Sunda Shelf, which geologically tie together much of Southeast Asia's islands, peninsulas, and mainland as a physical unit. The north-south trending mountains of mainland Southeast Asia, although physically related to the taller Himalayas to the north, have been heavily weathered and rounded in the tropical rainy climate. They are much lower (few are higher than 10,000 feet [3,048 meters]) than their counterparts in southern Tibet. The ranges parallel one another and separate the main drainage systems. Characteristically, active seismic forces such as faulting are not part of the physical landscape directly associated with the mainland highlands and the Sunda Shelf. Such forces are, however, very common along the archipelagic southern flank of the shelf and from the Philippines to New Guinea (Fig. 36–2).

Separating the major river basins that form the corelands of the five countries of mainland Southeast Asia are rugged highlands. A series of north-south parallel mountain ranges that extend southward from eastern Tibet composes the basic framework. These main ranges are, from east to west, the Annamite Chain of Vietnam, the Shan Highlands of western Thailand and eastern Burma that extend the length of the Malay Peninsula, and the Arakan Yoma of western Burma.

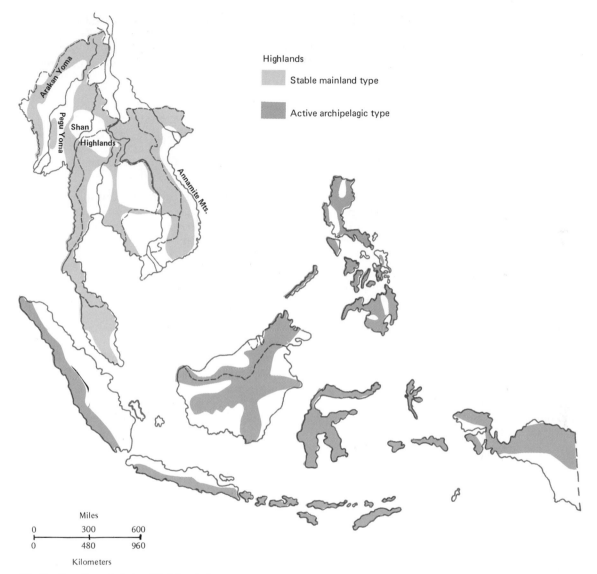

Highlands

▨ Stable mainland type

▨ Active archipelagic type

Miles

0   300   600

0   480   960

Kilometers

36–2   *Southeast Asia: Highland Areas*

Although accounting for less than one-half the total land area of Southeast Asia, the mainland component probably contains the greatest cultural fragmentation. For example, the groups that inhabit the highlands are made up of a number of diverse people who have long existed largely as tribes with cultural orientations and linguistic systems different from those of their more advanced, less isolated lowland cousins. Most of these highland people, by virtue of the territory they occupy, have found themselves politically attached to one or another state whose political, social, and religious ideals they may not share. Long ago, the stage was set for conflicts between the politically dominant majority that inhabit the major river basins and the groups that occupy largely interior, highland locations. This problem continues to exist and rankle and is serious in almost every country in mainland Southeast Asia.

Archipelagic Southeast Asia

Contrasted with the stability and regularity of the north-south trending axes of mountains and river valleys of mainland Southeast Asia is the younger active belt of volcanism associated with many of Southeast Asia's islands. A string of volcanic islands stretch from Sumatra and Java east to the Celebes and Moluccas and north to the Philippines. Not only is this area one of the most geologically active regions on earth, but it is also a highly diverse land surface, a good reflection of the recency of the processes of landscape formation. As one section of the circum-Pacific belt of volcanism known as "The Pacific Ring of Fire," the landforms of this part of Southeast Asia contrast sharply with the more stable mainland. Moreover, the influence of volcanic materials on local soils has been profound enough to affect fertility levels and cropping patterns and adds still another complicating factor to the region's physical outlines.

The island realm of Southeast Asia, to which Malaya is added, shares a common ethnic heritage. The term "Malay realm" has been used to describe the ethnic makeup of the bulk of its inhabitants, although the presence of large numbers of Chinese and other significant minorities makes this description inappropriate except for the most general discussion. Indigenous peoples of ancient Australoid or negroid stock are also present in certain locations, and in New Guinea, native peoples are predominantly negroid. Moreover, despite the common historic ethnic and linguistic affinities of the predominant Malays in Indonesia, Malaysia, and the Philippines, the Filipinos are largely Christians, whereas most other Malays are Moslem, a fact that inhibits understanding and interchange among these otherwise closely related peoples.

Southeast Asia as a Region

Perhaps the most common unifying features Southeast Asian people have are their shared location and a sense of belonging in their corner of the world. Together, the insular and mainland components share a tropical region with similar environmental resources and obstacles along with a marine location.

It is these benefits that offer the greatest hope, for it is the resource base of tropical agriculture and forests as well as the extensive petroleum reserves that hold the greatest future potential for economic development and growth.

Tropical Environments as Specialized Ecosystems

Despite a diversity in landforms, the shared marine orientation and tropical climate give the region a common set of environmental characteristics. These commonalities promote similar economic patterns and provide the basis for a unified approach to economic growth and development.

Effects of Equatorial and Tropical Location

Within five or six degrees of the equator, high average humidity and temperature are common. Singapore (2°N. latitude), with little seasonal variation in rainfall or temperature, is a classic equatorial climatic station. Similar climatic regimes are found throughout much of the Indonesian and Malaysian archipelagos, although locally landforms and prevailing winds may alter the typical pattern somewhat.

Away from the equator, seasonality of rainfall with distinct wet and dry seasons is common. In general, these seasons follow the alternations of the monsoon wind system. During the northern hemisphere summer, the monsoon takes the form of a southerly wind from April through September, and during winter, a northerly wind from October through March. Although most rainfall is concentrated in the summer months, surface relief creates some variations. Examples of unusual and unexpected patterns are found in the autumn maximum of rainfall along Vietnam's central coast and the summer dry areas in the rain-shadow of Burma's southeastern mountains.

A Fragile Environment

The uniformly high temperatures and heavy rainfall of the tropical environment produce a very fragile ecological condition. Such a condition demands careful management if use of the environment is not to result in serious and often nearly irreversible damage to the natural resource base. Tropical rainforests, the original vegetative cover of much of Southeast Asia, are characteristic of most of the humid tropics of the world.

The rainforest appears luxuriant and prolific, and often the uninformed assume that a rich soil supports this thick forest cover. In fact, the soil is not rich but has been heavily leached of plant nutrients. The rich forest growth is supported in large part by the sparse accumulation and rapid decomposition of its own litter on the floor of the forest. The presence of the forest is based on the rapid recycling of the nutrients. Thus, the growing forest produces, in large part, its own food supply. Climatic conditions provide a favorable environment for chemical decomposition to change organic materials to usable nutrients. The process involves a great number of species of trees, plants, and vines. It is an ecological system extremely sensitive to any alteration.

Two exceptions to the generally infertile tropical rainforest soils are present in Southeast Asia. One exception is the fertile alluvial soils of the floodplains and deltas where plant nutrients are periodically added to the soil by flooding. The second exception is on the island of Java where a youthful soil of volcanic origin provides a soil of great fertility. The areas of these fertile soils are densely settled and provide the basis of livelihood for most of Southeast Asia's peoples.

Mankind is an active agent in landscape change. Human beings have utilized a number of different approaches in the exploitation and use of the tropical rainforest ecosystem. Sometimes the approach has been cautious and to some extent emulates natural conditions by growing many crops together. In other cases, increasing population or ignorance and inexperience have resulted in destruction of the fragile ecosystem.

Imperata Grasses: A Challenge of the Tropics

One of the most serious problems in Southeast Asia's rainforest region has been the replacement of the forest by tough, fire-resistant, fibrous grasses of the species *Imperata*. These grasses are known locally by a variety of names such as *cogonales, cogan,* or just plain elephant grass. Where a large area of the forest has been cleared and improperly maintained, or where small plots are continuously cultivated, these grasses invade the fields and once established are difficult to eradicate. *Imperata* grasses have taken over large areas in the Philippines, Thailand, and Indo-China.

The consequences of *Imperata* are serious, for present indigenous farming methods cannot control the grasses. Heavy equipment and machinery are required for *Imperata* removal—equipment the people do not have, nor do they have access to such equipment. In Southeast Asia where the population growth rate is high and agriculture an important means of livelihood, the loss of land to *Imperata* is a serious problem.

Interestingly, *Imperata* has not been a significant problem where land is farmed strictly in the traditional shifting (migratory) manner and where demographic stability is maintained. The small forest clearings are cultivated for only a few years and then abandoned. Forests around the edges of the clearings gradually encroach on the clearings and shade out the *Imperata*.

Shifting Cultivation: Early Use of the Humid Tropics

Over large areas of the humid tropics, both in Southeast Asia and in other tropical locations, shifting agriculture is widespread (Fig. 36–3). It is a low-yield system, and the use of hand tools is characteristic. The system involves brief periods of cropping on a small parcel of land (one to three years) followed by a lengthy period of fallow (twelve to eighteen years). During the fallow period, the forest (secondary forest) is allowed to regenerate. The cycle of clearing, burning, cultivating, and fallowing is a means whereby the forest and soil store plant nutrients that are then released over a short period when the land is cropped.

The release of the plant nutrients is rapid, and crop yields usually decline quickly after the first year. Eventually the field is abandoned, and the forest is allowed to reestablish itself and begin the cycle anew. In some cases the forest grows to maturity, but more commonly a slash-and-burn cycle is renewed after a dozen years or so. The objective is to use the land only until yields decline and before any serious soil destruction takes place.

Shifting Cultivation as a Conserving System

Shifting cultivation has a number of interesting properties. It is an adaptive, though perhaps primitive, method to re-create natural conditions. A tremendous number of different crops, vines, cover crops, tubers, and pulses are intertilled and replicate

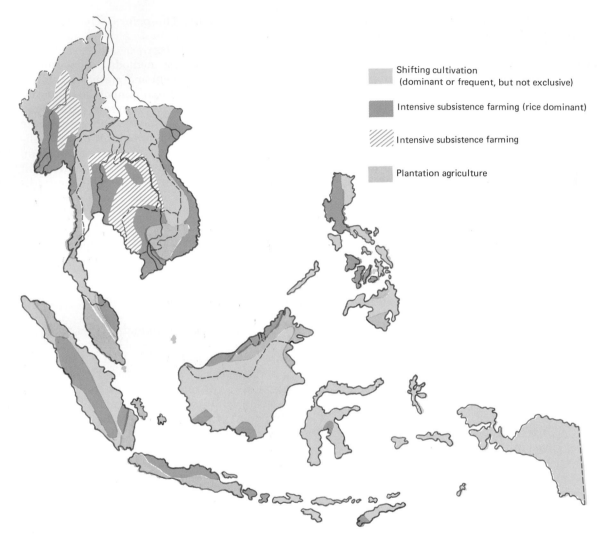

Shifting cultivation
(dominant or frequent, but not exclusive)

Intensive subsistence farming (rice dominant)

Intensive subsistence farming

Plantation agriculture

36–3 Southeast Asia: Major Agricultural Systems

in part the natural characteristics of the tropical ecosystem. The diversity of the shifting cultivator's cultigens adds strength to the crop system by providing a more balanced diet and a more assured food supply, and further protects the soil. The ability to grow things quickly and get the surface covered protects the soil from rain and sun and retards the process of leaching. All of these techniques indicate

that where populations are sparse and sufficient time is allowed the land in the fallow period, shifting cultivation is conserving rather than destructive. The system can support a sparse population indefinitely. Yet a balance between a sensitive environment and the population must be maintained. If the population grows or if too frequent cropping takes place, the end result may be degradation of the tropical rainforest and a rapid decline of soil nutrients with no replacement mechanism. If the balance is disturbed by too many people and too frequent cropping cycles, for example, large areas may be lost to elephant grasses. The land may then be useless to both shifting cultivators and other sedentary users who might desire the land for more intensive exploitation.

Contemporary Advantages

Tropical Resources

Southeast Asia has a considerable number of assets and advantages common to its location and environ-

36–4 Southeast Asia: Major Mineral Areas

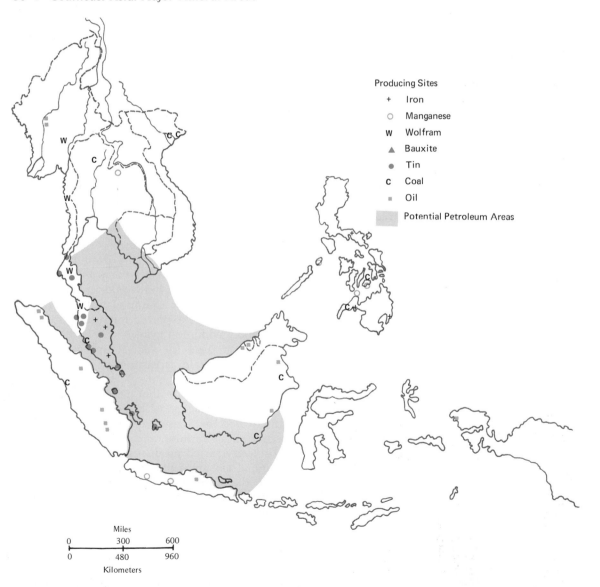

Producing Sites

+   Iron

○   Manganese

w   Wolfram

▲   Bauxite

●   Tin

c   Coal

▪   Oil

     Potential Petroleum Areas

Miles

| 0 | 300 | 600 |

| 0 | 480 | 960 |

Kilometers

ment. Among these are metallic minerals and fossil fuels as well as the forest and agricultural products of the tropics. Several minerals exist in large concentrations. Tin is probably the best known, and the major tin-producing belt, located in granitic ore and placer deposits of the Malay Peninsula and adjacent islands, is among the most productive and accessible in the world (Fig. 36–4). Other important minerals found in large quantities include iron ore (the Philippines and West Malaysia), manganese (the Philippines and Indonesia), and wolfram (Burma and Thailand). Significant minerals in lesser quantities include gold, bauxite, copper, zinc, and chromium. Tins leads all minerals in the value of its production, and this mineral has played an important role in the development and economic growth of the Malay Peninsula and the Indonesian islands of Banka and Billiton.

Of greater importance than the metallic wealth are the extensive fossil fuel resources, especially oil. Coal is found in exploitable quantities in several states (the Philippines, Indonesia, Thailand, and Vietnam), but only the deposits of Tonkin are ample enough for the requirements of industrialization. Petroleum, by contrast, exists in great quantities in the Indonesian archipelago and along the northern coast of Borneo. The existence of this resource is particularly fortunate, for Indonesia is among the poorest states of the region and greatly needs the foreign exchange from the sale of crude oil. Indonesia

A large Malay village in Brunei. Brunei possesses extensive oil resources that are now being exploited. (United Nations)

exports more than 1 billion dollars worth of oil yearly, mostly to Japan.

The Sultanate of Brunei, on the coast of north Borneo, is small and rich in oil. Its status is analagous to the small oil sheikdoms of the Persian Gulf, with a small population and a large resource of oil. It stayed out of the federation of Malaysia in order not to have its wealth diluted by joining a larger state.

Since 1970, almost every state in Southeast Asia has been involved in the search for oil. On the mainland several large sedimentary basins appear attractive as potential sources, but the major hope seems to lie offshore. The Sunda Shelf underlies the Gulf of Thailand, the Java Sea, and much of the South China Sea with an average depth of about 150 feet (46 meters). Offshore drilling is relatively simple in such shallow water, and the future may hold considerable promise for this particular resource. Although there is the question of political control and ownership of the offshore deposits, these deposits may be a means of generating capital and assisting economic development. If offshore oil exploration shows large reserves, faster rates of economic growth throughout the region may be anticipated.

The Environment as a Resource

One product that is especially prominent among the resources associated with the tropical environment is timber. Tropical hardwoods include several valuable species as, for example, teak, ebony, and mahogany which are exported for furniture, veneer, and plywood. In addition to saw timber are trees such as rattan palm used for special purposes such as carvings or household accessories.

The tropical environment offers much more than its forest species. Once that people experimented with and learned to cope with the special conditions, they discovered that these conditions could be adapted to continuous farming and exploited intensively. This discovery raised productivity levels far beyond the marginal existence of the shifting cultivator and permitted the support of dense populations. It is to these more intensive forms of agriculture that attention is now directed.

Paddy Rice Agriculture

Within Southeast Asia, the most common form of agriculture is the cultivation of paddy rice (Fig. 36–3). In all but one country (Malaysia), paddy rice accounts

The tropical rainforest of Malaysia covers over 75 percent of the land. Lumber and forest products rank third in the country's list of exports. Forest research programs have been started to improve wood yields by introducing pine and other fast-growing trees. (United Nations)

for more than half the agricultural area; moreover, it supports most of the population as the primary food grain. Traditionally, rice has also been a major export commodity for Burma, Thailand, Cambodia, and Vietnam. In recent years these states have not produced up to their potentials, but the outlook for increased production is bright if political conditions stabilize.

Paddy rice is grown is several ways in Southeast Asia; all depend on the presence of water to sustain plant growth. The main difference is the source and control of water. In some cases, as in the Irrawaddy Delta of Burma and parts of the Philippines, rainwater is the source. More common is the use of supplementary water. In the floodplain of the Menam Chao Phraya in Thailand, and along the lower Mekong in Cambodia and Vietnam, the irrigation of wet rice depends on the annual flooding of the rivers. In other locations, as in Java, water is provided and controlled through irrigation systems. Irrigation comprises the most sophisticated type of wet rice cultivation and also the highest yield per land unit.

Wet rice is a special crop. It is very sensitive to heat, sunlight, and applications of water. The nature of water-supply systems and the reliability of the control and amount of water largely determine the

Wet rice is the principal crop of Southeast Asia agriculture. Rice is usually cultivated and harvested by hand. (United Nations/FAO/H. Null)

yields. Where water is carefully controlled and where impoundment takes place through the construction of diked fields, growth is fastest and most prolific. It is important to keep water at a carefully controlled depth in the field for most of the growing cycle. The fields are commonly composed of 2 to 3 feet (0.6 to 1 meter) of unconsolidated alluvium underlain by impervious clay. The impervious clay keeps the water from percolating downward, and most water loss is through evapotranspiration. Water not only provides moisture for the rice plants but also nourishes them, for the water brings with it mineral nutrients; and these are vital to the plants in the absence of concentrated applications of chemical or organic fertilizer.

Wet rice cultivation in this part of the world is closely attuned to the conditions of the natural environment. To the extent that rice farming has relied more on primitive techniques than on capital substitutes in the form of machines and chemical fertilizers, it has been a relatively low-yield system. Yet this more primitive system means less reliance on external suppliers and outside interference. It also raises the prospect of a great potential for increased production, for with better systems of irrigation and fertilization coupled with new seed strains, great increases in yield seem likely and the feeding of a much larger population easily possible. One such example is provided by the controlled water supply in Java which may suggest a development approach for other parts of Southeast Asia.

Agricultural Innovation, Involution, and Population Growth[2]

Java is an exception to many of the generalizations that are commonly posed about Southeast Asia, and the reasons are of interest to anyone who examines the nature of economic development. First, Java and its adjacent island, Madura, are probably the most densely populated part of Southeast Asia and contrast sharply with other large islands in the Indonesian archipelago.

In the nineteenth century, as Dutch colonials set about exploiting their tropical colony, they decided to concentrate on sugar cane as a cash crop and to improve the growing conditions and technologies. In so doing, they invested in new and sophisticated systems of irrigation (a form of capital input). The system yielded well, and sugar cane production increased dramatically. More important, the Dutch allowed the irrigation system to be used to control water for wet rice cultivation on a part-time basis. Here an unforeseen consequence resulted. Rice yields increased rapidly, and at the end of the nineteenth century, the vagaries of sugar prices led to a decline in demand for sugar.

[2]An excellent description, from which this discussion draws heavily, is found in Clifford Geertz, *Agricultural Involution: The Process of Ecological Change in Indonesia* (Berkeley and Los Angeles: University of California Press, 1963), 176pp.

Consequently the irrigation system designed to commercialize agriculture based on a cash crop increasingly affected a subsistence food crop. This effect permitted the subsistence farming system to increase yields rapidly and to support an increasing population. The growing population provided a low-cost and increasing supply of labor which was used to increase output. In this way the agriculture system turned in on itself. The pattern of production based on an increased food supply ultimately led to a rapidly growing population and increased labor supply. Unfortunately, improvements in living levels and average incomes failed to materialize, for the rapidly growing population drained off the increase in output, a vicious circle of a total growth spiral without per capita improvement or related innovations in cropping techniques. This situation is similar to that experienced in China, a high-level equilibrium trap (Chapter 34).

Commercialization of Agriculture

Two agricultural systems common throughout Southeast Asia have been described, and each has its distinctive characteristics. Shifting cultivation, although it does not support many people, involves a large area, most of which is fallow at any given time. Where population is sparse in the hilly and mountainous sections of the region, shifting cultivation is the common form of land use. In the adjacent alluvial lowlands, wet rice cultivation dominates agriculture and supports dense populations, a feature that is particularly distinctive when mapped (Figs. 36–3 and 36–5). The third major agricultural system neither supports the dense populations of wet-rice cropping nor is as consumptive of land as is shifting cultivation. Yet in some ways this third type of agriculture, estate or plantation agriculture, is the most important, for it is an agricultural system that involves the extensive use of capital and is commercialized. Plantation agriculture is the sector most strongly associated with money and international trade linkages. To that extent this form of agriculture is a dynamic force for modernization and economic progress and plays a special role in the development of the region.

Estate Crop Cultivation and Cash Cropping

The growing of estate crops in Southeast Asia has changed markedly, and these changes reflect, in a large part, the developmental process under way. Originally derived from the system of shifting cultivators and from crops indigenous to Southeast Asia (pepper, tapioca, and gambier), export cropping was a marginal enterprise initially, at least for European planters. It was only in the late nineteenth century after the establishment of the joint-stock company that Europeans took advantage of the capital resources available and expanded greatly the production of crops such as coffee, coconuts, and rubber. At that point, estate agriculture became successful.

Crops and Cropping Practices

A number of different crops have been utilized in estate cropping, and some have succeeded better than others. Moreover, there is considerable variation in where particular crops are grown. In Java and the Philippines, sugar has long dominated, but until recently special marketing arrangements and guaranteed prices for certain quantities may have resulted in stagnation for the agricultural systems rather than in progress. In contrast, rubber production, concentrated mainly in Malaysia and Sumatra and grown both in estates and on small holdings, has been very successful and accounts for over one-half of the area under estate crops in the region. Other important cash crops include coconuts, coffee, and recently vast plantings of oil palm. All of these crops are remarkably well suited to the local environments, although the rubber tree, imported from the New World via England in the late nineteenth century, is probably the most adaptable and can thrive under a much greater variety of conditions.

Costs of cultivation and harvesting, processing costs, assembly and transport cost to market, and market price all affect the success of commercial crop cultivation. A favorable combination of these factors determines the degree of success and the location of particular plantings. Rubber in the past has succeeded, for it can be cultivated and harvested cheaply and is relatively easy to process and transport. As a result, natural rubber producers have managed to remain competitive with synthetic producers.

Earlier in this century, the British government, whose colonies is Southeast Asia produced most of the world's rubber, attempted to control the price of rubber on the world market. The British curtailed rubber sales. Since other sources could not meet the demand, rubber prices rose. Such a practice is called valorization, and can only be effective if most producing areas work together to control supplies carefully. In the case of the British valorization program, the high prices stimulated rubber cultivation elsewhere, particularly on the nearby Dutch island of Sumatra. Within a few years Dutch rubber undercut the British attempts to control supplies and thus price. Later, both the British and Dutch joined together in a second valorization attempt. The Japanese invasion of Southeast Asia in World War II and the subsequent development of synthetic rubber spelled the end to producer attempts to control natural rubber prices.

A more recent valorization program is that of the Organization of Petroleum Exporting Countries (OPEC), which controls 85 percent of the oil entering international trade. As long as the OPEC controls such a large percentage of the oil trade and no other energy sources are developed, these producing countries can fix prices at levels of their choice. Valorization programs, however, have a history of impermanence.

Other crops such as coconuts or oil palm may require more sophisticated forms of processing; often only a large estate or firm can be financially successful with these crops by producing in large quantities. The foreign-owned estate with its greater capitali-

36–5 *Southeast Asia: Population Distribution*

One dot represents 100,000 people

zation has distinct advantages. Yet, as Fryer has pointed out, the foreign-owned estate, despite its many advantages, and its potential success as a moneymaker and development device in Southeast Asia, has a political problem.[3] The bulk of Southeast Asia's population resents these vestiges of the colonial period, and many of the voices of Asian nationalism strongly disapprove the continuation of what they view as a form of Western European economic colonialism and exploitation.

Immigrant Chinese also proved adept at growing cash crops. With their special closely knit social organization, the Chinese were able to meet labor and marketing requirements within their ethnic group and thus possessed special advantages over all other groups. Native smallholders were unable to develop the marketing and transport linkages so vital to Chinese success. Thus, cash cropping of several varieties existed; some of it was estate based, and some was focused on smallholders.

In sum, cash cropping accounts for about one-half the estimated cultivated area in Southeast Asia (20 million acres or 8 million hectares). The contributions of these crops are much greater than are obvious at first glance in terms of area or the number of people employed. The greatest contribution of the estate crop sector is not simply the capital earned through the sale of crops; rather, the nature and operation of the commercial structure and apparatus, so fundamental to large estate cropping, may be powerful tools in economic modernization. Modern estate cropping can promote economic growth and change through increased demands for modern techniques of cultivation, processing, transportation, and marketing.[4]

Alternative points of view regarding estate cropping are not so optimistic.[5] Some authorities assert that estates, especially when controlled by foreign investors, impede rather than promote economic growth. Their argument is that a dual economy is

created which can lead to friction within the producing area and a maldistribution of wealth. Moreover, characteristically, estates production is limited to one of two crops that are sold to only a small number of nations. Such a situation can lead to increasing dependency on a small resource base and on a limited market outlet.[6]

Population: Modest but Growing Rapidly

When compared with the rest of Monsoon Asia, Southeast Asia is not densely populated. There are, however, exceptions. For example, the population in parts of the islands of Java and Luzon, the lower delta regions of the Irrawaddy, the Menam Chao Phraya, the Mekong, and the Red River, and the city-state of Singapore, achieve densities common along the lower Yangtze and Ganges rivers. All the areas noted are the core regions of their respective countries. With these exceptions and a few lesser ones, Southeast Asia is not densely populated. The total population of the region is estimated at slightly over 320 million. The problem, however, lies in the extremely rapid rates of population growth in all of these states. With annual growth rates of 2.7 percent, many Southeast Asian countries will double their populations in about twenty-five years, a prospect that soon will propel these states into the demographic dilemma that faces modern India or China. Perhaps the major implication of the current situation is that Southeast Asia has a great opportunity for economic growth and development, yet this opportunity must be viewed in the context of rapid and potentially dangerous population growth.

The Time Cushion

In view of Southeast Asia's present resource base and available arable land, there is sufficient time for most countries to promote rapid economic growth and development. Yet without some constraints on population growth, the gap between poverty and higher incomes may not be bridged. The rapidly growing populations may continue to dilute and erode the gains of economic progress, a dilemma common in most less developed lands. Indonesia, the Philippines, and Thailand are good illustrations of this dilemma. All three have experienced sound progress

---

[3]Donald Fryer, *Emerging Southeast Asia, A Study in Growth and Stagnation* (New York: McGraw-Hill, 1970), pp. 225–256.

[4]Hla Myint, "The 'Classical Theory' of International Trade and the Underdeveloped Countries," *Economic Journal*, 68 (1958), 317–337, argues persuasively for the positive role of export crop production in developing nations.

[5]Several of these points of view are summarized and outlined in Benjamin Higgins, *Economic Development*, rev. ed., (New York: Norton, 1968), pp. 267–295.

[6]See also Chapter 26 for additional thoughts on plantations and other forms of export-oriented agricultural production.

in their economies in recent years, and all have significant potential for improving agricultural output. Yet their rapid population growth rates continue to offset the economic gains made through planning and investment. In these nations and elsewhere, the results of rapid population growth are poverty, social unrest, and political instability. These, then, are the common problems that both impede the growth process and reflect its incapacity to solve the multitude of problems that confront contemporary Southeast Asia.

Malaysia and Singapore, by comparison, offer a more optimistic prospect. In Malaysia, although the growth rate is high, population density is modest, and there is little prospect of overpopulation in the near future. Singapore is small and incredibly crowded; yet its growth rate is modest, and it has demonstrated an ability to deal with its economic problems through intelligent and innovative planning. This planning is both social and economic in nature. Family planning is pushed vigorously, and good housing is provided for most of the population. The government has been successful in promoting industrialization and providing reasonable economic opportunities and adequate jobs for most people. The result is that Singapore has one of the highest average income levels in Asia, a success story that suggests that Singapore has already advanced out of the less developed stage of ecomomic growth and is well into that stage Rostow has termed "the drive to maturity." Singapore unfortunately is an exceptional case. As a city-state and former showpiece colony of the British, it can hardly serve as a development model for the remainder of Southeast Asia. Yet it does demonstrate that Asians in Southeast Asia have the capacity for modernization and economic success.

Further Readings

CHANG, JEN-HU, "The Agricultural Potential of the Humid Tropics," *Geographical Review*, 58 (1968), 335–361. Chang's thesis, as this brief article suggests, is that under present technologies the humid tropics may be producing near their unit maximum yields. This argument is based on the climate characteristics of the region and associated photosynthetic properties of food grain.

DOBBY, E. H. G., *Southeast Asia* (London: University of London, 1960), 415 pp. Dobby's general geography of Southeast Asia, although a bit old, remains one of the standard texts.

FISHER, CHARLES A., *South-East Asia*, rev. ed. (London: Methuen, 1966), 831 pp. Fisher's geography of Southeast Asia, although oriented toward political topics, remains one of the best general statements on the region.

FRYER, DONALD W., *Emerging Southeast Asia: A Study in Growth and Stagnation* (New York: McGraw-Hill, 1970), 486 pp. Fryer's work is one of the most recent regional geographies on Southeast Asia. This economically oriented text explores the contrast between the stagnant and progressive economic systems that operate in the various states.

FURNIVALL, J. S., *Colonial Policy and Practice: A Comparative Study of Burma and Netherland Indies* (London: Cambridge University Press, 1948), 568 pp. A comparative study of colonialism under the British (Burma) and Dutch (Indonesia) forms the theme of this work. The comparison is penetrating and the explanation sound.

GEERTZ, CLIFFORD, *Agricultural Involution: The Process of Ecological Change in Indonesia* (Berkeley and Los Angeles: University of California Press, 1963), 176 pp. This volume presents an ecological approach to the developmental history of Java. The explanation of the development of Javanese agriculture under Dutch colonialism is especially penetrating and insightful.

GRIST, D. H., *Rice* (London: Longmans, Green, 1959), 548 pp. Grist's text on rice cultivation, with many examples from tropical Southeast Asia, is perhaps the best single explanation of cultivation practices and ecology of wet rice.

PELZER, KARL J., "Man's Role in Changing the Landscape of Southeast Asia," *Journal of Asian Studies*, 27 (1968), 269–279. This essay interprets broadly the role of man in modifying and adapting to the specialized environments of Southeast Asia.

PURCELL, VICTOR, *The Chinese in Southeast Asia*, 2nd ed. (London: Oxford University Press, 1965), 623 pp. Purcell outlines the distribution of Chinese in the "Nanyang" and traces the historical developments that led to their present situation in each country in the region.

RICHARDS, PAUL W., *The Tropical Rainforest* (London: Cambridge University Press, 1957), 450 pp. A full explanation of the ecology of the tropical rainforest with many examples from Southeast Asia forms the main theme of this work.

SPENCER, JOSEPH E., *Shifting Cultivation in Southeast Asia* (Berkeley and Los Angeles: University of California Publications in Geography, 1966), 247 pp. Spencer, a cultural geographer, examines this important economic system common in much of Southeast Asia.

Southeast Asia: Problems and Prospects

37

ALTHOUGH SOUTHEAST ASIA has extensive mineral resources, a potentially productive agricultural environment, and a relatively low population density, problems do exist. These problems—economic, political, and social—are serious enough to raise fundamental questions about the ability of the region to achieve many of the goals of economic growth and national development. It is to these problems and the forces underlying them that attention is now directed.

The Colonial Past and Regional Disunity

In some ways the contemporary map of Southeast Asia resembles the map of Europe. The region is fragmented into a number of independent states, most of which are relatively small in area if not in population. In the preceding chapter it was posited that the formation of these states, at least those on the mainland, to a large extent reflected the geographical outlines of the main drainage basins. Colonialism also has played a major role in shaping the evolution of political areas in this part of the world.

Political Forces in Mainland Southeast Asia

With the exception of Thailand, all states in Southeast Asia were colonies of European powers. Colonialism reinforced the geographical elements present and added a political legacy to the region, for colonial control was most frequently associated with port cities and focused on large rivers and river basins. This legacy has not always been happy, and internal strife and intraregional conflict have been common. Although the forces of Asian nationalism threaten the existence of several states and many boundaries within the region, the imprint of a colonial past lingers over the contemporary political map. The extent to which this imprint persists may hinge on the outcome of military conflict.

Nowhere is this conflict so apparent as in Indo-China, a zone of former French colonialism now divided among the countries of Khmer Republic (Cambodia), Laos, and Vietnam (formerly subdivided into a northern and a southern part). At least two primary purposes were involved in forming these separate nations. On the one hand were the political interests of the big powers—the French, Chinese, and American—and the question of political control of Vietnam. On the other hand were the local nationalistic interests of self-government with political divisions focused on ethnic-tribal lines. The Khmer people dominated the Khmer Republic, and the interests of Laotian groups centered on what is today Laos. The resolution of both external and internal interests as they related both to nation formation and political orientation, after three decades of fighting, remains uncertain.

To the west, Burma was formerly a British colony. Although Burmese ethnics dominate the core of the Irrawaddy drainage basin, a number of tribal people live in the surrounding horseshoe of hills, plateaus, and mountains. Along the northern and eastern boundaries with China and Thailand, an insurgency

673

exists. In one locale it involves some remnants of Chinese Nationalist forces left from the 1949 Communist takeover in China. This insurgency has created a serious border problem with China. Indeed, the proximity of a united and powerful China is a military reality that affects the political behavior of all of Southeast Asia.

Thailand was never formally a European colony. Today, it continues its independence, but it too must walk an uneasy tightrope of shifting allegiances as power shifts among larger states outside the region. Its border areas are not secure, and the various tribes inhabiting its highland regions form a potentially serious fragmenting element.

Political Stability in the Archipelagos

Malaysia, composed of former British Malaya (the Malay Peninsula) and the Bornean states of Sarawak and British North Borneo (Sabah), has fewer prob-

lems (Fig. 37–1). Serious political unrest existed in the 1950s. The British played a key role in controlling this emergency. Although some problems persist, they appear manageable. The core of the revolt involved a distinctive ethnic minority, the Chinese. Probably the most troubling result of this earlier insurgency was the hardening of the ethnic division between the native Malays who compose about one-half of the population and the large Chinese minority who make up 36 percent of the population. Groups from the Indian subcontinent account for another 10 percent. This division, which also is to a large extent urban and rural (with the Chinese located mostly in cities), continues to darken the political horizons and growth potential of this state.

The island states have in some ways a more cohesive political situation. At least their inhabitants are ethnically more homogeneous. Yet Indonesia, with political animosities between Communist and

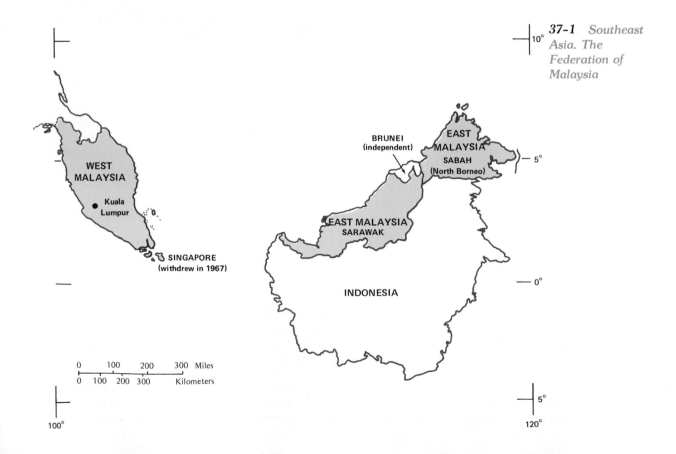

37–1 *Southeast Asia. The Federation of Malaysia*

non-Communist, has already witnessed one costly and brutal domestic confrontation in 1965. The violence associated with every election in the Philippines bears testimony to the fragility of American-inspired political democracy there. In the southern Sulu Islands and Mindanao of the Philippines, a religious split between Christians and Moslems is a continuing source of trouble aand bloodshed.

Nowhere in Southeast Asia is there political peace and stability. Singapore, a city-state, may be the sole claimant to an orderly political climate. Yet even in this model of Asian progress and modernization, there are many complaints that the government is repressive and allows little freedom to its citizens. It is against this background of political unrest, recent revolution, and warfare that the potentials for growth and change in Southeast Asia must be evaluated. Certainly there is no easy route to economic growth and social progress. Some political freedoms may have to be sacrificed in order to achieve stability so necessary to the orderly progression of economic and social growth.

Insurgency in Southeast Asia

Among the most serious problems that face Southeast Asia is the question of the political legitimacy of various groups and governments. World War II and the end of European colonialism yielded to the rising power of popular native governments which in turn ushered in a new chapter in the history of self-government in the region. These local governments represent several different political perspectives.

Alternative Political Systems and the Rise of Insurgency

Of those political movements available in Southeast Asia, the Communist party was among the best organized. The stress on sound and effective organization and discipline among the party members was obviously a strength. Marxist movements had become popular among young European-educated Southeast Asians in the 1920s and 1930s. Marxist groups were organized, and many programs and ideas for the future of Southeast Asia were outlined. Once that World War II demonstrated the

vulnerability of European colonial administrations, the stage was set for political movements, such as the Communists, to use the surge of Asian nationalism to achieve their goals.

By contrast, most former colonizing powers did not want to see Communist governments seize control in their former colonies. To that end, most of these European states worked vigorously to install political regimes acceptable to the West as they prepared to grant independence to their colonies. Malaya and the Philippines are good examples of this policy. In Vietnam, a war resulted over the nature of the succeeding political regime, and a settlement was reached whereby the territory was partitioned between competing political systems, a Communist north and a Western-oriented south. Part of the difficulty for some of the Western powers resulted when the South Vietnamese government did not prove popular and strong. The challenge of well-organized Marxist-inspired and -supported insurgency movements became a serious threat to the political survival of several states and the stability of the entire region.

The Nature of Insurgency

As portrayed in the classic examples in Southeast Asia, insurgency movements commonly arise in remote base areas associated with new revolutionary political ideas that have been coupled with native nationalism. Insurgencies are often tied to local political problems in impoverished rural locations and may focus on one ethnic group. Gradually the base area expands, and more support is available for the paramilitary guerrilla troops. Finally, guerrilla insurgents are strong enough to challenge the military units of the established government. Several levels or stages of growth are involved, and these are associated with several different units of territory (Fig. 37–2). From rural hamlets to districts to provinces spread the propaganda and activity and finally the control of the insurgents.[1]

Virtually every country in Southeast Asia has been affected by insurgent political-military efforts. Malaya and the Philippines had serious insurgency problems in the 1950s, but sharp reactions coupled with im-

[1]A good introduction and description of this process may be reviewed in Robert W. McColl, "The Insurgent State: Territorial Bases of Revolution," *Annals, Association of American Geographers,* 59 (1969), 613–631.

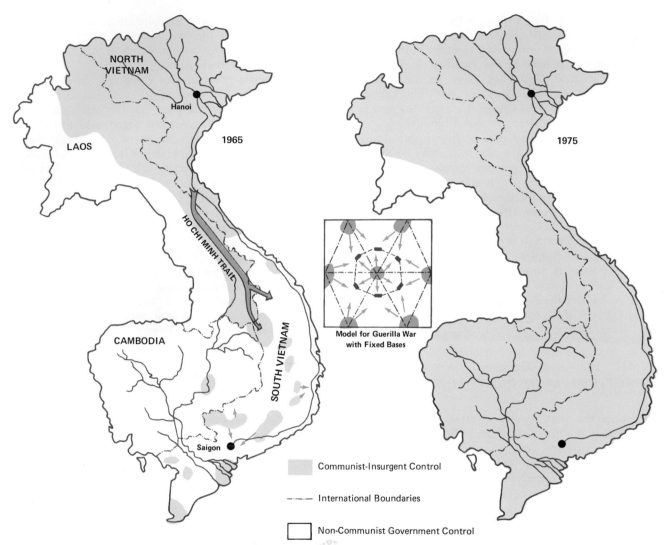

NORTH VIETNAM

Hanoi

1965

LAOS

HO CHI MINH TRAIL

CAMBODIA

SOUTH VIETNAM

Saigon

Model for Guerilla War with Fixed Bases

1975

Communist-Insurgent Control

—·—·— International Boundaries

Non-Communist Government Control

37–2   Indo-China: Expansion of Insurgency

proved social conditions and popular support of the government appear to have curbed insurgent movements in both countries. In Indonesia, the Communist party was legalized and in the early 1960s was the third largest in the world. A bloody coup d'état resulted in outlawing the party, a policy that continues today.

Indo-China has experienced more military trauma over Communist–non-Communist division than any other part of the region. This trauma no doubt ac-

counts for the enormous difficulties in promoting economic progress in Indo-China. The French and American involvements in Vietnam and adjacent states have not resolved the issue, and the political and military struggle in Indo-China has consumed valuable resources and energies. In Indo-China's case, the insurgency long ago progressed beyond the base area stage and for years involved main forces fighting in Vietnam, the Khmer Republic, and Laos.

Until political stability is attained, there is likely

to be little if any real opportunity for meaningful economic growth and progress. Political stability and control are a prerequisite of meaningful economic development. Only in those states that have achieved at least a credible level of peace and stability and an outward-looking government (Malaysia, Singapore, the Philippines, and Thailand) has economic progress been significant. It may be that Indonesia, after a decade of domestic strife and revolutions, should also be considered one of Southeast Asia's progressive states, but it is still too early to gauge accurately how significant the economic growth in Indonesia has been and how successful is the development program.

Efforts to Forge Regional Unity

Not all of the events of the past thirty to forty years in Southeast Asia have been fragmenting, disruptive, and pessimistic. Counterpoised to the problem and difficulties posed by insurgent movements, revolution, and military actions have been the efforts to forge regional unity in Southeast Asia. Some efforts have had political objectives. For example, the formation of the Southeast Asia Treaty Organization (SEATO) in 1954 was primarily a political-military arrangement aimed at containing Asian communism. Others are developmental organizations such as the Association of Southeast Asian Nations (ASEAN) and the Mekong River Basin Development Scheme. ASEAN, formed in 1967, is a political-economic organization organized to promote cooperation among Southeast Asian states. Its aim is to accelerate the pace of regional development and growth through cooperative trade and shared projects. The Mekong River Project, first conceived in 1954, is a Southeast Asian version of the American TVA (Tennessee Valley Authority) plan.

SEATO was the first regional organization in Southeast Asia. Supported by the United States, Great Britain, Australia, and New Zealand, its main goal was defense of the region against aggression, especially Communist aggression. It also has certain objectives for promoting economic cooperation. The Vietnam conflict hurt this alliance, and France and Pakistan, two nonregional members, have not participated in recent years. Changing policies and the withdrawal of Thailand, a long-time staunch supporter of SEATO, have rendered the organization moribund. Whether this organization could function for collective defense in the future is questionable.

ASEAN is specifically interested in promoting economic growth through regional integration and cooperation. Thailand, the Philippines, and Malaysia were the original members, and Indonesia and Singapore joined in 1967.

Another important agency supporting the goal of economic growth in the region is the Asian Development Bank, which is headquartered in Manila. Similar in concept and operation to the World Bank and other regional financial institutions that promote development, this bank is financed through deposits of wealthy states and lends money for specific projects among the participating states primarily in Monsoon Asia. Projects range from irrigation and agricultural schemes to major transportation and industrial investments. The bank's staff is composed largely of Asians from the member states.

The Mekong River Basin Project

The Mekong River Basin Development Project may be among the most significant projects in the region as a result of the far-reaching political implications involved. This project, which includes a comprehensive river basin development scheme (flood control, navigation, hydropower production, and irrigation), is focused on the Mekong and its tributaries (Fig. 37–3). Although none of the construction projects would lie in northern Vietnam, power could be supplied to that area. Thailand, Laos, Khmer Republic, and southern Vietnam, whose territory contains the Mekong watershed, would be the main beneficiaries.

The scope and benefits of this project could be enormous. An area of 200,000 square miles (518,000 square kilometers) with a population of more than 50 million would be affected. The project was first outlined by a United Nations commission in 1957, and work on some of the smaller tributaries has already been completed. This work, however, is a small part of the total project, a project that could promote progress and development throughout the region. Following the fall of South Vietnam to Communist forces, however, United States interest in the Mekong project has waned. It is difficult to

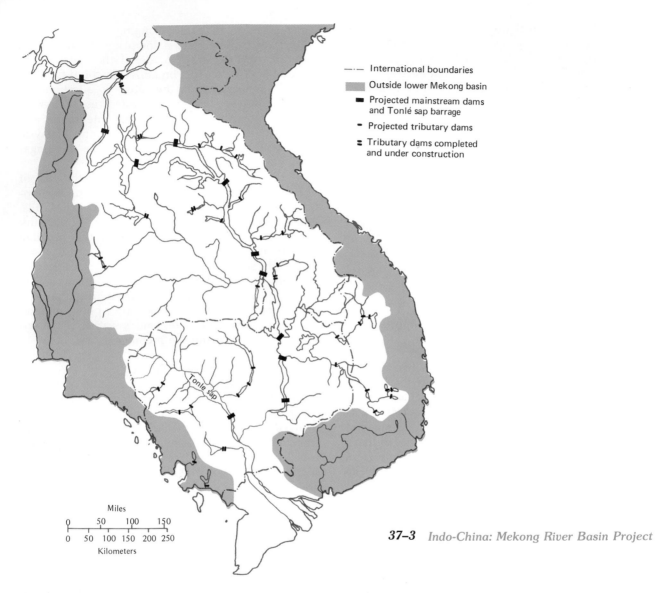

- —·— International boundaries
- ▨ Outside lower Mekong basin
- ■ Projected mainstream dams and Tonlé sap barrage
- ▬ Projected tributary dams
- ≈ Tributary dams completed and under construction

Tonle sap

Miles
0   50   100   150
0 50 100 150 200 250
Kilometers

37–3 *Indo-China: Mekong River Basin Project*

predict the future of the project at this time in view of United States disengagement from Indo-China. It may be that funding and technical assistance will be furnished by various United Nations' agencies and the Asian Development Bank.

Advantages of Regional Unity

Southeast Asia, like Europe, is broken politically into small fragments. Southeast Asians have suffered many of the same difficulties as have Europeans.

World power and influence are difficult to wield where small states argue and bicker among themselves. Cooperation and union make for strength. As Fryer has noted, political fragmentation is probably the strongest argument for Southeast Asians to cooperate with each other and join together in common economic and political alliances.[2]

[2]Donald Fryer, *Emerging Southeast Asia: A Study in Growth and Stagnation* (New York: McGraw-Hill, 1970), pp. 436–449.

Perhaps more important than the potential political advantages in regional unity are the possibilities for improved economic growth rates. Indo-China is a sad example of the effects of wartime conditions on economic growth and progress. During such periods, improvements in cities, ports, industries, transportation and power systems, irrigation networks, and other sectors are negated by widespread destruction. National incomes decline, and people get poorer with their savings lost and their activities and life styles disrupted. The normal and orderly processes and inducements to investment are not available, and external assistance, with all of its international political implications, is the only alternative available.

Animosities and lack of cooperation among states that fall short of actual fighting also are damaging. Inability to establish regional organizations that function effectively—for example, to achieve meaningful customs reductions—may impede the economic growth rates of all states in the region. Competition among producers of similar products weakens the ability of all to negotiate and trade effectively with the outside market. The example of oil-producing states in the Middle East is a striking illustration of the potential rewards that may accrue in marketing scarce resources. Perhaps the Southeast Asian states can bury their mutual jealousies and suspicions to cooperate and reap such rewards. Cooperative marketing cartels for tin, vegetable oil, and rubber might provide better control over price and production. Whether such union and cooperation are possible within the context of bitter animosities among neighbors remains to be seen. During the early part of the twentieth century, British and Dutch interests in Southeast Asia did join together in a valorization program to control rubber and tin prices. The precedent exists, and the great success of the oil-exporting nations' cartel (OPEC) may encourage the Southeast Asian nations to bury disagreements for economic gain.

Cultural Pluralism

Contemporary Southeast Asia exhibits diverse political and ethnic qualities. To a large extent, society and polity are closely related, and many of the region's political problems stem from the strains of multiethnic societies. Cultural pluralism is evident in economic organizations, for there exists a close association between specific occupations and particular ethnic groups. Certain ethnic groups are heavily concentrated in urban locations, whereas others are focused in rural areas or around mining sites. Pluralism and separatism rather than integration have characterized the economies, societies, and politics of most states in the region.

Natives, Chinese, and Indians

In most of Southeast Asia, there exist large Chinese and Indian minorities (Fig. 37–4). The number of these minorities varies considerably from country to country. In some cases, as in Malaysia, the Chinese comprise about 35 percent of the population and the Indians another 10 percent. In the Philippines and Indonesia, however, the Chinese represent only 2 or 3 percent of the total population, and the number of Indians is too small to be of any consequence. Some Chinese immigrants have been integrated and absorbed into the native population. They have been urged or forced to assume local identities, to use the local language, and to take local names. In Vietnam and Thailand, integration has been reasonably successful. Chinese have been present in large numbers for a long time. Much intermarriage has taken place, with resulting cultural assimilation. Yet a range of problems inherent in a multiethnic society exists, and no approach has proved entirely satisfactory in coping with the difficulties involved in dealing with several different groups. Racial and ethnic divisions persist and occasionally result in violent political outbursts and activities.

The Chinese in Southeast Asia

The Chinese have been involved in Southeast Asia for many centuries. The history of Vietnam, for example, is in part a history of Chinese attempts to extend their influence southward. Yet it was in the nineteenth century, in part associated with British colonialism, that extensive Chinese migration to the region took place. The British initially were interested in developing a tractable and industrious labor force to exploit tin deposits in the Malay Peninsula and to assist in agricultural development programs in northern Borneo. Inasmuch as the British held the native Malays in low regard for purposes of organized labor, they sought workers from the out-

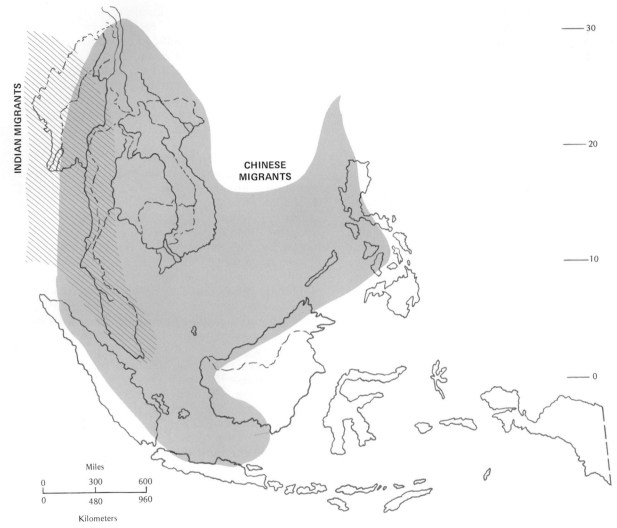

37–4 *Southeast Asia: Nonnative Ethnic Groups*

side and encouraged the immigration of Chinese from South China.

Over the years, large numbers of Chinese emigrated to Malaya and other parts of Southeast Asia. Typically they were not content to remain long as indentured servants or hired laborers. Moreover, the nature of Chinese society, with its closely knit family, clan, and dialect group associations, led to the formation of a broad network of ties between China and the migrants in Southeast Asia. These ties were fundamentally social and cultural but were rapidly extended to trade, production, and shipping. Chinese in Malaya quickly acquired land and began to produce cash crops such as coconuts, sugar, spices, and rubber. They also established small tin mines with modest capital requirements. Their social system provided them with the linkages to assure a constant supply of cheap labor and ready access to markets, both local and international. Their society ensured that they had friends or relatives in business, shipping, and production, and the home country provided a steady supply of inexpensive, hardworking laborers.

Over time, the Chinese carved out a large niche in the economy, not just of Malaya but of all states in Southeast Asia. In some cases they were little involved in primary production, but in every country they concentrated in cities and specialized in trade, shipping, and finance. In some cases they gained control of the rice milling industry, a key aspect of most of the agricultural economies of Southeast Asia. In other cases they dominated virtually all forms of local commerce, partly because of their ability to play the role of middleman in linking the colonial administrations of the European powers with the native populace. The European colonials were unwilling to perform these functions, and the native populations had no acquaintance with the particular crafts of commercial entrepreneurship. The Chinese filled a vacuum. In the process they established themselves in a position of privilege in the various Southeast Asian societies.

In time, the Chinese enlarged their traditional economic system to the point that their position was powerful and could not be controlled by external forces. Locally, their power, prestige, money, and increasing numbers became a potent political force and a source of resentment among native populations. It has been a general characteristic of Chinese who have migrated to foreign areas that they have maintained their customs, traditions, and linguistic patterns and were reluctant to integrate fully into the local society. The more Chinese there were in a concentration, the more vigorously they resisted acculturation and integration. So powerful has been this force that the customs and traditions of old China were much better preserved among the Southeast Asian communities than in China itself.[3] It is this cultural distinctiveness, coupled with a privileged economic position, that has created enormous resentment against the Chinese in Southeast Asia. Their clannishness has made them an easy and convenient

target for the frustration and anger of impoverished masses of Southeast Asians and their political spokesmen.

The Primate City: A Southeast Asian Dilemma

Nowhere in Southeast Asia are the problems of a fragmented society, a dual economy, and an uncertain political system so conspicuous as in the great primate cities of the region. In these cities are found the largest concentrations of wealth juxtaposed with the most dense clusters of poverty. Here the main political, administrative, educational, and medical institutions are located "cheek-by-jowl" with the greatest squatter concentrations and the most vocal and often radical politicians, the most modern stores and factories, and the largest number of itinerant peddlers and vendors. Indeed the term "primate" was coined to describe those cities in which were found disproportionately large concentrations of the administrative, economic, educational, and service activities and wealth of the state along with a large percentage of the urbanized population.

Such a phenomenon is not peculiar or unique to Southeast Asia. In Southeast Asia, however, it occurs in almost every country. The great capitals—Bangkok, Rangoon, Djakarta, and Manila—are examples (Fig. 37-5). These are the largest cities of their countries, the greatest ports and commercial trade centers. They contain the major international airports, and they account for a major share of the industrial output of the various states. Here too are found the best universities, hospitals, research centers, newspapers, and other cultural institutions. In some cases, as in Thailand or Burma, there are virtually no other cities with more than 100,000 inhabitants.

Primate Cities and Modernization: The Fragment Hypothesis

Primate cities in Southeast Asia dominate the political and economic life of their respective countries. Yet it is not at all clear if the consequences of this dominance are positive or negative. The division between city and countryside has raised a serious

[3]Victor Purcell, *The Chinese in Southeast Asia,* 2nd ed. (London: Oxford University Press, 1966), pp. 30–39, presents a good summary of Chinese society and customs in Southeast Asia.

● Cities with more than 1 million inhabitants

○ Cities with less than 1 million inhabitants

National capitals are underlined.

37–5 *Southeast Asia: Major Cities*

question about the productive benefits of these cities. It has been suggested that this urban-rural division is also a division between a modernizing economy and society and a traditional economy and folk society. Most serious among the consequences of this division are the association of the modern urban sector with one ethnic group, the Chinese, and the traditional or folk sector with another ethnic group, the native populations. The two sectors have been labeled "fragments," for they are seen as fragmenting or divisive elements.[4]

In such cases, many have criticized the primate city and asserted that it is not an effective agent of growth

[4]Allen Goodman of Harvard University first developed the fragment hypothesis as described here. See Allen E. Goodman, "The Political Implications of Urban Development in Southeast Asia: The 'Fragment' Hypothesis," *Economic Development and Cultural Change,* 20 (1971), 117–130.

and change in a developing land. The contention is that much of the wealth and many of the productive facilities of the country are concentrated in the one great city, so that few benefits diffuse throughout the remainder of the country. Only a small hinterland adjacent to or near the primate city is able to secure the full benefits of economic growth and progress generated in the large city. Large areas are far removed, and many people remain unaffected by the productive influence of the primate city. Where such spatially unequal growth takes place, it is argued, the seeds of discontent are sown. The wealth of the state is not spread equally, and great contrasts of wealth and poverty emerge that are most apparent by their location.

If a division between the primate city and the rest of the nation exists, the stage is set for a more serious consequence: the emergence of two distinct and antagonistic political camps: one urban-oriented, modern and often Chinese, the other rural-based, traditional and native. Under such conditions, one school of thought has emerged that claims the fragmenting of the Southeast Asian region along these lines will impede other growth and modernization processes. The primate city is seen as the leading agency in the divisive situation, and the implications are clear. Until the dominance of these cities is broken, modernization and productive change in the region will be slow and difficult.

It may be that there is a powerful logic in this argument, and it is very appealing. On the other hand, what solution exists that would result in a greater dispersal of economic benefits throughout a state? Would it be wise to attempt to stimulate growth away from the great cities? It might be that such attempts are economically unsound and would actually retard the rate of growth. It is these kinds of questions that confront planners and politicians in Southeast Asia as they seek to promote more rapid economic growth and modernization in their region.

Examples of Stagnation, Transformation, and Growth

A variety of approaches to development have been attempted in Southeast Asia. These range from Com-

munist to socialist to free enterprise. No one approach has succeeded, and it is possible to evaluate the achievements of different approaches in specific countries. Such an evaluation may permit some conclusions on the relative merits of these development strategies in the context of the Southeast Asian region.

Burma

Despite the fact that Burma is a large state and has a resource base more than sufficient to support its present population, its growth record has been very poor. Its per capita output of goods and services actually declined between 1939 and 1969, a situation that speaks poorly of Burma's particular approach to development. The approach, however, is unique, a sort of authoritarian isolationism and locally derived socialism. A military coup d'état in 1958 was followed by a vacillating policy of partial democracy. In 1962, the army, on the pretext that non-Burman ethnic subversion threatened the country, took over the government and dismissed the parliamentary government.

Since 1962, the path to modernization has been difficult. Not only has the government been authoritarian in its policies and socialist in its economic planning but it has also shut itself off from the outside world. This self-imposed isolation may well be the most shortsighted policy of all. Burma's failure to borrow and learn from the rest of the world, Communist, socialist, and capitalist, in planning economic growth and in soliciting external assistance and investment has been a disaster. In 1973, after years of stagnation, and perhaps even declining economic well-being, the military government yielded slightly. In addition to promulgating a new constitution, the government relaxed its policy of isolation and began to encourage more foreign contacts. Although it is still too early to estimate the consequences of this new policy, it suggests a recognition that the policy of isolated self-sufficiency is unworkable. For modern Southeast Asia, the goals of progress and growth appear to depend on cooperation and exchange on a global as well as regional scale.

Indonesia

Following independence from Dutch colonialism in 1949, Indonesia emerged as an important voice among the nonaligned states of the world. Led by the charismatic but troubled Sukarno, the island republic in the 1950s and early 1960s attempted to portray itself as a leader among those states that were following a new form of independent socialism. In Indonesia, the political rhetoric and the ambitions of the leaders far outdistanced economic performance and the ability of the state to support its claim that Indonesia was the best model for other less developed areas. In some years in the 1950s and early 1960s, there were declines in the per capita output of goods and services. This situation quickly led to severe social and political stress that finally resulted in a 1965 coup d'état and the overthrow of Sukarno.

Since 1966, the new military regime has set its major sights on repairing the economic neglect and damage of the Sukarno regime. One aspect of this policy was to seek assistance from outside sources such as Japan and the United States. The performance of the military government has not been showy, nor has the economic growth rate been spectacular; but a new, sober emphasis has been placed on solid achievements in economic planning and growth. Completion of *Repelita I* (the first five-year development plan) in 1974 was one such achievement in which annual economic growth averaged about 7 percent. The goals of *Repelita II* (the second five-year plan) are more ambitious and are focused on faster rates of growth in the agricultural sector.

Indonesia is still a poor country. Its per capita GNP is only about $90. It is a good example, however, of the fact that economic conditions can change and improve even where the situation is highly pessimistic and unstable. Although the path to economic transformation and modernization is difficult, painful, and slow, the goals are attainable, and Indonesia justifiably may point with pride to its recent record of accomplishment and look to the future with confidence. One uncertain aspect of Indonesia's achievement is the extent of distributing the fruits of the new growth broadly among the masses of peasants. Unless some improvements are made available to this large group, the outlook for future political and social stability is poor.

The Philippines: Progress from the Green Revolution

Although political events and developments in the Philippines during recent years have been un-

happy and cause for concern about the true nature of Philippine democracy, some sectors of the Philippine economy show considerable vigor and promise. One of the most successful examples is found in the agricultural sector as seen in the experience of certain rural areas in growing newly developed strains of high-yielding rice. New seed strains are not, however, the only innovation. A broad process of modernizing Philippine agriculture has been under way, and it includes a variety of supporting procedures, practices, and materials. These include increased use of insecticides, chemical fertilizers, and mechanical threshing machines; better systems and methods of irrigation; and more efficient use of labor gangs to provide adequate labor during periods of peak demand.

Such new and improved cropping practices, especially when coupled with the new seed strains developed at the International Rice Research Institute in Los Banos, have resulted in dramatic increases in yield, farm output, and farm income in some rural areas.[5] For example, in a four-year period coinciding with the introduction of new rice strains (1965–1969), yields nearly doubled; such gains are becoming more common throughout the Philippines as the new rice strains and supporting technologies are being diffused. Yet without a concomitant reduction in the rate of population growth, the Philippines may discover in another generation or two that the technical breakthrough in agricultural production does not lead to per capita gains in income with long-term improvement in levels of living. If the time cushion provided by the new agricultural breakthrough is not used wisely, the Philippines may fall into the same high-level equilibrium trap as did Java almost a century earlier.

Malaysia and Singapore

Probably the two richest countries in Southeast Asia are the progressive twins, Singapore and Malaysia. Upon independence in 1963, the two were politically integrated, but the Chinese-Malay dichotomy was severe enough to disrupt the political alliance. Despite political separation, both the city-state of Singapore and the Federation of Malaysia

---

[5]Robert Huke, "San Bartolome and the Green Revolution," *Economic Geography*, 50 (1974), 47–58.

Throughout Southeast Asia the Chinese form an important ethnic group. Many Chinese are concentrated in the cities working as store owners and traders. Shown here is a street scene in Bangkok, Thailand. (United Nations)

continue to thrive economically and stand as solid examples of prosperity in Southeast Asia. Despite the partially authoritarian nature of politics in the two countries, both have vigorous and growing private sectors in their economies. These dynamic free market forces, coupled with an outward-looking attitude, may account for the steady economic growth of these countries. Certain important aspects of Singapore's economy, however, have more state control than exists in Malaysia. For example, roughly one-half the population of the city-state is now housed in public projects, a case where the private sector has not been as successful as in other forms of enterprise.

Singapore, with its per capita GNP of roughly $1,300, stands next to Japan and Hong Kong as the most prosperous place in Monsoon Asia. Despite the physical constraints and problems of its small area, the city-state has made good use of its location and traditional economic functions of trade, shipping, and service. Building on this base, it has recently turned to sophisticated manufacturing in which the skills and industry of its talented and disciplined labor force are employed. Cameras, watches, and electronics goods are some of the products. Oil refining, which depends on Indonesian and Persian Gulf crude, is also a major industry for Singapore. Singapore in many ways has succeeded. The island republic is a "have" among "have nots," an island of prosperity in a larger sea of poverty. Singapore demonstrates, as does Japan, that Asians even under adverse conditions can achieve living levels comparable to those of Europeans, and they can do it in a relatively short time.

Malaysia's growth rate has been more modest than Singapore's; nevertheless, the federation has progressed steadily in recent years. Annual economic growth rates of 6 to 7 percent have been maintained consistently. Malaysia has been especially fortunate in the exploitation of its natural and agricultural resources. Tin, rubber, timber, coconuts, and more recently palm oil are the main products, and in recent years commodity prices have risen. It is difficult, however, to predict whether high demand levels for these commodities will continue.

Perhaps the most impressive thing about Malaysia's progress is that the division in Malaysian society between Chinese and Malay natives has not stifled or interrupted the growth. Despite their ethnic and political differences, the two groups have managed to coexist and cooperate to promote their common well-being. If this lesson can be learned and applied throughout the region, the prospects for rapid economic growth seem likely.

Implications for Future Growth

There is no single path to rapid economic growth and social change in Southeast Asia any more than there is a single solution to all of the region's problems. Despite the absence of a unitary approach to modernization, some lessons may be learned from past experience. Those states that have become preoccupied with their own problems or ambitions and have disregarded cooperation and assistance from a community of states have performed poorly. Burma and Indonesia are good examples. At the same time, it seems clear that the potentials for growth and progress in Southeast Asia are enormous. Moderate policies that emphasize the goals of growth and take advantage of international assistance appear to be the most successful. The shift in Indonesian policies and political and economic behavior after Sukarno's ouster suggests the desirability of a moderate policy. Moreover, the turnabout in Indonesia is an excellent indication of the effect that political policies have on the economy and how quickly things can change for the better—a hopeful outlook for many other less developed, impoverished areas of the world.

Vietnam

The 1973 Paris Peace Agreement that ended direct United States participation in the fighting in Vietnam brought a short-term peace to what, until the 1954 Geneva accords, was Indo-China. A year later fighting erupted again, and many observers were astonished at the speed of the collapse of South Vietnam. Since 1954, the bitterness among the Vietnamese people with respect to which group would guide the political destiny of Vietnam has sharpened, and the Vietnam War of the 1960s and 1970s was a painful expression of this bitterness.

The dispute between the two Vietnams in part stemmed from the desire of the northern half to incorporate the southern half and forge one unified Vietnam. The two halves complement each other economically, for the north is more industrial and is a food-deficient region. The south traditionally is a rice surplus area and of value to balance off the shortages. There was, however, an argument for separating the two halves, for the Annamite Mountains, Indo-China's most prominent physical feature, touch the sea midway down the coast very near the former demilitarized zone. Historically, the north and south have not always been politically integrated.

The South

The traditional rice surpluses of southern Vietnam disappeared during the wartime difficulties of the 1960s. The problem of peasant insurgency and rural insecurity became so great that much productive land was simply abandoned or little cultivated. The war had many other consequences for the predominantly agrarian economy of southern Vietnam. Some of these consequences accelerated the pace of economic growth; for example, United States aid, among other things, provided training programs for Vietnamese civilians, made huge investments in capital improvements (roads, harbors, cities, power grids, and irrigation systems), and drew large numbers of people into urban centers to contribute to the industrial and commercial sectors of the economy. The abrupt withdrawal of United States soldiers and civilians from Indo-China in 1971–1972, and the cutbacks in United States military and economic aid, created problems of unemployment, inflation, and instability in the economy. These problems continued and, coupled with insecurity in rural areas, created

Singapore is an island of prosperity in Southeast Asia. Levels of living often are comparable to those of Europe. Almost half of Singapore's inhabitants live in public housing similar to the Toa Payoh Housing Estates shown here. (United Nations)

687

an atmosphere of uncertainty about South Vietnam's future. This uncertainty may have played a part in the quick demise of the government in early 1975 and its replacement by a Marxist government.

The potential for economic growth, following resolution of the political and military problems, is great. Most promising is the enormous potential for increased agricultural production. Restoring abandoned land to cultivation and opening new land in the Mekong delta marshlands are two methods that could promote greater output. Increased yields on lands already cropped through improved irrigation, new seed strains, and more chemical fertilizer and insecticides are equally promising. It is this latent possibility for surplus food production that would appear to explain the protracted desire of northern

Vietnam to incorporate the south into a larger political union.

The North

Northern Vietnam's heartland is focused on the Red River drainage basin and the Tonkin Delta (Fig. 36–1). In many ways this region, with its dense population and intensive double-cropping rice complex, resembles Chinese occupancy patterns found farther north. The northern Vietnamese, however, have not been able to replicate the agricultural achievements and yields of the Chinese, and northern Vietnam has long suffered from food shortages. These shortages have, to some extent, been offset by the presence of metallic minerals and coal. Based on these resources, the former French colonial regime had initiated a

Saigon, largest city in southern Vietnam, received large numbers of refugees from outlying areas. Many of these new urban dwellers live in densely crowded slums. (United Nations / P. Teuscher)

number of industrial development projects. The Communist government has continued and expanded these industries. During the decade of the 1960s, both the Soviet Union and China contributed heavily to the north and countered United States aid in the south. The war, and especially heavy bombing, however, left much of the industrial complex in ruins.

Since the war, Vietnam has embarked on a three-year crash program to rebuild the industrial establishment in the north. The overall, long-term economic growth goals of the centrally planned economy were interrupted and delayed by the war. Agricultural deficits are likely to continue and will impede further economic growth. It is uncertain about how much the Russians and Chinese will continue to contribute. Although the nature and direction of political events in Indo-China remain unclear, one truth appears to have emerged since 1974. No matter what the exact name or legal status of the several countries, one political force will play an important role in the affairs of Indo-China, be it Vietnam, Laos, or the Khmer Republic: that political force is communism. The Communist party has long dominated the Tonkin heartland of northern Vietnam. Recent events suggest that it can be expected to influence directly the political destiny of an increasing part of mainland Southeast Asia.

Further Readings

BUTWELL, RICHARD, *Southeast Asia: A Political Introduction* (New York: Praeger, 1975), 242 pp. Butwell's book focuses on the politics of the region and the great complexity of the struggle to forge meaningful political union and compromise throughout Southeast Asia.

DUTT, ASHOK, *Southeast Asia: Realm of Contrasts* (Dubuque, Iowa: Kendall-Hunt, 1974), 194 pp. This volume, the work of ten regional specialists, is the most recent regional geography on Southeast Asia. Although there is considerable variation among the various chapters, up-to-date coverage and current topics are included.

FALL, BERNARD, *Street Without Joy,* 4th ed. (Harrisburg, Pa.: Stackpole, 1964), 408 pp. A French journalist with long-established and close ties to Vietnam examines the nature of the colonial past and the resulting conflict. He assesses the prospects for future stability.

FISHER, CHARLES A., "The Vietnamese Problem in Its Geographical Context," *Geographical Journal,* 131 (1965), 502–515. A geographer interprets the division of Vietnam into two parts on the basis of history and geographic factors.

GOODMAN, ALLEN, "The Political Implication of Urban Development in Southeast Asia: "The 'Fragment' Hypothesis," *Economic Development and Cultural Change,* 20 (1971), 117–130. Goodman interprets the multiethnic quality of Southeast Asian cities as disruptive to normal political processes and development.

McCOLL, ROBERT W., "The Insurgent State: Territorial Bases of Revolution," *Annals, Association of American Geographers,* 59 (1969), 613–631. McColl's emphasis on the insurgent state examines the process of its territorial expansion. Many of the examples are drawn from Southeast Asia.

McGEE, T. G., *The Southeast Asian City* (London: G. Bell, 1967), 204 pp. McGee focuses his attention on the great primate cities of the region and attempts to develop a general characterization of them.

POLLARD, VINCENT K., "ASA and ASEAN, 1961–1967: Southeast Asian Regionalism," *Asian Survey,* 10 (1970), 244–255. Difficulties of forging regional union and unity in Southeast Asia form the main theme in this essay.

SEWELL, W. R. DERRICK, "The Mekong Scheme: Guideline for a Solution to Strife in Southeast Asia," *Asian Survey,* 8 (1968), 448–455. Prospects for resolving the conflicts in Indo-China are suggested through river basin development of the Mekong and many of its tributaries.

SHAPLEN, ROBERT, *Time Out of Hand: Revolution and Reaction in Southeast Asia* (New York: Harper & Row, 1969), 465 pp. An American journalist traces the political history of Southeast Asia. Much of the material previously appeared in *The New Yorker.*

TAYLOR, ALICE (ed.), *Focus on Southeast Asia* (New York: Praeger, 1972), 229 pp. This set of short essays on various countries and geographical topics in Southeast Asia was originally published as *Focus* essays by the American Geographical Society.

Glossary

Agricultural calendar The chronological sequence of farming operations throughout the year including the period of land preparation, sowing, cultivation, and harvesting different crops.

Agricultural Revolution A period beginning 7,000 to 10,000 years ago characterized by domestication of plants and animals and development of farming.

Air mass A large body of the atmosphere that has uniform or nearly uniform temperature and moisture characteristics.

Alluvium Material that has been transported and deposited by water, often very fertile.

Andean Group A principal subgroup of the Latin American Free Trade Association (LAFTA) comprising Venezuela, Colombia, Ecuador, Peru, Bolivia, and Chile; these countries are committed to several joint efforts to speed economic integration.

Apartheid The policy of the South African government of strict white-nonwhite segregation.

Appalachia That part of the Appalachian Highlands of the United States designated as a poverty area.

Arable Land that can be cultivated profitably.

Association of Southeast Asian Nations (ASEAN) A political-economic organization formed in 1967 to promote cooperation among member nations.

Autarky Self-sufficiency, or the policy of economic independence.

Autonomous Region The highest-level political unit in China by which ethnic minorities are theoretically allowed some local self-rule and preservation of customs.

Aymara The principal Indian group and language of Indians living in the Lake Titicaca Basin; this group has maintained its integrity both in Inca times and during the subsequent Spanish intrusions.

Aztec One of the four high civilizations of pre-Columbian Latin America centered on the area around present-day Mexico City.

Bank An area of the ocean where the sea is relatively shallow. Fish tend to congregate on the banks. Banks thus have become important fishing areas.

Bauxite The ore from which aluminum is obtained.

Bedouin An Arab who lives by nomadic herding in the deserts of North Africa and the Middle East.

Benelux The nations of Belgium, the Netherlands, and Luxembourg considered as a group.

Berber A group of North African people living in the Sahara Desert and along the Mediterranean coast.

Biomass The total weight per unit area of all organisms in an ecosystem.

Biosphere That part of the earth occupied by various forms of life.

Boers Literally Dutch farmers, specifically the Dutch living in the Republic of South Africa.

Boreal Forest The large coniferous forest extending across Canada and part of northern United States.

Boserupian thesis The theory developed by Ester Boserup that rapid population growth can speed economic development.

Boxer Rebellion 1899–1900 political uprising promoted by a Chinese secret society, in part, directed against the penetration of China by foreign interests and foreign missionaries.

Brahmin The highest group in the hierarchy of the Hindu caste system comprising basically Aryans who are teachers and religious leaders.

Buffer state A small relatively weak country between two large nations; the buffer state serves to reduce conflict between the larger nations.

By-product A secondary or incidental material obtained in the mining or manufacturing process; for example, the primary mineral being mined may be lead, but silver is also recovered out of the lead ore, or in refining petroleum for gasoline, heavy hydrocarbons are also obtained that can be used in the chemical industry.

Caballero Spanish term for gentleman and horseman.

Calcification A soil-forming process in dry areas; resulting soil is characterized by the presence of soluble minerals.

Campo cerrado A vegetative type found in the Central West of Brazil and covering 75 percent of the region, used principally for cattle grazing (mixed grassland and woodland).

Campo limpo A grassland vegetative type found in the Central West of Brazil; agricultural potential of campo limpo land is considered very low.

Campo sujo A grassland with scattered woods found in the Central West of Brazil; agricultural potential of campo sujo land is considered low.

Cape Colored A term used in South Africa designating persons of mixed Caucasian and non-Caucasian parentage.

Capital Any form of wealth that can be used to produce more wealth; resources.

Cardinal temperatures The temperature range for a specific plant within which it can grow; the upper and lower temperature limits for plant growth.

Caribbean Common Market (CARICOM) A new organization comprising some of the Caribbean Free Trade Association (CARIFTA) members with a goal of economic integration.

Caribbean Free Trade Association (CARIFTA) A group of British and former British colonies in and around the Caribbean Sea formed to reduce trade barriers among member states.

Carrying capacity The maximum number of animals or people an area can support; population-carrying capacity has many variables such as nutrition level, level of living, and trade.

Caste A rigid system of social stratification based upon occupation with a person's position passed on by inheritance; derived from the Hindu culture.

Central American Common Market (CACM) A group of five Central American nations (Guatemala, El Salvador, Honduras, Nicaragua, and Costa Rica) who strive for economic integration.

Centrally planned economies Nations with a productive system controlled by the national government and based on communistic and socialistic principles.

Cerradão A scrub woodland vegetative type found in the Central West of Brazil; agricultural potential of *cerradão* land is considered low.

Chernozemic A type of soil formed by the calcification process and inherently one of the most fertile soils of the world (from Russian meaning "black earth").

Chibcha One of the four high civilizations of pre-Columbian Latin America centered on the area around present-day Bogota, Colombia.

Ching Dynasty The last imperial dynasty of China, A.D. 1644–1911, and founded by the Manchus.

Circular causation A development theory based on the idea of an upward or downward spiraling effect.

City-state A sovereign country consisting of a dominant urban unit and surrounding tributary areas.

Climax vegetation The concept that if an area is without interference by mankind for a long enough period, an assemblage of plants will eventually dominate that reflects climatic, soil, and moisture conditions.

Coke Coal that has undergone partial combustion and that is almost pure carbon; used for smelting iron ore.

Cold War The 1945 to mid-1960s period of hostility, just short of open warfare, between the Soviet Union and the United States and its allies.

Collectivization The process of forming collective or communal farms, especially in Communist countries; nationalization of private landholdings.

Command economy A centrally controlled and planned livelihood system as, for example, the socialist and Communist systems.

Commercialism The system of interchange of goods; the activities that lead to sale and exchange of products.

Common market A group of nations that have joined together to form a customs union with no or reduced tariff walls among members and a uniform tariff system with the outside world.

(The) Common Market The European Economic Community (EEC), the European Common Market.

Commune The basic socioeconomic unit of present-day China; the Chinese rural collective community.

Comparative advantage The idea that if there are several alternative uses of resources in a region, the most advantageous use tends to be selected.

Conference of Berlin A meeting of most European states, plus the United States, convened in 1885 by the German chancellor Bismarck. It was held to establish the ground rules for occupation of African territory by European states and to confirm the international status of the Congo territories controlled by the Belgian king Leopold.

Consumer goods Commodities produced for use by an individual or family such as a car, radio, and clothing; designed for consumption; they do not assist in creating additional wealth.

Conurbation A network of urban centers that have grown together.

Coreland The area of dense population that forms the urban-industrial heart of a nation; also the cultural, economic, political center of a nation or group of people.

Cottage industry (household) A system of manufacture in which raw materials are processed in the worker's home.

Council of Mutual Economic Assistance (CMEA or COMECON) The organization formed in 1949 to regulate economic relations between the Soviet Union and Eastern Europe.

Creole A person born in Latin America of European, usually Spanish, heritage.

Crop rotation The practice of using the same parcel of land for a succession of different cultigens in order to maintain or improve yields.

Crude birth rate The number of live births per year per 100 people (also commonly given per 1,000 people).

Crude death rate The number of deaths per year per 100 people (also commonly given per 1,000 people).

Cultigen A culturally improved plant; a domesticated plant.

Cultural determinism The theory that a person's range of action is limited largely by the society within which he or she lives; for example, food preference, desirable occupation, and laws.

Cultural landscape The mankind-modified environment including fields, houses, highways, planted forests, and mines, as well as weeds, pollution, and garbage dumps.

Cultural norm A standard of conduct sanctioned by a society.

Cultural pluralism The presence of two or more groups within the same area who follow different ways of life.

Cultural Revolution The upheaval in China during the 1960s when old cultural patterns were condemned and new Maoist patterns were strongly enforced.

Culture The ways of life of a population which are transmitted from one generation to another.

Culture complex A group of culture traits that are employed together as, for example, how clothing is made and distributed to consumers.

Culture hearth The source area for particular traits and complexes.

Culture realm The area within which the population possesses similar traits and complexes as, for example, the Chinese realm or Western society.

Culture trait A single element or characteristic of a group's culture as, for example, dress style.

Dark Age The period in Europe from about A.D. 500 to 1000, during which Europe was in political, economic, and cultural decline.

Debt peonage A system formerly used in Latin America to hold the labor force in bondage; debts accumulated by one generation were passed on to the next.

Demographic transformation A theory, based on Western European experience, of the relationship between birth and death rates and urbanization and industrialization.

Developed Used in this book to identify nations or regions that have a high level of economic production per person; *rich, advanced,* and *modern* are synonymous terms.

Development Decade The United Nations' designation of the 1960s to emphasize and encourage modernization of poor nations.

Diffusion The spread of an idea or material object over space.

Double-cropping The raising of two crops in succession in the same field during one growing season.

Dravidian One of the earliest inhabitants of India; dark-skinned Caucasoids of peninsular India. Also the language spoken by such groups as Tamil, Kanarese, and Telugu.

Dual economy The presence of two separate economic systems within a region, common to the poor world and characterized by one system geared for local needs and another for the export market.

Dulab See Noria.

Economic dualism See Dual economy.

Economies of scale The decrease in production costs brought about by high volume production; mass production techniques are commonly used.

Ecosystem The assemblage of plants and animals in a particular environment and their interdependence.

Ejido A form of land tenure in Mexico by which land is given to a farming community. The community may allocate parcels of land to individuals. Title to the land, however, remains with the community.

Entrepôt A center, usually a port, where goods are collected for redistribution; often a transshipment point.

Entrepreneur The "organization man"; a person who organizes economic activities.

Environment That which surrounds; the setting, both natural and cultural, within which a group lives.

Environmental determinism A general geographical theory, now largely discredited, based on the premise that the physical environment controls, directs, or influences what mankind does; variations in the physical environment should therefore be associated with different levels of economic well-being.

Environmental maintenance The care of physical surroundings that is necessary for sustained productivity.

Erosion The picking up and transport of earth materials by moving water and other natural agents.

Estáncia A Spanish term used in Latin America to describe a large rural landholding usually devoted to stock raising, especially horses and cattle; similar to a ranch.

Estate The term characteristically used in parts of Africa and Asia for a large export-oriented agricultural enterprise usually owned by a foreign company and using local labor; a plantation.

Ethnic group A group of people who share a common and distinctive culture.

Ethnic religions Religions associated with a particular group and area such as Shintoism, Taoism, and Hinduism.

European Coal and Steel Community (ECSC) A common market type organization set up after World War II to facilitate movement of coal and iron ore among the Benelux nations and France and West Germany.

European Economic Community (EEC) The common market structure consisting of nine full members with a goal of economic integration.

European explosion The spread of Western culture traits and complexes worldwide.

European Free Trade Association (EFTA) A group of seven European nations (Austria, Denmark, Norway, Portugal, Sweden, Switzerland, and United Kingdom) who joined together in 1959 to reduce tariffs among member nations. The transfer of Denmark and the United Kingdom to the European Common Market led to abolishment of the EFTA.

Evapotranspiration The water loss of plants by processes of evaporation and transpiration.

Exploitive agriculture Farming with the idea of maximizing short-term income with no consideration of long-term effects.

Extensive agriculture A farming system characterized by small inputs of labor or capital per land unit.

Factory A manufacturing unit based on quantity production, a distinct division of labor, and use of inanimate power-driven machinery.

Fault A fracture of the earth's crust associated with vertical or horizontal movement.

Fazenda A Portuguese term used in Brazil to describe a large rural landholding; these units may be devoted either to crop or animal production or to both.

Feed grain Cereals such as maize (corn), oats, and sorghum used as provender for poultry and livestock.

Ferroalloy A mineral such as manganese mixed with iron to make steel; ferroalloys give steel a variety of properties such as rust-resistance or strength.

Field rotation The movement of cropping from one parcel of land to another; a practice common in tropical areas of low population density.

Fiord (Fjord)   Glacially eroded valleys that subsequently have been invaded by the sea.

Flemish   Belgians who speak a Dutch dialect and occupy northern Belgium; one of the nation's two principal national groups.

Flow resources   A renewable resource such as trees, grass, rivers, and animals.

Folding   The process by which sedimentary rock strata are bent to form linear hills and valleys.

Food grain   Cereals such as wheat, rye, rice, maize (corn) used for human consumption.

Form utility   The change in shape, constitution, or character of a material to increase its usefulness and hence its value as, for example, smelting iron ore or processing cloth into clothing.

Fossil fuels   Organic energy sources formed in past geologic time such as coal, petroleum, and natural gas.

Fragment Hypothesis   The idea that the growing urban-rural differences in Southeast Asia have created two separate culture groups; a divisive process.

Free trade   The unrestricted movement of goods among independent nations.

Fund resources   Nonrenewable resources such as minerals.

General Agreement on Tariffs and Trade (GATT)   A multinational agreement of the mid-1960s by which signatory nations agreed to reduce tariffs, and, in general, improve international trade practices.

Ghana Empire   A West African trading state that flourished from the ninth to the thirteenth century but dates from earlier times; the modern nation of Ghana has taken its name from this empire.

Golden Horde   The Mongol state established west of the Ural Mountains in the Soviet Union in the thirteenth century and lasting to the beginning of the sixteenth century.

Golden horseshoe   An industrial district in Canada extending from Toronto to the western end of Lake Ontario to Hamilton.

Gosplan   The Soviet Union state agency charged with making national development plans; usually has five-year goals.

Great Leap Forward   The 1958–1961 attempt in China to socialize agriculture and increase production both in farming and industry.

Green Revolution   The use of new high-yielding hybrid plants—mainly rice, corn, and wheat—to increase food supplies and the infrastructure necessary to obtain greater production and distribute them to the consumer.

Gross national product (GNP)   The total value of all goods and services produced and provided during a single year.

Growing season   The period during the year that crops can be grown without artificial heat.

Growth pole theory   The concentration of development efforts on selected sites, usually urban, with the expectation that the improvements made will spread outward from the state.

Guarani   The major Indian group of Paraguay and adjacent Argentina and Brazil; the Guarani language is spoken throughout Paraguay.

Habitat   The environment; the place where an organism lives.

Hacienda   Spanish term used in Latin America for large rural landholding usually devoted to crop and animal production; these units formerly had a high degree of internal self-sufficiency and operated under the patron system.

Han Dynasty   Chinese dynasty, 206 B.C. to A.D. 220, during which large areas were brought under Chinese control; Han bureaucracy served as a model for succeeding dynasties.

Han people   The name used by the Chinese to refer to themselves to distinguish "cultured" Chinese from others; most of China proper is populated by the Han.

Hardwood   Angiosperm trees, commonly called broadleaf, and usually deciduous.

Heavy industry   A term applied to manufactures that use large amounts of raw materials such as coal, iron ore, and sand and that have relatively low value per unit weight.

High-level equilibrium trap   The condition of increasing population growth tied to increasing production so that levels of living remain constant; this situation apparently existed in traditional China.

Hinterland   The tributary area of a port of city; the size and productivity of the hinterland are usually reflected in size and wealth of the port or city.

Hsia Fang   The practice in China of transferring

urban elites to rural areas to perform manual labor; a practice designed to reeducate and maintain the group's revolutionary spirit.

Humus Partially decayed organic material that is an important constituent of soils; improves water-holding capacity, provides some plant nutrients, and makes the soil easier to cultivate.

Hybridization The selective crossing of different varieties or species of plants or animals to produce offspring possessing certain desired characteristics.

Hydrosphere The water bodies that cover the earth's surface.

Igneous rock Rock formed from molten material; igneous rock formed within the earth's surface tends to solidify very slowly, allowing minerals to agglomerate. Most metallic mineral deposits are of igneous rock origin.

Imperata A Southeast Asian type of grass that invades cleared areas and is difficult to control; elephant grass.

Import substitution The policy to encourage local production of needed goods rather than importing them; subsidies, loans, and protective tariff regulations are often the means of assuring local production.

Inca One of the four high civilizations of pre-Columbian Latin America centered on the city of Cuzco, Peru, and extending in the Andes of South America from southern Colombia to central Chile.

Indo-Aryan One of the earliest inhabitants in India who came from the northwest and spoke a Sanskrit language; light-skinned Caucasoids of northern India.

Industrial Revolution The period of rapid technologic change and innovation beginning in the mid-eighteenth century in England and which subsequently spread worldwide. The principal attribute of the revolution has been the development of cheap and massive amounts of controlled inanimate energy.

Inertia Used in this book in the sense of continuation of an economic activity in an area, especially manufacturing, after the original factors favoring the location have ceased.

Infrastructure The services and supporting activities necessary for a commercial economy to function, such as roads, banking, schools, hospitals, and government.

Intensive agriculture A farming system characterized by large inputs of labor or capital per land unit.

Intercropping The raising of two or more crops in the same field at the same time.

Iron Curtain A popular expression for the boundary between Eastern and Western Europe signifying the difficulty of movement of goods and people into Eastern Europe.

Isohyet A line along which all points have the same precipitation value; usually designated in average yearly amount.

Isotherm A line along which all points have the same temperature; usually designated in average monthly or yearly values.

Jihad A holy war, traditionally declared by the spiritual and secular leader (caliph) of Islam, against unbelievers.

Job out Subcontracting by a manufacturer to a specialized firm or person for production of a specific item or component.

Jute A hard fiber used in bagging and cordage produced primarily in Bangladesh and adjacent parts of India; obtained from a species of *Corchorus.*

Kibbutz An Israeli collective farming community.

Kolkhoz A Soviet collective farm. Members of the *Kolkhoz* lease the land from the government.

Koryo Dynasty The dynasty lasting from A.D. 918 to 1392 during which Korean culture matured.

Kraft process A method of processing resinous woodpulp that yields a strong fibrous paper used for wrapping paper and cardboard. Development of the process allowed the use of pine and other coniferous trees previously of little utility.

Kulaks The more wealthy Russian peasant farmers; this group was exiled, imprisoned, or killed in the collectivization of agriculture in the 1920s and 1930s.

Lake Chad Basin Commission An organization of states with territory in the Lake Chad drainage basin which promotes development projects within the area.

La Plata Group A subgroup of the Latin American Free Trade Association comprising Argentina, Brazil, Paraguay, and Uruguay oriented toward cooperation in development of the watershed of the Rio de la Plata.

Laterization   A soil-forming process in warm and humid areas in which the soils are heavily leached of water-soluble nutrients (calcium, phosphorus, potash, nitrogen) leaving iron and aluminum compounds.

Latifundia   Literally large landholdings; the control of most of the land by a small percentage of landowners.

Latin America   That part of the New World south of the United States comprising a cultural region, largely but not totally of former Spanish or Portuguese colonies.

Latin American Free Trade Association (LAFTA)   A loosely organized multinational grouping of Latin American nations that have joined together for economic integration.

Leaching   The movement of material downward in the soil by water; the material may be deposited at a lower level in the soil or may be flushed from the soil.

Least-cost point   The location(s) where costs of assembling materials, their processing, and shipment to market are lowest.

Legumes   Plants of the family *leguminosae*, especially those used agriculturally, such as beans, peas, alfalfa, and soybeans. Bacteria on the underground nodules of these plants have the ability to remove nitrogen from the air and fix it in a form usable by plants.

Levee   A raised bank along a watercourse formed by deposition of water-carried material during flooding; along parts of some rivers, that is, Mississippi and Hwang Ho, the natural levees have been built up further to protect surrounding areas from flooding.

Level of living   The actual material well-being of a person or family as measured by diet, housing, and clothing.

Life style   The mode or manner of behavior of an individual or group; the way of life as evidenced in dress, material possessions, and diet.

Light industry   A term applied to manufactures that use small amounts of raw materials and employ small or light machines.

Lingua franca   An auxilliary language used among peoples of different speech, commonly used for trading and political purposes.

Linkage   The connection or interdependence among industries either horizontally (over area) or vertically (at different stages of production); examples of vertical linkages are wheat from the farmer to miller to baker to store to customer.

Lithosphere   The solid portion of the earth.

Local relief   The difference in elevation between the highest and lowest points of an area.

Loess   Deposits of wind-transported, fine-grained material that are usually easily tilled and quite fertile.

Machine tools   Any power-operated instrument used to shape, form, or cut materials as, for example, a drill press or lathe.

Machine Tractor Station (MTS)   The Soviet agency in charge of all mechanized agricultural equipment and whose workers operate the equipment on *Kolkhoz* and *Sovkhoz* land.

Malthusian Doctrine   A theory advanced by Thomas Malthus that human populations tend to increase more rapidly than the means to care for the population.

Mandingo (Malinke) Empire   One of the largest West African states that endured several centuries and was centered in the area of present-day southern Mali.

Manioc (cassava)   A woody plant of the genus *Manihot* having tuberous roots that are used extensively in tropical areas as a food source; source of tapioca.

Maoism   The philosophy and behavior patterns extolled by the Chinese Communist leader Mao Tse-tung.

Maori   The Polynesian people of New Zealand; presently account for about 8 percent of New Zealand's population.

Mare Nostrum   Literally our sea; the Roman term for the Mediterranean Sea signifying Rome's control.

Marginal land   An area in which the production costs are almost equal to income. Little or no savings are possible.

Market economies   Nations with a productive system based upon capitalistic principles.

Marshall Plan   The system of economic and technological aid given by the United States to Western Europe after World War II.

Mata da primeira class   A vegetative type found in the Central West of Brazil and indicative of excellent land; first-class forest.

Mata da segunda class   A vegetative type found

in the Central West of Brazil and indicative of good agricultural land; second-class forest.

Materialism Devotion to tangible objects as opposed to spirtual needs and thoughts.

Mau Mau Africans of Kenya who practiced terrorism against European settlers and their supporters as part of the revolutionary movement that eventually led to national independence.

Maya One of the four high civilizations of pre-Columbian Latin America situated in southern Mexico and northern Central America.

Megalopolis Originally the continuous urban zone between Boston and Washington, D.C.; now used to describe any region where urban areas have coalesced to form a single massive urban zone.

Meiji period The period dating from 1867, during which the Japanese society and economy were transformed from feudal to modern.

Mekong River Basin Development Project A proposed Southeast Asian version of the Tennessee Valley Authority by which the nations of the basin would cooperate in various social and economic programs.

Mercantilism The philosophy by which most colonizing nations controlled the economic activities of their colonies: in essence, the colony existed for the benefit of the mother country.

Mestizo A person of mixed European and Indian blood in Latin America.

Mezzogiorno Southern Italy; one of the poorest regions of Western Europe.

Middle Ages The period in European history from about A.D. 500 to 1350; the first part of the Middle Ages also called the Dark Ages.

Middle Kingdom China, especially the Han Chinese area.

Millet (dura) A widely cultivated cereal grass (*Setaria*), particularly in parts of Africa and Asia, used as both a feed and food grain.

Mineral deposit A naturally occurring concentration of one or more earth materials.

Minifundia Literally, small landholding; most of the landowners combined control only a small percentage of the total area in farms.

Miracle grains The new hybrid varieties of maize, wheat, and rice that produce a greater quantity of useful calories than traditional varieties.

Mixed forest Woods composed of both deciduous and evergreen trees; a transitional woodland.

Monsoon The seasonal reversal in surface wind direction over the southeast quarter of Asia.

Moraine A ridge or mound of unconsolidated rock material that has been deposited by glaciers; moraines are indicators of the location of a glacier's outer edge.

Moslem One who surrenders to the will of God (Allah) as revealed by the Prophet Mohammed; a follower of Islam (submission).

Mulatto A person of mixed European and African blood in Latin America.

Multiplier effect The idea that for each new worker employed in an industry, other jobs are created to support and service the workers.

Nation-state A political grouping of people who occupy a definite area and who share a common set of beliefs and values.

Nationalism The emotional attachment by an individual or group to a country or region.

Nationalization The expropriation of an enterprise by the government and the operation of the enterprise by the government.

Naxalite A radical Indian Marxist rural revolutionary.

Neolithic period The latter part of the Old World Stone Age characterized by well-developed stone implements and some food raising.

New World The American continents of North and South America.

Night soil Human excreta used as a fertilizer, a practice common in parts of Monsoon Asia and Europe.

Noria A waterwheel used in parts of Spain, Latin America, North Africa and the Middle East, and eastern Asia to raise water.

North Atlantic Treaty Organization (NATO) A multinational military alliance of Western nations founded in 1949.

Obstacle Anything that inhibits a person from achieving a want or a desire or a goal.

Old World The continents of Europe, Asia, and Africa.

Operations Bootstraps The government of Puerto Rico's development plan emphasizing manufacturing, agricultural reform, and tourism.

Opium War The 1839–1842 British-initiated war designed to protect British commercial interests in China. The pretext was the refusal of the British to cooperate with the Chinese in halting opium traffic; the Chinese then sought to restrict British commercial activities.

Ore A mineral deposit that is economically profitable to mine; the material mined.

Organization for African Unity A group of African states concerned with the problems of political and economic relations among African nations and between Africa and other parts of the world.

Organization for Economic Cooperation and Development Originally established to assist the Marshall Plan fund distribution; in 1961 reorganized and now strives for expanded international trade and poor world development.

Organization of Petroleum Exporting Countries (OPEC) A thirteen-nation group of oil-producing countries that controls 85 percent of all the petroleum entering international trade. These nations have joined together to regulate oil production and prices; a valorization scheme.

Ottoman Empire A Turkish sultanate that controlled a large area from southern Eastern Europe through the Middle East and into North Africa; the empire survived for about six centuries until it collapsed after World War I.

Paddy (padi) rice A term used in Monsoon Asia referring to the rice plant; also to rice grown in flooded fields.

Paleolithic Age The earlier part of the Old World Stone Age during which time mankind lived by hunting, fishing, and gathering.

Palestine Liberation Organization (PLO) An umbrella organization created by the Arab states in 1964 to control and coordinate Palestinian opposition to Israel. Originally subservient to the wishes of the Arab states, the PLO is now dominated by the independent, nationalist ideology of El-Fatah, the largest and most moderate of the Palestinian resistance groups.

Pampas A Spanish term for grasslands also used as a proper noun to identify the most important region of Argentina.

Patron system The economic and social relationships between the large landowner in Latin America and his workers; a paternalistic relationship.

Pattern The distributional arrangement of a phenomenon as, for example, the distribution of population, climate, or cropping.

Paulista A person from São Paulo, the economic core area of Brazil; *Paulista* migrations in the past covered much of modern-day Brazil.

Permafrost Permanently frozen ground.

Petrochemical industries Manufacturing plants whose source of raw materials is petroleum or coal; examples of such industries are fertilizer, synthetic rubber and fiber, medicines, and plastics.

Photoperiod Length of daylight or the active period of photosynthesis.

Place utility The change in location of a product usually to increase its value as, for example, the movement of oil from the Arctic to eastern United States.

Placer deposits A concentration of one or more minerals in alluvial materials.

Plantation A large agricultural-producing unit emphasizing one or two crops that are sold and having distinct labor and management groups.

Plural society See Cultural pluralism.

Podzolization A soil-forming process in cool and humid areas in which both soluble minerals and iron and aluminum are leached from the soil.

Polder A tract of land reclaimed from the sea and protected by dikes; about 40 percent of the Netherlands is polders.

Population explosion The rapid natural increase in human numbers within a short time period, generally within the last 100 years.

Population pressure The strain or demands a population places on an area's resources; the ratio of number of animals or people to carrying capacity.

Poverty Material deprivation of such severity as to affect biologic and social well-being. The lack of income or its equivalent necessary to provide an adequate level of living.

Primary activities Economic pursuits involving production of natural or culturally improved resources, such as agriculture, livestock raising, forestry, fishing, and mining.

Primary processing Manufacturing using the products of primary activities such as wheat, iron ore, fish, and trees; the commodity of primary processing may be sold to consumers, that is, fish, or used as a raw material for further processing, that is, steel bars.

Primate city An urban center more than twice the size of the next largest city; primate cities have a high proportion of the nation's economic activity. Such cities are most obvious in the poor world.

Principle of the use of best resources first If accessible, use of high-quality resources will yield a greater economic return than resources of lower quality; therefore, high quality resources tend to be used first.

Producer goods A product used to create income, such as a tractor, newspaper press, or processing machine.

Productive capacity The total amount of resources that can be marshaled in an area under a given level of technology. Although it is a useful concept, productive capacity cannot be accurately measured.

Proven reserves The amount of a given mineral known to exist and is economically feasible to mine.

Pulse Certain cultivated leguminous plants used for food, especially peas and beans.

Push-Pull migration A theory used to explain the movement of people from rural areas to urban centers; the migrants are forced out of one area by limited opportunity and attracted to cities by perceived advantages.

Qanat (kanat) An underground tunnel, sometimes several miles long, tapping a water source; the water flows through the tunnel by gravity.

Quechua A major Indian language of Andean South America; the principal Inca language.

Region A portion of the earth that has some internal feature of cohesion or uniformity as, for example, the trade area of a city or an area of similar climate.

Regional specialization The division of production among areas; each area produces those goods or provides those services in which it has some advantage and trades for goods that can be produced more cheaply elsewhere.

Renaissance The reawakening of arts, letters, and learning in Europe from the fourteenth to sixteenth centuries; the transition period between medieval and modern Europe.

Residual soils Soils formed in place and in time reflect local environmental forces such as slope, climate, and vegetation.

Resource Anything a person can use to satisfy a need.

Rift A portion of the earth's crust characterized by subsidence of the crust between two parallel faults.

Sahel The semiarid grassland area along the southern margin of the Sahara Desert in western and central Africa.

Secondary activities The processing of material to add form utility, manufacturing.

Secondary processing Manufacturing using the products of primary processing such as flour and steel bars and producing a commodity that is more valuable such as bread and automobiles.

Sedentary agriculture A farming system based on continual cropping or use of the same fields.

Sertão A Portuguese term used in Brazil to refer to the frontier spirit and life; the backlands.

Settled agriculture See Sedentary agriculture.

Shaduf An irrigation method and means of raising water indigenous to North Africa consisting of a bucket pivoted on a long pole.

Shang Dynasty The period from about 1765 to 1123 B.C. in China, during which most of the cultural attributes of modern China were established.

Shatter belt A politically unstable region, especially Eastern Europe before Soviet domination.

Shifting agriculture A farming system based on periodic change of cultivated area, land rotation.

Silla The dominant tribal group that began the first Korean dynasty A.D. 700–918.

Sinic Chinese or Chinese related.

Soccer War A brief armed encounter between El Salvador and Honduras on the surface over a soccer match between teams of the countries but actually over a long-standing border dispute and Salvadorean settlement in the disputed area.

Sodium nitrate A source of nitrogen, one of the three major plant foods; also used in explosives. Chile's sodium nitrate deposits formerly were the only significant source of nitrogen; today, however, most nitrates are extracted from the atmosphere.

Softwood Gymnosperm trees, commonly called needle-leafed and usually evergreen.

Songhai Empire A west African trading state that flourished in the fifteenth and sixteenth centuries and centered along the middle Niger River.

Southeast Asia Treaty Organization (SEATO) A multinational military alliance of nations allied with the West in Southeast Asia and including Australia, New Zealand, the United States, and the United Kingdom.

Soviet Socialist Republic (SSR) The highest level

in the political-territorial division of the USSR, in which major ethnic groups are permitted some semblance of self-identity.

Sovkhoz A Soviet state farm operated by the state with paid employees.

Spatial organization The structures and linkages of human activities in an area.

Stages of economic growth A theory developed by Walter Rostow in which five stages of economic organization are recognized: traditional society, preconditions for take-off, the take-off, drive to maturity, and high mass consumption.

Standard of living The material well-being judged to be adequate by a society or societal subgroup as measured by diet, housing, and clothing.

Standard Metropolitan Statistical Area (SMSA) A governmental designation in the United States for a central city of at least 50,000 inhabitants and the surrounding counties.

Steppe A large grassland region in southwestern Soviet Union. Any area of short grass; also an arid type of climate being more moist than a desert but still water-deficient.

Subsistence agriculture A farming system in which the farmer and his family consume most or all of the production; noncommercial.

Suq The traditional urban marketplace in North Africa.

Suspended culture A society whose behavioral patterns undergo little or no change.

Sustained yield forestry Harvesting timber only at its annual growth rate.

Taconite A low-grade iron ore formerly considered worthless but now an important resource brought about by technological advances.

Taiga The large coniferous forest extending across northern Soviet Union.

Taiping Rebellion One of China's major uprisings (1850–1864) that proved almost fatal to the Manchu dynasty. The rebellion arose out of local conflicts, but quickly grew and spread throughout most of the country. Embracing vague principles of equality and religion, the rebellion failed, but it demonstrated the weaknesses of Manchu control.

Tariff The system of duties or customs imposed by a nation on imports and exports.

Technocratic theory A theory that asserts technology increases at a rate greater than population.

Tennessee Valley Authority (TVA) A United States regional development commission charged with the planning and execution of development projects in the Tennessee River Valley.

Terra roxa A latosol found in Brazil and other tropical areas that is especially suitable for coffee cultivation.

Tertiary activities Economic pursuits in which a service is performed, such as retailing, wholesaling, government, teaching, medicine, repair, and recreation.

Third World The countries aligned with neither Communist nor Western nations; generally these are poor nations.

Threshold The number of customers needed in order to support a particular activity or business.

Tierra caliente The hot land; the lowest of the highland climate possessing the characteristics of either the tropical rainy or tropical wet and dry climates.

Tierra fria The cold land; a highland climate located between the *tierra templada* and *tierra helada* and forms the upper limit of crop cultivation.

Tierra helada The ice land; the uppermost highland climate characterized by nightly frost, and at higher elevations permanent ice and snow.

Tierra templada The temperate land; a highland climate with mid-latitude climatic characteristics located between the *tierra templada* and *tierra fria.*

Till plain An undulating area with surface materials of glacial deposition origin.

Time cushion A period made available by technological advances to bring population growth under control before population food demands become greater than the supply; a Neo-Malthusian concept. Technologic advances include the Industrial Revolution, and more recently, the Green Revolution.

Tokugawa period The period from A.D. 1615 to 1867, during which the focus of power in Japan shifted to the Kanto Plain area; during this period many of the modern Japanese characteristics were firmly fixed in the culture.

Trans-Amazon Highway An ambitious project to construct an east-west highway the length of the Brazilian Amazon; part of a larger project to develop the Amazon Basin.

Transhumance The practice of moving animals seasonally between summer alpine and winter lowland pastures.

Treaty of Rome The agreement signed in 1957 by the Benelux nations, France, West Germany, and Italy that created the European Economic Community.

Tribalism Allegiance primarily to the local group and continued observation of the group's customs and life style.

Tsetse fly A member of the genus *Glossina*, common in parts of Africa, that transmits a number of diseases harmful to humans and domestic animals.

Tubers Food plants with edible roots or other subterranean parts such as the white potato, sweet potato, manioc, and yams.

Tundra Originally a vast treeless plain in the Soviet Union; now the treeless area polarward of the limit of forest and equatorward of polar ice cap; also the climate of this type region.

Ujamaa A Swahili word meaning "familyhood" that expresses a feeling of community and of cooperative activity. The word is used by the Tanzanian government to indicate a commitment to rapid economic development according to principles of socialism and communal solidarity.

Underdeveloped Used in this book to identify the nations or regions that have a low level of economic production per person; terms such as "poor," "undeveloped," "emerging," "backward," "less developed," and "developing" have also been used to designate such areas.

Underemployment The incomplete use of labor either because a person works only part-time or seasonally or because individuals are used inefficiently as, for example, the use of five people to perform a task that two could do, or a person working below his or her skill level.

Universalizing religions Religions considered by their adherents as appropriate for all mankind. Proselytizing is often common.

Untouchables The lowest group in the hierarchy of the Hindu caste system comprising basically Dravidian people who have the most menial occupations.

Urbanism The social-cultural aspects of city living; the way of life of those who live in cities.

Urbanization The agglomeration of people, the process of becoming city dwellers.

U-turn migration The movement from large urban centers to the suburbs and middle-sized cities, commonly in the United States, Japan, and Western Europe.

Valorization The attempt to control a commodity's price by a nation or a group of nations that produces the item.

Value added The residual value of a product or service after all production costs or other charges have been subtracted.

Viceroyalty The largest Spanish colonial unit in the New World. At independence, new nations formed themselves along the colonial political boundary lines of viceroyalties and their subparts.

Virgin and idle land program The attempt to expand Soviet Union agriculture to the east into Siberia and central Asia.

Walking on Two legs The present Chinese development approach stressing both traditional and modern production methods, city and rural areas, and large and small production units.

Walloons French-speaking Belgians, who are concentrated in southern Belgium; one of the nation's two principal national groups.

Wood pulp Ground wood used for newsprint and book paper; usually softwood.

Xerophytic Plants that are able to withstand moisture deficiency, vegetation common to desert areas.

Yang-shao An early stage of Chinese culture dating from 3000 to 5000 B.C. and characterized by dry field agriculture.

Yayoi The culture group located in Kyushu, from which many Japanese traits and complexes evolved.

Yi Dynasty The period during which the Korean peninsula was unified under a single government and lasted from A.D. 1392 to 1910.

Zaibatsu A large Japanese financial enterprise similar to a conglomerate in the United States but generally more integrated horizontally and vertically.

Zero population growth The maintenance of a stable population in which births and in-migration are balanced by deaths and out-migration.

Zulu A large African tribe of relatively high cultural attainment located along the southeastern coast.

Selected National Statistics

Appendix

Nation	Population (millions)	Area Sq. Mi. (1,000s)	Area Sq. Km. (1,000s)	Birth Rate (%)	Death Rate (%)	Infant Mortality (less than age 1) (%)
Anglo-America						
United States	216.7	3676	9520	1.5	0.9	1.6
Canada	23.5	3852	9976	1.6	0.7	1.5
Western Europe						
Austria	7.5	32	84	1.2	1.3	2.1
Belgium	9.9	12	31	1.2	1.2	1.6
Denmark	5.1	17	43	1.4	1.0	1.2
Finland	4.8	130	337	1.4	0.9	1.0
France	53.4	210	544	1.4	1.1	1.2
Germany (West)	61.2	96	249	1.0	1.2	2.0
Greece	9.1	51	132	1.6	0.9	2.4
Iceland	0.2	40	103	2.1	0.7	1.1
Ireland	3.2	27	70	2.2	1.1	1.7
Italy	56.5	116	301	1.5	1.0	2.1
Luxembourg	0.4	1	3	1.1	1.2	1.5
Netherlands	13.9	16	41	1.3	0.8	1.1
Norway	4.0	125	324	1.4	1.0	1.1

Population Under Age 15 (%)	Life Expectancy at Birth (years)	Population Annual Growth Rate (%)	Per Capita GNP ($)	Daily Food Supply (cal)	Annual Per Capita Energy Consumption (gals. of oil)	Percent Labor Force in Agriculture
26	72	0.6	7060	3330	2320	3.7
27	73	0.8	6650	3180	1940	8.2
24	71	0.0	4720	3310	795	14.8
23	71	0.0	6070	3380	1353	4.8
23	74	0.4	6920	3240	1045	11.2
23	71	0.5	5100	3050	908	21.3
24	72	0.4	5760	3210	880	13.7
22	71	−.2	6610	3220	1192	7.5
24	72	0.7	2360	3190	417	43.1
30	74	1.4	5620	2900	1147	17.7
31	71	1.1	2420	3410	705	26.5
24	72	0.5	2940	3180	653	18.8
20	70	−.1	6050	3380	3218	7.7
26	74	0.5	5590	3320	1246	8.1
24	74	0.4	6540	2960	1014	11.9

Nation	Population (millions)	Area Sq. Mi. (1,000s)	Sq. Km. (1,000s)	Birth Rate (%)	Death Rate (%)	Infant Mortality (less than age 1) (%)
Western Europe (*cont.*)						
Portugal	9.2	36	92	2.0	1.1	3.8
Spain	36.5	195	505	1.8	0.8	1.4
Sweden	8.2	174	450	1.3	1.1	0.8
Switzerland	6.2	16	41	1.2	0.9	1.2
United Kingdom	56.0	94	244	1.2	1.2	1.6
Eastern Europe						
Albania	2.5	11	29	3.3	0.8	8.7
Bulgaria	8.8	43	111	1.7	1.0	2.3
Czechoslovakia	15.0	49	128	2.0	1.2	2.1
Germany (East)	16.7	42	108	1.1	1.4	1.6
Hungary	10.7	36	93	1.8	1.2	3.3
Poland	34.7	121	313	1.9	0.9	2.5
Romania	21.7	92	238	2.0	0.9	3.5
Yugoslavia	21.8	99	256	1.8	0.9	4.0
Soviet Union	259.0	8600	22275	1.8	0.9	2.8
Japan	114.2	144	372	1.7	0.6	1.0
Australia	13.9	2968	7687	1.7	0.8	1.6
New Zealand	3.2	104	269	1.8	0.8	1.6
Latin America						
Argentina	26.1	1072	2777	2.3	0.9	5.9
Belize	0.1	9	23	4.2	1.1	6.3
Bolivia	4.8	424	1099	4.4	1.8	10.8
Brazil	112.0	3286	8512	3.7	0.9	8.2
Chile	11.0	292	757	2.4	0.8	7.7
Colombia	25.2	440	1139	3.3	0.9	9.7
Costa Rica	2.1	20	51	2.9	0.5	3.8
Cuba	9.6	44	115	2.2	0.6	2.9
Dominican Republic	5.0	19	49	4.6	1.1	9.8
Ecuador	7.5	109	284	4.2	1.0	7.8
El Salvador	4.3	8	21	4.0	0.8	5.8
Guatemala	6.4	42	109	4.3	1.2	8.0
Guyana	0.8	83	215	3.2	0.7	4.0
Haiti	5.3	11	28	3.6	1.6	15.0
Honduras	3.3	43	112	4.9	1.5	11.7
Jamaica	2.1	4	11	3.0	0.7	2.6
Mexico	64.4	762	1973	4.2	0.7	6.6
Nicaragua	2.3	50	130	4.8	1.4	12.3

Population Under Age 15 (%)	Life Expectancy at Birth (years)	Population Annual Growth Rate (%)	Per Capita GNP ($)	Daily Food Supply (cal)	Annual Per Capita Energy Consumption (gals. of oil)	Percent Labor Force in Agriculture
27	69	0.9	1610	2900	202	33.3
28	72	1.0	2700	2600	411	26.0
21	75	0.2	7880	2810	1193	8.3
23	73	0.4	8050	3190	777	7.8
24	72	0.1	3840	3190	1131	2.8
40	67	2.5	600	2390	145	66.3
22	71	0.6	2040	3290	855	46.6
23	71	0.8	3710	3180	1407	16.9
22	71	−.4	4230	3290	1475	13.0
20	70	0.6	2480	3280	718	25.0
24	70	1.0	2910	3280	940	39.0
25	69	1.1	1300	3140	709	56.0
26	68	1.0	1480	3190	377	49.8
26	69	0.9	2620	3280	1056	25.7
24	74	1.1	4460	2510	762	19.6
28	71	0.9	5640	3280	1189	8.1
30	72	1.0	4680	3200	673	11.8
29	68	1.3	1590	3060	369	16.4
39	60	2.3	—	—	—	30.0
42	47	2.6	320	1900	67	55.5
42	61	2.8	1010	2620	124	45.6
36	63	1.6	760	2670	278	23.7
43	61	2.5	550	2200	125	37.9
44	68	2.4	910	2610	92	42.1
37	70	1.6	800	2700	230	30.6
48	58	3.5	720	2120	82	61.3
45	60	3.2	550	2010	70	50.9
46	58	3.2	450	1930	46	56.2
45	53	3.1	650	2130	46	61.0
44	68	2.4	560	2390	186	27.9
42	50	2.0	180	1730	5	77.0
47	54	3.5	350	2140	41	66.4
46	68	2.3	1290	2360	283	29.5
46	63	3.5	1190	2580	237	45.2
48	53	3.4	720	2450	84	49.1

Nation	Population (millions)	Area Sq. Mi. (1,000s)	Area Sq. Km. (1,000s)	Birth Rate (%)	Death Rate (%)	Infant Mortality (less than age 1) (%)
Latin America (*cont.*)						
Panama	1.8	29	76	3.1	0.5	4.0
Paraguay	2.8	157	407	4.0	0.9	6.5
Peru	16.6	496	1285	4.1	1.2	11.0
Puerto Rico	3.2	3	9	2.3	0.6	2.4
Surinam	0.4	63	163	3.7	0.7	3.0
Trinidad	1.0	2	5	2.4	0.6	3.4
Uruguay	2.8	69	178	2.1	1.0	4.5
Venezuela	12.7	352	912	3.7	0.6	4.9
Africa						
Algeria	17.8	920	2382	4.8	1.5	14.2
Angola	6.3	481	1247	4.7	2.4	20.3
Benin	3.3	43	113	5.0	2.3	18.5
Botswana	0.7	232	600	4.6	2.3	9.7
Burundi	3.9	11	28	4.1	2.0	15.0
Cameroon	6.7	184	475	4.0	2.2	13.7
Central African Republic	1.9	241	623	4.3	2.2	19.0
Chad	4.2	496	1284	4.4	2.4	16.0
Congo	1.4	132	342	4.5	2.1	18.0
Egypt	38.9	387	1002	3.6	1.2	11.6
Ethiopia	29.4	472	1222	4.3	1.8	18.1
Gabon	0.5	103	268	3.2	2.2	17.8
Gambia	0.6	4	11	4.2	2.1	16.5
Ghana	10.4	92	239	4.7	2.0	15.6
Guinea	4.7	95	246	4.7	2.3	17.5
Guinea-Bissau	0.5	14	36	4.0	2.5	20.8
Ivory Coast	7.0	125	322	4.6	2.1	16.4
Kenya	14.4	225	583	4.9	1.6	11.9
Lesotho	1.1	12	30	3.9	2.0	11.4
Liberia	1.7	43	111	5.0	2.1	15.9
Libya	2.7	679	1760	4.8	0.9	13.0
Malagasy Republic	8.0	227	587	5.0	2.1	17.0
Malawi	5.3	46	118	4.8	2.4	14.2
Mali	5.9	479	1240	5.0	2.6	18.8
Mauritania	1.4	398	1031	4.5	2.5	18.7
Morocco	18.3	172	447	4.8	1.6	13.0
Mozambique	9.5	203	525	4.3	2.0	16.5
Niger	4.9	489	1267	5.2	2.6	20.0
Nigeria	66.6	357	924	4.9	2.3	18.0
Rhodesia	6.8	151	391	4.8	1.4	12.2
Rwanda	4.5	10	26	5.1	2.2	13.3
Senegal	5.3	76	197	4.6	2.1	15.9
Sierra Leone	3.2	28	72	4.5	2.1	13.6

Population Under Age 15 (%)	Life Expectancy at Birth (years)	Population Annual Growth Rate (%)	Per Capita GNP ($)	Daily Food Supply (cal)	Annual Per Capita Energy Consumption (gals. of oil)	Percent Labor Force in Agriculture
43	66	2.6	1060	2580	157	41.5
45	62	3.1	570	2740	32	52.7
45	56	2.9	810	2320	124	44.8
37	72	1.7	2300	2530	837	8.1
50	66	3.0	1180	2450	604	22.2
40	66	1.8	1900	2380	868	18.8
28	70	1.1	1330	2880	201	15.2
45	65	3.1	2220	2430	552	25.6
48	53	3.2	780	1730	95	60.7
42	38	2.3	680	2000	39	64.0
46	41	2.7	140	2260	8	49.8
48	56	2.3	330	2040	—	86.6
44	40	2.1	100	2040	3	87.1
40	41	1.8	270	2410	17	85.0
42	41	2.1	230	2200	11	91.2
40	38	2.0	120	2110	3	90.2
42	44	2.4	500	2260	41	41.8
41	52	2.3	310	2500	62	54.4
44	42	2.5	100	2160	6	84.1
32	41	1.0	2240	2220	231	81.8
41	44	2.1	190	2490	14	82.1
47	48	2.7	460	2320	35	58.4
43	41	2.4	130	2200	18	84.7
37	38	1.5	390	—	8	21.0
43	44	2.5	500	2430	52	84.5
46	50	3.3	220	2360	33	82.1
40	46	1.9	180	—	—	89.7
42	45	2.9	410	2170	87	75.6
49	55	3.9	5080	2570	175	32.1
45	44	2.9	140	2530	14	86.0
35	43	2.4	150	2210	11	89.3
49	38	2.4	90	2060	5	91.0
42	38	2.0	310	1970	22	87.3
46	53	3.2	470	2220	49	56.8
43	44	2.3	310	2050	28	73.9
43	38	2.7	130	2080	6	92.8
45	41	2.7	310	2270	18	62.2
48	52	3.4	540	2660	120	63.9
44	41	2.9	90	1960	2	93.3
43	44	2.5	370	2370	28	79.7
43	44	2.4	200	2280	21	71.5

Nation	Population (millions)	Area Sq. Mi. (1,000s)	Sq. Km. (1,000s)	Birth Rate (%)	Death Rate (%)	Infant Mortality (less than age 1) (%)
Africa (*cont.*)						
Somalia	3.4	246	638	4.7	2.2	17.7
South Africa	26.1	472	1222	4.0	1.5	11.7
Sudan	16.3	968	2506	4.8	1.8	14.1
Swaziland	0.5	7	17	4.9	2.2	14.9
Tanzania	16.0	365	945	4.7	2.2	16.2
Togo	2.3	22	56	5.1	2.3	12.7
Tunisia	6.0	63	164	3.4	1.1	12.5
Uganda	12.4	91	236	4.3	1.6	16.0
Upper Volta	6.4	106	274	4.8	2.6	18.2
Zaire	26.3	906	2345	4.5	2.0	16.0
Zambia	5.2	291	753	5.0	1.9	16.0
Southwest Asia						
Iran	34.8	636	1648	4.4	1.6	13.9
Iraq	11.8	168	435	4.4	1.1	9.9
Israel	3.6	8	21	2.8	0.7	2.2
Jordan	2.9	38	98	4.8	1.5	9.7
Kuwait	1.1	6	16	4.4	0.5	4.4
Lebanon	2.8	4	10	4.0	1.0	5.9
Oman	.8	82	212	5.0	1.9	13.8
Qatar	.1	9	22	5.0	1.9	13.8
Saudi Arabia	7.6	830	2150	5.0	2.0	15.2
Syria	7.8	71	185	4.5	1.5	9.3
Turkey	41.9	301	781	3.9	1.2	11.9
United Arab Emirates	.2	32	84	5.0	1.8	13.8
Yemen Arab Republic	5.6	75	195	5.0	2.1	15.2
Yemen People's Republic	1.8	111	288	5.0	2.1	15.2
South Asia						
Afghanistan	20.0	250	648	4.3	2.1	18.2
Bangladesh	83.3	55	143	4.7	2.0	13.2
Bhutan	1.2	18	47	4.4	2.0	—
India	622.7	1226	3177	3.4	1.3	12.2
Nepal	13.2	54	141	4.3	2.0	16.9
Pakistan	74.5	346	896	4.4	1.5	12.1
Sri Lanka	14.1	25	66	2.8	0.8	4.5
East Asia						
China	900.0	3692	9561	2.7	1.0	5.5
Hong Kong	4.5	.4	1	2.0	0.5	1.7
Korea (North)	16.7	47	121	3.6	0.9	—
Korea (South)	35.9	38	98	2.4	0.7	4.7

Population Under Age 15 (%)	Life Expectancy at Birth (years)	Population Annual Growth Rate (%)	Per Capita GNP ($)	Daily Food Supply (cal)	Annual Per Capita Energy Consumption (gals. of oil)	Percent Labor Force in Agriculture
45	41	2.6	100	1830	8	84.7
41	52	2.5	1320	2740	609	30.9
45	51	3.0	290	2160	27	82.0
48	44	2.7	470	—	—	81.3
47	44	2.5	170	2260	14	86.2
46	41	2.7	270	2330	13	73.3
45	55	2.3	760	2250	81	49.9
44	50	2.7	250	2130	9	85.9
43	38	2.3	90	1710	3	86.8
44	44	2.5	150	2060	14	79.8
46	46	3.1	540	2590	105	72.8
47	51	2.8	1440	2300	243	46.0
48	53	3.2	1280	2160	171	46.6
33	72	2.1	3580	2960	552	9.7
48	53	3.3	460	2430	73	33.7
44	69	3.9	11510	—	1771	1.6
41	64	3.0	1070	2280	221	19.6
45	47	3.1	2070	—	49	—
45	47	3.1	8320	—	3388	—
45	45	2.9	3010	2270	231	66.0
49	57	3.0	660	2650	111	51.1
42	57	2.7	860	3250	119	67.7
45	47	3.2	10480	—	2996	—
45	45	2.9	210	2040	7	79.2
48	45	2.9	240	2070	68	64.5
44	40	2.2	130	1970	13	81.7
43	47	2.7	110	1840	6	85.9
42	44	2.3	70	—	—	94.4
40	50	2.1	150	2070	39	69.3
40	44	2.3	110	2080	2	93.9
46	51	2.9	140	2160	36	58.9
39	68	2.0	150	2170	28	55.1
33	62	1.7	350	2170	124	67.8
31	71	1.4	1720	2370	241	4.3
42	61	2.6	430	2240	515	54.7
40	65	1.7	550	2520	185	51.0

Nation	Population (millions)	Area		Birth Rate (%)	Death Rate (%)	Infant Mortality (less than age 1) (%)
		Sq. Mi. (1,000s)	Sq. Km. (1,000s)			
East Asia (*cont.*)						
Mongolia	1.5	604	1565	4.0	1.0	—
Taiwan	16.6	14	36	2.3	0.5	2.6
Southeast Asia						
Burma	31.8	262	678	4.0	1.6	12.6
Indonesia	136.9	753	1951	3.8	1.4	13.7
Khmer Republic	8.1	70	181	4.7	1.9	12.7
Laos	3.5	91	237	4.5	2.3	12.3
Malaysia	12.6	128	333	3.5	0.7	7.5
Philippines	44.3	116	300	3.5	0.8	7.4
Singapore	2.3	.2	1	1.8	0.5	1.4
Thailand	44.4	199	514	3.5	1.1	8.9
Vietnam	47.3	128	333	4.2	2.0	4.3

SOURCES: *Population Data Sheet* (Washington, D.C.: Population Reference Bureau); *Statistical Yearbook; Demographic Yearbook* (New York: United Nations, 1974); *Production Yearbook* (Rome: Food and Agriculture Organization, 1975).

Population Under Age 15 (%)	Life Expectancy at Birth (years)	Population Annual Growth Rate (%)	Per Capita GNP ($)	Daily Food Supply (cal)	Annual Per Capita Energy Consumption (gals. of oil)	Percent Labor Force in Agriculture
44	61	3.0	700	2380	202	61.9
43	69	1.8	890	2620	—	42.0
40	50	2.4	110	2210	11	59.6
44	48	2.4	180	1790	31	66.3
45	45	2.8	120	2430	—	—
42	40	2.2	70	2110	12	78.8
45	63	2.8	720	2460	89	55.5
43	58	2.7	370	1940	60	53.7
34	68	1.3	2510	—	410	3.4
45	58	2.4	350	2560	57	79.9
41	50	2.1	160	2340	28	76.4

Index

Italic page numbers indicate figures; n = note; t = table.

A

713

M

N

Y

Z

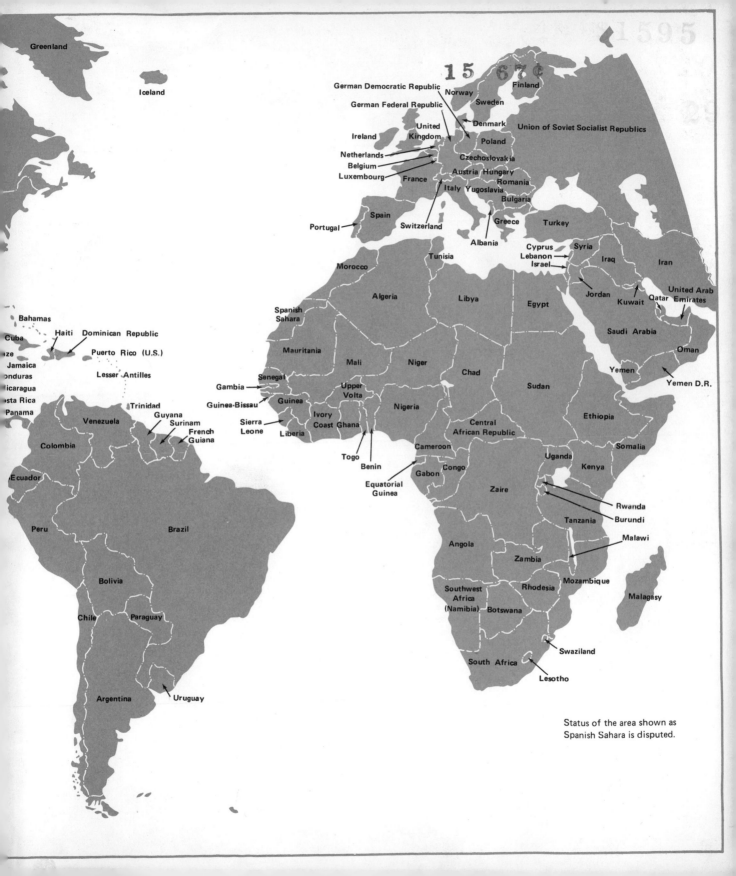

Status of the area shown as
Spanish Sahara is disputed.